69

Methods in Membrane Lipids

METHODS IN MOLECULAR BIOLOGY™

John M. Walker, SERIES EDITOR

419. **Post-Transcriptional Gene Regulation,** edited by *Jeffrey Wilusz, 2008*
418. **Avidin–Biotin Interactions:** *Methods and Applications,* edited by *Robert J. McMahon, 2008*
417. **Tissue Engineering, Second Edition,** edited by *Hannsjörg Hauser and Martin Fussenegger, 2007*
416. **Gene Essentiality:** *Protocols and Bioinformatics,* edited by *Andrei L. Osterman, 2008*
415. **Innate Immunity,** edited by *Jonathan Ewbank and Eric Vivier, 2007*
414. **Apoptosis in Cancer:** *Methods and Protocols,* edited by *Gil Mor and Ayesha Alvero, 2008*
413. **Protein Structure Prediction, Second Edition,** edited by *Mohammed Zaki and Chris Bystroff, 2008*
412. **Neutrophil Methods and Protocols,** edited by *Mark T. Quinn, Frank R. DeLeo, and Gary M. Bokoch, 2007*
411. **Reporter Genes for Mammalian Systems,** edited by *Don Anson, 2007*
410. **Environmental Genomics,** edited by *Cristofre C. Martin, 2007*
409. **Immunoinformatics:** *Predicting Immunogenicity In Silico,* edited by *Darren R. Flower, 2007*
408. **Gene Function Analysis,** edited by *Michael Ochs, 2007*
407. **Stem Cell Assays,** edited by *Vemuri C. Mohan, 2007*
406. **Plant Bioinformatics:** *Methods and Protocols,* edited by *David Edwards, 2007*
405. **Telomerase Inhibition:** *Strategies and Protocols,* edited by *Lucy Andrews and Trygve O. Tollefsbol, 2007*
404. **Topics in Biostatistics,** edited by *Walter T. Ambrosius, 2007*
403. **Patch-Clamp Methods and Protocols,** edited by *Peter Molnar and James J. Hickman, 2007*
402. **PCR Primer Design,** edited by *Anton Yuryev, 2007*
401. **Neuroinformatics,** edited by *Chiquito J. Crasto, 2007*
400. **Methods in Membrane Lipids,** edited by *Alex Dopico, 2007*
399. **Neuroprotection Methods and Protocols,** edited by *Tiziana Borsello, 2007*
398. **Lipid Rafts,** edited by *Thomas J. McIntosh, 2007*
397. **Hedgehog Signaling Protocols,** edited by *Jamila I. Horabin, 2007*
396. **Comparative Genomics,** *Volume 2,* edited by *Nicholas H. Bergman, 2007*
395. **Comparative Genomics,** *Volume 1,* edited by *Nicholas H. Bergman, 2007*
394. **Salmonella:** *Methods and Protocols,* edited by *Heide Schatten and Abe Eisenstark, 2007*
393. **Plant Secondary Metabolites,** edited by *Harinder P. S. Makkar, P. Siddhuraju, and Klaus Becker, 2007*
392. **Molecular Motors:** *Methods and Protocols,* edited by *Ann O. Sperry, 2007*
391. **MRSA Protocols,** edited by *Yinduo Ji, 2007*
390. **Protein Targeting Protocols, Second Edition,** edited by *Mark van der Giezen, 2007*
389. **Pichia Protocols, Second Edition,** edited by *James M. Cregg, 2007*
388. **Baculovirus and Insect Cell Expression Protocols, Second Edition,** edited by *David W. Murhammer, 2007*

387. **Serial Analysis of Gene Expression (SAGE):** *Digital Gene Expression Profiling,* edited by *Kare Lehmann Nielsen, 2007*
386. **Peptide Characterization and Application Protocols,** edited by *Gregg B. Fields, 2007*
385. **Microchip-Based Assay Systems:** *Methods and Applications,* edited by *Pierre N. Floriano, 2007*
384. **Capillary Electrophoresis:** *Methods and Protocols,* edited by *Philippe Schmitt-Kopplin, 2007*
383. **Cancer Genomics and Proteomics:** *Methods and Protocols,* edited by *Paul B. Fisher, 2007*
382. **Microarrays, Second Edition:** *Volume 2, Applications and Data Analysis,* edited by *Jang B. Rampal, 2007*
381. **Microarrays, Second Edition:** *Volume 1, Synthesis Methods,* edited by *Jang B. Rampal, 2007*
380. **Immunological Tolerance:** *Methods and Protocols,* edited by *Paul J. Fairchild, 2007*
379. **Glycovirology Protocols,** edited by *Richard J. Sugrue, 2007*
378. **Monoclonal Antibodies:** *Methods and Protocols,* edited by *Maher Albitar, 2007*
377. **Microarray Data Analysis:** *Methods and Applications,* edited by *Michael J. Korenberg, 2007*
376. **Linkage Disequilibrium and Association Mapping:** *Analysis and Application,* edited by *Andrew R. Collins, 2007*
375. **In Vitro Transcription and Translation Protocols:** *Second Edition,* edited by *Guido Grandi, 2007*
374. **Quantum Dots:** *Applications in Biology,* edited by *Marcel Bruchez and Charles Z. Hotz, 2007*
373. **Pyrosequencing® Protocols,** edited by *Sharon Marsh, 2007*
372. **Mitochondria: Practical Protocols,** edited by *Dario Leister and Johannes Herrmann, 2007*
371. **Biological Aging:** *Methods and Protocols,* edited by *Trygve O. Tollefsbol, 2007*
370. **Adhesion Protein Protocols,** *Second Edition,* edited by *Amanda S. Coutts, 2007*
369. **Electron Microscopy:** *Methods and Protocols, Second Edition,* edited by *John Kuo, 2007*
368. **Cryopreservation and Freeze-Drying Protocols,** *Second Edition,* edited by *John G. Day and Glyn Stacey, 2007*
367. **Mass Spectrometry Data Analysis in Proteomics,** edited by *Rune Matthiesen, 2007*
366. **Cardiac Gene Expression:** *Methods and Protocols,* edited by *Jun Zhang and Gregg Rokosh, 2007*
365. **Protein Phosphatase Protocols:** edited by *Greg Moorhead, 2007*
364. **Macromolecular Crystallography Protocols:** *Volume 2, Structure Determination,* edited by *Sylvie Doublié, 2007*
363. **Macromolecular Crystallography Protocols:** *Volume 1, Preparation and Crystallization of Macromolecules,* edited by *Sylvie Doublié, 2007*
362. **Circadian Rhythms:** *Methods and Protocols,* edited by *Ezio Rosato, 2007*
361. **Target Discovery and Validation Reviews and Protocols:** *Emerging Molecular Targets and Treatment Options, Volume 2,* edited by *Mouldy Sioud, 2007*

METHODS IN MOLECULAR BIOLOGY™

Methods in Membrane Lipids

Edited by

Alex M. Dopico

*Department of Pharmacology,
The University of Tennessee Health Science Center,
Memphis, TN*

HUMANA PRESS ✳ TOTOWA, NEW JERSEY

© 2007 Humana Press Inc.
999 Riverview Drive, Suite 208
Totowa, New Jersey 07512

www.humanapress.com

All rights reserved. No part of this book may be reproduced, stored in a retrieval system, or transmitted in any form or by any means, electronic, mechanical, photocopying, microfilming, recording, or otherwise without written permission from the Publisher. Methods in Molecular Biology™ is a trademark of The Humana Press Inc.

All papers, comments, opinions, conclusions, or recommendations are those of the author(s), and do not necessarily reflect the views of the publisher.

This publication is printed on acid-free paper. ∞
ANSI Z39.48-1984 (American Standards Institute)

Permanence of Paper for Printed Library Materials.
Cover illustration: Fig. 3 (top far left panel) from Chapter 34, "Lipid Domains in Supported Lipid Bilayer for Atomic Force Microscopy".

Production Editor: Rhukea J. Hussain
Cover design by Donna Niethe

For additional copies, pricing for bulk purchases, and/or information about other Humana titles, contact Humana at the above address or at any of the following numbers: Tel.: 973-256-1699; Fax: 973-256-8341; E-mail: orders@humanapr.com; or visit our Website: www.humanapress.com

Photocopy Authorization Policy:
Authorization to photocopy items for internal or personal use, or the internal or personal use of specific clients, is granted by Humana Press Inc., provided that the base fee of US $30.00 per copy is paid directly to the Copyright Clearance Center at 222 Rosewood Drive, Danvers, MA 01923. For those organizations that have been granted a photocopy license from the CCC, a separate system of payment has been arranged and is acceptable to Humana Press Inc. The fee code for users of the Transactional Reporting Service is: [978-1-58829-662-7/07 $30].

Printed in the United States of America. 10 9 8 7 6 5 4 3 2 1

e-ISBN: 978-1-59745-519-0

Library of Congress Control Number: 2007925516

Preface

My interest in membrane lipids has to be traced back to the Medical School years at the University of Buenos Aires. In 1980 I was taking my first course in Biochemistry, in which students had to go through the pages of Biochemistry by *Lehninger*, or Principles of Biochemistry by *White et al.,* (thank God, no handouts were available) to master the complicated structure of lipids. I was struck by the contrast between the structural complexity of membrane lipids and the simple, almost exclusive function commonly attributed to them; that lipids served as an undifferentiated sea where proteins moved and translated cell signaling was consensus. Paraphrasing the late Carl Sagan when asked about the possibility of mankind being the only intelligent life form in the vast and complex universe: *"If that's the case, what a waste!"* After a few years of following the consequences of lipid intake on human health at the Hypertension Division of the Cardiology Hospital, Natl. Acad. Sci., my interest in vascular reactivity brought me to the world of ion channel proteins. Naturally, when I started postdoctoral work, the choice was to study membrane lipid modulation of ion channel function. In a later career move, scientific interests only slightly shifted to the modulation of ion channel function by small amphiphilic drugs and the relevance of the membrane lipid microenvironment in such modulation. In a way, the organization of this volume and the topics contained herein reflect this personal journey: the book starts with methodologies used to address lipid distribution and organization, and ends with methods used to study interactions among membrane lipids, bilayer-spanning proteins, and drugs.

The guiding principle of this book is the presentation of methodologies developed by both experimentalists and theoreticians to study structure and/or function of membrane lipids. The volume has been organized in sections that nucleate chapters not around a given set of related techniques but around more or less defined topics of study in Membrane Lipidology. Thus, while each chapter is distinctly defined by the laws and principles of Mathematics, Computational Science, Physicochemistry, Optics, etc., the division of the volume into section is driven by Biology. The book sequentially presents methods to study (i) lipid distribution, structure, and lipid-lipid interactions, (ii) lipid phases, (iii) mono- and bilayer lipid curvature and stress, (iv) lipid domains, and (v) membrane lipid-protein/drug interactions. This organization offers the reader the possibility of comparing (and evaluating any possible complementation among) different techniques to tackle a given biological problem, let's say, membrane lipid diffusion. Naturally, as any attempt to classify and organize human knowledge, the organization of the book is rather artificial, and the overlapping of some topics across sections is recognized. The reader may also find with ease an alternative location of a given chapter.

Most chapters written by experimentalists follow the typical format of the MiMB Humana Series, that is, they start with an Introduction, where the biological question and the basic principles of the technique are described and followed with Materials, which list all chemicals, reagents and devices required to conduct the experiment; Methods, which contains a step-by-step description of each protocol, and the Notes section, which highlights experimental difficulties, pitfalls, and ways to circumvent them. While each chapter presents an Introduction to the particular topic and method, I felt it was necessary to also include in this book two brief introductory chapters dealing with the diversity of membrane lipids and lipid polymorphism. It is the expectation and hope of both authors and myself that the present volume will represent not only a valuable reference manual for experienced biophysicists and biochemists specialized in research on membrane lipids, but also appeal to investigators in scientific areas, whether theoreticians or experimentalists.

I would like to thank, first and foremost, the authors of all chapters, whose contributions are the foundation of this volume. It has been an honor to work with you. I want to express my deepest gratitude to Prof. John M. Walker, the MiMB Series Editor, for his encouragement and guidance throughout the editing process. I also want to thank NIAAA and NHLBI for their continuous support of my research on the interactions among small amphiphilic agents, ion channel proteins and membrane lipids. Finally, my gratitude goes to my family and my lab personnel, for their patience and their understanding of what "time constraints" mean.

Alex M. Dopico
Memphis, TN
July 12, 2006

Contents

Preface v
Contributors xi
Color Plates xvi

1. A Glance at the Structural and Functional Diversity of Membrane Lipids
   Alex M. Dopico and Gabor J. Tigyi 1

2. Membrane Lipid Polymorphism: Relationship to Bilayer Properties and Protein Function
   Richard M. Epand 15

3. Acrylodan-Labeled Intestinal Fatty Acid-Binding Protein to Measure Concentrations of Unbound Fatty Acids
   Jeffrey R. Simard, Frits Kamp, and James A. Hamilton 27

4. Measuring Molecular Order and Orientation Using Coherent Anti-Stokes Raman Scattering Microscopy
   Hilde A. Rinia, George W. H. Wurpel, and Michiel Müller ... 45

5. Preparation of Oriented, Fully Hydrated Lipid Samples for Structure Determination Using X-ray Scattering
   Stephanie A. Tristram-Nagle 63

6. Nuclear Magnetic Resonance Investigation of Oriented Lipid Membranes
   Olivier Soubias and Klaus Gawrisch 77

7. Molecular Dynamics Simulations as a Complement to Nuclear Magnetic Resonance and X-Ray Diffraction Measurements
   Scott E. Feller 89

8. Use of Inverse Theory Algorithms in the Analysis of Biomembrane NMR Data
   Edward Sternin 103

9. Statistical Thermodynamics Through Computer Simulation to Characterize Phospholipid Interactions in Membranes
   Mihaly Mezei and Pál Jedlovszky 127

10. Fluorometric Assay for Detection of Sterol Oxidation in Liposomal Membranes
    Parkson Lee-Gau Chong and Michelle Olsher 145

11 Fluorescence Detection of Signs of Sterol Superlattice Formation in Lipid Membranes
    Parkson Lee-Gau Chong, Berenice Venegas, and Michelle Olsher 159

12 Differential Scanning Calorimetry in the Study of Lipid Phase Transitions in Model and Biological Membranes: Practical Considerations
    Ruthven N. A. H. Lewis, David A. Mannock, and Ronald N. McElhaney 171

13 Pressure Perturbation Calorimetry
    P. D. Heiko Heerklotz 197

14 Fourier Transform Infrared Spectroscopy in the Study of Lipid Phase Transitions in Model and Biological Membranes: Practical Considerations
    Ruthven N. A. H. Lewis and Ronald N. McElhaney 207

15 Optical Dynamometry to Study Phase Transitions in Lipid Membranes
    Rumiana Dimova and Bernard Pouligny 227

16 Fluorescence Assays for Measuring Fatty Acid Binding and Transport Through Membranes
    Kellen Brunaldi, Jeffrey R. Simard, Frits Kamp, Charu Rewal, Tanong Asawakarn, Paul O'Shea, and James A. Hamilton 237

17 Measurement of Lateral Diffusion Rates in Membranes by Pulsed Magnetic Field Gradient, Magic Angle Spinning-Proton Nuclear Magnetic Resonance
    Klaus Gawrisch and Holly C. Gaede 257

18 Using Fluorescence Recovery After Photobleaching to Measure Lipid Diffusion in Membranes
    Conrad W. Mullineaux and Helmut Kirchhoff 267

19 Single-Molecule Fluorescence Microscopy to Determine Phospholipid Lateral Diffusion
    Michael J. Murcia, Sumit Garg, and Christoph A. Naumann 277

20 Modeling 2D and 3D Diffusion
    Michael J. Saxton 295

21 Measurement of Water and Solute Permeability by Stopped-Flow Fluorimetry
    John C. Mathai and Mark L. Zeidel 323

22 Fluorescence Microscopy to Study Pressure Between Lipids
    in Giant Unilamellar Vesicles
    Anna Celli, Claudia Y. C. Lee, and Enrico Gratton 333

23 X-Ray Scattering and Solid-State Deuterium Nuclear Magnetic
    Resonance Probes of Structural Fluctuations in Lipid Membranes
    Horia I. Petrache and Michael F. Brown 341

24 Determination of Lipid Spontaneous Curvature from X-Ray
    Examinations of Inverted Hexagonal Phases
    Michael M. Kozlov 355

25 Shape Analysis of Giant Vesicles With Fluid Phase Coexistence
    by Laser Scanning Microscopy to Determine Curvature,
    Bending Elasticity, and Line Tension
    **Samuel T. Hess, Manasa V. Gudheti, Michael Mlodzianoski,
    and Tobias Baumgart** 367

26 Laser Tweezer Deformation of Giant Unilamellar Vesicles
    Cory Poole and Wolfgang Losert 389

27 Measurement of Lipid Forces by X-Ray Diffraction and Osmotic Stress
    **Horia I. Petrache, Daniel Harries,
    and V. Adrian Parsegian** 405

28 Micropipet Aspiration for Measuring Elastic Properties
    of Lipid Bilayers
    Marjorie L. Longo and Hung V. Ly 421

29 Langmuir Films to Determine Lateral Surface Pressure
    on Lipid Segregation
    Antonio Cruz and Jesús Pérez-Gil 439

30 Detergent and Detergent-Free Methods to Define Lipid Rafts
    and Caveolae
    Rennolds S. Ostrom and Xiaoqiu Liu 459

31 Near-field Scanning Optical Microscopy to Identify Membrane
    Microdomains
    Anatoli Ianoul and Linda J. Johnston 469

32 Fluorescence Microscopy to Study Domains
    in Supported Lipid Bilayers
    Jonathan M. Crane and Lukas K. Tamm 481

33 Fluorescence Resonance Energy Transfer to Characterize
    Cholesterol-Induced Domains
    Luís M. S. Loura and Manuel Prieto 489

34 Lipid Domains in Supported Lipid Bilayer for Atomic Force Microscopy
Wan-Chen Lin, Craig D. Blanchette, Timothy V. Ratto, and Marjorie L. Longo 503

35 Nuclear Magnetic Resonance Structural Studies of Membrane Proteins in Micelles and Bilayers
Xiao-Min Gong, Carla M. Franzin, Khang Thai, Jinghua Yu, and Francesca M. Marassi 515

36 Laurdan Studies of Membrane Lipid-Nicotinic Acetylcholine Receptor Protein Interactions
Silvia S. Antollini and Francisco J. Barrantes 531

37 Single-Molecule Methods for Monitoring Changes in Bilayer Elastic Properties
Olaf S. Andersen, Michael J. Bruno, Haiyan Sun, and Roger E. Koeppe II 543

38 Ion-Channel Reconstitution
Francisco J. Morera, Guillermo Vargas, Carlos González, Eduardo Rosenmann, and Ramon Latorre 571

39 The Use of Differential Scanning Calorimetry to Study Drug-Membrane Interactions
Thomas M. Mavromoustakos 587

40 Atomic Force Microscopy to Study Interacting Forces in Phospholipid Bilayers Containing General Anesthetics
Zoya V. Leonenko, Eric Finot, and David T. Cramb 601

Index 611

Contributors

OLAF S. ANDERSEN • *Department of Physiology and Biophysics, Weill Medical College of Cornell University, New York, NY, USA*

SILVIA S. ANTOLLINI • *Instituto de Investigaciones Bioquímicas de Bahía Blanca, Bahía Blanca, Argentina*

TANONG ASAWAKARN • *Faculty of Veterinary Science, Chulalongkorn University, Bangkok, Thailand, and Cell Biophysics Group, School of Biology, University of Nottingham, Nottingham, UK*

FRANCISCO J. BARRANTES • *Instituto de Investigaciones Bioquímicas de Bahía Blanca, Bahía Blanca, Argentina*

TOBIAS BAUMGART • *Chemistry Department, University of Pennsylvania, Philadelphia, PA, USA*

CRAIG D. BLANCHETTE • *Biophysics Graduate Group, Division of Biological Sciences, University of California, Davis, CA, USA, and Molecular Biophysics and Functional Nanostructures Group, Chemistry and Materials Science, Lawrence Livermore National Laboratory, Livermore, CA, USA*

MICHAEL F. BROWN • *Departments of Chemistry and Physics, University of Arizona, Tucson, AZ, USA*

KELLEN BRUNALDI • *Departments of Physiology and Biophysics, and of Pharmacology and Experimental Therapeutics, Boston University School of Medicine, Boston, MA, USA*

MICHAEL J. BRUNO • *Department of Physiology and Biophysics, and Graduate Program in Biochemistry and Structural Biology, Weill Medical College of Cornell University, New York, NY, USA*

ANNA CELLI • *Laboratory of Fluorescence Dynamics Department of Physics, University of Illinois, Urbana-Champaign, IL, USA*

PARKSON L-G. CHONG • *Department of Biochemistry, Temple University School of Medicine, Philadelphia, PA, USA*

DAVID T. CRAMB • *Department of Chemistry, University of Calgary, Calgary, Alberta, Canada*

JONATHAN M. CRANE • *Department of Molecular Physiology and Biological Physics, University of Virginia, Charlottesville, VA, USA*

ANTONIO CRUZ • *Departamento de Bioquimica, Facultad de Biologia, Universidad Complutense, Madrid, Spain*

RUMIANA DIMOVA • *Max Planck Institute of Colloids and Interfaces, Potsdam, Germany*

ALEX M. DOPICO • *Department of Pharmacology, The University of Tennessee Health Science Center, Memphis, TN, USA*

RICHARD M. EPAND • *Department of Biochemistry and Biomedical Sciences, McMaster University, Hamilton, Ontario, Canada*

SCOTT E. FELLER • *Department of Chemistry, Wabash College, Crawfordsville, IN, USA*

ERIC FINOT • *Department of Chemistry, University of Calgary, Calgary, Alberta, Canada*

CARLA M. FRANZIN • *The Burnham Institute, La Jolla, CA, USA*

HOLLY C. GAEDE • *Laboratory of Membrane Biochemistry and Biophysics, NIAAA, National Institutes of Health, Bethesda, MD, USA*

SUMIT GARG • *Department of Chemistry and Chemical Biology, Indiana University-Purdue University, Indianapolis, IN, USA*

KLAUS GAWRISCH • *Laboratory of Membrane Biochemistry and Biophysics, NIAAA, National Institutes of Health, Bethesda, MD, USA*

XIAO-MIN GONG • *The Burnham Institute, La Jolla, CA, USA*

CARLOS GONZÁLEZ • *Centro de Estudios Cientificos, Valdivia, Chile, and NINDS, National Institutes of Health, Bethesda, MD, USA*

ENRICO GRATTON • *Laboratory of Fluorescence Dynamics, Department of Physics, University of Illinois, Urbana-Champaign, IL, USA*

MANASA V. GUDHETI • *Chemistry Department, University of Pennsylvania, Philadelphia, PA, USA*

JAMES A. HAMILTON • *Department of Physiology and Biophysics, Boston University School of Medicine, Boston, MA, USA*

DANIEL HARRIES • *Laboratory of Physical and Structural Biology, NICHHD, National Institutes of Health, Bethesda, MD, USA*

HEIKO HEERKLOTZ • *Division of Biophysical Chemistry, Biozentrum of the University of Basel, Basel, Switzerland*

SAMUEL T. HESS • *Chemistry Department, University of Pennsylvania, PA, USA*

ANATOLI IANOUL • *Department of Chemistry, Carleton University, Ottawa, Ontario, Canada*

LINDA J. JOHNSTON • *Steacie Institute for Molecular Sciences, National Research Council Canada, Ottawa, Ontario, Canada*

PÁL JEDLOVSZKY • *Department of Colloid Chemistry, Eötvös Loránd University, Budapest, Hungary*

FRITS KAMP • *Ludwig Maximilian University, Munich, Germany*

HELMUT KIRCHHOFF • *Institut für Botanik, Westfälische-Wilhelms-Universität Műnster, Műnster, Germany*

ROGER E. KOEPPE II • *Department of Chemistry and Biochemistry, University of Arkansas, Fayetteville, AR, USA*

Contributors

MICHAEL M. KOZLOV • *Department of Physiology and Pharmacology, Sackler Faculty of Medicine, Tel Aviv University, Tel Aviv, Israel*

RAMON LATORRE • *Centro de Estudios Científicos, Valdivia, Chile*

CLAUDIA Y. C. LEE • *Laboratory of Fluorescence Dynamics, Department of Physics, University of Illinois, Urbana-Champaign, IL, USA*

ZOYA V. LEONENKO • *Department of Chemistry, University of Calgary, Calgary, Alberta, Canada*

RUTHVEN N. A. H. LEWIS • *Department of Biochemistry, University of Alberta, Alberta, Canada*

WAN-CHEN LIN • *Biophysics Graduate Group, Division of Biological Sciences, University of California, Davis, CA, USA*

XIAOQIU LIU • *Department of Pharmacology and the Vascular Biology Center of Excellence, The University of Tennessee Health Science Center, Memphis, TN, USA*

MARJORIE L. LONGO • *Department of Chemical Engineering and Materials Science, University of California, Davis, CA, USA*

WOLFGANG LOSERT • *Institute for Research in Electronics and Applied Physics, Institute for Physical Science and Technology, Department of Physics, University of Maryland, College Park, MD, USA*

HUNG V. LY • *Department of Chemical Engineering and Materials Science, University of California, Davis, CA, USA*

LUÍS MIGUEL SANTOS LOURA • *Centro de Química-Física Molecular, Instituto Superior Técnico Lisboa, Portugal, and Centro de Química and Departamento de Química, Universidade de Évora, Évora, Portugal*

DAVID A. MANNOCK • *Department of Biochemistry, University of Alberta, Alberta, Canada*

FRANCESCA M. MARASSI • *The Burnham Institute, La Jolla, CA, USA*

JOHN C. MATHAI • *Renal-Electrolyte Division, Department of Medicine, University of Pittsburgh School of Medicine, Pittsburgh, PA, USA*

THOMAS MAVROMOUSTAKOS • *Institute of Organic and Pharmaceutical Chemistry, National Hellenic Research Foundation, Athens, Greece*

RONALD N. MCELHANEY • *Department of Biochemistry, University of Alberta, Alberta, Canada*

MIHALY MEZEI • *Department of Physiology and Biophysics, Mount Sinai School of Medicine, New York University, New York, NY, USA*

MICHAEL MLODZIANOSKI • *Chemistry Department, University of Pennsylvania, Philadelphia, PA, USA*

FRANCISCO J. MORERA • *Centro de Estudios Cientificos, Valdivia, Chile, and Universidad Austral de Chile, Valdivia, Chile*

MICHIEL MÜLLER • *Biophysics and Microscopy Group, Swammerdam Institute for Life Sciences, University of Amsterdam, Amsterdam, The Netherlands*

CONRAD W. MULLINEAUX • *School of Biological and Chemical Sciences, Queen Mary, University of London, London, UK*

MICHAEL J. MURCIA • *Department of Chemistry and Chemical Biology, Indiana University-Purdue University Indianapolis, IN, USA*

CHRISTOPH A. NAUMANN • *Department of Chemistry and Chemical Biology, Indiana University-Purdue University, Indianapolis, IN, USA*

MICHELLE OLSHER • *Department of Biochemistry, Temple University School of Medicine, Philadelphia, PA, USA*

PAUL O'SHEA • *Cell Biophysics Group, School of Biology, University of Nottingham, Nottingham, UK*

RENNOLDS S. OSTROM • *Department of Pharmacology and the Vascular Biology Center of Excellence, The University of Tennessee Health Science Center, Memphis, TN, USA*

V. ADRIAN PARSEGIAN • *Laboratory of Physical and Structural Biology, NICHHD, National Institutes of Health, Bethesda, MD, USA*

JESÚS PÉREZ-GIL • *Departamento de Bioquímica, Facultad de Biología, Universidad Complutense, Madrid, Spain*

HORIA I. PETRACHE • *Department of Physics, Indiana University-Purdue University, Indianapolis, IN, USA, and Laboratory of Physical and Structural Biology, NICHHD, National Institutes of Health, Bethesda, MD, USA*

CORY POOLE • *Institute for Research in Electronics and Applied Physics, Institute for Physical Science and Technology, Department of Physics, University of Maryland, College Park, MD, USA*

BERNARD POULIGNY • *Centre de recherche Paul-Pascal, CNRS, France*

MANUEL JOSÉ ESTEVEZ PRIETO • *Centro de Química-Física Molecular, Instituto Superior Técnico, Lisboa, Portugal*

TIMOTHY V. RATTO • *Molecular Biophysics and Functional Nanostructures Group, Chemistry and Materials Science, Lawrence Livermore National Laboratory, Livermore, CA, USA*

CHARU REWAL • *Department of Physiology and Biophysics, Boston University School of Medicine, Boston, MA, USA*

HILDE A. RINIA • *Biophysics and Microscopy Group, Swammerdam Institute for Life Sciences, University of Amsterdam, Amsterdam, The Netherlands*

EDUARDO ROSENMANN • *Centro de Estudios Científicos, Valdivia, Chile*

MICHAEL J. SAXTON • *Department of Biochemistry and Molecular Medicine, University of California, Davis, CA, USA*

JEFFREY R. SIMARD • *Departments of Physiology and Biophysics and of Pharmacology and Experimental Therapeutics, Boston University School of Medicine, Boston, MA, USA*

OLIVIER SOUBIAS • *Laboratory of Membrane Biochemistry and Biophysics, NIAAA, National Institutes of Health, Bethesda, MD, USA*
EDWARD STERNIN • *Department of Physics, Brock University, St. Catharines, Ontario, Canada*
HAIYAN SUN • *Department of Chemistry and Biochemistry, University of Arkansas, Fayetteville, AR, USA. Current affiliation: Departments of Physiology and Neuroscience, Johns Hopkins University School of Medicine, Baltimore, MD, USA*
LUKAS K. TAMM • *Department of Molecular Physiology and Biological Physics, University of Virginia, Charlottesville, VA, USA*
KHANG THAI • *The Burnham Institute, La Jolla, CA, USA*
GABOR J. TIGYI • *Department of Physiology, The University of Tennessee Health Science Center, Memphis, TN, USA*
STEPHANIE TRISTRAM-NAGLE • *Biological Physics Group, Physics Department, Carnegie Mellon University, Pittsburgh, PA, USA*
GUILLERMO VARGAS • *Centro de Estudios Científicos, Valdivia, Chile*
BERENICE VENEGAS • *Department of Biochemistry, Temple University School of Medicine, Philadelphia, PA, USA*
GEORGE W. H. WURPEL • *FOM Institute for Atomic and Molecular Physics (AMOLF), Amsterdam, The Netherlands*
JINGHUA YU • *The Burnham Institute, La Jolla, CA, USA*
MARK L. ZEIDEL • *Department of Medicine, Beth Israel Deaconess Medical Center and Harvard Medical School, Boston, MA, USA*

Color Plates

Color plates follow p. 294

COLOR PLATE 1   Modeled structure of ADIFAB with **(left)** and without **(right)** FA bound *(26)*. (Fig. 1, Chapter 3; *see* complete caption on p. 28.)

COLOR PLATE 2   Fluorescence assays utilizing the ADIFAB probe to detect unbound concentrations of FA. (Fig. 2, Chapter 3; *see* complete caption on p. 31.)

COLOR PLATE 3   Effects of various buffers on ADIFAB titration curves. (Fig. 6, Chapter 3; *see* complete caption on p. 37.)

COLOR PLATE 4   Fluorescence approaches developed to study the mechanism of FA transport in membranes. (Fig. 1, Chapter 16; *see* complete caption on p. 238.)

COLOR PLATE 5   The effect of external pyranine on measurements of FA diffusion through membranes. (Fig. 10, Chapter 16; *see* complete caption on p. 252.)

1

A Glance at the Structural and Functional Diversity of Membrane Lipids

Alex M. Dopico and Gabor J. Tigyi

Summary

In the postgenomic era, spatially and temporally regulated molecular interactions as signals are beginning to take center stage in the understanding of fundamental biological events. For years, reductionism derived from the "fluid mosaic" model of the cell membrane has portrayed membrane lipids as rather passive molecules that, whereas separating biologically relevant aqueous phases, provided an environment so that membrane proteins could fulfill the specificity and selectivity required for proper cell signaling. Whereas these roles for membrane lipids still stand, the structural diversity of lipids and their complex arrangement in supramolecular assemblies have expanded such limited, although fundamental roles. Growing developments in the field of membrane lipids help to understand biological phenomena at the nanoscale domain, and reveal this heterogeneous group of organic compounds as a long underestimated group of key regulatory molecules. In this introductory chapter, brief overviews of the structural diversity of membrane lipids, the impact of different lipids on membrane properties, the vertical organization of lipids into rafts and caveolae, and the functional role of lipids as mediators of inter- and intracellular signals are provided. Any comprehensive review on membrane lipids, whether emphasizing structural or functional aspects, will require several volumes. The purpose of this chapter is to provide both introduction and rationale for the selection of topics that lie ahead in this book. For this reason, the list of references primarily includes reviews on particular issues dealing with membrane lipids wherein the reader can find further references.

Key Words: Caveolae; compressibility; entropy; leaflet asymmetry; lipid rafts; lipid signaling; membrane lipids; phase separation; spontaneous curvature.

1. Structural Diversity of Membrane Lipids

In contrast to other major classes of biologically relevant organic molecules (nucleotides, amino acids, and carbohydrates), which can be grouped around a clear-cut chemical definition, the term *lipid* engulfs a vast number of molecules that merely share "a much better solubility in organic solvents than in water" *(1–3)*. It might be argued that almost all lipids ultimately are "biological amphiphiles" *(3)*, largely resulting from the addition of polar groups to hydrocarbon chains. However, the polar regions and, thus, the amphiphilic character of some molecules like triglycerides, have in practical terms a negligible effect in attenuating the hydrophobicity of the hydrocarbon chains. The strong hydrophobicity of triglycerides sharply contrasts with the amphiphilic nature of inositol triphosphate (IP_3), a lipid molecule with multiple negative charges at physiological pH. In brief, the rather imprecise chemical definition that was coined with the term "lipids" reveals little but structural heterogeneity.

When applied to lipids, the constraint "membrane" does not do much to limit structural heterogeneity, and the number of lipid species and corresponding functions that are being discovered in biological membranes keep increasing. As examples, let us just mention that about 100 lipid species exist in the very simple red blood cell, and more than 600 lipid species are considered part of most biological membranes *(4–6)*.

Structural heterogeneity of lipid molecules is the foundation of membrane lipid diversity. The latter, however, is buttressed by the uneven distribution of lipid species spatially as well as temporally, which results in complex multidimensional organization of membrane leaflets. Let us consider for a moment the differential steady-state composition and distribution of lipid species across natural membranes. To begin with, lipids are differentially distributed across bilayer leaflets, with preferred location of negatively charged lipids, such as phosphatidylserine (PS) and phosphatidylinositol in the inner leaflet. In contrast, sphingomyelin is much more abundant in the outer leaflet *(6–8)*. A differential lipid composition across leaflets of naturally occurring bilayers is essential for distinct lipid species to be able to interact with different components of cell membranes: glycocalix and protein receptor sites in the outer side, and cytoskeletal, anchoring, and internal signaling molecules in the inner side of the membrane; for example, the membrane translocation of protein kinase C (PKC) requires specific recognition of negatively charged phospholipids, with PS and phosphatidic acid being usually involved *(9)*.

On the other hand, differential lipid composition across membrane leaflets leads to different *physical* properties across leaflets. For example, the predominance of negatively charged lipid species in the inner leaflet leads to a high density of negative charge in the cytosolic side of cell membranes (~0.6 Q/nm^2), which in the absence of significant counterions, leads to a surface potential. The surface potential affects local ion concentration and, thus, membrane protein conformation and function (of ion channels and pumps, in particular). Notably, this surface potential is critically controlled by local variations in membrane lipids *(10)*. Moreover, this type of membrane polarity may change temporally during specific cell processes; for example, PS is exposed to the cell membrane outer leaflet during sperm cell capacitation *(11)*.

The physicochemical asymmetry existing across membrane leaflets is critical to modulate specific functions performed by asymmetrically oriented domains/functional units in membrane-associated proteins (e.g., adenylate cyclase *[12]*, and most ion-channel proteins). The critical role of physicochemical asymmetry across leaflets in maintaining cell life requires tight modulation. A change in lipid polarity and asymmetry is a hallmark of apoptotic death. Indeed, the increase of PS in the outer leaflet can be used to determine that a cell is entering apoptosis *(13,14)*. Given the wide range of cell functions that depend on lipid asymmetry, it is not surprising that cells use a variety of enzymes (flippases, floppases, and scramblases) *(15,16)* to actively maintain lipid asymmetry across membrane leaflets.

In addition to lipid asymmetry across bilayer leaflets, complexity in membrane lipid distribution is provided by the existence of vertical lipid domains. These domains, spanning either one or both bilayer leaflets, are characterized by the predominance of distinct lipid species, resulting in a local lipid composition different from that of the "bulk" bilayer (*see* **Subheading 3.**).

Membrane lipid distribution and composition vary among kingdoms and species. To illustrate this point, hopanoids and other terpenes are particularly rich in the cytoplasmic membranes of some bacteria, whereas sterols are generally absent in prokaryotes *(17,18)*. Within eukaryotes, further, differences can be found: cholesterol and sitosterol are the primary steroids in membranes of animals and plants, respectively. Even within a species, membrane lipid composition varies across organelles, with the cell membrane being abundant in cholesterol (15–50% of total lipid) and internal organelle membranes being scarce in this steroid (e.g., ≤5% in mitochondrial membranes) *(6)*. Lipid composition heterogeneity across organelles is furthered by differential distribution of proteins involved in vectorial transfer of specific lipids across organelles *(19)*; a well-studied example is the vectorial transport of different ceramides along the endoplasmic reticulum, Golgi apparatus, and the cell membrane *(20)*. Finally, membrane lipid composition and distribution can be influenced by environmental factors and/or disease. For example, changes in n-3 and n-6 polyunsaturated fatty acids in the diet are reflected in the acyl chain composition of membranes in higher animals *(21)*.

2. Some Common Themes—Lipid Assemblies

In spite of the astonishing diversity of membrane lipids highlighted in the previous section, a few generalizations can be made. All membrane lipid molecules interact with each other by means of relatively weak chemical forces *(22–24)*. With the exception of "lipid adducts," in which lipids are covalently bound to proteins (e.g., sterol–Hedgehog interaction, isoprenylation, palmitoylation, myristoylation, etc.) *(25–28)*, lipids also interact with membrane proteins through weak chemical forces *(24,29–31)*. Weak forces, together with the high flexibility of hydrocarbon chains in glycero- and sphingolipids, characteristically provide the properties of soft matter to the bilayer that results from lipid self-assembly. It is this combination of structural heterogeneity and interacting weak forces among lipid molecules that makes entropy play a key role in membrane structure and function. Given the relationship between entropy and temperature, modifications of the latter might lead to *trans–gauche* C–C bond interconversions in the acyl chains of bilayer lipids *(32)*, rendering another possible source of variability in lipid structure. Indeed, only lipids with a limited degree of disorder in terms of range and rate of movement and, thus, a given "effective shape," will be able to self-assemble into the largely flat bilayer, two-dimensional structure (lamellar phase) that prevails in natural membranes *(18,24,33)*. The determinant role of molecular structure in lipid self-assembly, phases, and polymorphism is discussed in Chapter 2 of this volume. Moreover, it is the combination of structural heterogeneity, weak interacting forces, and "limited disorder" what allows inserted peptides and proteins to modify the physicochemical environment of surrounding lipids *(34–38)*. Conversely, physicochemical properties determined by the lipid microenvironment composition determine structure and function of associated peptides/proteins *(39–42)*. The bilayer physicochemical environment is particularly important in modulating the function of embedded, integral proteins. For example, integral protein conformation and function have been demonstrated to be controlled by bilayer curvature-derived stress, thickness, and lateral pressure profile (*see* below under this **Subheading**). Bilayer thickness forces the hydrophobic spanning region of the protein to match the length of the hydrophobic region of the bilayer; the lipid species that provides a better hydrophobic match will likely prevail at the protein–lipid interface (although hydrophobic mismatch is possible) *(31,36,43–45)*.

Lipid fractions in close vicinity with bilayer-spanning proteins/peptides may show dynamics much slower than those of the bulk lipids, leading to the concept of "vicinal lipids" or "lipid annulus," a relevant site for pharmacological modulation of membrane protein function *(46,47)*. Thus, embedded proteins will "recruit" certain lipids, and "repel" others altogether, leading to membrane lipid sorting. Chemical diversity of lipids does allow for a wide variety of choices and adjustments, to exquisitely match the properties of the individual membrane proteins.

Hydrophobic matching coupled with membrane polarity (and associated changes in bilayer curvature) modulates the activity of integral membrane proteins, a classical example given by ion-channel proteins/peptides. Simple channel-forming peptides (gramicidin A and alamethicin in particular), have served as excellent models to understand how changes in the physicochemical nature of lipids surrounding the ion-channel affect distinct aspects of channel function such as gating and permeation *(48–51)*. Moreover, modification of channel function through changes *in the physical properties of the channel lipid microenvironment* has been increasingly recognized for a variety of complex, multisubunit channel-forming proteins, such as nicotinic acetylcholine and N-methyl-D-aspartate receptors, volume-activated Cl^-, voltage-gated Ca^{2+}, voltage-gated K^+, and large-conductance, Ca^{2+}-dependent K^+ channels, and connexons *(46,50,52–56)*.

Another common feature is that most naturally occurring membranes contain nonlamellae-forming lipids (disregarding structural particulars). These lipids may be required to stabilize functionally relevant conformational states of some proteins *(57–59)*. In addition, shifts between nonlamellar and lamellar structures, often occurring with changes in bilayer curvature (*see* Chapter 2 in this volume and **refs.** *33* and *60*), are involved in defined physiological processes, such as membrane fusion *(59,61,62)*. The fusogenic behavior of a bilayer that depends on the equilibrium between the lamellar and the hexagonal phases can be decreased by the so-called bilayer stabilizers, which include steroids (e.g., cholesterol sulfate) and drugs (e.g., some antivirals). The activity of a variety of membrane-bound enzymes, PKC in particular, is usually decreased with an increased amount of bilayer stabilizer in the biological membrane *(63)*.

As mentioned earlier, the existence of a "limited disorder" and the characteristics of soft matter that most lipids provide to the bilayer are essential for membrane structure and function. This commonality in natural membranes is related to the particular condition (phase) in which the vast majority of natural lipid bilayers exist, i.e., the "liquid-ordered" lipid phase *(18,33)*. In general, a longitudinal axis (more or less perpendicular to the bilayer plane) can be recognized in bilayer-forming lipid species, rendering another source of structural variation and level of organization based on whether the direction of these longitudinal axes is more or less ordered. Thus, different phases, termed "mesophases," exist between liquids and solids *(64,65)*. In eukaryotic membranes, it is largely up to cholesterol to stretch and order the hydrocarbon chains of glycero- and sphingolipids, leading to the liquid-ordered phase *(18,33)*. In simple terms, this phase is not "disordered enough" to allow ions and other small molecules to easily leak across the lipid bilayer, yet it is not "ordered enough" to prevent lipid and protein movements (translation in particular) within the bilayer. Both the separation of distinct ionic media by natural membranes, and the movement within the lipid bilayer of lipid species and membrane proteins involved in signaling are essential for life as we know it.

Whereas the different types of movements performed and allowed by membrane lipids and the methods to study them will be considered later within specific chapters, let us just mention in this introductory chapter one practical example of studying lipid mobility. The construction of membrane protein microarrays, of paramount importance for human health, as membrane proteins represent the majority of drug targets, is dependent on an understanding of protein diffusion, and its control by the surrounding lipid lateral fluidity and movement *(66)*. The fact that membrane lipids differ in shape not only makes lipid phases possible, but also allow lipid monolayers acquire spontaneous curvature, another common feature in naturally occurring lipid bilayers. However, under identical conditions (pH, ionic strength, etc.) if both monolayers have the same lipid composition, the bilayer will have no net tendency to curve. Thus, a most obvious mechanism to induce bilayer curvature is provided by differential lipid composition between the two membrane leaflets. This seems to be the general rule, not the exception, as considered in the **Subheading 1**.

Finally, variations in both rate and range of acyl/alkyl chain motion, and the existence of different lipid effective shapes, are not only responsible for the existence of lipid phases but also lipid polymorphism. Polymorphism and the behavior of lipids in water will be considered in Chapter 2 (*see also* **refs. 5,60**). For now, let us just mention that the liquid interface of the lipid bilayer allows bilayer bending and compression/expansion, another common feature of naturally occurring membranes. Yet again, it is lipid heterogeneity (in this case primarily through variations in the acyl chain composition of membrane glycerolipids) what exerts a powerful force controlling bilayer rigidity and compressibility *(18,33,67)*. Moreover, increasing the abundance of a single lipid species in a natural membrane may dramatically alter its physicochemical properties. For example, increasing cholesterol in animal membranes increases both bending rigidity and compressibility area of their constituting lipid bilayers *(18,68)*.

Control of bilayer rigidity and/or compressibility is required as these bilayer properties need to adjust to perform distinct physiological tasks in different cells/membrane domains within the cell membrane. For example, a "soft" membrane is critical for red blood cells to permeate capillaries having an internal diameter narrower than the erythrocyte diameter *(69)*. In addition, because lipids in bilayers form "a two-dimensional liquid interface in a three-dimensional space," the shape of the interface is also controlled by the Gaussian curvature modulus *(33,70)*. It should be clear to the reader that in spite of their softness, disorder, and diversity of constituent lipid species, *natural lipid bilayers are characterized by a stratified design*, with a layer of kinetically restricted water, an "amphiphilic region" (polar head groups of sphingo- and glycerolipids and their glycerol backbone), a relatively ordered acyl chain segment, and a hydrophobic core where disordered acyl chain segments prevail *(71,72)*. This organization keeps acyl chains away from the aqueous medium. It also generates a "lateral pressure profile," as pressures across the bilayer are unevenly distributed: repulsion (positive pressure) prevails between polar headgroups, interfacial tension (negative pressure) generates in the amphiphilic region, and repulsion prevails again between disordered acyl chains. This lateral pressure profile largely determines the spontaneous curvature of the bilayer and modifies membrane protein conformation and function *(73,74)*.

Modification of bilayer pressure profile induced by short-chain alkanols and other "interfacially active compounds" appears to be a major contributor to the action of these drugs on membrane protein function *(75)*. In a more general context, amphiphilic hormones (e.g., steroids), long chain alcohols, and other drugs can easily partition within lipid interfaces and also shift the equilibrium from one lipid phase to another. By doing so, these interfacially active compounds may modify conformation and function of membrane proteins that distribute differentially across lipid domains.

3. Vertical Domains, Rafts, and Caveolae

As found with other molecules, lipids that are alike attract each other more strongly than lipid molecules of different chemical nature *(76)*, which leads to lipid phase separation. Indeed, the chemical heterogeneity of membrane lipid species multiplies the possibility of local phase separation within the bilayer. Thus, membrane lipids are laterally organized in the bilayer plane in a nonrandom fashion, which leads to the existence of "lipid domains" (whether horizontal or vertical) in cell membranes *(77–80)*. However, given the movement of lipids in a given phase, it should be stressed that these lipid domains always carry a temporal constraint relative to the lipid in the "bulk" phase.

In addition to the differential chemical affinities among membrane lipids and local phase separation, membrane domains are favored by the bilayer insertion of peptides and proteins, in particular those associated with the cytoskeleton *(34,81–83)*. In addition, domain formation appears to be favored by endocytic lipid trafficking between the cell membrane and other organelles *(84)*. Notably, some ion channel populations and most membrane proteins involved in the recognition of neurotransmitters, hormones, or other messenger signals brought to the cell from the extracellular medium seem to be localized in special "vertical" domains existing in the cell membrane *(85–90)*. When compared with the rest of the membrane, these domains appear to be enriched in specific lipid species, with cholesterol and sphingolipids being highly abundant in the outer leaflet of the vertical domain *(77,78,80)*. These proteolipid membrane domains have received collectively the term of "rafts," yet discrepancies still exist on the specific chemistry that should define them, as well as their dimensions *(78,80,91,92)*. Furthermore, raft-like structures have been identified in lipid mixtures in the absence of peptide/protein. However, it is clear that domain (raft) formation leads to the segregation of components and dynamic compartmentalization of the cell membrane, which is critical for membrane-associated cell signaling. The dynamic, temporal nature of these vertical domains is underscored by the fact that "rafts" redistribute rapidly and even disappear in response to signaling molecules *(81)*. Defined vertical lipid domains in cell membranes have been demonstrated to directly control diffusion of molecules through and within the membrane, regulate membrane-associated protein activity *(34,88)*, and participate in physiological process that involve cell recognition of its environment, such as cell motility and adhesion *(81,93)*. Finally, the existence of lateral lipid domains in membranes necessarily modifies the physicochemical properties of the bulk bilayer, such as compressibility, bending, and permeability, as well as membrane lipid–protein interactions *(34,78,88,94)*.

The concept of "caveolae" is also subject of much debate. The simplest point of view states that caveolae result from the association of rafts and some "caveolin" proteins, caveolin-1 in particular. However, some investigators claim that caveolae may exist in the absence of caveolin proteins. Furthermore, whereas a flask-shaped invagination of the cell membrane was initially considered a defining theme of caveolae, flat membrane domains in the bilayer consisting of raft-like domains enriched in caveolin-type proteins might also be considered "caveolae" *(95–97)*. At any rate, it is abundantly clear that both lipid "rafts" and "caveolae" nucleate and integrate critical cell signaling molecules, such as growth-factor receptors, several G protein-coupled receptors, a wide variety of ion channels, protein and lipid kinases, and some other key enzymes. Colocalization of signaling molecules in distinct lipid domains likely facilitates the transduction of a message from one signaling molecule to the other (saving energy to the cell), and facilitates cross-talk among relevant signaling pathways.

Given their wide involvement in clustering and regulation of signaling molecules, it is not surprising that disruption of raft and/or caveolae may lead to disease, including muscular dystrophy, neoplasia-associated processes, and cardiovascular pathology *(98–100)*. Indeed, rafts and caveolae are increasingly receiving pharmacological and therapeutic attention. For example, caveolin-1 is known to bind and activate endothelial nitric oxide synthase (eNOS). In endothelial cells, eNOS makes NO, which is critical in vasodilation, angiogenesis, and wound-repair, among other physiological processes, making the caveolin-1/eNOS dyad a target of therapeutic intervention *(97,101)*. Caveolae may also be involved in the transport into and within the cell of diverse molecules including cholesterol and folate receptors. The possibility that caveolae could be used to incorporate and import into the cell therapeutic agents has begun to be actively explored in recent years *(97,102)*.

4. Membrane Lipid Function

Structural diversity necessarily brings functional diversity. Biological functions of lipids can be found in most biochemistry and cell biology textbooks (e.g., *see* **refs.** *1–3*), and comprehensive, recent reviews on membrane lipids are available *(33,103–110)*. Lipidologists have traditionally made the distinction between "structural" and "signaling" lipids. The main functions of the former (typical examples include triglycerides, phospholipids, and cholesterol) is to store energy or provide insulation between two conducting media (extracellular fluid and cytosol, or cytosol and intraorganelle fluid) and/or an appropriate environment for the functioning of embedded proteins. It is within this function as boundary (whether as part of a lipid bilayer or a mixed micelle) that the amphiphilic character of certain lipids becomes more obvious (i.e., interfacially active compounds). On one hand, hydration and desorption of membrane lipids appear to be critical in controlling bilayer organization and protein function. On the other hand, bilayer lipids are in position to "organize" near water molecules, which results in less dense and more structured water molecules than those of bulk water. This more structured net of water molecules help to control molecular movements, such as slowing down diffusion *(23,33)*.

The main function of "signaling lipids" (IP_3, steroid hormones, prostaglandins, sphingosine-1-phosphate, etc.) is to mediate a biological message brought by a hormone, paracrine regulator, or neurotransmitter from the cell surface to the interior of the cell/nucleus. However, the distinction between signaling and structural lipids has become increasingly blurred. To begin with, a given lipid type may serve as structural lipid in one body organ/tissue and as signaling lipid in another. A typical example of this duality is found in ceramides, which are the key impermeabilizing structure in the stratum corneum of the skin on one hand, and participate in apoptosis signaling on the other *(111,112)*.

In addition, typical "signaling" lipids whose action largely resides in the cytosol (e.g., prostaglandins, leukotrienes, and lysophospholipids) are usually enzymatic products of "structural" lipids residing in the cell membrane. Thus, it is the cell membrane with its "structural" elements (synthetizing enzymes and precursors lipids) and their dynamic lipid microenvironment, the level at which a cascade leading to signaling lipids is primarily regulated. Accordingly, the targeting of lipid cascades by the pharmaceutical industry shows no sign of slowing down. Starting with aspirin and prostaglandin synthetase inhibition, more recent research has given us the development of selective Cox-2 inhibitors to treat inflammation, pain, and fever *(97)*.

Moreover, membrane lipids that are precursors of signaling molecules, and thus, were classically considered as structural lipids, have been increasingly found to regulate biological processes by themselves. Phosphatidylinositol 4,5-bisphosphate (PIP_2) represents a notorious example of a membrane lipid long sought to merely represent a precursor of signaling lipids (IP_3 and diacylglycerol) that is now recognized as a key signaling molecule (for a review on PIP_2 as modulator of membrane ion-channel protein function, *see* **ref.** *113*). An even more striking example of the blurred limit between structural and signaling lipids is provided by PS. Long considered an structural membrane lipid, PS (and other negatively charged phospholipids) is critical in membrane translocation of PKC, and displays selective binding to MARCKS, GAP43, gravin, GRK5, ErbB family of proteins, N-methyl-D-aspartate receptors, scramblases, and others *(29)*.

The lack of a clear distinction between structural and functional lipids probably helped to originate terms such as "bioactive lipids" and the like. This terminology leads us to conclude that "biopassive lipids" and/or "inorganically active lipids" must exist, which does not seem to make much sense. Lysophospholipids received their name from the detergent (lysing) properties they possess, some at low and others at high micromolar concentrations. However, lysophospholipid receptors show binding affinities for these lipids that are in the low nanomolar range *(114)*. The best example for this is the platelet-activating factor receptor, which is activated by picomolar concentrations of the lipid *(115)*. Thus, once again, partial knowledge of membrane lipids gave rise to misleading terminology.

In synthesis, it is abundantly clear that the function of membrane lipids is not properly described by the concept of inert building blocks of membrane bilayers. On the contrary, these structurally heterogeneous molecules thrive as dynamic determinants of membrane function and precursors of extra- and intracellular mediators. Interestingly, in some cases the same product of a membrane lipid can function as a mediator on both sides of the lipid bilayer: for example, sphingosine-1-phosphate is a ligand of G protein-coupled receptors as well as a second messenger mediating Ca^{2+} release from the endoplasmic reticulum *(116)*. Moreover, lysophosphatidic acid plays triple functions: it is an intermediate of membrane phospholipid biosynthesis, a ligand of G protein-coupled receptors, and also a ligand of a lipid-regulated nuclear transcription factor (i.e., the peroxisome proliferator-activated receptor γ) *(114)*.

5. Conclusion

Membrane lipid heterogeneity was largely overlooked owing to the long acceptance of the fluid-mosaic model of membrane. This model, although widely successful, emphasizes that it is up to membrane proteins to provide structural heterogeneity and, consequently, specificity of function; membrane lipids are left to the passive role of a fluid ocean in which proteins float, move (mainly laterally) and, thus, perform their tasks. Together, the concepts of membrane lipid vertical domains, differential lipid movement in the vicinity of embedded membrane proteins/peptides, asymmetric distribution of lipids across membrane leaflets, heterogeneity in lipid macroscopic organization (polymorphism) and phases, and the existence of mono- and bilayer curvature (not to mention cytoskeletal-membrane lipid associations), as we know them today, are no longer compatible with the overall homogeneity and randomness of membrane lipids that are implied from the fluid-mosaic model.

It should be clear that the complexities of membrane lipid organization, movements, function, and interactions with other molecules critical for life, all recognize a common starting point: the structural heterogeneity of lipid molecules. This heterogeneity poses a problem in Lipidology, i.e., finding a clear-cut structurally based definition of "lipid," but at the same time also guarantees a bright future to lipid research.

Acknowledgments

This work was supported by National Institutes of Health (NIH) grants HL77424 and AA11560 Alejandro M. Dopico (AMD), and HL79004, HL61469, and CA92160 Gabor J. Tigyi (GJT).

References

1. Nelson, D. L. and Cox, M. M. (eds.) (2004) *Lehninger Principles of Biochemistry.* Freeman, New York, NY.
2. Alberts, B., Johnson, A., Lewis, J., Raff, M., Roberts, K., and Walter, P. (eds.) (2002) *Molecular Biology of the Cell.* Garland Science, New York, NY.
3. Smith, E. L., Hill, R. L., Lehman, I. R., Lefkowitz, R. J., Handler, P., and White, A. (eds.) (1983) *Principles of Biochemistry-General Aspects.* McGraw Hill Book Company, New York, NY.
4. Wolf, C. and Quinn, P. J. (2004) Membrane lipid homeostasis. *Subcell Biochem.* **37**, 317–357.
5. Dowhan, W. (1997) Molecular basis for membrane phospholipids diversity: why are there so many lipids? *Annu. Rev. Biochem.* **66**, 199–232.
6. Sackmann, E. (1995) Biological membranes. Architecture and function, in *Handbook of Biological Physics: Structure and dynamics of membranes. From cells to vesicles*, vol. 1 (Lipowsky, R. and Sackmann, E., eds.), Elsevier Science B. V., Amsterdam, pp. 1–63.
7. Bevers, E. M., Comfurius, P., Dekkers, D. W., Harmsma, M., and Zwaal, R. F. (1998) Transmembrane phospholipids distribution in blood cells: control mechanisms and pathophysiological significance. *Biol. Chem.* **379**, 973–986.
8. Op den Kamp, J. A. F. (1979) Lipid asymmetry in membranes. *Ann. Rev. Biochem.* **48**, 47–71.
9. Bolsover, S. R., Gómez, D., Fernández, J. C., and Corbalán-García, S. (2003) Role of the Ca^{2+}/phosphatidylserine binding region of the C2 domain in the translocation of protein kinase C alpha to the plasma membrane. *J. Biol. Chem.* **278**, 10,282–10,290.
10. Murray, D., Arbuzova, A., Hangyas-Mihalyne, G., et al. (1999) Electrostatic properties of membranes containing acidic lipids and adsorbed basic peptides: theory and experiment. *Biophys. J.* **77**, 3176–3188.
11. de Vries, K. J., Wiedmer, T., Sims, P. J., and Gadella, B. M. (2003) Caspase-independent exposure of aminophospholipids and tyrosine phosphorylation in bicarbonate responsive human sperm cells. *Biol. Reprod.* **68**, 2122–2134.
12. Stubbs, C. D. (1983) Membrane fluidity: structure and dynamics of membrane lipids. *Essays Biochem.* **19**, 1–39.
13. Williamson, P. and Schlegel, R. A. (2004) Hide and seek: the secret of the phostidylserine receptor. *J. Biol.* **3**, 14.
14. Kagan, V. E., Fabisiak, J. P., Shvedova, A. A., et al. (2000) Oxidative signaling pathway for externalization of plasma membrane phosphatidylserine during apoptosis. *FEBS Lett.* **477**, 1–7.
15. Daleke, D. L. (2003) Regulation of transbilayer plasma membrane phospholipid asymmetry. *J. Lipid Res.* **44**, 233–242.
16. Bevers, E. M., Comfurius, P., Dekkers, D. W., and Zwaal, R. F. (1999) Lipid translocation across the plasma membrane of mammalian cells. *Biochim. Biophys. Acta* **1439**, 317–330.
17. Sahm, H., Rohmer, M., Bringer-Meyer, S., Sprenger, G. A., and Welle, R. (1993) Biochemistry and physiology of hopanoids in bacteria. *Adv. Microb. Physiol.* **35**, 247–273.
18. Bloom M., Evans, E., and Mouritsen, O. G. (1991) Physical properties of the fluid lipid-bilayer component of cell membranes: a perspective. *Quart. Rev. Biophys.* **24**, 293–397.
19. Holthuis J. C. and Levine, T. P. (2005) Lipid traffic: floppy drives and a superhighway. *Nat. Rev.* **6**, 209–220.
20. Perry, R. J. and Ridgway, N. D. (2005) Molecular mechanisms and regulation of ceramide transport. *Biochim. Biophys. Acta* **1734**, 220–234.
21. Hulbert, A. J., Turner, N., Storlien, L. H., and Else, P. L. (2005) Dietary fats and membrane function: implications for metabolism and disease. *Biol. Rev. Cambridge Phil. Soc.* **80**, 155–169.
22. Mukherjee, S. and Maxfield, F. R. (2000) Role of membrane organization and membrane domains in endocytic lipid trafficking. *Traffic* **1**, 203–211.

23. Rand, R. P. and Parsegian, V. A. (1989) Hydration forces between phopholipid bilayers. *Biochim. Biophys. Acta* **988**, 371–376.
24. Israelachvili, J. N., Marčelja, S., and Horn, R. G. (1980) Physical principles of membrane organization. *Quar. Rev. Biophys.* **13**, 121–200.
25. Mann, R. K. and Beachy, P. A. (2004) Novel lipid modifications of secreted protein signals. *Annu. Rev. Biochem.* **73**, 891–923.
26. Resh, M. D. (2004) Membrane targeting of lipid modified signal transduction proteins. *Subcell Biochem.* **37**, 217–232.
27. Marsh, D., Horvath, L. I., Swamy, M. J., Mantripragada, S., and Kleinschmidt, J. H. (2002) Interactions of membrane-spanning proteins with peripheral and lipid-anchored membrane proteins: perspectives from protein-lipid interactions. *Mol. Membr. Biol.* **19**, 247–255.
28. Ingham, P. W. (2001) Hedgehog signaling: a tale of two lipids. *Science* **294**, 1879–1881.
29. McLaughlin, S., Hangyas-Mihlyne, G., Zaitseva, I., and Golebiewska, U. (2005) Reversible-through calmodulin-electrostatic interactions between basic residues on proteins and acidic lipids in the plasma membrane. *Biochem. Soc. Symp.* **72**, 189–198.
30. de Planque, M. R. and Killian, J. A. (2003) Protein-lipid interactions studied with designed transmembrane peptides: role of hydrophobic matching and interfacial anchoring. *Mol. Membr. Biol.* **20**, 271–284.
31. Lee, A. M. (2003) Lipid-protein interactions in biological membranes: a structural perspective. *Biochim. Biophys. Acta* **1612**, 1–40.
32. Seelig, J. and Seelig, A. (1980) Lipid conformation in model membranes and biological membranes. *Quart. Rev. Biophys.* **13**, 19–61.
33. Mouritsen, O. G. (ed.) (2005) *Life-As a matter of fact. The emerging science of lipidomics.* Springer-Verlag, Berlin, Heildelberg, Germany.
34. Epand, R. M. (2004) Do proteins facilitate the formation of cholesterol-rich domains? *Biochim. Biophys. Acta* **1666**, 227–238.
35. Marsh, D. and Pali, T. (2004) The protein-lipid interface: perspectives from magnetic resonance and crystal structures. *Biochim. Biophys. Acta* **1666**, 118–141.
36. Marsh, D. (2003) Lipid interactions with transmembrane proteins. *Cell Mol. Life Sci.* **60**, 1575–1580.
37. Bechinger, B. (2000) Biophysical investigations of membrane perturbations by polypeptides using solid-state NMR spectroscopy. *Mol. Membr. Biol.* **17**, 135–142.
38. Epand, R. M. and Epand, R. F. (2000) Modulation of membrane curvature by peptides. *Biopolymers* **55**, 358–363.
39. Kazlauskaite, J. and Pinherio, T. J. (2005) Aggregation and fibrillization of prions in lipid membranes. *Biochem. Soc. Symp.* **72**, 211, 212.
40. Tamm, L. K., Hong, H., and Liang, B. (2004) Folding and assembly of beta-barrel membrane proteins. *Biochim. Biophys. Acta* **1666**, 250–263.
41. Booth, P. J. and Curran, A. R. (1999) Membrane protein folding. *Curr. Opin. Struct. Biol.* **9**, 115–121.
42. Epand, R. M. and Lester, D. S. (1990) The role of membrane biophysical properties in the regulation of protein kinase C activity. *Trend Pharmacol. Sci.* **11**, 317–320.
43. Jensen, M. O. and Mouritsen, O. G. (2004) Lipids do influence protein function-the hydrophobic matching hypothesis revisited. *Biochim. Biophys. Acta* **1666**, 205–226.
44. de Planque, M. R. and Killian, J. A. (2003) Protein-lipid interactions studied with designed transmembrane peptides: role of hydrophobic matching and interfacial anchoring. *Mol. Membr. Biol.* **20**, 271–284.
45. Dumas, F., Lebrun, M. C., and Tocanne, J. F. (1999) Is the protein/lipid hydrophobic mismatch principle relevant to membrane organization and functions? *FEBS Lett.* **458**, 271–277.

46. Barrantes, F. J. (2002) Lipid matters: nicotinic acetylcholine receptor-lipid interactions. *Mol. Membr. Biol.* **19,** 277–284.
47. Horvath, L. I., Arias, H. R., Hankovszky, H. O., Hideg, K., Barrantes, F. J., and Marsh, D. (1990), Association of spin-labeled local anesthetics at the hydrophobic surface of acetylcholine receptor in native membranes from *Torpedo marmorata*. *Biochemistry* **29,** 8707–8713.
48. Andersen, O. S., Koeppe, R. E., 2nd., and Roux, B. (2005) Gramicidin channels. *IEEE Trans. Nanobiosci.* **4,** 10–20.
49. Aguillella, V. M. and Bezrukov, S. M. (2001) Alamethicin channel conductance modified by lipid charge. *Eur. Biophys. J.* **30,** 233–241.
50. Sansom, M. S. (1998) Models and simulations of ion channels and related membrane proteins. *Curr. Opin. Struct. Biol.* **8,** 237–244.
51. Bechinger, B. (1997) Structure and functions of channel-forming peptides: magainins, cecropins, melittin and alamethicin. *J. Membr. Biol.* **156,** 197–211.
52. Cascio, M. (2005) Connexins and their environment: effects of lipid composition on ion channels. *Biochim. Biophys. Acta* **1711,** 142–153.
53. Romanenko, V. G., Rothblat, G. H., and Levitan, I. (2004) Sensitivity of volume-regulated anion current to cholesterol structural analogues. *J. Gen. Physiol.* **123,** 77–87.
54. Crowley, J. J., Treistman, S. N. and Dopico A. M. (2003) Cholesterol antagonizes ethanol potentiation of human brain BK_{Ca} channels reconstituted into phospholipid bilayers. *Mol. Pharmacol.* **64,** 365–372.
55. Casado, M. and Ascher, P. (1998) Opposite modulation of NMDA receptors by lysophospholipids and arachidonic acid: common features with mechanosensitivity. *J. Physiol.* **513,** 317–330.
56. Lundbæk, J. A. and Andersen, O. S. (1994) Lysophospholipids modulate ion channel function by altering the mechanical properties of lipid bilayers. *J. Gen. Physiol.* **104,** 645–673.
57. Brown, M. F. (1997) Influence of nonlamellar-forming lipids on rhodopsin. *Curr. Topics Membr.* **44,** 285–356.
58. Epand, R. M. (1996) The properties and biological roles of non-lamellar forming lipids. *Chem. Phys. Lipids* **81,** 101–264.
59. Lohner, K. (1996) Is the high propensity of ethanolamine plasmalogens to form non-lamellar lipid structures manifested in the properties of biomembranes? *Chem. Phys. Lipids* **81,** 167–184.
60. Seddom, J. M. and Templer, R. H. (1995) Polymorphism of lipid-water systems, in *Handbook of Biological Physics: Structure and dynamics of membranes. From cells to vesicles, vol. 1* (Lipowsky, R. and Sackmann, E., eds.), Elsevier Science B. V., Amsterdam, pp. 97–160.
61. Cherezov, V., Siegel, D. P., Shaw, W., Burgess, S. W., and Caffrey, M. (2003) The kinetics of non-lamellar phase formation in DOPE-Me: relevance to biomembrane fusion. *J. Membr. Biol.* **195,** 165–182.
62. Hafez, I. M. and Cullis, P. R. (2001) Roles of lipid polymorphism in intracellular delivery. *Adv. Drug Deliv. Revs.* **47,** 139–148.
63. Epand, R. M. (1990) Relationship of phospholipids hexagonal phases to biological phenomena. *Biochem. Cell Biol.* **68,** 17–23.
64. Peng, J. B., Barnes, G. T., and Gentic, I. R. (2001) The structures of Langmuir-Blodgett films of fatty acids and their salts. *Adv. Colloid. Interface Sci.* **25,** 163–219.
65. Cevc, G. (ed.) (1993) *Phospholipids Handbook*. Marcel Dekker Inc., New York, NY.
66. Fang, Y., Frutos, A. G., and Lahiri, J. (2002) Membrane protein microarrays. *J. Am. Chem. Soc.* **124,** 2394–2395.
67. Nagle, J. F. and Tristram-Nagle, S. (2000) Structure of lipid bilayers. *Biochim. Biophys. Acta* **1469,** 159–195.

68. Cruzeiro-Hansson, L., Ipsen, J. H., and Mouritsen. (1989) Intrinsic molecules in lipid membranes change the lipid-domain interfacial area: cholesterol at domain interfaces. *Biochim. Biophys. Acta* **979**, 166–176.
69. Evans, E. A. (1989) Structure and deformation properties of red blood cells: concepts and quantitative methods. *Methods Enzymol.* **173**, 3–35.
70. Siegel, D. P. and Kozlov, M. M. (2004) The Gaussian curvature elastic modulus of N-monomethylated dioleoylphosphatidylethanolamine: relevance to membrane fusion and lipid phase behavior. *Biophys. J.* **87**, 366–374.
71. Vereb, G., Szollosi, J., Matko, J., et al. (2003) Dynamic, yet structured: the cell membrane three decades after the Singer-Nicolson model. *Proc. Natl. Acad Sci. USA* **100**, 8053–8058.
72. Tieleman, D. P., Marrink, S. J., and Berendsen, H. J. C. (1997) A computer perspective of membranes: molecular dynamics studies of lipid bilayer systems. *Biochim. Biophys. Acta* **1331**, 235–270.
73. Cantor, R. S. (1999) Lipid composition and the lateral pressure profile in bilayers. *Biophys. J.* **76**, 2625–2639.
74. Marsh, D. (1996) Lateral pressure in membranes. *Biochim. Biophys. Acta* **1286**, 183–223.
75. Cantor, R. S. (1997) The lateral pressure profile in membranes: a physical mechanism of anesthesia. *Biochemistry* **36**, 2339–2344.
76. Regen, S. L. (2002) Lipid-lipid recognition in fluid bilayers: solving the cholesterol mystery. *Curr. Opin. Chem. Biol.* **6**, 729–735.
77. Deveaux, P. F. and Morris, R. (2004) Transmembrane asymmetry and lateral domains in biological membranes. *Traffic* **5**, 241–246.
78. Simmons, K. and Vaz, W. L. C. (2004) Model systems, lipid rafts, and cell membranes. *Annu. Rev. Biophys. Biomol. Struct.* **33**, 269–295.
79. Vereb, G., Szollosi, J., Matko, J., et al. (2003) Dynamic, yet structured: the cell membrane three decades after the Singer-Nicolson model. *Proc. Natl. Acad. Sci. USA* **100**, 8053–8058.
80. Brown, D. A. and London, E. (1998) Structure and origin of ordered lipid domains in membranes. *J. Membr. Biol.* **164**, 103–114.
81. Golub, T. and Pico, C. (2005) Spatial control of actin-based motility through plasmalemmal PtdIns(4,5)P2-rich raft assemblies. *Biochem. Soc. Symp.* **72**, 119–127.
82. Meiri, K. F. (2004) Membrane/cytoskeletal communication. *Subcell Biochem.* **37**, 247–282.
83. Luna, E. J. and Hitt, A. L. (1992) Cytoskeleton-plasma membrane interactions. *Science* **258**, 955–963.
84. Mukherjee, S. and Maxfield, F. R. (2000) Role of membrane organization and membrane domains in endocytic lipid trafficking. *Traffic* **1**, 203–211.
85. Holowka, D., Gosse, J. A., Hammond, A. T., et al. (2005) Lipid segregation and IgE receptor signaling: a decade of progress. *Biochim. Biophys. Acta* **1746**, 252–259.
86. Golub, T., Wacha, S., and Caroni, P. (2004) Spatial and temporal control of signaling through lipid rafts. *Curr. Opin. Neurobiol.* **14**, 542–550.
87. Martens, J. R., O'Connell, K., and Tamkun, M. (2004) Targeting of ion channels to membrane microdomains: localization of KV channels to lipid rafts. *Trends Pharmacol. Sci.* **25**, 16–21.
88. Szabo, I., Adams, C., and Gulbins, E. (2004) Ion channels and membrane rafts in apoptosis. *Pflügers Arch.* **448**, 304–312.
89. Alonso, M. A. and Millan, J. (2001) The role of lipid rafts in signaling and membrane trafficking in T lymphocytes. *J. Cell Sci.* **114**, 3957–3965.
90. Simmons, K. and Ikonen, E. (1997) Functional rafts in cell membranes. *Nature* **387**, 569–572.
91. Grzybek, M., Kozubek, A., Dubiclecka, P., and Sikorski, A. F. (2005) Rafts-the current picture. *Folia Histochem. Cytobiol.* **43**, 3–10.
92. Allende, D., Vidal, A., and McIntosh, T. J. (2004) Jumping to rafts: gatekeeper role of bilayer elasticity. *Trends Biochem. Sci.* **29**, 325–330.

93. Harris, T. J. and Siu, C. H. (2002) Reciprocal raft-receptor interactions and the assembly of adhesion complexes. *Bioessays* **24,** 996–1003.
94. O'Shea, P. (2005) Physical landscapes in biological membranes: physico-chemical terrains for spatio-temporal control of biomolecular interactions and behaviour. *Phil. Trans. Royal Soc. Lond. Series A* **363,** 575–588.
95. Stan, R. V. (2005) Structure of caveolae. *Biochim. Biophys. Acta* **1746,** 334–348.
96. Ostrom, R. S. and Insel, P. A. (2004) The evolving role of lipid rafts and caveolae in G protein-coupled receptor signaling: implications for molecular pharmacology. *Br. J. Pharmacol.* **143,** 235–245.
97. Marx, J. (2001) Caveolae: a once-elusive structure gets some respect. *Science* **294,** 1861–1866.
98. Li, X. A., Everson, W. V., and Smart, E. J. (2005) Caveolae, lipid rafts, and vascular disease. *Trends Cardiovasc. Med.* **15,** 92–96.
99. Bouras, T., Lisanti, M. P., and Pestell, R. G. (2004) Caveolin-1 in breast cancer. *Cancer Biol. Ther.* **3,** 931–941.
100. Cohen, A. W., Hnasko, R., Schubert, W., and Lisanti, M. P. (2004) Role of caveolae and caveolins in health and disease. *Physiol. Rev.* **84,** 1341–1379.
101. Chicani, G., Zhu, W., and Smart, E. J. (2004) Lipids: potential regulators of nitric oxide generation. *Am. J. Physiol. Endocrinol. Metab.* **287,** E386–E389.
102. Bathori, G., Cervenak, L., and Karadi, I. (2004) Caveolae-an alternative endocytotic pathway for targeted drug delivery. *Crit. Rev. Ther. Drug Carrier Syst.* **21,** 67–95.
103. Lee, A. G. (2004) How lipids affect the activities of integral membrane proteins. *Biochim. Biophys. Acta* **1666,** 62–87.
104. Cristea, I. M. and Degli Esposi, M. (2004) Membrane lipids and cell death: an overview. *Chem. Phys. Lipids* **129,** 133–160.
105. Futerman, A. H. and Hannun, Y. A. (2004) The complex life of simple sphingolipids. *EMBO Rep.* **5,** 777–782.
106. Bankaitis, V. A. and Morris, A. J. (2003) Lipids and the exocytotic machinery of eukaryotic cells. *Curr. Opin. Cell Biol.* **15,** 389–395.
107. Edidin, M. (2003) Lipids on the frontier: a century of cell-membrane bilayers. *Nat. Rev. Mol. Cell Biol.* **4,** 414–418.
108. Gruenberg, J. (2003) Lipids in endocytic membrane transport and sorting. *Curr. Opin. Cell Biol.* **15,** 383–388.
109. Tillman, T. S. and Cascio, M. (2003) Effects of membrane lipids on ion channel structure and function. *Cell Biochem. Biophys.* **38,** 161–190.
110. Wallis, J. G. and Browse, J. (2002) Mutants of Arabidopsis reveal many roles for membrane lipids. *Prog. Lipid Res.* **41,** 254–278.
111. Goni, F. M., Contreras, F. X., Montes, L. R., Sot, J., and Alonso, A. (2005) Biophysics (and sociology) of ceramides. *Biochem. Soc. Symp.* **72,** 177–188.
112. Bouwstra, J. A., Honewell-Nguyen, P. L., Gooris, G. S., and Ponec, M. (2003) Structure of the skin barrier and its modulation by vesicular formulations. *Prog. Lipid Res.* **42,** 1–36.
113. Suh, B. C. and Hille, B. (2005) Regulation of ion channels by phosphatidylinositol 4,5-bisphosphate. *Curr. Opin. Neurobiol.* **15,** 370–378.
114. Tigyi, G. and Parrill, A. L. (2003) Molecular mechanisms of lysophosphatidic acid action. *Prog. Lipid Res.* **42,** 498–526.
115. Ishii, S., Nagase, T., and Shimizu, T. (2002) Platelet-activating factor receptor. *Prostaglandins Other Lipid Mediat.* **68–69,** 599–609.
116. Chalfant, C. E. and Spiegel, S. (2005) Sphingosine 1-phosphate and ceramide 1-phosphate: expanding roles in cell signaling. *J. Cell Sci.* **118,** 4605–4612.

2

Membrane Lipid Polymorphism

Relationship to Bilayer Properties and Protein Function

Richard M. Epand

Summary

Bilayers are the most familiar arrangement of phospholipids. However, even as bilayers, phospholipids can arrange themselves in a variety of morphologies from essentially flat structures found in large liposomes or when adhered to a flat solid support, to the curved structures found in small liposomes or as bicontinuous cubic phases. Phospholipids can also arrange themselves as curved monolayers, such as in the hexagonal phase, and they can even form spherical or ellipsoid-shaped micelles. A number of factors will determine the final morphology of a lipid aggregate including the structure of the lipid, the nature of the lipid headgroup and its degree of hydration, and the temperature. In addition to being interesting in its own right, the property of lipid polymorphism can be applied to understand how fundamental intrinsic curvature properties of a membrane alter the physical properties of a membrane bilayer. This, in turn, will affect the functional characteristics of membrane proteins, with several possible mechanisms explaining the coupling of membrane properties with protein function.

Key Words: Bilayer; cubic phase; curvature strain; hexagonal phase; interfacial enzyme catalysis; lateral pressure profile; lipid polymorphism.

1. Introduction

In the past couple of years, there has been an increased interest in the role of lipids in biology. There are several reasons for this. In part, it is a component of the "systems biology" approach in which classes of biochemical molecules are considered together, so that one takes into account the interactions among components. This has led to the establishment of several fields such as genomics, proteomics and, more recently, lipidomics *(1)*, among others. Several groups are now attempting to coordinate approaches to the study of lipidomics, perhaps the first of which was the Lipidmaps Consortium, which led to a suggested new classification system for lipids *(2)*. There has also been technical advances in the applications of mass spectroscopy to lipid analysis *(3)*. Combined with these approaches, there has been a greater appreciation for the roles of certain lipids as secondary messengers as well as the importance of lipids in modulating the functional properties of membranes. There has also been a long-standing interest in the role of dietary lipids on human health. This includes the popular concern about the impact of cholesterol and triglycerides in the diet and their possible relationship to atherosclerosis.

More recently, there has been increased attention to the detrimental health consequences of *trans*-fatty acids *(4–8)*, as well as the beneficial effects of ω-3 fatty acids *(9–12)*. Many of the biological effects of dietary lipids are a consequence of altering the biophysical properties of the host's cell membranes (nonspecific effect), rather than to a more specific effect. The relationship between lipid polymorphic properties and membrane function that will be analyzed in this article is an example of a nonspecific effect.

2. Lipid Phases

2.1. Normal vs Inverted

Pure phospholipids are capable of undergoing transformations from one shape or morphology to another. This is termed lipid polymorphism, i.e., the ability of lipids to take on structures of different shapes. Some other recent reviews have pointed to the relationship between lipid polymorphism and membrane function *(13,14)*. Lipid phases are divided into two general types: normal and inverted phases. Normal phases are those in which the polar moiety of the lipid faces outward from the lipid structure, whereas the nonpolar portion of the molecule makes up the structure core. The lipid arrangement is opposite in inverted phases: the polar groups face inward and the nonpolar portion occupies the exterior of the structure. Normal and inverted phases are also referred to as type I and type II phases, respectively. Specific examples will be illustrated below.

2.2. Bilayer/Lamellar Phase

Lipids in biological membranes are arranged primarily, if not exclusively, as bilayers. This is sometimes referred to as a lamellar phase, to distinguish it from bicontinuous cubic phases (*see* **Subheading 2.3.**) in which the lipid is also arranged as a bilayer. The difference is that the cubic phase is a three-dimensional structure, whereas the common kind of bilayer is essentially a flat, two-dimensional structure. Phospholipids that form flat bilayers spontaneously stack when hydrated, and form repeating lamellae of multilamellar vesicles (MLVs). Hence, these types of flat bilayers are often referred to as lamellar phases to distinguish them from bilayers that form cubic phases.

A flat bilayer that is not infinitely large would have hydrophobic edges, which would be exposed to water. This would be a markedly destabilizing feature, although ruptured cell membranes are suggested to contain fragments of bilayer flat pieces with exposed edges. Because these fragments are large, the edges include only a small fraction of the total surface area. A nascent high-density lipoprotein is also made up of a flat disk of phospholipid bilayer protected at the edges by plasma apolipoproteins. However, these are exceptions, and a much more common situation is that lipids in the lamellar phase form spherical vesicles that do not have exposed edges. When these vesicles are stacked, one inside the other, like an onion, the structure is called an MLV. These structures are so large (on the order of microns) that on a molecular scale they are locally almost perfectly flat.

MLVs can be converted into unilamellar vesicles. Unilamellar vesicles have advantages for studying membrane functions such as transport across the bilayer or membrane fusion, because these processes would not be complicated by several events occurring simultaneously in stacks of bilayers. One way to make unilamellar vesicles is by extrusion through polycarbonate filters having pores of uniform size. This process of extrusion produces single wall, large unilamellar vesicles (LUVs). The size of LUVs can be varied by choosing polycarbonate filters having different pore diameters. However, LUVs with diameters more than about 200 nm are generally contaminated with vesicles that have more than one lamella. It is also difficult to make vesicles smaller than 50 nm in diameter by this method. The smallest vesicles that can form are usually made by sonication, producing Small Unilamellar Vesicles

(SUV). These have diameters of the order of 20 nm. The bilayer in SUVs has greater curvature than that found in larger vesicles, simply because of the morphology of the particle. The nature of this curvature in SUV is opposite for the two monolayers of the bilayer. The outer monolayer is curved with an expanded headgroup cross-sectional area, whereas the inner monolayer has a contracted headgroup cross-sectional area. The opposite is the case for the hydrocarbon, with the inner monolayer having more expanded hydrocarbon space and a more contracted one for the outer monolayer. SUVs are intrinsically unstable because of these curvature effects. To distinguish the two kinds of curvatures of monolayers, the curvature of the type found in the outer monolayer is called positive curvature and that of the inner monolayer, negative curvature. The definitions of positive and negative are arbitrary, but they do reflect the opposite curvatures for the two kinds of bending.

2.3. Cubic Phases

The term cubic phase refers to the symmetry elements in the arrangement of the unit cells. There are many types of cubic phases. Cubic phases can be divided into two general classes: inverted micellar and bicontinuous. In an inverted micellar cubic phase the lipids are packed into spherical aggregates, with each sphere representing a unit cell that packs together with other unit cells in a cubic array. Each of the micelles is "inverted" in the sense that the polar head groups are pointing inward, toward the center of the sphere, whereas the surface of the sphere is hydrophobic. Because of their hydrophobic exterior, inverted micelles in water are more likely to form ordered aggregates of cubic phase, when compared with normal phase micelles; these will disperse in water as single micelles.

The other type of cubic phase is the bicontinuous cubic phase. The term bicontinuous refers to the fact that both the lipid phase and the water channels are continuous in space (i.e., one can move along the aqueous channels from one location in the structure to another without having to cross over a lipid barrier; similarly, one can move along with lipid for large distances without having to enter the water phase). This bicontinuous arrangement is accomplished with several different morphologies, all of which have cubic symmetry but belong to different space groups (i.e., they have different symmetry elements). An excellent and detailed discussion of the various types of cubic phases can be found in the book by Hyde et al. *(15)*. This book also discusses the possible biological relevance of the cubic phase, as do several review articles *(16–18)*.

2.4. Hexagonal Phase

In the hexagonal phase, lipids are packed together to form hollow cylinders. The most common kind of hexagonal phase formed in water is the inverted-hexagonal phase, referred to as the H_{II} phase. In this phase, the methyl ends of the acyl chains form the exterior of the lipid cylinders, and the headgroups face the central core of the cylinder, which is filled with water. These cylinders are packed together with hexagonal symmetry in which each cylinder is surrounded by six other cylinders, hence the name hexagonal phase. Being a type II inverted phase structure that includes a bundle of cylinders, the exterior of the bundle would be hydrophobic and in contact with water. In addition, a hexagonal array of straight cylinders would have the entire acyl chains exposed to water at the ends of the cylinders. Nevertheless, H_{II} phases are stable in excess water. Three factors may contribute to this. First, the cylinders may not be straight but have sufficient bending so that they can form a toroidal structure, thus avoiding the exposure of blunt ends. There is some evidence from electron microscopy for this type of structure. The second factor is that these are very large aggregates, so that the surface-to-volume ratio is small. Thus, destabilizing interactions between water and the exterior of the hexagonal phase aggregate may be compensated for by favorable interactions within the hexagonal phase structure. Finally, there may be a monolayer of lipid surrounding the H_{II} aggregate and exposing the headgroups of the monolayer. Although there is no evidence for such a monolayer, its existence would be difficult to detect because it would represent a very small fraction of the total lipid in the structure.

3. Factors Determining Phase Preference

Although it was mentioned that the bilayer was the most common type of organization of lipid molecules in biological systems, many of the major lipid components of biological membranes, in isolated purified form, will form nonlamellar phases under ordinary conditions of room temperature and excess water. For example, phosphatidylethanolamine (PE) from natural sources will spontaneously form an H_{II} phase in excess water *(19,20)*. Normal micelles have detergent-like properties and are formed by bile salts, which function biologically as detergents, as well as by lysolipids and gangliosides. Some of the factors that determine what phase will form under a particular set of conditions are considered.

3.1. Lipid Structure

The molecular structure and properties of the lipid molecule will be a major factor in determining the type of phase a lipid will form. This is determined in part by steric factors. Steric interactions among headgroups as well as the acyl chains are important. For example, PE and phosphatidylcholine (PC) are both zwitterionic lipids with similar chemical structures. The additional three methyl groups on the nitrogen of PC contribute a steric component to prevent this lipid from forming an inverted phase. The acyl chains also have repulsive steric interactions that are greater for unsaturated than for saturated lipids, and also greater for longer chain length or branched acyl chains. These factors can become sufficiently large that even a lipid such as PC, which normally favors a bilayer arrangement, will readily convert into a hexagonal phase *(21)*.

Acyl chain double bonds can exist as one of two geometric isomers; either the *cis*, commonly found in nature, or the *trans*-isomer, with its relation to health risks, as mentioned in the introduction. The presence of a *cis*-double bond will cause the acyl chain to have a kink at that position, resulting in increased steric repulsion toward the methyl terminus; this increased repulsion can be relieved, at least in part, by converting the bilayer into a hexagonal phase. The H_{II} phase will be more stable because its curved monolayer will have a larger cross-sectional area on the outer, hydrophobic surface, and can therefore accommodate the steric repulsion caused by the presence of a *cis*-double bond. Such an increase in the steric repulsion toward the methyl terminus occurs to a much smaller degree with *trans*-double bonds in the acyl chain because this geometry causes less change in the direction of the acyl chain. These findings have given rise to the shape concept of lipid polymorphism, with cone-shaped lipids forming structures with positive curvature, such as micelles, whereas inverted cone-shaped lipids would form inverted phases, such as the H_{II} phase.

However, steric factors are not the only properties that will determine the interaction forces among adjacent lipid molecules. In addition, attractive interactions, such as hydrogen bonding among headgroups, can affect the phase preference. This is likely to be an additional factor in the preference of PE for inverted phases. Charge repulsion among headgroups will have the opposite effect and will inhibit the formation of inverted phases. There are many examples of biological lipids that are anionic. In general, these lipids do not form inverted phases because of electrostatic repulsion among headgroups. However, addition of a cation, such as Ca^{2+} or lowering the pH to protonate the lipid, in many cases causes the lipid to convert from the lamellar to the hexagonal phase. A prime example of this is cardiolipin, a lipid with four acyl chains that readily forms inverted phases, yet only after the negative charge of the headgroup is neutralized, for example, by binding cations *(22)*.

3.2. Hydration

Intermolecular hydrogen bonding between lipid headgroups competes with hydrogen bonding between a lipid headgroup and water. Thus, in cases where there is more bonding between headgroups, such as PE, there is less hydrogen bonding with water and, therefore a lower degree of hydration of the lipid. Lower hydration of the lipid will promote inverted phase formation. One way to lower the effective hydration is with salting-out salts that promote hexagonal phase formation *(23)*. Another way to reduce the hydration is simply to decrease the amount of water added to the lipid sample. However, there is an additional factor; there has to be sufficient water to fill the core of the hexagonal phase cylinders. This leads to the "re-entrant phenomenon:" as the hydration is lowered the hexagonal phase becomes the more stable phase until a point is reached (at even lower hydration) at which the lamellar phase reappears *(24)*.

3.3. Temperature

Temperature is an important factor in regulating lipid polymorphism. This arises as a consequence of acyl chain splay. Carbon–carbon single bonds between adjacent methylene groups in the acyl chains can undergo *trans*- to *gauche*-isomerization. This is different from the interconversion between *trans*- and *cis*-double bonds that require rearrangement of covalent bonds. Conversion of a *trans*- to *gauche*-form simply requires free rotation around a single bond. The *trans*-rotamer is more stable because the extensions of the acyl chain coming off at opposite sides of the C–C bond result in less steric repulsion. At low temperature, most of the C–C bonds are *trans* and the acyl chain is fully extended, approximating a linear rod. As the temperature increases, entropy drives the conversion of some *trans*-bonds to *gauche*. This will result in a deviation of the acyl chain from its linear direction, with the result that the methyl terminal end of the acyl chain will have a larger deviation from the bilayer normal. As lipid polymorphism is usually studied in the presence of water, to be more biologically relevant, this usually restricts the range of temperatures to roughly 0–100°C. Within this temperature range, some lipids form only one phase and do not exhibit thermotropic phase transitions. However, for those lipids that do exhibit thermotropic phase transitions, the lamellar phase is favored at lower temperatures and the hexagonal phase at higher temperatures.

The cubic phase is often thought to be an intermediate between the lamellar and the H_{II} phase *(25)*. However, the cubic phase exhibits very slow kinetics, both in formation and conversion to other phases. Therefore, it is often uncertain what the stability of the cubic phase is, relative to other phases, under any particular condition. It has been suggested that the cubic phase is formed by a pathway different from that of the H_{II} phase *(26)*, rather than as an intermediate in the pathway for formation of the hexagonal phase.

4. Lipid Polymorphism and Membrane Properties

There are few examples of nonlamellar structures forming in biological systems; there are examples of lipid micelles that perform important biological functions, such as micelles formed by bile salts functioning to disperse triglycerides in the intestine. There also have been some reports about the presence of cubic phases in cells *(16,27)*. The formation of cubic phases in cells would be particularly intriguing because it could provide directed paths and barriers for translocating substances that could quickly form and dissipate. In addition to these limited, and in the case of cubic phases, speculative roles for nonlamellar phases, there is also the modulation of biological functions resulting from the presence of nonlamellar-forming lipids. The mechanism(s) by which bilayer physical properties are modulated by the presence of such lipids is discussed in **Subheading 4.1**.

4.1. Curvature Strain

It is useful to consider the curvature properties of a lipid monolayer. A bilayer in a LUV or MLV will have almost no physical curvature on a molecular scale (i.e., locally, the bilayer is flat). However, the constituent monolayers of the bilayer may have an intrinsic curvature that is not flat. Despite the fact that its physical shape is essentially flat, if each monolayer could bend to its preferred shape, without a change in the polarity of the environment of any of its groups, it would then achieve its intrinsic curvature. This cannot happen in a symmetrical bilayer because each monolayer would bend in opposite directions, leaving a void between the ends of the acyl chains. However, the arrangement of lipid in the H_{II} phase can approximate that of a monolayer that has attained its intrinsic curvature. The extent of this curvature can readily be calculated from the lattice spacing of the H_{II} phase measured by diffraction. To be accurate, this measured intrinsic curvature requires a correction, which results from the fact that the H_{II} phase does not fill all space if it remains as perfectly rounded cylinders. There will be voids between the cylinders that have to be filled by *gauche* to *trans*-isomerization of the acyl chains that point toward these voids. This will require energy, and as a consequence, the H_{II} cylinder diameter will decrease to lower the extent of this hydrocarbon packing problem. There are experimental ways to correct for this factor, and thus an accurate value of the intrinsic curvature can still be obtained. This curvature is usually expressed as an intrinsic radius of curvature, which is defined as the distance from the center of the cylinder to the pivotal plane; the pivotal plane is the position in the lipid structure whose cross-sectional area does not change when the monolayer bends *(28)*.

A monolayer organized in a structure in which its physical curvature is equal to its intrinsic curvature will not possess any curvature strain. It will not have any driving force to change shape. Lipids that spontaneously form highly curved monolayers will have instability because of curvature strain when they form a planar bilayer. The curvature energy associated with each of the monolayers depends on two factors: the intrinsic curvature and the elastic-bending modulus. In other words, the curvature strain of a monolayer will be equal to the energy required to unbend it from the form in which it has achieved its intrinsic curvature to the flat structure of the bilayer. This energy per unit area of interface is given by:

$$\frac{0.5K_C}{R_0^2}$$

where K_c is the elastic bending modulus of the lipid monolayer and R_0 is the lipid monolayer's spontaneous radius of curvature in excess water. The elastic-bending modulus is a measure of the stiffness of the monolayer. The easier it is to bend the monolayer, the less curvature strain will be acquired by forming a structure whose curvature is different from the intrinsic curvature. The earlier discussion, to be more precise, refers to the mean curvature. There is a second curvature modulus called the Gaussian curvature. This refers to an average of the sum of positive and negative curvatures. Thus, there can be morphologies, such as "saddle points" (structures shaped like a saddle) that have zero mean curvature but they have Gaussian curvature, which is the sum of negatively and positively curved surfaces, irrespective of their sign. The importance of Gaussian curvature in the stability of membrane fusion intermediates has been recognized *(29)*.

4.2. Lateral Pressure Profile

An alternative formulation of curvature-related instability has been proposed by Cantor *(30,31)*. This approach focuses on the variation of lateral pressure as a function of the position in the bilayer. Often lateral pressure profiles and curvature strain are alternative ways of describing changes in bilayer properties resulting from membrane curvature strain. If there is a lateral pressure at the methyl terminus of the acyl chains higher than in that present in other regions of the bilayer, the scenario is equivalent to the bilayer having negative curvature strain.

Each of the two formulations has their own advantages. Curvature strain is a simpler idea and is more amenable to direct experimental measurement. However, the concept of lateral pressure profile provides a more detailed molecular description and can be more informative to interface, providing information about the structure and location of substances within the membrane *(32)*. One of the lines of evidence supporting the concept of the lateral pressure profile is that it correctly predicts the variation of anesthetic potency as a function of chain length for different classes of molecules *(33,34)*.

4.3. Tilt Modulus

In addition to monolayer bending and lateral pressure at a particular depth in the membrane, lipid monolayers can also be subjected to changes in acyl chain tilt *(35)*. It has been calculated that the tilt modulus involves two major contributions. One contribution arises from the stretching of the hydrocarbon chains on tilt deformation, which also results in loss of chain conformational flexibility. The second contribution is purely entropic, arising from the constraints imposed by tilt deformation. The two factors have comparable energies. This formulation represents an alternative way of taking into account the factors that lead to curvature strain.

5. Biological Roles of Membrane Curvature

There are many specific interactions as well as bulk physical properties that modulate biological function. Some bulk properties are intrinsic monolayer curvature, motional properties within the membrane (as reflected, e.g., by order parameter gradients), and properties of the membrane interface, including polarity, penetration of water, and charge. In addition, there is domain formation, which reflects the nonuniform distribution of molecules in the plane of the membrane.

In this regard, there is an interesting example in which bending of a lipid monolayer causes demixing of membrane lipid components *(36)*. This section focuses specifically on the role of nonlamellar-forming lipids in modulating certain membrane-related functions.

5.1. Homeostasis

Nonbilayer-forming lipids have been suggested to play an important role in biological membranes by establishing an environment with an optimal balance between stability and plasticity *(37)*. In particular, studies with bacteria have demonstrated that changing the growth temperature of the organism leads to an alteration in the lipid composition of the membrane, both in *Escherichia coli* *(38)* as well as in *Acholeplasma laidlawii* *(39)*, to maintain a constant curvature instability. For the latter organism, it has been suggested that there is a specific enzyme, diglucosyldiacylglycerol transferase, whose activity is sensitive to the balance between lamellar and nonlamellar lipids *(40)*. This is likely to be a factor in explaining the origin of curvature instability under a variety of conditions. However, the fact that homeostasis of curvature instability is maintained suggests that this physical property has important biological consequences.

5.2. Membrane Fusion

Membrane fusion is a process in which two planar bilayers must undergo rearrangement into a nonlamellar structure. Intermediates that have been proposed in membrane fusion involve different changes in membrane curvature for the joining of opposing monolayers *(41,42)*. Changes in membrane curvature required to form fusion intermediates can be facilitated by the insertion of protein segments. There has been much interest in the so-called "fusion peptides," which are segments of proteins that promote membrane fusion. Several examples of fusion peptides are known from studies of the fusion of enveloped viruses to target membranes. Several of these peptides have been shown to promote membrane fusion by themselves and also increase the negative curvature strain of membranes *(43)*. Another contributing factor for the tendency of viral fusion peptides to promote nonlamellar phases is suggested to be their partitioning into hydrocarbon voids and thereby stabilizing highly curved fusion intermediates *(44)*.

Regarding lipid polymorphism, the formation of cubic phases has attracted interest because of its similarities to the formation of a membrane fusion pore *(45)*. The lipid-lined aqueous channel that connects one unit cell with another in a bicontinuous cubic phase has the same structure as a fusion pore. A lowering of the rupture tension of the membrane by hydrophobic peptides can accelerate the conversion of an intermediate to the fusion pore as well as to a cubic phase *(46)*. Of course, the bicontinuous cubic phase is a three-dimensional lipid phase with many unit cells, whereas a fusion pore is formed as an isolated structure. Nevertheless, similar changes in membrane properties may promote both processes. In addition, recent evidence proposes to determine the structure of the initial hemifusion intermediate using diffraction methods *(47)*.

5.3. Activity of Membrane-Bound Enzymes

The activity of several integral membrane and amphitropic proteins (proteins that exchange between membrane and aqueous compartments) has been shown to be sensitive to membrane curvature strain. These proteins include rhodopsin *(48)*, G proteins *(49–51)*, the proapoptotic Bcl-2 proteins, *t*-Bid *(52)*, and Bax *(53)* among others. In addition, the assembly of lactose permease has been shown to be dependent on the presence of PE, a lipid imparting negative curvature stress *(54)*. More specifically, curvature stress has been suggested to modulate the free energy and folding of integral membrane proteins *(55)*. The focus will be on the properties of two amphitropic enzymes whose activity is modulated by membrane curvature, yet apparently through different mechanisms. These enzymes are protein kinase C (PKC) and phosphocholine cytidylyltransferase (CT).

The activities of both PKC *(56)* and CT *(57)* are enhanced by increasing the negative curvature strain of the membrane. However, in the case of PKC there are two lines of evidence that indicate that there is not a direct relationship between curvature strain and enzyme activity *(58)*. One indication comes from studies with a series of PEs containing 18:1 acyl chains but differing in the position of the C–C double bond. The curvature properties of this series of lipids have been determined *(59)*, and the activity of PKC does not correlate with the curvature strain measured with membranes containing different members of this homologous series of lipids *(60)*. However, the activity of PKC does correlate with properties of an interfacial fluorescent probe. The situation with CT is different, and indicates that a change in curvature strain is the mechanism by which nonlamellar-forming lipids modulate the activity of this enzyme *(61,62)*. The other indication that the activity of PKC is not directly modulated by membrane curvature strain comes from studies of the activity of PKC in the presence of lipids in the cubic phase *(63)*. Spontaneous conversion of bilayers to the cubic phase will result in the relief of negative curvature strain. In order to compare the activity of PKC in the cubic phase with the activity in the lamellar phase, the cubic phase of mono-olein was converted to a lamellar phase by the addition of progressively larger amounts of phosphatidylserine. In addition, the activity of PKC using bilayer membranes made up of dielaidoyl PE that were converted into a bicontinuous cubic phase was compared with the addition of a small amount of alamethicin *(64)*. In both systems, the activity of PKC was greater in the cubic phase than in the lamellar phase. Furthermore, it was shown that this difference was not because of the small change in membrane composition but rather to the change of phase. Hence, despite the relief of negative curvature strain, the cubic phase is more potent in activating PKC. Therefore, it is concluded that although the activity of CT appears to be directly coupled with membrane curvature, the activity of PKC is modulated by nonlamellar-forming lipids by a less direct mechanism.

6. Summary

There is substantial evidence that the presence of nonlamellar-forming lipids is important for the functioning of cells. There are a number of processes including membrane fusion and the activity of membrane-bound enzymes that are affected by the presence of these lipids. The mechanism by which these lipids alter membrane properties appears to differ for different systems. In addition, other factors such as "fluidity" and lipid domain formation are also important for some membrane functions.

References

1. Mouritsen, O. G. (2005) *Life as a matter of fat: The emerging science of lipidomics*. Springer, Berlin.
2. Fahy, E., Subramaniam, S., Brown, H. A., et al. (2005) A comprehensive classification system for lipids. *J. Lipid. Res.* **46,** 839–861.
3. Murphy, R. C., Fiedler, J., and Hevko, J. (2001) Analysis of nonvolatile lipids by mass spectrometry. *Chem. Rev.* **101,** 479–526.

4. King, I. B., Kristal, A. R., Schaffer, S., Thornquist, M., and Goodman, G. E. (2005) Serum trans-fatty acids are associated with risk of prostate cancer in beta-Carotene and Retinol Efficacy Trial. *Cancer Epidemiol. Biomarkers Prev.* **14,** 988–992.
5. Lopez-Garcia, E., Schulze, M. B., Meigs, J. B., et al. (2005) Consumption of trans fatty acids is related to plasma biomarkers of inflammation and endothelial dysfunction. *J. Nutr.* **135,** 562–566.
6. Mozaffarian, D., Rimm, E. B., King, I. B., Lawler, R. L., McDonald, G. B., and Levy, W. C. (2004) trans fatty acids and systemic inflammation in heart failure. *Am. J. Clin. Nutr.* **80,** 1521–1525.
7. Roach, C., Feller, S. E., Ward, J. A., Shaikh, S. R., Zerouga, M., and Stillwell, W. (2004) Comparison of cis and trans fatty acid containing phosphatidylcholines on membrane properties. *Biochemistry* **43,** 6344–6351.
8. Stender, S. and Dyerberg, J. (2004) Influence of trans fatty acids on health. *Ann. Nutr. Metab.* **48,** 61–66.
9. Lim, G. P., Calon, F., Morihara, T., et al. (2005) A diet enriched with the omega-3 fatty acid docosahexaenoic acid reduces amyloid burden in an aged Alzheimer mouse model. *J. Neurosci.* **25,** 3032–3040.
10. Peet, M. and Stokes, C. (2005) Omega-3 fatty acids in the treatment of psychiatric disorders. *Drugs* **65,** 1051–1059.
11. Seo, T., Blaner, W. S., and Deckelbaum, R. J. (2005) Omega-3 fatty acids: molecular approaches to optimal biological outcomes. *Curr. Opin. Lipidol.* **16,** 11–18.
12. Wu, M., Harvey, K. A., Ruzmetov, N., et al. (2005) Omega-3 polyunsaturated fatty acids attenuate breast cancer growth through activation of a neutral sphingomyelinase-mediated pathway. *Int. J. Cancer* **117,** 340–348.
13. Epand, R. M. (1998) Lipid polymorphism and protein-lipid interactions. *Biochim. Biophys. Acta* **1376,** 353–368.
14. de Kruijff, B. (1997) Lipid polymorphism and biomembrane function. *Curr. Opin. Chem. Biol.* **1,** 564–569.
15. Hyde, S., Andersson, S., Larsson, et al. (1997) *The Language of Shape: The Role of Curvature in Condensed Matter: Physics, Chemistry and Biology.* Elsevier, Amsterdam.
16. Landh, T. (1995) From entangled membranes to eclectic morphologies: cubic membranes as subcellular space organizers. *FEBS Lett.* **369,** 13–17.
17. Lindblom, G. and Rilfors, L. (1992) Nonlamellar phases formed by membrane lipids. *Adv. Colloid Interface Sci.* **41,** 101–125.
18. Luzzati, V. (1997) Biological significance of lipid polymorphism: the cubic phases. *Curr. Opin. Struct. Biol.* **7,** 661–668.
19. Cullis, P. R. and de Kruijff, B. (1979) Lipid polymorphism and the functional roles of lipids in biological membranes. *Biochim. Biophys. Acta* **559,** 399–420.
20. Gruner, S. M. (1992) *Nonlamellar Lipid Phases,* (Yeagle, P., ed.), Boca Raton, FL, pp. 211–250.
21. Epand, R. M., Epand, R. F., Decicco, A., and Schwarz, D. (2000) Curvature properties of novel forms of phosphatidylcholine with branched acyl chains. *Eur. J. Biochem.* **267,** 2909–2915.
22. Rand, R. P. and Sengupta, S. (1972) Cardiolipin forms hexagonal structures with divalent cations. *Biochim. Biophys. Acta* **255,** 484–492.
23. Epand, R. M. and Bryszewska, M. (1988) Modulation of the bilayer to hexagonal phase transition and solvation of phosphatidylethanolamines in aqueous salt solutions. *Biochemistry* **27,** 8776–8779.
24. Kozlov, M. M., Leikin, S., and Rand, R. P. (1994) Bending, hydration and interstitial energies quantitatively account for the hexagonal-lamellar-hexagonal reentrant phase transition in dioleoylphosphatidylethanolamine. *Biophys. J.* **67,** 1603–1611.
25. Jordanova, A., Lalchev, Z., and Tenchov, B. (2003) Formation of monolayers and bilayer foam films from lamellar, inverted hexagonal and cubic lipid phases. *Eur. Biophys. J.* **31,** 626–632.
26. Siegel, D. P. (1999) The modified stalk mechanism of lamellar/inverted phase transitions and its implications for membrane fusion. *Biophys. J.* **76,** 291–313.

27. Deng, Y., Marko, M., Buttle, K. F., Leith, A., Mieczkowski, M., and Mannella, C. A. (1999) Cubic membrane structure in amoeba (*Chaos carolinensis*) mitochondria determined by electron microscopic tomography. *J. Struct. Biol.* **127,** 231–239.
28. Rand, R. P. and Parsegian, V. A. (1997) *Hydration, curvature, and bending elasticity of phospholipid monolayers*, (Epand, R. M., ed.), San Diego, CA, pp. 167–189.
29. Siegel, D. P. and Kozlov, M. M. (2004) The gaussian curvature elastic modulus of N-monomethylated dioleoylphosphatidylethanolamine: relevance to membrane fusion and lipid phase behavior. *Biophys. J.* **87,** 366–374.
30. Cantor, R. S. (1999) The influence of membrane lateral pressures on simple geometric models of protein conformational equilibria. *Chem. Phys. Lipids* **101,** 45–56.
31. Cantor, R. S. (1999) Lipid composition and the lateral pressure profile in bilayers. *Biophys. J.* **76,** 2625–2639.
32. Cantor, R. S. (1997) The lateral pressure profile in membranes: a physical mechanism of general anesthesia. *Biochemistry* **36,** 2339–2344.
33. Cantor, R. S. (2001) Breaking the Meyer-Overton rule: predicted effects of varying stiffness and interfacial activity on the intrinsic potency of anesthetics. *Biophys. J.* **80,** 2284–2297.
34. Mohr, J. T., Gribble, G. W., Lin, S. S., Eckenhoff, R. G., and Cantor, R. S. (2005) Anesthetic potency of two novel synthetic polyhydric alkanols longer than the n-alkanol cutoff: evidence for a bilayer-mediated mechanism of anesthesia? *J. Med. Chem.* **48,** 4172–4176.
35. May, S., Kozlovsky, Y., Ben Shaul, A., and Kozlov, M. M. (2004) Tilt modulus of a lipid monolayer. *Eur. Phys. J. E. Soft Matter* **14,** 299–308.
36. Ding, L., Weiss, T. M., Fragneto, G., Liu, W., Yang, L., and Huang, H. W. (2005) Distorted hexagonal phase studied by neutron diffraction: lipid components demixed in a bent monolayer. *Langmuir* **21,** 203–210.
37. Garab, G., Lohner, K., Laggner, P., and Farkas, T. (2000) Self-regulation of the lipid content of membranes by non-bilayer lipids: a hypothesis. *Trends Plant Sci.* **5,** 489–494.
38. Morein, S., Andersson, A., Rilfors, L., and Lindblom, G. (1996) Wild-type *Escherichia coli* cells regulate the membrane lipid composition in a "window" between gel and non-lamellar structures. *J. Biol. Chem.* **271,** 6801–6809.
39. Osterberg, F., Rilfors, L., Wieslander, A., Lindblom, G., and Gruner, S. M. (1995) Lipid extracts from membranes of *Acholeplasma laidlawii A* grown with different fatty acids have a nearly constant spontaneous curvature. *Biochim. Biophys. Acta* **1257,** 18–24.
40. Vikstrom, S., Li, L., and Wieslander, A. (2000) The nonbilayer/bilayer lipid balance in membranes. Regulatory enzyme in *Acholeplasma laidlawii* is stimulated by metabolic phosphates, activator phospholipids, and double-stranded DNA. *J. Biol. Chem.* **275,** 9296–9302.
41. Chernomordik, L. V. and Kozlov, M. M. (2005) Membrane hemifusion: crossing a chasm in two leaps. *Cell* **123,** 375–382.
42. Zimmerberg, J. and Chernomordik, L. V. (1999) Membrane fusion. *Adv. Drug Deliv. Rev.* **38,** 197–205.
43. Epand, R. M. and Epand, R. F. (2003) Fusion peptides and the mechanism of viral fusion. *Biochim. Biophys. Acta* **1614,** 116–121.
44. Haque, M. E., Koppaka, V., Axelsen, P. H., and Lentz, B. R. (2005) Properties and Structures of the Influenza and HIV Fusion Peptides on Lipid Membranes: Implications for a Role in Fusion. *Biophys. J.* **89,** 3183–3194.
45. Cherezov, V., Siegel, D. P., Shaw, W., Burgess, S. W., and Caffrey, M. (2003) The kinetics of non-lamellar phase formation in DOPE-Me: Relevance to biomembrane fusion. *J. Membr. Biol.* **195,** 165–182.
46. Siegel, D. P. and Epand, R. M. (2000) Effect of influenza hemagglutinin fusion peptide on lamellar/inverted phase transitions in dipalmitoleoylphosphatidylethanolamine: implications for membrane fusion mechanisms. *Biochim. Biophys. Acta* **1468,** 87–98.

47. Yang, L. and Huang, H. W. (2003) A rhombohedral phase of lipid containing a membrane fusion intermediate structure. *Biophys. J.* **84,** 1808–1817.
48. Brown, M. F. (1994) Modulation of rhodopsin function by properties of the membrane bilayer. *Chem. Phys. Lipids* **73,** 159–180.
49. Escriba, P. V., Ozaita, A., Ribas, C., et al. (1997) Role of lipid polymorphism in G protein-membrane interactions: nonlamellar-prone phospholipids and peripheral protein binding to membranes. *Proc. Natl. Acad. Sci. USA* **94,** 11,375–11,380.
50. Vogler, O., Casas, J., Capo, D., et al. (2004) The Gbetagamma dimer drives the interaction of heterotrimeric Gi proteins with nonlamellar membrane structures. *J. Biol. Chem.* **279,** 36,540–36,545.
51. Yang, Q., Alemany, R., Casas, J., Kitajka, K., Lanier, S. M., and Escriba, P. V. (2005) Influence of the membrane lipid structure on signal processing via G protein-coupled receptors. *Mol. Pharmacol.* **68,** 210–217.
52. Epand, R. F., Martinou, J. C., Fornallaz-Mulhauser, M., Hughes, D. W., and Epand, R. M. (2002) The apoptotic protein tBid promotes leakage by altering membrane curvature. *J. Biol. Chem.* **277,** 32,632–32,639.
53. Basañez, G., Sharpe, J. C., Galanis, J., Brandt, T. B., Hardwick, J. M., and Zimmerberg, J. (2002) Bax-type apoptotic proteins poraPe pure lipid bilayers through a mechanism sensitive to intrinsic monolayer curvature. *J. Biol. Chem.* **277,** 49,360–49,365.
54. Zhang, W., Bogdanov, M., Pi, J., Pittard, A. J., and Dowhan, W. (2003) Reversible Topological Organization within a Polytopic Membrane Protein Is Governed by a Change in Membrane Phospholipid Composition. *J. Biol. Chem.* **278,** 50,128–50,135.
55. Hong, H. and Tamm, L. K. (2004) Elastic coupling of integral membrane protein stability to lipid bilayer forces. *Proc. Natl. Acad. Sci. USA* **101,** 4065–4070.
56. Epand, R. M. and Lester, D. S. (1990) The role of membrane biophysical properties in the regulation of protein kinase C activity. *Trends Pharmacol. Sci.* **11,** 317–320.
57. Attard, G. S., Templer, R. H., Smith, W. S., Hunt, A. N., and Jackowski, S. (2000) Modulation of CTP: phosphocholine cytidylyltransferase by membrane curvature elastic stress. *Proc. Natl. Acad Sci. USA* **97,** 9032–9036.
58. Mosior, M. and Epand, R. M. (1999) Role of the membrane in the modulation of the activity of protein kinase C. *J. Liposome Res.* **9,** 21–41.
59. Epand, R. M., Fuller, N., and Rand, R. P. (1996) Role of the position of unsaturation on the phase behavior and intrinsic curvature of phosphatidylethanolamines. *Biophys. J.* **71,** 1806–1810.
60. Giorgione, J. R., Kraayenhof, R., and Epand, R. M. (1998) Interfacial membrane properties modulate protein kinase C activation: role of the position of acyl chain unsaturation. *Biochemistry* **37,** 10,956–10,960.
61. Davies, S. M., Epand, R. M., Kraayenhof, R., and Cornell, R. B. (2001) Regulation of CTP: phosphocholine cytidylyltransferase activity by the physical properties of lipid membranes: an important role for stored curvature strain energy. *Biochemistry* **40,** 10,522–10,531.
62. Drobnies, A. E., Davies, S. M., Kraayenhof, R., Epand, R. F., Epand, R. M., and Cornell, R. B. (2002) CTP: phosphocholine cytidylyltransferase and protein kinase C recognize different physical features of membranes: differential responses to an oxidized phosphatidylcholine. *Biochim. Biophys. Acta* **1564,** 82–90.
63. Giorgione, J. R., Huang, Z., and Epand, R. M. (1998) Increased activation of protein kinase C with cubic phase lipid compared with liposomes. *Biochem.* **37,** 2384–2392.
64. Keller, S. L., Gruner, S. M., and Gawrisch, K. (1996) Small concentrations of alamethicin induce a cubic phase in bulk phosphatidylethanolamine mixtures. *Biochim. Biophys. Acta* **1278,** 241–246.

3

Acrylodan-Labeled Intestinal Fatty Acid-Binding Protein to Measure Concentrations of Unbound Fatty Acids

Jeffrey R. Simard, Frits Kamp, and James A. Hamilton

Summary

The concentration of long-chain (14–18 carbons) fatty acids (FA) free in solution (unbound) is difficult to measure directly because of the low aqueous solubility of these common dietary FA. One indirect and convenient way to measure the concentration of unbound FA is a method using the fluorescent-(acrylodan) labeled intestinal FA-binding protein (ADIFAB). Under appropriate conditions, ADIFAB fluorescence measures unbound FA, regardless of any third phase such as albumin, FA-binding proteins, or membranes. With knowledge of the total amount of FA in the system and the assumption that the amount of FA bound to ADIFAB is negligible, equilibrium constants or partition coefficients for FA in equilibrium with the third phase can be calculated. Herein, the use of ADIFAB is described to measure unbound FA concentration using oleic acid as a typical long-chain FA. Attempts were not made to calibrate the accuracy of ADIFAB for FA concentration, but to investigate its reliability and reproducibility under differing buffer conditions. It is shown that ADIFAB fluorescence is sensitive to biologically prevalent ions and that calibration curves must be constructed for conditions that do not closely match those previously published. The results with in vitro systems suggest that there will be caveats with the application of ADIFAB to measure FA concentrations in vivo, where the precise environment of the probe is not known or cannot be tightly controlled.

Key Words: Fatty acid transport; fatty acids; flip–flop; fluorescence; membranes; ADIFAB.

1. Introduction

Glucose and fatty acids (FA) are the two most important dietary nutrients available for energy in the body. Glucose serves as the primary fuel source for the brain and skeletal muscles, whereas FA is primarily utilized by cardiac muscle *(1,2)*. FA not immediately oxidized for energy are delivered to the adipose tissue to be stored as triglycerides (TG) until energy stores are depleted. Whereas the mechanism by which glucose enters the cell is well established, the mechanism by which FA cross the cell membrane is still widely debated.

The laboratory previously developed a series of novel dual-fluorescence assays to monitor directly the movement of FA through model and biological membranes (*see* Chapter 16). These assays use a variety of fluorescent probes to discriminate between the adsorption and transbilayer movement of FA in membranes *(3–10)*. Acrylodan-labeled intestinal FA-binding protein (ADIFAB) can be used to measure the binding of FA to the outer membrane leaflet (adsorption) in the same experiment in which a pH-sensitive fluorophore detects protons released by FA as they diffuse through the membrane (transmembrane movement) *(7)*. Studies performed with this assay suggest that FA can diffuse rapidly across model lipid membranes *(4–10)* and also across the plasma membranes of isolated rat adipocytes *(6,11)*,

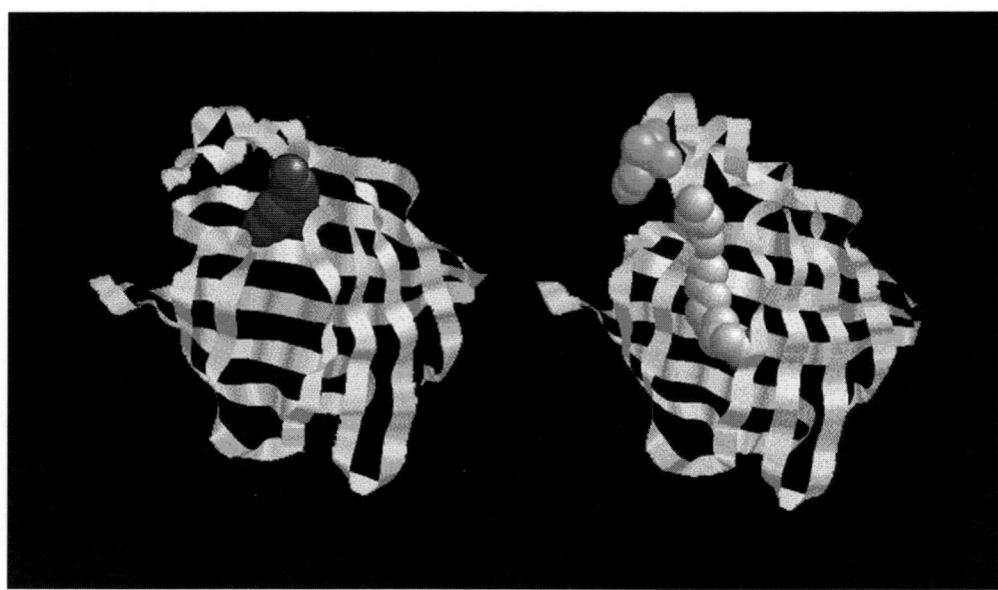

Fig. 1. Modeled structure of ADIFAB with **(left)** and without **(right)** FA bound *(26)*. Protein structure is depicted as a yellow ribbon diagram. Acrylodan is shown in blue (without FA) and green (with FA) to illustrate a change in fluorescence. FA ligand is shown in a space-filling form and atoms are color-coded (carbon, gray; oxygen, red).

undifferentiated and differentiated 3T3-L1 adipocytes (J. A. Hamilton, unpublished), and HepG2 cells *(12,13)*. These studies also reveal that FA diffusion into isolated rat adipocytes is rapid enough to supply the intracellular metabolism of FA into TG *(11)*.

ADIFAB is a derivative of the rat FA-binding protein (FABP) found in the intestine (I-FABP). Rat I-FABP and human I-FABP are structurally well characterized, as complete X-ray and nuclear magnetic resonance structures are available *(14–17)*. The protein binds FA in vitro, and is thought to function in vivo as a cytosolic transporter for FA shuttling between the plasma membrane and other intracellular membranes. ADIFAB is made by covalently labeling rat I-FABP with a fluorescent acrylodan molecule (**Fig. 1**) that is sensitive to changes in polarity *(18)*. Specifically, the acrylodan molecule is affixed to a lysine residue in the helical "lid" covering the β-barrel structure *(14,19)* that forms a predominantly hydrophobic-binding pocket in which a single FA molecule binds with high affinity *(20)*. The helical "lid" region is believed to be mobile and serve as the portal through which FA enter and exit the protein *(14)*. ADIFAB senses the concentration of unbound FA through movements of the acrylodan-labeled "lid" as the molecule enters and exit the binding pocket. When a FA binds to ADIFAB, the "lid" is believed to move into an open conformation, which exposes the acrylodan moiety to a more polar aqueous environment with changes in fluorescence *(14,18)*.

Technical guidelines for applications of ADIFAB have been presented *(21,22)*. However, as the probe becomes more widely used and is applied in complex environments *(23)*, it is important to evaluate more carefully the limitations of the probe and potential pitfalls. ADIFAB binds biomolecules other than FA in a manner that affects its fluorescence spectrum. These include reduced enzyme A (CoA-SH), acetyl CoA and oleoyl CoA *(24)*. Such effects are likely to lead to inaccuracies of quantifying FA in the cytoplasm of cells and, possibly, in biological fluids.

2. Materials

2.1. FA Stock Solutions (K⁺-Salt)

1. Pure (99%) FA in the form of powder (saturated) or oil (mono- and polyunsaturated) (Sigma, St. Louis, MO). Store at room temperature (saturated) or at –20°C (mono- and polyunsaturated).
2. Reagent-grade chloroform.
3. 0.1 N Potassium hydroxide solution.

2.2. Preparation of Small Unilamellar Vesicles

1. Phosphatidylcholine isolated from chicken egg (eggPC) and dissolved in chloroform (20 mg/mL) (Avanti Polar Lipids, Alabaster, AL). Tightly seal the cover of the bottle with Parafilm and store at –20°C between uses (*see* **Note 1**).
2. 50-mL Glass round-bottom flasks and 3-mm glass beads.
3. HEPES/KOH buffer: 50 mM HEPES titrated with KOH, pH 7.4 (*see* **Note 2**). Buffer is stable at 4°C for up to 3 wk. Warm to the working temperature of 25°C before fluorescence experiments.
4. Sephadex G-25 (Amersham Biosciences/GE Healthcare, Piscataway, NJ).
5. 50-mL Size-exclusion column with a stopcock to regulate the flow of eluate.
6. Open-ended UltraClear™ centrifuge tubes (½ × 2½ in.²; 13 × 64 mm²) (Beckman, Fullerton, CA).
7. Sonicator (Sonifier® Cell Disruptor 350, Branson Sonic Power Company, Danbury, CT).

2.3. Preparation of Large Unilamellar Vesicles

1. Materials as described in **Subheading 2.2., steps 1–5**.
2. High-pressure extruder with a 25-mm diameter-wide sample compartment. Flow regulator attachment for gas tank.
3. 0.1-µm Nucleopore track-etch polycarbonate membranes and drain disks (25-mm diameter) (Whatman, Florham Park, NJ).

2.4. Isolation and Preparation of Rat Adipocytes

1. Two to three male Sprague-Dawley rats weighing 225–275 g (Charles River Laboratories, Wilmington, MA).
2. Surgical scalpel and scissors.
3. Modified Krebs-Ringer bicarbonate (MOPS-KRB) buffer (bicarbonate-free): 20 mM MOPS, 118 mM NaCl, 5 mM KCl, 1.1 mM MgSO$_4$, 1.1 mM KH$_2$PO$_4$, 2.5 mM CaCl$_2$, and 5.1 mM glucose, pH 7.4. Typical KRB buffers also include 25 mM bicarbonate (NaHCO$_3$) An equal volume of MOPS-KRB buffer supplemented with 3% (w/v) fraction V bovine serum albumin (BSA) must also be prepared for the initial steps of adipocyte isolation. BSA-free MOPS-KRB buffer is stable at 4°C for up to 3 wk. Add BSA to the buffer the day of the experiment. Warm to the working temperature of 37°C before tissue preparation and fluorescence experiments.
4. Collagenase B (Boehringer-Manheim, Indianapolis, IN).
5. Nitex Nylon mesh (250-µm pore-size) (3–300/50, Tetko, Briacliff Manor, NY) and a Swinnex filter holder (47-mm diameter) (Millipore, Billerica, MA).
6. Separation funnel with stopcock.

2.5. Stock Solution of ADIFAB

1. ADIFAB (Molecular Probes, Eugene, OR/FFA Sciences, San Diego, CA) is dissolved at 100 µM by adding 134 µL of storage buffer per 200 µg of lyophilized ADIFAB. ADIFAB is stable for more than 3 mo when stored at 4°C in storage buffer *(25)*. Storage buffer: 50 mM Tris-HCl, 1 mM ethylenediaminetetraacetic acid (EDTA), and 0.05% sodium azide, pH 8.0.

2.6. Fluorescence Measurements

1. Polystyrene cuvets with four clear sides and spinbars (spectral cell stir bars) (Fisher Scientific, Pittsburgh, PA).
2. Gel-loading pipet tips (1–200 µL) (VWR Scientific, Bridgeport, NJ).

2.7. The Pitfalls of Using ADIFAB to Measure Unbound FA Concentrations

1. Materials for the experiment are as described in **Subheading 2.6**.
2. HEPES/KOH buffer as described in **Subheading 2.2., step 3**.
3. Measuring buffer: 20 mM HEPES, 150 mM NaCl, 5 mM KCl, and 1 mM Na$_2$HPO$_4$, pH 7.4.
4. HEPES/KRB buffer: 20 mM HEPES, 118 mM NaCl, 5 mM KCl, 2.5 mM CaCl$_2$, 1.1 mM MgSO$_4$, 1.1 mM Na$_2$HPO$_4$, and 5.1 mM D-glucose, pH 7.4.
5. MOPS/KRB buffer: 20 mM MOPS, 118 mM NaCl, 5 mM KCl, 2.5 mM CaCl$_2$, 1.1 mM MgSO$_4$, 1.1 mM Na$_2$HPO$_4$, and 5.1 mM D-glucose, pH 7.4.
6. Hepes-EDTA-sucrose (HES) buffer: 10 mM HEPES, 250 mM sucrose, and 1 mM EDTA, pH 7.4.
7. Measuring buffer containing varying amounts of HEPES, NaCl, CaCl$_2$, and D-glucose.
8. HEPES/KOH buffer containing variable amounts of NaCl, Na$_2$SO$_4$, and Na$_2$HPO$_4$.

3. Methods

ADIFAB measures the concentration of unbound FA in an aqueous solution. Thus, one can add aliquots of FA to ADIFAB in buffer and measure changes in ADIFAB fluorescence up to the point wherein the FA begins to aggregate at its solubility limit *(21)*. As one knows how much FA was added, a relationship between the ADIFAB fluorescence and the concentration of unbound FA can be established. Using that information, one can then add a FA-binding moiety such as albumin *(26)*, FABP *(27)*, or membranes *(21,22,28)* into solution to determine the concentration of unbound FA in equilibrium with the third moiety. If one knows the total amount of FA added to the system and assumes that the amount of FA bound to ADIFAB is negligible, then one can evaluate equilibrium constants or the partition coefficients of different FA in equilibrium with albumin or membranes. This approach has also been applied to measure the relative amounts of different FA in the plasma *(29)*.

The laboratory uses ADIFAB in the presence of vesicles or cells in suspension to measure the *unbound* concentration of FA in equilibrium with the membrane following the partitioning of most FA molecules into the lipid bilayer **(Fig. 2)**. ADIFAB can also be used to measure indirectly the internalization/metabolism of FA in cells. As FA at the inner membrane leaflet is esterified into acyl-CoA, which is unable to diffuse back across the membrane, an inwardly directed FA concentration gradient is maintained across the membrane. This gradient results in the movement of FA from the outer to inner leaflet, and a net loss of FA from the external buffer with time. In the absence of an internal FA "sink" inside protein-free lipid vesicles ADIFAB fluorescence increases to a stable value following the addition of uncomplexed FA. However, the initial rapid increase in ADIFAB fluorescence is followed by a slow return to baseline (minutes) in adipocyte studies, which was shown to correlate with the metabolism of FA into TG *(11)*.

Herein, the effect of buffer constituents on the efficacy of ADIFAB for measuring the concentration of unbound FA is investigated. The accuracy of ADIFAB determinations of the concentration of unbound FA can be questioned because they have not been validated by independent methods and, in fact, differ from the values determined from other methods *(24)*. However, as the authors are more interested in comparing relative concentrations and changes in the partitioning of FA between different experiments, than in determining absolute concentrations of unbound FA, they do not independently validate the accuracy of the method. The results show that ADIFAB is highly sensitive to salts found in biological systems, raising questions about the reliability of this probe for measuring FA in systems that cannot be tightly controlled, although strategies are presented to allow reliable measurement in well-controlled in vitro systems.

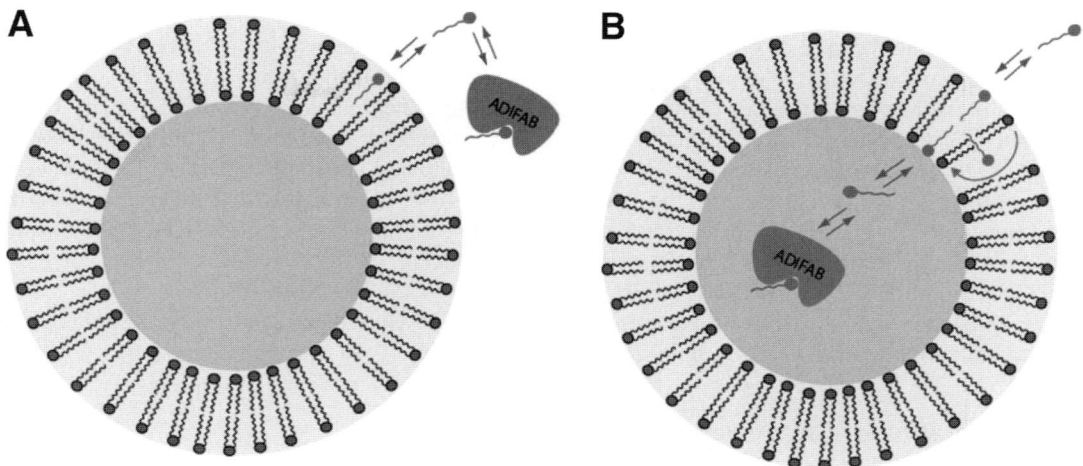

Fig. 2. Fluorescence assays utilizing the ADIFAB probe to detect unbound concentrations of FA. Upon addition of FA to a suspension of lipid vesicles, ADIFAB reports the concentration of FA in the external buffer that is in equilibrium with the membrane (**A**). ADIFAB can also be trapped inside vesicles to measure the concentration of FA in equilibrium with the inner leaflet (**B**). In this assay, kinetic changes in ADIFAB fluorescence reflect FA adsorption, flip–flop, and desorption. In the assay in panel A, kinetic changes in ADIFAB fluorescence reflect FA binding while the assay shown in panel B reflect the combined steps of adsorption, flip–flop and desorption.

3.1. FA Stock Solutions (K^+-Salt)

1. Weigh out approx 250 mg of pure FA (*see* **Note 3**).
2. Place the FA inside a glass scintillation vial and dissolve them in enough chloroform to establish a FA concentration range of 20–30 mg/mL. Quickly place a screw cap on the vial to prevent evaporation of the chloroform and consequent changes in the concentration of the dissolved FA.
3. Determine the dry weight of the dissolved FA in order to calculate the exact concentration of FA. Using a small aluminum-weighing dish, tare a microgram-scale microbalance. Use a 600 Series Hamilton syringe (Hamilton company, Reno, NV) to deliver exactly 5 µL of the chloroform/FA mixture to an aluminum-weighing NV dish. Using a hotplate, dry away the chloroform rapidly. Weigh and record the mass of the dried FA. Repeat this step at least five times and calculate the average mass of FA dried for each trial. Using the average mass and the volume of FA dried in each case, calculate the exact concentration of FA in the chloroform.
4. Dry the FA/chloroform mixture completely under a stream of N_2 gas and lyophilize the FA for 1 h to remove all traces of chloroform.
5. Dissolve the lyophilized FA in enough 0.1 *N* KOH to make a 10 m*M* stock solution of FA (K^+-salt). Adjust the pH of each stock to >10 to ensure complete dissolution of the FA into micelles (*see* **Note 4**).

3.2. Preparation of Small Unilamellar Vesicles

1. Place 50 mg of eggPC into a 50-mL round-bottom flask and dry under a vacuum in a rotary evaporator (*see* **Note 5**).
2. Place the dried lipid film in a lyophilizer for 1 h to remove all traces of chloroform.
3. Remove the lipids from the lyophilizer and place five to ten glass beads and the 2 mL of HEPES/KOH buffer into the round-bottom flask. Swirl gently the mixture by hand until all lipids have been removed from the sides of the flask. The lipid mixture will appear opaque.
4. Allow the lipids to hydrate either (1) overnight at 4°C or (2) for 1 h at room temperature (*see* **Note 6**).

5. Transfer the hydrated lipids to an open-ended polypropylene centrifuge tube. Place the tip of the sonicator into the lipid mixture and center it approx 1 mm from the bottom of the tube (*see* **Note 7**). A light flow of N_2 gas is delivered into the centrifuge tube through a small rubber tube connected to a pressurized gas tank (*see* **Note 1**). The rubber tubing is held in place by a piece of $\frac{1}{2}$ in. Teflon tape (Fisher Scientific Co., Pittsburgh, PA), which is stretched around the sonicator tip (*see* **Note 8**).
6. Partially submerge the sample tube into an ice water bath and pulse sonicate for 1 h (output control 3, 30% (duty cycle) or until the solution changes from being highly turbid to nearly clear in appearance (*see* **Note 9**). Following sonication, small unilamellar vesicles (SUVs) are formed when the lipid mixture has become completely translucent.
7. Centrifuge the SUV suspension for 30 min at approx 5000*g* (*see* **Note 10**).
8. Estimate the final eggPC concentration in the form of SUV based on the quantity of eggPC dried and the volume of SUV collected (*see* **Note 11**).

3.3. Preparation of Large Unilamellar Vesicles

1. Prepare the lipids by performing **Subheading 3.2., steps 1–4**.
2. Place liquid N_2 into a small durer and prepare a beaker full of warm/hot water. Place the round-bottom flask containing the lipids into the liquid N_2 while constantly swirling the lipids by hand. Do not remove the lipids from the liquid N_2 until the mixture is completely frozen. Transfer the round-bottom flask to the warm/hot water bath again swirling the lipid mixture by hand. Do not remove the lipids from the water until the mixture is completely melted. Repeat this freeze–thaw cycle a total of five times (*see* **Note 12**). The lipid mixture will appear opaque and turbid.
3. Place a drain disk and a single 0.1-μm pore size membrane inside a high-pressure extruder. Attach the extruder to a high-pressure gas tank containing N_2 gas.
4. Wash the membrane by placing deionized water into the extruder. Force the water through the membrane two times before adding the lipid mixture.
5. Add the lipid mixture to the extruder and force the sample through the membrane and drain disk two times, each time collecting the lipids in a 15-mL conical tube. Remove the drain disk and membrane and replace it with two new 0.1-μm pore size membranes. Wash these membranes with water as described in **step 4**. Repeat the lipid extrusion through these membranes 10 times. The lipid mixture will become completely translucent as large unilamellar vesicles (LUVs) are formed in suspension (*see* **Note 13**).
6. Estimate the final eggPC concentration in LUV as described in **Subheading 3.2., step 8**.

3.4. Isolation and Preparation of Rat Adipocytes

1. Anesthetize and sacrifice two to three adult male Sprague-Dawley rats weighing 225–275 g.
2. Surgically remove the epididymal and perirenal fat pads and place the tissue into a 50-mL conical tube containing 10–15 mL of warm MOPS/KRB buffer supplemented with 3% (w/v) BSA.
3. Cut the tissue into small pieces using surgical scissors. This is accomplished by using the scissors to rapidly blend the mixture of fat and buffer inside the conical tube for approx 5 min.
4. Incubate the chopped adipose tissue in a water bath shaking at 100–120 oscillations/min for 30 min at 37°C with collagenase B. Add 10–15 mg of collagenase B per 5 g of fat tissue to complete the digestion.
5. Filter the isolated cells through nylon mesh into a new 50-mL conical tube.
6. Wash the cells three times in a separation funnel with warm MOPS/KRB buffer warmed to 37°C (BSA-free). After each wash, allow 2–3 min for the cells to float to the top and form an oil-like suspension (*see* **Note 14**). Place the washed cells in a 50-mL conical tube.
7. Prepare the isolated adipocytes for fluorescence measurements by removing excess buffer from beneath the oil-like cell suspension, until the final volume of the cell suspension to buffer is 25% (v/v).
8. Place the cells in a 37°C still water bath and allow them to "relax" for 30 min.

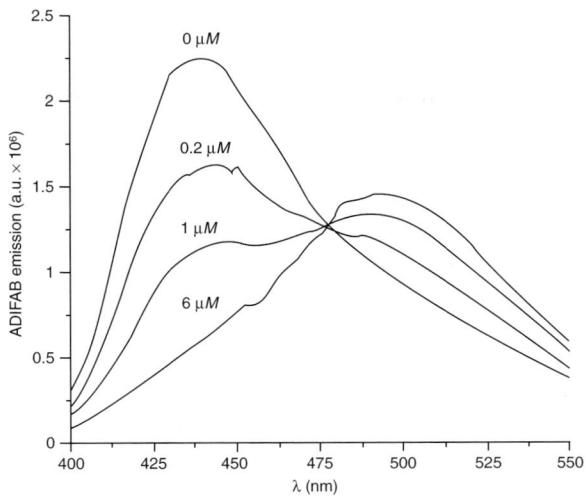

Fig. 3. Emission spectra of ADIFAB (0.2 μM) in 20 mM measuring buffer taken in the presence of increasing amounts of oleate. Unbound oleate was delivered into the cuvet and the emission spectrum was measured. ADIFAB fluorescence is typically measured as a ratio of emission intensities at 505 and 432 nm. Increasing concentrations of FA change the ADIFAB ratio by reducing emission at 432 nm and increasing emission at 505 nm.

3.5. Fluorescence Measurements—Calibration of ADIFAB Fluorescence With Unbound FA Concentration

1. Add ADIFAB to a polystyrene cuvet containing 3 mL of HEPES/KOH buffer to a final concentration of 0.2 μM.
2. Place the cuvet into a fluorimeter (*see* **Note 15**) to monitor ADIFAB fluorescence.
3. Open a fluorescence software program (Datamax for Windows) and select the desired fluorescence protocol for measuring either the emission spectrum or the dual-emission fluorescence ratio (R) of ADIFAB. Open the excitation and emission slits to at least 5 nm. The emission spectrum of ADIFAB is measured by exciting the probe at 386 nm. In the absence of FA, the ADIFAB spectrum exhibits a major peak with an emission maximum at 432 nm and a minor peak with an emission maximum near 485–490 nm (**Fig. 3**).
4. Add small amounts of FA (0.2–1 μM increments) to the cuvet using a Hamilton syringe (5 μL capacity) and measure the emission spectrum of ADIFAB after each addition of FA. With increasing FA concentration, the emission signals at 432 and 505 nm should decrease and increase, respectively. Typical spectra for this type are shown in **Fig. 3**.
5. Calculate the ADIFAB emission ratio using the following equation: $R = 505/439$ nm.
6. Plot a calibration curve using ADIFAB R on the y-axis and FA concentration on the x-axis (*see* **Note 16**).

3.6. Using ADIFAB to Measure the Binding of FA to Vesicles and Isolated Rat Adipocytes

1. For vesicles, add a desired amount of SUV or LUV to a polystyrene cuvet containing a mini magnetic stirrer; bring to a total volume of 3 mL with HEPES/KOH buffer. For isolated rat adipocytes, add 3 mL of the prepared cell suspension (in MOPS/KRB buffer) to a polystyrene cuvet containing a mini magnetic stirrer (*see* **Note 17**).
2. Add ADIFAB to the cuvet to a final concentration of 0.2 μM. Place the cuvet into the fluorimeter and select the desired fluorescence protocol (*see* **Note 15**) for measuring the dual-emission fluorescence ratio (R) of ADIFAB as described in **Subheading 3.5., step 3**.

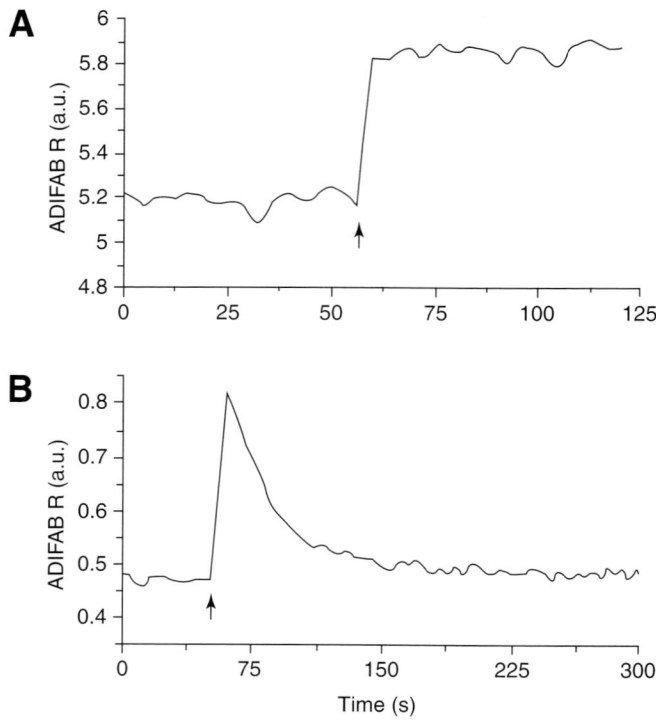

Fig. 4. Typical ADIFAB data obtained upon adding unbound oleate to vesicles and isolated rat adipocytes. FA rapidly bind to the external leaflet of vesicles (**A**) and adipocytes (**B**) until equilibrium is reached between FA in the membrane and external buffer. ADIFAB fluorescence increases to a maximal value in cell experiments, but slowly returns to baseline as FA is metabolized intracellularly. Loss of intracellular FA results in the inward flux of FA across the membrane, which causes the concentration of FA in the external buffer to decrease.

3. Deliver the desired volume of FA stock solution into the sample using a micropipetor equipped with a gel-loading pipet tip. Insert the micropipetor into the injection port of the fluorimeter while monitoring changes in the ADIFAB R. ADIFAB fluorescence rapidly increases to a stable level when FA is added to vesicles, but slowly decreases after reaching a maximal value in rat adipocyte experiments (*see* **Note 18**). Typical data for vesicles and cells are shown **Fig. 4**.
4. Determine the concentration of FA remaining in the external buffer at equilibrium by extrapolation from an ADIFAB calibration curve that was plotted using data obtained under the same experimental conditions. ADIFAB calibration curves are described in **Subheading 3.5., step 7**.
5. Determine the concentration of FA remaining in the external buffer at equilibrium by another method. Using obtained ADIFAB R values, the unbound concentration of FA is given by the following equation (*see* **Note 19**):

$$FA = \left[K_d Q (R - R_0)\right] / \left(R_{max} - R\right)$$

6. Using the total amount of FA added to cells or vesicles and the calculated value for unbound FA, calculate the amount of FA bound to the membrane (*see* **Note 20**).

3.7. The Pitfalls of Using ADIFAB to Measure Unbound FA Concentrations

1. For the experiments described in this subheading, add 0.2 µM ADIFAB to polystyrene cuvets containing buffers of varying composition, pI, HEPES concentration, and NaCl concentration.
2. Place a cuvet into the fluorimeter and select the desired fluorescence protocol for measuring the emission spectrum of ADIFAB as described in **Subheading 3.5., step 3**.
3. Measure the emission spectrum in the absence of FA.

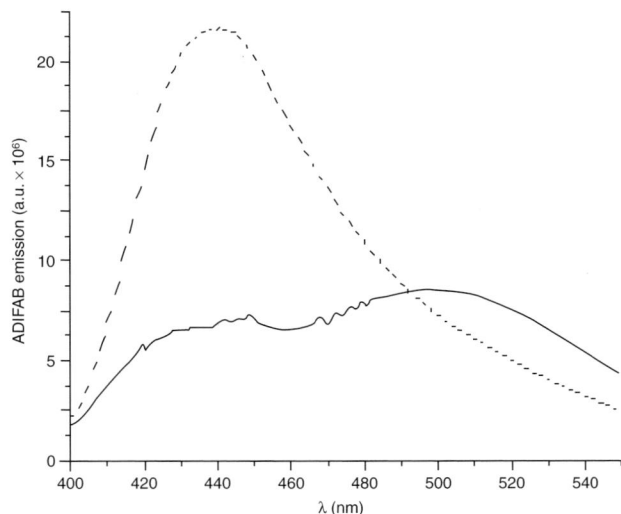

Fig. 5. Effects of different buffers on the emission spectra of ADIFAB taken in the absence of FA. These spectra were obtained from ADIFAB (0.2 μM) placed in 20 mM measuring buffer containing high concentrations of salts (dashed line; *see* **Subheading 3.7.**) or 100 mM HEPES/KOH buffer, which contains no added salts (solid line, *see* **Subheading 3.7.**). Emission at 432 nm is significantly decreased in the HEPES/KOH buffer such that the ADIFAB ratio ($R = \lambda_{505}/\lambda_{432}$) is effectively increased despite the absence of FA.

3.7.1. The Effect of Different Buffers on the ADIFAB Emission Spectrum

1. Prepare 1 L of 100 mM HEPES/KOH buffer and 20 mM measuring buffer. Adjust the pH to 7.4 using KOH.
2. Place 3 mL of 100 mM HEPES/KOH buffer and 3 mL of 20 mM measuring buffer into two separate cuvets and measure the ADIFAB emission spectrum in the absence of FA. (*see* **Note 21**) Spectra obtained for these buffers are shown in **Fig. 5**.
3. *Test the effect of HEPES concentration on ADIFAB emission:* place 5 mL of the 100 mM HEPES/KOH buffer into two 50-mL conical tubes. Using deionized water, dilute the 100 mM HEPES/KOH to make 250 mL of 20 mM and 50 mM HEPES/KOH buffer. Place 3 mL of HEPES/KOH buffer (20, 50, or 100 mM) into separate cuvets and measure the ADIFAB emission spectra (*see* **Note 22**).
4. *Test the effect of NaCl concentration on ADIFAB emission:* place 100 mL of 50 mM HEPES/KOH buffer into a 100-mL glass beaker. Add NaCl to the 50 mM HEPES/KOH buffer to a final concentration of 150 mM. Place 10 mL of this buffer into two 50-mL conical tubes. Use the 50 mM HEPES/KOH (without NaCl) to dilute the buffer in each tube until the NaCl concentration is reduced to 100 and 50 mM. Place 3 mL of each 50 mM HEPES/KOH buffer (containing 50, 100, or 150 mM NaCl) into separate cuvets and measure the ADIFAB emission spectra (*see* **Note 23**).
5. *Test the effect of pI and salt composition on ADIFAB emission by replacing NaCl with an isotonic amount of Na_2SO_4* (*see* **Note 24**): place 100 mL of 50 mM HEPES/KOH buffer into a 100-mL glass beaker. Add Na_2SO_4 to the 50 mM HEPES/KOH buffer to a final concentration of 50 mM Na_2SO_4. Place 3 mL of this buffer into a cuvet and measure the ADIFAB emission spectrum (*see* **Note 25**).

3.7.2. The Effect of Different Buffers on ADIFAB Sensitivity to FA

1. *Test the effect of pI on ADIFAB sensitivity to FA by using a HEPES/KOH buffer containing a low concentration of NaCl:* place 3 mL of 50 mM HEPES/KOH buffer (containing 50 mM NaCl) into a cuvet and measure the ADIFAB emission spectrum in the presence of varying concentrations of FA (*see* **Note 26**).

Table 1
Base Composition of the Buffers Used to Study ADIFAB Sensitivity to Changing Concentrations of Oleate

Buffer components (mM)	HEPES/KRB buffer	MOPS/KRB buffer	HES buffer	Measuring buffer
HEPES	20	–	10	20
MOPS	–	20	–	–
NaCl	118	118	–	150
KCl	5	5	–	5
MgSO$_4$	1.1	1.1	–	–
CaCl$_2$	2.5	2.5	–	–
Na$_2$HPO$_4$	–	–	–	1
KH$_2$PO$_4$	1.1	1.1	–	–
Glucose	5.1	5.1	–	–
Sucrose	–	–	250	–
EDTA	–	–	1	–

2. *Prepare various buffers that are commonly used in fluorescence studies of FA-binding to vesicles and cells:* prepare 300 mL of 50 mM HEPES/KOH, 20 mM measuring buffer, 20 mM HEPES/KRB buffer, 20 mM MOPS/KRB buffer, and 10 mM HES buffer (*see* **Note 27**). Adjust the pH of all buffers to 7.4 using KOH. Buffer compositions are listed in **Table 1**.
3. *Test the effect of pI and salt composition on the ADIFAB calibration plot by supplementing HEPES/KOH buffer containing low and high concentrations of NaCl with a fixed concentration of Na$_2$SO$_4$:* place 100 mL of 50 mM HEPES/KOH buffer (containing 50 or 150 mM NaCl) into a 100-mL glass beaker. Add Na$_2$SO$_4$ to the buffer to a final concentration of 40 mM Na$_2$SO$_4$. Place 3 mL of each 50 mM HEPES/KOH buffer (containing 50 or 150 mM NaCl together with 40 mM Na$_2$SO$_4$) into separate cuvets and measure ADIFAB emission spectra in the presence of increasing concentrations of FA (*see* **Note 28**). Record the ADIFAB R values for each concentration of FA tested.
4. *Test the effect of phosphate anions on the ADIFAB calibration plot by supplementing HEPES/KOH buffer containing a high concentration of NaCl with a fixed concentration of Na$_2$HPO$_4$:* place 100 mL of 50 mM HEPES/KOH buffer (containing 150 mM NaCl) into a 100-mL glass beaker. Add Na$_2$HPO$_4$ to the buffer to a final concentration of 1 mM Na$_2$HPO$_4$. Place 3 mL of the 50 mM HEPES/KOH buffer (containing 150 mM NaCl together with 1 mM Na$_2$HPO$_4$) into a cuvet and measure ADIFAB emission spectra in the presence of increasing concentrations of FA (*see* **Note 29**). Record the ADIFAB R values for each concentration of FA tested.
5. *Test the effect of increased HEPES concentration on the ADIFAB calibration plot by supplementing measuring buffer with additional HEPES:* Place 100 mL of 20 mM measuring buffer into a 100-mL glass beaker. Add HEPES to the buffer to increase the HEPES concentration to 50 mM. Place 3 mL of the 50 mM measuring buffer into a cuvet and measure ADIFAB emission spectra in the presence of increasing concentrations of FA (*see* **Note 30**). Record the ADIFAB R values for each concentration of FA tested.
6. *Test the effect of glucose on the ADIFAB calibration plot by supplementing measuring buffer with D-glucose:* place 100 mL of measuring buffer (20 and 50 mM) into two 100-mL glass beakers. Add D-glucose to each buffer to a final concentration of 5 mM glucose. Place 3 mL of each buffer into separate cuvets and measure ADIFAB emission spectra in the presence of increasing concentrations of FA (*see* **Note 31**). Record the ADIFAB R values for each concentration of FA tested.

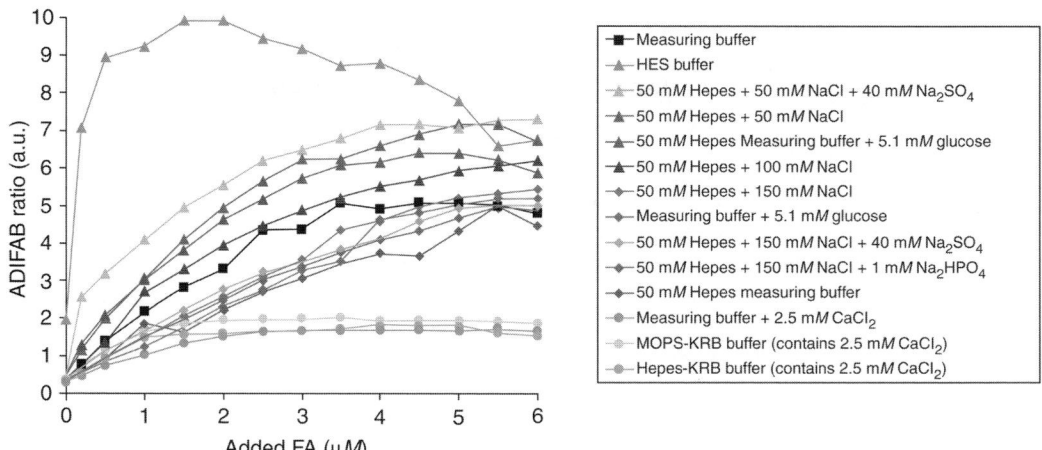

Fig. 6. Effects of various buffers on ADIFAB titration curves. ADIFAB fluorescence ($R = \lambda_{505}/\lambda_{432}$) was measured in the presence of increasing amounts of unbound oleate (0–6 µM). The titration curve of ADIFAB in 20 mM measuring buffer was chosen as the standard response for ADIFAB in these experiments (black). ADIFAB was more sensitive to FA in lower pI buffers, which produced more hyperbolic titration curves (blue color shades). HES buffer was the lowest pI buffer and produced the most hyperbolic curve. ADIFAB fluorescence increased more linearly over a higher range of oleate concentrations in higher pI buffers (green color shades). The addition of $CaCl_2$ to any buffer saturated the titration curves at approx 2 µM oleate (orange color shades). These data reveal that the standard measuring buffer for ADIFAB does not accurately report FA concentration more than 4 µM. Under such conditions, high pI buffers composed principally of Nacl should be used and Ca^{+2} ions should be avoided.

7. *Test various buffers that are commonly used in fluorescence studies of FA-binding to vesicles and cells:* place 3 mL of 20 mM measuring buffer, 20 mM HEPES/KRB buffer, 20 mM MOPS/KRB buffer, and 10 mM HES buffer into separate cuvets, and measure ADIFAB emission spectra in the presence of increasing concentrations of FA (*see* **Note 32**). Record the ADIFAB R values for each concentration of FA tested.
8. *Test the effect of Ca^{2+} on the ADIFAB calibration plot by supplementing measuring buffer with $CaCl_2$:* place 100 mL of 50 mM measuring buffer into a 100-mL glass beaker. Add $CaCl_2$ to the buffer to a final concentration of 2.5 mM $CaCl_2$. Place 3 mL of the buffer into a cuvet and measure ADIFAB emission spectra in the presence of increasing concentrations of FA (*see* **Note 33**). Record the ADIFAB R values for each concentration of FA tested.
9. Plot a calibration curve for each buffer using ADIFAB R and FA concentrations as described in **Subheading 3.5., step 6** (*see* **Note 34**). The curves for various buffers are shown in **Fig. 6**.

4. Notes

1. To prevent oxidation of unsaturated lipid chains, trap either N_2 or Ar gas inside the bottle of eggPC before storage at –20°C.
2. HEPES concentration can range between 20 and 100 mM. Modify the buffer concentration to optimize signal-to-noise depending on the fluorescence probes being used in the dual-fluorescence assay; *see* Chapter 16 for more information.
3. Density must be used to calculate the amount of pure mono- and polyunsaturated FA, which are liquid at room temperature.
4. To produce FA stock solutions in ethanol or dimethyl sulfoxide, add the desired amount of solvent to pure FA.

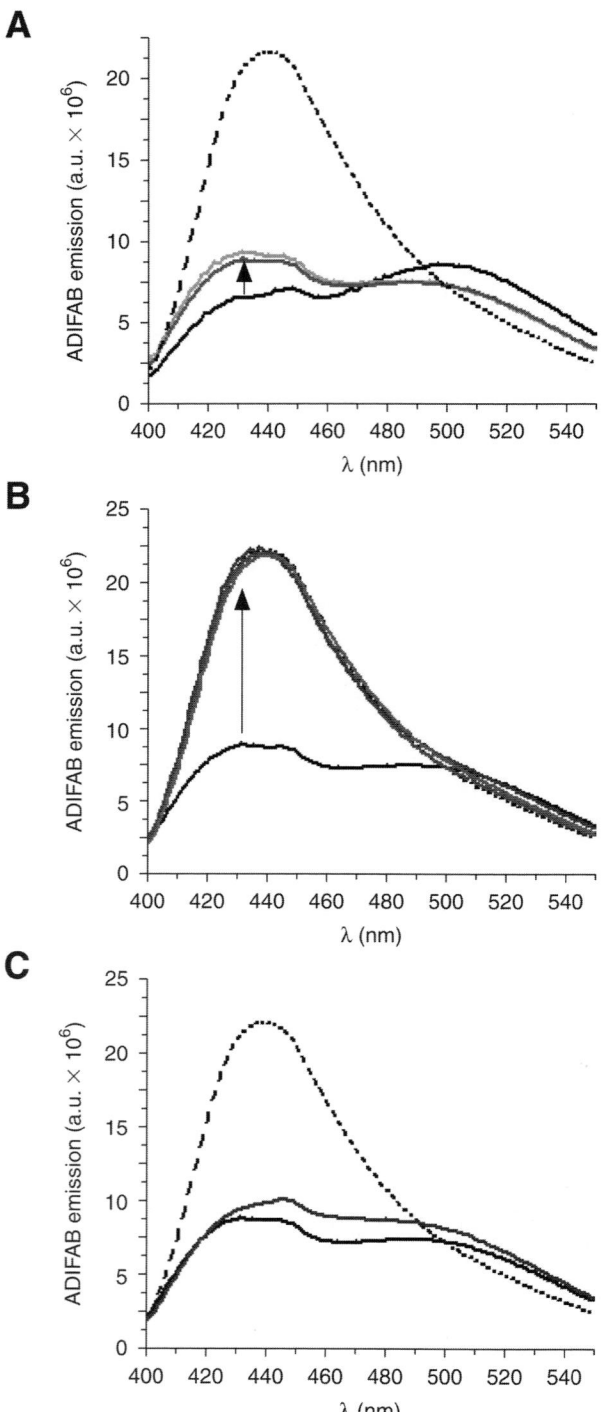

Fig. 7. Effects of buffer components on the emission spectra of ADIFAB, taken in the absence of FA. These spectra were obtained to determine the buffer components required to optimize the emission profile of ADIFAB. Emission spectra of ADIFAB in 100 mM HEPES buffer (solid line) and 20 mM measuring buffer (dashed line) are shown in each panel for reference. ADIFAB emission spectra were measured in HEPES/KOH buffers of decreasing HEPES concentration (**A**), increasing NaCl concentration (**B**), and with 50 mM Na$_2$SO$_4$ (**C**). Emission at 432 nm increased as the HEPES

5. During this time, the round-bottom flask containing the lipids is spun over a heated water bath to facilitate evaporation of the chloroform and create a thin film of dried eggPC.
6. The lipid hydration must be carried out above the gel–liquid crystalline transition temperature (T_m). The T_m of eggPC is approx 0°C.
7. It has been found that this placement of the sonicator tip maximizes the dispersal of energy throughout the entire sample and minimizes the loss of sample through splashing.
8. Care must be taken not to wrap the Teflon tape around the sonicator tip too tightly so that the N_2 gas can escape.
9. The sonication must be carried out above the T_m of the lipid. The ice water bath prevents overheating of the sample, the consequent breakdown of lipid acyl chains and the generation of free FA.
10. Centrifugation removes any metal particulates released from the sonicator tip during the sonication step.
11. The exact eggPC concentration is determined by assaying total phosphorous content (30).
12. The repetitive freeze–thaw cycles assist the formation of unilamellar lipid membranes. If this step is not performed, the subsequent extrusion of the lipids will be slow and difficult.
13. The extrusion must be carried out above the T_m of the lipid.
14. The TG stored in these cells make them float to the top of any aqueous suspension.
15. The laboratory uses a Spex® Fluoromax-2 fluorimeter (Jobin Yvon, Edison, NJ) equipped with Datamax for windows software.
16. This ratio increases linearly with the concentration of unbound FA and saturates at the solubility limit of the particular FA being measured. This curve can be used to estimate the concentration of unbound FA that is in equilibrium with lipid membranes following addition of FA to a suspension of cells or vesicles.
17. For experiments involving cells, the base of the sample holder is warmed to 37°C by a flow of water pumped into the sample compartment from an external temperature-controlled water bath.
18. In SUV and LUV, there is no internal sink to drive the influx of additional FA across the membrane and the FA are poorly soluble in the aqueous compartment. As FA are metabolized in the adipocyte, they are replaced by FA from the external buffer. Thus, the concentration of FA in equilibrium with ADIFAB decreases, as does the fluorescence of ADIFAB.
19. R_0 is the ADIFAB ratio measured in the absence of FA including background subtraction at each wavelength. R is the ADIFAB ratio measured in the presence of FA, including background subtraction at each wavelength. R_{max} is the ADIFAB ratio when ADIFAB is saturated and typically has a value of 11.5 (22). Q is the ratio of ADIFAB emission intensities measured at 432 nm in the presence of zero and saturating concentrations of FA and typically has a value of 19.5 (22). K_d is the ADIFAB dissociation constant for the FA being studied. The ADIFAB K_d for some common FA have been published previously (21,22).
20. Applications of these calculations and the use of ADIFAB for measuring the binding of FA to membranes have been reviewed elsewhere (5).
21. For fluorescence experiments with vesicles containing an entrapped pH dye, the laboratory typically uses a simple 100 mM HEPES/KOH buffer containing no salts. The difference in the ADIFAB emission spectrum when using this buffer results from higher HEPES concentration and or the absence of one or more salts found in the measuring buffer.
22. Typical data are shown in **Fig. 7**. Dilution of HEPES concentration has a minor effect on ADIFAB emission by increasing the signal observed at 432 nm to a small extent. Decreasing the concentration to less than 50 mM does not have any additional benefit.

Fig. 7. *(Continued)* concentration was reduced to 50 mM (dark green) and 25 mM (light green). Emission at 432 nm increased to expected values (dashed line) in the presence of 50 mM (light blue), 100 mM (light purple), and 150 mM NaCl (dark purple). Replacement of 150 mM NaCl with and isotonic concentration of Na_2SO_4 (dark red) did not restore emission at 432 nm to normal intensity suggesting a role for Cl^- in mediating ADIFAB emission at this wavalength.

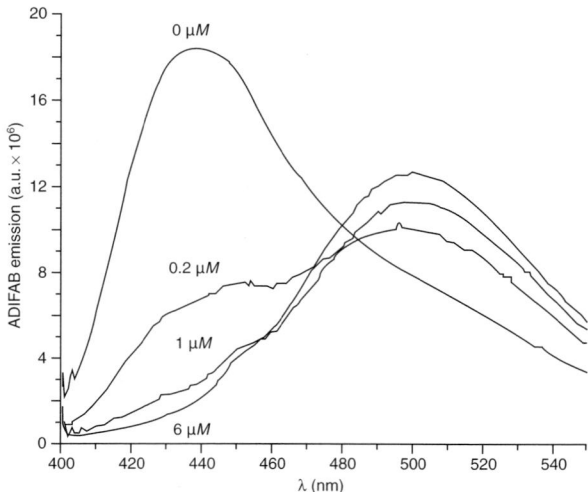

Fig. 8. Emission spectrum of ADIFAB (0.2 mM) in 50 mM HEPES/KOH buffer (+50 mM Nacl [low salt]) and measured in the presence of increasing amounts of oleate. In the presence of low salt, emission at 432 nm is more sensitive to lower amounts and less sensitive to higher amounts of FA.

23. Typical data are shown in **Fig. 7**. NaCl modestly increases the emission signal at 432 nm relative to 505 nm to a level reported previously *(21,22)*. Although NaCl restores the ADIFAB emission spectrum to what is expected (compare with **Fig. 3**, zero micromolar oleate); this effect does not appear to be strongly dependent on the concentration of NaCl because 50 mM and 150 mM NaCl produce similar results.
24. The pI of a solution is a quantity representing the interactions of ions with water molecules and other ions in solution and is calculated according to the following equation:

$$I = \tfrac{1}{2} \Sigma\, m_i(z_i)^2$$

where m_i is the molar concentration of ion (i), z_i is the valence factor for ion (i), and I is the total pI.
25. Typical data are shown in **Fig. 7**. The emission maximum at 432 nm is slightly higher than that observed with no salts, but significantly lower than that observed with only NaCl present. These results suggest that Cl$^-$ ions may be required for optimal performance of the ADIFAB probe when reporting FA concentrations. Thus, the correct emission profile of ADIFAB is not dependent on the total composition of the "measuring buffer" used to first describe the function of the ADIFAB probe, but on the presence of certain ions.
26. This buffer has an ionic strength of $I = 50$ mM (compare with **Fig. 3**, ionic strength of $I = 150$ mM). Typical spectra obtained in the presence of 0, 0.2, 0.5, and 1 µM oleate are shown in **Fig. 8**. There are significant changes in the spectrum as oleate concentration increases from 0 to 0.2 µM, with smaller changes occurring with higher amounts of added FA. These findings suggest that ADIFAB is overly sensitive to very low levels of FA in low pI buffers.
27. Because effects were observed on ADIFAB emission spectra obtained in the absence of FA, it was hypothesized that it was possible for pI, NaCl concentration, and overall buffer composition to have an effect on the sensitivity of ADIFAB to changing concentrations of unbound FA. As measuring buffer was used to obtain the ADIFAB K_d for several common FA *(22)*, it was used as the standard for ADIFAB fluorescence titration. If these K_d values are used to calculate the concentration of unbound FA, a buffer that performs similar to measuring buffer must be chosen.
28. The addition of 40 mM Na$_2$SO$_4$ to a low pI buffer containing 50 mM Cl$^-$ ions increases the sensitivity of ADIFAB to low concentrations of oleate (<0.2 µM). In this buffer, the calibration

curve is more hyperbolic between 0 and 1 μM oleate when compared with the same buffer in the absence of Na_2SO_4. The addition of Na_2SO_4 to a high pI buffer containing 150 mM Cl⁻ "ions" decreases the sensitivity of ADIFAB to FA. In this buffer, the calibration curve is nearly linear over the range of 0–6 μM oleate. Because both salts contain Na⁺ ions, this effect is likely caused by the relative amounts of Cl⁻ and SO_4^{-2} ions. The concentrations of unbound FA used in this series of studies were 0, 0.2, 0.5, 1, 1.5, 2, 2.5, 3, 3.5, 4, 4.5, 5, 5.5, and 6 μM.

29. The concentration of Na_2HPO_4 added to the buffer in this experiment was chosen to match the concentration found in measuring buffer (the standard buffer used for measuring ADIFAB fluorescence). This experiment suggests that HPO_4^{-2} and PO_4^{-3} ions do not have an effect on the sensitivity of ADIFAB to changing concentrations of FA.

30. Increasing the HEPES concentration in the measuring buffer causes the ADIFAB calibration curve to become more linear at higher concentrations of FA when compared with 20 mM measuring buffer.

31. Adding glucose to 20 mM measuring buffer causes the ADIFAB calibration curve to become more linear at higher concentrations of FA. However, adding glucose to 50 mM measuring buffer causes ADIFAB to become more sensitive to changes in FA concentration.

32. Measuring buffer is the standard buffer used to calibrate changes in ADIFAB R with changes in unbound FA concentration. The ADIFAB calibration plot saturates at 6 μM oleate, the solubility limit of oleate, as shown in a previously published study *(22)*. However, the ADIFAB calibration plot saturates at 2 μM oleate when adding FA to HEPES/KRB and MOPS/KRB buffers. The low pI of HES buffer caused ADIFAB to be hypersensitive to less than 1 μM concentrations of oleate. In this buffer, the ADIFAB calibration plot saturates at 1 μM oleate. To estimate the solubility limit of FA or measure unbound FA concentrations in vivo, buffers containing physiological concentrations of specific salts should be used.

33. Addition of 2.5 mM $CaCl_2$ prevents the normal reduction of emission signal at 432 nm in response to less than 2 μM oleate. This causes the ratiometric fluorescence of ADIFAB to saturate at 2 μM FA, which is much lower than any other buffer that does not contain Ca^{+2}. This is supported by the finding that the solubility of oleate is decreased in the presence of calcium *(21)*. This experiment suggests that the early saturation of ADIFAB R observed in HEPES/KRB and MOPS/KRB buffers is because of the presence of 2.5 mM $CaCl_2$ in these buffers.

34. ADIFAB sensitivity to less than 0.2 μM FA increases in low pI buffers. In these buffers, the calibration curves are more hyperbolic at less than 2 μM oleate. Although the slope of each curve varies between 0 and 1 μM oleate, ADIFAB fluorescence increases nearly linearly with increasing FA concentration in several buffers. Buffers containing at least 100 mM NaCl produce the most linear calibration curves, and despite some variance, are similar to the standard measuring buffer. The linear response of ADIFAB allows quantitative measurements of unbound FA concentrations to be made accurately. Changes in these titration curves reflect the effects of pI on the ADIFAB probe and the solubility of the FA being measured. Because FA solubility changes with pI, the K_d and Q from ADIFAB (*see* **Note 19** above) will change with buffer composition. Therefore, it is necessary to calibrate ADIFAB fluorescence in the particular experimental buffer being used in any study.

References

1. Brinkmann, J. F., Abumrad, N. A., Ibrahimi, A., van der Vusse, G. J., and Glatz, J. F. C. (2002) New insights into long-chain fatty acid uptake by heart muscle: a crucial role for fatty acid translocase/CD36. *Biochem. J.* **367**, 561–570.
2. Zhang, X., Fitzsimmons, R. L., Cleland, L. G., et al. (2003) CD36/fatty acid translocase in rats: distribution, isolation from hepatocytes, and comparison with the scavener receptor SR-B1. *Lab. Invest.* **83**, 317–332.
3. Civelek, V. N., Hamilton, J. A., Tornheim, K., Kelly, K. L., and Corkey, B. E. (1996) Intracellular pH in adipocytes: effects of free fatty acid diffusion across the plasma membrane, lipolytic agonists, and insulin. *Proc. Natl. Acad. Sci. USA* **93**, 10,139–10,144.
4. Hamilton, J. A. (1998) Fatty acid transport: difficult or easy? *J. Lipid Res.* **39**, 467–481.

5. Hamilton, J. A. and Kamp, F. (1999) How are free fatty acids transported in membranes? Is it by proteins or by free diffusion through the lipids? *Diabetes* **48,** 2255–2269.
6. Hamilton, J. A., Johnson, R. A., Corkey, B., and Kamp, F. (2001) Fatty acid transport: the diffusion mechanism in model and biological membranes. *J. Mol. Neurosci.* **16,** 99–107.
7. Hamilton, J. A., Guo, W., and Kamp, F. (2002) Mechanism of uptake of long-chain fatty acids: do we need cellular proteins? *Mol. Cell Biochem.* **239,** 17–23.
8. Kamp, F. and Hamilton, J. A. (1992) pH gradients across phospholipid membranes caused by fast flip-flop of unionized fatty acids. *Proc. Natl. Acad. Sci. USA* **89,** 11,367–11,370.
9. Kamp, F., Zakim, D., Zhang, F., Noy, N., and Hamilton, J. A. (1995) Fatty acid flip-flop in phospholipid bilayers is extremely fast. *Biochemistry* **34,** 11,928–11,937.
10. Kamp, F. and Hamilton, J. A. (1993) Movement of fatty acids, fatty acid analogues, and bile acids across phospholipid bilayers. *Biochemistry* **32,** 11,074–11,085.
11. Kamp, F., Guo, W., Souto, R., Pilch, P., Corkey, B. E., and Hamilton, J. A. (2003) *J. Biol. Chem.* **278,** 7988–7995.
12. Hamilton, J. A. and Guo, W. (2001) Fatty acid uptake and metabolism in HepG2 cells. Abstract 365A, 45th annual Biophysical Society meeting, Baltimore, MD.
13. Guo, W., Huang, N., Cai, J., Xie, W., and Hamilton, J. A. (2006) Fatty acid transport and metabolism in HepG2 cells. *Am. J. Physiol. Gastrointest. Liver Physiol.* **290,** 528–534.
14. Zhang, F., Lücke, C., Baier, L. J., Sacchettini, J. C., and Hamilton, J. A. (1997) Solution structure of human intestinal fatty acid binding protein: implications for ligand entry and exit. *J. Biomol. NMR* **9,** 213–228.
15. Hodsdon, M. E., Ponder, J. W., and Cistola, D. P. (1996) The NMR solution structure of intestinal fatty acid-binding protein complexed with palmitate: application of a novel distance geometry algorithm. *J. Mol. Biol.* **264,** 585–602.
16. Lücke, C., Gutierrez-Gonzalez, L. H., and Hamilton, J. A. (2003) Intracellular lipid binding proteins: evolution, structure, and ligand binding. Wiley-VCH GmbH & Co. KGaA, Weinheim, Germany, 95–118.
17. Sacchettini, J. C., Gordon, J. I., and Banaszak, L. J. (1988) The structure of crystalline *Eschericia coli*-derived rat intestinal fatty acid-binding protein at 2.5-A resolution. *J. Biol. Chem.* **263,** 5815–5819.
18. Prendergast, F., Meyer, M., Carlson, G., Iida, S., and Potter, J. (1983) Synthesis, spectral properties, and use of 6-acryloyl-2-dimethylaminonaphthalene (Acrylodan). A thiol-sensitive, polarity-sensitive fluorescent probe. *J. Biol. Chem.* **258,** 7541–7544.
19. Scapin, G., Gordon, J. I., and Sacchettini, J. C. (1992) Refinement of the structure of recombinant rat intestinal fatty acid-binding apoprotein at 1.2-A resolution. *J. Biol. Chem.* **267,** 4253–4269.
20. Hamilton, J. A. (2002) How fatty acids bind to proteins: the inside story from protein structures. *Prostaglandins Leukot Essent. Fatty Acids* **67,** 65–72.
21. Richieri, G. V., Ogata, R. T., and Kleinfeld, A. M. (1992) A fluorescently labeled intestinal fatty acid binding protein: interactions with fatty acids and its use in monitoring free fatty acids. *J. Biol. Chem.* **267,** 23,495–23,501.
22. Richieri, G. V., Ogata, R. T., and Kleinfeld, A. M. (1999) The measurement of free fatty acid concentration with the fluorescent probe ADIFAB: a practical guide for the use of the ADIFAB probe. *Mol. Cell Biochem.* **192,** 87–94.
23. Kampf, J. P. and Kleinfeld, A. M. (2004) Fatty acid transport in adipocytes monitored by imaging intracellular free fatty acid levels. *J. Biol. Chem.* **279,** 35,775–35,780.
24. McArthur, M. J., Atshaves, B. P., Frolov, A., Foxworth, W. D., Kier, A. B., and Schroeder, F. (1999) Cellular uptake and intracellular trafficking of long chain fatty acids. *J. Lipid Res.* **40,** 1371–1383.
25. FFA Sciences website, http:www.ffasciences.com. Last accessed April 2007.
26. Richieri, G. V., Anel, A., and Kleinfeld, A. M. (1993) Interactions of long-chain fatty acids and albumin: determination of free fatty acid levels using the fluorescent probe ADIFAB. *Biochemistry* **32,** 7574–7580.

27. Richieri, G. V., Ogata, R. T., and Kleinfeld, A. M. (1996) Kinetics of fatty acid interactions with fatty acid binding proteins from adipocyte, heart, and intestine. *J. Biol. Chem.* **271,** 31,068–31,074.
28. Richieri, G. V. and Kleinfeld, A. M. (1995) Continuous measurement of phospholipase A2 activity using the fluorescent probe ADIFAB. *Anal. Biochem.* **229,** 256–263.
29. Richieri, G. V. and Kleinfeld, A. M. (1995) Unbound free fatty acid levels in human serum. *J. Lipid Res.* **36,** 229–240.
30. Bartlett, G. R. (1959) Phosphorous assay in column chromatography. *J. Biol. Chem.* **234,** 466–468.

4

Measuring Molecular Order and Orientation Using Coherent Anti-Stokes Raman Scattering Microscopy

Hilde A. Rinia, George W. H. Wurpel, and Michiel Müller

Summary

Coherent anti-Stokes Raman scattering microscopy develops rapidly into a powerful technique to image both the chemical composition and physical state in complex samples from biophysics, biology, and the material sciences. This nonlinear vibrational technique increases the signal relative to spontaneous Raman scattering and does not require labeling of the specimen. A theoretical description of the technique is provided and the two major modes of operation: picosecond- and multiplex coherent anti-Stokes Raman scattering are discussed. The potential of the technique is demonstrated with examples of direct measurement of acyl chain order and orientation in lipid monolayers, bilayers, and lipid vesicles.

Key Words: Coherent anti-Stokes Raman scattering (CARS); lipid biophysics; lipid chain order; lipid chain orientation; microscopy; multiplex CARS; Raman; vibrational spectroscopy.

1. Introduction

Enabled by recent short-pulse laser technology developments, coherent anti-Stokes Raman scattering (CARS) microscopy is being used increasingly as a unique microscopic tool over the last few years. Contrary to light and fluorescence microscopy, CARS microscopy provides direct chemical and physical image contrast. Although already demonstrated in 1982 by Duncan et al. *(1)*, high-resolution microscopy was only shown to be feasible with Ti:Sapphire-based laser technology *(2)*. Since then, the field has grown rapidly, demonstrating the use of various laser sources, excitation, and detection schemes and—most importantly—unique applications in biology, biophysics, and the material sciences *(3–8)*.

Vibrational spectroscopy probes the intra- and intermolecular vibrational structure of molecules and molecular assemblies, providing detailed information on both the chemistry and physics of the sample. Unique to this area of spectroscopy is that many of the spectral features remain resolvable even at room temperature and in complex samples such as live cells and that it does not require any form of labeling. A number of features make CARS especially suitable for high-resolution vibrational microscopy:

1. It is a nonlinear optical technique that provides signal levels that are many orders of magnitude (typically ≥3–4) stronger than attainable for spontaneous Raman scattering.
2. Near-infrared wavelengths can be used that ensure submicron optical resolution and minimal sample heating and damage.

3. The nonlinear dependence of the CARS signal on input laser intensity provides inherent optical sectioning and three-dimensional (3D) imaging capability.
4. Finally, the signal propagates coherently and is of shorter wavelength than all the input fields making it readily detectable in the presence of a fluorescence background.

Many of the CARS microscopy experiments undertaken in recent years have focused on studies of lipids both in model systems *(3,7,9)* and in cellular membranes *(5,10,11)*. The reason for this lies in the relatively high Raman scattering cross-section for the acyl chain C–H stretch vibrations and the relatively high concentrations of these species in membranes and lipid bodies. Furthermore, vibrational spectroscopy, is uniquely positioned for biophysical studies of lipids: chemical selectivity can be used to discriminate between saturated and unsaturated lipid species or to identify the presence of cholesterol, whereas the physical selectivity offered by the vibrational spectrum may be used to determine the phase of, and intermolecular interactions, among lipid species *(12,13)*.

This chapter is organized as follows. In **Subheading 2.**, the technique of CARS microscopy is introduced, both in theory and experimental design and considerations. Special attention is paid to explain the differences between two modes of operation: picosecond (ps)-CARS and multiplex CARS. In **Subheading 3.**, a specific experimental setup for multiplex CARS microscopy is introduced and the procedures for the preparation of lipid model membranes for biophysical studies are described. In **Subheading 4.**, methods to extract the relevant information from complicated CARS spectra, issues to consider to improve the sensitivity in CARS microscopy experiments, and application of CARS microscopy to study lipid acyl chain order and orientation in model membranes are presented. Finally, in **Subheading 5.**, some of the inherent problems and limitations associated with the technique are discussed.

2. The Technique
2.1. CARS Microscopy

CARS is a third-order nonlinear process in which three laser beams coherently interact with the sample to generate a coherent signal beam (*see* **Fig. 1**). More specifically, a *pump* beam with frequency ω_{pu}, and a *Stokes* beam with frequency ω_s, setup a grating, which a third laser beam (denoted by *probe*, with frequency ω_{pr}) undergoes a Bragg diffraction, generating an *anti-Stokes* signal beam at frequency $\omega_{as} = \omega_{pu} - \omega_s + \omega_{pr}$. The nonlinear process is resonantly enhanced when the difference frequency between pump and Stokes matches a vibrational energy level in the sample ($\omega_{pu} - \omega_s = \omega_{vib}$). In a plane wave approximation and for a nonabsorbing medium, the CARS signal strength is given by *(14)*:

$$I_{CARS} \propto \left|\chi^{(3)}\right|^2 I_{pu} I_s I_{pr} \mathrm{sinc}^2\left(\Delta \vec{k} \cdot d/2\right) \tag{1}$$

where I denotes the intensity of the lasers and CARS signal, d is the thickness of the scattering volume, $\mathrm{sinc}(x) = (\sin x)/x$, and $\Delta \vec{k} = \vec{k}_{as} - \vec{k}_{pu} + \vec{k}_s - \vec{k}_{pr}$ describes the phase mismatch between the wave vectors k. It has been shown *(2,15)* that for high numerical aperture focusing conditions the phase matching condition is less stringent. Therefore, a collinear beam geometry ($\vec{k}_{as} = \vec{k}_{pu} = \vec{k}_s = \vec{k}_{pr}$) is commonly used to achieve maximum spatial resolution in microscopy applications. Also for practical reasons *pump* and *probe* are generally derived from the same laser source ($\omega_{pu} = \omega_p$ and $I_{pu} = I_{pr}$).

CARS Microscopy

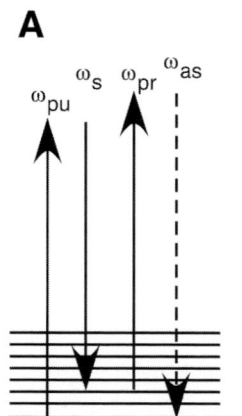 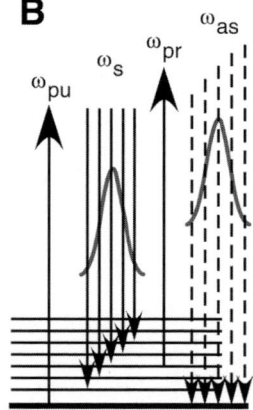

Fig. 1. Energy diagram for **(A)** ps- and **(B)** multiplex CARS. The thick line denotes the electronic ground state and the thin lines denote vibrational levels. Solid arrows denote input laser fields and dashed arrows denote the generated coherent signal field. In multiplex CARS the width of the *Stokes* laser beam determines the frequency extent of vibrational levels probed simultaneously, whereas the spectral width of the *pump* and *probe* beam determine the spectral resolution.

For a sample consisting of one or more molecular species, the CARS signal is proportional to:

$$I_{CARS} \propto \left| \sum_k N_k \chi_k^{(3)} \right|^2 \tag{2}$$

where N_k denotes the number of molecules of type k. The nonlinear susceptibility ($\chi^{(3)}$) for each molecular species consists of a resonant (R) and a nonresonant (NR) contribution (*see* **Fig. 2**).

$$\chi^{(3)} = \chi_R^{(3)} + \chi_{NR}^{(3)} \tag{3}$$

Far away from one-photon resonances, the resonant contribution to the nonlinear susceptibility can be written as:

$$\chi_R^{(3)} \propto \sum_j \frac{A_j}{\delta_j - i\Gamma_j} \tag{4}$$

where A_j is a real constant containing the vibrational scattering cross-section of the vibrational mode j, $\delta_j = \Omega_j - \omega_L + \omega_S$ denotes the detuning from the vibrational resonance Ω_j, Γ_j is the half width at half maximum (HWHM) of the spontaneous Raman scattering vibrational transition, and the summation runs over all vibrational resonances. It follows from these expressions that the CARS spectral lineshape is given by:

$$I_{CARS} \propto \sum_{k,l} N_k \chi_{k,NR}^{(3)} N_l \chi_{l,NR}^{(3)} + 2\sum_{k,l}\left[N_k N_l \chi_{k,NR}^{(3)} \sum_i \frac{\delta_{li} A_{li}}{\delta_{li}^2 + \Gamma_{li}^2} \right] + \sum_{k,l}\left[N_k N_l \sum_{i,j} \frac{A_{ki} A_{lj}\left(\delta_{ki}\delta_{lj} + \Gamma_{ki}\Gamma_{lj}\right)}{\left(\delta_{ki}^2 + \Gamma_{ki}^2\right)\left(\delta_{lj}^2 + \Gamma_{lj}^2\right)} \right] \tag{5}$$

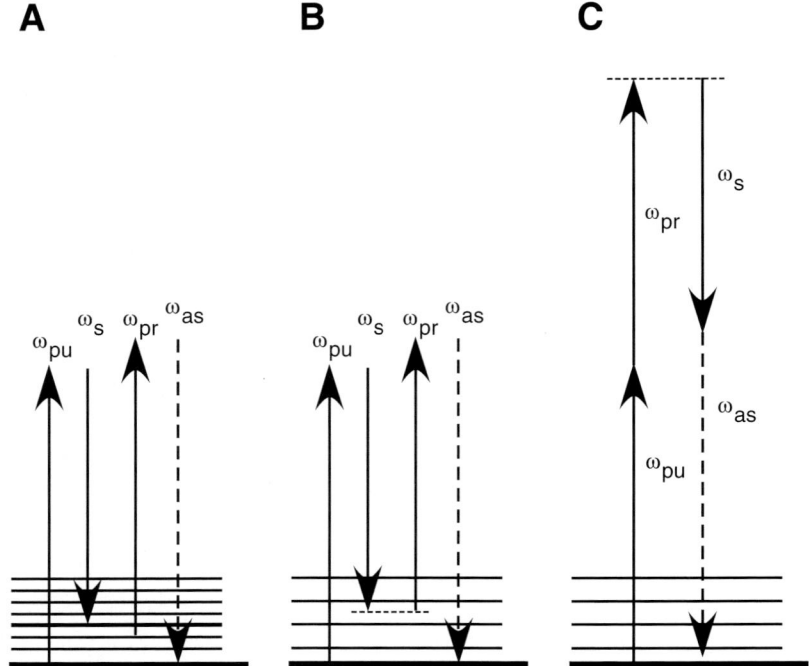

Fig. 2. Energy diagrams for the (vibrationally) resonant (**A**) and NR (**B,C**) contributions to the CARS signal. The thin dotted horizontal line denotes a so-called virtual energy level.

where the subscripts k, l run over the different molecular species and i, j run over the vibrational resonances of each molecular species. In comparison, the spontaneous Raman scattering lineshape is described by a sum of Lorentzian contributions:

$$I_{\text{Raman}} \propto \sum_{k}\left\{\sum_{i}\frac{N_k A_{ki}}{\delta_{ki}^2 + \Gamma_{ki}^2}\right\} \qquad (6)$$

It should be noted that when the first electronic excited state is far away from one-photon resonances, spontaneous Raman and CARS spectra can directly be compared if they are both acquired under parallel excitation and detection polarization conditions *(16)*.

2.2. Experimental Procedures

2.2.1. ps-CARS

Duncan et al. *(1,17)* reported the first—relatively low resolution—CARS microscopy investigations. CARS microscopy was discovered again in the late 1990s with the first successful CARS microscopy experiments at high spatial resolution *(2,18)*. These experiments were based on using picosecond laser sources for both *pump (probe)* and *Stokes*. This mode will be denoted by ps-CARS in what follows. In ps-CARS a single frequency in the vibrational spectrum is probed for a certain wavelength setting of the lasers. The spectral resolution is determined by the bandwidth of the lasers. Thus, in order to acquire a full CARS spectrum, the *Stokes* laser has to be tuned to different wavelengths. In terms of signal generation, this mode of operation is the most efficient *(19)*, and it permits fast scanning in case of strong vibrational scatterers *(20)*.

As various points in the vibrational spectrum are acquired subsequently, the spectral signal-to-noise ratio (SNR) for this mode of operation is strongly affected by laser-induced fluctuations (pulse-to-pulse power fluctuation, timing jitter between *pump* and *Stokes* laser pulses, and so on). Generally, the NR background dominates the signal in the forward direction, laser-induced noise in this background may easily "drown" the resonant signal. It is for this reason that the strong NR background is generally thought to be a drawback in CARS microscopy, and various methods have been developed to minimize its contribution to the total signal.

2.2.2. Suppressing the NR Background

The magnitude of the NR background contribution to the total CARS signal (and thus relative to the resonant contribution), can be suppressed using the different properties of the two contributions. For example, depolarization affects the resonant and NR contribution to the CARS signal in a different way. Specific polarization settings for *pump*, *Stokes*, and *anti-Stokes* can be used for maximum suppression of the NR contribution, whereas retaining a significant part of the resonant contribution *(21,22)*. Alternatively, the difference in temporal response of the resonant and NR contribution to the CARS signal can be utilized to isolate the resonant component in time-resolved measurements *(23)*. Also, combined phase and polarization-control experiments have been shown to considerably reduce the NR background contribution *(24,25)*.

However, today the most common method to suppress the NR background in CARS microscopy experiments, is the use of *epi*-mode detection. It has been shown *(26,27)* that, whereas for extended samples the CARS signal is generated predominantly in the forward direction, the CARS signal for samples with a thickness less than the wavelength is generated equally in the forward and backward direction. Thus, for a thin vibrationally resonant feature in a thick bulk sample, the strong bulk NR contribution is mainly generated in the forward direction, whereas the CARS signal in the backward direction is primarily because of the resonant (and NR) contribution of the feature of interest. This method of the detection is easily incorporated in standard confocal microscopes and can readily be combined with laser beam scanning. The only disadvantage is that the signal strength is determined in part by the thickness of the feature of interest.

In case of complete suppression of the NR contribution, the CARS signal for a single vibrationally resonant moiety is given by:

$$I_{CARS}^{resonant\ only} \propto \left| N_k \chi_{R,k} \right|^2 \tag{7}$$

The SNR in this case is given by *(16)*:

$$SNR_{CARS}^{resonant\ only} = \frac{\left| N_k \chi_{R,k} \right|^2 \Delta t}{\sqrt{\left(I_D \Delta t \right)^2 + \left| N_k \chi_{R,k} \right|^2 \Delta t}} \tag{8}$$

where Δt is the total measuring time and I_D is the dark count rate (including dark current, read-out noise, and so on). Thus, for decreasing concentrations the CARS signal decreases quadratically with the number of vibrationally resonant molecules and the detection sensitivity is ultimately limited by the dark count rate of the experimental setup.

2.2.3. Multiplex CARS

In multiplex CARS *(3,28–32)*, a significant range of vibrational frequencies is addressed simultaneously using a combination of a spectrally narrow-band *pump* (and *probe*) laser and a broad-band *Stokes* laser. Because the *anti-Stokes* signal is generated concurrently over a significant part of the vibrational spectrum, the spectral SNR for this mode of operation is determined only by Poisson noise inherent to the detection process. This yields CARS spectra of unprecedented quality that permit detailed vibrational analysis.

As an example (*see also* **Subheading 4.**), consider the detection of the CARS signal from a lipid bilayer in water. The total—forward emitted—multiplex CARS signal from the resonant lipid (L) within the NR bulk water (W) is given by:

$$I_{CARS}^{multiplex} \propto \left| N_W \chi_{NR,W} + N_L \chi_{NR,L} + N_L \chi_{R,L} \right|^2 \tag{9}$$

$$= \left| N_W \chi_{NR,W} + N_L \chi_{NR,L} \right|^2 + 2(N_W \chi_{NR,W} + N_L \chi_{NR,L}) \text{Re}(N_L \chi_{R,L}) + \left| N_L \chi_{R,L} \right|^2$$

where N_W and N_L are the molarities of water and lipid, respectively. Because for a single bilayer $N_L \ll N_W$, the resonant term becomes negligible with respect to the cross term, and the total CARS signal can be approximated by:

$$I_{CARS}^{multiplex} \sim \left| N_W \chi_{NR,W} \right|^2 + 2 N_W N_L \chi_{NR,W} \text{Re}(\chi_{R,L}) \tag{10}$$

It follows that at low concentrations the CARS signal strength depends linearly on the concentration of the resonant feature. For multiplex CARS, this implies that the detection sensitivity is solely determined by Poisson noise and a large NR background signal can actually be used to increase the detection sensitivity.

3. Materials
3.1. Experimental Setup

A number of different experimental setups for CARS microscopy have now been put forward. The main differences among CARS microscopy configurations lie in the laser sources used, the method of scanning, and the type of detection. A combination of two Ti:Sapphire lasers has been used most often for CARS microscopy. One of the lasers serves as both *pump* and *probe*, the other as *Stokes*. In ps-CARS, both lasers produce ps laser pulses (typically in the order of 3–5 ps). In multiplex CARS, the *Stokes* laser operates in femtosecond mode, providing a broad bandwidth. A technological challenge for the Ti:Sapphire-based systems is temporal synchronization between the two lasers. Any time jitter between the *pump* and *Stokes* results in signal strength fluctuations. Advanced schemes have been developed to secure tight synchronization *(33–35)*.

Various alternative laser sources have been recently developed and successfully used in CARS microscopy applications. These include the use of optical parametric oscillators in ps-CARS *(20)* and broad bandwidth generation in optical fibers *(36–38)* in multiplex CARS. In forward-detected CARS microscopy, generally the sample is moved relative to the laser beam focus in all 3D. In contrast, in *epi*-detected CARS microscopy, laser beam scanning is often used. The main detection differences among these methodologies result from the use of either ps- or multiplex CARS. In the former, a single detector such as a photomultiplier

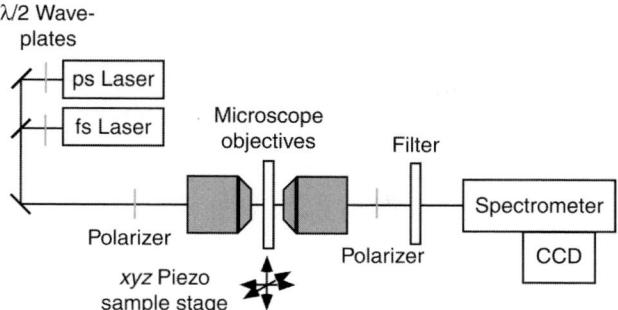

Fig. 3. Schematic of the experimental setup for multiplex CARS microscopy experiments.

tube or avalanche photodiode can be used, whereas in multiplex CARS a CCD camera in combination with a spectrometer is used.

As an illustration, **Fig. 3** shows the experimental setup for multiplex CARS experiments on lipid vesicles and lipid mono- and bilayers as described below. Two tunable mode-locked Ti:Sapphire lasers, with a repetition rate of 80 MHz and pumped by an Nd:YVO4 laser (Millenium, Spectra Physics, USA), are made collinear and synchronized with a "lok-to-clock" (Spectra Physics, USA) system. An additional home-built, long-term feedback system ensures a timing jitter between the lasers of <1 ps. The *pump/probe* laser operates in picosecond mode, with pulses of 3 ps (bandwidth approx 5/cm·fwhm), centered around 710 nm. The *Stokes* laser produces 80 fs pulses, corresponding to a bandwidth of approx 180/cm·fwhm. The center wavelength can be tuned between 750 and 950 nm corresponding to Raman shifts in the range of approx 750–3500/cm. The polarization of the two lasers is controlled by achromatic $\lambda/2$ plates. An additional polarizer in the combined beam path ensures equal polarization for both lasers. The lasers are focused with a microscope objective (1.25 NA, ×63, oil immersion) onto the sample. The effective probing volume is given by the lateral and axial resolution of 0.31 ± 0.04 and 1 ± 0.05 μm, respectively *(39)*. The sample is piezo-scanned in 3D (PZT P-611.3S, Physik Instrumente Germany). The CARS signal is collected in the forward direction with a second microscope objective (0.65 NA, ×40), passed through a polarizer, and filtered by a 710-nm short-wave pass filter (Omega, USA). Multiplex CARS spectra are recorded on a CCD camera coupled to a spectrograph (MS 257, Oriel, USA). The spectral resolution of approx 5/cm is determined by the resolving power of the spectrograph.

3.2. Lipid Sample Preparation

For the preparation of multilamellar vesicles (MLV), the lipids are premixed in the appropriate molar ratio in chloroform, which is subsequently evaporated under a nitrogen stream. Traces of organic solvent are removed by placing the sample *in vacuo* for at least 2 h. MLV (typically 2 mM) are prepared by hydrating the lipids in phosphate-buffered saline buffer (pH = 7.4) at a temperature more than T_m, whereas vortexing for at least 5 min. The vesicles are freeze–thawed five times to ensure proper mixing of the components. A microscopic sample is made by placing the vesicles on a coverslip coated with 0.5 wt% agarose gel. A second coverslip is placed on top and the sample is sealed.

Supported lipid mono- and bilayers on glass are made using a Langmuir trough (Kibron Inc., Finland). Lipids dissolved in chloroform are spread onto a water surface and compressed to a

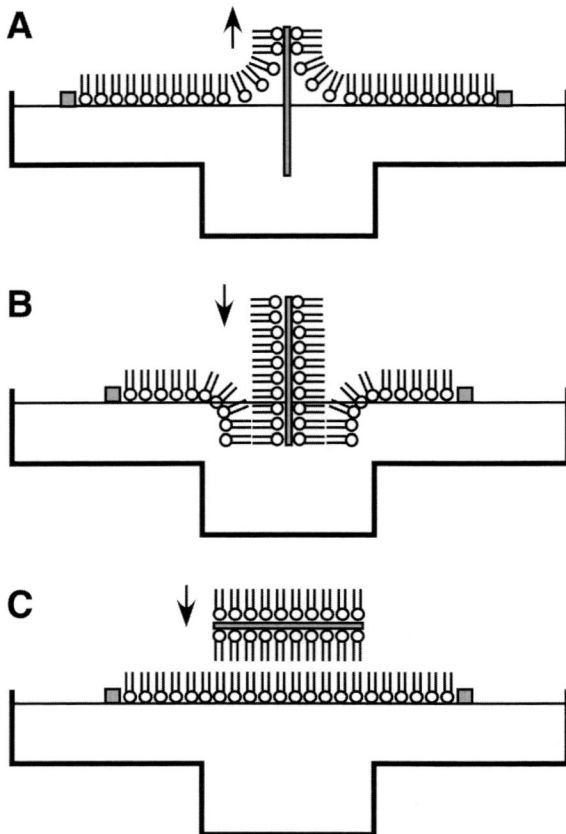

Fig. 4. Schematic of the preparation of supported lipid mono- and bilayers on glass. (**A**) Langmuir–Blodgett transfer of a lipid monolayer to a hydrophilic glass surface. (**B**) Langmuir–Blodgett transfer of a second lipid monolayer. (**C**) Langmuir–Schäfer transfer of a second lipid monolayer.

surface pressure of 30 mN/m. The first lipid monolayer is transferred to a thoroughly cleaned (hydrophilic) cover glass using the Langmuir–Blodgett technique (i.e., pulling it from the water subphase through the lipid monolayer on water). Transfer efficiency is checked and always found to be close to 100%. The second monolayer is applied by using either the Langmuir–Blodgett technique again (i.e., pushing the lipid covered glass back into the water subphase), or the Langmuir–Schäfer technique. In the latter case, the lipid-covered glass is oriented parallel to the water surface when pushed through the lipid layer (*see* **Fig. 4**). The sample is closed with a second cover glass and sealed with epoxy. No differences were observed in the CARS spectra of lipid bilayers made with either the Langmuir–Blodgett or Langmuir–Schäfer technique.

Single lipid monolayers can be studied in CARS microscopy by making an asymmetric bilayer, wherein only one of the lipid monolayers is deuterated. Through the isotope substitution the vibrational spectrum shifts from approx 2900 to 2100/cm. Thus, when studying the C–H stretch vibrational region from one lipid monolayer, the other (deuterated) lipid monolayer only adds a NR contribution to the CARS signal.

4. Methods
4.1. Data Analysis

Owing to the coherent addition of both resonant and NR contributions (*see* **Eq. 5**), interpretation of the CARS spectrum generally requires extensive curve fitting. Such curve fitting requires high SNR spectra, which can be obtained using multiplex CARS. A unique relation exists between the spontaneous Raman spectrum and the CARS spectrum if both are obtained under "all-parallel" polarization conditions *(16)*. Thus, for the spontaneous Raman spectrum the signal should be detected through an analyzer oriented parallel to the polarization of the excitation laser. Similarly, for the CARS spectrum *pump*, *Stokes*, and *probe* should have parallel polarization, and the *anti-Stokes* signal should be detected through an analyzer with an orientation parallel to the incident laser fields.

All CARS spectra are analyzed by least-squares fitting, using the general description for CARS signals far away from one-photon resonances:

$$\frac{I_{CARS}}{I_{CARS}^{ref}}(\omega_{pu} - \omega_s) = \frac{1}{\left|\chi_{NR}^{glass}\right|^2} \left| \chi_{NR} + \sum_j \frac{A_j}{\Omega_j - (\omega_{pu} - \omega_s) + i\Gamma_j} \right|^2 \quad (11)$$

Here $\chi_{NR}/\chi_{NR}^{glass}$ is a frequency-independent parameter that describes the total NR background in the focal volume relative to the glass NR background, and $(\omega_{pu} - \omega_s)$ is the frequency difference of the *pump* and *Stokes* laser. The vibrational resonances are characterized by their amplitudes (A_j), linewidths (Γ_j), and spectral positions (Ω_j). For the C–H stretch region of the lipids eight peaks are needed to faithfully reproduce the experimental data. In addition, as the lipid signal from a single lipid mono- or bilayer is extremely weak, the resonant vibrational contribution of water (around 3200/cm) also needs to be included in the fit. The initial guesses to the fits are taken from Raman spectra of the lipid and H_2O. Because of the *Stokes* laser spectral intensity profile, the noise is not constant over the whole measured vibrational spectrum. Thus, CARS frequencies corresponding to low-intensity Stokes-laser frequencies appear with relatively higher noise. This is taken into account by weighing the fit with the Stokes profile. The parameters in the fit are constrained to positive amplitudes and peak positions and widths that do not deviate by more than 15% from their spontaneous Raman values.

4.2. Sensitivity

An example of the sensitivity and spectral SNR that can be realized with multiplex CARS microscopy is given in **Fig. 5**. **Figure 5A** shows the CARS spectrum of a 1,2-dipalmitoyl-*sn*-glycero-3-phosphocholine (DPPC) lipid bilayer supported on glass. The CARS signal strength is given as I_{CARS}/I_{CARS}^{ref}, where I_{CARS}^{ref} is the CARS signal obtained at the same axial position as I_{CARS}, but in the absence of lipid. Note that this dimensionless quantity is independent of the laser powers or any other experimental parameters, and directly reflects the strength of the resonant signal relative to the NR background. Thus, the CARS signal strength can be related in absolute terms to local lipid density. A measure for this density is given by $A_{-2845}/(NR \cdot \Gamma_{-2845})$, where A_{-2845} and Γ_{-2845} are the amplitude and linewidth of the –2845/cm ($\upsilon_s[CH_2]$ symmetric methylene stretch) vibrational resonance as obtained from the fit. For the DPPC bilayer a value of $A_{-2845}/(NR \cdot \Gamma_{-2845}) = 0.007 \pm 0.0009$ was obtained. **Figure 5B**

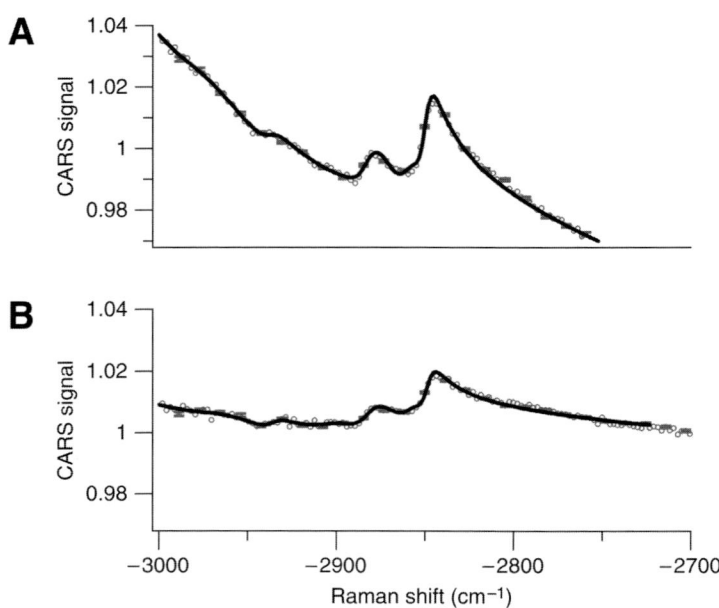

Fig. 5. Multiplex CARS spectrum of a DPPC lipid bilayer (**A**) and DPPC lipid monolayer (**B**). Error bars denote one standard deviation. The solid line is a least-square weighted fit with eight lipid and one water vibrational resonances. The CARS signal is given relative to the NR background and can be related directly to the absolute local lipid density.

shows the CARS spectrum of a monolayer of DPPC on a monolayer of d62-1,2-dipalmitoyl-D62-*sn*-glycero-3-phosphocholine (d62-DPPC) supported on glass. Because of the deuteration of one of the leaflets of the lipid bilayer, the CARS signal results from a single (nondeuterated) lipid monolayer. The observed CARS signal strength is $A_{-2845}/(NR \cdot \Gamma_{-2845}) = 0.003 \pm 0.004$, in agreement with half of the lipid bilayer of **Fig. 5A**.

4.3. Acyl Chain Order

Given the sensitivity of multiplex CARS microscopy as demonstrated in **Subheading 4.2**, it is possible to directly measure the acyl chain order in single-supported lipid bilayers. **Figure 6** shows typical room temperature, spontaneous Raman spectra of the C–H stretch region of DPPC, and 1,2-dioleoyl-*sn*-glycero-3-phosphocholine (DOPC). At room temperature, DPPC is in the gel phase with the acyl chains highly ordered in a fully stretched configuration and with a minimum of *gauche* kinks. In this state, the Raman spectrum is dominated by the $\upsilon_s(CH_2)$ symmetric methylene stretch and $\upsilon_a(CH_2)$ asymmetric methylene stretch at 2847 and 2882/cm, respectively. The Raman spectrum for DOPC in the liquid-crystalline phase at room temperature shows some marked changes: most notably, the $\upsilon_a(CH_2)$ loses intensity relative to the $\upsilon_s(CH_2)$, which shifts to higher wavenumbers. The ratio $R = I_{2935}/I_{2880}$ (where 2935/cm means a Fermi resonance) is often used as a phenomenological parameter that describes the acyl chain order in lipid bilayers (*12,40,41*).

Because of the general very low cross-section of spontaneous Raman scattering, it is extremely difficult to obtain spontaneous Raman spectra of a single bilayer (*42*). The nonlinear interaction in CARS strongly amplifies the Raman signal-reducing acquisition times to the

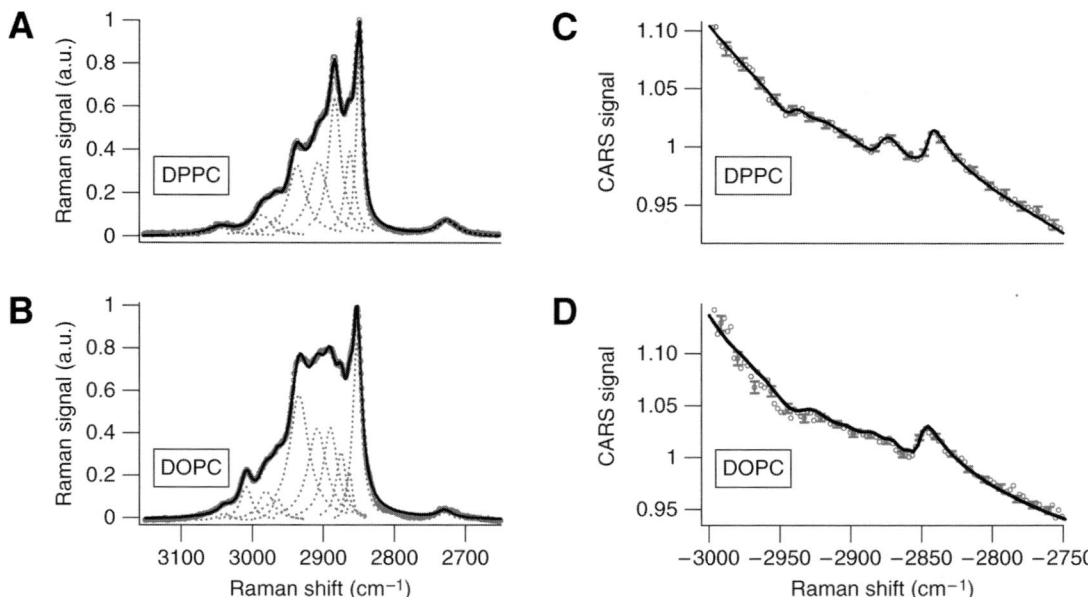

Fig. 6. Spontaneous Raman spectrum of (**A**) DPPC and (**B**) DOPC. The solid line is a fit of the experimental data to a sum of Lorentzian lines, depicted with dotted lines. Multiplex CARS spectrum of a single-supported bilayer of (**C**) DPPC and (**D**) DOPC. The solid line is a fit of the experimental data and error bars denote one standard deviation. All spectra are taken at room temperature. (Data from **ref. 7**).

milisecond–second regime, permitting the evaluation of acyl chain order parameters in supported bilayers. As an example, **Fig. 6C,D** show the CARS spectra of single bilayers of DPPC and DOPC, respectively. It is clear that the signal from the $\upsilon_a(CH_2)$ significantly decreases from gel-to-liquid crystalline phase also for a single lipid bilayer. Using the fitting procedure described in **Subheading 4.1**, acyl chain order parameters can be determined for the supported DPPC and DOPC bilayers and compared with order parameters obtained for MLV (7).

4.4. Acyl Chain Orientation

By using the polarization properties of CARS microscopy, specific molecular orientations can be probed. As an example, the multiplex CARS analysis of acyl chain orientational order in MLV is shown. The principal concept of this method follows. The strength of the generated CARS signal field depends on the relative orientation of the molecular transition dipole moments with respect to the polarization of the laser fields. For instance, if the polarization of the *pump*, *Stokes*, and *probe* lasers are all parallel, then, the generated *anti-Stokes* signal from the symmetric methylene stretching vibrations of the acyl chain of a lipid molecule has a maximum when this vibration is oriented parallel to the incident polarization. In this case, the polarization of the signal is also parallel to that of the rest of the fields. In other words, in an all-parallel configuration the generated CARS signal shows a maximum for acyl chains oriented perpendicular to the plane of polarization. On the other hand, for the skeletal mode C–C acyl chain vibrations (around 1100/cm), which are oriented mainly along the acyl chain, a maximum signal is found for acyl chains oriented parallel to the incident laser fields. In a first approximation, lipid molecules in a MLV are all

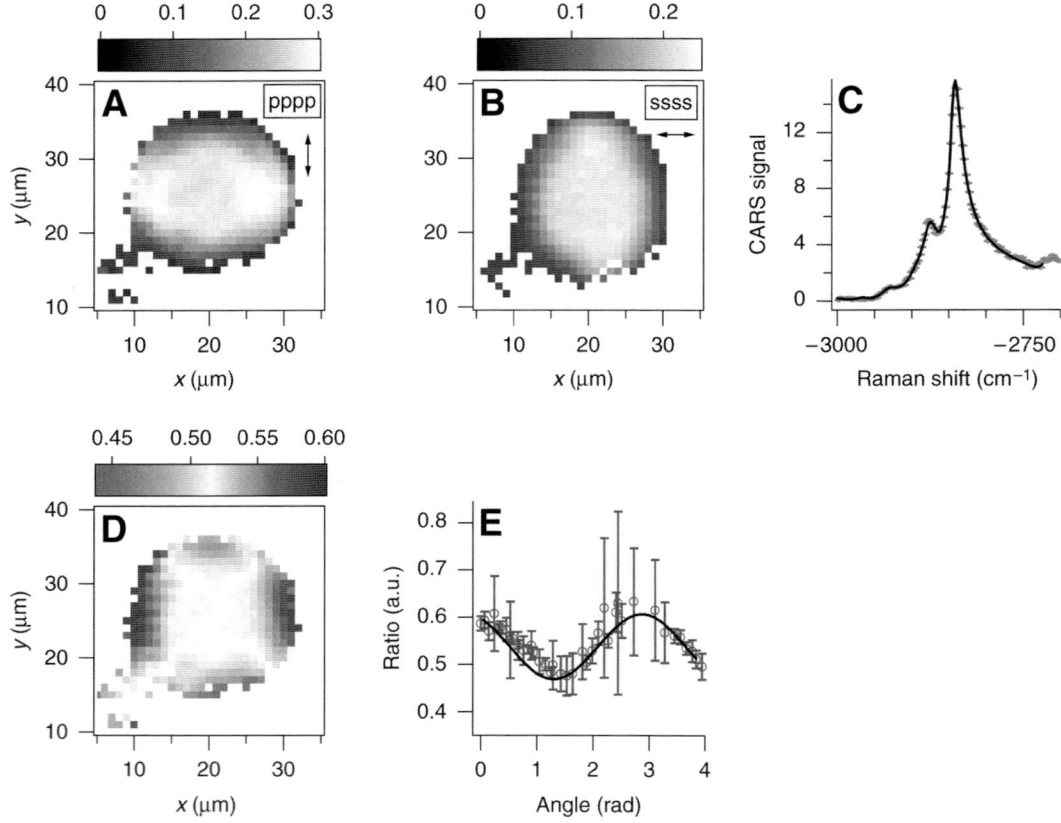

Fig. 7. Multiplex CARS images of the $\upsilon_s(CH_2)$ (–2845/cm) mode of a SPM MLV with the polarization of all lasers and signal fields in (A) *pppp*- and (B) *ssss*-configuration. The two-sided arrow denotes the orientation of the polarization. (C) Typical multiplex CARS spectrum from the image in **7A** at position $x = 28$ and $y = 26$ μm. The solid line is a least-squares fit to the data. (D) Ratio image of **7A** and **7B** as defined by **Eq. 14**. Values less than 0.5 are found in the top and bottom quadrant, whereas values more than 0.5 are found in the left and right quadrants. (E) Variation of the ratio of **7D** along the rim of the vesicle and a $\cos^2\varphi$ fit to the data (solid line). Error bars denote one standard deviation.

oriented with their acyl chains parallel to the radial vector. Thus, acyl chain orientational order can be studied by making several CARS images of an MLV for different orientations of the laser and signal polarizations.

Consider a multiplex CARS image (**Fig. 7A**) of the equatorial plane of a perfectly spherical MLV of egg-sphingomyelin (SPM). At room temperature, SPM is in the gel phase with most of the acyl chains in an all *trans*-configuration. The CARS image is derived from measured CARS spectra in the region of –3100–2700/cm at every pixel position in the image. From a least-squares fit of the spectrum to **Eq. 11**, the amplitude of the $\upsilon_s(CH_2)$ symmetric methylene stretching mode at –2845/cm is determined, and this value is plotted as a function of spatial position. The polarization of both the incident and generated fields is indicated by the arrows (**Fig. 7A**). Clearly, the CARS signal shows a maximum for orientations of the acyl chains perpendicular to the polarization of the fields.

The orientation of the lipid acyl chain of each molecule can be described by the angle φ_i^0 between the radial vector of the MLV and the laboratory frame. The orientational dependence of the CARS signal field from the symmetric C–H stretching vibration of the methylene units of lipid acyl chains then follows *(43)*:

$$\bar{E}_{CARS}(\varphi_i) \propto \sum_i^N \rho(\varphi_i)\cos^4\varphi_i \qquad (12)$$

where φ_i describes the angle between the *pump* polarization and the molecular transition dipole moment. In practice, various influences (e.g., local acyl chain tilt, *gauche*-kinks, membrane ruffling, and axial averaging over the focal region) will lead to a local distribution of lipid chain angles relative to φ^0. This local orientational disorder in the orientation of the acyl chains can be incorporated in the analysis by assuming a Gaussian distribution:

$$\rho(\varphi_i) = \exp\left(\frac{-(\varphi_i - \varphi_i^0)^2}{2\sigma^2}\right) \qquad (13)$$

where σ is the standard deviation of the orientational distribution. In the experiments, two multiplex CARS images are acquired from the equatorial plane of a MLV. In each image the polarization of *pump*, *Stokes*, *probe*, and *anti-Stokes* are parallel. All polarizations are flipped by 90° between the two images, with the two polarization conditions being denoted by *pppp* and *ssss*. By plotting the ratio:

$$R = \sqrt[4]{I_{v_s(CH_2)}^{pppp}} \Big/ \left(\sqrt[4]{I_{v_s(CH_2)}^{pppp}} + \sqrt[4]{I_{v_s(CH_2)}^{ssss}}\right) \qquad (14)$$

an image of the orientational order with negligible contribution from differences in lipid density is obtained. **Eq. 14** varies between 0 and 1 in case of perfect orientational order. The amplitude of the variation in CARS signal strength as a function of φ^0 decreases with increasing lipid chain orientational disorder, approaching 0.5 for an isotropic orientation of methylene stretching vibrations.

For SPM the width of the orientational distribution of the acyl chains is found to be $\sigma = 0.90 \pm 0.02$, slightly higher than found for DPPC. In this study *(43)*, it was also found that unsaturated lipids such as DOPC and 1-palmitoyl-2-oleoyl-*sn*-glycero-3-phosphocholine (POPC) show a significantly decreased orientational order. The addition of cholesterol, on the other hand, significantly increases orientational order of the acyl chains.

5. Notes

CARS microscopy is a powerful technique for probing both the chemical composition and the physical state of complex samples. Its vibrational specificity provides unique contrast without the requirement of labeling. However, as any technique, CARS microscopy is not without limitations. Although in principle a CARS signal can be generated from a single molecule, it derives its large increase in signal yield, relative to spontaneous Raman, from its nonlinear coherent interaction with the sample. This works best at high concentrations wherein the quadratic resonant term in **Eq. 5** dominates. At lower concentrations, CARS encounters its limitations. In epi-detection ps-CARS, the signal drops quadratically with the number of

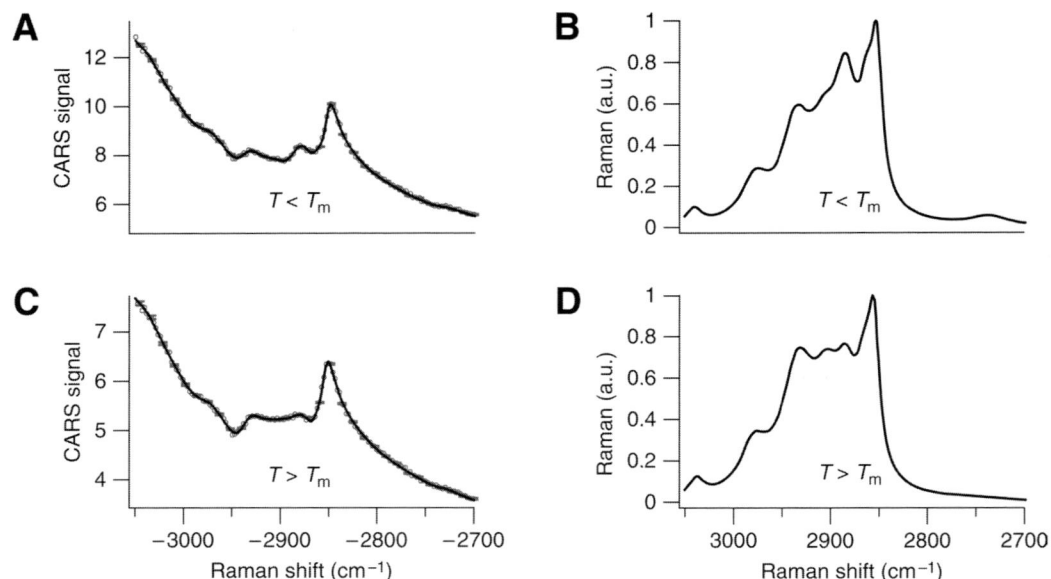

Fig. 8. Multiplex CARS spectra for DMPC small unilamellar vesicles below (**A**) and above (**C**) the main phase-transition temperature. The spectral parameters (line amplitude, positions, and widths) from the fitted CARS spectra are used to calculate the corresponding spontaneous Raman spectrum for DMPC vesicles below (**B**) and above (**D**) the phase-transition temperature.

vibrationally resonant modes. In multiplex CARS the signal drop is linear, but a high overall signal count is required to counteract Poisson noise from the NR background. In practice, sensitivity levels in the order of 5–10 mM have been realized *(7,16,44)*. Other techniques, such as surface-enhanced Raman scattering (*see* **refs. 45,46**) and tip-enhanced nonlinear Raman scattering *(47)*, have been developed to increase the sensitivity for Raman scattering to the single molecule level.

The vibrational spectrum is rich in information. It contains highly specific chemical information (contained in the so-called "finger-print" region) and may also shed light on intermolecular interactions. However, especially for biological applications, wherein the variation in chemical-building blocks is limited, spectral differences between different moieties or physical states are often subtle. In addition, many regions of the vibrational spectrum are highly congested. For example, a faithful representation of the C–H stretch vibrational spectrum of a single lipid species generally requires at least eight overlapping vibrational modes. In CARS, the vibrational spectrum results from interference of all these vibrational resonances. This interference makes the interpretation of CARS spectra complex and poses stringent requirements on the SNR of the spectral acquisition.

An example of the influence of congested spectral bands on the CARS spectrum is shown in **Fig. 8**. A multiplex CARS spectrum of 1,2-dimyristoyl-*sn*-glycero-3-phosphocholine (DMPC) small unilamellar vesicles is measured both below and above the main phase-transition temperature (T_m = 24°C). The high SNR of the multiplex CARS spectra permits resolving the small differences in the spectrum for the gel phase (**Fig. 8A**) and liquid-crystalline phase (**Fig. 8C**) vesicles. When the parameters for the vibrational modes (amplitudes, positions, and widths) determined from the fit of the multiplex CARS data are used to calculate the corresponding spontaneous Raman spectrum, the influence of chain melting on the vibrational spectrum becomes even more apparent **(Fig. 8B,D)**. The ratio of the Raman intensity at 2935 and 2880/cm is often used as a phenomenological chain order parameter that corresponds to the chain melting in this phase transition *(12)*. Clearly, CARS spectra with excellent SNR are required to permit such data analysis in a complex vibrational band such as the C–H stretch region of lipids. As demonstrated above, selective deuteration can be used to enhance the spectral selectivity of multiplex CARS microscopy and spectroscopy. Although applications of CARS microscopy have thus far focused mainly on lipids, the technique has interesting potential for the study of small molecules, such as water and urea *(4,39)*, which cannot readily be labeled for fluorescence microscopy research.

Acknowledgments

This research was in part supported with financial aid from the Netherlands Organization for Scientific Research and a research programmes of the Stichting voor Fundamenteel Onderzoek der Materie and Aard- en Levenswetenschappen under grant no. 1520378 and 1520383.

References

1. Duncan, M. D., Reijntjes, J., and Manuccia, T. J. (1982) Scanning coherent anti-Stokes microscope. *Opt. Lett.* **7,** 350–352.
2. Zumbusch, A., Holtom, G. R., and Xie, X. S. (1999) Three dimensional vibrational imaging by coherent anti-Stokes Raman scattering. *Phys. Rev. Lett.* **82,** 4142–4145.
3. Müller, M. and Schins, J. M. (2002) Imaging the thermodynamic state of lipid membranes with multiplex CARS microscopy. *J. Phys. Chem. B* **106,** 3715–3723.
4. Cheng, J. X., Pautot, S., Weitz, D. A., and Xie, X. S. (2003) Ordering of water molecules between phospholipid bilayers visualized by coherent anti-Stokes Raman scattering microscopy. *Proc. Natl. Acad. Sci. USA* **100,** 9826–9830.
5. Nan, X., Cheng, J. X., and Xie, X. S. (2003) Vibrational imaging of lipid droplets in live fibroblast cells with coherent anti-Stokes Raman scattering microscopy. *J. Lipid Res.* **44(11),** 2202–2208.
6. Potma, E. O., Xie, X. S., Muntean, L., et al. (2004) Chemical Imaging of Photoresists with Coherent Anti-Stokes Raman Scattering (CARS) Microscopy. *J. Phys. Chem. B* **108,** 1296–1301.
7. Wurpel, G. W. H., Schins, J. M., and Müller, M. (2004) Direct measurement of chain order in single lipid mono- and bilayers with multiplex CARS. *J. Phys. Chem. B* **108,** 3400–3403.
8. Kennedy, A. P., Sutcliffe, J., and Cheng, J. X. (2005) Molecular Composition and Orientation in Myelin Figures Characterized by Coherent Anti-Stokes Raman Scattering Microscopy. *Langmuir* **21,** 6478–6486.
9. Potma, E. O. and Xie, X. S. (2003) Detection of single lipid bilayers with coherent anti-Stokes Raman scattering (CARS) microscopy. *J. Raman Spectrosc.* **34,** 642–650.
10. Holtom, G. R., Thrall, B. D., Chin, B., and Colson, D. (2001) Achieving molecular selectivity in imaging using multiphoton Raman spectroscopy techniques. *Traffic* **2,** 781–788.
11. Cheng, J., Jia, Y. K., Zheng, G., and Xie, X. S. (2002) Laser-Scanning Coherent Anti-Stokes Raman Scattering Microscopy and Applications to Cell Biology. *Biophys. J.* **83,** 502–509.
12. Levin, I. W. (1984) Vibrational spectroscopy of membrane assemblies, in *Advances in Infrared and Raman Spectroscopy, vol. 11* (Clark, R. J. H. and Hester, R. E., eds.), Wiley, pp. 1–48.
13. Snyder, R. G., Tu, K., Klein, M. L., Mendelsohn, R., Strauss, H. L., and Sun, W. (2002) Acyl chain conformation and packing in dipalmitoylphosphatidylcholine bilayers from MD simulation and IR spectroscopy. *J. Phys. Chem. B* **106,** 6273–6288.
14. Eesley, G. L. (1981) *Coherent Raman Spectroscopy*. Pergamon Press, New York.

15. Bjorklund, G. C. (1975) Effects of focusing on third-order nonlinear processes in isotropic media. *IEEE J. Quant. Electr.* **11,** 287–296.
16. Müller, M., Schins, J. M., and Wurpel, G. W. H. (2004) Shot-noise limited detection sensitivity in multiplex CARS microscopy. *SPIE* **5323,** 195–204.
17. Duncan, M. D. (1984) Molecular discrimination and contrast enhancement using a scanning coherent anti-Stokes Raman microscope. *Opt. Commun.* **50,** 307–312.
18. Müller, M., Squier, J., de Lange, C. A., and Brakenhoff, G. J. (2000) CARS microscopy with folded BoxCARS phasematching. *J. Microsc.* **197,** 150–158.
19. Cheng, J. X. and Xie, X. S. (2004) Coherent Anti-Stokes Raman Scattering Microscopy: Instrumentation, Theory, and Applications. *J. Phys. Chem. B* **108,** 827–840.
20. Potma, E. O. and Xie, X. S. (2004) CARS Microscopy for Biology and Medicine. *Opt. Photon. News* **15,** 40–45.
21. Akhmanov, S. A., Bunkin, A. F., Ivanov, S. G., and Koroteev, N. I. (1978) Polarization active Raman spectroscopy and coherent Raman elipsometry. *Sov. Phys. JETP* **47,** 667–678.
22. Cheng, J., Book, L. D., and Xie, X. S. (2001) Polarization coherent anti-Stokes Raman scattering microscopy. *Opt. Lett.* **26,** 1341–1343.
23. Volkmer, A., Book, L. D., and Xie, X. S. (2002) Time-resolved coherent anti-Stokes Raman scattering microscopy: imaging based on Raman free induction decay. *Appl. Phys. Lett.* **80,** 1505–1507.
24. Oron, D., Dudovich, N., Yelihn, D., and Silberberg, Y. (2002) Quantum control of coherent anti-Stokes Raman processes. *Phys. Rev. A* **65,** 43408.
25. Oron, D., Dudovich, N., and Silberberg, Y. (2003) Femtosecond Phase-and-Polarization Control for Background-Free Coherent Anti-Stokes Raman Spectroscopy. *Phys. Rev. Lett.* **90,** 213902.
26. Volkmer, A., Cheng, J., and Xie, X. S. (2001) Vibrational imaging with high sensitivity via epi-detected coherent anti-Stokes Raman scattering microscopy. *Phys. Rev. Lett.* **87,** 23901.
27. Cheng, J., Volkmer, A., Book, L. D., and Xie, X. S. (2001) An epi-detected coherent anti-Stokes Raman scattering (E-CARS) microscope with high spectral resolution and high sensitivity. *J. Phys. Chem. B* **105,** 1277–1280.
28. Cheng, J., Volkmer, A., Book, L. D., and Xie, X. S. (2002) Multiplex coherent anti-Stokes Raman scattering microspectroscopy and study of lipid vesicles. *J. Phys. Chem. B* **106,** 8493–8498.
29. Müller, M., Schins, J. M., Nastase, N., Wurpel, G. W. H., and Brakenhoff, G. J. (2002) Imaging the chemical composition and thermodynamic state of lipid membranes with multiplex CARS microscopy. *Biophys. J.* **82,** 175a.
30. Müller, M. and Wurpel, G. W. H. (2004) Measurement of chain order in single lipid mono- and bilayers with multiplex coherent anti-Stokes Raman scattering. *Biophys. J.* **86,** 333a.
31. Chen, P. C., Joyner, C. C., and Burns-Kaurin, M. (1999) Multiplex coherent anti-Stokes spectroscopy by use of a nearly degenerate broadband optical parametric oscillator. *Appl. Opt.* **38,** 5894–5898.
32. Otto, C., Voroshilov, A., Kruglik, S. G., and Greve, J. (2001) Vibrational bands of luminescent zinc(II)-octaethyl-porphyrin using a polarization-sensitive 'microscopic' multiplex CARS technique. *J. Raman Spectrosc.* **32,** 495–501.
33. Jones, D. J., Potma, E. O., Cheng, J. -X., et al. (2002) Synchronization of two passively mode-locked, picosecond lasers within 20 fs for coherent anti-Stokes Raman scattering microscopy. *Rev. Sci. Instrum.* **73,** 2843–2848.
34. Potma, E. O., Jones, D. J., Cheng, J. -X., Xie, X. S., and Ye, J. (2002) High-sensitivity coherent anti-Stokes Raman scattering microscopy with two tightly synchronized picosecond lasers. *Opt. Lett.* **27,** 1168–1170.

35. Kee, T. W. and Cicerone, M. T. (2004) Simple approach to one-laser, broadband coherent anti-Stokes Raman scattering microscopy. *Opt. Lett.* **29,** 2701–2703.
36. Paulsen, H. N., Hilligse, K. M., Thgersen, J., Keiding, S. R., and Larsen, J. J. (2003) Coherent anti-Stokes Raman scattering microscopy with a photonic crystal fiber based light source. *Opt. Lett.* **28,** 1123–1125.
37. Yakovlev, V. V. (2003) Advanced instrumentation for non-linear Raman microscopy. *J. Raman Spectrosc.* **34,** 957–964.
38. V. V. Yakovlev, "Broadband cost-effective nonlinear Raman microscopy," SPIE 5323, 214–222 (2004).
39. Wurpel, G. W. H., Schins, J. M., and Müller, M. (2002) Chemical specificity in 3D imaging with multiplex CARS microscopy. *Opt. Lett.* **27,** 1093–1095.
40. Snyder, R. G., Hsu, S. L., and Krimm, S. (1978) Vibrational spectra in the C-H stretching region and the structure of the polymethylene chain. *Spectrochim. Acta* **34A,** 395–406.
41. Orendorff, C. J., Ducey, M. W., and Pemberton, J. E. (2002) Quantitative Correlation of Raman Spectral Indicators in Determining Conformational Order in Alkyl Chains. *J. Phys. Chem. A* **106,** 6991–6998.
42. Lhert, F., Capelle, F., Blaudez, D., Heywang, C., and Turlet, J. -M. (2000) Raman spectroscopy of phospholipid black films. *J. Phys. Chem. B* **104,** 11,704–11,707.
43. Wurpel, G. W. H., Rinia, H. A., and Müller, M. (2005) Imaging orientational order and lipid density in multilamellar vesicles with mulitplex CARS microscopy. *J. Microsc.* **218,** 37–45.
44. Potma, E. O. and Xie, X. S. (2005) Direct Visualization of Lipid Phase Segregation in Single Lipid Bilayers with Coherent Anti-Stokes Raman Scattering Microscopy. *Chem. Phys. Chem.* **6,** 77–79.
45. Kneipp, K., Wang, Y., Kneipp, H., et al. (1997) Single molecule detection using surface-enhanced Raman scattering (SERS). *Phys. Rev. Lett.* **78,** 1667–1670.
46. Nie, S. and Emory, S. R. (1997) Probing single molecules and single nanoparticles by surface-enhanced Raman scattering. *Science* **275,** 1102–1106.
47. Ichimura, T., Hayazawa, N., Hashimoto, M., Inouye, Y., and Kawata, S. (2004) Application of tip-enhanced microscopy for nonlinear Raman spectroscopy. *Appl. Phys. Lett.* **84,** 1768–1770.

5

Preparation of Oriented, Fully Hydrated Lipid Samples for Structure Determination Using X-Ray Scattering

Stephanie A. Tristram-Nagle

Summary

This chapter describes a method of sample preparation called "the rock and roll method," which is basically a solvent evaporation technique with controlled manual sample movement during evaporation of solvent from lipid/solvent mixtures that produces well-oriented thick stacks of about 2000 lipid bilayers. Many lipid types have been oriented using different solvent mixtures that balance solubilization of the lipid with uniform deposition of the lipid solution onto solid substrates. These well-oriented thick stacks are then ideal samples for collection of both X-ray diffraction data in the gel phase and X-ray diffuse scattering data in the fluid phase of lipids. The degree of orientation is determined using visual inspection, polarizing microscopy, and a mosaic spread X-ray experiment. Atomic force microscopy is used to compare samples prepared using the rock and roll method with those prepared by spin-coating, which produces well-oriented but less homogeneous lipid stacks. These samples can be fully hydrated through the vapor provided that the hydration chamber has excellent temperature and humidity control.

Key Words: AFM; hydration; mosaic spread; oriented lipid bilayers; polarizing microscopy; rock and roll method; X-ray scattering.

1. Introduction

Lipid bilayers form the underlying structure of every plant and animal cell membrane wherein, in general, they occur in a fluid, fully hydrated state. Whereas X-rayed, multilamellar arrays of lipid bilayers produce sharp Bragg diffraction peaks only in the lower-temperature crystalline phases due to their well-ordered structure *(1)*, fluid phase lipid bilayers produce diffuse scattering in the fully hydrated state due to thermal fluctuations in the water space between fluctuating bilayers (for a review, *see* **ref. 2**). The Nagle laboratory pioneered a diffuse scattering technique that collects and analyzes two-dimensional (2D) diffuse X-ray scattering from fully hydrated, fluid-phase lipid stacks oriented onto a solid substrate *(3–5)*. These data yield structural and material parameters about specific lipids, but require thick samples (~4–10 µm) since diffuse scattering is weaker than sharp, Bragg diffraction peaks. This chapter will present the step-by-step procedures of how to prepare oriented, thick bilayer sample films and how to hydrate them.

The name "rock and roll method" derives from the popular culture; it was borrowed since it describes well the procedure for making oriented stacks of lipid bilayers first published by the lab in 1993 *(6)*. The appropriate amount of dry lipid to form a 10-µm thick film (*see* **Note 1** and **Subheading 3.1., step 1**) is dissolved in the appropriate solvent(s). The choice of solvent(s) depends on the lipid chain length, unsaturation, backbone, and headgroup type, and is crucial for the success of the films (*see* **Subheading 3.1., step 3**). The lipid solution is poured onto one of the substrates (*see* **Subheading 2., step 3**), which is hand-*rocked* in four

From: *Methods in Molecular Biology, vol. 400: Methods in Membrane Lipids*
Edited by: A. M. Dopico © Humana Press Inc., Totowa, NJ

directions (N, S, E, and W), forcing the lipid solution to *roll* out to the edges of the substrate. These actions subject the lipid bilayers to shear force, which helps them orient during evaporation of the solvent. The rock and roll procedure requires a large glove box to produce defect-free films, because a solvent-rich vapor slows down the evaporation of solvent from the films (*see* **Subheading 3.3.** and **Note 2**).

The rock and roll procedure is compared herein with two other methods that were previously used to form thin films (~20 bilayers):

1. A simple stationary solvent evaporation technique described in detail by Seul and Sammon *(7)* (*see* **Subheading 3.4.**).
2. A "high-tech" method called spin-coating *(8)* (*see* **Subheading 3.5.**).

A third method published recently *(9)* that codissolves solid naphthalene with the solvents and lipid, renders well-aligned and easily hydrated thick lipid films. However, it was deemed unsuitable for the analysis because the resulting films contain many small holes, which preclude an accurate estimation of the sample thickness.

2. Materials

1. *Lipids*: most lipids are purchased from Avanti Polar Lipids (Alabaster, AL) due to their high purity and, more recently, low acyl-chain migration in mixed chain lipids. Occasionally, novel lipids are synthesized for the Nagle lab by collaborators; these are checked by electrospray ionization mass spectrometry to verify their purity. Lyophilized lipids are stored at –20°C in their original vials in dessicators before use. The original vials are equilibrated at room temperature for at least 15 min before opening to avoid condensation of water into the lipid, which can lead to hydrolysis of chains. When stored and opened this way, the saturated chain lipids do not degrade for many years, as verified using thin-layer chromatography (46:15:3, chloroform:methanol:7 N NH$_4$OH [v:v:v]), stained with molybdic acid stain *(10)*.
2. *Solvents*: solvents are of high-pressure liquid chromatography grade; purity of the solvents and cleanliness in forming the sample films are essential. Solvents are poured from the original solvent container into disposable 20-mL glass scintillation vials, which are discarded after 1 d of sample preparation.
3. *Substrates*: usually a 1-mm thick silicon wafer (100 face) cut to 1.5 × 3 cm^2 is used. The size of the wafer is cut to be the same as the size of the Peltier cooler (Melcor, Trenton, NJ) that supports the silicon wafer during the hydration experiment, so that the sample is cooled evenly (*see* **Subheading 3.7.**). The silicon wafers can be used with or without dopant. The first substrate that was used with some success for gel phase samples was a glass microscope slide. Microscope slides have the advantage that orientation can be checked easily using a polarizing microscope (*see* **Subheading 3.4.**); however, the diffuse background from this substrate is large and not easily subtracted from the fluid phase diffuse data. Alternatively, the substrate is a 3 × 3 cm^2 piece of freshly cleaved 35-µm thick mica (Mica New York Corp, NY), which can be curved and glued onto a cut, glass beaker. Another type of substrate is a thin (70 µm) 3 × 3 cm^2 00 glass cover slip (Precision Glass and Optics, Santa Ana, CA).
4. *Water*: nanopure water is used to hydrate the lipids from the vapor phase in a freshly cleaned hydration chamber (described in **Subheading 3.7.**). Water purity and cleanliness of the chamber are essential to reach full hydration.

3. Methods

3.1. Preparation of Samples

1. To calculate the mass of lipid required to make a 10-µm film, consider a 10-µm thick column of many copies of a single lipid pair in a bilayer stack. Convert to number of molecular bilayers by dividing by the thickness of a dried lipid bilayer (compare with ~60 Å for dried

dipalmitoylphosphatidylcholine [DPPC]). Determine the total number of lipids in the film by first multiplying the aforementioned number of bilayers by two, then dividing by the surface area of a dried lipid (~50 Å^2), and then multiplying by the total surface area of the substrate. Then divide by Avogradro's number to get moles, and multiply by the gram molecular weight of the lipid to get grams of lipid. The actual thickness of these films measured by atomic force microscopy (AFM) is slightly smaller than the calculated thickness, presumably because of some collection of lipid near the edges of the substrate *(11)*.
2. Weigh lyophilized lipid using an analytical balance and place the lipid (~4–10 mg, calculated as aforementioned) into a disposable, small glass test tube. Disposable test tubes are used to assure cleanliness. Add 100–200 μL solvent to the weighed lipid in the test tube in the glove box. Agitate the tube to solubilize the lipid. The solvents required for each lipid were determined by the method of trial and error, within the general guidelines that a hydrophobic solvent is needed to dissolve the lipid, whereas an amphipathic solvent is needed to decrease the contact angle of the lipid solution with the hydrophilic substrate, thus causing the solution to spread. Often one solvent is not sufficient to both solubilize the lipid and cause the solution to spread on the silicon, mica, or glass surface. Listed below are the solvents or solvent mixtures (shown as volume ratios), which were successful for each lipid in the lab. Results may vary across laboratories due to relative humidity, temperature, and size (or lack) of the glove box (*see* **Note 2**). If the solvents below do not successfully form a smooth-looking film to the naked eye on the first attempt, then a different ratio may be used to redissolve the film.
3. Suggested solvents for specific lipids:
   a. Dioleoylphosphatidylcholine (DOPC)—easiest lipid to orient. 2:1, 2.5:1, or 3:1 chloroform: methanol, or 1:1 chloroform: trifluoroethanol (TFE).
   b. Dilauroylphosphatidylcholine (DLPC)—difficult to orient. 1:1:1 chloroform:methanol:TFE.
   c. Dimyristoylphosphatidylcholine (DMPC)—chloroform:TFE 1:1 or 2:1, chloroform:methanol 1:1 or 1.5:1, and neat isopropanol.
   d. DiC15phosphatidylcholine—chloroform:methanol 2.5:1.
   e. Dipalmitoylphosphatidylcholine (DPPC)—chloroform:methanol 2–3.5:1.
   f. D,L-DPPC—chloroform:methanol 2:1.
   g. DiC17phosphatidylcholine—chloroform:methanol 3:1.
   h. Distearoylphosphatidylcholine (DSPC)—chloroform:methanol 2:1 or 3:1.
   i. DiC19phosphatidylcholine—chloroform:methanol 2.5:1, methanol:carbon tetrachloride 5:1.
   j. DiC20phosphatidylcholine—methanol:carbon tetrachloride 4:1.
   k. DiC22phosphatidylcholine—methanol:carbon tetrachloride 4:1.
   l. DiC24phosphatidylcholine—difficult to orient. Methanol:carbon tetrachloride 3:1 and chloroform:methanol 2:1.
   m. DSPC-Br$_4$—chloroform:TFE, 1:1.
   n. Dihexadecylphosphatidylcholine (DHPC)—chloroform:methanol 2:1.
   o. Dierucoylphosphatidylcholine DiC22:1PC—chloroform:TFE, 1:1.
   p. Palmitoyloleoylphosphatidylcholine (POPC)—chloroform:TFE, 1:1.
   q. Dimyristoylphosphatidylserine (DMPS)—difficult to orient. TFE:toluene, 4:1, or neat toluene.
   r. Dioleolyphosphatidylserine (DOPS)—chloroform:TFE, 2:1.
   s. Stearoyldocosahexaenoylphosphatidylcholine—methylene chloride:methanol, 3:1.
   t. Stearoyldocosapentaenoylphosphatidylcholine—methylene chloride:methanol, 3:1.
   u. Dilaurylphosphatidylethanolamine (DLPE)—neat hexafluoroisopropanol in the hood (*see* **Note 3**).
   v. Myristoylpalmitoylphosphatidylcholine (MPPC)—chloroform:methanol, 2:1.
   w. Myristoylstearoylphosphatidylcholine (MSPC)—chloroform:methanol, 2.5:1.
   x. 1,2-Di-*O*-myristoyl-3-N,N,N-trimethylaminopropane (DM-TAP)—chloroform:methanol, 10:1.
   y. Most mixtures combining DOPC, DPPC, DLPC, brain sphingomyelin, and cholesterol (raft and super lattice mixtures)—chloroform:TFE, 1:1 (*see* **Note 4**).

z. Mixtures with alamethicin and DOPC or DOPC-Br$_4$—chloroform, then redissolved in chloroform:TFE, 1:1.
aa. Mixtures with human immunodeficiency virus fusion peptide FP-23 and DOPC or diC22:1PC—neat hexafluoroisopropanol, in the hood (*see* **Note 3**).

Salts can be added to any of the aforementioned lipid–solvent mixtures. Salts crystallize on drying, yet they diffuse readily between bilayers on subsequent hydration (*see* **Subheading 3.7.**) (*see also* **ref. 12**). In the glove box, close the test tube prepared with above-containing solvent and lipid, with a silicone or neoprene stopper during the following preparation of the substrate, so that the ratio of solvents does not change due to unequal evaporation. Rubber stoppers should not be used as they are degraded by the solvents.

3.2. Preparation of the Substrates

The normal procedure is to simply bathe the Si wafers in chloroform in a glass Petri dish inside the glove box; then, the wafers are rubbed with a cotton swab (*see* **Note 5**). This procedure not only cleans the wafer, but also fills the glove box with a chloroform-saturated atmosphere. The Si wafers are sometimes rubbed with a Kimwipe in the final step to remove any remaining residue, and any lint is carefully blown away with a small rubber bulb. When mica is used, the 3 × 3 cm^2 piece is freshly cleaved using a technique shown to the author by Jacob Israelachvilli; this consists of wedging a fine gauge needle or pin into the edge of a square of grade 2 mica to separate a thin layer of mica from the original slab, which is then slowly teased away using a tweezer. The thickness should be about 35 µm (i.e., flexible enough to become slightly curved, yet thick enough to contain lipid and solvent), with no cracks or holes. The mica should be used within 1 or 2 h after cleaving to avoid dust collecting on its surface.

The third substrate, i.e., 3 × 3 cm^2 thin glass cover slips, should be cleaned with chloroform swabbing as for the silicon wafer. At just 70-µm thick, these cover slips break easily, especially during cleaning. The thin glass cover slip is useful for X-raying lipid samples through the back of the glass at a 45° angle of incidence; the scattering from the sample is then totally unobstructed by the substrate, which is not the case for a grazing angle of incidence on the silicon wafer or curved mica. These substrates are then attached to the cap of a 20-mL disposable glass scintillation vial using either sticky tack or clay (*see* **Fig. 1**). This method of attachment leaves the sides of the substrate completely free, which is essential to allow surface tension and hold the solvent solution on the top of the substrate. If the substrates are placed directly onto a larger substrate, such as a microscope slide, the lipid–solvent mixture tends to run off at the edges, due to a small contact angle of the lipid solution with the substrate.

3.3. Rock and Roll Method

When the author first began making films in the early 1990s, she tried to evaporate the solvents in the hood, but rapid evaporation of the solvent caused many surface defects (a frosted appearance) that were poorly oriented. Even working on the open lab bench close to an open window or air conditioner caused the lipid surface to become "frosty" during evaporation of the solvent. Therefore, investment in a suitable glove box is key for the success of this method. The Nagle lab glove box is 8 ft^3 made from 3/16 in. thick Plexiglas, with rubber, gloveless ports for entry of hands. Humidity is monitored with a probe (Rotronics, Dallas, TX), but not controlled. The humidity generally increases from about 40 to 70% relative

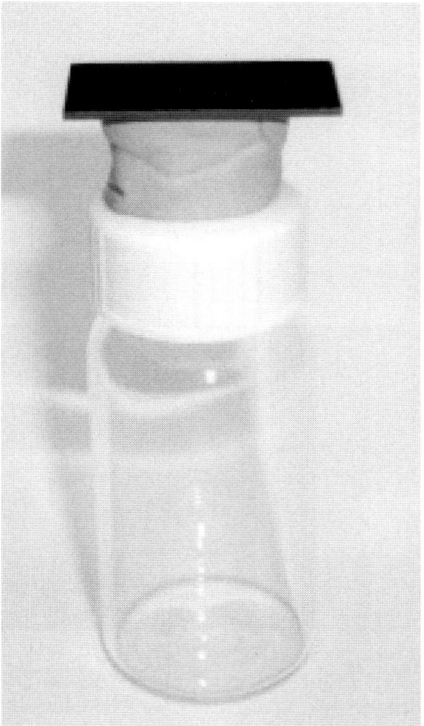 

Fig. 1. A silicon substrate (15 × 30 mm^2) is attached through sticky tack (synthetic clay) to a 20-mL scintillation vial. **(Left)** Side view, **(Right)** top view. The highly polished silicon surface appears black in this photograph.

humidity (RH) during the film formation, because of perspiration from human hands in the glove box (*see also* **Note 2**).

In the glove box, the lipid in solvent is poured onto the substrate attached as described in **Subheading 3.2.**, with rocking and rolling (*see* **Subheading 1.**). As the sample dries, a higher angle of rocking (up to 90°) may be required in order to induce the now-viscous solution to move down the wafer. The rocking continues until the rolling of lipid in solvent over the surface stops; at this point the substrate is removed from the sticky tack and placed onto a flat surface in the glove box, which is why flat substrates work best (although the author has also used a hanging drop rock and roll method on an inverted curved beaker). The entire process takes about 5 min. Rotating the substrate on the sticky tack before detaching it avoids breaking the thinner substrates. The film is left to dry for 1 d in the glove box and an additional day in air on the lab bench or in a fume hood to assure complete removal of solvent (*see* **Note 6**).

3.4. Assessment of Orientation

After the sample on mica or glass cover slip has dried for at least 1 d on the lab bench, it is first examined by eye; gross misorientation appears as large holes, crusted lipid, and a frosted appearance. If there is no evidence of gross misorientation, the sample is examined using an optical microscope under crossed polarizers as described by Asher and Pershan *(13)*.

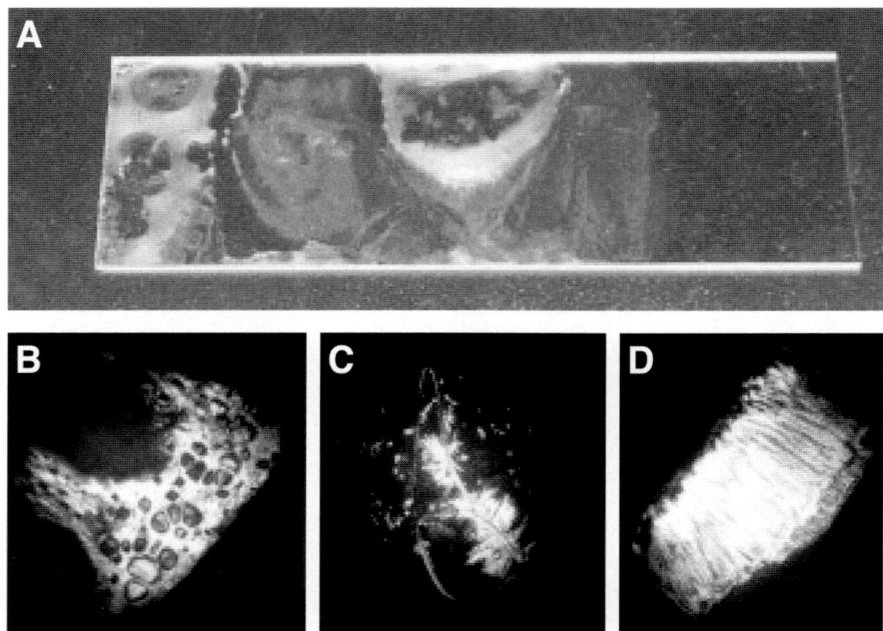

Fig. 2. (**A**) Visual appearance of a sample of DPPC dried from chloroform:methanol (3:1) onto a cleaned, stationary glass microscope slide. (**B–D**) Defects observed using crossed polars in the optical microscope on the far left side of the sample shown in **A**. The lipid in the center of the slide appeared dark gray under crossed polars, i.e., fairly well oriented but not uniform. Lipid was not deposited on the far right side of the slide. Results like these were the motivation for the rock and roll procedure.

If the film is completely smooth, well oriented, and homogeneous, there will be a uniform black field under crossed polars. Any defects retard the light along the defect, which results in loss of total cancellation of crossed polarized light, giving a bright spot or streak. Many curious and astonishing structures have been observed by the author and her students (*see also* **Note 7**). The sample in **Fig. 2** was prepared by pouring DPPC solubilized in 3:1 chloroform:methanol onto a cleaned glass microscope slide; this was rocked briefly to evenly distribute the lipid solution over the substrate. It was then left to dry on the lab bench without any further movement, similar to the procedure of Seul and Sammon (*7*), except that the amount of lipid in this sample was for a 10-μm thick film. The visual appearance of the film after drying on the lab bench is shown in **Fig. 2A**. The various parts of the same film visualized under crossed polars in the optical microscope are shown in **Figs. 2B–D**. A photograph of a sample prepared using the rock and roll procedure is shown in **Fig. 6**.

Once the sample appears uniform under crossed polars, it can be X-rayed either in the dried or hydrated state using the lab-rotating anode or during the synchrotron experiment. The measurement of mosaic spread is obtained by rotating the flat sample in 0.005° steps through a chosen Bragg θ angle. The Bragg peak intensity is measured and plotted vs angle of rotation. This is fit with a Gaussian function, and the sigma of the Gaussian function is reported in degrees as the mosaic spread. An example of a rocking scan to determine mosaic spread is shown in **Fig. 3**.

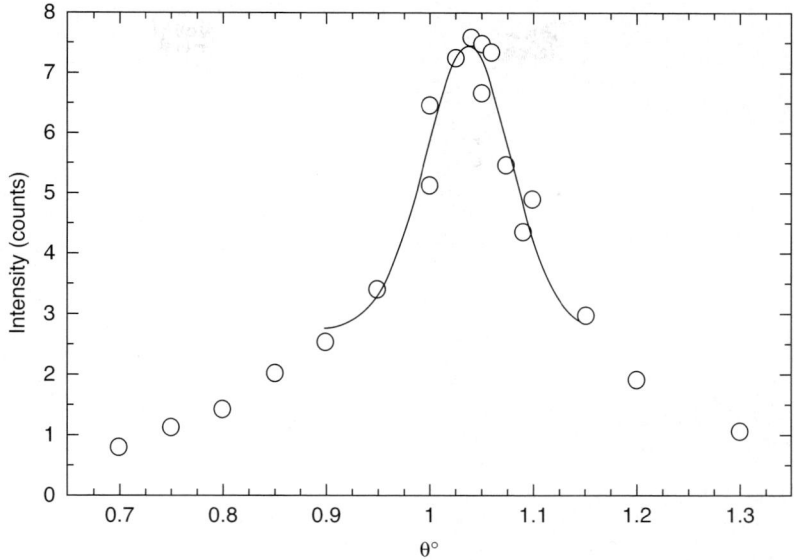

Fig. 3. Mosaic spread of a 10-μm thick DMPC sample oriented onto a silicon wafer. This rocking scan was centered on the second order Bragg reflection from a lamellar stack of hydrated bilayers. Each data point was collected at a separate angle of incidence. The half-width of the Gaussian fit (solid line) is 0.08°.

3.5. AFM to Characterize Uniformity and Smoothness of Sample Films

With the help of Justin Legleiter, a talented graduate student in Chemistry at Carnegie Mellon University, the author used AFM to probe microscopic differences between lipid films prepared using the rock and roll procedure, with those using spin-coating. Details of the tapping mode AFM are presented in **ref. 11**. Many variables such as spinning speed, method of application of the lipid in solvent (dropwise or stationary), concentration of lipid in solvent, and choice of solvent were attempted. The samples shown below in this comparison study were deposited on 3×3 cm² pieces of freshly cleaved mica and the thicknesses were approx 200 bilayers. The AFM 3D and the phase images of a 1.1 mg DMPC sample dissolved in 100 μL TFE:chloroform (2:1), and prepared using the rock and roll method are shown in **Fig. 4A,B**, respectively. Three steps of one bilayer thickness are shown in the lower right-hand corner of both **Fig. 4A,B**. These are sometimes called edge dislocations. The corresponding images for a sample prepared by spin-coating are shown in **Fig. 4C,D**. The sample was 0.94 mg of DMPC in 100 μL TFE:chloroform (2:1), added drop-by-drop as the sample was spinning at approx 20g. The main difference between the images in **Fig. 4** is that in the spin-coated sample there are many large holes, and several protrusions of one or more bilayers. Although the lipid appears to be flat and well oriented in **Fig. 4C,D**, the film is less homogeneous and extensive than that in **Fig. 4A,B**.

Another comparison of these two methods was carried out using the neat solvent, isopropanol, first reported by Seul and Sammon *(7)*. **Figure 5A,B** show the 3D and phase images of a DMPC sample (1.2 mg DMPC dissolved in 100 μL isopropanol) prepared by the rock and roll method. The figure shows that the surface is quite smooth, with four extensive steps of one

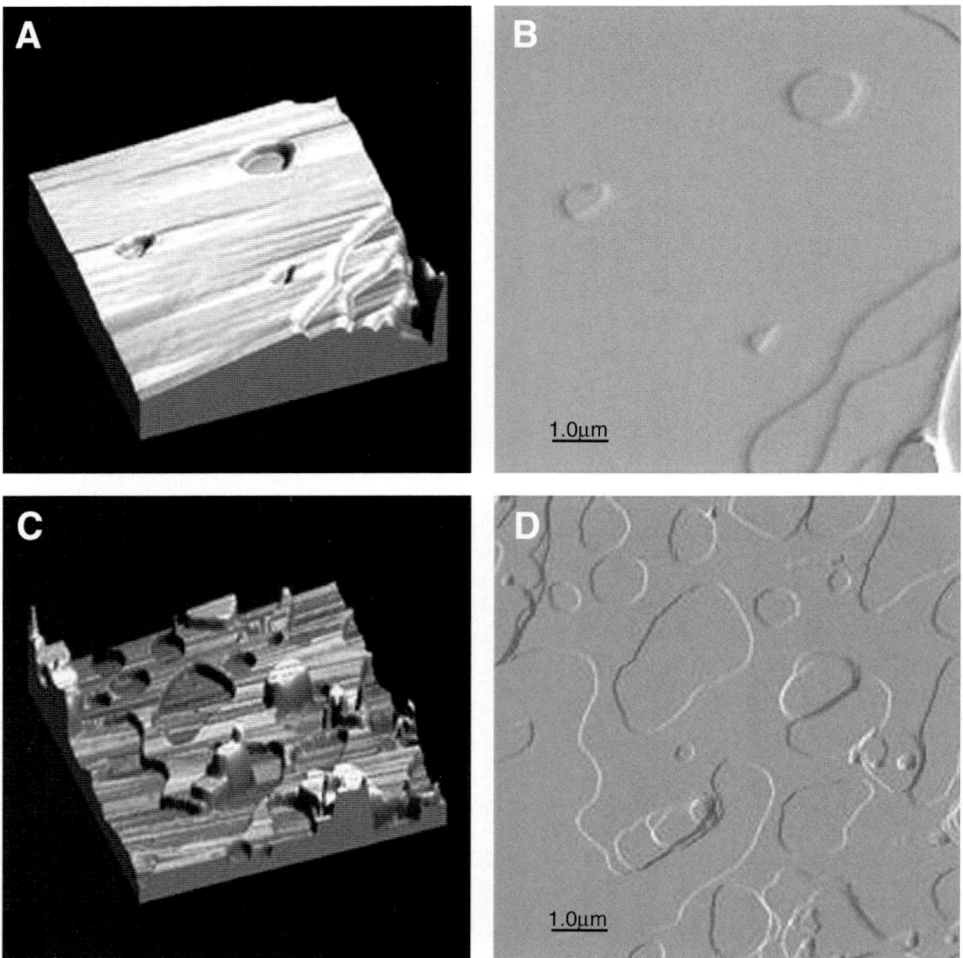

Fig. 4. Tapping mode AFM of DMPC deposited from TFE:chloroform spin-oating (2:1) onto freshly cleaved mica either by the rock and roll method (**A** and **B**) or by spin-coating (**C** and **D**). The 3D images have a 25-nm height range.

bilayer thickness each. For **Fig. 5C,D**, the DMPC was dissolved in 200 μL isopropanol, placed onto the freshly cleaved mica and then spun at approx 20g until no more solvent remained (about 2 min). As in the case of TFE:chloroform, the film prepared by the rock and roll method is shown to be smoother, more homogenous, and more uniform. Another comparison using neat TFE gave the same result. A combined technique of rock and roll followed by spin-coating during the final drying step resulted in less loss of material, but did not produce films of significantly better orientation than the rock and roll procedure alone, as judged by polarizing microscopy.

An additional benefit of the rock and roll procedure, when compared with spin-coating, is that no material is lost because of spinning; this result is particularly important when expensive peptides are added to the films. Thus, these AFM studies show that (at least for films of thickness ~200 bilayers) the rock and roll method produces better-oriented, smoother, and more uniform films.

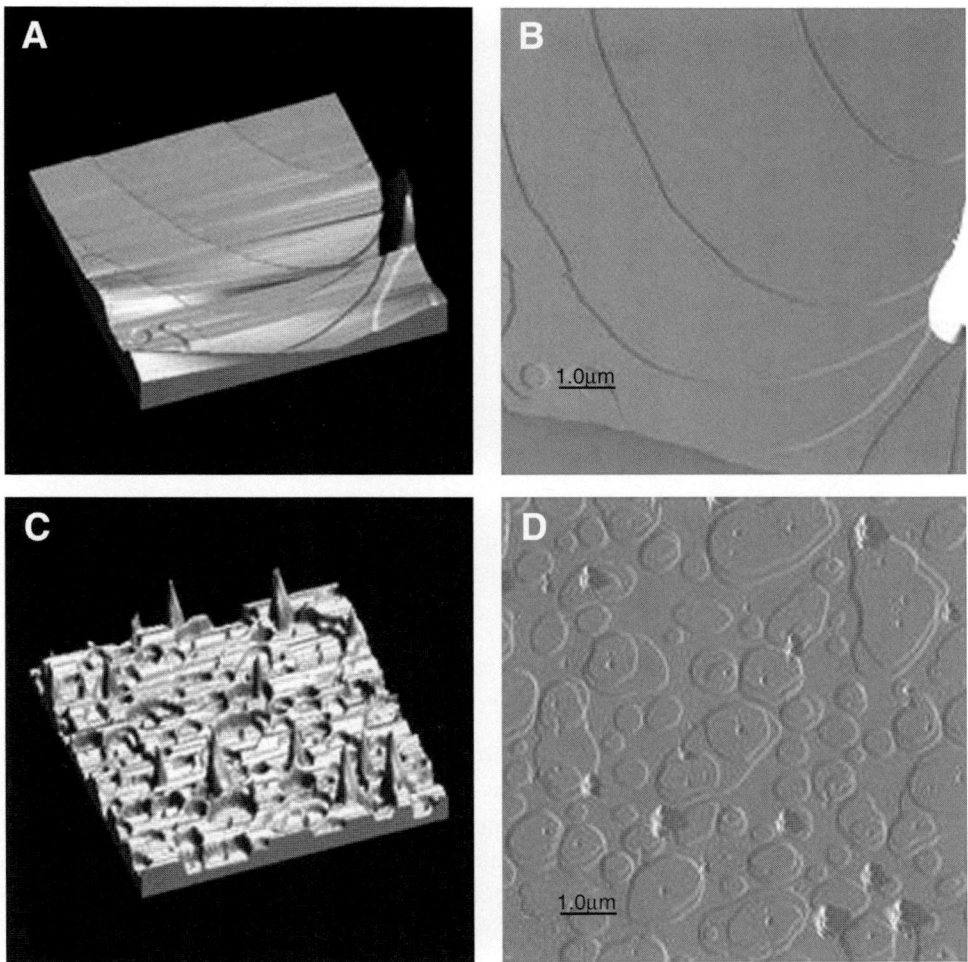

Fig. 5. Tapping mode AFM of DMPC from isopropanol deposited onto freshly cleaved mica either by the rock and roll method (**A** and **B**) or by spin-coating (**C** and **D**). Height ranges were 25 and 50 nm for **5A,C**, respectively.

3.6. Packing Samples

The sample on the 1.5 × 3 cm² silicon wafer is then trimmed to a thin strip of 4 or 5 × 30 mm² long, by removing sample along both long edges of the silicon wafer using a fresh single-edge razor blade, followed by wiping off any remaining lipid with a dry cotton swab. One reason for trimming the samples is to remove lipid that may be less well oriented near the edges of the wafer. The other reason concerns the X-ray experiment (5). The trimmed samples are stored in multicompartment plastic boxes within a large glass dessicator at 4°C. The samples are usually prepared 1–3 wk before a synchrotron trip, but samples on all three substrates are quite stable for several months when stored in a dessicator at 4°C. They are carried to the Cornell high-energy synchrotron source (CHESS) in the dessicator in an ice-filled container. A picture of a sample prepared in this way is shown in **Fig. 6**.

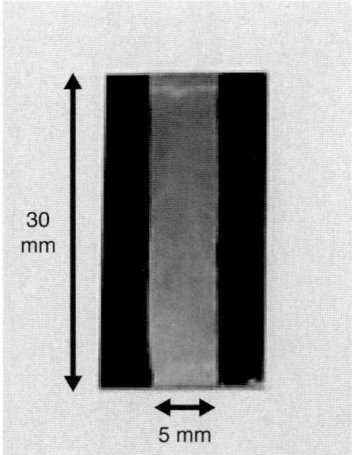

Fig. 6. DPPC dissolved in chloroform:methanol (3.5:1) was oriented onto a silicon substrate using the rock and roll procedure and trimmed as described in **Subheading 3.6.**

3.7. Sample Hydration Through the Vapor

Hydration of lipid bilayers through the vapor phase has been difficult *(14)*. To summarize briefly, although it was possible to fully hydrate gel phase bilayers through the vapor *(6,15,16)*, it was at first impossible to hydrate through the vapor in the more biologically relevant fluid phase *(17,18)*. The seeming paradox (i.e., that the vapor above pure water is not at 100% relative humidity even though it is in equilibrium with pure water) generated controversy in several labs *(19)*. The trick that worked in the gel phase was to cool the sample relative to the surrounding vapor using a Peltier cooler under the sample substrate *(6)*; however, this solution was not sufficient to condense the considerably higher water content into a fluid phase lipid. The key to successfully hydrating a sample through the vapor was excellent temperature control; when this is done, the vapor pressure paradox does not exist *(19,20)*. The researchers have been fortunate to be able to use two X-ray chambers for hydrating lipids in the fluid phase. The details and pictures of these chambers are described in **refs. 5** and **21** and will not be repeated in this chapter. One important design feature is a rapid flow to the chamber of the coolant that is temperature-controlled by an external water bath. In addition, thick or double walls (double 6-µm thick mylar windows for entry and exit of X-rays) and additional foam insulation on the outside of the chamber for extreme temperatures (>50°C and <10°C) are needed. A final important feature is to add a water-filled piece of filter paper (suggestion of Peter Rand) to the inside top of the chamber with fingers that extend into the hydrating water reservoir. This increases the water evaporation surface and helps to decrease hydration equilibration times. When the Peltier current under the sample is set to cool the sample by 0.1°C relative to the hydration chamber temperature *(5)*, hydration at 30°C occurs in about 30 min. If a slower hydration is desired (as when collecting data at many D-spacings within 10 Å of full hydration), a smaller current is used. The current can also be reversed to slow down the hydration or slightly dry out the sample by heating the surface of the Peltier in contact with the sample. Controlling the hydration speed and extent this

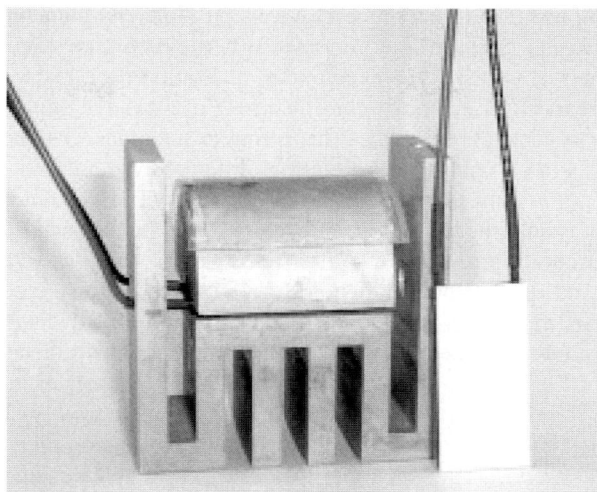

Fig. 7. DPPC dissolved in 2:1 chloroform:methanol is deposited onto mica, which is glued to a cut, glass, 50-mL beaker. This is in contact with a curved aluminum holder; Dow heat sink compound is normally spread under the beaker for better thermal contact during the hydration experiment. There is a Peltier cooler under the sample (not visible); a second Peltier cooler is shown to the right of the holder as a demonstration. An alternate sample holder for flat silicon substrates does not have the semicylindrical aluminum piece.

way was a huge achievement, allowing data collection of many samples during a single trip to CHESS.

The chamber uses flat samples, which are rotated during the X-ray data collection. Alternatively, mica can be used as substrate (described in **Subheading 2., step 3**). The mica with lipid sample is curved and glued onto a cut, 50-mL glass beaker. Devcon 5 min epoxy and clamps are used to hold the mica in place during the drying of the epoxy. Since the sample is curved, all the angles of incidence that are obtained by rotating a flat sample can be obtained in one scan at a grazing angle of incidence. The cut, glass beaker sits on a semicylinder of solid aluminum, which is in contact with the Peltier cooler (*see* **Fig. 7**). One small drawback to using this sample is that there are several sharp and intense mica reflections that are difficult to subtract out with a background scan, resulting in their inclusion in the final data. However, as the diffuse background scattering from mica is very low, these sharp peaks were found to be acceptable.

4. Notes

1. Thicker films up to 40 µm were attempted, but the mosaic spread, or degree of misorientation, increased. Thinner films were tried (2 µm), but the intensity of diffuse scattering was reduced substantially when compared with the reflectivity from the substrate and general background. Films between 4 and 10 µm are a compromise between satisfactory X-ray signal-to-noise and low mosaic spread. However, films less than 8 µm are sometimes not successful because there is not enough lipid to cover the surface homogeneously.
2. When teaching the rock and roll method to colleagues, the author reached success only when a suitable glove box was available. The use of a small, plastic disposable glove box was not successful because the atmosphere in the glove box rapidly became too humid due to perspiration

from the hands. The use of thin neoprene gloves is possible, yet somewhat awkward, given the necessary manipulations. Even in a large glove box, the humidity sometimes rises above 80% RH when gloves are not worn. If this occurs, then the ratio of the hydrophobic/hydrophilic solvent should be increased: for example, chloroform:TFE, from 1:1 to 2:1.

3. Hexafluoroisopropanol's label warns that it is dangerous to all organs. Wear neoprene gloves and work with this solvent in the hood only. This procedure is different from use of the glove box described in **Subheading 3.3.**, but this solvent is unique in that it can produce quite uniform films while drying quickly. This solvent is only successful for some lipids and should be avoided if possible.

4. Films formed from cholesterol mixtures are usually not as well oriented as phospholipids, but the orientation of mixtures of lipids and cholesterol is improved by annealing at 10–20°C above the main phase-transition temperature of the highest melting lipid. This is carried out in a humid environment in an annealing chamber with a water-filled sponge attached above and below the lipid sample for 6–12 h.

5. For cleaning the silicon wafers, the author has tried a chromic acid wash, followed by copious rinsing with Barnstead/Thermolyne (Dubuque, IA) nanopure water, then an HCl wash, followed by copious rinsing with Barnstead nanopure water, with or without a final step of swabbing with high-pressure liquid chromatography chloroform using a cotton swab. The results are similar with or without the two acid washes before the chloroform swabbing. The final step of chloroform swabbing imparts a slightly hydrophobic surface to the wafers, which is beneficial. On one occasion the author tried to save time by using a solution of hot, concentrated Contrad (Decon Labs, King of Prussia, PA). This is a basic surfactant solution that is used for washing glassware, which was recommended by a colleague. This was nearly disastrous: the Contrad etched the silicon wafer surface so that it lost its shine; this was not restored with a hydrofluoric acid dip, as used for the removal of a silicon oxide layer. The mosaic spread from lipid samples prepared on these wafers was worse than that with the normal chloroform swabbing technique, and there was a large diffuse scatter from these substrates, similar to that from glass substrates. Therefore, it is best not to aggressively clean the polished silicon wafer surface.

6. Some care is needed to master the rock and roll technique. The rocking should be slow and shallow at first, when there is still much of the solvent on the substrate. Excessive rocking at this time can cause the lipid solution to fall off the edges of the substrate. The angle and speed of rocking can be increased as the sample dries. Rocking too slowly at this time, or not forcing the lipid solution to roll out to the edges of the substrate, produces a poorly oriented sample.

7. Defects can be described as volcanoes, feathers, tubules, squigglies, simple crosses, and fatty streaks. If there is water present, a familiar "Maltese cross" appears because of the circular shape of the multilamellar vesicles that retard the light in a radial fashion *(22)*. If Maltese crosses are observed, the sample is not well oriented and should be discarded. Simple crosses are more prominent if the methanol content of the solvent mixture is too high, and squigglies and feathers are more prominent when the chloroform content of the solvent mixture is too high. Squigglies and feathers are more disruptive than a surface layer of methanol-induced simple crosses. Another feature, termed "Grandjean terraces" and described nearly 100 yr ago *(23)*, looks like steps or plateaus (*see* **Figs. 4A,B** and **5A,B**). These edge dislocations, unlike in the thin films of Seul and Sammon *(7)*, do not degrade the orientation of the sample, probably because they are largely confined just to the outer surface of thick films.

Acknowledgments

I would like to thank my husband and colleague, Prof. John Nagle, for continued collaboration and support, and Dr. Norbert Kučerka for supplying the mosaic spread graph. I would also like to thank Dr. Horia Petrache for help with construction of the hydration chamber described in ref. 5. Supported by National Institutes of Health grant no. GM44976 (JFN).

References

1. Luzzati, V. (1968) X-Ray Diffraction Studies of Lipid-Water Systems, in *Biological Membranes*, (Chapman, D., ed.), Academic Press, London, pp. 71–123.
2. Tristram-Nagle, S. and Nagle, J. F. (2004) Lipid Bilayers: Thermodynamics, structure, fluctuations, and interactions. *Chem. Phys. Lipids* **127,** 3–14.
3. Lyatskaya, Y., Liu, Y., Tristram-Nagle, S., Katsaras, J., and Nagle, J. F. (2001) Method for obtaining structure and interactions from oriented lipid bilayers. *Phys. Rev. E* **63,** 011907(1–9).
4. Liu, Y. and Nagle, J. F. (2004) Diffuse scattering provides material parameters and electron density profiles of biomembranes. *Phys. Rev. E* **69,** 040901(R).
5. Kučerka, N., Liu, Y., Chu, N., Petrache, H. I., Tristram-Nagle, S., and Nagle, J. F. (2005) Structure of fully hydrated fluid phase DMPC and DLPC bilayers using X-ray scattering from oriented multilamellar arrays and from unilamellar vesicles. *Biophys. J.* **88,** 2626–2637.
6. Tristram-Nagle, S., Zhang, R., Suter, R. M., Worthington, C. R., Sun, W. -J., and Nagle, J. F. (1993) Measurement of chain tilt angle in fully hydrated bilayers of gel phase lecithins. *Biophys. J.* **64,** 1097–1109.
7. Seul, M. and Sammon, M. J. (1990) Preparation of surfactant multilayer films on solid substrates by deposition from organic solution. *Thin Solid Films* **185,** 287–305.
8. Mennicke, U. and Salditt, T. (2002) Preparation of solid-supported lipid bilayers by spin-coating. *Langmuir* **18,** 8172–8177.
9. Hallock, K. J., Wildman, K. H., Lee, D. -K., and Ramamoorthy, A. (2002) An innovative procedure using a sublimable solid to align lipid bilayers for solid-sate NMR studies. *Biophys. J.* **82,** 2499–2503.
10. Dittmer, J. C. and Lester, R. L. (1964) A simple, specific spray for the detection of phospholipids on thin-layer chromatograms. *J. Lipid Res.* **5,** 126–127.
11. Tristram-Nagle, S., Liu, Y., Legleiter, J., and Nagle, J. F. (2002) Structure of gel phase DMPC determined by X-ray diffraction. *Biophys. J.* **83,** 3324–3335.
12. Petrache, H. I., Tristram-Nagle, S., Harries, D., Kučerka, N., Nagle, J. F., and Parsegian, V. A. (2006) Swelling of Phospholipids by Monovalent Salt. *J. Lipid Res.* **47,** 302–309.
13. Asher, S. A. and Pershan, P. S. (1979) Alignment and defect structures in oriented phosphatidylcholine multilayers. *Biophys. J.* **27,** 393–421.
14. Rand, R. P. and Parsegian, V. A. (1989) Hydration forces between phospholipid bilayers. *Biochim. Biophys. Acta* **988,** 351–376.
15. Levine, Y. K. (1973) X-Ray Diffraction Studies of Membranes. *Prog. Surf. Sci.* **3,** 279–352.
16. Katsaras, J., Yang, D. S. C., and Epand, R. M. (1992) Fatty-acid chain tilt angles and directions in dipalmitoylphosphatidylcholine bilayers. *Biophys. J.* **63,** 1170–1175.
17. Jendrasiak, G. L. and Hasty, J. H. (1974) Hydration of phospholipids. *Biochim. Biophys. Acta* **337,** 79–91.
18. Smith, G. S., Safinya, C. R., Roux, D., and Clark, N. A. (1987) X-ray study of freely suspended films of a multilamellar lipid system. *Mol. Cryst. Liq. Cryst.* **144,** 235–255.
19. Nagle, J. F. and Katsaras, J. (1999) Absence of a vestigial vapor pressure paradox. *Phys. Rev. E* **59,** 7018–7024.
20. Katsaras, J. (1998) Adsorbed to a rigid substrate, dimyristoyl-phosphatidylcholine multibilayers attain full hydration in all mesophases. *Biophys. J.* **75,** 2157–2162.
21. Katsaras, J. and Watson, M. J. (2000) Sample cell capable of 100% relative humidity suitable for x-ray diffraction of aligned lipid multibilayers. *Rev. Sci. Inst.* **71,** 1737–1739.
22. Tristram-Nagle, S. and Wingert, L. M. (1990) A thermotropic study of 1-deoxy-1-(N-methyloctanamido)-D-glucitol (MEGA-8) using microscopy, calorimetry and x-ray diffraction. *Mol. Cryst. Liq. Cryst.* **188,** 41–56.
23. Grandjean, F. (1916) The orientation of anisotropic liquids on the surface of crystals. *Bull. Soc. Franc. Min.* **39,** 164–213.

6

Nuclear Magnetic Resonance Investigation of Oriented Lipid Membranes

Olivier Soubias and Klaus Gawrisch

Summary

^{31}P-and ^{2}H-nuclear magnetic resonance are widely used to study order and dynamics, but also orientation of lipid bilayers at solid interfaces. Herein, the various techniques of orienting lipid bilayers at interfaces, the requirements for acquisition of nuclear magnetic resonance spectra, and the basic principles for spectra interpretation are described. Preparation of two types of oriented bilayers is presented: multilamellar bilayers oriented at flat solid interfaces and unilamellar tubular bilayers inside porous substrates. The latter bilayers are accessible from a bulk water phase, which permits adjustment of pH, as well as delivery of proteins and ligands to the bilayer surface.

Key Words: AAO; ^{2}H NMR; lipids; membranes; mosaic spread; oriented bilayers; ^{31}P NMR; porous aluminum oxide; solid state NMR.

1. Introduction

Solid-state nuclear magnetic resonance (NMR) spectroscopy provides valuable information on structure, dynamics, and the phase state of lipid/water dispersions. Orienting membranes with respect to the magnetic field direction yields considerable narrowing of the NMR resonances equivalent to an increase in NMR sensitivity by one order of magnitude. Spatially anisotropic NMR parameters like chemical shifts and quadrupolar interactions are easily measured. The orientational dependence of these parameters is used to determine bilayer alignment, and the effective magnitude of anisotropic interaction contains valuable information on the lipid phase state, as well as of structure and dynamics of bilayers. In addition, preparation of oriented bilayers is a necessity for many NMR experiments on peptide and protein-containing membranes. Preparation methods for oriented samples, parameters for the acquisition of NMR spectra, and instructions for interpretation of ^{31}P- and ^{2}H NMR spectra are given.

Over the last 30 yr, a large number of laboratories have experimented with NMR on oriented membranes (e.g., *see* the publications from the laboratories of K. Arnold, M. Bloom, B. Bechinger, K. Beyer, M. F. Brown, E. J. Dufourc, K. Gawrisch, R. G. Griffin, G. Gröbner, J. A. Killian, F. M. Marassi, A. Milon, S. J. Opella, A. Ramamoorthy, J. Seelig, F. Separovic, E. Sternin, A. S. Ulrich, and A. Watts, just to mention a few). However, information on sample preparation is typically written in fine print without much detail. Although the information in this chapter is primarily based on the authors' experience with NMR sample preparation, they have benefited greatly from communicating with other laboratories on the subject matter. The readers are also referred to Chapter 5.

2. Materials

1. Highly purified lipids dissolved in organic solvent or as dry powder. Polyunsaturated lipids that are prone to oxidation must be dissolved in organic solvent with the antioxidant butylated hydroxytoluene (BHT) added at a lipid/BHT molar ratio of 250/1 and sealed in ampoules. All lipids must be stored in an ultracold freezer.
2. Use only ultrapure solvents: methanol, ethanol, chloroform, methylene chloride, 2-propanol, 1,1,1,3,3,3-hexafluoro-2-propanol, and fuming nitric acid.
3. Water for ^{2}H NMR experiments must be deuterium-depleted (2–4 ppm ^{2}H). For ^{31}P NMR experiments deionized water is sufficient.
4. Ultrathin microscope cover slips cut to the size of the flat coil NMR probe, for example, $9 \times 8 \times 0.07$ mm^3 (Paul-Marienfeld GmbH and Co., KG, Lauda-Königshofen, Germany).
5. Porous anodic aluminum oxide (AAO) filters, for example, 13-mm diameter, 0.2-µm nominal pore size (Whatman International Ltd., Maidstone, England).
6. 100-µL Hamilton syringe.
7. Ultrasonification equipment (optional).
8. Solid-state NMR spectrometer with a proton resonance frequency of 300 MHz or higher, equipped with a flat coil or large solenoidal coil probe. For ^{31}P NMR experiments a dual resonance ^{1}H/^{31}P probe is required; for ^{2}H NMR on lipids with perdeuterated hydrocarbon chains a single resonance, ^{2}H NMR probehead is sufficient. Experiments on selectively deuterated molecules may benefit from using a dual resonance ^{1}H/^{2}H NMR probe that permits proton decoupling. ^{1}H- and X-band amplifiers (^{1}H: 0.1–1 kW, X-band: 0.3–1 kW). Preamplifiers rated for high-power NMR experiments. A low-pass filter that stops the ^{1}H decoupling frequency before the preamplifier of the X-channel. The radio-frequency channels must have sufficient bandwidth to guarantee a phase and amplitude distortion-free detection of spectral width of at least 200 kHz.

3. Methods
3.1. Preparation of Glass Slides

1. The cleaning must be performed in a fume hood. The person conducting the cleaning must wear proper protective clothing and eye protection. Place 20–60 glass slides as single layer on the bottom of a glass Petri dish or use a rack made of Teflon that holds the slides upright, and cover them with fuming nitric acid. Cover the container and store the slides in acid for several hours. Remove the slides from the acid and flush them repeatedly with deionized, 0.22 µm filtered water to remove any traces of acid.
2. If a lipid–water dispersion is applied to the glass, the slides should be kept in deionized water until used. If lipids are applied from organic solvent, slides should be dried in a dust-free environment.
3. Align the slides flat in a suitable tray that limits the spread of liquids beyond slide boundaries, for example, the cover of well plates.

3.2. Preparation of Oriented Bilayers From Organic Solvent

1. Dissolve the lipids in a suitable organic solvent (*see* **Note 1**) at a concentration of 10–20 mg/mL. If lipids are stored in a solvent that is inappropriate for preparation of oriented bilayers, the solvent must be removed in a stream of argon or very pure nitrogen gas generated from liquid nitrogen. For the removal of solvents use a rotating glass tube to spread the drying lipid as a thin film over the glass surface. Very thin lipid films may dry well enough without application of vacuum. Alternatively, samples must be further dried by application of high vacuum for 1 h. Use a cold trap filled with dry ice or liquid nitrogen to trap solvent and pump oil vapors.
2. A total of 0.1–2 mg of lipid per cm^2 of slide surface may be applied. Lower amounts of lipid usually result in less mosaic spread of bilayer orientations. The solvent should be applied in

20–50 µL/cm² increments, for example, with a glass microliter syringe that has a bent needle such that the solvent can be spread over the surface evenly. Dry the sample in air and repeat solvent application until the desired amount of lipid per slide has been deposited. Lipids that are prone to oxidation must be applied to the glass slides inside a glove box filled with an oxygen-free atmosphere. This is conveniently achieved by flushing the glove box at high rate with nitrogen gas from a liquid nitrogen tank for gas use.
3. Depending on lipid layer thickness and solvent properties, it might be necessary to place the sample into high vacuum for 1–2 h to remove remaining traces of solvent.
4. Hydrate the bilayers in an inert gas atmosphere with a relative humidity adjusted to 60–95%, for example, by using concentrated salt solutions. For deuterium NMR experiments, salt solutions must be prepared using deuterium-depleted water to avoid a strong ^{2}HHO resonance.
5. Various procedures have been recommended to reduce mosaic spread of bilayer orientations, for example, drying and hydrating samples repeatedly or annealing of samples at elevated temperature.
6. Stack an appropriate number of glass slides such that sensitivity requirements of the NMR equipment are met. Membranes with very low mosaic spread, very sensitive NMR equipment, and long acquisition times may afford as little as 0.1 mg of lipid per sample to achieve a reasonable signal-to-noise ratio in the spectra. Lipid quantities in the low-milligram range yield good signal-to-noise at acquisition times from minutes to 1 h even at somewhat increased mosaic spread. Whereas a single slide may be sufficient for experiments on submilligram quantities of lipid, larger quantities require stacking of slides. Difficulties with sample preparation increase with the number of stacked slides.
7. Insert the stack of slides into an appropriate container that fits into the NMR coil, for example, a flat glass cell with a silicon rubber plug that permits adjustment of humidity. Humidity may be controlled by a concentrated salt solution in a separate region of the container or by strips of filter paper that were soaked in water. If an even pressure is applied to the stack of slides, for example, by wrapping them with Teflon band, samples may be immersed in water. As a general rule, mosaic spread increases with increasing bilayer hydration. Other types of containment have been used as well, for example, samples were wrapped into several layers of parafilm or samples were sealed in small polyethylene bags. However, the seal of any container may not prevent slow evaporation of water, in particular, in experiments conducted over an extended period of time.

3.3. Preparation of Oriented Bilayers From Lipid Water Dispersions

There are frequent occasions when preparation of oriented samples from organic solvents is not practical or undesirable. Under these circumstances samples may be prepared from a dispersion of lipids in water.

1. Prepare a homogenous, monodisperse batch of liposomes, for example, by ultrasonication. The following instructions were written for use of a Branson 250 sonifier (Branson Ultrasonics, Eemnes, The Netherlands) coupled to a microsonifier tip: sonicate the vortexed dispersion of membrane material at a duty cycle of 50% and power levels 4–5 for 15 min to produce a nearly transparent solution of small unilamellar vesicles. Ultrasonication must be conducted at a temperature slightly above the gel–fluid phase transition of the lipids. Because the dispersion heats severely during sonication, this usually requires cooling of the dispersion, for example, in an ice bath. Sonication is not advised for membranes prone to oxidation (e.g., unsaturated lipids), and membranes containing certain proteins and peptides. In such cases, homogeneous preparations may be produced by repeated extrusion of the dispersion through polycarbonate filters with pore sizes ranging 0.4–0.1 µm. Homogenous preparations of proteoliposomes can be also achieved by diluting micellar solutions of lipids and protein with buffer below the critical micellar concentration of the detergent and consecutive removal of the detergent by dialysis.
2. Remove large membrane particles and metallic shavings from the sonifier tip by gentle centrifugation, for example, 5 min at 1000g.

3. Spread an appropriate volume of the lipid–water dispersion on the slides to apply 0.1–2 mg of lipid per square centimeter as described for the use of organic solvents. One mg of a typical phosphatidylcholine per square centimeter yields a stack of approx 3000 bilayers.
4. Dry the slides in a gentle stream of dry nitrogen gas generated from a liquid nitrogen tank for gas use.
5. Hydrate and pack samples as described in **Subheading 3.2.**

Oriented samples have been successfully prepared from very concentrated multilamellar lipid water dispersions containing 50% water by weight. However, mosaic spread of such preparations is likely to be higher. The homogenous lipid paste is evenly applied to a clean slide, covered with a second slide, and the two slides are gently moved against each other while applying moderate pressure. Excess lipid is removed from the edges. Samples may contain a significant fraction of multilamellar liposomes with bilayers that are oriented at random. Mosaic spread is reduced by cycles of hydration/dehydration and by annealing of samples.

3.4. Preparation of Single, Tubular Lipid Bilayers From Liposomes Using Porous Substrates

1. Place two to five 13-mm diameter, porous AAO filters into a mini extruder (Avanti Polar Lipids, Alabaster, AL) equipped with two 1-mL syringes or into a stainless steel thermobarrel extruder (Lipex Biomembranes, Inc., Vancouver, BC, Canada) connected to a pressurized N_2 tank *(1)*.
2. Rinse the filters repeatedly with 1 mL of water at a rate of 10–20 mL/min. Flush syringes with buffer.
3. Pass 1 mL of lipid/water dispersion (0.5–5 mg of lipid/mL) through the filters 5–15 times. Flush syringes with buffer.
4. Flush the filters with 1–5 mL of buffer at a rate of 3–12 mL/min. For reproducibility and convenience this can be done with an infusion pump. The flushing removes any remaining liposomes as well as multiple bilayers except for the last one at the cylindrical surface of AAO pores.
5. The outer filter may have some buildup of membrane material at the surface and is better not used for NMR experiments. Every filter contains from 0.1 to 0.5 mg of lipid in tubular arrangement. The long axis of tubules is oriented perpendicular to the filter surface. An appropriate number of filters may be stacked and packed into a suitable container filled with buffer. The mosaic spread of tubule orientations can be reduced by repeated freeze-thaw cycles.

3.5. ^{31}P NMR

3.5.1. NMR Acquisition and Processing Parameters

1. Set up the solid-state NMR instrument in double resonance configuration with detection on the ^{31}P NMR channel and decoupling on the ^1H channel.
2. Orient sample with its bilayer normal parallel to the magnetic field, adjust temperature, tune, and match both ^{31}P- and ^1H NMR channels. Samples with tubular bilayers in porous aluminum oxide should be oriented with their tubule axis parallel to the magnetic field (filter normal parallel to field).
3. Acquire the free induction decay (FID) with a Hahn echo pulse sequence $[(\pi/2) - \tau_1 - (\pi) - \tau_2 -$ acquire] *(2)*. The purpose of using an echo sequence is to delay acquisition of the ^{31}P NMR resonance signal past the decay of electronic ringing of the resonance coil and preamplifier circuits after the strong radio-frequency pulses. This ensures registration of resonance signals without baseline distortions. Typical acquisition parameters are:
   - Spectral width 500 ppm or higher.
   - Spectral filter width 250 kHz or higher.

- Acquisition with 2048 data points per FID or higher.
- Set the carrier frequency slightly outside the spectral region of resonances.
- Set τ_1 to 20–100 µs to suppress the ringing of resonance circuits (shorter delay times are desirable to avoid attenuation of signal components with short spin–spin relaxation times).
- Chose amplifier power level settings such that the $\pi/2$ pulses length is 5 µs or less to ensure homogeneous excitation over the entire bandwidth.
- The delay time τ_2 should be kept sufficiently short to start acquisition of data points well before the echo maximum, which occurs approximately at τ_1 after the π-pulse.
- Set the recycle delay time to at least three times the spin-lattice relaxation time (typically 1–5 s).
- Adjust the proton decoupling frequency to the resonances of lipid glycerol and polar head methylene groups (~4.2 ppm).
- Adjust 5–25 kHz of proton continuous wave decoupling beginning with the $\pi/2$ pulse. A higher decoupling field strength improves spectral resolution because of better suppression of ^1H–^{31}P dipolar interactions but it raises sample temperature.

4. For processing, adjust data points in the FID by left-shifting, such that the signal starts precisely at the echo maximum. This is easier to achieve when the data set has been acquired with a very short dwell time, for example, 1 µs. Alternatively, the delay time τ_2 may be used to fine adjust the delay such that a data point coincides with the echo maximum; special software may be used to interpolate the FID between data points to shift all data points of the FID by a fraction of the dwell time. Such software is available from this laboratory on request.
5. Multiply the FID with an exponential window function to reduce noise, for example, by multiplication with an exponential function equivalent to a line broadening of 10–100 Hz, and convert the spectrum into the frequency domain by a Fourier transformation. To improve spectral resolution, the number of data points of the acquired signal may be increased up to fourfold by "zero-filling."
6. Phase-correct the spectrum using zero-order phase correction only. This requires that the FID was properly left-shifted to begin exactly at the echo maximum and that any intensity from ringing of resonance circuits has fully decayed by the time of the echo maximum. Success is determined by the flatness of the spectral baseline.
7. The chemical shift scale of ^{31}P NMR spectra is typically reported relative to the signal of 85% phosphoric acid. If an absolute calibration of spectra is required, a signal of phosphoric acid in a capillary should be acquired using the same experimental setup. However, even with such precautions precision is usually ±1 ppm only.

3.5.2. Interpretation of Spectra

1. *Quality of orientation—mosaic spread:* the ^{31}P isotope is a spin-1/2 nucleus with a natural abundance of 100%. Its gyromagnetic ratio is approx 40% of the value of protons yielding a good sensitivity. The shape of ^{31}P spectra is dominated by a large anisotropy (orientation dependence) of chemical shift *(3)*. Therefore, ^{31}P NMR is a very convenient tool for the investigation of lipid alignment. In fluid (liquid crystalline) bilayers, lipids perform rapid diffusional motions about the normal to the membrane surface. This motion generates a tensor of chemical shift with effective axial symmetry *(4)*. Lipid bilayers orient with their bilayer normal parallel to the field resonate at $\sigma_{//}$, whereas bilayers orient with their normal perpendicular to the magnetic field resonate at $\sigma_\perp$. The difference $(\sigma_{//} - \sigma_\perp)$, called anisotropy of chemical shift, is typically 45 ± 5 ppm for fluid lipid bilayers, and depends somewhat on the lipid species. Lipid dispersions with a totally random orientation of lipid bilayers yield so-called powder spectra with low intensity at $\sigma_{//}$ and high intensity at $\sigma_\perp$, as shown in **Fig. 1A**. The spectrum of oriented bilayers consists of a well-resolved resonance with an orientation-dependent chemical shift that depends on bilayer orientation according to:

$$\sigma(\theta) = \sigma_{iso} - \frac{2}{3}(\sigma_\perp - \sigma_{//}) \cdot \left(\frac{3\cos^2\theta - 1}{2} \right)$$

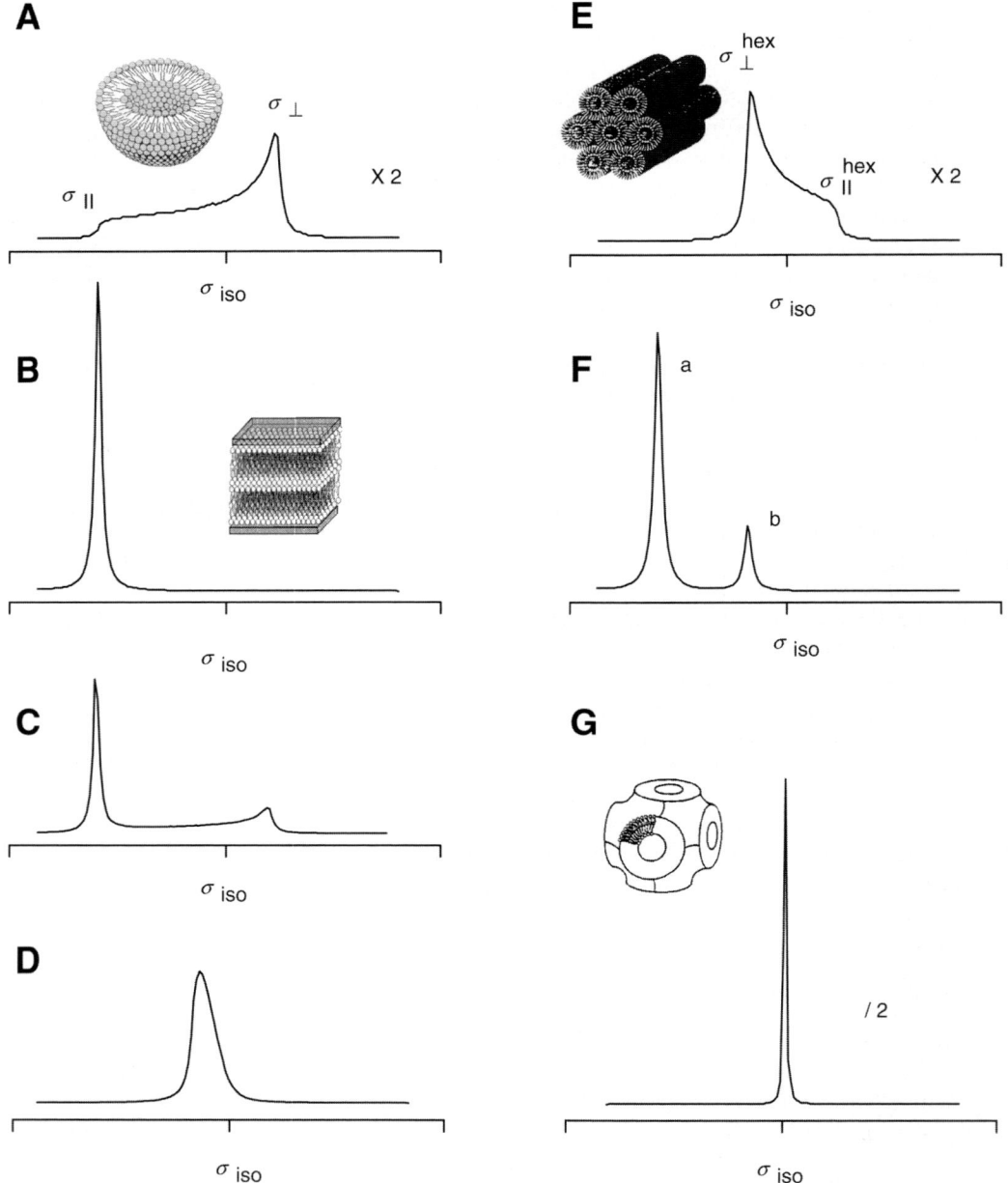

Fig. 1. (**A**) ^{31}P NMR powder pattern for a spherical distribution of bilayer orientations as observed for nonoriented multilamellar liposomes. (**B**) Perfectly oriented sample with the bilayer normal parallel to the magnetic field: the linewidth is determined entirely by spin–spin relaxation, T_2 (line broadening = 0.7 ppm). (**C**) Same as in **Fig. 1B** but with a superimposed signal from randomly oriented bilayers. The integral intensity (amount of lipid) in oriented and nonoriented bilayers is equal. The weak resonance at high field is a good indicator for the presence of nonoriented material. (**D**) Resonance signal of a sample oriented with the bilayer normal at 45° to the magnetic field. The sample orientation has a mosaic spread described by a Gaussian distribution with a width of σ = 2.5°. (**E**) ^{31}P NMR powder pattern representative of lipids in a nonbilayer, inverse-hexagonal H_{II} phase. (**F**) Same as in **Fig. 1B** with 80% of lipids in a lamellar phase (signal a) and 20% in a H_{II} phase (signal b). (**G**) ^{31}P NMR spectrum of lipids in a cubic phase. In this phase, the anisotropy of chemical shift is

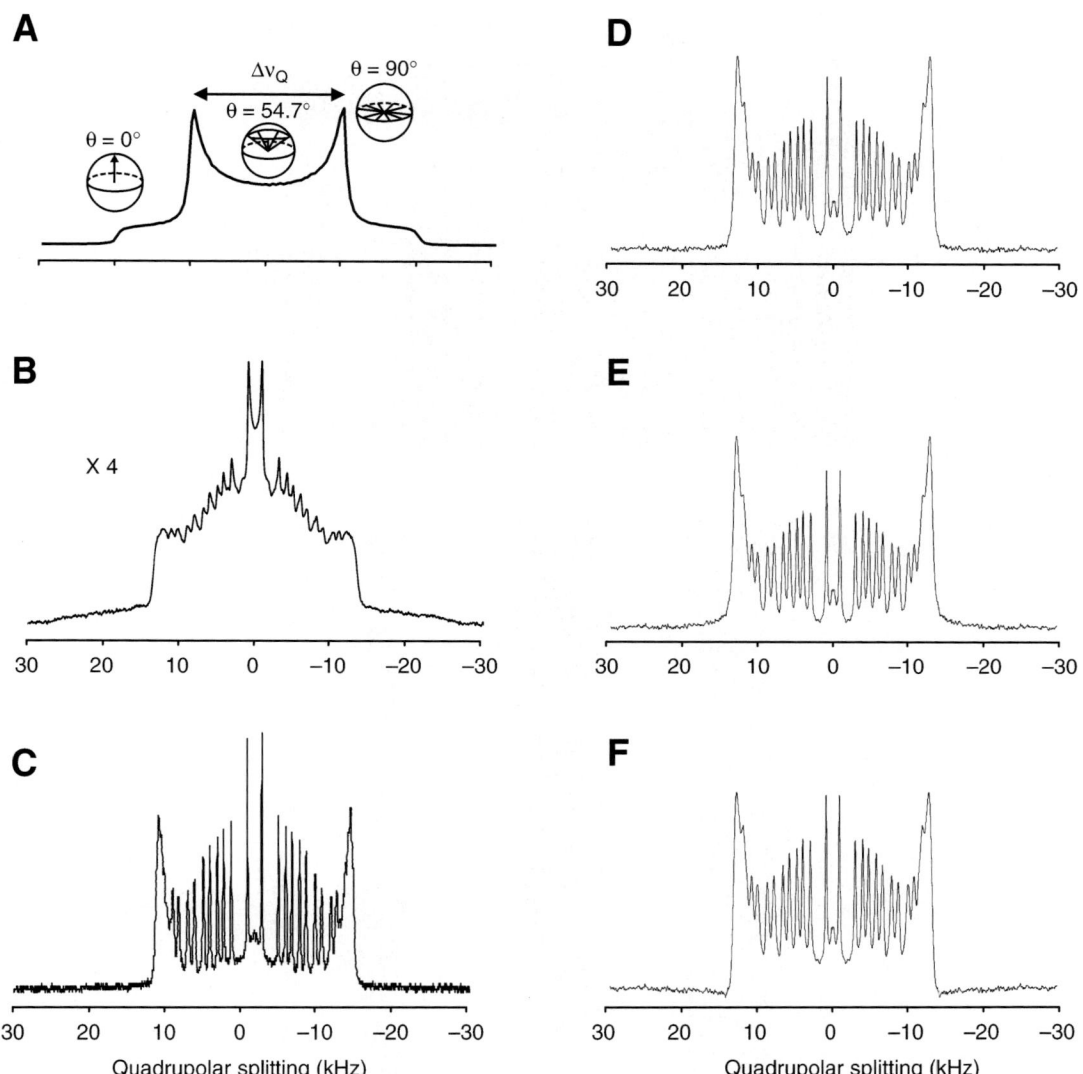

Fig. 2. (**A**) ^{2}H NMR spectrum of a single deuteron in bilayers oriented at random to the magnetic field. The distance in Hertz between the two maxima is called quadrupolar splitting, Δv_Q. (**B**) ^{2}H NMR spectrum of nonoriented, multilamellar liposomes is made up of 18:0(d35)-22:6n3PC with a perdeuterated *sn*-1 chain at 35°C in 50 wt% deuterium depleted water. (**C**) ^{2}H NMR spectrum of oriented bilayers of 18:0(d35)-22:6n3PC with the bilayer normal perpendicular to the magnetic field. Temperature and hydration conditions are the same as in **Fig. 2B**. (**D**) ^{2}H NMR spectrum of oriented, single bilayers of 18:0(d35)-22:6n3PC at 35°C in AAO filters. The normal-to-filter surface is oriented parallel to the magnetic field. (**E,F**) Simulation of a "bad" adjustment of the echo maximum before Fourier transformation. In (**E**) and (**F**) the echo maximum was shifted by –1 and +1 µs, respectively. Even such a small deviation has a significant influence on the appearance of spectra. Please note the upward and downward tilts of the baseline.

and θ is the angle between the normal to the lipid bilayer surface and the magnetic field B_0. In fluid lipid bilayers, the maximal possible splitting of 250 kHz is typically reduced to values of less than 70 kHz by fast lipid motions. This reduction is expressed in terms of an order parameter, S_{CD} (*12*). **Figure 2B** shows a typical ^{2}H spectrum of lipids with perdeuterated

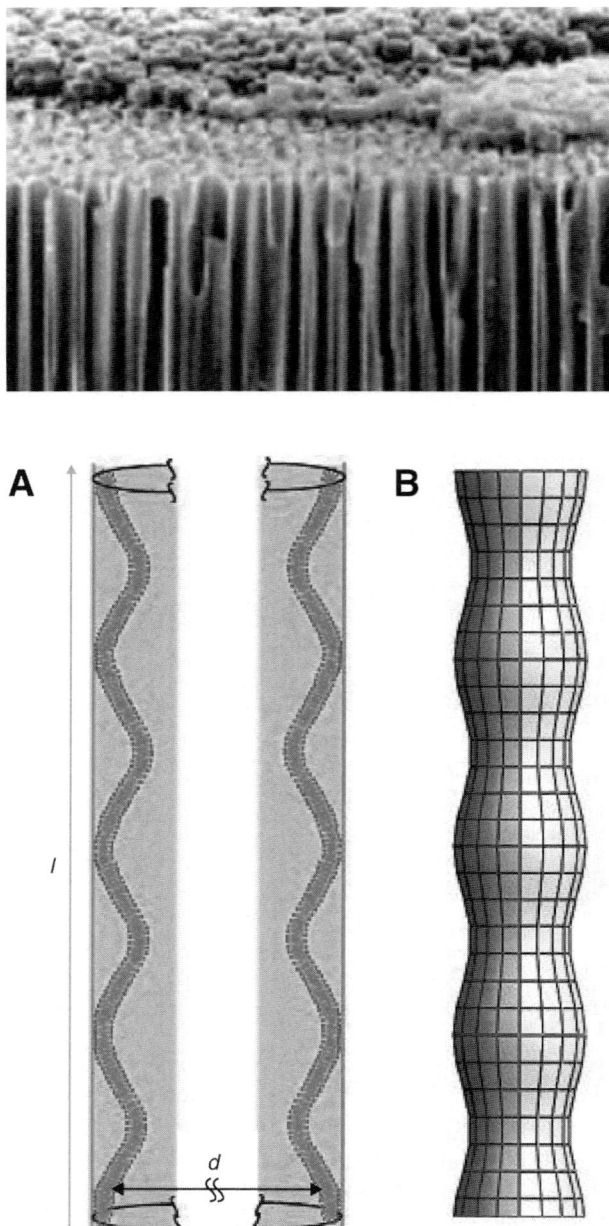

Fig. 3. Top-scanning electron micrograph of highly porous AAO filters. The average pore diameter is 200 nm. **(Bottom A)** A model for lipid adsorption consistent with the NMR data shown in **Fig. 1D**. The lipids adsorb as wavy tubular bilayers. According to the ^{2}H NMR and ^{1}H-MAS NMR experiments, these tubules possess undulations with a radius of curvature of 100–400 nm. Pockets of water with an average thickness of 3 nm are trapped between the tubules and the AAO surface. **(Bottom B)** Illustration of the shape of the lipid bilayer tubules inside the AAO pores. Such lipid cylinders with variable diameter have lowest free energy when they belong to the family of surfaces with constant total curvature called Delaunay surfaces (*see* **ref. 1** for details).

hydrocarbon chains in fluid bilayers that are oriented at random to the magnetic field. The spectrum is a superposition of orientation-dependent quadrupolar splittings, one for each carbon of the acyl chains. The ^{2}H NMR resonances of oriented bilayers with low mosaic spread are a superposition of well-resolved doublets, one doublet for each deuterium labeled carbon **(Fig. 2C,D)**.

Phospholipids are conveniently deuterated at specific sites in the polar head group or at one or both hydrocarbon chains. The magnitude of headgroup-order parameters is a sensitive measure of headgroup orientation *(13)*, whereas chain-order parameters are very sensitive measures of effective chain length and lateral area per lipid molecule *(14)*. ^{2}H NMR is used to probe not only the level of mosaic spread of bilayer orientations, but also the changes in bilayer structure and dynamics from addition of cholesterol, amphiphilic drugs, peripheral and integral membrane peptides, and proteins. Phase transitions of the lipid matrix are easily detected by ^{2}H NMR: order parameters (S_{CD}) increase drastically on the transition into gel- and crystalline phases and resolution decreases. In contrast, a transition to nonlamellar fluid phases reduces order parameters (*see* **Note 4**) *(10)*.

4. Notes

1. Solvent properties have a strong influence on the spreading and morphology of lipid films at solid interfaces. The differences arise from changes in interface energies between the lipid solution/suspension and air as well as the lipid solution/suspension and the solid support. Solvents are selected according to their ability to solubilize lipids as well as their vapor pressure, which is responsible for their rate of evaporation.
2. Integral intensity of the randomly oriented liposomes in this example is equal to the integral intensity of oriented bilayers, but signal height of well-oriented bilayers exceeds severalfold the signal height of nonoriented bilayers. Therefore, the presence of substantial amounts of nonoriented bilayers in the spectra is frequently overlooked. Detection of the nonoriented fraction of bilayers requires a flat baseline and a good signal-to-noise ratio of the spectra.
3. Other arrangements of lipids may be responsible for appearance of an isotropic ^{31}P NMR resonance as well (e.g., spontaneous formation of liposomes with diameters of 100 nm or less, formation of micelles, or lipid hydrolysis).
4. ^{2}H-labeling of lipids does not change lipid conformation and motional properties. However, there is a small effect from deuteration on the strength of hydrogen bonds and on phase-transition temperatures. For example, chain perdeuterated DPPC has a gel–fluid phase transition at T_m = 37°C, whereas protonated DPPC has a T_m = 41.5°C. As rule of thumb, the main phase-transition temperature of lipids is lowered by 2°C per deuterated hydrocarbon chain.

Acknowledgments

This work was supported by the Intramural Research Program of the National Instituted on Alcohol Abuse and Alcoholism, NIH.

References

1. Gaede, H. C., Luckett, K. M., Polozov, I. V., and Gawrisch, K. (2004) Multinuclear NMR studies of single lipid bilayers supported in cylindrical aluminum oxide nanopores. *Langmuir* **20,** 7711–7719.
2. Rance, M. and Byrd, R. A. (1983) Obtaining high fidelity spin-1/2 powder spectra in anisotropic media: phase-cycled Hahn echo spectroscopy. *J. Magn. Reson.* **52,** 221–240.

3. Griffin, R. G. (1976) Observation of the effect of water on the 31P nuclear magnetic resonance spectra of dipalmitoyllecithin. *J. Am .Chem. Soc.* **98,** 851–853.
4. Kohler, S. J. and Klein, M. P. (1976) ^{31}P nuclear magnetic resonance chemical shielding tensors of phosphorylethanolamine, lecithin, and related compounds: Applications to head-group motion in model membranes. *Biochemistry* **15,** 967–974.
5. Seelig, J. (1978) 31 P nuclear magnetic resonance and the head group structure of phospholipids in membranes. *Biochim. Biophys. Acta* **515,** 105–140.
6. Arnold, K., Gawrisch, K., and Volke, F. (1979) 31 P NMR investigations of phospholipids I. Dipolar interactions and the 31 P NMR lineshape of oriented phospholipid/water dispersions. *Stud. Biophys.* **75,** 189–197.
7. Soubias, O., Saurel, O., Reat, V., and Milon, A. (2002) High resolution 13C NMR spectra on oriented lipid bilayers: from quantifying the various sources of line broadening to performing 2D experiments with 0.2-0.3 ppm resolution in the carbon dimension. *J. Biomol. NMR* **24,** 15–30.
8. Cullis, P. R. and de Kruijff, B. (1978) The polymorphic phase behaviour of phosphatidylethanolamines of natural and synthetic origin. A 31 P NMR study. *Biochim. Biophys. Acta* **513,** 31–42.
9. Davis, J. H., Jeffrey, K. R., Bloom, M., Valic, M. I., and Higgs, T. P. (1976) Quadrupolar echo deuteron magnetic resonance spectroscopy in ordered hydrocarbon chains. *Chem. Phys. Lett.* **42,** 390–394.
10. Davis, J. H. (1983) The description of membrane lipid conformation, order and dynamics by ^{2}H-NMR. *Biochim. Biophys. Acta* **737,** 117–171.
11. Seelig, J. (1977) Deuterium magnetic resonance: theory and application to lipid membranes. *Q. Rev. Biophys.* **10,** 353–418.
12. Seelig, J. and Seelig, A. (1980) Lipid conformation in model membranes and biological membranes. *Q. Rev. Biophys.* **13,** 19–61.
13. Macdonald, P. M. and Seelig, J. (1988) Anion binding to neutral and positively charged lipid membranes. *Biochemistry* **27,** 6769–6775.
14. Schindler, H. and Seelig, J. (1975) Deuterium order parameters in relation to thermodynamic properties of a phospholiped bilayer. A statistical mechanical interpretation. *Biochemistry* **14,** 2283–2287.

7

Molecular Dynamics Simulations as a Complement to Nuclear Magnetic Resonance and X-Ray Diffraction Measurements

Scott E. Feller

Summary

Advances in the field of atomic-level membrane simulations are being driven by continued growth in computing power, improvements in the available potential energy functions for lipids, and new algorithms that implement advanced sampling techniques. These developments are allowing simulations to assess time- and length scales wherein meaningful comparisons with experimental measurements on macroscopic systems can be made. Such comparisons provide stringent tests of the simulation methodologies and force fields, and thus, advance the simulation field by pointing out shortcomings of the models. Extensive testing against available experimental data suggests that for many properties modern simulations have achieved a level of accuracy that provides substantial predictive power and can aid in the interpretation of experimental data. This combination of closely coupled laboratory experiments and molecular dynamics simulations holds great promise for the understanding of membrane systems. In the following, the molecular dynamics method is described with particular attention to those aspects critical for simulating membrane systems and to the calculation of experimental observables from the simulation trajectory.

Key Words: Correlation function; force field; lipid bilayer; molecular dynamics; simulation; modeling.

1. Introduction

The molecular dynamics (MD) method, as most typically applied to lipid bilayers, is based on the solution to the equations of motion for a set of particles representing each of the atoms in the system. The most straightforward implementation is to solve the Newtonian equations of motion,

$$\ddot{x} = \frac{F_x}{m} \qquad F_x = \left(\frac{\partial U}{\partial x}\right) \tag{1}$$

where m and F_x are the mass and x component of force, respectively. Equivalent expressions for y and z are also included for each of the N atoms. Starting from an initial set of coordinates and velocities, the forces on each atom are calculated from the derivatives of the empirical potential energy function (U) described in the **Subheading 2**.

Equation 1 cannot be solved analytically but its numerical solution is straightforward, at least when compared with typical coupled nonlinear differential equations encountered in the physical sciences. A particularly simple algorithm, obtained from a Taylor's series expansion around $x(t)$, is the Verlet algorithm *(1)*

$$x(t + \Delta t) = 2x(t) - x(t - \Delta t) + \frac{F_x(t)}{m}\Delta t^2 \tag{2}$$

which gives the position of each atom based on the current position and force along with the previous position. Notice that the velocities $v_x(t) = \dot{x}(t)$ are not needed to calculate the trajectory. However, they are used for the calculation of the kinetic energy and the pressure. There are several variations on the basic Verlet algorithm; examples are "leapfrog Verlet" and "velocity Verlet" *(2)*. All generate identical trajectories in the microcanonical ensemble, but differ in the definition of the velocities. In the simulation literature, the statistical mechanical ensemble that is generated in the simulation is typically referred to by a shorthand notation indicating those variables that are held constant. For example, the microcanonical ensemble is often termed "*NVE*" to designate that the particle number (*N*), volume (*V*), and internal energy (*E*) are constant. The definition of velocity becomes important when implementing algorithms for MD in the canonical (*NVT*) or isothermal–isobaric (*NPT*) ensembles. These methods, which have gained widespread use in the simulation of bilayer systems, modify the equations of motion given by **Eq. 1** to maintain the temperature and/or pressure at a fixed value. Most groups have adopted the "extended system" formalism, originally developed by Andersen *(3)* for constant pressure MD. An additional degree of freedom (df) is added to the simulation, corresponding to the volume of the cell, with the force acting on this added df being determined by the difference between instantaneous and applied (target) pressures. For a cubic simulation cell, the equations of motion become

$$\dot{x} = v + \tfrac{1}{3}\frac{\dot{V}}{V}x \qquad \dot{v}_x = \frac{F_x}{m} - \tfrac{1}{3}\frac{\dot{V}}{V}v_x \qquad \dot{V} = \frac{1}{W}[P(t) - P_{\text{ext}}] \tag{5}$$

where W is the "mass" parameter for the extra df (this df is often referred to as the piston). The value of the mass does not influence equilibrium properties but does affect the rate at which the system responds to the pressure imbalance. As a practical matter, the mass must be sufficiently small that the system can respond on the MD time-scale; however, very small masses may lead to unphysical oscillations with frequency proportional to $W^{1/2}$. Extensions to Andersen's method have been described, allowing the use of noncubic simulation cells *(4)*, the addition of a constant temperature df *(5)*, and the control of oscillations by means of a Langevin equation *(6)*.

The application of constant pressure MD to lipid bilayers is complicated by the anisotropic nature of the system. In contrast to a simulation of pure water or a neat alkane where only the size of the cell need varies, the dimensions in the directions, lateral (*x* and *y*) and normal to the membrane (*z*) must change independently because of the large differences in compressibility. Macroscopic membrane thermodynamics suggest that bilayer tension, which is proportional to the difference between pressure in the normal and lateral directions, must be zero at equilibrium because the system is at its free energy minimum with respect to surface area *(7)*. Thus, it has been argued that an isotropic pressure tensor (e.g., $P_{xx} = P_{yy} = P_{zz} = 1$ atm) should be applied in a lipid bilayer simulation with the lateral and normal simulation cell dimensions allowed to adjust independently. In such a simulation, the molecular surface area adjusts freely and can be compared directly with experimental estimates from X-ray or nuclear magnetic resonance (NMR) measurements. Unfortunately, several complications have arisen in the application of this approach. Several groups have observed that the membrane tension/molecular surface area is influenced by the number of lipid molecules in the simulation cell *(8,9)*. An explanation for this finite-size effect in terms of artificial periodicity imposed by the boundary conditions has been presented *(8,10)*.

An alternative explanation for the failure of most force fields to give the correct molecular surface area in a constant pressure simulation is that the potentials are simply not accurate enough to give the correct lateral pressure. There are two challenges in developing a force field for constant pressure simulations of lipids. The first is a general problem inherent even in isotropic systems, namely that the pressure is computed as a small difference between two large terms,

$$P = \rho k_B T - \sum \vec{f} \cdot \vec{r} \qquad (6)$$

where the first term gives the kinetic energy contribution to the pressure, whereas the summation referred to as the virial *(2)*, describes the interactions between atoms. Using water as an illustration, the kinetic energy contribution is approx 4000 atm at room temperature. Ideally, this huge term is almost entirely balanced out by the virial contribution to give a total pressure of approx 1 atm when the simulation is carried out at the correct density. However, to determine the forces with such accuracy is a challenging task and small errors in the force contributions to the virial by just a few percents can lead to pressures that are tens or hundreds of atmospheres in error. Subtle differences in simulation methodology are sufficient to induce these large changes in the pressure; for example, one of the motivations behind reparameterization of the popular TIP4P water model was that, the widespread adoption of accurate Ewald summation methods for the calculation of the Coulombic force resulted in calculated pressures that were significantly different from those calculated using truncation methods as had been used in the original parameter development *(11)*. Although these uncertainties in the calculated pressure calculation appear huge, their effect on molecular volumes is modest, because of the relative incompressibility of liquids (e.g., a 200 atm change in pressure produces only a ~1% shift in density).

A second challenge for constant pressure simulations of lipid bilayers is the nature of the local tangential pressure. The pressure in the plane of the membrane (P_L) is not uniform, but a function of position along the membrane normal: $P_L = P_L(z)$. Thus, the lateral pressure computed from the simulation is an average over a strongly varying function of position. Recent computations of the lateral pressure profile show variations of thousands of atmospheres over the thickness of the membrane *(12–14)*. An example profile (**Fig. 1**) shows that a near-zero bilayer surface tension is obtained from the cancelation of large terms of opposing signs. The large negative pressure, arising from the water–alkane interfacial tension is balanced out by the positive chain pressure, and to a lesser extent, by headgroup repulsions. A decomposition of the lateral pressure into entropic and enthalpic components by Lindahl and Edholm *(12)* showed a similar cancelation of large terms of opposing sign.

2. Force Fields

The energy function (U) is central to the method of MD simulation. All interactions, both intra- and intermolecular, must be faithfully represented by an empirical function of the atomic coordinates. In principle, U may take on any form; however, as a practical matter it should be differentiable (to analytically determine forces) and must contain a sufficient number of parameters to accurately describe the potential energy surface without having more parameters than those that can be reasonably estimated. Popular programs for biomolecular simulation such as chemistry at HARvard molecular mechanics (CHARMM) *(15)*, assisted model building with energy refinement (AMBER), and GROningen molecular simulation (GROMOS) typically contain intramolecular terms for bond lengths, bond angles, and torsion angles, and intermolecular terms representing Coulombic and dispersion interactions. In the following,

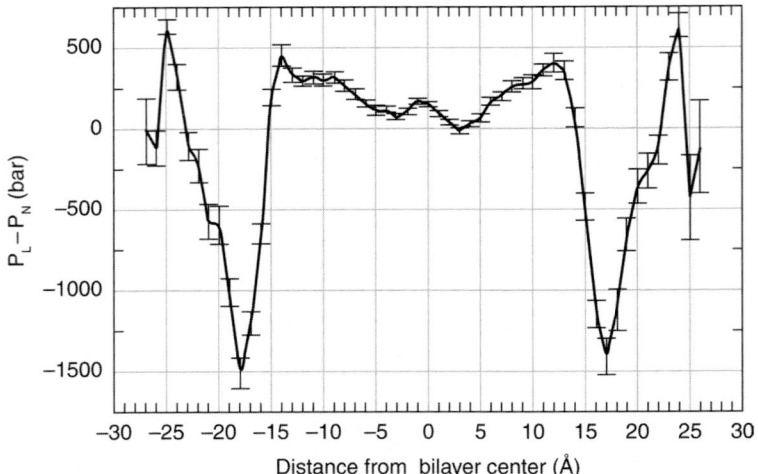

Fig. 1. Lateral pressure as a function of position along the bilayer normal.

a brief description of the energy function used within CHARMM is given; the format of the specific force field used in a simulation is usually given (or should be given) within the methods sections of an article. For a more complete description of phospholipid parameter set development, the reader is referred to **ref. 16**.

The bond energy is represented by a harmonic function of the difference between instantaneous and equilibrium bond length.

$$U(l) = \sum_{bonds} k_b (l - l_0)^2 \qquad (7)$$

Notice that in CHARMM a nonstandard definition of the force constant is used to eliminate multiplications, such subtle differences often make it difficult for a direct comparison of force field parameters. Values for l_0 and k_b are typically fit to reproduce equilibrium geometries and vibrational frequencies of simple model compounds. For example, parameters for the alkyl chains in the lipid bilayer might be determined by fitting the parameters to the experimental geometry and frequencies for butane. The bond angle energy, also typically represented by a quadratic function, depends on the difference between the instantaneous and equilibrium values of the bond angle.

$$U(\theta) = \sum_{angles} k_\theta (\theta - \theta_0)^2 \qquad (8)$$

As with the bond length, parameters for the bond angles are fit to reproduce equilibrium geometries and vibrational frequencies. Within the CHARMM potential energy function an additional function, the Urey–Bradley term,

$$U(r^{1-3}) = \sum_{angles} k_{1-3} (r^{1-3} - r_0^{1-3})^2 \qquad (9)$$

is included to obtain a satisfactory description of the vibrational modes. This contribution is a quadratic function of the distance between atoms separated by two chemical bonds; thus, it depends on the two bond lengths and the bond angle between the atoms.

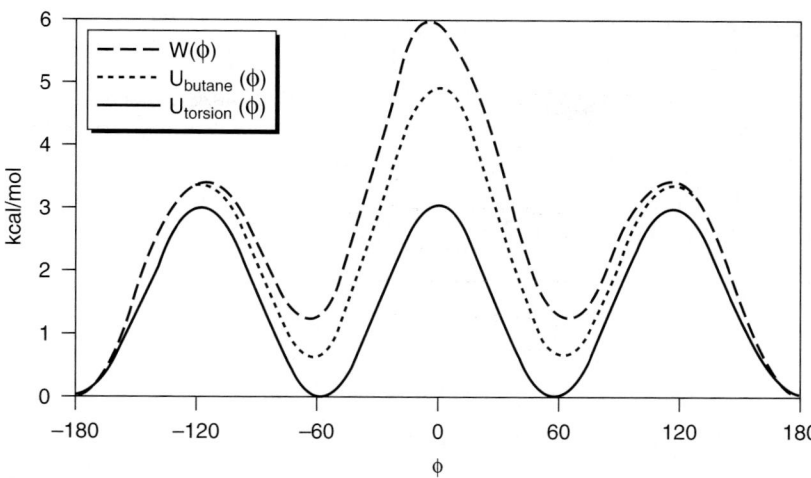

Fig. 2. Torsional energy surface for rotation around a single bond in a saturated chain.

One of the most critical components of the potential energy function for chain molecules such as lipids is the torsional potential. The torsional potential is often represented by a Fourier series in the dihedral angle (ϕ):

$$U(\varphi) = \sum_{\text{dihedrals}} k_\varphi [\cos(n\varphi - \varphi_0) + 1] \qquad (10)$$

Two terms, one with $n = 1$ and a second with $n = 3$ (both with $\phi_0 = 0$), are sufficient to produce a torsional potential energy surface, typical of alkanes, with two local minima corresponding to the *gauche* conformations and a global minimum at the *trans* position. Appropriate choices of ϕ_0 and n can similarly produce energy surfaces for rotation around double bonds. However, the barriers to rotation are typically too large to be sampled in an MD simulation making energy differences between conformers less important. In addition to the bare torsional potential (**Eq. 10**), the potential energy surface is also influenced by Coulombic and dispersion interactions between atoms separated by three or more chemical bonds. These "1–4" interactions are omitted or scaled from their full strength in many parameter sets but are not modified in the CHARMM lipid parameter set. The dihedral parameters are generally fit to reproduce conformational energy differences, for example, between *gauche* and *trans*, and rotational barrier heights. These energies are obtained from a combination of experimental data and *ab initio* quantum mechanical calculations. **Figure 2** gives the potential of mean force for rotation about the central dihedral of the palmitic acid chain of dipalmitoylphosphatidylcholine (DPPC). This parameterization includes four cosine terms fit to high-level *ab initio* results *(17)*. The bare torsional potential (**Eq. 10**) is included in **Fig. 2** for comparison along with the complete butane potential energy surface. From this figure, it is clear that the 1–4 interactions contribute significantly to the potential energy surface, and that parameterizations neglecting these interactions will, by necessity, use a very different bare torsional potential energy function. From the butane energy surface the difference between *trans* and *gauche* states is 0.63 kcal/mol, and from the DPPC distribution a gauche fraction of 23% is obtained. Both are in good agreement with experimental estimates from IR spectroscopy. The current CHARMM lipid force field utilizes an improved alkane torsional potential, fit by considering results (well depths and barrier heights) from a number of short alkanes *(18,19)*. Their demonstration of the influence of the conformational state of adjoining dihedrals on the alkane torsional surface points out that the force field development process is a continual exercise, ideally leading to a constant improvement arising from additional goal data against which the models can be tested.

The remaining terms in the potential energy function describe intermolecular forces. The dispersion or van der Waals interaction is modeled using the Lennard–Jones 6–12 potential:

$$U = \sum_i \sum_{j<i} \varepsilon_{ij} \left[\left(\frac{\sigma_{ij}}{r_{ij}} \right)^{12} - \left(\frac{\sigma_{ij}}{r_{ij}} \right)^{6} \right] \tag{11}$$

where ε is a measure of the attraction between atoms and σ is determined by the atom sizes. Lennard–Jones parameters are typically determined by carrying out condensed phase simulations and adjusting the parameters to reproduce experimental densities and heats of vaporization. Experimental data from crystal structures of lipids (or lipid fragments) as well as liquid state properties of neat alkanes, have been frequently used in developing appropriate parameters for lipid sets. The double summation in **Eq. 11** is formally over all intermolecular pairs, and over intramolecular pairs separated by more than two chemical bonds. However, in practice the summation is truncated, so it only includes atom pairs separated by a distance smaller than the cutoff radius (r_c). The short-range nature (r^{-6}) of these interactions allows cutoff radii in the range of 8–14 Å to be used with relatively small errors introduced into the forces. To further reduce these errors, it is common practice to use termination functions that smoothly bring the interaction to zero over the outermost 1–2 Å of the cutoff sphere, thus removing discontinuities in the potential. Energies and pressures can be corrected for the neglect of long-range van der Waals attraction *(2)* using a continuum approximation for the region outside (r_c). However, until recently these were rarely applied in membrane simulations because the anisotropic structure makes the use of these methods awkward. A recent article describing the implementation of a long-range correction to the pressure addresses these issues *(20)*.

The final term in the CHARMM energy function accounts for Coulombic interactions between nonbonded atoms,

$$U = \sum_i \sum_{j<i} \frac{q_i q_j}{r_{ij}} \tag{12}$$

where q is the partial charge assigned to each atom. This term is crucial for a successful description of lipid bilayer structure because of the strong headgroup–headgroup, headgroup–solvent, and solvent–solvent interactions, all of which contain a large electrostatic component. Assignment of atomic partial charges is often based on *ab initio* calculations. However, some investigators have found it necessary to scale the charges up or down in order to obtain stable bilayer simulations.

In addition to experimental dipole moments and aqueous heats of solvation, the CHARMM force field charges are also determined by *ab initio* calculations, but in a very different manner. Complexes formed from small model compounds (e.g., tetramethylammonium) and water were studied through quantum mechanics to determine the intermolecular potential energy surface. The charges were subsequently adjusted to fit the empirical energy surface to the quantum mechanical result. This approach emphasizes the importance of lipid–solvent interactions in the parameterization. Simulations of the alkane–water interface tested the hydrophobic interaction between water and the hydrocarbon core of the membrane. These simulations allow direct evaluation of the surface tension and provide an especially sensitive test of intermolecular potentials. Simulations of water/octane and water/1-octene using the CHARMM force field produce surface tensions in good agreement with experiment and are able to reproduce the surface tension lowering because of unsaturation *(21)*.

The evaluation of Coulombic interactions is an important methodological issue in membrane simulation. In the periodically replicated simulation cell generally used for bilayer simulations, the double summation in **Eq. 12** formally includes all atoms and their periodic images (an infinite sum). In contrast to the Lennard–Jones potential with its rapid r^{-6} convergence, truncation of the r^{-1} electrostatic potential can lead to severe artifacts in an MD simulation. The most common effect is an increased structure induced by force truncation: for example, the decay length for orientational ordering of water at the DPPC membrane interface increased 50% when spherical truncation at $r_c = 12$ Å was used *(22)*. More recently, a severe dependence of the surface area on the method of electrostatic force calculation was shown by Patra et al. *(23)*.

Many techniques have been developed to minimize the errors using termination functions, whereas some researchers have used algorithms that modify the interaction pair list. For example, the intermolecular interactions can be evaluated based on distance between molecules, rather than atoms, to insure that forces are calculated between electrically neutral groups. Membrane simulations have also been designed such that hydrocarbon–hydrocarbon interactions are truncated at relatively short distances (~9 Å), whereas headgroup–water interactions were calculated over a much longer range (20 Å). However, the most reliable approach appears to be the use of Ewald summation techniques. The Ewald method breaks the summation in **Eq. 11** down into two summations: a short ranged term that is summed in real space to $r_c \sim 10$ Å, and a second term that is calculated in reciprocal space (taking advantage of the periodic nature of the system). The high accuracy of the Ewald summation comes with a large computational expense that scales approximately as N^2 (making it costly for large systems such as hydrated bilayer membranes). An improved algorithm now used in almost all membrane simulations, the Particle Mesh Ewald method *(24)*, relies on fast Fourier transform methods to determine the reciprocal space summation, and is sufficiently fast for use in bilayer simulations.

3. Experimental Observables

The product of the MD computer experiment is a set of atomic coordinates as a function of time, typically stored to disk with a frequency of approximately once per picosecond. Thus, a 10-ns simulation produces an ensemble of 10,000 system configurations. Various properties can be obtained from this trajectory; for example, positions and functions of positions such as distances or orientations, properties of the entire system such as the volume or surface area, or energetic quantities such as the van der Waals attraction between two groups of atoms. If the quantity of interest is denoted by $A(t)$, to make explicit that the properties sampled in the MD simulation are functions of time, then assuming that the simulation is sampling the correct equilibrium distribution:

$$A_{obs} = \langle A(t) \rangle$$

where A_{obs} is the macroscopic observable that would be measured in the laboratory and the brackets denote an average over time. An example of a time average from the simulation that could be compared directly with experiment is the deuterium order parameter (S_{CD}). The ergodic hypothesis will be fulfilled so long as the simulation is sufficiently long to explore the phase space of the system, i.e., to sample a sufficient number of representative conformations. How long might this be? The answer will depend on the property one wishes to measure. As an example, transitions between torsional states occur as often as 100 times per nanosecond for chain dihedrals near the terminal methyl end of the lipid, but less than once per nanosecond for certain torsions in the glycerol regions.

To quantify how rapidly $A(t)$ is fluctuating it is useful to calculate the autocorrelation function defined as:

$$c(t) = \langle A(t') \cdot A(t'+t) \rangle$$

where the average is over all t'. The values of the correlation time at $t = 0$ and $t = \infty$ give the average of the squares of the time series elements and the average of the elements themselves, respectively:

$$c(0) = \langle A^2 \rangle$$

and

$$c(\infty) = \langle A \rangle^2$$

It is often more convenient to work with the normalized autocorrelation function, denoted here as $C(t)$. To obtain the normalized autocorrelation function, first the plateau value is subtracted $\langle A \rangle^2$, resulting in a function, which goes to zero at long times. After subtracting this quantity, the zero point in the function is:

$$\langle A^2 \rangle - \langle A \rangle^2 = \sigma^2[A]$$

Thus, if the correlation function is divided by the square of the standard deviation after subtracting the square of the mean, the normalized autocorrelation function is obtained:

$$C(t) = \frac{\langle A(t') \cdot A(t+t') \rangle - \langle A(t') \rangle^2}{\langle A(t')^2 \rangle - \langle A(t') \rangle^2} \qquad (13)$$

Examples of normalized correlation functions, for the state of a dihedral angle in a saturated fatty acid and in a polyunsaturated fatty acid, are given in **Fig. 3**.

The rate of decay of the correlation function can be further quantified by computing the relaxation time (τ_R) from the integral of the normalized correlation function:

$$\tau_R = \int_0^\infty C(t)\,dt.$$

If the correlation function is described by a single exponential with a decay time (τ), i.e., $C(t) = exp(-t/\tau)$, then the decay time and the relaxation time are identical. However, in

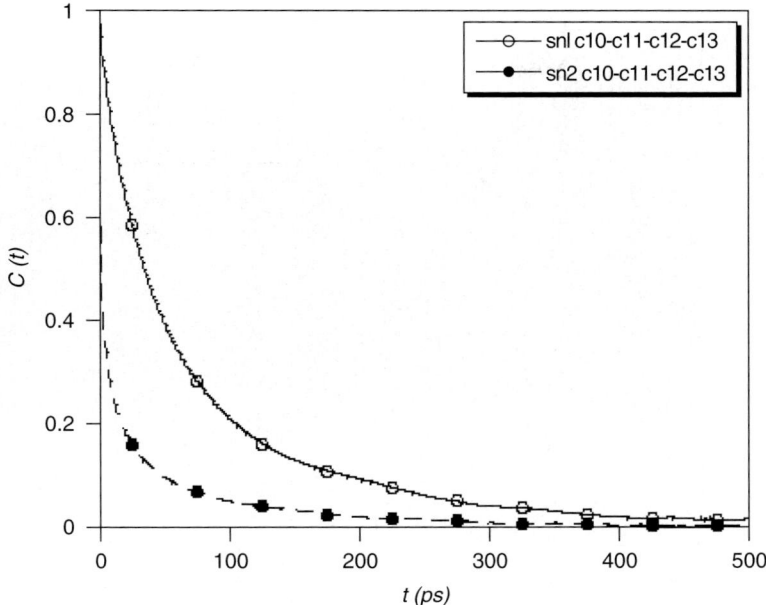

Fig. 3. Dihedral autocorrelation function calculated from a simulation of a 1-stearoyl λ-2-docosa-hexaenoyl λ-3-phosphatidylcholine bilayer.

general the decay of the correlation function will be complex, and not necessarily described by a single exponential function. The relaxation time is an important quantity both as a tool for analyzing the simulation and as a quantity that can be related to certain experimental observables. In analyzing the simulation, it provides a measure of the time required for the system to lose memory of its previous state. Thus, for quantities having short relaxation times, the computer experiment can provide many independent samples, and the statistical error in the quantity being measured can be estimated by breaking the simulation trajectory up into blocks (each block with a length more than the relaxation time) and a confidence level computed from the resulting independent samples.

To demonstrate this with a concrete example, the bilayer repeat spacing (D) determined from an MD simulation of fully hydrated dioleylphosphatidylcholine is considered. The repeat spacing, which is obtained experimentally from an X-ray diffraction experiment, is determined from the simulation as the length of the unit cell normal to the membrane surface in a constant pressure MD simulation. Then, the time series, $D(t)$, can be plotted over the entire course of the simulation (*see* **Fig. 4**). The mean value $<D(t)>$ is 62.783 Å and the standard deviation, $\sigma[D(t)]$, is 0.240 Å. From the time series, it is clear that points very nearby in time are highly correlated, whereas on a longer time-scale the data seems stochastic. To determine the relaxation time, the normalized autocorrelation function for the repeat spacing was calculated (*see* **Fig. 5**) and integrated to obtain $\tau_R \approx 25$ ps.

As is often the case, the relaxation occurs over a range of time-scales with some very fast relaxation removing most of the correlation between time-points, but with slower processes required to completely decorrelate the data. The data in **Fig. 5** suggests that if the trajectory is broken up into 40 pieces (D_i), each 1 ns in length, the mean values computed from each section can be safely

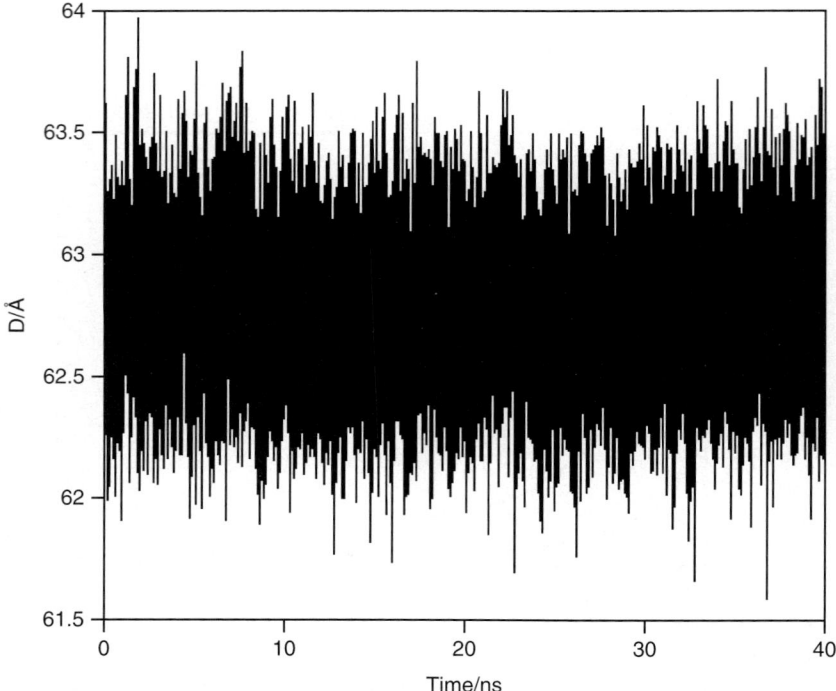

Fig. 4. Repeat spacing (*D*) as a function of time from a simulation of a dioleylphosphatidylcholine bilayer.

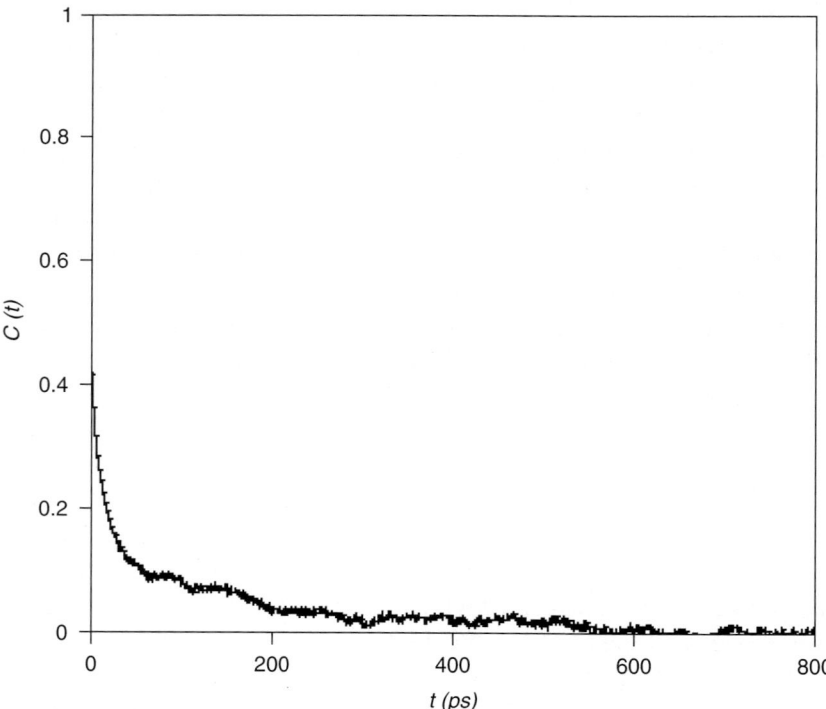

Fig. 5. Autocorrelation function of the time series presented in **Fig. 4**. The relaxation time computed from the integral of the curve is approx 25 ps.

treated as independent experiments. This analysis gives $\langle \bar{D}_i \rangle$ = 62.783 Å, $\sigma[\bar{D}_i]$ = 0.047. Thus, the standard error of the mean is estimated as $\sigma/\sqrt{N}$ where $N = 40$ in the present case. The final result for the simulated D spacing is then 62.783 ± 0.007 Å. The high precision with which this quantity can be determined may be considered atypical, as it is more commonly the case that the relaxation time is comparable with the simulation length, leading to much greater statistical uncertainty.

To conclude, consider an example wherein the correlation function itself is directly related to the experimental observable, as is the case for many quantities measured in NMR relaxation measurements. A powerful tool for the description of dynamics in the lipid bilayer are ^{13}C and ^{2}H spin-lattice relaxation NMR experiments that probe the reorientational dynamics of the C–H bond vectors in the lipid acyl chains. For a system of unoriented bilayers, the relevant correlation function is the second Legendre polynomial of the dot product between the vector at time t' and the vector a time t later:

$$C(t) = \left\langle P_2 \left[\vec{\mu}(t') \cdot \vec{\mu}(t'+t) \right] \right\rangle$$

Figure 6 shows the reorientational correlation function for the $C7$ carbon position in a dimyristoylphosphatidylcholine bilayer. The decay of the correlation function is complex, occurring over the time-scale from 1 ps to 100 ns as shown nicely by transforming the data to a log–log scale **(Fig. 6, inset)**. The NMR experiments do not probe the relaxation time directly, but rather are related to the reorientational correlation function through the spectral density, $J(\omega)$, which is the real component of the Fourier transform of $C(t)$,

$$J(\omega) = \int_0^\infty C(t) \cos(\omega t)\, dt$$

by the relation:

$$\frac{1}{T_1} = \frac{3}{16} \left(\frac{e^2 qQ}{\hbar} \right)^2 \left[J(\omega_D) + 4J(2\omega_D) \right] \tag{14}$$

Figure 7 gives the spectral density computed from the correlation function in **Fig. 6**. Current MD simulations, on the order of hundreds of nanoseconds in length, are now capable of sampling lipid dynamics on time-scales that nicely match those probed by high-field NMR spectrometers (*see* **Fig. 7** wherein the deuteron Larmor frequency is indicated). The possibility of combined NMR/MD approaches to problems in membrane biophysics is thus a promising area of investigation.

4. Conclusions

Several methodological issues have been discussed here: potential energy function and its parameterization, numerical integration method and length of time step, treatment of long-range electrostatic forces, and choice of statistical mechanical ensemble. An ever-present concern is the length of the simulation; for example, have the properties of interest been sufficiently sampled so that the results are statistically significant? Another issue is the size of the simulation cell. To date, the number of molecules included in the simulation has been

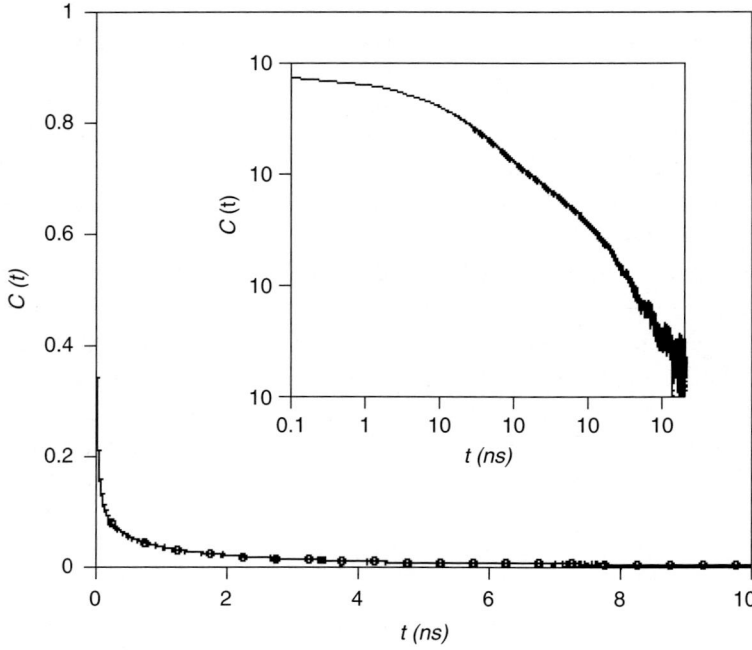

Fig. 6. Reorientational correlation function for the C–H bond vector at the seventh carbon position in dimyristoylphosphatidylcholine.

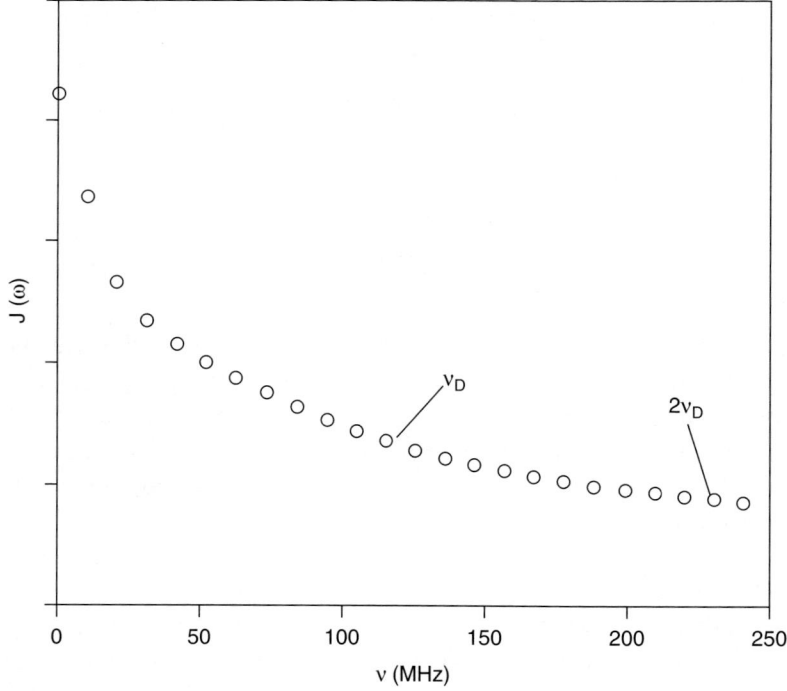

Fig. 7. Spectral density calculated from the Fourier transform of the correlation function plotted in **Fig. 6**. The value of the deuteron Larmor frequency in a high-field NMR spectrophotometer (750-MHz proton) is indicated in the plot.

determined by hardware limitations, and has been in the range of 10,000–100,000 atoms. However, further research may show that the small length scales inherent in these simulations are not sufficient for some problems.

Many of these issues are even more critical as simulations of complex model membranes are attempted. For example, simulations of pure lipid membranes benefit from the increased precision available by averaging over the properties of many indistinguishable molecules. However, a simulation of a single-ion channel in a membrane may need to be followed for a length 50–100 times greater to obtain equivalent statistics. Additionally, systems with low solute/lipid ratios may require many lipid molecules to faithfully reproduce the macroscopic conditions. The recent history of hardware and algorithmic advances, along with the rapid growth of parallel computing, suggest these problems will be addressed early in the coming decade.

Acknowledgments

The author is grateful to the many experimentalist colleagues with whom he has discussed relating simulation results to laboratory observables, in particular Michael Brown, Klaus Gawrisch, Daniel Huster, John Nagle, and Steve Wassall. Financial support is acknowledged from the National Science Foundation, the National Institutes of Health, and the Dreyfus Foundation.

References

1. Verlet, L. (1967) Computer experiments on classical fluids. I. Thermodynamical properties of Lennard-Jones molecules. *Phys. Rev.* **159**, 98–103.
2. Allen, M. P. and Tildesley, D. J. (1987) *Computer Simulation of Liquids*, Clarendon, Oxford.
3. Andersen, H. C. (1980) Molecular dynamics simulations at constant temperature and/or pressure. *J. Chem. Phys.* **72**, 2384–2393.
4. Nose, S. and Klein, M. L. (1983) Constant pressure molecular dynamics for molecular systems. *Mol. Phys.* **50**, 1055–1076.
5. Hoover, W. G. (1985) Canonical dynamics: Equilibrium phase-space distributions. *Phys. Rev. A* **31**, 1695–1697.
6. Feller, S. E., Zhang, Y., Pastor, R. W., and Brooks, B. R. (1995) Constant pressure molecular dynamics simulation: The Langevin piston method. *J. Chem. Phys.* **103**, 4613.
7. Jahnig, F. (1996) What is the surface tension of a bilayer membrane? *Biophys. J.* **71**, 1348–1349.
8. Feller, S. E. and Pastor, R. W. (1996) On simulating lipid bilayers with an applied surface tension: periodic boundary conditions and undulations. *Biophys. J.* **71**, 1350–1355.
9. Lindahl, E. and Edholm, O. (2000) Mesoscopic undulations and thickness fluctuations in lipid bilayers from molecular dynamics simulations. *Biophys. J.* **79**, 426–433.
10. Feller, S. E. and Pastor, R. W. (1997) Length scales of lipid dynamics and molecular dynamics. *Pac. Symp. Biocomput.* 142–150.
11. Horn, H. W., Swope, W. C., Pitera, J. W., et al. (2004) Development of an improved four-site water model for biomolecular simulations: TIP4P-Ew. *J. Chem. Phys.* **120**, 9665–9679.
12. Lindahl, E. and Edholm, O. (2000) Spatial and energetic-entropic decomposition of surface tension in lipid bilayers from molecular dynamics simulations. *J. Chem. Phys.* **113**, 3882–3893.
13. Gullingsrud, J. and Schulten, K. (2004) Lipid bilayer pressure profiles and mechanosensitive channel gating. *Biophys. J.* **86**, 3496–3509.
14. Carrillo-Tripp, M. and Feller, S. E. (2005) Evidence for a Mechanism by Which Omega-3 Polyunsaturated Lipids May Affect Membrane Protein Function. *Biochemistry* **44**, 10,164–10,169.
15. Brooks, B. R., Bruccoler, R. E., Olafson, B. D., States, D. J., Swaminathan, S., and Karplus, M. (1983) CHARMM: A program for macromolecular energy, minimization, and dynamics calculations. *J. Comp. Chem.* **4**, 187–217.

16. Schlenkrich, M., Brickmann, J., MacKerell, A. D., Jr., and Karplus, M. (1996) Empirical potential energy function for phospholipids: criteria for parameter optimization and applications, in *Biological Membranes: A Molecular Perspective from Computation and Experiment* (Merz, K. and Roux, B., eds.), Birkhauser, Boston, pp. 31–81.
17. Feller, S. E. and MacKerell, A. D., Jr. (2000) An improved empirical potential energy function for molecular simulations of phospholipids. *J. Phys. Chem. B* **104,** 7510–7515.
18. Klauda, J. B., Brooks, B. R., and Pastor, R. W. (2005) Adjacent gauche stabilization in linear alkanes: implications for polymer models and conformational analysis. *J. Phys. Chem. B* **109,** 15,684–15,686.
19. Klauda, J. B., Brooks, B. R., MacKerell, A. D., Jr., Venable, R. M., and Pastor, R. W. (2005) An ab Initio Study on the Torsional Surface of Alkanes and Its Effect on Molecular Simulations of Alkanes and a DPPC Bilayer. *J. Phys. Chem. B* **109,** 5300–5311.
20. Laguee, P., Pastor, R. W., and Brooks, B. R. (2004) Pressure-Based Long-Range Correction for Lennard-Jones Interactions in Molecular Dynamics Simulations: Application to Alkanes and Interfaces. *J. Phys. Chem. B* **108,** 363–368.
21. Feller, S. E., Yin, D., Pastor, R. W., and MacKerell, A. D., Jr. (1997) Molecular dynamics simulation of unsaturated lipid bilayers at low hydration: parameterization and comparison with diffraction studies. *Biophys. J.* **73,** 2269–2279.
22. Feller, S. E., Pastor, R. W., Rojnuckarin, A., Bogusz, S., and Brooks, B. R. (1996) Effect of Electrostatic Force Truncation on Interfacial and Transport Properties of Water. *J. Phys. Chem.* **100,** 17,011–17,020.
23. Patra, M., Karttunen, M., Hyvonen, M. T., Falck, E., Lindqvist, P., and Vattulainen, I. (2003) Molecular Dynamics Simulations of Lipid Bilayers: Major Artifacts Due to Truncating Electrostatic Interactions. *Biophys. J.* **84,** 3636–3645.
24. Essmann, U., Perera, L., Berkowitz, M. L., Darden, T., Lee, H., and Pedersen, L. G. (1995) A smooth particle mesh Ewald method. *J. Chem. Phys.* **103,** 8577.

8

Use of Inverse Theory Algorithms in the Analysis of Biomembrane NMR Data

Edward Sternin

Summary

Treating the analysis of experimental spectroscopic data as an inverse problem and using regularization techniques to obtain stable pseudoinverse solutions, allows access to previously unavailable level of spectroscopic detail. The data is mapped into an appropriate physically relevant parameter space, leading to better qualitative and quantitative understanding of the underlying physics, and in turn, to better and more detailed models. A brief survey of relevant inverse methods is illustrated by several successful applications to the analysis of nuclear magnetic resonance data, yielding new insight into the structure and dynamics of biomembrane lipids.

Key Words: Alignment; de-Pakeing; hexagonal; lamellar; model membranes; order parameter; phase transitions; phospholipid; relaxation rate; Tikhonov regularization.

1. Introduction to the Mathematics of Ill-Posed Problems

Some of the consequences of widely available and ever-increasing computer power are obvious: faster and more detailed data acquisition, more elaborate display capabilities, and an ability to search through vast arrays of data. However, the ubiquity of computer power can also influence the conceptual aspects of data analysis, changing the way new problems are approached. Enhanced numerical ability to transform the observed experimental data into forms that are better suited for conceptual analysis yields a profound change in the way the data is modeled and analyzed. In turn, this better footing for the conceptual thinking leads to better models, and through making it obvious what experimental aspects need improvement, to better data.

In a traditional research paradigm, experimental data is kept scrupulously separate from the theoretical model, to avoid biasing the observer. However, converting measured data into another, more convenient form is acceptable; the Fourier transform (FT) being an excellent example. Inverse theory methods extend this idea to the possibility of the nature of this transformation itself being a "parameter of the fit," in some sense. In many situations, the regularizing influence of real physical constraints is formally added to the inverse algorithm, and ensures that a physically meaningful transformation is obtained in the end. The process of finding and testing the optimal transform algorithm can be long and difficult, and it may not always work. However, it has become clear that for a wide range of problems that one encounters in the interpretation of spectroscopic nuclear magnetic resonance (NMR) data, this works reliably and without bias.

The resulting enrichment of the kind of questions one can ask about the data has had a great impact on several significant problems of current interest in the study of biological and model membrane systems. In this section, a brief introduction to the mathematics of the inverse problems is presented. Many excellent books on inverse problems exist; the goal here is not to reproduce their material but to introduce the language of the inverse problems and to illustrate the effectiveness of inverse methods. To this end, a conceptually simple example of multiexponential analysis of simulated relaxation data (signals decaying in time) is used initially. In subsequent sections, practical biomembrane examples are considered, and a successful extraction of a parameter distribution function is used as a stepping-stone toward a better understanding of the physical properties of each system.

1.1. Indirectly Observed Data

Solving an integral equation*

$$f(y) = \int g(x) K(y,x) dx \tag{1}$$

where $f(y)$ is the experimentally measured quantity (e.g., for an NMR spectrum, y would have the physical meaning of frequency), $g(x)$ is the desired set of parameters to be determined (e.g., chemical shifts), and $K(y, x)$ is a *kernel* function that describes the way the experimental data depends on these parameters is, in general, a mathematically *ill-posed* problem. Only a few select kernel functions allow for a complete unambiguous inverse calculation,

$$f(y) \rightarrow g(x)$$

Again, FT is an example. For others, including most of the kernels relevant to NMR, there is no analytical solution to the inverse problem, and numerical approximate solutions are the only practical alternatives.

Ill-posedness in fact has a formal definition *(1)*; briefly, the three essential conditions of ill-posedness are existence, uniqueness, and stability. However, strategies have been developed to combat all three.

Existence: a solution $g(x)$ may not exist at all; the way out is to be explicit about what is the question that one is asking, and to ask not for the "true" $g(x)$, but for "a reasonable approximation $\tilde{g}(x)$." Here, "reasonable" is in the least-squares sense, a minimum of misfit norm

$$\Psi(\tilde{g}) = \left\| f(y) - \int \tilde{g}(x) K(y,x) dx \right\|^2 \rightarrow \min \tag{2}$$

This misfit norm, the "distance" between the measured $f(y)$ and the approximation calculated as an integral over $\tilde{g}(x)$, is sometimes called the least-squares error norm. A minimum of misfit ensures compatibility of the fit with the measured data.

Uniqueness: many $\tilde{g}(x)$ may satisfy **Eq. 1** equally well; the best strategy is to bring in additional physical input. Of all compatible solutions, choose the one that best satisfies this additional constraint. Depending on the problem the optimal constraint criteria may take different forms, but several important criteria are quite universal and apply on very basic physical grounds. For example, a reasonable physical function should be "smooth" (i.e., locally

---

*When the limits of integration are fixed as appropriate for the majority of problems of practical interest, this is the so-called Fredholm integral equation (FIE) of the first kind.

well-approximated by a linear function), and minimizing the norm of the second derivative $\|g''(x)\|$ in addition to the misfit Ψ, provides an universal and very useful criterion for the selection of the best of all possible solutions.

Stability: a small perturbation in $f(y)$ (e.g., experimental noise) may cause a large change in $\tilde{g}(x)$. This aspect of ill-posedness turns out to be the most difficult one to deal with. The solution requires *regularization* of the problem, which enforces local stability of the $f(y) \rightarrow g(x)$ mapping.

Regularization is the key to finding the inverse solution and it deserves a special detailed discussion, but first it should be noted that so far only the continuous spaces of both the experimental variable y and of the parameter-space variable x have been dealt with. In practice, a set of experimental measurements is typically taken as a discrete set of data points, and so it is useful at this point to formulate a discrete version of the inverse problem. In the language of matrices,

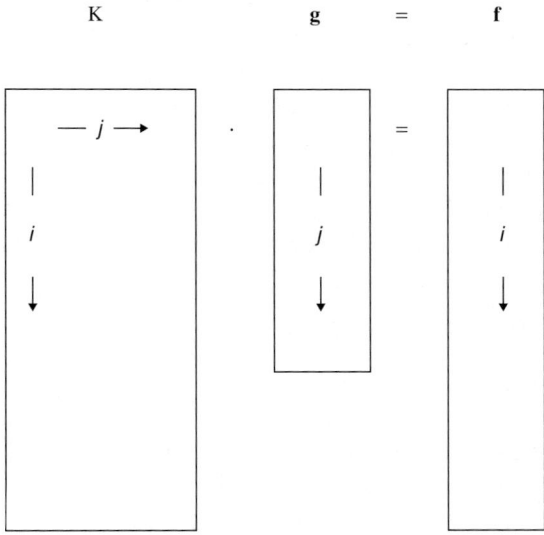

where vectors $\mathbf{f} = \{f_i\} = \{f(y_i), i = 1 \ldots m\}$ and $\mathbf{g} = \{g_j\} = \{g(x_j), j = 1 \ldots n\}$, are related through the matrix $\mathbf{K} = \{K_{ij}\} = \{K(y_i, x_j), i = 1 \ldots m, j = 1 \ldots n\}$. Note how the dimensions of the problem need not be equal (an even-determined problem); depending on the number of available data points f_i and the number of "parameters of the fit" g_j, the inverse problem may end up being over- or underdetermined. For an experimentalist, it is somewhat startling to think of a fit to *more* parameters than there are measured data points, but this simply underscores the fact that regularized solutions of inverse problems are fundamentally different from least-squares fits.

In the matrix context, the inverse problem is equivalent to calculating the inverse matrix, $\mathbf{K}^{-1}$ such that $\mathbf{g} = \mathbf{K}^{-1}\mathbf{f}$. If such an inverse does not exist (a majority of cases), the problem becomes the calculation of the *pseudoinverse*, the subject of a large and well-developed field of mathematics. Unfortunately, its significant mathematical complexity represents a formidable challenge to a nonspecialist, and many day-to-day experimental difficulties may dramatically influence the choice of numerical strategies. A small part of the field of inverse problems will be explored with a very specific goal of implementing some of these techniques as suitable for the problems of biomembrane NMR. As it turns out, selecting the right kind of numerical strategy is as important as the specifics of a linear algebra algorithm, and much depends on the experience and judgment of the person performing the calculation. In short, "successful *inverse problem solving is strongly dependent on the analyst*" *(2)*.

1.2. Regularization

The key to finding a pseudoinverse solution is regularization, and to examine various ways in which a pseudoinverse problem can be regularized, the following illustrative example will be introduced:

$$f(t) = \int_{r_{min}}^{r_{max}} g(r)\, e^{-rt}\, dr \qquad (3)$$

where the generalized x, y variables have been replaced with r, t as appropriate for a multi-exponential analysis of a decay curve; t represents time and r has the physical meaning of a relaxation rate. The data $f(t)$ is a decay curve in the linearly sampled time domain, and the desired inverse solution $g(r)$ is a distribution of relaxation rates. Multiexponential fits are notoriously difficult, and rarely extend beyond a superposition of just one or two terms; systematic misfit is often tolerated or accommodated by the use of *ad hoc* corrections (e.g., stretched exponentials). In this section an example of a broad asymmetric distribution of relaxation rates simulated in the logarithmically sampled range of $r_{min} = 10^{-3}$ to $r_{max} = 10^0$ will be used. This simulated $g(r)$ is used to calculate $f(t)$ according to **Eq. 3**; a random normally distributed noise at 1% of the maximum $f(t)$ intensity is added; and a pseudoinverse solution is sought from the resulting noisy dataset. This is repeated throughout this section as various aspects of regularization are examined.

1.2.1. Regularization Through Discretization

The first effect to consider is the discretization itself, already introduced in the previous section. The act of choosing a good set of points in the parameter space for which to ask about the value of the distribution function, by itself has a significant regularizing influence on the inverse problem. The deterministic relationship between the time- and frequency-domain point distribution is that a feature of FT is not available in a general inverse problem, and thus the best strategy for selecting a proper level of discretization is not a hard-number prescription similar to a Nyquist theorem, but a set of guidelines to follow:

- Select enough points in the grid of parameter values to allow for the entire physically relevant range to be covered with sufficient resolution to reproduce all of the essential features in the data; and no more.
- Acquire enough data points, sufficiently spread out in the observation domain to resolve contributions from different parameter values.
- Overdetermined problems, by at least a factor of two, are easier to solve.
- Higher parameter grid densities require better signal-to-noise ratio in the data.

Ultimately, numerical simulation and testing, and trial-and-error are required to establish a good discretization scheme for a given problem. Near the optimum, the results of the calculation should be largely independent of the exact choice of the discretization grids.

Figure 1 illustrates the regularizing effects of discretization alone, as applied to the example of a distribution of relaxation rates. Level of discretization is varied by changing the total number of points across the parameter space. At just the right level of regularization (middle plot) the recalculated $\tilde{g}$ is a faithful representation of the true g.

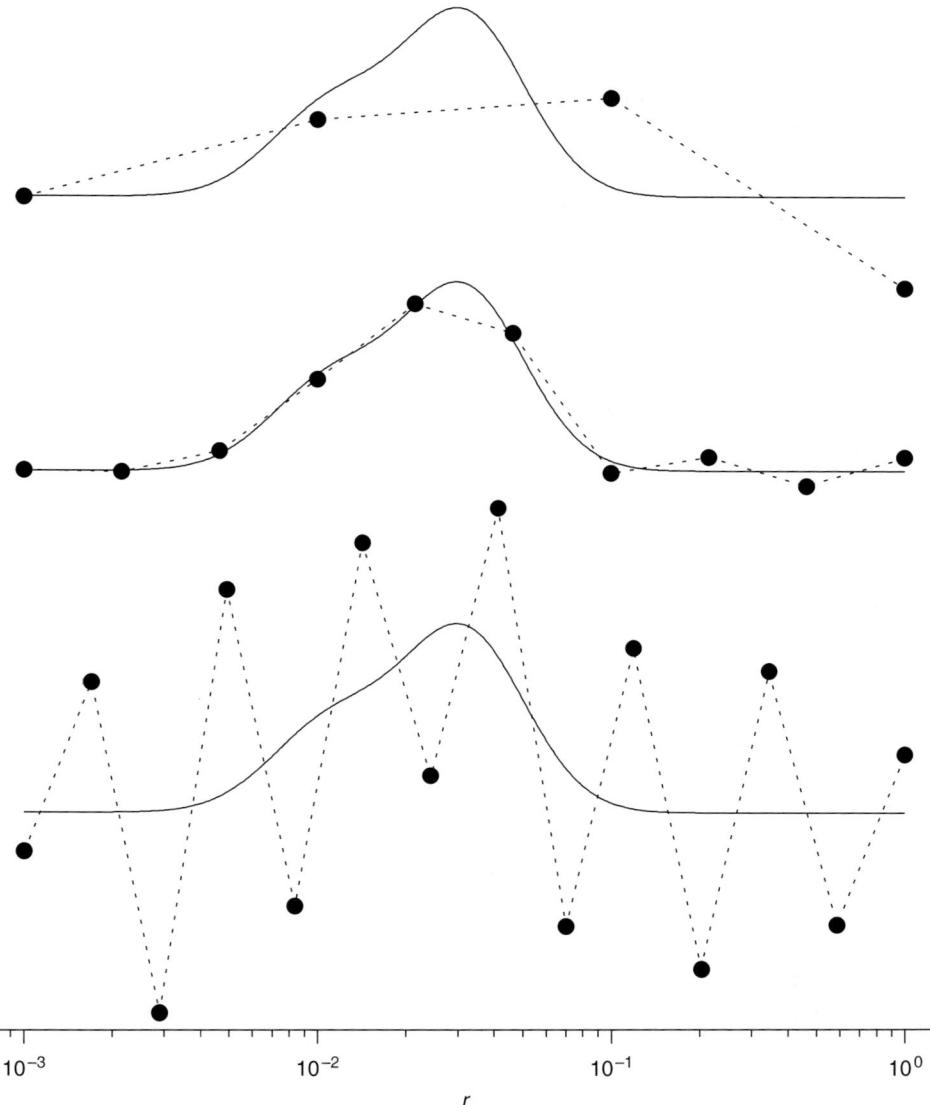

Fig. 1. Regularization through proper discretization. Overly smooth pseudoinverse solution (**top**) fails to reproduce some of the features of the true g, and insufficient regularization produces an unstable pseudoinverse (**bottom**). The optimum discretization (**middle**) faithfully reproduces the true g, shown with a solid line in all three graphs.

1.2.2. The L-Curve

It is easy to recognize the optimal level of regularization when the true $g(x)$ is known; in a real experimental situation this is not the case. Fortunately, a convenient and very revealing way to formally test any valid regularization mechanism is through the so-called L-curve, which is a plot of the norm of the regularized solution $\|\tilde{g}\|$ vs the corresponding misfit norm $\|\mathbf{f} - \mathbf{K}\tilde{g}\|$, as the regularization is varied *(3,4)*. As the name implies, this curve has a characteristic L-shape illustrated in **Fig. 2** for a number of grid densities in addition to the three shown in **Fig. 1**. The corner of the curve represents the optimum regularization level, a point

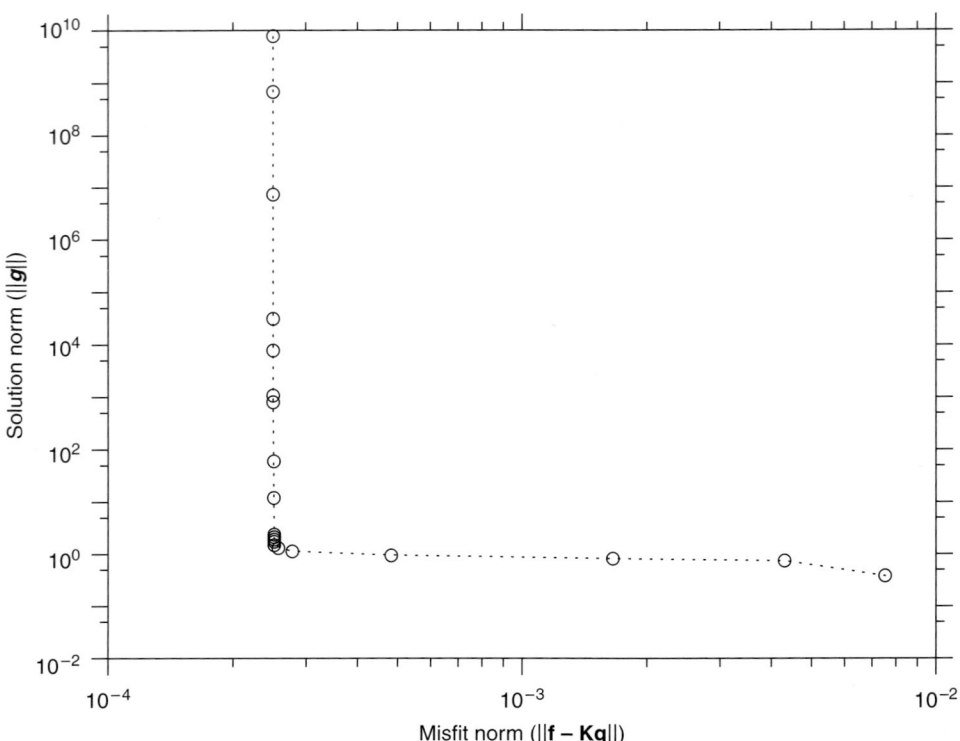

Fig. 2. L-curve obtained by systematically varying grid density. The three graphs of **Fig. 1** correspond to three points on this curve, one near the corner (the optimum, middle plot of **Fig. 1**) and two on the branches of the L. Upper-left region corresponds to loss of stability in $\tilde{g}$, and the lower-right corner represents a loss of compatibility with the data.

of compromise between the two key quantities to be minimized if inverse solution is successful. Note how several points are crowded together near the corner of the L-curve; as mentioned before, near the optimum the exact choice of discretization becomes less important.

1.2.3. Regularization Through Singular-Value Decomposition Truncation

The second key to a successful pseudoinverse solution is to properly truncate the singular-value decomposition (SVD) of the transformation matrix. Many discrete ill-posed problems exhibit the property of a gradual decrease in the size of their singular values. The following qualitative statement can be made for many physical problems: as the kernel matrix is expanded in its singular values, the lower-valued ones tend to magnify the effects of noise in the measured data. Thus, an effective way to regularize the inverse solution is to truncate the SVD expansion, in a manner quite similar to the truncation of the Taylor series expansion of a function. This has to be done in the region free of rapid changes in the size of the singular values. Yet again, there is no rigid prescription as to what is the appropriate level of truncation, and multiple trial-and-error attempts might be required.

If SVD truncation is an appropriate regularization mechanism for a given problem, a systematic examination of the misfit norm again generates an L-curve, with the optimal truncation level near the corner of it. This optimal level of SVD-truncation both provides numerical stability and

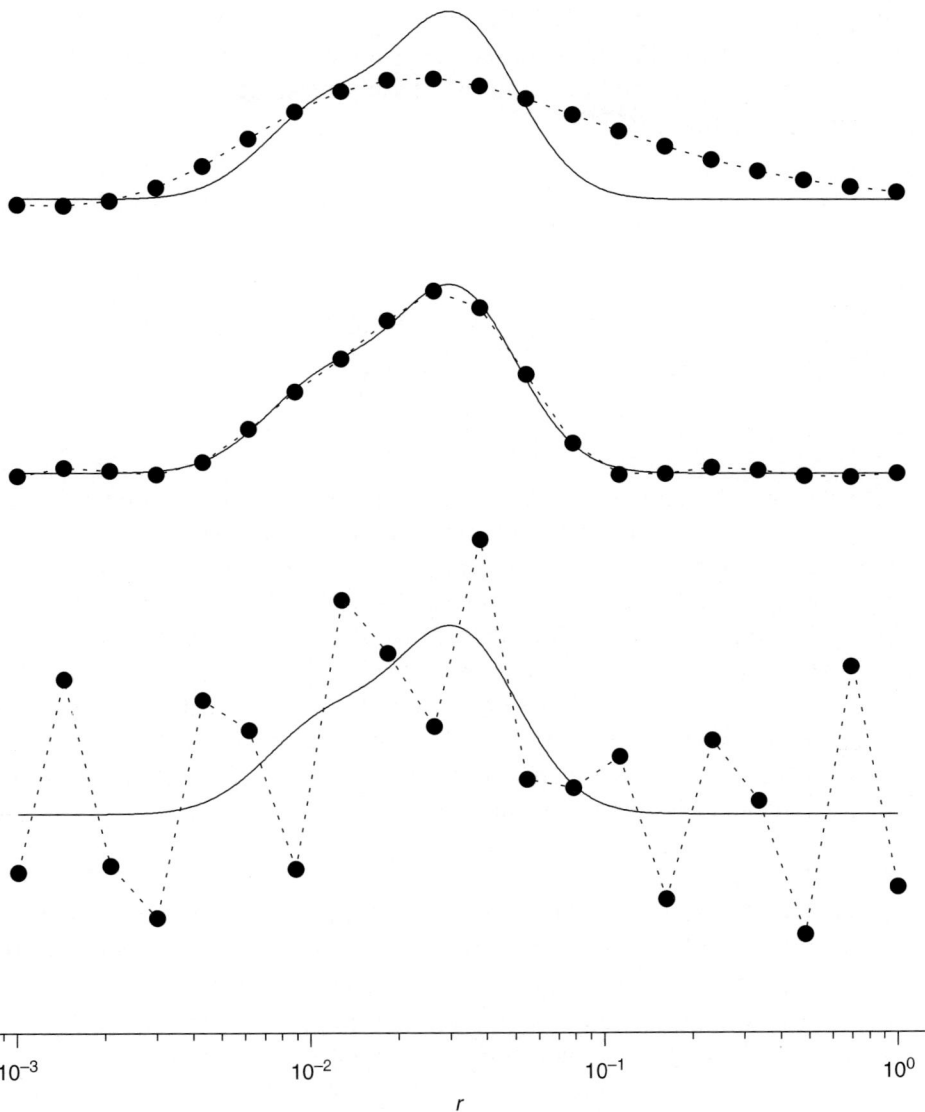

Fig. 3. Regularization through SVD truncation. Overly smooth pseudoinverse solution **(top)** fails to reproduce some of the features of the true g, and insufficient regularization produces an unstable pseudoinverse **(bottom)**. The optimum SVD truncation **(middle)** faithfully reproduces the true g, shown with a solid line in all three graphs.

ensures compatibility with the data. Near the optimum, the inverse solution is stable with respect to changes in the level of truncation. This is illustrated in **Fig. 3** wherein, as before, the three graphs illustrate the effects of (top to bottom): excessive, optimal, and insufficient regularization through SVD truncation. The full L-curve is similar to **Fig. 2** and is not shown.

1.2.4. Tikhonov Regularization

Whereas the previous two methods of regularizing a pseudoinverse arise from general aspects of signal sampling, the true key to a successful inverse solution lies in the incorporation of additional physical constraints through Tikhonov regularization *(5)*. The procedure is to minimize a modified misfit functional, with the least-squares term ensuring compatibility with the data as usual (*see* **Eq. 2**), and the additional terms enforcing some essential property of the resulting inverse solution:

$$\Psi\{g\} = \left\| f(y) - \int g(x) K(x,y) dx \right\|^2 + \lambda T\{g\} \tag{4}$$

where

$$T\{g\} = \begin{cases} \|g\|^2 & (\text{Tikhonov}) \\ \|g''\|^2 & (\text{Philips}) \\ -\int g \log g \, dx & (\text{maximum entropy}) \end{cases} \tag{5}$$

For each form of regularization functional (three common ones are shown),[*] the dimensionless parameter λ controls the balance between compatibility with the data and the regularizing effects of $T\{g\}$. Clearly, the right choice for λ is crucial: if λ value is too small, a stable solution will not be found; if λ is too large, essential physical features in the solution will be obscured. It should be noted that other forms of the regularization functional are possible and for any given problem and finding the best one may represent a significant challenge; the three popular ones shown in **Eq. 5** are conceptually simple, have been well-studied theoretically, and are known to work for a wide variety of physical data containing random noise. **Figure 4** illustrates the dependence of the inverse on the value of λ; in the interests of brevity, the corresponding L-curve is not shown.

1.2.5. Example: Distributions of Relaxation Rates

As an illustration of a successful regularization of a difficult problem, consider the previously introduced example of multiexponential analysis of a decay curve. The inherent difficulty of fitting a noisy signal measured on a linear scale with an exponentially varying function means that it is extremely difficult to answer even the most basic questions, for example, whether the "spectrum" of relaxation rates is a broad continuous distribution or a sum of several narrow processes, each with its own relaxation rate. Traditionally, each of the possibilities would be modeled separately, and the fits compared with the data. Choosing the right kind of a model function would be an essential qualitative step, left to the discretion of the investigator. Regularizing the inverse solution allows us instead to remap the data into a distribution of relaxation rates, in a model-free manner. Examination of the resulting distribution immediately reveals the underlying physical mechanisms and the nature of model that should be pursued, as shown in **Fig. 5**.

Two simulated relaxation rate distribution functions, $g_a(r)$ and $g_b(r)$, are used to generate two time-domain signals according to **Eq. 3**; 1% random noise is added, and the resulting $f_a(t)$ and $f_b(t)$ are then inverted over the grid of 100 logarithmically distributed relaxation rate values. The two time-domain signals appear similar and do not provide clues to the nature of the underlying relaxation-rate distribution. On the other hand, mapping the data into the domain of

---

[*]The umbrella name of Tikhonov regularization is often used to refer to the calculational scheme defined by **Eqs. 4** and **5**, and covers all forms of the regularization functional.

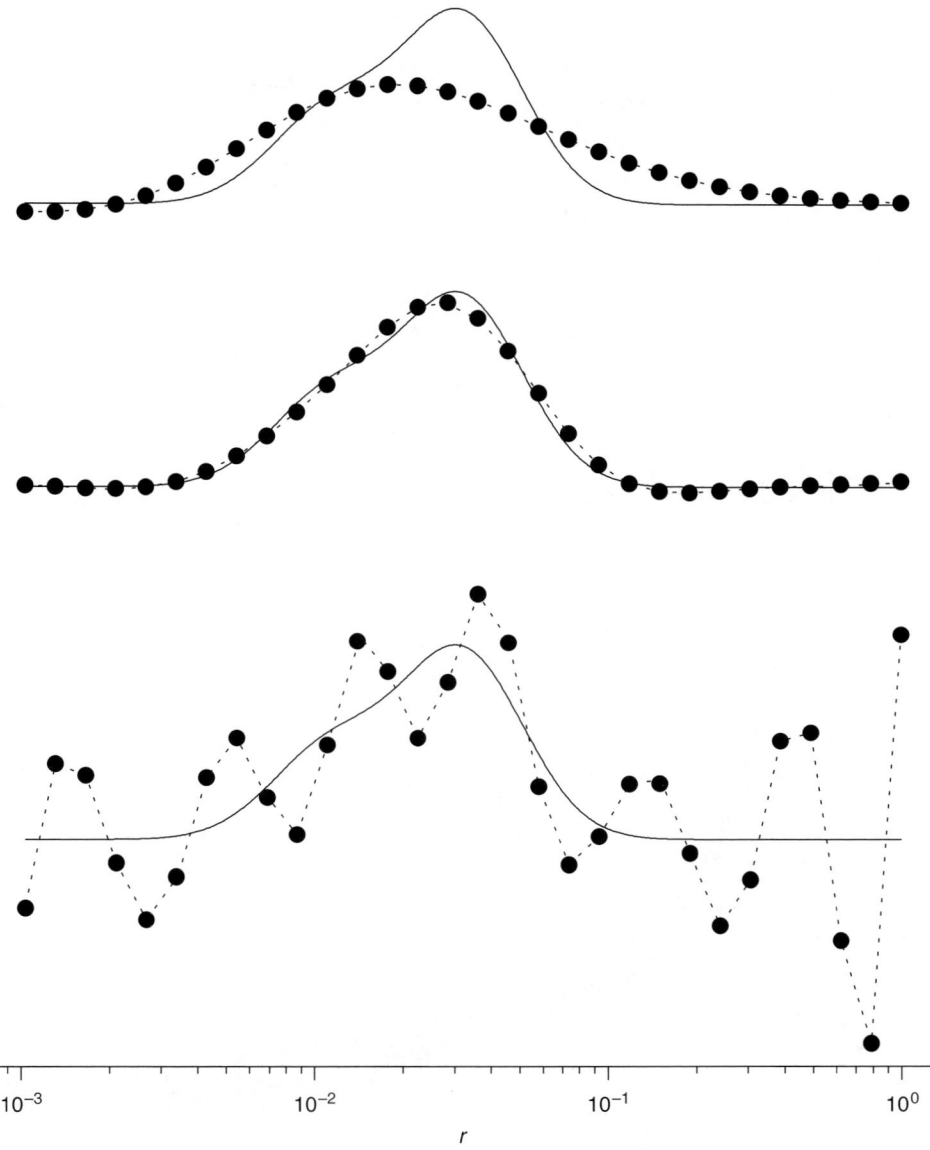

Fig. 4. Tikhonov regularization. Overly smooth pseudoinverse solution **(top)** fails to reproduce some of the features of the true g, and insufficient regularization produces an unstable pseudoinverse **(bottom)**. The optimum choice of λ **(middle)** faithfully reproduces the true g, shown with a solid line in all three graphs.

relaxation rates through a regularized inverse solution immediately differentiates between a "broad asymmetric hump" in $g_a(r)$ and a "superposition of three distinct lines" in $g_b(r)$. To be sure, there are discrepancies between the true and the recalculated distribution functions; these arise as a consequence of the noise, and can only be reduced by improving the quality of the input data. Nevertheless, the essential qualitative nature of the underlying processes is clearly revealed. This is a simulated example; however, the same method was used to establish that photochromic reaction spiropyran ↔ merocyanine occurs not in a distribution of

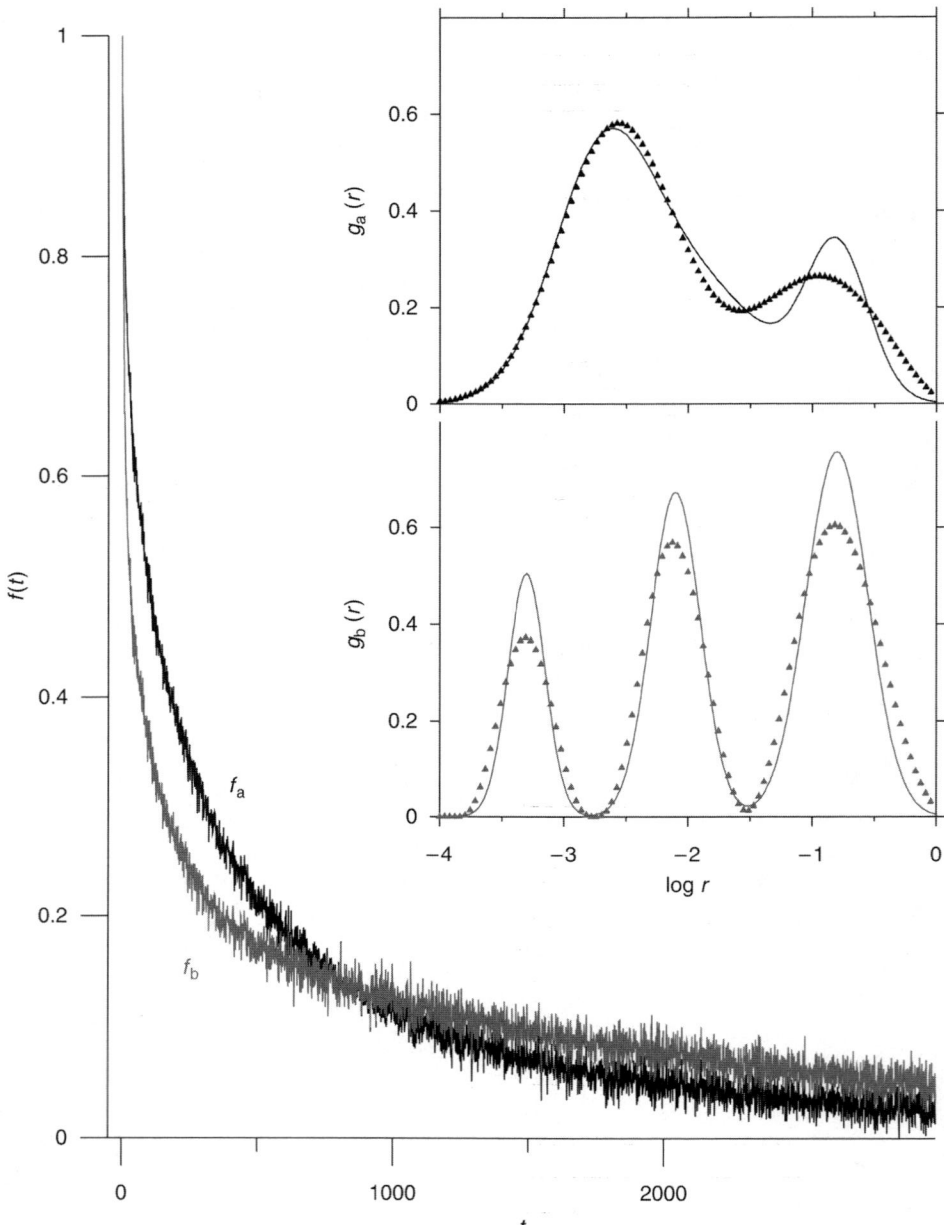

Fig. 5. Example of a successful inverse solution: multiexponential decay. Solid lines in the insert represent true distributions of relaxation rates; discrete points are the inverse solutions over a grid of 100 logarithmically spaced relaxation rates.

local environments with a wide spectrum of site-dependent local reaction rates, but as a coexistence of several different isomeric reactions, which run in parallel *(6)*.

1.2.6. Self-Consistency

L-curve is an excellent graphical tool for reviewing effects of regularization. With the help of an L-curve, proper choice of discretization grids and level of SVD truncation can be established fairly quickly for many cases of interest. However, explicit optimization of the regularization parameter λ may require additional numerical work, and the method of choice for a broad range of experimental situations is the so-called self-consistency method *(7,8)* that allows the optimal value for λ to be determined solely from the data itself. The numerical details are beyond the scope of this discussion, but the essence of the method is to monitor not just the total misfit norm, but the statistical *distribution* of the misfit function, and to insist that this distribution is identical to the distribution of experimental noise, determined independently, typically from a baseline area of the experimental data wherein no known signal exists or from a separate null experiment. Conceptually simple, this idea is of fundamental importance; it allows one to state that the obtained approximate solution reflects all of the information from the measured data, without overfitting the noise. In this sense, the pseudoinverse becomes not just one possible solution, but *the best* solution within the constraints of the experimental errors, and can be thought of as a faithful mapping of the experimental data into the parameter space.

2. Case Studies
2.1. De-Pakeing: Distribution Functions in Biomembrane NMR

Anisotropy of molecular motions in biological and model membranes is responsible for partial motional averaging of the orientation-dependent second-rank tensor interactions, such as anisotropic chemical shift, nuclear dipole–dipole, and nuclear quadrupolar interactions. Axial symmetry of fast reorientational motions produce the following fundamental scaling relationship

$$\omega(\theta) = x P_2(\cos\theta) = x \frac{3\cos^2\theta - 1}{2} \tag{6}$$

between the spectroscopic observable ω and the inherent motionally averaged anisotropy parameter x. Although both have the dimensions of frequency, it is important to note that only ω domain is accessible experimentally. Here θ is the angle between the symmetry axis of fast molecular motions and the external magnetic field. The system may exhibit more than a single inherent time-averaged anisotropy, giving rise to an *anisotropy distribution function* $g(x)$, for example, the order parameter profile $S_{CD}(n)$ (where n labels different molecular sites along the hydrocarbon chain) extracted from the quadrupolar splittings of an ^{2}H NMR spectrum; a set of isotropic chemical shifts or chemical shift anisotropies from a ^{13}C NMR spectrum, and so on.

On the other hand, experimental spectra may contain contributions from domains of different orientations, giving rise to an *orientational distribution function* $p(\theta)$. Individual domains remain static on the time-scale of the experiment, but each anisotropy x makes a spectral contribution associated with every orientation present in the sample. This gives rise to a continuous lineshape function, $s_x(\omega)$, for each x. Because $P_2(\cos\theta)$ varies from +1 to −1/2, each $s_x(\omega)$ contributes to the total $S(\omega)$ in the range from x to $-x/2$. The other contributing function, $p(\theta)$, is the probability of encountering a domain oriented at an angle θ with respect to the external magnetic field, for example, for a single crystal at 0°, $p(\theta) = \delta(\theta)$ and thus $s_x(\omega) = \delta(\omega - x)$; for a perfectly random powder $p(\theta) = \sin\theta$, and from **Eq. 6** it immediately follows that $s_x(\omega) = [3x(x + 2\omega)]^{-1/2}$.

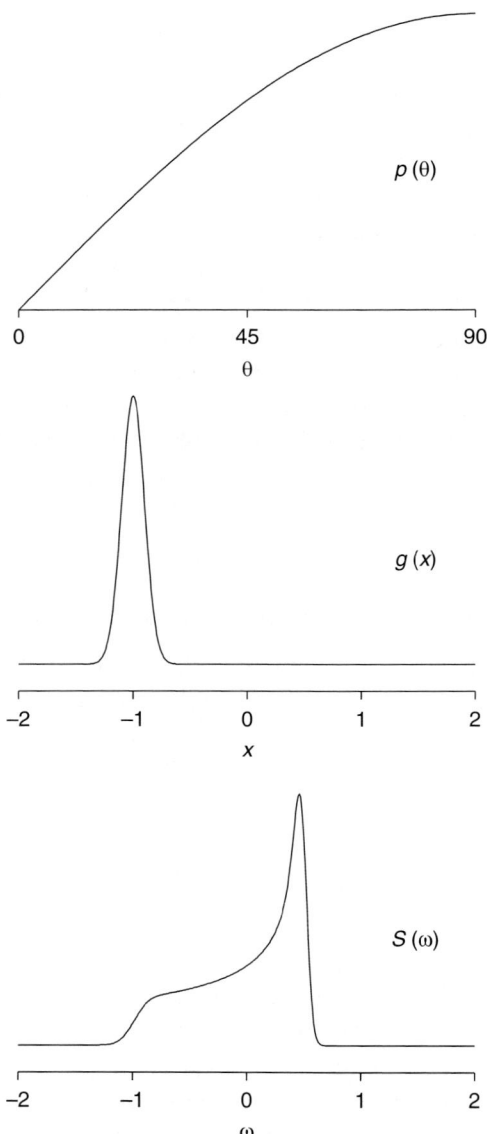

Fig. 6. Both orientational and anisotropy distribution functions, $p(\theta)$ and $g(x)$, contribute to the formation of the powder spectrum, as described by **Eqs. 7–8**. Here, $p(\theta) = \sin \theta$ is used as appropriate for a random distribution of domain orientations.

Both $g(x)$ and $p(\theta)$ matter, and the spectral lineshape $S(\omega)$ can be written in two equivalent ways:

$$S(\omega) = \int g(x) \left[p(\theta) \frac{\partial \theta}{\partial \omega} \right] dx \tag{7}$$

$$= \int p(\theta) \left[g(x) \frac{\partial y}{\partial \omega} \right] d\theta \tag{8}$$

with **Eq. 7** implying a $g(x)$-weighted sum of lineshapes, and **Eq. 8** implying a $p(\theta)$–weighted sum of oriented spectra. **Figure 6** illustrates the relationship between $g(x)$, $p(\theta)$, and $S(\omega)$.

Clearly, **Eqs. 7** and **8** describe an inverse FIE problem, where either of the two distribution functions $g(x)$ and $p(\theta)$ need to be extracted from the experimentally measured spectrum $S(\omega)$ that now plays the role of $f(y)$ from **Eq. 1**. The case of $S(\omega) \to g(x)$ (**Eq. 7**) implies that the form of $p(\theta)$ is known, for example a fully random distribution of orientations is $p(\theta) = \sin\theta$. This is the so-called *de-Pake-ing** (9–11) that is widely used for the analysis of solid-state NMR spectra. For example, "de-Pake-ing" ^{2}H NMR spectra of chain-deuterated phospholipids in a model membrane represent directly the orientational-order parameter profile $S_{CD}(n)$ that reports on the molecular motions in the bulk of the membrane. Monitoring changes in this order parameter profile caused by structural phase changes, interactions with other membrane constituents, and so on, is a sensitive and powerful experimental tool (12).

The case of **Eq. 8** wherein the spectral anisotropies $g(x)$ are known and the orientational distribution $p(\theta)$ needs to be measured is a separate inverse FIE problem. It has been used, for example, to examine the aligning properties of a porous media filled with a liquid crystalline material (13). However, in biomembrane NMR, de-Pakeing is the inverse problem of primary interest. Unfortunately, in this case one needs to know what $p(\theta)$ is, and it is well known that the assumption of $p(\theta) = \sin\theta$ is often inadequate, because of the effects of partial alignment of membrane bilayers by an external magnetic field.

2.2. Magnetic Alignment: Extracting Both g(x) and p(θ)

Lipid bilayers have anisotropic magnetic susceptibility,

$$\Delta\chi = \chi_\parallel - \chi_\perp < 0 \qquad (9)$$

which gives rise to an interaction between an induced magnetic moment of each small domain of area A, thickness d, and bilayer normal $\vec{n}$, with the external magnetic field $\vec{H}$; the resulting torque

$$\vec{\tau} = \Delta\chi\, A\, d\, (\vec{H}\cdot\vec{n})(\vec{H}\times\vec{n}) \qquad (10)$$

orients the domain preferentially so that $\vec{n} \perp \vec{H}$. Physically, this corresponds to spherical vesicles being deformed into ellipsoids with their long axes preferentially aligned along the magnetic field. **Figure 7** illustrates a typical example of a spectral distortion caused by such partial magnetic alignment; the distortion is in the suppressed size of the spectral "shoulder" associated with $\theta = 0$ orientation, relative to the undistorted lineshape shown in **Fig. 6**. The overall lineshape is dominated by the strong $\theta = 90°$ peak, and thus the distortion appears irrelevant. However, de-Pakeing is extremely sensitive to small lineshape changes, and the distribution of anisotropies it extracts is strongly affected by such distortions. This data was acquired at a relatively low field strength of $7T$, the effects are more pronounced at higher fields.

Therefore, it is essential to de-Pake in the presence of an unknown orientational distribution. This is probably impossible for a completely arbitrary $p(\theta)$, as either anisotropy or orientation may be responsible for a spectral contribution at any given observation frequency (*see* **Eq. 6**). However, one can test and compare a parameterized family of models that describe a physically reasonable $p(\theta)$, de-Pake repeatedly, and select the best model parameters in the same way, by finding a global minimum of the misfit norm. Fortunately, partial magnetic alignment

---

*After G. E. Pake who first reported lineshapes like that shown in **Fig. 6**.

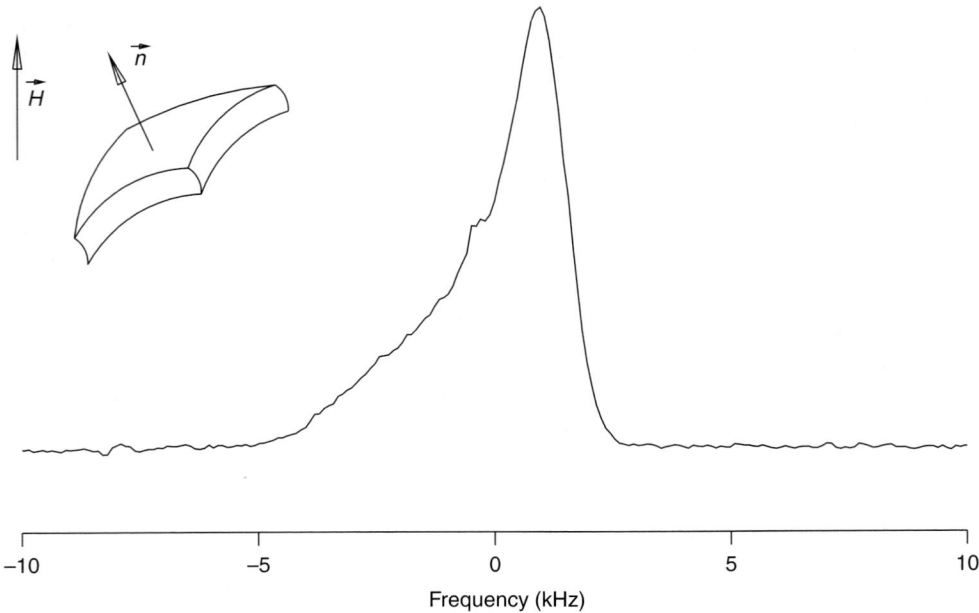

Fig. 7. Partial magnetic alignment of biomembrane domains by an external magnetic field is seen in a ^{31}P NMR spectrum of 83 wt% 1-palmitoyl-2-oleoyl-*su*-glycero-3-phosphoethanolamine (POPE) + 17 wt% 1-palmitoyl-2-oleoyl-*su*-glycero-3-phosphoglycerol (POPG) mixture in water at 25°C, in a moderate field of 7T.

can be described as a relatively well-behaved continuous deformation of $p(\theta)$ from its random form, $\sin \theta$. As a result, several one-parameter models for $p(\theta)$ are physically reasonable *(14,15)*:

- Ellipsoidal (high correlation between orientations of adjacent domains):

$$p_E(\theta) \propto \sin\theta \left[1 - (1-\kappa_E)\cos^2\theta\right]^{-2}$$

- Boltzmann (adjacent-domain orientations are uncorrelated):

$$p_B(\theta) \propto \sin\theta \, \exp\left[\kappa_B \cos^2\theta\right]$$

- Legendre polynomial expansion (general):

$$p_L(\theta) \propto \sin\theta \sum_{i=1}^{\infty} A_i P_i(\cos\theta) \approx \sin\theta(1 - A\kappa_L \cos^2\theta)$$

when κ_E, κ_B, or κ_L, as appropriate, are varied, the orientational distribution function undergoes a continuous deviation from the random $p(\theta) = \sin\theta$. Use of these models for a simultaneous extraction of $g(x)$ and *some* limited information about the form of $p(\theta)$ requires a slight modification to the inverse procedure:

- The misfit function is modified to include the appropriate κ:

$$\Psi\{g(x);\kappa\} = \left\| S(\omega) - \int g(x) K(\omega,x;\kappa) dx \right\|^2 + \lambda T\{g(x)\} \qquad (11)$$

- To deal with the highly nonlinear dependence on κ, an appropriate range of κ-values is swept for each model in turn and the misfit norm is monitored.

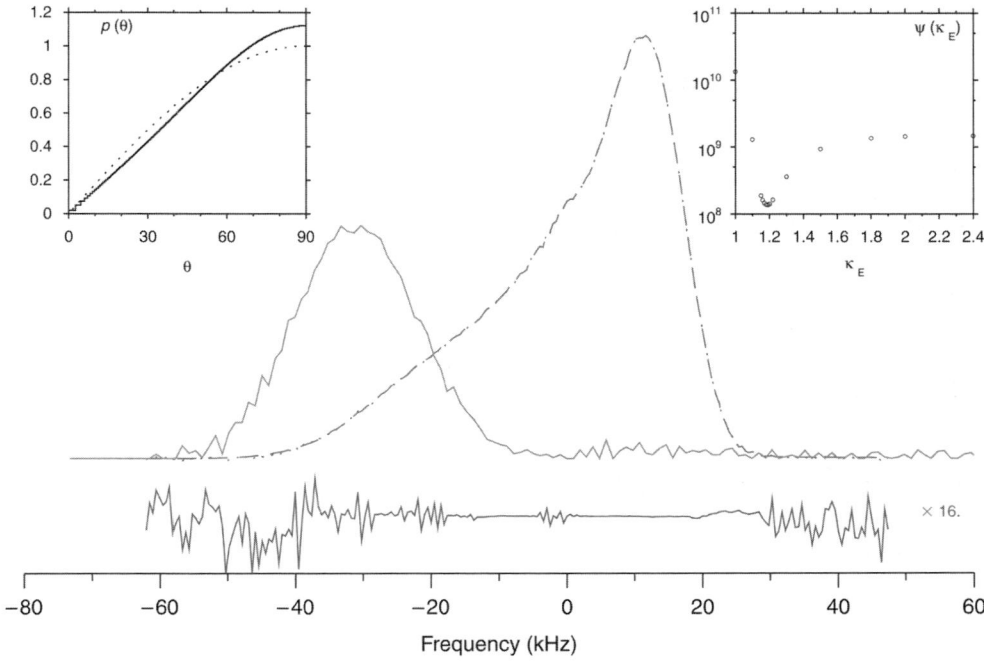

Fig. 8. Simultaneous extraction of $\tilde{g}(x)$ and $\tilde{p}(\theta)$ from a partially magnetically aligned powder spectrum: min $\Psi \to \tilde{\kappa} \to \tilde{p}(\theta) \to \tilde{g}(x)$. Dashed line is the same ^{31}P NMR spectrum shown in Fig. 7, whereas the solid line is $\tilde{g}(x)$, the result of de-Pakeing using TR. The dotted line, visually indistinguishable from the input data, is the powder spectrum recalculated for the found $\tilde{g}(x)$; the difference between the two (the misfit) is shown below on a (×16) scale. The right insert shows that for the optimum value $\kappa_E = 1.19 \pm 0.005$ the misfit norm is 100 times lower than for $\kappa_E = 1$, i.e., when a random $p(\theta) = \sin \theta$ is *assumed*. The left insert shows the $p(\theta)$ that corresponds to this optimum κ_E; the random $\sin \theta$ is shown with a dotted line for comparison.

- The lowest of all minima in misfit norm corresponds to: (a) the best model and its optimal $\tilde{\kappa}$-value, which in turn determines the best estimate $\tilde{p}(\theta)$, and (b) the best inverse $\tilde{g}(x)$.

Figure 8 presents an example whereby a ^{31}H NMR spectrum from a partially magnetically aligned model membrane sample was successfully de-Paked using Tikhonov regularization (TR), yielding not only the orientational order parameter profile, but also an estimate of the degree of alignment, corresponding to an ellipsoidal-like deformation of a spherical orientational distribution function. The optimal value $\kappa_E = 1.19$ has a simple physical meaning of the square of the ratio of the semiaxes of an ellipsoid of rotation, so in this case the alignment is very mild. However, even this mild effect would obscure the emergence of the second broad spectral component at about +15 KHz; this spectrum was measured near a $L_\alpha \leftrightarrow H_{II}$ phase boundary, and this signal is from a very small amount of the second (H_{II}) component. Without the refinement that the inverse solution afforded herein, a small rise in the spectrum, which is in fact, the peak of the H_{II} powder pattern, would likely to have been dismissed as an artifact. In magnetic fields typical of modern NMR spectrometers hydrated biomembranes at biologically relevant temperatures exhibit a degree of alignment that typically corresponds to $\kappa_E \approx 2$ to 8.

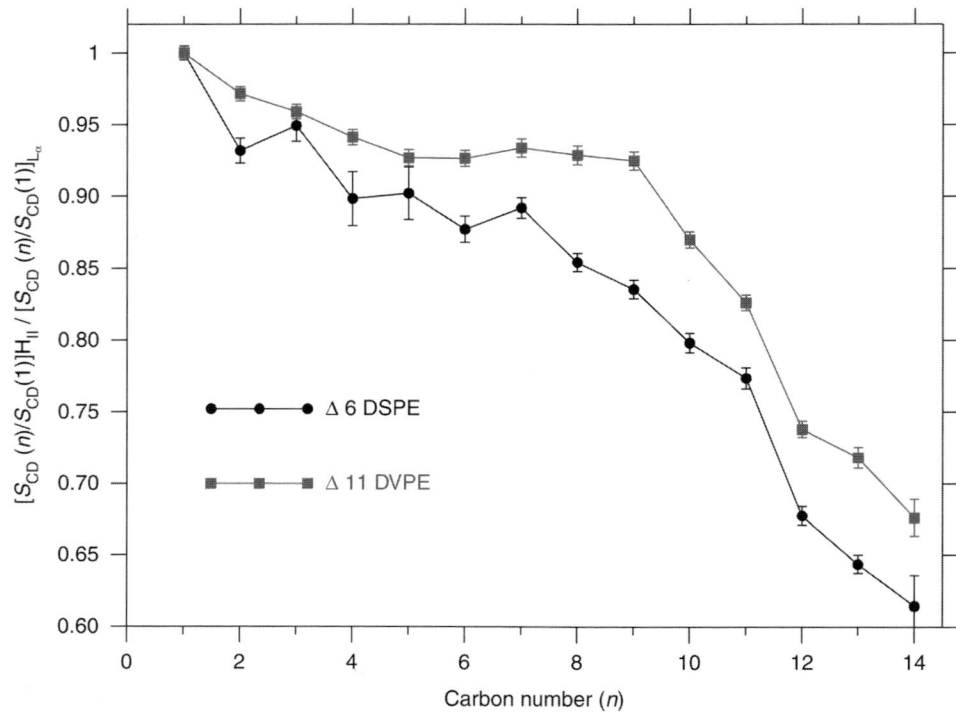

Fig. 9. Changes in the order parameter profiles of monounsaturated lipids with the double bond at different positions along the chain, monitored through a small amount of d_{29}-tetradecanol. $S_{CD}(n)$ are extracted through de-Pakeing in the presence of partial magnetic alignment, separately for L_α and H_{II} phases, normalized by their respective $S_{CD}(1)$, and their ratio calculated. Δ6 DSPE: 1,2-dipetroseliuoyl-*sn*-glycero-3-phosphoethanolamine; Δ11 DVPE: 1,2-divacceiioyl-*sn*-glycero-3-phosphoethanolamine.

2.2.1. Chain Orientational Order Near $L_\alpha \leftrightarrow H_{II}$ Transition

Being able to calculate $g(x)$ in the presence of alignment means that spectral contributions from different structural phases can be precisely measured, improving the quality of determining phase boundaries in complex structural mixtures *(15)*. It is also essential for obtaining orientational order parameters from the ^{2}H NMR of chain-deuterated biomembrane samples. TR-based de-Pakeing allows highly reliable $S_{CD}(n)$ determination *(15,16)*, leading to a measurement of effects previously thought undetectable. For example, thermodynamic properties of model membranes containing monounsaturated phospholipids depend strongly on the position of the double bond along the hydrocarbon chain, with very slight changes in chemical structure causing dramatic shifts in phase-transition temperatures. To examine the chain orientational order for any anomalies associated with the position of the double bond a small amount of deuterated tetradecanol can be added. It is known to localize in the bilayer and act as a reliable reporter of the order parameter profiles *(17)*, and this can be exploited to monitor subtle changes in the order parameter profiles as the sample undergoes $L_\alpha \leftrightarrow H_{II}$ structural phase transition *(18)*. The result is shown in **Fig. 9** wherein a local modulation clearly shifts with the position of the double bond. Plotted on the vertical axis is a dimensionless ratio of order parameters representing, in essence, the extent to which average local order remains "bilayer-like" as the sample undergoes a dramatic change in local monolayer curvature associated with the formation of the H_{II} phase. The effect is delicate, and comparable in magnitude with the distortions in the order parameter profile caused by magnetic alignment; the ability to extract $\tilde{g}(x)$ simultaneously with $\tilde{p}(\theta)$ is essential in making it visible.

2.2.2. Alignment and Structural Organization

Figure 9 shows how to overcome a "distortion" resulting from a partial alignment; the parameters of optimal $\tilde{p}(\theta)$ gathered in the process of obtaining an inverse solution yield additional information about the sample. However, it is interesting to note that an inverse solution may be as instructive when it fails as when it succeeds. For example, one-parameter models for orientational distribution functions may fail to produce a satisfactory inverse. This suggests presence of additional complexity in the sample that necessitates exploring higher-order terms in the Legendre polynomial expansion, or some other combination of nonoverlapping orientation-dependence functions. With some increase in processing power and extra care taken in the acquisition of high-fidelity NMR data, one can study samples containing a superposition of multiple contributions from different structural arrangements, for example, a mixture of randomly oriented small micelles or multilamellar vesicles, and partially aligned giant unilamellar vesicles or even strongly aligned bicelles. The search for a minimum of misfit norm now extends over a two-parameter space: κ of the aligned fraction and the fraction of total sample that is in the aligned phase. **Figures 10** and **11** test the "golden standard" of aligned biomembrane samples, namely, a glass-aligned "sandwich" preparation of a model membrane system; d_{35}-SDhPC[*] in this case. Sharp peaks typical of aligned samples tend to obscure the broad powder patterns of the unaligned fraction and with some adjustment to the phasing of the spectrum the latter may blend into the baseline. The resulting spectrum may appear perfectly aligned, and de-Pakeing it—unnecessary. However, **Fig. 10** shows an inverse solution obtained through de-Pakeing and using a single-parameter ellipsoidal model. The data is indistinguishable from the recalculated "fit" to the spectrum; however, a careful examination alerts us to a systematic misfit in the wings of the spectrum, making this inverse solution inadequate. On the other hand, the inverse shown in **Fig. 11** is free of systematic misfit, and a good de-Paked spectrum is obtained. This inverse solution is consistent with the sample containing a 70/30% (±2%) aligned/randomly oriented mixture of structural phases, with the oriented one having the mean distribution of angles of about ±3°. This spectrum was selected as being one of the best glass-aligned examples; "fully aligned" samples that in fact contain 40% or more of unaligned phase are quite typical. More than a third of the sample goes essentially unnoticed.

More than a diagnostic tool to monitor the quality of sample preparation, the structural phase information obtained in the process is valuable by itself; for example, such two-parameter fits have been applied to the spectra of a so-called bicellar short/long-chain lipid mixture (1,2-dimyristoyl-*sn*-glycero-3-phosphocholine:1,2-dicaproyl-*sn*-glycero-3-phosphocholine in a 4.1:1 molar ratio). Such spectra, similar in appearance to the one shown in **Figs. 10** and **11** have been interpreted as evidence of "perfect" alignment in bicelles *(19)*. The information obtained using mixed-model de-Pakeing, allows construction of a rich, structural-phase diagram of a bicellar mixture *(15)*; throughout most of the temperature range bicelles turn out to be mixed structural phases. This has been confirmed independently in a recent detailed study *(20)*.

---

[*]1-steroyl-2-docosahexaenoyl-*sn*-glycero-3-phosphocholine.

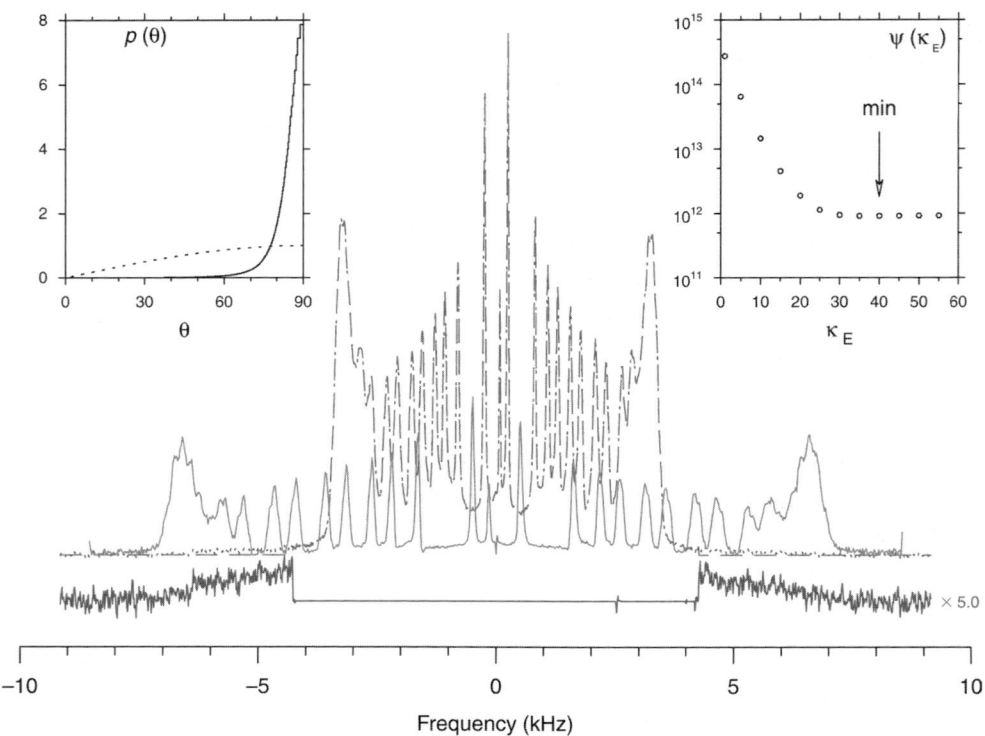

Fig. 10. NMR spectrum of a glass-aligned bilayer sample of d_{35}-SDhPC, de-Paked using a single-parameter ellipsoidal model. Notation and the arrangement of inserts is identical to **Fig. 8**. The data (dotted line) and the fit (dashed lines) are essentially indistinguishable, but a systematic misfit in the wings of the spectrum reveals the inadequacy of this model.

Figures 10 and **11** demonstrate the need for high-fidelity experimental data and the care that must be used in the data processing: the spectral features that yield such detailed information are small and easily obscured by excessive filtering or the artifacts of other common processing methods that are used to improve the visual appearance of the spectra. These figures also illustrate how an improved, albeit computationally costly, data analysis leads to a realization of a need for a better model, to an introduction and use of such a model, and finally, to a better and more complete understanding of the story told by the data. As mentioned earlier, the model itself becomes a "fit parameter," and is improved on in the process. This is the true strength of regularization.

2.3. Regularization in Two Dimensions

The challenge in the examples presented here lies primarily in the quality of data acquisition; if a reliable data set is available and the kernel function is reasonably well behaved, the inverse is likely to succeed. Most frequently, the data is one-dimensional (1D), although some successful inverse solutions have been performed to obtain 2D distribution functions as well. For example, 2D DOQSY and DECODER data obtained from spider dragline silk was inverted to yield a Ramachandran map, i.e., a 2D distribution of backbone torsion angles (ϕ, ψ) *(21,22)*. Tikhonov regularization functional (**Eq. 5**) does not explicitly depend on the (possibly multidimensional) parameter-space variable x, and so the minimization problem

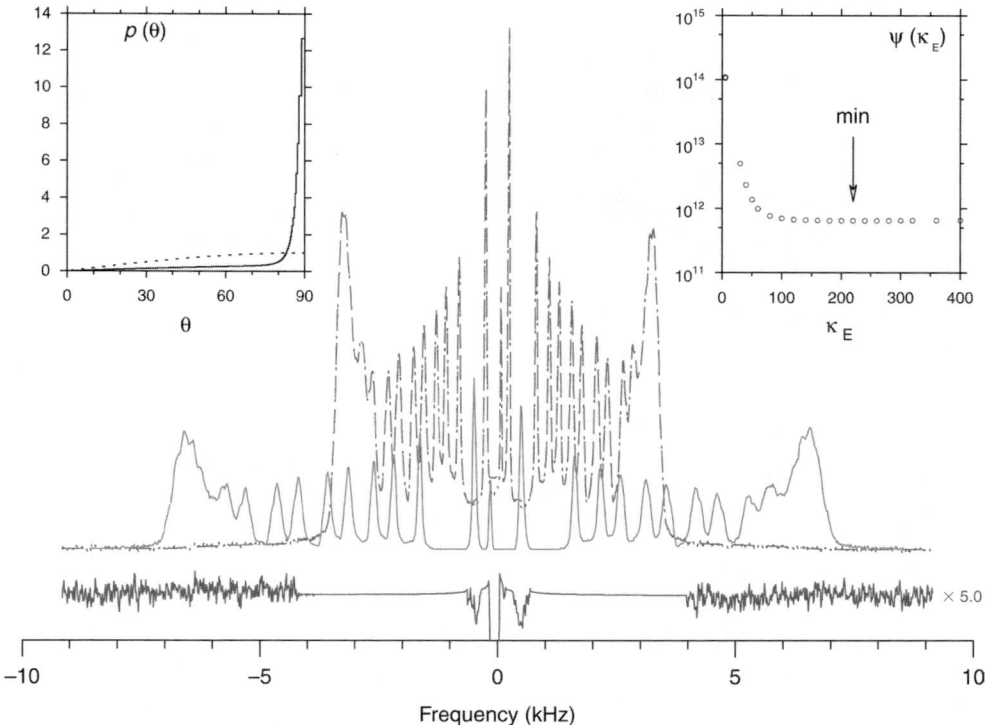

Fig. 11. NMR spectrum of a glass-aligned bilayer sample of d_{35}-SDhPE, de-Paked using a 70/30% aligned/randomly oriented mixed model. Notation and the arrangement of inserts is identical to **Fig. 10**. The minimum in $\Psi(\kappa)$ is relatively shallow, especially seen on the logarithmic scale, partly because of a small systematic misfit near zero associated with an isotropic line, likely from residual HDO that has a different isotropic chemical shift. This artifact does not influence the de-Paked spectrum of the lipid. Note the absence of a systematic misfit in the wings of the spectrum, as in **Fig. 10**. A large value of the optimal $\kappa_E = 220 \pm 10$ corresponds to a narrow (about ±3°) distribution of the bilayer normals in the aligned fraction (the left insert).

can be reformulated as an equivalent 1D problem. Reindexing $\{g_{ij}\} \to \{g_l\}$, where $i = 1 ... m$, $j = 1 ... n$, and $l = m \times (j - 1) + i$, converts a 2D grid of (ϕ_i, ψ_j) values into an equivalent 1D distribution *(22)*. However, even this problem, although massive in scope, belongs to the same class as all of the examples in the preceding sections, namely, that of the overdetermined problems: the size of the dataset exceeds that of the parameter space.

Somewhat unexpectedly, this is not a necessary condition, although for underdetermined problems the difficulties of obtaining a successful inverse solution become greater, and the importance of choosing the best regularization functional for a given problem increases. In 2D NMR problems, in particular, there is considerable redundancy in the data, as the adjacent rows and columns of the 2D dataset typically overlap in their dependence on the adjacent values in the parameter distributions. Regularization terms that provide a certain amount of 2D coupling in the parameter space can help to overcome the difficulties of inverting underdetermined problems. One such example is shown in **Fig. 12**, wherein a distribution of backbone torsional angles $g(\phi,\psi)$ is simulated over a 5° × 5° grid, with two Gaussian peaks representing a certain mixture of near α-helical and near β-sheet motifs, a total of 2701 points

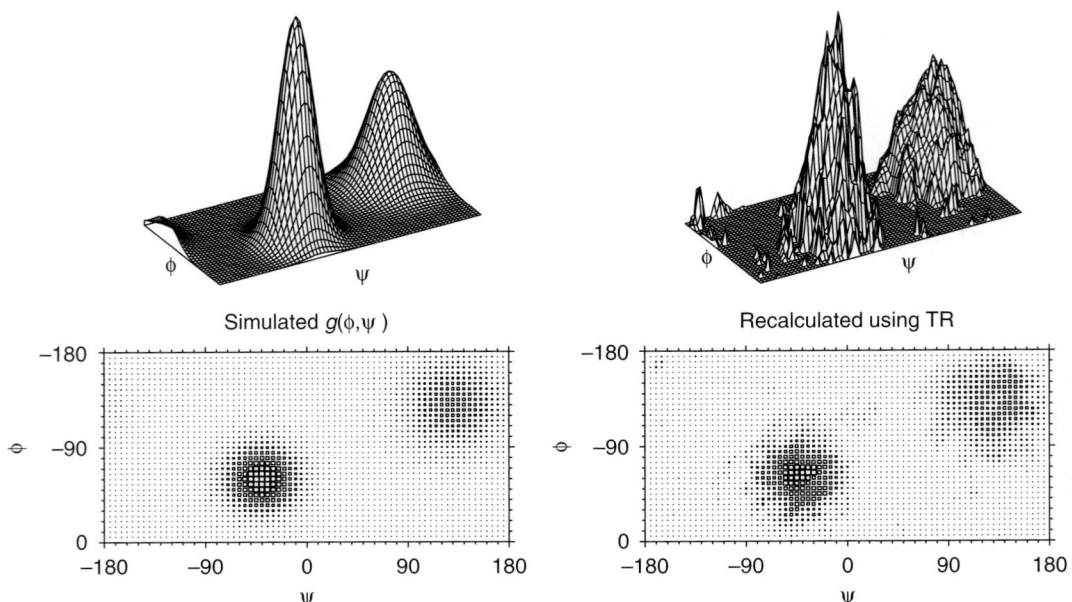

Fig. 12. Intensities of 190 CP MAS cross-peaks are calculated from a simulated 73 × 37 Ramachandran torsional angle distribution $g(\phi, \psi)$ (on the left) and 0.1% random noise added. The resulting dataset is inverted using Tikhonov regularization with Laplacian as the regularization functional; the result (on the right) faithfully reproduces the initial distribution function *(29)*. Inversion using only the least-squares minimization fails to recover the true $g(\phi, \psi)$, at any noise level (not shown).

in parameter space. A set of 190 ^{1}H–^{13}C CP MAS cross-peak intensities is then calculated from this distribution, as appropriate for Alzheimer's β-amyloid fibrils *(23)*. A 0.1% random noise is added to the simulated data, and this vastly underdetermined (190-by-2701) inverse functional, $T(g) = \left\| \nabla^2, g \right\|^2$, another well known form of regularization. The true distribution is successfully recovered **(Fig. 12)**; for an experimentally obtained data this would allow a determination of the relative amounts of different backbone motifs in the sample. The question of whether Alzheimer's fibrils pack in a parallel or an antiparallel fashion is of significant structural importance; this can now be tested using a fairly limited dataset that can be obtained indirectly in CP MAS and finite-pulse radio-frequency-driven recoupling at constant time experiments *(24)*. Without regularization only very limited qualitative fits can be performed; these suggest the presence of a fairly broad distribution of backbone angle pairs *(25,26)*.

3. A Regularization Primer

The computational "cost" of getting some first-hand experience with regularized inverse solutions is surprisingly low. In addition to many excellent implementations in the form of C and Fortran libraries (e.g., GENEREG[*] *[27]*), high-level macro language implementations have been gaining popularity (e.g., Regularization Tools for MatLab[†] *[28]*). The latter-style tools

---

[*]www.fmf.uni-freiburg.de/service/sg_wissinfo/Software/.
[†]www2.imm.dtu.dk/~pch/Regutools/.

```
//////////////////////////////////////// regularize()
function [g] = regularize(t,f,r,K,svd_n,lambda)
  m=length(f); n=length(r);
  if m < n then
    error("Not meant for underdetermined problems, need more data");
  end;

  [U,S,V]=svd(K);        //  SVD of the kernel matrix

  nt=n;
  if svd_cnt > 0 then    //  if requested, truncate singular values
    nt=min(n,svd_cnt);
    nt=max(nt,2);        //  but not too few!
  end;

  sl=S(nt,nt);           //  Tikhonov regularization
  for k=1:nt
    sl(k,k)=S(k,k)/(S(k,k)^2+lambda);
  end;

  g=V(1:n,1:nt)*sl*U(1:m,1:nt)'*f;    // return g(r)
endfunction;
/////////////////////////////////////////////////////////
//  multi-exponential inverse analysis of time decay curves
//      f(t) = \int g(r) exp(-r*t) dr
/////////////////////////////////////////////////////////
//      r = vector or r values
//      n = number of points in r
//      g = vector of unknowns, g(r)
//      t = vector of t values
//      m = number of points in t
//      f = vector of measured data, f(t)
//      K = (m x n) kernel matrix
//      svd_cnt = keep only this many singular values
//      lambda = Tikhonov regularization parameter
//      datafile = (noisy) data, in two columns: (t,f)
/////////////////////////////////////////////////////////
datafile='test.dat';
r_steps=30; // r-grid has this many points,
r_min=1e-3; // from this minimum,
r_max=1e0;  // to this maximum
svd_cnt=11; // set to 0 for no SVD truncation'
lambda=2e-4; // set to 0 for no Tikhonov regularization

//  read in the time-domain data
fd=mopen(datafile); [n,t,f]=mfscanf(-1,fd,"%f %f"); mclose(fd);

//  create a vector of r values, logarithmically spaced
r_inc = (log(r_max)-log(r_min)) / (r_steps-1);
r = exp([log(r_min):r_inc:log(r_max)]);

//  set up our kernel matrix, normalize by the step in r
K = exp(-t*r) * r_inc;

//  call the inversion routine
[g] = regularize(t,f,r,K,svd_cnt,lambda);

//  plot the original data f(t) and our misfit
scf(1); clf; plot(t,f,'-');
misfit = f - K*g ;
Psi = sum(misfit.^2)/(length(f)-1);
scale=0.2*max(f)/max(misfit);
plot(t,scale*misfit,'-r');
legend('input data, f(t)','misfit, x'+string(scale));
xtitle('LS error norm = '+string(Psi),'t','f(t)');

//  separately, plot the result of the inversion
scf(2); clf; plot(log10(r),g,':o');
xtitle('inverse solution','log r','g(r)');
```

Fig. 13. Example of SciLab code implementing inversion through Tikhonov regularization, as appropriate for the multiexponential analysis example shown in **Fig. 5**.

are particularly interesting because they allow a novice user to focus on the algorithm rather than on the computational details, as they provide a compact feature-rich vector-oriented syntax. The example shown in **Fig. 13** was used to solve the multiexponential example of the **Introduction**.

For clarity, this sample code uses a particular value for all relevant parameters (grid density, level of SVD truncation, regularization parameter λ); by adding appropriate loops, all figures of the **Introduction** can be obtained. The heart of the code is a very short function defined at the beginning; this function is called from the main routine, which contains mostly housekeeping calls. The only computational task in the main routine is to fill in the values of the discretized kernel matrix, a sum of exponential decay curves in this case; this portion of the code needs to be changed to use it for another inverse problem. The code is written in the syntax of SciLab[*], a public-domain alternative to MatLab[†]; using it with MatLab involves only minor syntactic changes.

Acknowledgments

Heartfelt thanks to Hartmut Schäfer for many illuminating discussions and for his help with software; Robert Tycko for the use of simulated CP MAS data; and Ivan Polozov for the glass-aligned d_{35}-SDhPC spectrum. Figures were prepared using the software developed at the Tri-University Meson Facility, Vancouver, Canada. This work was supported by Natural Sciences and Engineering Research Council of Canada.

References

1. Hadamard, J. (1923) *Lectures on the Cauchy problem in linear partial differential equations.* Yale University Press, New Haven, USA.
2. Santamarina, J. C. and Fratta, D. (2005) *Discrete signals and inverse problems. An Introduction for Engineers and Scientists.* p. 276, John Wiley & Sons Ltd., Chichester, England.
3. Lawson, C. L. and Hanson, R. J. (1974) *Solving linear least squares problems.* Prentice-Hall, Englewood Cliffs, New Jersey.
4. Hansen, P. C. (1998) *Rank-deficient and discrete ill-posed problems: numerical aspects of linear inversion,* in *Monographs on Mathematical Modeling and Computation, vol. 4,* Society for Industrial and Applied Mathematics Philadelphia PA.
5. Tikhonov, A. N. and Arsenin, V. Y. (1977) *Solutions of Ill-Posed Problems.* John Wiley & Sons, New York.
6. Schäfer, H., Albrecht, U., and Richert, R. (1994) Dispersive first-order reactions I: Data analysis. *J. Chem. Phys.* **182,** 53.
7. Honerkamp, J. and Weese, J. (1990) Tikhonov's regularization method for ill-posed problems. A comparison of different methods for the determination of the regularization parameter. *Contin. Mech. Thermodyn.* **2,** 17–30.
8. Weese, J. (1992) A reliable and fast method for the solution of Fredholm integral equations of the first kind based on Tikhonov regularization. *Comput. Phys. Commun.* **69,** 99–111.
9. Bloom, M., Davis, J. H., and MacKay, A. L. (1981) Direct determination of the oriented sample NMR spectrum from the powder spectrum for systems with local axial symmetry. *Chem. Phys. Lett.* **80,** 198–202.
10. Sternin, E., Bloom, M., and MacKay, A. L. (1983) De-Pake-ing of NMR spectra. *J. Magn. Reson.* **55,** 274–282.
11. Whittall, K., Sternin, E., Bloom, M., and MacKay, A. L. (1989) Time- and frequency-domain "dePakeing" using inverse theory. *J. Mag. Res.* **84,** 64–71.

---

[*]www.scilab.org.
[†]Registered name of The Mathworks, Inc., www.mathworks.com.

12. Davis, J. H. (1991) Deuterium nuclear magnetic resonance spectroscopy in partially ordered systems, in *Isotopes in the Physical and Biomedical Sciences, vol. 2*, (Buncel, E. and Jones, J. R., eds.), Elsevier, Amsterdam, pp. 99–157.
13. Schäfer, H. and Stannarius, R. (1995) Calculation of orientational distributions of partially ordered samples from NMR spectra. *J. Magn. Reson. B* **106,** 14–23.
14. Schäfer, H. Madler, B., and Sternin, E. (1998) Determination of Orientational Order Parameters from 2H NMR Spectra of Magnetically Partially Oriented Lipid Bilayers. *Biophys. J.* **74(2),** 1007–1014.
15. Sternin, E., Schäfer, H. Polozov, I., and Gawrisch, K. (2001) Simultaneous determination of orientational and order parameter distributions from NMR spectra of partially oriented model membranes. *J. Magn. Reson.* **149,** 110–113.
16. Sternin, E., Nizza, D., and Gawrisch, K. (2001) Temperature dependence of DMPC/DHPC mixing in bicelles and its structural implications. *Langmuir* **17,** 2610–2616.
17. Sternin, E., Fine, B., Bloom, M., Tilcock, C. P., Wong, K.F., and Cullis, P. R. (1988) Acyl chain orientational order in the hexagonal H_{II} phase of phospholipid-water dispersions. *Biophys. J.* **54(4),** 689–694.
18. Sternin, E., Zaraiskaya, T., Razavi, R., and Epand, R.M. (2006) Changes in molecular order across the lamellar-to-inverted hexagonal phase transition depend on the position of the double-bond in monounsaturated phospholipid dispersions. *Chem. Phys. Lipids* **140,** 98–108.
19. Sanders, C. R. and Schwonek, J. P. (1992) Characterization of magnetically orientable bilayers in mixtures of dihexanoylphosphatidylcholine and dimyristoylphosphatidylcholine by solid-state NMR. *Biochemistry* **31,** 8898–8905.
20. Triba, M. N., Warschawski, D. E., and Devaux, P. F. (2005) Reinvestigation by phosphorus NMR of lipid distribution in bicelles. *Biophys. J.* **88,** 1887–1901.
21. van Beek, J. D., Beaulieu, L., Schfer, H., Demura, M., Asakura, T., and Meier, B. H. (2000) Solid-state NMR determination of the secondary structure of *Samia cynthia ricini* silk. *Nature* **405,** 1077–1079.
22. van Beek, J. D., Meier, B. H., and Schäfer, H. (2003) Inverse methods in two-dimensional NMR spectral analysis. *J. Magn. Reson.* **162,** 141–157.
23. Tycko, R., Weliky, D. P., and Berger, A. E. (1996) Investigation of molecular structure in solids by two-dimensional NMR exchange spectroscopy with magic angle spinning. *J. Chem. Phys.* **105,** 7915–7930.
24. Bennett, A. E., Weliky, D. P., and Tycko, R. (1998) Quantitative conformational measurements in solid state NMR by constant-time homonuclear dipolar recoupling. *J. Am. Chem. Soc.* **120,** 4897–4898.
25. Petkova, A. T., Ishii, Y., Balbach, J. J., et al. (2002) A structural model for Alzheimer's β-amyloid fibrils based on experimental constraints from solid state NMR. *PNAS* **99,** 16742–16747.
26. Chan, J. C. C. and Tycko, R. (2003) Solid-state NMR spectroscopy method for determination of the backbone torsion angle psi in peptides with isolated uniformly labeled residues. *J. Am. Chem. Soc.* **125,** 11,828–11,829.
27. Roths, T. Marth, M., Weese, J., and Honerkamp, J. (2001) A generalized regularization method for nonlinear ill-posed problems enhanced for nonlinear regularization terms. *Comput. Phys. Commun.* **139,** 279–296.
28. Hansen, P. C. (1994) Regularization Tools: A Matlab package for analysis and solution of discrete ill-posed problems. *Numerical Algorithms* **6,** 1–35.
29. Keyvanloo, A. (2004). Extracting Ramachandran torsional angle distributions from 2D NMR data using Tikhonov regularization. M.Sc. thesis, Brock University, St. Catharines, Ontario.

9

Statistical Thermodynamics Through Computer Simulation to Characterize Phospholipid Interactions in Membranes

Mihaly Mezei and Pál Jedlovszky

Summary

This chapter describes the major issues that are involved in the statistical thermodynamics of phospholipid membranes at the atomic level. The ingredients going into models of lipid bilayers are summarized: force fields, representation of long-range interactions, and boundary conditions. Next, the choice of thermodynamic ensembles, and the two main options for the generation of a representative sample of configurations: molecular dynamics and Monte Carlo are discussed. The final issue that is dealt with describes the various ways the generated ensembles can be analyzed.

Key Words: Ewald sum; force field; free-energy profile; molecular dynamics; Monte Carlo; Voronoi tesselation.

1. Introduction

Statistical thermodynamic description of a system recognizes the fact that the behavior exhibited by the system cannot be explained by a single state. Instead, the system's behavior is the result of its sampling an ensemble of states. The fundamental result of statistical thermodynamics is the characterization of such ensembles in terms of the Boltzmann distribution. Thermodynamic description of a system can use different sets of independent variables. Once the independent variables are set, other variables are determined by various thermodynamic relations (e.g., the equation of state). Each choice defines a different set of such relationships. Corresponding to each choice of independent variables there is a statistical thermodynamic ensemble, with their respective formalism. Whereas in the infinite system size limit, the results are the same irrespective of the ensemble chosen, for finite sizes they can give answers that differ by an amount that is proportional to $1/N$, with N being the number of particles in the system. Because the formalism is different for each ensemble, the choice is usually governed by computational convenience.

Although analytical theories exist for the characterization of ensembles of simple systems, systems of the complexity of lipid membranes are not amenable to such treatment without extreme simplification. However, such complex systems are amenable to be modeled in full atomic detail using computer simulations. This section presents the various steps involved in characterizing a phospholipid membrane using computer simulation.

2. Construction of the System

The construction of a model for a phospholipid membrane involves several choices, each involving some trade-off. The first choice is the representation of intermolecular energies and/or forces. Although it would be desirable to use quantum-mechanical techniques, they are still prohibitively expensive. The currently favored choice is the use of molecular mechanical force fields treating nonbonded interactions in a pair-wise additive manner and describing the intramolecular interactions with bond stretching and bending as well as torsional terms, although there exist force fields that also include cooperative terms, usually represented as polarization energy. Statistical treatment in general assumes a large enough sample that the average behavior of the sample is representative of the full-size system. The larger the number of molecules in the model, the better the representation, yet the calculation is more expensive. The accuracy of a model with limited number of molecules is increased significantly by the use of periodic boundary conditions: a basic cell containing the system is surrounded by periodic replicas in all three dimensions. This device eliminates surface effects at the expense of introducing artifactual periodicity into the model.

2.1. Force Field

2.1.1. All-Atom Representation

Molecular mechanical force fields express the energy of the system $E(X^N)$ as a sum of several terms and—when needed—calculate the force acting on each atom as the gradient of this energy:

$$E(X^N) = E_{NB} + E_{14} + E_{BOND} + E_{ANG} + E_{TOR} \qquad (1)$$

where E_{NB} is the nonbonded energy, summed over all pairs of atoms X_i and X_j, separated by the distance r_{ij}, that are on different molecules or in the same molecule but separated by more than three bonds, E_{14} is the nonbonded energy summed over all pairs of atoms separated by exactly three bonds, and E_{BOND}, E_{ANG}, and E_{TOR} are the intramolecular energies summed over all bonds, bond angles, and torsions, respectively. The nonbonded term is usually given in the form

$$E_{NB}(X_i, X_j) = 4\varepsilon \left[\left(\frac{\sigma}{r_{ij}}\right)^{12} - \left(\frac{\sigma}{r_{ij}}\right)^{6} \right] + \frac{q_i q_j}{r_{ij}} \qquad (2)$$

where ε_{ij} and σ_{ij} are the so-called Lennard–Jones parameters, representing the depth of the attraction owing to dispersion forces and the extent of exchange repulsion, respectively; q_i and q_j are the partial charges assigned to atoms i and j to represent the electrostatic interaction between them. Although some force fields assign ε_{ij} and σ_{ij} values for different *pairs* of atom types, most assign values for each atom type and obtain ε_{ij} and σ_{ij} as a combination of the two. The combination rules either involve calculating the geometric mean of both (e.g., in the OPLS force field *[1]*) or the so-called Lorentz–Berthelot rule: arithmetic mean for σ_{ij} and geometric mean for ε_{ij} (e.g., in force fields CHARMM *[2]* and AMBER *[3]*). Partial charges are either obtained from empirical rules (*see* **ref. 4**) or from *ab-initio* calculations, using a fitting procedure that finds partial charges by ensuring the best reproduction of the electric field around a molecule. In general, the Lennard–Jones parameters are established independently of the molecule the atom is in; partial charges are assigned for each molecule.

It is important to keep in mind two facts that may eventually result in fundamental reparametrization of nonbonded interactions. First, this form, although well established (even entrenched), owes its existence to the necessity of saving computational time at the expense of introducing too steep repulsion. Second, whereas the three terms appear to neatly represent the physics of three different types of interactions (exchange repulsion, dispersion, and electrostatics), when performing a least-squares fit to actual data the matrix obtained is usually nearly singular, indicating that the functions proportional to r^{-12}, r^{-6}, and r^{-1} are nearly linearly dependent. Besides the practical problem of having to deal with nearly singular matrices this means that the coefficients derived will contain contributions from different types of interactions. As a result, the transferability of the parameters suffers. E_{14} is represented with the same functional form as E_{NB}, but either with a different set of nonbonded parameters (e.g., in CHARMM) or applying an overall correction factor to both the Lennard–Jones and the electrostatics part (e.g., in AMBER).

E_{BOND} and E_{ANG} are generally represented with harmonic terms:

$$E_{BOND} = k_{ij}^{b} \left(r_{ij} - r_{ij}^{0} \right)^{2}, \qquad (3)$$

and

$$E_{ANG} = k_{ij}^{a} \left(\alpha_{ijk} - \alpha_{ijk}^{0} \right)^{2} \qquad (4)$$

or

$$E_{ANG} = k_{ik}^{UB} \left(r_{ik} - r_{ik}^{0} \right)^{2}, \qquad (5)$$

where i and j are the atoms forming the bond of length r_{ij}, i, j, and k are the atoms forming the bond angle α_{ijk}, the superscript 0 refers to the equilibrium value, and the parameters k^a, k^b, and k^{UB} are the respective harmonic strengths. The second form of E_{ANG} is usually referred to as the Urey–Bradley term. E_{TOR} represents in general two types of terms. The contribution of the conformational state of a bond to the energy of the molecule is usually expressed as a trigonometric function of the torsion angle δ:

$$E_{TOR} = k_{ijkl} \, [1 + \cos(n_{ijkl} \delta_{ijkl} + \delta_{ijkl}^{0})], \qquad (6)$$

where the parameter k_{ijkl} represents the strength of the interaction, and the parameters n_{ijkl} and δ_{ijkl}^{0} depend on the type of the bond. The other type is called improper torsion and is used to enforce either the chirality of an atom or to keep a bond in a plane (e.g., in the case of an aromatic ring). For an atom k with bonded neighbors i, j, and l, the improper torsion is a harmonic function of the angle between the planes formed by atoms i, j, and k and by atoms j, k, and l. It cannot be emphasized enough that the various terms of each extensively used force field have evolved as a whole, and mixing terms from various sources is likely to lead to inferior results.

2.1.2. Simplified Lipid Representation

There have been efforts to reduce the computational expense by introducing simplifications into the all-atom representation in such a way that the essence of the interactions is conserved. Foremost among such simplification is the simplified treatment of hydrogens. Both CHARMM and AMBER have parameter sets wherein the apolar hydrogens have been mapped onto the carbon atom they are bonded to. In this treatment, there will be different carbon atom types depending on how many hydrogens are mapped. An intermediate solution was presented by the GROMACS force field *(5)* that introduced the concept of frozen groups: hydrogens do appear explicitly but their movement is not independent of the heavy atom they are bonded to.

As even without explicit hydrogens the time-scale available for simulating lipid bilayers is generally inadequate to study rare events, such as the exchange of lipids between bilayers (also called "flip–flop transitions"), further simplifications have been introduced *(6–9)*. These models concatenate the headgroup into a few hydrophilic centers and replace the hydrophobic tails with a few centers connected with a harmonic spring. Such models are able to reproduce even the spontaneous formation of the membrane bilayer within reasonable computer time *(7)*.

2.2. Periodic Boundary Conditions

The most important property required of a simulation cell used under the traditional periodic boundary conditions is that it can be used to pack the three-dimensional space by appropriately translated copies of itself without leaving void space. The conceptually simplest of such shape is the cube. However, as the periodic system is used only to avoid having to introduce a surface, the effect of periodicity should be minimized. In simulating general solutions, this calls for a shape that has the largest inscribed sphere for a given volume and led to the introduction of rhombic dodecahedron and truncated octahedron. In modeling lipid bilayers the distance between layer images should be kept as large as possible because the concerted effect of a bilayer is much stronger than the interaction between individual lipids in the same layer. This led to the use of prism shape cells for lipid bilayers wherein the axis of the prism is along the bilayer normal. Consistent with the use of prism, one can still use a cross-section that has the largest inscribed circle for a given area, leading to the choice of hexagonal prism *(10)*.

Simulations of lipid bilayers can take advantage of periodic systems wherein the replicas of the simulation cell filling the space seamlessly are generated by translation and rotation. Dolan et al. *(11)* have shown that using either $P2_1$ or Pc symmetry the neighboring box will contain an image of the *opposite* layer. Under these symmetries the two layers can end up exchanging lipids *without actually flipping over* because a lipid leaving the cell at one side will cause its image to re-enter the simulation cell in the opposite layer. This provides a computationally efficient way to equilibrate the two layers of a membrane (an important task if the two layers have different guests embedded into them). Without using one of these nontraditional periodic boundary conditions, such an equilibration requires either the direct exchange of lipids between the two layers or the separate determination of the requisite number of lipids based on the area/headgroup of the lipids and the guest. The first solution is computationally impractical, whereas the reliability of the second is questionable. Thus, it is somewhat of a surprise that use of these boundary conditions has not been widely adopted for the modeling of lipids with proteins embedded, as witnessed by a recent review on such simulations *(12)*.

2.3. Treatment of Long-Range Interactions

In general, the energy of interaction between two atoms decreases with the distance between them. Thus, significant savings in computer time can be achieved by treating interactions between distant atoms separately from interactions between pairs closer to each other. One option is to set them to zero when the distance exceeds some predefined threshold, usually called cutoff. The other option is to use a simplified representation of interactions between distant pairs. In the context of simulating lipid membranes this takes the form of using a formalism to obtain the interaction with simplified forms of all periodic cells, extending to infinity, realized by a construct called Ewald sum (*see* **Subheading 2.3.2.**).

2.3.1. Cutoffs

If interactions among all atoms of the system were to be calculated, the computation cost (even assuming pair-wise additive potentials) would be proportional to the *square* of the number of atoms. However, the interactions die off as the distance between them increases, so such a procedure would result in calculating many near-zero quantities. This observation prompted the introduction of cutoffs: a distance threshold beyond which all interaction energies are assumed to be zero. However, this procedure is not without pitfalls. Besides the obvious problem of being "greedy" and choosing too short of a threshold, there are the problems created by the discontinuity at the cutoff distance (causing artifactual heating during molecular dynamics runs) and the neglect of contributions from the small but numerous distant pairs that can add up to a significant amount even when the choice of threshold was not "greedy." The discontinuity, only affecting adversely molecular dynamics simulations, can be eliminated by the introduction of a so-called switching function that continuously changes the function to be cutoff at zero over a finite interval. The cumulative contribution of pairs beyond the distance R_C whose interaction is proportional to $1/r^k$ has the general form of

$$\int_{R_C}^{\infty} \frac{C}{r^k} r^2 dr = \frac{C}{k-3} \left[\infty^{-k+3} - R_C^{-k+3} \right] \tag{7}$$

For large enough R_C this provides significant contribution only if the interactions die off slower than $1/r^4$. For $k = 3$ (as is the case for dipole–dipole interactions) the integral will depend on the way the triple integration is carried out, i.e., on the shape of the system as it is extended to infinity. For $k < 3$ (as is the case of charge–charge and charge–dipole interactions) the integral diverges for sure. To avoid such problem it has been recognized early on *(13)* that, whenever possible, atoms should be grouped into neutral clusters and the cutoff between any two atoms should be based on the distance between the cluster centers.

2.3.2. Ewald Sum

In general the total dipole moment of a simulation cell is nonzero. Thus, the electrostatic interactions between a simulation cell and its periodic replicas can add up to a significant amount. However, the summation of these terms is nontrivial: the resulting infinite series is only conditionally convergent. As a consequence, the final sum depends on the order of summation, just as the integral of the distant dipolar contributions depend on the shape of the system being integrated to infinity. Ewald *(14)* introduced a technique that calculates the dipolar sum as two absolute convergent series, one of them in the reciprocal space. The relation between the Ewald sum and the summation order has been analyzed by Campbell *(15)*. Use of the Ewald sum has been facilitated by the introduction of the particle-mesh technique *(16)* that significantly reduced its computational complexity. However, note that its use corresponds to a system on infinite stack of bilayers (separated by water layers) instead of a single bilayer. To avoid this artifact, the Ewald technique has been extended to systems that are periodic in only two of the three spatial dimensions *(17)*.

3. Generation of Boltzmann Sample of Configurations

There are three different decisions that have to be made when establishing the procedure for generating a Boltzmann-weighted ensemble: (1) the choice of thermodynamic ensemble, (2) the method of sampling in the ensemble chosen, and (3) as all methods of sampling obtain successive configurations from the previous one, the generation of the initial configuration.

3.1. Thermodynamic Ensembles

The thermodynamic ensembles most frequently used include the canonical (N, V, T), microcanonical (N, V, E), isothermal–isobaric (N, p, T), and grand–canonical (μ, V, T) ensembles. For modeling membrane systems sometimes the surface tension is also included as an additional variable, leading to simulations in the ensemble (N, p, γ, T) *(18,19)*. The choice of the ensemble is made based on the importance of which thermodynamic property has to be guaranteed to give the right (i.e., experimental) value and the sampling advantage a particular ensemble offers. By setting V and N constant, the density (and for lipid bilayers, the area/headgroup) can be set to the desired value but the pressure will be obtained from the simulation, and because of the approximate nature of the force field, it cannot be guaranteed to turn out to be 1 atm. However, for heterogeneous systems the density is a complex function of the components, and assuming the incorrect value may lead to the appearance of large voids (bubbles) in the system.

Classical molecular dynamics corresponds to sampling in the microcanonical ensemble, but the technique has been generalized to other ensembles by including an additional fictitious degree of freedom. The choice of the ensemble also affects the sampling efficiency. Using constant p (and γ) requires periodic change in the volume (and in the cross-section of the cell), helping the system to cross barriers. Using constant μ requires the change in the number of particles (i.e., insertions and deletions) and this helps equilibration between different pockets or semipockets of the system in general and speeds up the penetration of the solvent by an order of magnitude *(10)*. Whereas successful insertions into condensed phase systems are generally rare, the cavity-biased technique made efficient use of the (μ, V, T) ensemble feasible *(20,21)*.

3.2. Method of Sampling

Currently, two major classes of methods are used for sampling configurations of condensed phases: molecular dynamics and Monte Carlo (MC). Molecular dynamics takes advantage of the fact that systems following Newton's law of motion will sample a Boltzmann-distributed ensemble, whereas MC methods use the mathematical construct called Markov chain that can also generate a Boltzmann-distributed ensemble. From a mathematical point of view, the MC approach solves a problem with weaker restrictions than molecular dynamics because satisfying Newton's law of motion is a sufficient but not necessary condition for the generated set to follow a Boltzmann distribution. However, current practice favors molecular dynamics as it was found to work well and the few realizations of MC attempted so far have not proven to be superior to it. However, it is our belief that the potential of the MC approach has not been fully exploited yet *(22)*. Note also that simulations can combine the two techniques to exploit the respective advantages of each *(23–25)*.

3.2.1. Molecular Dynamics

The large number of degrees of freedom in a system of solvated lipid bilayers implies that Newton's law of motion has to be solved by numerical quadratures. There are several of such quadratures developed (*see* **ref. 26**), but each is limited in the time step they can make in order to maintain conservation of energy. Some increase in the time step is possible if the highest frequency motions (i.e., the vibrations involving hydrogen atoms) are frozen. This is usually achieved by applying a constraint on the bond lengths (usually implemented by the SHAKE method *[27]*) involving hydrogens, allowing the increase of the time step from the customary 1–2 fs.

3.2.2. Monte Carlo

The MC technique used for simulation of atomic and molecular assemblies, usually referred to as the Metropolis method *(28)*, is based on the construction of a Markov chain whose limiting distribution π is the Boltzmann distribution in the ensemble under consideration. This requires the construction of a transition matrix P such that $\pi = P\pi$. P is constructed with the help of another matrix Q whose elements q_{ij} form a transition matrix of an irreducible Markov chain on the same states. The q_{ij} matrix elements are usually referred to as the *a priori* transition probabilities. The simulation proceeds from state i by selecting a candidate state j with probability q_{ij} that is accepted with probability *(26)*.

$$P_{acc} = \min\left(1, \frac{\pi_j \, q_{ij}}{\pi_i \, q_{ji}}\right) \tag{8}$$

The original Metropolis method uses the particular choice of $q_{ij} = q_{ji}$. In general, the shift from state i to state j changes a small part of the system, to avoid having too small (π_j/π_i) in the above equation. The aim of the sampling technique design is to select the largest possible change in the system that is still likely to be accepted. This is achieved by the judicious choice of coordinates for the change and for the magnitude and direction of change. For the sampling of polymer conformations, such as lipids, this usually means the use of torsional coordinates. The choice of sampling techniques is only limited by the practitioners' imagination. Here three such techniques that have been applied to lipid simulations or are considered having strong potentials are briefly described.

3.2.2.1. CONFIGURATIONAL BIAS MC

Configurational bias MC generates new elements of the Markov chain by growing (possibly only part of) a polymer chain unit by unit, adding each by considering the other atoms in the vicinity to minimize steric overlap; the bias introduced by these choices can be controlled and corrected for *(29,30)*. The most attractive feature of this technique is the significantly reduced correlation between successive accepted members of the Markov chain. However, as the system becomes denser, the probability of acceptance becomes progressively smaller.

3.2.2.2. EXTENSION-BIAS MC

Extension biasing is based on the observation that the maximum atomic displacement resulting from a given torsion angle change depends on the shape of the part of a molecule that is affected by the change in that torsion angle; it scales the torsion angle stepsize parameter with the inverse square root of the largest distance from the torsion axis *(10)*. So far it was applied to torsions that move the full length of the polymer chain, but it is equally applicable to local moves affecting only a polymer segment *(31,32)* (although it has not been done yet).

3.2.2.3. SCALED COLLECTIVE VARIABLES

Sampling in terms of the so-called scaled collective variables *(33)* is a well-established technique that finds a special linear combination of natural variables (e.g., torsion angles) that result in significantly better sampling. The coefficients are obtained from the eigenvectors and eigenvalues of the Hessian of the system, with a significant additional computational expense being involved. Although (to the authors knowledge) it has not been applied to lipid systems, a novel and efficient implementation could calculate the Hessians separately for each lipid instead of calculating a single Hessian for all of the lipids, as this reduces the computational complexity of calculating Hessians by a full order of magnitude.

3.3. Initial Configuration

With lipid simulations becoming more and more widespread, reasonably well-equilibrated initial configurations can be obtained from earlier simulations of the same or similar systems. Lacking such "crutch," condensed phase simulations start either from a randomly generated configuration or from a crystalline state (corresponding to a start at infinite or zero absolute temperature, respectively). For simulations of a lipid bilayer these choices are likely to be both extremely inefficient because the time-scales of both lateral diffusion and orientational relaxation in a bilayer indicate that providing a well-equilibrated system at ambient or physiological temperatures would require very long simulations. In spite of this dire prognosis such calculations have been performed successfully *(34)*. A well-established strategy to build a new bilayer *(35)* is random selection of phospholipids from a preequilibrated and prehydrated library of DPPC generated by MC simulations in the presence of a mean field *(36,37)*.

4. Analysis of the Generated Ensemble
4.1. Density Profiles

In characterizing the average structure of the membrane at different regions along its normal axis probably the most important tool is the density profile of various atoms or atomic groups. The calculation of density profiles is a rather straightforward task: the average occurrence of the atoms of interest per configuration has to be counted in different lateral slices of the membrane and divided by the volume of the slice. To obtain better statistics it is generally advised to average the obtained profiles over the two layers of the membrane. Conversely, the comparison of the density profiles in the two separate layers can provide information on the sampling efficiency of the equilibrium structure of the membrane layers.

To get an overall view about the distribution of the atoms across the membrane the mass and electron density profiles of the system are of particular importance. The relevance of the calculation of electron density profile in simulations is stemmed from the fact that it can also be measured in X-ray diffraction experiments, and hence it is one of the important properties of the system through which the quality of the simulation can directly be tested against experimental data. The general shape of the mass and electron density profiles in phospholipid membranes shows that the highest and lowest density part of the membrane is the region of the headgroups and the middle region of the chain terminal CH_3 groups, respectively. The distance of the density peaks corresponding to the two headgroup regions is a simple measure of the membrane thickness. Furthermore, the absence or presence of a thin but deep minimum in the middle of membranes of more than one component can provide information on whether all the components can extend to the middle of the bilayer or not *(38,39)*.

Besides the overall density profiles of the system, the density profile of various atomic groups can also provide valuable information on the organization of the membrane structure along its normal axis. Thus, the density profile of the water molecules shows how deeply water can penetrate into the bilayer. The evolution of this profile during the simulation is a rather sensitive measure of the equilibration of the system *(10)*. This profile can also easily be converted to the free energy profile of water up to the position beyond which the obtained water density reaches zero. Although any meaningful analysis of such density profiles usually requires their determination across the entire membrane, the distance range within which the rather computationally demanding calculation of the water free energy profile has to be performed can substantially be reduced in this way *(40)*.

The density profiles of various atomic groups can also be considered when the system investigated has to be divided into separate regions. Such a partitioning of the membrane can be useful in the analysis of various structural or dynamical features that show considerable variation along the membrane normal axis owing to the large inhomogeneity of the system. In this way, the properties of interest can be separately analyzed in the separate membrane regions. The usual partitioning of the membrane divides it to regions dominated by the hydrocarbon chains, the headgroups, and the hydrating water molecules, respectively. However, based on the density profiles of various atomic groups, more sophisticated yet physically still meaningful dividing schemes can also be derived *(40,41)*.

The comparison of the density distribution of different atoms or atomic groups along the membrane normal axis can also give some indication on the average alignment of various parts of the lipid molecules. Thus, for instance, the comparison of the density profiles of the P and N atoms *(18,24,38–40,42)*, or the negatively charged phosphate and positively charged choline groups *(43–47)* in pure membranes of phosphatidylcholine lipids has revealed that, although the N atom or choline group density peak is somewhat farther from the middle of the bilayer than that of the P atoms or phosphate groups, the two density peaks largely overlap with each other. This observation indicates that the dipole vector of the lipid headgroups (roughly pointing from the P to the N atom) is, on average, directed more likely toward the aqueous phase than toward the membrane interior; however, this preference is rather weak. This conclusion, drawn solely from the behavior of density profiles has also been confirmed by detailed analyses of the headgroup structure *(18,38,39,48–51)*.

4.2. Order Parameter

In the liquid crystalline (L_α) phase of the membrane the conformation of the hydrocarbon tails of the lipid molecules is disordered. This conformational disorder can be characterized by various different quantities, such as the average tilt angle of each of the C–C bonds along the hydrocarbon tails *(39,52)* or the ratio of the appearance of the *trans* and *gauche* alignments of the dihedral angles around these bonds *(42,53–55)*. However, the vast majority of the studies calculate the profile of the CH_2 group order parameter along the lipid tails for characterizing their conformational and orientational order. The importance of the determination of the order parameter profile in the simulation is that it can also be measured by nuclear magnetic resonance spectroscopy, and thus, it is another quantity through which the simulation results can be compared with experimental data.

The order parameter tensor of a CH_2 group is defined as

$$S_{ij} = \frac{\langle 3\cos\gamma_i \cos\gamma_j - \delta_{ij}\rangle}{2}, \qquad (9)$$

where indices i and j run through the x, y, and z axes of the local Cartesian frame fixed to the CH_2 group, γ is the angle formed by the corresponding axis of this frame with the membrane normal, δ_{ij} is the Kronecker δ, and $\langle....\rangle$ denotes ensemble averaging. The local frame is defined in such a way that its x-axis connects the two H atoms, the y-axis is the main symmetry axis of the CH_2 group, whereas the z-axis is perpendicular to the plane of the three atoms. In the case of using a force field of simplified lipid representation (i.e., when the entire CH_2 group is treated as a united atom) this frame has to be defined without knowing the orientation and geometry of the CH_2 group. However, a definition equivalent with the aforementioned one can still be given in this case. Thus, the z-axis connects the C atoms located before and after the CH_2 group of interest along the hydrocarbon chain, the y-axis is perpendicular to z and lays also in the plane containing the CH_2 group of interest and its two neighboring C atoms, whereas the x-axis is perpendicular to both y and z *(56)*.

The deuterium order parameter S_{CD} that is measurable by nuclear magnetic resonance spectroscopy is related to the elements of the order parameter tensor through the relation

$$S_{CD} = \frac{2S_{xx} + S_{yy}}{3} \qquad (10)$$

Because of the symmetry of the CH_2 group the S_{xy} and S_{xz} elements of the order parameter tensor are zero. Furthermore, isotropic rotation around the z-axis leads to the relations of $S_{yz} = 0$ and $S_{zz} = -2S_{xx} = -2S_{yy}$ *(57)* (the S_{zz} parameter is often referred to as S_{chain} as well.). In this case, the S_{CD} order parameter of a given CH_2 group can simply be calculated as

$$S_{CD} = \left\langle \frac{3}{2}\cos^2\alpha - \frac{1}{2}\right\rangle, \qquad (11)$$

where α is the angle formed by the C–H bond with the membrane normal. Conversely, a noticeable deviation of the obtained S_{zz} values from $-2S_{xx}$ or $-2S_{yy}$ indicates the existence of rotational anisotropy along the molecular axis joining two C atoms that are separated by two C–C bonds *(56)*.

4.3. Structure of the Headgroup Region

The structure of the dense headgroup region of the membranes, consisting of the polar part of the constituting lipid molecules, the waters penetrated deepest into the bilayer, and also fractions of the hydrocarbon chains is of key importance in determining the properties of the membrane. The headgroup structure is resulted from the delicate interplay between the lipid–lipid and lipid–water interactions. The overall organization of the polar lipid headgroups and waters hydrating them can be characterized by the electrostatic potential between the aqueous phase and the membrane interior, a quantity that can again be compared with experimental data, and its contributions because of the lipid and water molecules. The detailed description of the headgroup region structure includes the analysis of the lipid headgroups as well as the structure of the interfacial water.

4.3.1. Electrostatic Potential

The electrostatic potential difference between the aqueous phase and the interior of the membranes built up by neutral phospholipid molecules results from the microscopic separation of the center of the positive and negative charges owing to the orientational preferences of the lipid headgroups and water molecules. This electrostatic potential difference, often referred to as the dipole potential Φ, can be calculated using the Poisson equation at the distance Z from the middle of the bilayer along its normal axis as the double integral of the charge density profile $\rho_Q(Z)$:

$$\Phi(Z) = -\frac{1}{\epsilon_0} \int_0^Z dZ' \int_0^{Z'} \rho_Q(Z'') \, dZ'', \tag{12}$$

where ϵ_0 is the vacuum permittivity. It is not surprising that the contribution of the lipid and water molecules to this potential is of opposite sign: the orientation of the water dipole vectors is driven by the charge distribution of the lipid headgroups along the membrane normal axis. Owing to the fact that the positively charged group of the neutral phospholipid molecules is always located at the end of the headgroup chain, and hence they are, on average, farther from the bilayer center than the negatively charged phosphate group that is attached directly to the glycerol backbone, the lipid contribution to the electrostatic potential of the aqueous phase with respect to the membrane interior is positive. However, this positive potential is usually overcompensated by the negative potential contribution because of the preferential orientation of the water molecules, resulting in a net negative electrostatic potential value in the aqueous phase *(24,45,50,58,59)*, which is in agreement with existing experimental data *(60)*.

4.3.2. Lipid Headgroup Structure

Perhaps the most important property characterizing the structure of the lipid headgroups is the distribution of its tilt angle relative to the membrane normal (the orientation of the phosphorylcholine headgroup is often described by the vector pointing from its P to N atom, called the PN vector, see **refs. *18,38,39,48–51***). However, a detailed analysis of the headgroup structure requires a thorough investigation of the interactions acting between the headgroups as well. Because phospholipid molecules lack hydrogen bond donor H atoms, the most prevalent interactions acting between neighboring headgroups are charge pairing *(49)* and water bridging (i.e., when the two lipid headgroups are forming hydrogen bonds with the same, bridging water molecule) *(61)*. In addition, in the case of mixed membranes containing also H-donor molecules (e.g., cholesterol) hydrogen bonding can also occur between these molecules and phospholipids *(62)*. The presence and relative importance of these interactions as well as their details (e.g., participating atoms, equilibrium distance) can be analyzed in detail using the partial pair correlation function of appropriately chosen atom pairs *(40,42,47–49,61–66)*. Because the lateral packing of the molecules is mainly determined by the interactions acting between the headgroups, the detailed investigation of the local lateral structure (e.g., by Voronoi analysis) can also shed some light to the nature of the headgroup–headgroup interactions *(65,67,68)*. Furthermore, the relative arrangement of the neighboring headgroups can be described by the distribution of the angle formed by the vectors describing their orientation (usually the PN vector) *(66)*, whereas their spatial distribution around each other can well be characterized by the distribution of the angle formed by two neighboring headgroups (represented by the position of an appropriately chosen atom, e.g., P or N) around the central one.

4.3.3. Structure of the Interfacial Water

The change of the orientational order of the interfacial water molecules along the membrane normal axis can be characterized by the profiles (i.e., the average values obtained in different lateral membrane slices) of appropriately chosen orientational parameters *(51,58)*. Based on the behavior of these profiles the headgroup region can be divided into separate layers in which the full distribution of these orientational parameters can then be meaningfully analyzed *(58)*. In phospholipid membranes, as in other polar interfaces the most important orientational parameter in this respect is clearly the angle formed by the water dipole vector with the membrane normal axis. However, the description of the alignment of entire water molecules relative to the bilayer requires the introduction of other orientational parameters as well. In analyzing the interfacial orientation of water molecules it should be kept in mind that the unambiguous description of the orientational preferences of a rigid molecule relative to an external vector (e.g., the membrane normal) requires the calculation of the bivariate joint distribution of two independent orientational variables (e.g., the angular polar coordinates of the external vector in a local frame fixed to the individual molecules) *(69,70)*.

4.4. Analysis of Voids

The properties of the voids in lipid membranes are closely related to the key biological functions of the membranes. Thus, several small molecules of vital physiological importance (e.g., O_2, CO_2, NO, and so on) go through the membrane of the cells by passive transport. This diffusion process is obviously related to the properties of the voids in the membrane. Furthermore, some theories explain the phenomenon of anesthesia partly by changes in the void distribution of the membranes owing to the anesthetics that are dissolved in the membrane interior *(71)*. The properties of the voids to be calculated in order to thoroughly characterize the organization of the free volume in the system include the distribution of their size, shape, connectivity, and orientation in the different regions of the membrane. In defining voids, a distinction has to be made between the *empty* free volume (i.e., the entire space that is not covered by the atomic spheres) and the *accessible* free volume (i.e., the free volume pockets that are accessible for a spherical probe of a given size) *(72)*. Obviously, the determination of the accessible free volume requires the introduction of an extra parameter (i.e., the radius of the probe). In the limiting case of the probe radius of zero the accessible and empty free volumes become equivalent. Voids in the membrane can either be detected using a set of test points, or analytically using the Voronoi–Delaunay (VD) method *(73)*.

4.4.1. The Test Point Approach

In this method a large set of test points are generated (either randomly or along a grid) in the system, and the points that are farther from all the atoms of the system than a given limiting distance are marked. The ratio of these marked points and the total number of points generated provide immediately the fraction of the accessible free volume corresponding to the probe radius equal to the limiting distance used in the procedure *(45,68,72,74–76)*. However, the identification of the voids is a rather difficult task, because it has to be done on a system of marked and unmarked discrete points. A computationally efficient way of solving this problem involves union/find type algorithm that results in a tree structure containing the information on how the marked sites are connected to each other *(76)*. A more serious problem of the test point approach is that it introduces a certain numerical inaccuracy in the results. To keep this numerical inaccuracy sufficiently low, a large number of test points has to be used, which makes the entire analysis computationally rather costly. Furthermore, the computational cost of such calculations increases proportionally with the cube of the system size.

4.4.2. The VD Method

In analyzing voids in systems built up by large molecules, such as lipid membranes, a generalized variant *(77)* of the original VD method *(73)* has to be used. In this approach, the system is divided into cells, which are called the Voronoi S-regions. Each of these cells is associated with an atom, and covers the region of space every point of which is closer to the surface of this particular atom than to any other atom of the system. Obviously, these cells fill the space without gaps and overlaps. The faces, edges, and vertices of these cells are the loci of the spatial points that are equally close to the surface of two, three, and four atoms, respectively, and are closer to these atoms than to any other atom of the system. Thus, the vertices of the Voronoi S-regions are the centers of the empty interstitial spheres that can be inscribed between the atoms of the system. These spheres can be regarded as elemental cavities; any complex void present in the system is built up by them. Furthermore, each edge of a Voronoi S-region connecting two of its vertices represents the fairway passing through the narrow bottleneck between the nearest atoms, and can be characterized by the radius of this bottleneck. The set of the Voronoi S-network vertices and edges of the system are forming a network, called the Voronoi S-network that can be used to map the interatomic voids in the system. Thus, each void accessible by a given spherical probe can be represented by a connected cluster of the S-network edges, the bottleneck radius of all of which exceeds the radius of the probe. Knowing the position and radius of the atoms as well as of the elemental interatomic cavities (represented by the S-network vertices) the volume of the void can easily be calculated *(77)*. Although the determination of the Voronoi S-network vertices and edges requires rather sophisticated algorithms, this approach can detect the voids present in the system in an exact yet computationally less demanding way *(41,77,78)*.

4.5. Analysis of the Solvation

4.5.1. Solvation of Large Molecules

The ultimate goal of the lipid membrane simulations is to model the complex environment of the membranes of living cells, and thus, help in understanding their biological functions on the molecular level. However, presently available computer capacities only allow the simulation of a few solute molecules, for example, anesthetics *(79,80)*, coenzymes *(81)*, peptides *(82)*, oligonucleotides *(83)*, or a protein molecule *(12)* in a pure phospholipid membrane, or simulations of two component mixed membranes built up by phospholipid molecules as the main component and other natural amphiphils (e.g., cholesterol, *see* **refs.** *23,24,38,39,45,54,62,63,65,68,75,76,78*). Recently, Pandit et al. *(84)* have reported computer simulation of a three-component rafted membrane in which a domain of cholesterol and 18:0 sphingomyelin is embedded in the matrix of phospholipid molecules. These studies usually focus on the local *(23,65,76,82)* as well as overall *(24,38,39,45,68,76,78–84)* changes induced in the structure of the phospholipid membrane by the solutes, the specific (e.g., hydrogen bonding) interactions between the solute and phospholipid molecules *(62,63,65,83)*, as well as on the preferential position and diffusion of the solute in the membrane *(79–81)*.

4.5.2. Free Energy Profile of Small Molecules and Membrane Permeability

For a class of solvent molecules of biological relevance, i.e., small, neutral molecules of physiological importance (e.g., water, O_2, CO_2, NO, and so on), solvation in the membrane can be analyzed in considerably more detail than that of larger solutes. The biological role of these molecules requires their ability of passing through the membrane without the aid of any specific, membrane-bound proteins, and hence the profile of their solvation free energy across the membrane is of great importance. Such calculations *(40,74,75,85,86)* are usually performed by inserting the solute into a large set of test points in the system. Using the cavity insertion variant of the method *(74)*, when the test particle is only inserted into spherical cavities of the minimum radius of R_{cav}, the free energy profile $A(Z)$ along the membrane normal axis Z can be computed as

$$A(Z) = -k_B T \; [\ln\langle \exp(-U_{test}(Z))\rangle/k_B T + \ln\langle P_{cav}(Z)\rangle - 1] - pV/N, \qquad (13)$$

where k_B is the Boltzmann constant, N is the number of the particles, p, V, and T are the pressure, volume, and absolute temperature of the system, respectively, U_{test} is the interaction energy of the inserted test particle with the system, P_{cav} is the probability of finding a suitable empty cavity, and the brackets <….> denote ensemble averaging. P_{cav} is directly obtained in the calculation as the ratio of the number of gridpoint representing cavities to the total number of gridpoints in the cell. The calculation provides also the fraction of the accessible free volume for the probe of radius R_{cav} immediately. It should be noted that in the case of $R_{cav} = 0$ the P_{cav} probability is unity, and the original version of the particle insertion method *(87)* is given back. When the diffusion constant profile of the solute $D(Z)$ is also determined *(40,85,86)* (e.g., by the force correlation method, *see* **ref. 40**), the experimentally accessible permeability coefficient of the solute P can also be calculated *(40,85,86)* using the inhomogeneous solubility-diffusion model of Marrink and Berendsen *(85)*, as

$$\frac{1}{P} = \int_{Z_1}^{Z_2} \frac{\exp[A(Z)/k_B T]}{D(Z)} \, dZ. \qquad (14)$$

Acknowledgments

Pal Jedlovszky is supported by the Foundation for Hungarian Research and Higher Education as a Békésy György Fellow. P. J. acknowledges financial support of the Hungarian OTKA Foundation under Project no. T049673.

References

1. Jorgensen, W. L., Maxwell, D. S., and Tirado-Rives, J. (1996) Development and Testing of the OPLS All-Atom Force Field on Conformational Energetics and Properties of Organic Liquids. *J. Am. Chem. Soc.* **118,** 11,225–11,236.
2. Brooks, B. R., Bruccoleri, R. E., Olafson, B. D., States, D. J., Swaminathan, S., and Karplus, M. (1983) CHARMM: A program for macromolecular energy, minimization and dynamics calculation. *J. Comp. Chem.* **4,** 187–217, URL:http://www.charmm.org. April 1, 2007.
3. AMBER: Assisted model building with energy refinement, URL:http://www.amber.ucsf.edu/amber/amber.htm. April 1, 2007.
4. Gasteiger, J. and Marsali, M. (1980) Iterative partial equalization of orbital electronegativity—a rapid access to atomic charges. *Tetrahedron* **36,** 3219–3229.

5. GROMACS: The world's fastest molecular dynamics, URL:http://www.gromacs.org/. April 1, 2007.
6. Goetz, R. and Lipowsky, R. (1998) Computer simulation of bilayer membranes: Self-assembly and interfacial tension. *J. Chem. Phys.* **108**, 7397–7409.
7. Shelley, J. C., Shelley, M. Y., Reeder, R. C., Bandyopadhyay, S., and Klein, M. L. (2001) A Coarse Grain Model for Phospholipid Simulations. *J. Phys. Chem. B* **105**, 4464–4470.
8. Martí, J. (2004) A molecular dynamics transition path sampling study of model lipid bilayer membranes in aqueous environment. *J. Phys. Condens. Matter.* **16**, 5669–5678.
9. Marrink, S. J., de Vries, A. H., and Mark, A. E. (2004) Coarse Grained Model for Semiquantitative Lipid Simulations. *J. Phys. Chem. B* **108**, 750–760.
10. Jedlovszky, P. and Mezei, M. (1999) Grand canonical ensemble Monte Carlo simulation of a lipid bilayer using extension biased rotations. *J. Chem. Phys.* **111**, 10,770–10,773.
11. Dolan, E. A., Venable, R. M., Pastor, R. W., and Brooks, B. R. (2002) Simulations of membranes and other interfacial systems using $P2_1$ and Pc periodic boundary conditions. *Biophys. J.* **82**, 2317–2325.
12. Gumbart, J., Wang, Y., Aksimentiev, A., Tajkhorshid, E., and Schulten, K. (2005) Molecular dynamics simulations of proteins in lipid bilayers. *Curr. Opin. Struct. Biol.* **15**, 423–431.
13. Neumann, M. (1985) The dielectric constant of water. Computer simulations with the MCY potential. *J. Chem. Phys.* **82**, 5663–5672.
14. Ewald, P. (1924) Die Berechnung optischer und elektrostatischer Gitterpotentiale. *Ann. Phys.* **64**, 253–287.
15. Campbell, E. S. (1963) Existence of a "well defined" specific energy for an ionic crystal; justification of Ewald 's formulae and their use to deduce equations for multipole lattices. *J. Phys. Chem. Solids.* **24**, 197–208.
16. Essman, U., Perera, L., Berkowitz, M. L., Darden, T., Lee, H., and Pedersen, L. G. (1995) A smooth particle mesh Ewald method. *J. Chem. Phys.* **103**, 8577–8593.
17. Yeh, I. C. and Berkowitz, M. L. (1999) Ewald summation for systems with slab geometry. *J. Chem. Phys.* **111**, 3155–3162.
18. Chiu, S. W., Clark, M., Balaji, V., Subramaniam, S., Scott, H. L., and Jakobsson, E. (1995) Incorporation of Surface Tension into Molecular Dynamics Simulation of an Interface. A Fluid Phase Lipid Bilayer Membrane. *Biophys. J.* **69**, 1230–1245.
19. Tieleman, D. P. and Berendsen, H. J. C. (1996) Molecular dynamics simulation of a fully hydrated dipalmitoylphosphatidylcholine bilayer with different macroscopic boundary conditions and parameters. *J. Chem. Phys.* **103**, 4871–4880.
20. Mezei, M. (1980) A cavity-biased (T,V,μ) Monte Carlo method for the computer simulation of fluids. *Mol. Phys.* **40**, 901–906.
21. Mezei, M. (1987) Grand-canonical ensemble Monte Carlo simulation of dense fluids: Lennard-Jones, soft spheres and water. *Mol. Phys.* **61**, 565–582. Erratum (1989) **67**, 1207–1208.
22. Mezei, M. (2002) On the potential of Monte Carlo methods for simulating macromolecular assemblies, in *Third International Workshop for Methods for Macromolecular Modeling Conference Proceedings*, (Gan, H. H. and Schlick, T., eds.), Springer, New York, pp. 177–196.
23. Chiu, S. W., Jakobsson, E., and Scott, H. L. (2001) Combined Monte Carlo and Molecular Dynamics Simulation of Hydrated Lipid-Cholesterol Lipid Bilayers at Low Cholesterol Concentration. *Biophys. J.* **80**, 1104–1114.
24. Chiu, S. W., Jakobsson, E., and Scott, H. L. (2001) Combined Monte Carlo and molecular dynamics simulation of hydrated dipalmitoyl-phosphatidylcholine–cholesterol lipid bilayers. *J. Chem. Phys.* **114**, 5435–5443.
25. Ebersole, B. J., Visiers, I., Weinstein, H., and Sealfon, S. C. (2003) Molecular basis of partial agonism: orientation of indoleamine ligands in the binding pocket of the human serotonin 5-HT2A receptor determines relative efficacy. *Mol. Pharm.* **63**, 36–43.

26. Allen, M. P. and Tildesley, D. J. (1986) *Computer simulation of liquids*. Clarendon Press, Oxford.
27. Ryckaert, J. P., Ciccotti, G., and Berendsen, H. J. C. (1977) Numerical integration of cartesian equation of motion of a system with constraints; molecular dynamics of n-alkanes. *J. Comp. Phys.* **23**, 327–341.
28. Metropolis, N. A., Rosenbluth, A. W., Rosenbluth, M. N., Teller, A. H., and Teller, E. (1953) Equation of state calculation by fast computing machines. *J. Chem. Phys.* **21**, 1087–1092.
29. Rosenbluth, M. N. and Rosenbluth, A. W. (1955) Monte Carlo calculation of the average extension of molecular chains. *J. Chem. Phys.* **23**, 356–359.
30. Siepman, I. and Frenkel, D. (1992) Configurational bias Monte Carlo: a new sampling scheme for flexible chains. *Mol. Phys.* **75**, 59–70.
31. Dodd, L. R., Boone, T. D., and Theodorou, D. N. (1993) A concerted rotation algorithm for atomistic Monte Carlo simulation of polymer melts and glasses. *Mol. Phys.* **78**, 961–996.
32. Mezei, M. (2003) Efficient Monte Carlo sampling for long molecular chains using local moves, tested on a solvated lipid bilayer. *J. Chem. Phys.* **118**, 3874–3879.
33. Noguti, T. and Go, N. (1985) Efficient Monte Carlo method for simulation of fluctuating conformations of native proteins. *Biopolymers* **24**, 527–546.
34. Marrink, S. J., Tieleman, D. P., and Mark, A. E. (2000) Molecular Dynamics Simulations of the Kinetics of Spontaneous Micelle Formation. *J. Phys. Chem. B* **104**, 12,165–12,173.
35. URL:http://thallium.bsd.uchicago.edu/Roux Lab/; April 1, 2007.
36. De Loof, H. D., Harvey, S. C., Segrest, J. P., and Pastor, R. W. (1991) Mean field stochastic boundary molecular dynamics simulation of a phospholipid in a membrane. *Biochemistry* **30**, 2099–2113.
37. Pastor, R. W., Venable, R. M., and Karplus, M. (1991) Model for the structure of the lipid bilayer. *Proc. Natl. Acad. Sci. USA* **88**, 892–896.
38. Smondyrev, A. M. and Berkowitz, M. L. (1999) Structure of Dipalmitoylphosphatidylcholine/Cholesterol Bilayer at Low and High Cholesterol Concentrations: Molecular Dynamics Simulation. *Biophys. J.* **77**, 2075–2089.
39. Jedlovszky, P. and Mezei, M. (2003) Effect of Cholesterol on the Properties of Phospholipid Membranes. 1. Structural Features. *J. Phys. Chem. B* **107**, 5311–5321.
40. Marrink, S. J. and Berendsen, H. J. C. (1994) Simulation of Water Transport through a Lipid Membrane. *J. Phys. Chem.* **98**, 4155–4168.
41. Rabinovich, A. L., Balabaev, N. K., Alinchenko, M. G., Voloshin, V. P., Medvedev, N. N., and Jedlovszky, P. (2005) Computer simulation study of intermolecular voids in unsaturated phosphatidylcholine lipid bilayers. *J. Chem. Phys.* **122**, 084906. 12 pages.
42. Hyvönen, M. T., Rantala, T. T., and Ala-Korpela, M. (1997) Structure and Dynamic Properties of Diunsaturated 1-Palmitoyl-2-Linoleoyl-*sn*-Glycero-3-Phosphatidylcholine Lipid Bilayer from Molecular Dynamics Simulation. *Biophys. J.* **73**, 2907–2923.
43. Heller, H., Schaefer, M., and Schulten, K. (1993) Molecular Dynamics Simulation of a Bilayer of 200 Lipids in the Gel and Liquid-Crystal Phases. *J. Phys. Chem.* **97**, 8343–8360.
44. Tu, K., Tobias, D. J., and Klein, M. L. (1995) Constant Pressure and Temperature Molecular Dynamics Simulation of a Fully Hydrated Liquid Crystal Phase Dipalmitoylphosphatidylcholine Bilayer. *Biophys. J.* **69**, 2558–2562.
45. Tu, K., Klein, M. L., and Tobias, D. J. (1998) Constant-Pressure Molecular Dynamics Investigation of Cholesterol Effects in a Dipalmitoylphosphatidylcholine Bilayer. *Biophys. J.* **75**, 2147–2156.
46. Zubrzycki, I. Z., Xu, Y., Madrid, M., and Tang, P. (2000) Molecular dynamics simulation of a fully hydrated dimyristoylphosphatidylcholine membrane in liquid-crystalline phase. *J. Chem. Phys.* **112**, 3437–3441.
47. Saiz, L. and Klein, M. L. (2001) Structural Properties of a Highly Polyunsaturated Lipid Bilayer from Molecular Dynamics Simulations. *Biophys. J.* **81**, 204–216.

48. Husslein, T., Newns, D. M., Pattnaik, P. C., Zhong, Q., Moore, P. B., and Klein, M. L. (1998) Constant pressure and temperature molecular-dynamics simulation of the hydrated diphytanolphosphatidylcholine lipid bilayer. *J. Chem. Phys.* **109,** 2826–2832.
49. Pasenkiewicz-Gierula, M., Takaoka, Y., Miyagawa, H., Kitamura, K., and Kusumi, A. (1999) Charge Pairing of Headgroups in Phosphatidylcholine Membranes: A Molecular Dynamics Simulation Study. *Biophys. J.* **76,** 1228–1240.
50. Saiz, L. and Klein, M. L. (2002) Electrostatic interactions in neutral model phospholipid bilayer by molecular dynamics simulations. *J. Chem. Phys.* **116,** 3052–3057.
51. Åman, K., Lindahl, E., Edholm, O., Håkansson, P., and Westlund, P. O. (2003) Structure and Dynamics of Interfacial Water in an L_α Phase Lipid Bilayer from Molecular Dynamics Simulations. *Biophys. J.* **84,** 102–115.
52. Rabinovich, A. L., Ripatti, P. O., Balabaev, N. K., and Leermakers, F. A. M. (2003) Molecular dynamics simulations of hydrated unsaturated lipid bilayers in the liquid-crystal phase and comparison to self-consistent field modelling. *Phys. Rev. E* **67,** 011909. 14 pages.
53. Snyder, R. G., Tu, K., Klein, M. L., Mendelssohn, R., Strauss, H. L., and Sun, W. (2002) Acyl Chain Conformation and Packing in Dipalmitoylphosphatidylcholine Bilayers from MD Simulation and IR Spectroscopy. *J. Phys. Chem. B.* **106,** 6273–6288.
54. Hofsäß, C., Lindahl, E., and Edholm, O. (2003) Molecular Dynamics Simulations of Phospholipid Bilayers with Cholesterol. *Biophys. J.* **84,** 2192–2206.
55. Shinoda, W., Mikami, M., Baba, T., and Hato M. (2003) Molecular Dynamics Study on the Effect of Chain Branching on the Physical Properties of Lipid Bilayers: Structural Stability. *J. Phys. Chem. B* **107,** 14,030–14,035.
56. López-Cascales, J. J., García de la Torre, J., Marrink, S. J., and Berendsen, H. J. C. (1996) Molecular dynamics simulations of a charged biological membrane. *J. Chem. Phys.* **104,** 2713–2720.
57. van der Ploeg, P. and Berendsen, H. J. C. (1982) Molecular dynamics simulation of a bilayer membrane. *J. Chem. Phys.* **76,** 3271–3276.
58. Jedlovszky, P. and Mezei, M. (2001) Orientational Order of the Water Molecules Across a Fully Hydrated DMPC Bilayer. A Monte Carlo Simulation Study. *J. Phys. Chem. B* **105,** 3614–3623.
59. Pandit, S. A., Bostick, D., and Berkowitz, M. L. (2003) An algorithm to describe molecular scale rugged surfaces and its application to the study of a water/lipid bilayer interface. *J. Chem. Phys.* **119,** 2199–2205.
60. Gawrisch, K., Ruston, D., Zimmemberg, J., Parsegian, V. A., Rand, R. P., and Fuller, N. (1992) Membrane dipole potentials, hydration forces, and the ordering of water at membrane surfaces. *Biophys. J.* **61,** 1213–1223.
61. Pasenkiewicz-Gierula, M., Takaoka, Y., Miyagawa, H., Kitamura, K., and Kusumi, A. (1997) Hydrogen Bonding of Water to Phosphatidylcholine in the Membrane As Studied by a Molecular Dynamics Simulation: Location, Geometry, and Lipid-Lipid Bridging via Hydrogen-Bonded Water. *J. Phys. Chem. A* **101,** 3677–3691.
62. Pasenkiewicz-Gierula, M., Róg, T., Kitamura, K., and Kusumi, A. (2000) Cholesterol Effects on the Phosphatidylcholine Bilayer Polar Region: A Molecular Simulation Study. *Biophys. J.* **78,** 1517–1521.
63. Chiu, S. W., Jakobsson, E., Mashl, R. J., and Scott, H. L. (2002) Cholesterol-Induced Modifications in Lipid Bilayers: A Simulation Study. *Biophys. J.* **83,** 1842–1853.
64. Shinoda, K., Shinoda, W., Baba, T., and Mikami, M. (2004) Comparative molecular dynamics study of ether- and ester-linked phospholipid bilayers. *J. Chem. Phys.* **121,** 9648–9654.
65. Jedlovszky, P., Medvedev, N. N., and Mezei, M. (2004) Effect of Cholesterol on the Properties of Phospholipid Membranes. 3. Local Lateral Structure. *J. Phys. Chem. B* **108,** 465–472.
66. Jedlovszky, P., Pártay, L., and Mezei, M. (2007) The structure of the zwitterionic headgroups in a DMPC bilayer as seen from Monte Carlo simulation: Comparisons with ionic solutions. *J. Mol. Liquids,* **131–132,** 225–234.

67. Shinoda, W. and Okazaki, S. (1998) A Voronoi analysis of lipid area fluctuation in a bilayer. *J. Chem. Phys.* **109,** 2199–2205.
68. Falck, E., Patra, M., Karttunen, M., Hyvönen, M. T., and Vattulainen, I. (2004) Lessons of Slicing Membranes: Interplay of Packing, Free Area, and Lateral Diffusion in Phospholipid/Cholesterol Bilayers. *Biophys. J.* **87,** 1076–1091.
69. Jedlovszky, P., Vincze, Á., and Horvai, G. (2002) New insight into the orientational order of water molecules at the water/1,2-dichloroethane interface: a Monte Carlo simulation study. *J. Chem. Phys.* **117,** 2271–2280.
70. Jedlovszky, P., Vincze, Á., and Horvai, G. (2004) Full description of the orientational statistics of molecules near interfaces. Water at the interface with CCl_4. *Phys. Chem. Chem. Phys.* **6,** 1874–1879.
71. Cantor, R. S. (1999) Lipid Composition and the Lateral Pressure Profile in Bilayers. *Biophys. J.* **76,** 2625–2639.
72. Marrink, S. J., Sok, R. M., and Berendsen, H. J. C. (1996). Free volume properties of a simulated lipid membrane, *J. Chem. Phys.* **104,** 9090–9099.
73. Okabe, A., Boots, B., Sugihara, K., and Chiu, S. N. (2000) *Spatial Tessellations: Concepts and Applications of Voronoi Diagrams*. John Wiley, Chichester.
74. Jedlovszky, P. and Mezei, M. (2000) Calculation of the Free Energy Profile of H_2O, O_2, CO, CO_2, NO, and $CHCl_3$ in a Lipid Bilayer with a Cavity Insertion Variant of the Widom Method. *J. Am. Chem. Soc.* **122,** 5125–5131.
75. Jedlovszky, P. and Mezei, M. (2003) Effect of Cholesterol on the Properties of Phospholipid Membranes. 2. Free Energy Profile of Small Molecules. *J. Phys. Chem. B* **107,** 5322–5332.
76. Falck, E., Patra, M., Karttunen, M., Hyvönen, M. T., and Vattulainen, I. (2004) Impact of cholesterol on voids in phospholipid membranes. *J. Chem. Phys.* **121,** 12,676–12,689.
77. Alinchenko, M. G., Anikeenko, A. V., Medvedev, N. N., Voloshin, V. P., Mezei, M., and Jedlovszky, P. (2004) Morphology of Voids in Molecular Systems. A Voronoi-Delaunay Analysis of a Simulated DMPC Membrane. *J. Phys. Chem. B* **108,** 19,056–19,067.
78. Alinchenko, M. G., Voloshin, V. P., Medvedev, N. N., Mezei, M., Pártay, L., and Jedlovszky, P. (2005) Effect of Cholesterol on the Properties of Phospholipid Membranes. 4. Interatomic Voids. *J. Phys. Chem. B* **109,** 16,490–16,502.
79. Bassolino-Klimas, D., Alper, H. E., and Stouch, T. R. (1995) Mechanism of Solute Diffusion through Lipid Bilayer Membranes by Molecular Dynamics Simulation. *J. Am. Chem. Soc.* **117,** 4118–4129.
80. López-Cascales, J. J., Hernández Cifre, J. G., and García de la Torre, J. (1998) Anaesthetic Mechanism on a Model Biological Membrane: A Molecular Dynamics Simulations Study. *J. Phys. Chem. B* **102,** 625–631.
81. Söderhäll, J. A. and Laaksonen, A. (2001) Molecular Dynamics Simulation of Ubiquinone inside a Lipid Bilayer. *J. Phys. Chem. B* **105,** 9308–9315.
82. Duong, T. H., Mehler, E. L., and Weinstein, H. (1999) Molecular Dynamics Simulation of Membranes and a Transmembrane Helix. *J. Comp. Phys.* **151,** 358–387.
83. Bandyopadhyay, S., Tarek, M., and Klein, M. L. (1999) Molecular Dynamics Study of a Lipid–DNA Complex, *J. Phys. Chem. B* **103,** 10,075–10,080.
84. Pandit, S. A., Vasudevan, S., Chiu, S. W., Mashl, R. J., Jakobsson, E., and Scott, H. L. (2004) Sphingomyelin-Cholesterol Domains in Phospholipid Membranes: Atomistic Simulation. *Biophys. J.* **87,** 1092–1100.
85. Marrink, S. J. and Berendsen, H. J. C. (1996) Permeation Process of Small Molecules across Lipid Membranes Studied by Molecular Dynamics Simulations. *J. Phys. Chem.* **100,** 16,729–16,738.
86. Shinoda, W., Mikami, M., Baba, T., and Hato, M. (2004) Molecular Dynamics Study on the Effect of Chain Branching on the Physical Properties of Lipid Bilayers: 2. Permeability. *J. Phys. Chem. B* **108,** 9346–9356.
87. Widom, B. (1963) Some Topics in the Theory of Fluids. *J. Chem. Phys.* **39,** 2808–2812.

10

Fluorometric Assay for Detection of Sterol Oxidation in Liposomal Membranes

Parkson Lee-Gau Chong and Michelle Olsher

Summary

The authors have developed a fluorescence assay to measure the rate and extent of sterol oxidation in lipid bilayers. Dehydroergosterol (DHE), a fluorescent cholesterol analog, is used as a probe and at the same time as a membrane component. The assay can also be performed on bilayers containing a mixture of sterols including DHE and nonfluorescent sterols, such as cholesterol and ergosterol. The fluorescence intensity of DHE decreases on oxidation, so the rate and extent of free radical- or enzyme-induced sterol oxidation can be measured as a function of temperature and membrane composition. For the studies, two-component (e.g., phosphatidylcholine (PC)/DHE) and multicomponent (e.g., DHE/PC/bovine-brain sphingomyelin) large unilamellar vesicles were used, and sterol oxidation was initiated either by the peroxy radical generator 2,2′-azobis (2-amidinopropane) dihydrochloride or by the enzyme cholesterol oxidase. The data gathered from this assay may be used to examine the effects of water- and lipid-soluble antioxidants on membrane sterol oxidation produced by free radicals. This assay can be used to test the potency of antioxidants and pro-oxidants, and can be used to determine whether unknown substances demonstrate antioxidant activity against sterol oxidation. The assay can also be used as a tool to examine the effect of sterol lateral organization on sterol oxidation (in the presence or absence of antioxidants). In agreement with the sterol regular distribution model, it is found that both free radical- and enzyme-induced sterol oxidation vary with membrane sterol content in a well defined alternating manner.

Key Words: Antioxidants; cholesterol; dehydroergosterol; fluorescence; free radicals; membrane; sterol oxidation.

1. Introduction

The authors have developed a novel fluorescence assay that allows them to measure the rate and extent of sterol oxidation in both the presence and absence of antioxidant. *In vivo* oxidation of membrane cholesterol has important biomedical implications; the oxysterol products produced have been implicated in a broad list of pathophysiological changes in the human body, and are believed to be a major factor associated with many disease states. Cholesterol can be oxidized in the body by enzymes and/or reactive oxygen species. A proportion of oxysterol production in the body allows for removal of excess cholesterol as a component of bile acids. However, when the oxidative load becomes too great as a result of overproduction of reactive oxygen species and/or depletion of antioxidant protective mechanisms, oxysterol production can induce deleterious changes. For example, the accumulation of oxidized low-density lipoprotein (LDL) in the subendothelial space is correlated with many pathological changes in vascular cells and thought to lead to the formation of the atherosclerotic plaques that are seen in coronary heart disease *(1)*. Recent evidence points to the oxysterol contained within the oxidized-LDL particle as the cytotoxic agent *(2)*. Oxidation products of cholesterol are highly toxic to arterial smooth muscle cells *(3)* and vascular cells *(2)* as well as many other cell types *(4)*. Oxysterol production is associated with the etiology of many diseases, such as Alzheimer's Disease, diabetes, and cancer *(3,5,6)*. Elucidation of the mechanisms involved in the regulation of cholesterol oxidation would be useful in the development of therapeutic options for the prevention and treatment of many human diseases and conditions.

The conventional way to analyze the products of cholesterol oxidation requires: (1) the extraction of total lipids using organic solvents, (2) the enrichment, separation, and detection of oxycholesterols by thin-layer chromatography and high-performance liquid chromatography (HPLC), and (3) the confirmation of structural identity by mass and nuclear magnetic resonance spectrometry *(7)*. Although this approach is necessary for understanding the chemistry underlying the oxidation reaction, it also perturbs membrane organization and is too cumbersome for kinetic studies. Ultraviolet (UV) light has been used to detect particular cholesterol oxidation products in lipid vesicles, but its use is rather limited. This is because oxycholesterols do not absorb UV light with a reasonably high extinction coefficient, except for cholest-4-en-3-one *(8)* and 7-ketocholesterol *(9)*. The UV method is even less useful in intact cell membranes because of the presence of many other membrane constituents, which also absorb UV light.

The fluorescence assay described here is sensitive and can be used to measure the rate and extent of oxidation of sterol contained within a membrane without much membrane organization disruption. In this chapter, how to use a peroxy radical generator such as 2,2′-azobis [2-amidinopropane] dihydrochloride (AAPH) to induce sterol oxidation has been illustrated, but rates and extent of sterol oxidation induced by other pro-oxidants, such as 2,2′-azobis (2,4-dimethylvaleronitrile), a lipophilic peroxy radical generator, and by enzymes, such as cholesterol oxidase *(10)* can also be monitored.

It is also possible to get a relative measure of the degree of antioxidant protection afforded against sterol oxidation by looking at the length of the antioxidant-induced lag phase. The lag phase is a period of slow oxidation during which the antioxidant is scavenging some proportion of the free radicals being generated (*see* **Fig. 3**). Once the antioxidant has been depleted, the rate of sterol oxidation increases. By comparing the length of the lag phase along with the rate of sterol oxidation after the lag phase, it is possible to rank lipid- and water-soluble antioxidants according to their effects on sterol oxidation. When this assay is used to examine sterol oxidation by free radicals or by enzymes, either with or without the addition of antioxidant, only one sample preparation is required. The lipid system must include the fluorescent sterol, dehydroergosterol (DHE); however, other sterols (such as cholesterol or ergosterol) may be mixed with DHE (*see* **Note 1**).

The liposomes can be as simple as two-component vesicles, for example, 1,2-dimyristoyl-*sn*-glycero-3-phosphocholine (DMPC)/DHE, or as complex as multicomponent vesicles, LDL particles, or cells with incorporated DHE. These samples can be used to test the potency of antioxidants and pro-oxidants, and can be used to determine whether unknown substances demonstrate antioxidant activity. In addition to the aforementioned possibilities, the assay can be used as a tool to probe the relationship between membrane sterol lateral organization and sterol oxidation. The fluorescence assay is most suited for this application. For this specialized use of the assay, a more rigorous sample preparation is required.

The relationship between membrane lateral organization and sterol oxidation can be described by the sterol superlattice theory. This theory (*11,12*; reviewed in **refs. *13*** and ***14***) states that at certain mathematically predicted mole fractions of sterol, the sterol molecules are maximally organized into a type of regular distribution called a sterol superlattice. It is hypothesized that differences in sterol lateral organization can lead to large variations in membrane packing between components at a critical mole fraction compared with those at a noncritical mole fraction. Many membrane-associated properties, such as the activity of some surface-acting enzymes and the partitioning of nystatin into the bilayer, show multiple biphasic changes as a function of sterol mole percent *(10,15,16)*. The fluorescence assay can be used to examine the rate of sterol oxidation, which is another property showing multiple biphasic changes as a function of sterol mole fraction *(10,17)*. The rate of sterol oxidation should be proportional to sterol availability, and sterol is expected to be more accessible to the aqueous phase (containing water-soluble pro-oxidants or enzymes) at critical mole fractions (reviewed in **ref. 13**). The authors observed an increase in the rate of sterol oxidation at critical predicted mole fractions, and a decrease in the oxidation rate as mole percent deviated from critical *(10,17)*. When the assay was performed in the presence of antioxidants, a decrease in the length of the lag phase at or near predicted critical mole fraction as compared with the length of the lag phase at noncritical mole fractions was observed (Olsher and Chong, unpublished results). Thus, this application of the assay not only provides supporting evidence for the sterol superlattice theory, but also reveals a new type of regulation for sterol oxidation, that is, sterol oxidation varies with the extent of sterol superlattice in the membrane, and changes biphasically with sterol content at critical mole fractions.

The most important procedure involved in this application of the assay is sample preparation. A typical sample set contains between nine and fifteen independently prepared samples, each differing from the next by a very small increment (~0.3 mol%) in mole percent of sterol. The absolute amount of sterol remains constant from sample to sample; the mole fraction of sterol is varied by changing the amounts of the nonsterol lipid components in small increments from sample to sample. Thus, the end result is a series of samples in which the mole percent of sterol changes by small steps over an entire range of less than 3–4 mol%. Because such small differences in sterol mole fraction are used, all aspects of sample preparation must be carried out with extreme precision. Determination of stock solution concentrations, analysis of phosphorus concentration after extrusion and pipetting of lipids (or liposomes) must all be done accurately. The thermal history of each sample must be the same. Careful sample preparation insures that the differences observed between samples can only be attributed to differences in sterol mole percent (hence, differences in the lateral organization of sterol in the bilayer).

The assay protocol involves time trace monitoring of DHE fluorescence intensity after addition of a free radical generator (e.g., AAPH) or an oxidizing enzyme at desired temperature (37°C, in the study) (*see* **ref. 17**). The initial rate of sterol oxidation is then calculated from the decay of DHE fluorescence intensity. The assay may be performed in the presence or absence of antioxidants.

2. Materials
2.1. DHE Concentration Determination

1. DHE (Sigma, St. Louis, MO) is purified by HPLC before being dissolved in $CHCl_3$ to make a stock solution (*see* **Note 2**). Stock solution is sensitive to light, heat, and oxygen, and must be sealed with Teflon tape and parafilm and stored in darkness at −20°C.
2. 1,4-Dioxane (Burdick & Jackson, Morris Township, NJ).
3. Small glass culture tube with phenolic screw cap that has a chloroform-resistant Teflon interface (Pyrex 9826) (Corning, distributed by Fisher Scientific, Suwanee, GA).

4. Spectrophotometer and two-matched quartz cuvettes (also used in phospholipid and cholesterol determinations).
5. Microman pipettes (Gilson, Middleton, WI).

2.2. Phospholipid Concentration Determination

1. Phospholipids: 1-palmitoyl-2-oleoyl-*sn*-glycero-3-phosphocholine (POPC) and DMPC (Avanti Polar Lipids, Alabaster, AL) are prepared by dissolving lipids in $CHCl_3$ to produce a stock solution of approx 1 mM. Phospholipid stock solution is sealed with Teflon tape and parafilm and stored at −20°C.
2. Phosphorus standard solution (0.65 mM) (Sigma P3, 869).
3. 13 × 100 mm Disposable culture tubes.
4. 50% H_2SO_4, 30% H_2O_2, 5% ammonium molybdate, sodium bisulfite.
5. Fiske-Subbarow reagent (FSR) (Sigma).
6. Heating plate and heating blocks (20 wells).

2.3. Sample Preparation

1. DHE/$CHCl_3$ stock solution and phosphatidylcholine/$CHCl_3$ stock solution.
2. Storage tubes: Pyrex round bottom culture glass tubes that can be closed with a phenolic screw cap having a Teflon fluorocarbon-resin-faced rubber liner that is resistant to chloroform and many other organic solvents (*see* **Note 3**).
3. 50 mM Tris-HCl buffer, pH 7.0, 0.02% NaN_3, 0.1 mM ethylenediaminetetraacetic acid (EDTA) (Sigma, St. Louis, MO) (*see* **Note 4**), filtered through a syringe filter (0.2 μm).
4. Nitrogen (compressed gas) and Reacti-Vap III, Reacti-Therm III manifold (Rockford, IL).
5. Flexi-dry freeze-dryer (FTS Systems, Stone Ridge, NY).
6. Extruder (Lipex Biomembranes, Vancouver, Canada) and Nucleopore polycarbonate membranes (Whatman Int., Kent, UK).
7. Zetasizer HS-100 spectrometer (Malvern, Worce, UK).

2.4. Assay Protocol

1. Phosphatidylcholine/DHE large unilamellar vesicles (LUVs) are pipetted into reaction cuvette and mixed with calculated volume of Tris-HCl buffer.
2. AAPH (Aldrich, Milwaukee, WI) stock is prepared by mixing with Millipore (Billerica, MA) water. AAPH is stored at 4°C at all times; only remove from this temperature when it needs to be added to the reaction cuvette (*see* **Note 5**).
3. For assays including antioxidants, ascorbic acid is dissolved in (Millipore) water. Ascorbyl palmitate is dissolved in 100% ethanol.
4. Fluorometer (SLM 8000C) (SLM, Urbana, IL).

3. Methods
3.1. DHE Concentration Determination

1. Before pipetting DHE stock solution for concentration determination, DHE must be brought to room temperature and mixed gently to assure even distribution of molecules.
2. Pipette 2 mL of 1,4-dioxane (Burdick & Jackson) into small glass culture tube with chloroform-resistant screw cap.
3. Pipette 20 μL DHE stock solution into tube containing dioxane; mix well (overnight with rotation; cover tube with foil).
4. Wash out two matched quartz cuvettes (1-cm pathlength) and add DHE/dioxane mixture from screw-top tube.

5. Take absorbance measurements at 326 nm of DHE/dioxane mixture; collect data every 10 s; block light from spectrophotometer in between readings to prevent photodamage.
6. Collect for 50 s. Wait a few minutes and try again. Compare readings.
7. Use mean absorbance value A_{326} to calculate the concentration of DHE from: $[DHE] = (A_{326}/\varepsilon)$ where the extinction coefficient ε is equal to 10,600 $M^{-1}cm^{-1}$ in dioxane at 326 nm *(18)*; multiply by dilution correction factor (total volume divided by volume of DHE added to dioxane).

3.2. Phospholipid Concentration Determination

1. Phospholipid concentration in stock solution is determined by the method of Bartlett *(19)*, with some modifications.
2. Pipette phosphorus standard solution (0.65 m*M*) into 13 × 100 mm disposable culture tubes. Two tubes are left empty to serve as blanks; the remaining six tubes contain 13, 26, 39, 52, 65, and 78 nmols of phosphate, and will be used to produce a standard curve.
3. Pipette phospholipid stock solution for phosphate concentration determination. Calculate approximately how much sample would be required so that the unknown concentrations will fall within the standard curve values.
4. Add 200 µL of 50% H_2SO_4 to each tube and vortex gently.
5. Place tubes in preheated aluminum block (200°C) for 5 min.
6. Add 200 µL (*see* **Note 6**) of 30% H_2O_2 to each tube and char in aluminum heating block for 30 min.
7. Remove tubes from block and allow cooling.
8. Add 2.3 mL of H_2O (*see* **Note 7**) to each tube; then add 115 µL of 5% ammonium molybdate to each tube and vortex gently. If liquid becomes yellow after addition of ammonium molybdate, add a few crystals of sodium bisulfite and vortex until liquid becomes clear.
9. Prepare FSR according to manufacturer's instructions; add 115 µL of FSR to each tube and vortex gently.
10. Place tubes in a water bath at 100°C for 20 min or until samples turn blue.
11. Allow tubes to cool. Read absorbance at 660 nm (start with least concentrated samples).

3.3. Sample Preparation

When the assay is used for screening antioxidants or pro-oxidants, the sample preparation is simplified, because only one sample is required. However, the studies have shown that properties such as antioxidant protection and pro-oxidant potency may vary at different mole fractions of sterol because of differences in sterol lateral organization. This should be kept in mind when designing the experiment, so that the mole fraction of sterol remains constant for each sample being compared.

When this assay is used to examine the relationship between sterol oxidation and sterol lateral organization, then strict attention to all aspects of sample preparation is required. In each set, sterol mole fraction is increased by small increments (~0.3 mol%) for each sample. The absolute number of sterol molecules per sample does not vary; rather, sterol mole percent is changed by varying the amount of the other lipid components. *See* **Table 1** for a typical sample preparation design.

1. Before pipetting, DHE/$CHCl_3$ and phospholipid/$CHCl_3$ stock solutions must be brought to room temperature to assure even distribution of molecules.
2. Pipette calculated amounts of phospholipid and DHE stock solutions; stock solution tubes should remain in ice while pipetting to slow evaporation. All solutions (stock, vesicles, and so on) containing DHE must be protected from light (use opaque tubes or wrap tubes in foil).
3. Dry lipids under nitrogen gas stream until no visible solvent remains; then dry lipids overnight under vacuum.

Table 1
Typical Sample Preparation Scheme

Tube no.	Sterol (mol%)	DHE (nmol)	POPC (nmol)
1	18.4	125	554
2	18.8	125	540
3	19.1	125	529
4	19.4	125	519
5	19.7	125	510
6	20.0	125	500
7	20.3	125	491
8	20.6	125	482
9	20.9	125	473
10	21.2	125	465
11	21.5	125	456

A typical set of DHE/POPC mixtures shows increases in sterol mole percent by increments of 0.3 mol%. The absolute number of moles of sterol (DHE, in this case) remains constant from sample to sample, whereas the number of moles of the second component (POPC) varies from sample to sample. In this sample set, 20.0 mol% sterol is the only predicted critical sterol mole fraction for maximal superlattice formation *(23)*.

4. The next day, reconstitute lipids with 5 mL of filtered Tris-HCl buffer (pH 7.0) at the desired temperature (*see* **Note 8**). After reconstitution, flush samples with argon, cover tubes with parafilm and vortex for 2 min at the chosen temperature to form multilamellar vesicles (MLVs).
5. MLVs are then subjected to three cooling/heating cycles (for POPC/sterol or DMPC/sterol LUVs, 4°C for 30 min, and then 37°C for 30 min were used).
6. Let MLVs sit at room temperature for at least 4 d to allow them to come to thermal equilibrium or semiequilibrium for membrane lateral organization. Store vesicles under argon to avoid auto-oxidation of sterol before sterol oxidation measurements.
7. Prepare LUVs from MLVs by extrusion. Extrude at desired temperature 10 times through two stacked Nucleopore polycarbonate filters under nitrogen gas pressure to form homogeneous LUVs of desired size (*see* **Note 9**).
8. Store LUVs under nitrogen or argon gas at room temperature before activity measurements to avoid auto-oxidation of sterols.
9. Incubate at room temperature for 7 d. Measure the phospholipid concentration of each LUV sample by the method of Bartlett (*see* **ref. *19***) to determine the new lipid concentration after loss because of extrusion (*see* **Note 10**). Also, measure vesicle size by photon correlation spectroscopy (Malvern Zetasizer HS-1000 spectrometer) (*see* **Note 11**). The previous study *(10)* shows that the sterol mole percents in vesicles before and after extrusion differ by less than 0.2 mol%. Thus, for convenience, the sterol mole percents in MLVs were used to assess the relationship between sterol content and sterol oxidation in LUVs.

3.4. Assay Protocol

1. Turn on circulating water bath connected to fluorometer until it reaches the desired temperature (*see* **Note 12**). For the experiments on DHE/POPC and DHE/DMPC, the authors used a 15 min incubation period to allow samples to reach 37°C.
2. Turn on fluorometer (SLM 8000C) and set excitation wavelength to 325 nm with a bandpass of 0.5 nm. Observe emission through a monochromator set at 396 nm with a bandpass of 8 nm.
3. Add calculated amounts of phospholipid/DHE LUVs and Tris-HCl buffer to a fluorescence cuvette (total sample volume = 1.6 mL; final cuvette concentration = ~8 μM sterol and ~30 μM total lipid).
4. If antioxidant is to be used in the assay, it is added at this time.

Fluorometry and Sterol Oxidation

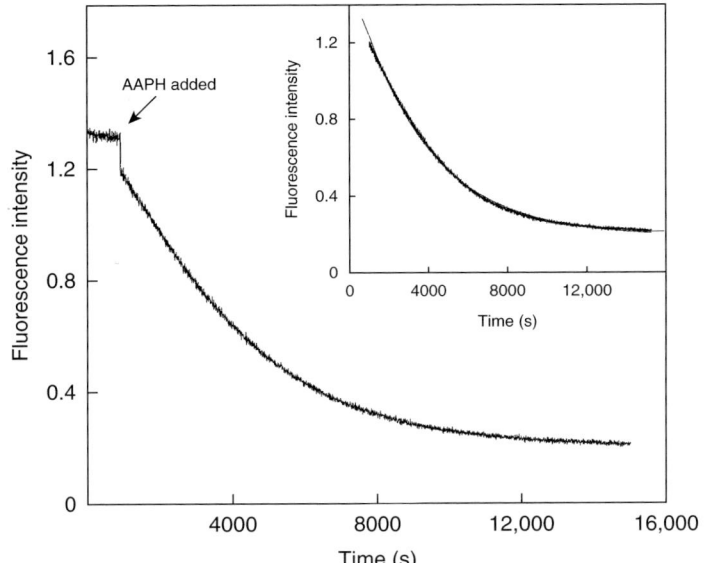

Fig. 1. Profile of a typical assay showing DHE fluorescence intensity decay. DHE (21.2 mol%)/POPC LUVs were incubated at 37°C; 30 μL of 300 mM AAPH was added at 1000 s (*see* **Note 15**) and rate of intensity decay was monitored at 395 nm. The sharp drop seen on addition of AAPH is because of energy transfer (*see* **Note 16**) and the asymptotic value at end of assay is referred to as "F_∞" (*see* **Note 17**). (**Inset**) Data of the AAPH-induced DHE fluorescence decay in 21.2 mol% DHE/POPC LUVs measured at 37°C are fitted well by first-order kinetics using the equation [DHE]/[DHE]$_0$ = F/F_0 = $(F_{obs} - F_\infty)/F_0$ = exp $(-kt + b)$. Here, b is a constant, and [DHE]$_0$ and F_0 denote the DHE concentration and fluorescence intensity, respectively, at reaction time zero (upon addition of AAPH), and a correlation coefficient R = 0.999 was obtained. k is the apparent rate constant of the reaction, and F_∞ is the DHE fluorescence intensity at a long time. For the data fitting, all the fluorescence intensities are normalized against the intensity at reaction time zero (reproduced from *[17]* with permission).

5. Place cuvette into sample compartment of the fluorometer and incubate under gentle magnetic stirring for chosen incubation period to allow sample to reach desired temperature.
6. After incubation, begin time trace monitoring. Monitor background fluorescence for at least 60 s. If background fluorescence is stable (i.e., no photobleaching is apparent, *see* **Note 13**), pause the data acquisition and add free radical generator or oxidizing enzyme by pipetting into center of reaction cuvette (*see* **Note 14**).

3.5. Data Analysis

3.5.1. Apparent Rate Constant

The authors have used data from this fluorescence assay to calculate three different parameters: initial rate of sterol oxidation, length of lag phase in the presence of antioxidant, and the apparent reaction rate constant (k). To calculate k, the reaction must run until DHE fluorescence intensity reaches an asymptotic value (**Fig. 1**).

In the absence of antioxidant, the steady decay of DHE fluorescence intensity induced by AAPH is best described by first order kinetics. A curve-fitting program (Kaleidagraph; v. 3.52; Synergy Software, Reading, PA) allows calculation of the apparent reaction rate constant (k) using equation *(1)*: $F = F_\infty + e^{(-kt + b)}$ (**Fig. 1, inset**). To calculate k in the presence of antioxidants, fit all data after lag phase has ended.

3.5.2. Initial Rate

Most of the assays performed to examine the rate of sterol oxidation focus on the parameter of initial rate rather than the apparent rate constant. As this kind of experiment is very tedious, and it is crucial to keep the sample thermal history identical for all the samples in the same sample set, the approach requiring a short operational time is preferred. An initial rate calculation requires substantially less time per assay than does solving for k. The initial rate is also particularly more useful for the purpose of examining the relationship between sterol oxidation and sterol lateral organization. When free radical oxidation occurs, the oxidized product is likely to cause some membrane perturbation and/or change in original membrane lateral organization because of the fact that the oxidized sterol product has a larger mean molecular cross-sectional area than native sterol *(20)*, and has a weaker interaction with the acyl chains of the neighboring phospholipids *(21)*. Initial rate of sterol oxidation will give us the most accurate snapshot of initial sterol organization in the bilayer (*see* **Note 18**).

Initial rate of sterol oxidation is calculated from the slope of DHE fluorescence intensity decrease *vs* time. By examining a typical plot (*see* **Fig. 1**), it is apparent that the first 15 min of sterol oxidation can be roughly fit to a straight line. However, a snapshot of the initial rate of oxidation at the very beginning of free radical production, before any change in lateral organization has occurred is needed. Each set of samples spanning a critical mole fraction must be analyzed separately in order to get an accurate picture for that particular set of LUVs. There are variations in the time used to define the initial rate calculation per sample set. One reason for these slight variations is because of the high sensitivity of fluorescence signals.

Fluorescence intensity is a relative value, depending on the amount of fluorophore in the cuvette, sample mixing, temperature fluctuations, and quenching because of impurities in the sample. For this reason, each complete set is analyzed separately, but all sets are normalized so that they can be compared with each other. Each sample set is analyzed by examining the first 5–10 min of fluorescence decay in a sequential manner. The first 3 min are plotted, and then new plots are created for the first 4, 5, 6, and so on minutes. The trend of the slopes of each sequential reaction profile shows the point at which the reaction rate first begins to slow. When the slope begins to become smaller in a systematic manner (meaning deviation from linearity), then the data points at shorter time should be used. By looking at the reaction profiles in this fashion, details are visible that would be overlooked if just the larger profile is examined. It is possible to visualize the beginning of deviation from linearity, which usually occurs between 4 and 7 min, depending on the sample set. Within a sample set, the deviation from linearity can be determined and, thus, the number of minutes that will be used to calculate the initial rate can be defined. An acceptable linear fit for that particular time span is usually a correlation coefficient (R) value of at least 0.8. However, the validity of a particular R value is based on a power analysis, which takes into account the sample size and calculates the lowest acceptable correlation coefficient that will still give a 95% confidence interval ($p < 0.05$). The correlation coefficient can be somewhat lower than normally would be accepted because its value is based not only on the spread of points, but on the number of points in the plot. If an initial rate is chosen for the first 5 min, the R values may be systematically lower than those for a set where the initial rate is defined as the first 6 min. The initial rate is defined as the smallest time frame that gives a linear plot and a reasonable correlation coefficient and power analysis.

Fluorometry and Sterol Oxidation

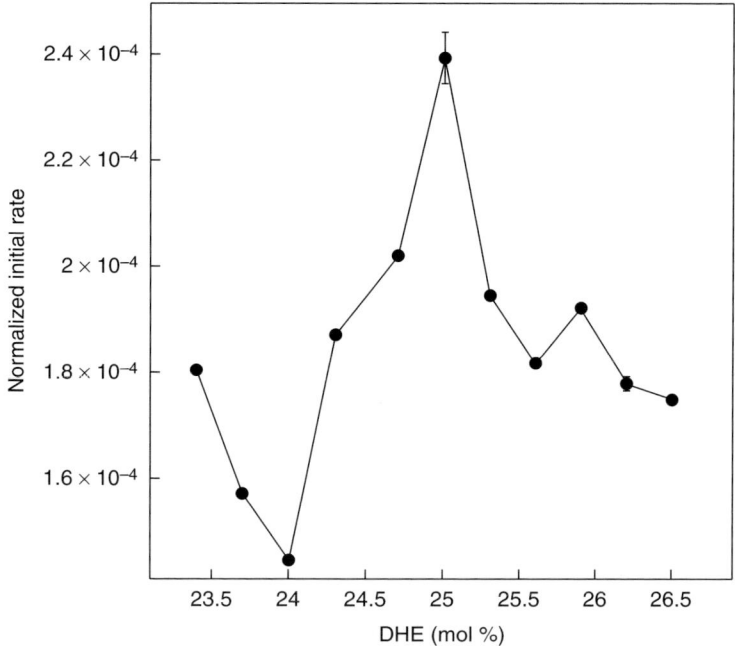

Fig. 2. Normalized initial rate of sterol oxidation in DHE/POPC LUVs (diameter ~200-nm). Assays were performed at 37°C; 30 μL of 300 mM AAPH was used to initiate oxidation. The vertical bars are the standard deviations of the measurements from three independently prepared samples. These data show that sterol oxidation rate undergoes a biphasic change at the critical sterol mole fraction 25.0 mol%, which is the only critical sterol mole fraction for maximal superlattice formation in this mole percent range examined *(23)*. Similar biphasic changes in AAPH-induced sterol oxidation were observed at other critical sterol mole fractions in DHE/POPC LUVs and in other membrane systems such as DHE/bovine-brain sphingomyelin/POPC LUVs *(17)*. Cholesterol oxidase-induced sterol oxidation exhibited similar multiple biphasic changes with membrane sterol content *(10)*.

3.5.3. Applications for This Assay

Any of the assay applications described can be extrapolated to more complex systems, such as multicomponent vesicles, lipoprotein particles, and cells. DHE can be incorporated into cells by using either the DHE-methyl-β-cyclodextrin conjugate, or by allowing spontaneous transfer of cholesterol and DHE between membranes through the aqueous phase *(22)*.

This fluorescence assay is an excellent tool for relating the rate of sterol oxidation to membrane sterol lateral organization, particularly because DHE is used not just as a fluorescent probe to monitor oxidation rate, but also serves as an actual membrane component. Thus, a view of a lipid bilayer that is undisturbed by extraneous probe molecules can be observed.

To look at sterol lateral organization, the initial rates of oxidation from samples can be compared within a sample set. Because all samples contain the same absolute amount of sterol, differences in initial rate of sterol oxidation must be a result of changes in the lateral organization of sterol molecules in the membrane at different sterol mole fractions. The plot above (*see* **Fig. 2**) shows a biphasic change in initial rate of sterol oxidation, with a maximum oxidation rate at one of the critical mole percents (25.0 mol%), as predicted by the sterol

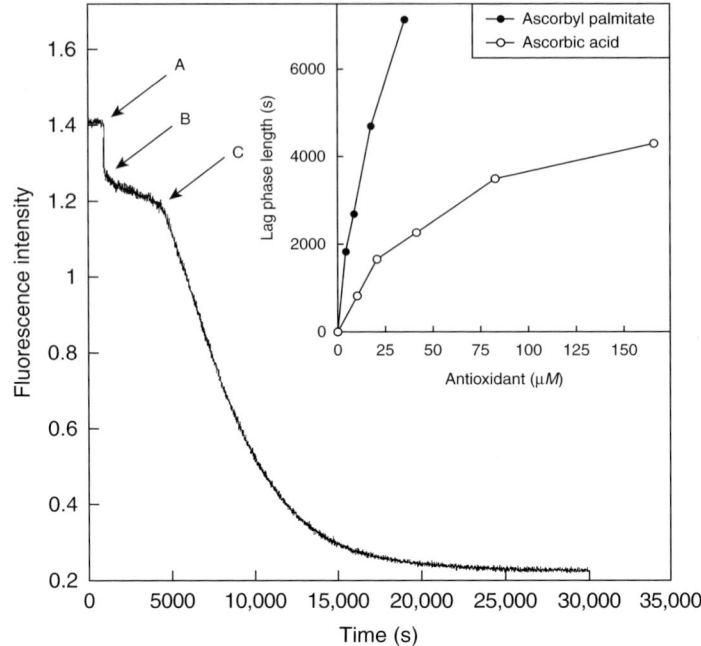

Fig. 3. Profile of a typical assay showing DHE fluorescence intensity decay after incubation with an antioxidant. DHE (19.1 mol%)/POPC LUVs were incubated with 22.5 μM ascorbyl palmitate for 15 min at 37°C before beginning of assay. Point A: addition of 30 μL of 300 mM AAPH; points A to B: energy transfer; points B to C: lag phase (slow oxidation step); after point C: beginning of late phase (fast oxidation step). **(Inset)** Plot showing comparison between length of lag phase produced by a water-soluble antioxidant (ascorbic acid) and a lipid-soluble antioxidant (ascorbyl palmitate) (Olsher and Chong, unpublished results). It is clear from the inset that, at the same given apparent dose, ascorbyl palmitate is much more potent in protecting against sterol oxidation than its water-soluble counterpart, ascorbic acid.

superlattice theory. Also the rate of sterol oxidation can be examined in the presence of an antioxidant. When the assay includes an antioxidant, the typical profile contains an additional lag phase (*see* **Fig. 3**).

Both lag phase and initial rate data immediately after lag can be fitted to separate linear lines; the cross point of these fitted lines minus the time before addition of AAPH equals the lag phase length. Generally, the end of lag phase is clearly visible and easily defined even without using the intersection of two fitted lines. By using the same DHE concentration for each sample, the length of the lag phase can be used to describe the relative protection from sterol oxidation given by different antioxidants (*see* **Fig. 3**, **inset**), or by a specific dose of one antioxidant. This procedure can be useful in ranking the potency of different antioxidants, or in examining the effect of water- and lipid-soluble antioxidants on sterol lateral organization. In addition, by assaying bilayers in the presence of a lipid-soluble antioxidant, such as ascorbyl palmitate, one can explore issues relating to membrane perturbation. If we choose to relate lateral organization to the rate of sterol oxidation in the presence of an antioxidant, we would follow the protocol described previously for examining sterol lateral organization, except that we incubate the sample with an antioxidant before adding pro-oxidant to the sample cuvette.

4. Notes

1. DHE is structurally similar to and behaves in the same fashion as cholesterol in terms of membrane incorporation, spontaneous transfer, and membrane lateral organization *(22–25)*. For this reason, multicomponent vesicles can be made with cholesterol or ergosterol as the main sterol component, provided that a small amount (e.g., 1 mol%) of DHE is added as a probe.
2. Results produced by using unpurified DHE are often quite different from those produced using HPLC-purified DHE incorporated into LUVs. This may be because of the presence of oxidized DHE in the unpurified stock. To purify DHE, a Waters Symmetry C_{18} column is used (5 μm; 3.9 × 150 mm) (Waters Corp., Milford, MA); mobile phase: methanol: acetonitrile 2:1, v/v; flow rate 1.5 mL/min; detection of absorbance at 326 nm. It is possible that the presence of oxidized DHE introduces error by affecting membrane packing in a nonsystematic way, and hence interferes with the measurement of the rate of sterol oxidation and the length of the lag phase (if antioxidants are used).
3. All glassware used for lipid storage and vesicle preparation should first be washed with polar and nonpolar solvents. For the vesicle preparation, the authors use one wash with chloroform:methanol (2:1 v/v), followed by one wash with ethanol, and a final washing with deionized water. Glassware must be dried completely. Gilson Microman pipettes are used to transfer organic solvents or lipid samples dissolved in organic solvents.
4. EDTA is used to chelate metal ions in solution. It was found that metal chelation is required for the measurement of the antioxidant capacity of ascorbyl palmitate. In the presence of free metal ions, ascorbyl palmitate either does not produce a lag phase at all or produces a greatly attenuated lag phase, and therefore does not provide typical antioxidant protection. Water-soluble vitamin C (ascorbic acid) demonstrates antioxidant protection both in the absence and presence of free metal ions, but EDTA was added to all buffer preparations.
5. Oxidizing agents, such as water- or lipid-soluble free radical generators or enzymes, can be used in this assay to oxidize sterol. AAPH is a water-soluble peroxy radical generator that undergoes thermal decomposition to yield N_2 plus two carbon radicals. Most of the carbon radicals then react rapidly with oxygen molecules to yield peroxy radicals *(26)*. The rate of AAPH decomposition is determined primarily by temperature and pH. At 37°C in neutral water, the half-life of AAPH decomposition is about 175 h *(27)*. Based on these calculations, the rate of peroxy radical formation is essentially constant for the duration of the assay, even when a significant lag phase is observed. However, if the sample set is large and must be run over the course of four or more days, there is some loss of potency of radical production. The potency is tested by comparing initial rates of sterol oxidation using the same sample (POPC/DHE LUVs; 19.1 mol% DHE; 30 μL of 300 m*M* AAPH per assay; 8 nmols DHE per assay) on four consecutive days, and calculating the percent decrease per day. Between daily assays, AAPH is stored at 4°C. After 4 d of storage, the initial rate of sterol oxidation is about 20% less than the original measurement. All samples comprising a single sample set should be assayed within 3 d.
6. The original procedure calls for a smaller amount of H_2O_2; however, it was found that the samples would sometimes turn brown after 30 min in the heating block if there was an insufficient amount of H_2O_2. To avoid this, add extra H_2O_2; 100–200 μL is sufficient to prevent this color change. If the samples are not clear on removal from heating block, the final absorbance readings will be affected.
7. The original assay calls for 2 mL of H_2O, 100 μL of ammonium molybdate and 100 μL of FSR reagent. The volume may be adjusted to allow for differences in the final volume required for a particular spectrophotometer, as long as the proportions remain consistent.
8. The temperature required for vortexing step (*see* **Subheading 3.3.**, **step 4**), heating/cooling step (*see* **Subheading 3.3.**, **step 5**) and the temperature at which the assay is performed (*see* **Subheading 3.4.**, **step 1**) all depend on the membrane system used. The temperature chosen is usually about 10°C more than the main transition temperature of the matrix lipid. For the studies, 37°C was used.

9. If the LUVs are used to study sterol lateral organization, then the Nucleopore filter used for extrusion should have pores that are more than 60 nm *(28)*. When vesicles are too small, the radius of curvature is large enough to disturb the lateral organization of the lipid molecules. For other studies, LUVs of any size can be used.
10. Lipid loss because of extrusion process was calculated to be about 40% *(10)*.
11. Size was determined before and immediately after extrusion, as well as after the postextrusion 7-d incubation period. Vesicle size was also measured before and after some fluorescence assays, both in the absence and presence of antioxidants. It was found that there was no significant change in vesicle size because of addition of AAPH or any antioxidant used in the assay.
12. The recommended operation temperature range for this assay is 15–47°C. Below 15°C, it is necessary to flush the sample compartment of the fluorometer with nitrogen to get rid of moisture condensation on optical components. If the assay is performed at more than 47°C, an additional factor must be considered. At higher temperatures, sterol oxidation occurs at a faster rate, and the kinetic trace of the AAPH-induced DHE oxidation usually reaches an asymptotic value within 1 h. At this point, both initial rate of oxidation and apparent rate constant (k) can be calculated as described (*see* **Subheading 3.5.**). However, continuation of the time trace monitoring past this point (at temperatures >47°C) would result in the appearance of a sudden steep rise in fluorescence intensity over time (Yoon and Chong, unpublished results). This is probably owing to the formation of either an additional compound or an excited state complex. In either case, the initial rate of AAPH-induced sterol oxidation and the k value calculated from the DHE fluorescence intensity decay (once the profile has reached an asymptotic value) are meaningful, and can be interpreted in the same way as data retrieved at lower temperatures. Even at higher temperatures (>47°C), HPLC analysis clearly shows that, on addition of AAPH, the initial decay of DHE fluorescence intensity is because of the oxidation of DHE (Yoon and Chong, unpublished results).
13. Photobleaching is the irreversible destruction of an excited fluorophore on exposure to light. To measure the contribution of decreased fluorescence intensity by photobleaching, fluorescence intensity was measured for more than 5 h under the experimental conditions (37°C) in the absence of AAPH. The contribution to intensity loss by photobleaching was found to be approximately 0.5%. This is a contribution small enough to be ignored.
14. Thirty microliters of 300 m*M* AAPH was used for each reaction. AAPH solution was freshly prepared for each sample set and used within 3 d. Prepare AAPH by measuring appropriate amount and mixing with (Millipore) water. Store at 4°C in between assays; remove from refrigerator only when it is time to add AAPH to reaction cuvette and replace when finished. The final ratio in the reaction cuvette of DHE:AAPH is approx 1:1.
15. For purposes of illustration, AAPH is added here after 1000 s. In a typical assay, background intensity can be monitored for a much shorter time period to verify that there is no appreciable photobleaching or other artifacts that would result in DHE oxidation in the absence of peroxyfree radical.
16. On addition of AAPH, there is an immediate sharp drop in DHE fluorescence intensity, followed by a steady decrease in fluorescence intensity over time (*see* **Fig. 1**). The sharp drop in fluorescence intensity that occurs on addition of free radical generator is owing to resonance energy transfer, which is the transfer of excited state energy from an excited donor (DHE) to an acceptor (AAPH). As AAPH is nonfluorescent, the DHE fluorescence is thus reduced. Energy transfer does not affect the calculation of initial rate of sterol oxidation or the determination of lag phase length because energy transfer occurs much faster than sterol oxidation. The steady decrease in intensity over time results from AAPH-induced sterol oxidation.
17. The fluorescence intensity measured at a long time after addition of AAPH is designated as F_∞, which is typically 10% of the initial intensity. The existence of a nonzero F_∞ value is probably a result of either unreacted DHE in an inaccessible area or production of a weakly fluorescent product.
18. When comparative studies are conducted, the initial rates from different samples must first be normalized to unity.

Acknowledgments

The authors gratefully acknowledge the support from American Heart Association (0255082N, 0425617U and 0655418U) and American Chemical Society Petroleum Research Fund (PRF No. 38205-AC7).

References

1. Aviram, M. (2000) Review of human studies on oxidative damage and antioxidant protection related to cardiovascular diseases. *Free Radic. Res.* **33(Suppl),** S85–S97.
2. Panini, S. R. and Sinensky, M. S. (2001) Mechanisms of oxysterol-induced apoptosis. *Curr. Opin. Lipidol.* **12,** 529–533.
3. Peng, S., Hu, B., and Morin, R. J. (1992) Effects of cholesterol oxides on atherogenesis, in *Biological effects of cholesterol oxides*, p. 167–189 (Peng, S. and Morin, R. J., eds.), CRC Press, Boca Raton, FL.
4. Chang, J. Y., Phelan, K. D., and Liu, L. Z. (1998) Neurotoxicity of 25-OH-cholesterol on NGF-differentiated PC12 cells. *Neurochem. Res.* **23,** 7–16.
5. Steinberg, D. (1991) Antioxidants and atherosclerosis. A current assessment. *Circulation* **84,** 1420–1425.
6. Bjorkhem, I. (2002) Do oxysterols control cholesterol homeostasis? *J. Clin. Invest.* **110,** 725–730.
7. Park, P. S. W. and Addis, P. B. (1992) Methods of analysis of cholesterol oxides, in *Biological effects of cholesterol oxides*, p. 33–70 (Peng, S. and Morin, R. J., eds.), CRC Press, Boca Raton, FL.
8. Smith, A. G. and Brooks, C. J. (1977) The substrate specificity and stereochemistry, reversibility and inhibition of the 3-oxo steroid delta 4-delta 5-isomerase component of cholesterol oxidase. *Biochem. J.* **167,** 121–129.
9. Park, P. S. W. and Addis, P. B. (1985) HPLC determination of C-7 oxidized cholesterol derivatives in foods. *J. Food Sci.* **50,** 1437–1441.
10. Wang, M. M., Olsher, M., Sugar, I. P., and Chong, P. L. -G. (2004) Cholesterol superlattice modulates the activity of cholesterol oxidase in lipid membranes. *Biochemistry* **43,** 2159–2166.
11. Chong, P. L. -G. (1994) Evidence for regular distribution of sterols in liquid crystalline phosphatidylcholine bilayers. *Proc. Natl. Acad. Sci. USA* **91,** 10,069–10,073.
12. Virtanen, J. A., Ruonala, M., Vauhkonen, M., and Somerharju, P. (1995) Lateral organization of liquid-crystalline cholesterol-dimyristoylphosphatidylcholine bilayers. Evidence for domains with hexagonal and centered rectangular cholesterol superlattices. *Biochemistry* **34,** 11,568–11,581.
13. Chong, P. L. -G. and Olsher, M. (2004) Fluorescence studies of the existence and functional importance of regular distributions in liposomal membranes. *Soft Mater.* **2,** 85–108.
14. Chong, P. L. -G. and Sugar, I. P. (2002) Fluorescence studies of lipid regular distribution in membranes. *Chem. Phys. Lipids* **116,** 153–175.
15. Wang, M. M., Sugar, I. P., and Chong, P. L. - G. (1998) Role of the sterol superlattice in the partitioning of the antifungal drug nystatin into lipid membranes. *Biochemistry* **37,** 11,797–11,805.
16. Liu, F. and Chong, P. L. -G. (1999) Evidence for a regulatory role of cholesterol superlattices in the hydrolytic activity of secretory phospholipase A2 in lipid membranes. *Biochemistry* **38,** 3867–3873.
17. Olsher, M., Yoon, S. I., and Chong, P. L. -G. (2005) Role of sterol superlattice in free radical-induced sterol oxidation in lipid membranes. *Biochemistry* **44,** 2080–2087.
18. Muczynski, K. A. and Stahl, W. L. (1983) Incorporation of danslated phospholipids and dehydroergosterol into membranes using a phospholipid exchange protein. *Biochemistry* **22,** 6037–6048.
19. Bartlett, G. R. (1959) Phosphorus assay in column chromatography. *J. Biol. Chem.* **234,** 466–468.
20. MacLachlan, J., Wotherspoon, A. T., Ansell, R. O., and Brooks, C. J. (2000) Cholesterol oxidase: sources, physical properties and analytical applications. *J. Steroid Biochem. Mol. Biol.* **72,** 169–195.

21. Moore, N. F., Patzer, E. J., Barenholz, Y., and Wagner, R. R. (1977) Effect of phospholipase C and cholesterol oxidase on membrane integrity, microviscosity, and infectivity of vesicular stomatitis virus. *Biochemistry* **16,** 4708–4715.
22. Bar, L. K., Chong, P. L. -G., Barenholz, Y., and Thompson, T. E. (1989) Spontaneous transfer between phospholipid bilayers of dehydroergosterol, a fluorescent cholesterol analog. *Biochim. Biophys. Acta* **983,** 109–112.
23. Liu, F., Sugar, I. P., and Chong, P. L. -G. (1997) Cholesterol and ergosterol superlattices in three-component liquid crystalline lipid bilayers as revealed by dehydroergosterol fluorescence. *Biophys. J.* **72,** 2243–2254.
24. Schroeder, F., Barenholz, Y., Gratton, E., and Thompson, T. E. (1987) A fluorescence study of dehydroergosterol in phosphatidylcholine bilayer vesicles. *Biochemistry* **26,** 2441–2448.
25. Chong, P. L. -G., Liu, F., Wang, M. M., Truong, K., Sugar, I. P., and Brown, R. E. (1996) Fluorescence evidence for cholesterol regular distribution in phosphatidylcholine and in sphingomyelin lipid bilayers. *J. Fluoresc.* **6,** 221–230.
26. Krasowska, A., Rosiak, D., Szkapiak, K., Oswiecimska, M., Witek, S., and Lukaszewicz, M. (2001) The antioxidant activity of BHT and new phenolic compounds PYA and PPA measured by chemiluminescence. *Cell Mol. Biol. Lett.* **6,** 71–81.
27. Niki, E. (1990) Free radical initiators as source of water- or lipid-soluble peroxyl radicals. *Methods Enzymol.* **186,** 100–108.
28. Chong, P. L. -G., Tang, D., and Sugar, I. P. (1994) Exploration of physical principles underlying lipid regular distribution: effects of pressure, temperature, and radius of curvature on E/M dips in pyrene-labeled PC/DMPC binary mixtures. *Biophys. J.* **66,** 2029–2038.

11

Fluorescence Detection of Signs of Sterol Superlattice Formation in Lipid Membranes

Parkson Lee-Gau Chong, Berenice Venegas, and Michelle Olsher

Summary

There is a significant amount of experimental data, obtained predominantly from fluorescence studies, showing that sterol-containing liposomes can exhibit multiple biphasic changes in membrane properties at specific critical mole fractions of sterol such as 20.0, 22.2, 25.0, 33.3, 40.0, and 50.0 mol%. This can be understood in terms of the sterol regular distribution (e.g., superlattice) model. Here, the authors use excitation generalized polarization of 6-lauroyl-2-dimethylamino-naphthalene fluorescence in fluid 1-palmitoyl-2-oleoyl-*sn*-glycero-3-phosphocholine/cholesterol unilamellar vesicles to illustrate the experimental procedures and conditions that are required to detect multiple biphasic changes at predicted sterol mole fraction values in liposomal membranes. For this detection, the use of small sterol increments over a wide sterol mole fraction range is essential. Lipid concentration, incubation time, thermal history, and degree of sterol oxidation of liposomal membranes are critical factors. The principles and methodologies described here can be extended to other probes or bioactive molecules, such as enzymes, and can be applied to study sterol lateral organization in multicomponent lipid membranes.

Key Words: Fluorescence; lateral organization; liposomes; membranes; sterols; superlattices.

1. Introduction

The importance of lateral heterogeneity of lipid membranes became more widely acknowledged after the postulation of the raft hypothesis in 1997 *(1)*. Although the coexistence of stable lipid domains in model membranes was observed long before this date, the presence of phase coexistence in cell membranes is still controversial and the understanding of such domains at the molecular level, even in model membranes, is poor. Experimental evidence is being accumulated supporting the presence of lateral regular distributions in sterol-containing liposomal membranes (reviewed in **refs. 2–4**). Abrupt biphasic profiles in physical properties were observed at specific critical sterol mole fractions (C_r) (e.g., 15.4, 20.0, 22.2, 25.0, 33.3, 40.0, and 50.0 mol% sterol) *(5–14)*. At present, this behavior can only be explained by a sudden lateral reorganization of the lipids because of the maximum formation of sterol superlattice in the plane of the membrane at C_r (*see* **Fig. 1**).

According to this model, regularly distributed sterol superlattices and irregularly distributed lipid areas coexist in sterol-containing membranes. The ratio of irregular to regular regions reaches a local minimum at C_r. In the regular regions, sterol molecules can be distributed into either hexagonal or centered-rectangular superlattices *(5,6)* (*see* **Fig. 1**). The model also proposes that membrane-free volume varies with sterol content in an alternating manner, reaching a local minimum at C_r *(7,8,15)* (*see* **Fig. 1**) because sterol molecules in the regular regions are more

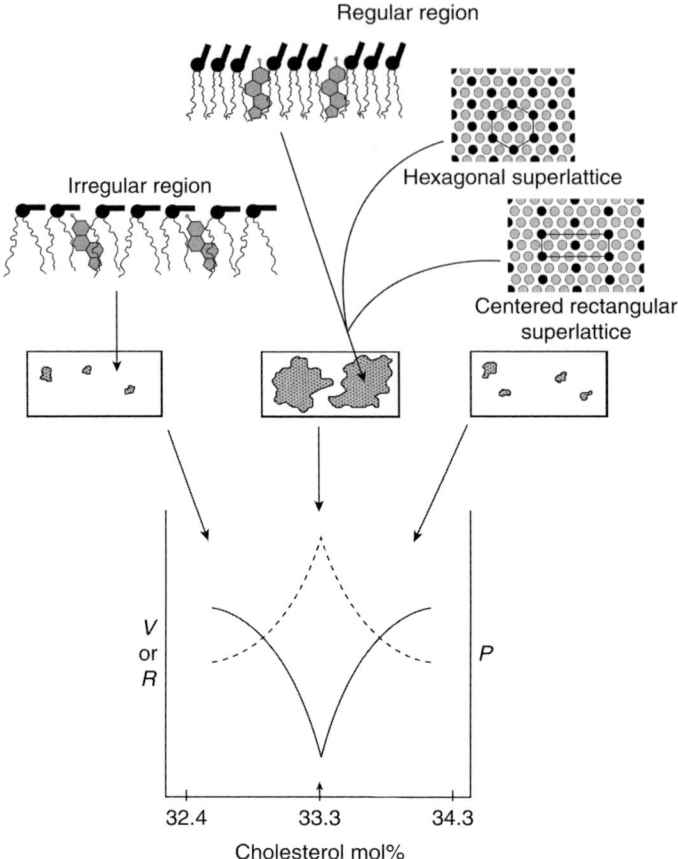

Fig. 1. Schematic diagram showing the current concepts underlying the sterol regular distribution model. The rectangle-like objects represent a lipid membrane, where regular (*shaded areas*) and irregular (*blank areas*) regions coexist. In regular regions, sterol molecules are regularly distributed into either hexagonal or centered rectangular superlattices within the host lipid matrix. There is a biphasic change in proportion of irregular region to regular region (*R, solid line*), membrane free volume (*V, solid line*), and the perimeter of regular region (*P, dashed line*) with membrane cholesterol content in the neighborhood of a critical sterol mole fraction C_r (e.g., 33.3 mol% cholesterol as indicated by arrow) theoretically predicted for maximal superlattice formation. (reproduced from **ref. 24** with permission)

tightly packed than those in the irregular regions *(7,9)*. Sterol exposure to water or water soluble materials, on the other hand, reaches a local maximum at C_r (*see* **Fig. 1**) (*7, 12, 14*), mainly because of the increase in the interfacial area between regular and irregular regions (*see* **Fig. 1**).

The molecular interactions that can simulate such regular distributions include a small, unfavorable sterol–sterol repulsion and a large, favorable sterol–phospholipid attraction as a result of the hydrophobic effect *(13,16)*. Another major factor is the multibody interactions of sterol with nearest neighbors. Sterol has a condensing effect on neighboring acyl chains, reducing the number of *gauche* configurations along the polymethylene units. This causes an unfavorable entropy loss. The fine balance between the unfavorable interactions (entropy loss and sterol–sterol repulsion) and the overall favorable hydrophobic effect drives the sterol to form superlattice-like regular distributions in the membrane.

The detection of multiple biphasic profiles at C_r can be reasonably accepted as evidence for sterol superlattice formation. In this chapter, excitation generalized polarization (GP_{ex}) *(17)* of 6-lauroyl-2-dimethylamino-naphthalene (LAURDAN) fluorescence in fluid 1-palmitoyl-2-oleoyl-*sn*-glycero-3-phosphocholine (POPC)/cholesterol large unilamellar vesicles (LUVs) is used to illustrate the experimental procedures and conditions that are required to detect multiple biphasic changes at predicted C_r values in liposomal membranes. The factors that may hinder or obliterate the detection of signs for superlattice formation are also discussed.

In short, use of small sterol increments (~0.3 mol%) over a wide sterol mole fraction range is essential. The concentrations of lipid stock solutions must be determined precisely and accurately. Proper use of pipettes, choice of lipid concentration in vesicle dispersions, vesicle incubation time, vesicle thermal history, and degrees of sterol oxidation are critical factors. The principles and methodologies described here can be extended to study membrane properties using other fluorescent probes *(5,6,8,18)* or bioactive molecules such as enzymes and free radicals *(10,12,14)*, and can be applied to study sterol lateral organization in multicomponent lipid membranes *(14)*.

2. Materials

2.1. Formation of LUV

1. Storage tubes: Pyrex round bottom glass tubes that can be closed with a phenolic screw cap having a Teflon liner (Fisher Scientific, Pittsburgh, PA) that is resistant to chloroform and many other organic solvents (*see* **Note 1**).
2. POPC (*see* **Note 2**) and cholesterol (*see* **Note 3**) are stored at −20°C (use −70°C for long-term storage). Stock solutions of POPC and cholesterol are made in chloroform at a concentration around 1 mM.
3. 50 mM Tris-HCl buffer, pH 7.2, 0.02% NaN_3, 10 mM ethylenediaminetetraacetic acid (EDTA), filtered through a syringe filter (0.2 μm).
4. Gilson Microman pipettes (Gilson, Middleton, WI).
5. Nitrogen (compressed gas) and Reacti-Vap III plus Reacti-Therm III manifold (Rockford, IL).
6. Flexi-dry freeze-dryer (FTS Systems, Stone Ridge, NY).
7. A vacuum chamber made by connecting a dessicator to the laboratory vacuum line.
8. Extruder (Lipex Biomembranes, Vancouver, Canada) and Nucleopore polycarbonate membranes (Whatman, Brentford, UK).
9. Malvern Zetasizer HS-1000 spectrometer (Malvern Inst., Ltd., UK); light source: 10 mW He–Ne laser set at 633 nm.

2.2. Phosphorus Assay

1. Phosphorus standard (0.65 mM) (Sigma Chemical Co., St. Louis, MO) stored at 4°C.
2. 50% Sulfuric acid, 30% hydrogen peroxide (stored at 4°C), 5% ammonium molybdate, and sodium bisulfite.
3. Fiske–Subbarow reducer (FSR) (Sigma Chemical Co., St. Louis, MO) stored at room temperature and protected from light.
4. Fisher isotemp water bath (also used in cholesterol assay).
5. Heating plate and heating blocks (20 wells).
6. Spectrophotometer (Model Lambda-25, Perkin-Elmer Instruments, Waltham, MA) (also used in cholesterol assay).

2.3. Cholesterol Assay

1. Cholesterol E kit (Wako Chemicals, Richmond, VA).
2. 0.1% (w/v) Triton X-100.

2.4. Fluorescence Measurements of LAURDAN

1. LAURDAN (Molecular Probes Inc., Eugene, OR) stock solution: LAURDAN dissolved in dimethyl sulfoxide (DMSO) (high-performance liquid chromatography [HPLC] grade) stored at −20°C in dark (*see* **Subheading 3.4.**).
2. Polypropylene screw cap amber vials (size: 2 mL) with O-ring (Perfector Scientific, Atascadero, CA).
3. Fluorometer (ISS, Inc., Champaign, IL) small stirring bar, and quartz cuvettes.

3. Methods

3.1. Preparation of LUVs

The observation of biphasic behavior as a function of sterol mole fraction strongly relies on vesicle preparation. To collect empirical evidence of this kind of profile, the sample set must be planned and prepared very carefully. A sample set is made up of several tubes of LUVs (*see* **Table 1**); each sample has a slightly different sterol mole fraction. For good resolution of the profile, it is desirable to vary the sterol mole fraction in small increments (0.2–0.4 mol%), especially when the C_r are very close to one another. This calls for highly accurate pipetting of the lipid stock solutions and careful determination of lipid concentrations in stock solutions. In addition, a biphasic membrane behavior at a given C_r has to be studied with a sample set spanning approx a 3 mol% region around the critical point (*see* **Note 4**). At least one sample should be prepared in triplicate to allow for the calculation of error; however, it is better to choose two sterol mole fractions (one at C_r, and the other far from C_r).

The preparation of all samples in the set must be completed within a short time frame, in order to ensure that the thermal history of each sample is virtually the same. For this reason, all vesicle samples in a sample set should be prepared on the same day. If the thermal history is not preserved, then the lateral organization equilibration process may be different from one sample to the next. In this case, observation of biphasic change is still possible, but a much longer incubation time may be required to observe the change. Thus, if the biphasic profile was not observed, it would be impossible to distinguish between samples that have not yet reached thermal equilibrium for lateral organization and samples that were improperly prepared. If attention is given to pipetting and to the thermal history of the samples, the observation of biphasic phenomena is more likely to be successful. **Table 1** describes the preparation of a typical sample set in which the absolute amount of sterol remains constant, whereas the amount of POPC is varied by small increments.

1. In a Pyrex test tube with a screw cap having a Teflon liner, dissolve an appropriate amount of POPC (or cholesterol) in HPLC-grade chloroform, making a stock solution of POPC (or cholesterol) at a concentration close to 1 mM. Use Gilson Microman pipettes to transfer organic solvents or lipid samples dissolved in organic solvents. Stock solutions are sealed with Teflon tape and parafilm and stored at −20°C.
2. The cholesterol and phospholipid concentrations in stock solutions are determined by a Cholesterol E kit from Wako and the method of Bartlett *(19)*, respectively.
3. Plan sample set as illustrated in **Table 1**; follow the table to pipette calculated amounts of POPC and cholesterol stock solutions into clean Pyrex test tubes (*see* **Note 5**). Stock solution tubes should remain on ice during the pipetting step and should be closed immediately when not in use, to slow evaporation.
4. Evaporate the chloroform in each tube of the same sample set at the same time under nitrogen gas stream using a manifold (Reacti-Vap III plus Reacti-Therm III) until no visible solvent remains. Then, dry lipids overnight under high vacuum (Flexi-Dry, FTS Systems).
5. The next day (before tubes are removed from vacuum), preheat a water bath and an incubator (large enough to contain a vortex) to 40°C, so that when the incubator door is opened for vortexing of samples, a temperature of 40°C is maintained (*see* **Note 6**).

Table 1
Example of Sample Set Around Cholesterol Critical Mole Fraction 33.3 mol%

Tube number	Cholesterol (mol%)	1 mM Cholesterol stock (μL)	1.3 mM POPC stock (μL)	Difference in volume (μL)	Cholesterol (nanomoles per mL buffer)	POPC (nanomoles per mL buffer)
1	31.3	500	843	0	100	219.3
2a	31.7	500	828	15	100	215.2
2b	31.7	500	828	15	100	215.2
2c	31.7	500	828	15	100	215.2
3	32.1	500	813	15	100	211.3
4	32.5	500	798	15	100	207.5
5	32.9	500	784	14	100	203.8
6a	33.3	500	769	15	100	200.0
6b	33.3	500	769	15	100	200.0
6c	33.3	500	769	15	100	200.0
7	33.7	500	756	13	100	196.5
8	34.1	500	743	13	100	193.1
9	34.5	500	729	14	100	189.5
10	34.9	500	717	12	100	186.4
11	35.3	500	704	13	100	183.0

The concentration of the stock solution is 1 mM for cholesterol and 1.3 mM for POPC. The final volume of the MLV sample is 5 mL, which gives [POPC] = 200 μM at 33.3 mol%. Two triplicates were made to generate typical error bars.

6. Warm buffer to 40°C in water bath. Remove tubes from vacuum and place them in the water bath.
7. Reconstitute lipids using the following procedure: pipette warm buffer into tubes, flush with argon, cap and seal with Teflon tape and parafilm. Vortex the lipid dispersions at 40°C for a certain amount of time (e.g., 5 min) until the pellet is completely resuspended to form multilamellar vesiles (MLVs). All the vesicle samples in the same set should be vortexed for the same amount of time. After reconstitution, flush samples again with argon, recap and reseal tubes with Teflon tape and parafilm.
8. Incubate the sample set at 40°C for 1 h. MLVs are then subjected to three cooling/heating cycles (each cycle equals a period of 30 min at 4°C followed by another 30 min at 40°C). Then, incubate MLVs at room temperature for 4 d in the vacuum chamber (a dessicator connected to the vacuum line) (*see* **Note 7**) to allow vesicles to come to thermal equilibrium for lateral organization (*see* **Note 8**).
9. On day five, prepare LUVs from MLVs by extrusion (*see* **Note 9**). It is advisable to extrude all samples on the same day. Preheat the extruder to 40°C. It is also necessary to have some hot deionized water (~40–60°C) that will be used to clean the extruder between samples. Set the nitrogen pressure (100–150 psi for 200-nm pore size in the polycarbonate membrane), which must remain constant throughout the extrusion of the sample set.
10. Incubate first tube at 40°C for 10 min, then extrude sample 10 times through two stacked Nucleopore polycarbonate membranes with the desired membrane pore size to form LUVs (*see* **Note 10**). After extrusion, flush tube with argon, seal with parafilm, and wash extruder 10 times with hot water. Repeat this for all tubes.
11. After extrusion, measure vesicle size with the Malvern Zetasizer (Malvern, UK). In a disposable cuvette, pipette 1.5 mL of filtered and degassed water and add 20 μL of the vesicle suspension; then, begin measurement.
12. Store LUVs for at least 4 d *(5)* under argon at room temperature in a vacuum chamber (*see* **Subheading 2.1.7.**) to avoid oxidation of sterols and phospholipids (*see* **Note 11**).

13. During the incubation period, it is also necessary to measure the phospholipid concentrations of each LUV sample by the method of Bartlett *(19)* to determine the new phospholipid concentration after loss as a result of extrusion. After phosphorus determination, dilute the stock solution to desired POPC concentration (*see* **Note 12**).

3.2. Phosphorus Assay

These instructions are modified from the method of Bartlett *(19)*.

1. Pipette phosphorus standard solution (0.65 m*M*) into test tubes. Two tubes are left empty to serve as references; the remaining six tubes contain 13, 26, 39, 52, 65, and 78 nmols of phosphorus, and will be used to produce a standard curve.
2. Pipette phospholipid stock solution for phosphorus concentration determination. Calculate approximately how much sample would be required so that the unknown concentrations will fall within the standard curve values.
3. Evaporate the organic solvent of the sample. If sample is in aqueous solution there is no need to do this step.
4. Add 200 µL of 50% sulfuric acid to each tube and vortex gently.
5. Place tubes in preheated aluminum block (200°C) for 5 min.
6. Add 200 µL (*see* **Note 13**) of 30% hydrogen peroxide to each tube and char in aluminum heating block for 30 min.
7. Remove tubes from block and allow cooling.
8. Add 2.3 mL of water (*see* **Note 14**) to each tube; then add 115 µL of 5% ammonium molybdate to each tube and vortex gently; if liquid becomes yellow after addition of ammonium molybdate, add a few crystals of sodium bisulfite, and vortex until liquid becomes clear.
9. Prepare FSR according to manufacturer's instructions; add 115 µL of freshly made FSR to each tube and vortex gently.
10. Place tubes in a water bath at 100°C for 20 min or until samples turn blue.
11. Allow tubes to cool. Read absorbance at 660 nm (start with least concentrated samples).

3.3. Cholesterol Assay (Modified From Manufacturer's Instructions)

1. Dilute the cholesterol standard solution (20 mg/mL, from the Cholesterol E kit (Wako Chemicals USA, Inc., Richmond, VA) 10 times.
2. Prepare 4 mL of the color reagent (from Wako) using specified proportions of reagent and buffer.
3. Pipette 20, 40, 60, 80, 100, and 120 µL of the diluted cholesterol standard into separate test tubes; two empty tubes will serve as blanks. Pipette the sample (cholesterol stock solution, ~1 m*M*) in suitable amounts in triplicate. Evaporate the solvent of the sample by nitrogen.
4. Add 300 µL of 0.1% Triton X-100. Vortex until the cholesterol pellet is resuspended.
5. Add 300 µL of the color reagent and vortex.
6. Place tubes in a water bath at 37°C for 5 min.
7. Take out the tubes and allow them to cool. Then add water so that all tubes contain a total volume of 1.3 mL.
8. Measure absorbance at 600 nm.

3.4. Measurements of LAURDAN GP_{ex}

1. Prepare LAURDAN stock solution by dissolving a very small amount of LAURDAN powder in 3 mL of DMSO (HPLC grade). Pipette an aliquot (e.g., 25 µL) of LAURDAN stock solution into 3 mL of methanol (Optima, Fisher Scientific, Pittsburgh, PA) in a cuvette and mix well. Estimate the concentration of the stock by measuring the absorbance at 364 nm (A_{364}) using a spectrophotometer. The LAURDAN concentration is calculated using the equation: $[LAURDAN]_{stock} = (A_{364}/\varepsilon) \times$ (dilution factor), where ε is the extinction coefficient of LAURDAN in methanol at 364 nm (20,000 M^{-1} cm^{-1}). The LAURDAN stock solution used in the GP_{ex} measurements (*see* **Figs. 2–4**) was 40 µ*M*.

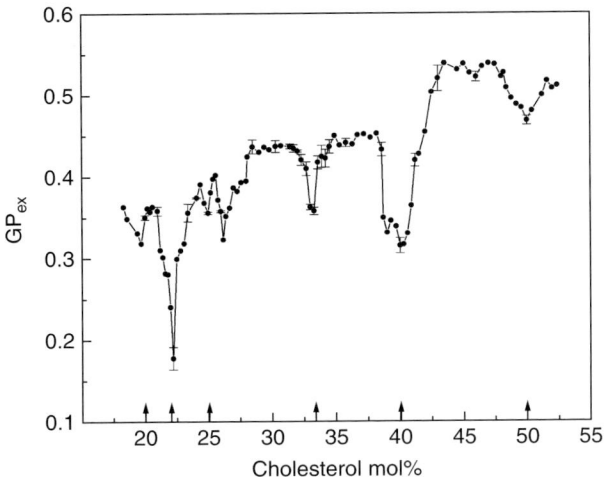

Fig. 2. Laurdan's GP_{ex} as a function of cholesterol content in cholesterol/POPC LUV (diameter ~160–180 nm). GP_{ex} was measured at 24°C. GP_{ex} minima (GP_{ex} dips) appear at 19.7, 22.2, 25.0, 26.2, 33.3, 40.3, and 50.0 mol%. Arrows indicate the theoretically predicted critical sterol mole fractions (C_r) for maximal superlattice formation in the mole percent range examined. Within the experimental errors (~0.1–0.4 mol% sterol; *see* **ref. 8**), the GP_{ex} dip values (except for 26.2 mol%) agree with the theoretical C_r values (i.e., 20.0, 22.2, 25.0, 33.3, 40.0 and 50.0 mol%). The vertical bars are the standard deviations of GP_{ex} obtained from three independently prepared vesicle samples. [POPC] was 40-60 μM and vesicles were incubated 7 days or more prior to GP_{ex} measurements. (reproduced from **ref. 24** with permission)

2. From sample number one (tube no. 1 in **Table 1**), add a volume of POPC that corresponds to 40 nmols into each of two small amber vials. Add buffer to give a final volume of 1.8 mL. Do the same for the rest of the sample set to generate two sets of vials.
3. To each of the vials in one of the two sets, add 5 μL of 40 μ*M* LAURDAN stock solution (LAURDAN-to-POPC molar ratio 1:200). Add the same amount of DMSO used in the first vial to the second set of vials (used as reference).
4. Incubate for at least 1 h.
5. Scan fluorescence emission spectrum from 400 to 520 nm for each vial containing LAURDAN (excitation wavelength: 340 nm); subtract the readings from the corresponding reference vial.
6. Calculate GP_{ex} *(17)* using the equation: $GP_{ex} = (I_{435} - I_{500})/(I_{435} + I_{500})$, where I_{435} and I_{500} are the fluorescence intensity at 435 nm and 500 nm, respectively.
7. Make a graph of the GP_{ex} values of all samples as a function of the cholesterol mol%. **Figure 2** illustrates the multiple biphasic changes in LAURDAN GP_{ex} observed using this method. In cholesterol/POPC LUVs, GP_{ex} reaches a local minimum at 19.7, 22.2, 25.0, 26.2, 33.3, 40.3 and 50.0 mol%; these values (except for 26.2 mol%) are at or near the theoretically predicted critical mole fractions (i.e., 20.0, 22.2, 25.0, 33.3, 40.0, and 50.0 mol%) in the mole percent ranges examined. The observation of GP_{ex} dips (rather than peaks) at C_r is consistent with the superlattice model in that sterol is more exposed to the aqueous phase at C_r than at non-C_r (*see* **Note 15**). It is known that LAURDAN GP_{ex} decreases when the water content near the chromophore of LAURDAN is high *(17)*. LAURDAN's chromophore in monopolar diester liposomes is located in the bilayer near the water–membrane interfacial region *(20)*.
8. **Figure 3** illustrates the importance of keeping sterols from auto-oxidation (*see* **Note 16**), which may obliterate the biphasic change in LAURDAN GP_{ex} at 50.0 mol% cholesterol in cholesterol/POPC LUVs.
9. **Figure 4** demonstrates that both vesicle incubation time and phospholipid concentration in the vesicle dispersion are also critical factors for the detection of biphasic changes in membrane

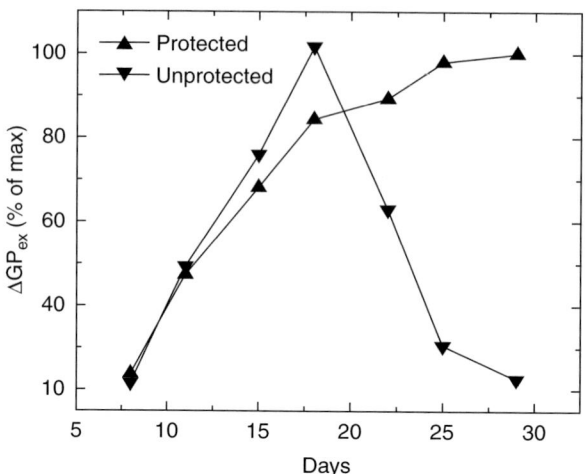

Fig. 3. Comparison of the effect of vesicle incubation time on the depth of the LAURDAN GP_{ex} dip (ΔGP_{ex}) detected at the critical sterol mole fraction 50.0 mol% cholesterol in cholesterol/POPC LUVs (diameter ~160–180 nm) for two sets of samples ([POPC] = 40 μM): (▲) protected (*see* **Subheading 3.1.12.**) and (▼) unprotected (neither placed in the vacuum chamber nor flushed with argon) from sterol oxidation. Vesicles were incubated and GP_{ex} was measured at 24°C. (reproduced from **ref. 24** with permission)

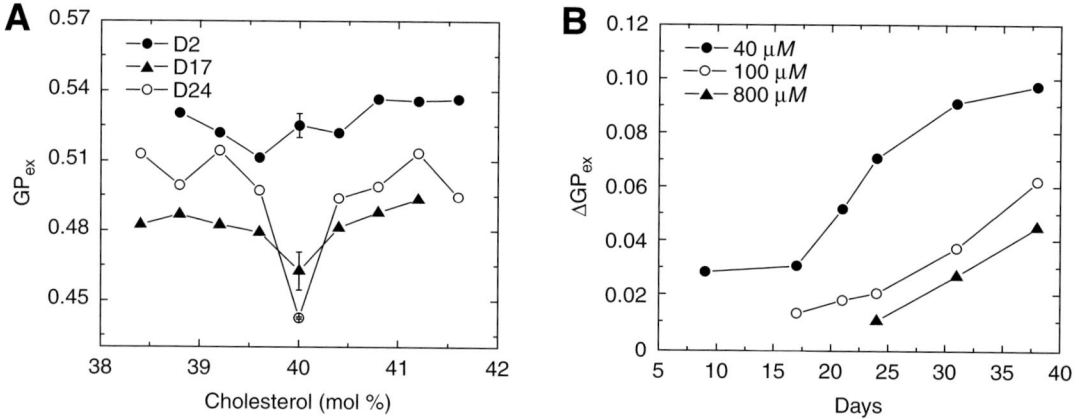

Fig. 4. **(A)** The evolution of the GP_{ex} dip profile over time for a set of samples in the neighborhood of 40.0 mol% cholesterol in cholesterol/POPC LUVs (diameter ~160–180 nm) with [POPC] = 40.0 μM. **(B)** Effects of vesicle incubation time and POPC concentration in the vesicle dispersions on the depth of the LAURDAN GP_{ex} dip (ΔGP_{ex}) detected at the critical mole fraction 40.0 mol% cholesterol in cholesterol/POPC LUVs. Vesicles were incubated, and GP_{ex} was measured at 24°C. All the samples were protected from sterol oxidation using the method described in the text. (reproduced from **ref. 24** with permission)

properties at C_r. If the sample set is carefully prepared and well protected from oxidation and the biphasic profile is not obvious, it may be possible that the sample set needs a longer incubation time (*see* **Note 17**).

4. Notes

1. For storing lipid stock solutions and making lipid vesicles, Pyrex round bottom test tubes are used that can be closed with a phenolic screw cap having a Teflon liner. All glassware used for

lipid storage and vesicle preparation should first be washed with polar and nonpolar solvents. For the vesicle preparation, the authors use one wash with chloroform:methanol (2:1 v/v), followed by one wash with ethanol, and a final washing with deionized water. Glassware must be dried completely.

2. Sphingolipids and phosphoacylglycerides other than POPC can also be used as matrix lipid(s). The purity of the phospholipids to be used can be checked by thin-layer chromatography.

3. In addition to cholesterol, the methodology presented in this chapter can be applied to other sterols such as ergosterol and the fluorescent cholesterol analog, dehydroergosterol *(8)*. Before use, the purity of sterols should be checked. These sterols can be purified by recrystalization in ethanol (or methanol) or by HPLC *(21)*.

4. Three to six sample sets should be prepared to cover some or all the C_r values (20.0, 22.2, 25.0, 33.3, 40.0 and 50.0 mol%) in the range of 18–55 sterol mol%. This would enable us to assess whether the biphasic points experimentally observed are correlated with the predicted C_r values. Here, it is suggested that a sample set spans a region that covers a single critical point C_r. Actually, a sample set may span a sterol mole fraction range much larger than 3 mol% such that more than one critical point can be covered in the same set. Selection of sample size in each set depends on whether the measurements of physical or biological properties of all the vesicle samples in the same set can be completed in a short period of time so the measured membrane properties are not a result of changes in sample thermal history.

5. Many factors determine the size of the sample set. All steps must be performed on each sample on the same day. For greater accuracy, use a POPC stock concentration that allows a large difference in the POPC volume to be pipetted into each sample tube. The minimum difference between consecutive tubes should be approx 10 µL (*see* **Table 1**). This volume can be measured accurately using Gilson Microman pipettes. In this way, small sterol mole percent increments such as 0.3 mol% can be readily achieved.

6. Samples must be vortexed at a temperature at least 10°C higher than the main phase transition temperature of the constituent lipids.

7. Lipids are susceptible to oxidation. Flushing samples with nitrogen or argon can protect from oxidation, but storing the sealed samples under vacuum will provide long-term protection.

8. These heating/cooling cycles and long incubation periods have frequently been used in liposome preparation for calorimetric studies *(22)*. This protocol provides a means of evenly distributing membrane components among lipid multilayers and attaining an equilibrium distribution of molecules within each monolayer.

9. Multiple biphasic changes in membrane properties with cholesterol content can be detected in MLVs *(5)* and LUVs *(12)*. The limitation is that when the vesicle is smaller than approx 60 nm in diameter, biphasic changes in membrane properties may be attenuated or abolished because of large curvature stress *(23)*. LUVs (around 200–800 nm in diameter) are preferred because the vesicle size distribution of extruded LUVs is much more narrow than that of MLVs. LUVs provide less scattered light, which might interfere with fluorescence measurements, and they do not precipitate easily.

10. The size distribution of the vesicles will be determined by the pore size of the polycarbonate membrane used and the number of times the sample is extruded. However, if small vesicles are present in the MLV dispersions, they will pass through the membrane pores without forming LUVs. The result will be a heterogenous distribution of particle sizes. To prevent this, freeze–thaw cycles are performed in order to increase the size of the MLV's before extrusion. Put dry ice in acetone, and place the sample in the dry-ice–acetone mixture for 2–3 min or until completely frozen. Then, incubate the sample at 40°C for 15 min. For good results, this cycle should be repeated at least 10 times. Moreover, a sample should not be extruded with a membrane in which the pore is less than half the size of the vesicles to be extruded. Otherwise, a large amount of lipid may be lost in the polycarbonate membrane and the extrusion will be very slow. Normally, at higher lipid concentrations, larger MLVs are formed, so if one wants a 200 nm LUV one should extrude first with a membrane having higher pore size, such as 400 nm, and then extrude with a 200 nm membrane pore.

11. Long incubation used here is to allow vesicles to come to thermal equilibrium for lateral organization, as mentioned in **Note 8**. The incubation time that is required to detect a biphasic change in membrane properties varies with vesicle lipid concentration (*see* **Note 17**), depends on the critical mole fraction region being examined, and changes with lipid membrane systems. An incubation of 12–24 h (or less) at 4°C is sufficient to detect a biphasic change in the ratio of pyrene excimer to monomer fluorescence intensity at the critical mole fraction in pyrene-labeled phosphatidylcholine/dimyristoylphosphatidylcholine LUVs ([lipid] ~2 μM) (*23*), a lipid membrane system also believed to form lipid superlattices (*2,4*). For sterol-containing vesicles with [lipid] less than 100 μM, the average minimum time to detect a biphasic change is about 4 d, but the actual incubation time needed for a biphasic change varies with C_r.

12. In principle, the cholesterol concentration of the LUVs has to be determined as well; thus, the sterol mole percent in each sample after extrusion can be known. However, in a previous study (*12*) it was shown that the sterol mole percents in vesicles before and after extrusion differed little (<0.2 mol%). Thus, for convenience, the sterol mole percents in MLVs can be used to assess the relationship between sterol content and membrane properties in LUVs. The phospholipid concentration of the LUVs determined here is needed to calculate the volume of the LUV dispersions to be pipetted into the sample cuvette so that LAURDAN GP_{ex} for each sample from the same set will be determined based on the same amount of POPC (*see* **Subheading 3.4.**).

13. The original procedure calls for a small amount of hydrogen peroxide; however, it was found that the samples would sometimes turn brown after 30 min in the heating block. If the samples are not clear on removal from the heating block, the final absorbance readings will be affected. To avoid this problem, the Bartlett assay was modified by adding 200 μL of hydrogen peroxide.

14. The original assay calls for 2 mL of water, 100 μL of ammonium molybdate, and 100 μL of FSR reagent. The volume may be adjusted to allow for differences in the final volume required for a particular spectrophotometer, as long as the proportions remain consistent.

15. The direction (either a dip or a peak) and the extent (depth of the dip or peak) of a biphasic change in membrane properties with sterol content at C_r vary with the fluorescent probes used and the type of enzymes or bioactive agents, depending on where in the membrane they reside and what membrane properties these extrinsic molecules are sensing. However, in all cases studied, the direction of the biphasic change in membrane properties at C_r can be explained by the principles of sterol superlattice formation (for review, *see* **ref. 3**).

16. A critical factor for generating multiple biphasic changes in membrane properties at C_r is the prevention of sterol oxidation. If the samples are well protected against oxidation of sterol, the depth of the GP_{ex} dip increases with increasing incubation time; if enough time is allotted, a maximum depth will be reached, which means that the maximum area covered by sterol superlattice has been achieved (*see* **Fig. 3**). However, if the sample is not well protected against sterol oxidation (as revealed by HPLC), then the GP_{ex} dip becomes hindered and even obliterated after a certain amount of incubation time (*see* **Fig. 3**). Sterol oxidation becomes a critical issue when the sample needs long incubation times to reach the lateral organization equilibrium. To protect vesicle samples from oxidation, vesicle samples should be flushed with argon every 2–3 d during the incubation time, and the sample tubes should be sealed with parafilm and stored in a low vacuum chamber. Alternatively, the samples can be stored inside a glovebox in an inert gas atmosphere.

17. The phospholipid concentration in vesicle dispersions (or the concentration of lipid vesicles) and the time for vesicle incubation (the time between the extrusion and the measurement of membrane properties) are also critical factors for generating biphasic changes in membrane properties at C_r. Frequently, a biphasic change in the GP_{ex} profile was only detected after the sample set was incubated for a certain amount of time (*see* **Fig. 4A**) so that the lateral organization of membrane lipids could reach or move toward equilibrium. This equilibration process strongly depends on lipid concentration. In **Fig. 4B**, it can be seen that the detection of the GP_{ex} dip at 40.0 mol% cholesterol takes a longer time for a more concentrated vesicle sample, while for all the lipid concentrations examined, the GP_{ex} dip becomes deeper over time. Note that the incubation time for the detection of a biphasic change in GP_{ex} and the lipid concentration dependence

of the dip depth (as shown in **Fig. 4**) may vary from one specific critical sterol mole fraction to another, but the general trend applies to all the critical sterol mole fractions.

Acknowledgments

The authors gratefully acknowledge the support from American Heart Association (0255082N, 0425617U, and 0655418U) and American Chemical Society Petroleum Research Fund (PRF no. 38205-AC7).

References

1. Simons, K. and Ikonen, E. (1997) Functional rafts in cell membranes. *Nature* **387,** 569–572.
2. Chong, P. L. -G. and Sugar, I. P. (2002) Fluorescence studies of lipid regular distribution in membranes. *Chem. Phys. Lipids* **116,** 153–175.
3. Chong, P. L. -G. and Olsher, M. (2004) Fluorescence studies of the existence and functional importance of regular distributions in liposomal membranes. *Soft Matter.* **2,** 85–108.
4. Somerharju, P., Virtanen, J. A., and Cheng, K. H. (1999) Lateral organisation of membrane lipids: the superlattice view. *Biochim. Biophys. Acta* **1440,** 32–48.
5. Chong, P. L. -G. (1994) Evidence for regular distribution of sterols in liquid crystalline phosphatidylcholine bilayers. *Proc. Natl. Acad. Sci., U.S.A.* **91,** 10,069–10,073.
6. Virtanen, J. A., Ruonala, M., Vauhkonen, M., and Somerharju, P. (1995) Lateral organization of liquid-crystalline cholesterol-dimyristoylphosphatidylcholine bilayers. Evidence for domains with hexagonal and centered rectangular cholesterol superlattices. *Biochemistry* **34,** 11,568–11,581.
7. Chong, P. L. -G., Liu, F., Wang, M. M., Truong, K., Sugar, I. P., and Brown, R. E. (1996) Fluorescence evidence for cholesterol regular distribution in phosphatidylcholine and in sphingomyelin lipid bilayers. *J. Fluoresc.* **6,** 221–230.
8. Liu, F., Sugar, I. P., and Chong, P. L. -G. (1997) Cholesterol and ergosterol superlattices in three-component liquid crystalline lipid bilayers as revealed by dehydroergosterol fluorescence. *Biophys. J.* **72,** 2243–2254.
9. Wang, M. M., Sugar, I. P., and Chong, P. L. -G. (1998) Role of the sterol superlattice in the partitioning of the antifungal drug nystatin into lipid membranes. *Biochemistry* **37,** 11,797–11,805.
10. Liu, F. and Chong, P. L. -G. (1999) Evidence for a regulatory role of cholesterol superlattices in the hydrolytic activity of secretory phospholipase A2 in lipid membranes. *Biochemistry* **38,** 3867–3873.
11. Wang, M. M., Sugar, I. P., and Chong, P. L. -G. (2002) Effect of double bond position on dehydroergosterol fluorescence intensity dips in phosphatidylcholine bilayers with saturated sn-1 and monoenoic sn-2 acyl chains. *J. Phys. Chem.* **106,** 6338–6345.
12. Wang, M. M., Olsher, M., Sugar, I. P., and Chong, P. L. -G. (2004) Cholesterol superlattice modulates the activity of cholesterol oxidase in lipid membranes. *Biochemistry* **43,** 2159–2166.
13. Parker, A., Miles, K., Cheng, K. H., and Huang, J. (2004) Lateral distribution of cholesterol in dioleoylphosphatidylcholine lipid bilayers: cholesterol-phospholipid interactions at high cholesterol limit. *Biophys. J.* **86,** 1532–1544.
14. Olsher, M., Yoon, S. I., and Chong, P. L. -G. (2005) Role of sterol superlattice in free radical-induced sterol oxidation in lipid membranes. *Biochemistry* **44,** 2080–2087.
15. Chong, P. L. -G. (1996) Membrane-free volume variation with bulky lipid concentration by regular distribution: a functionally important property explored by pressure studies of phosphatidylcholine bilayers in *High-Pressure Effects in Molecular Biophysics and Enzymology,* (Markley, J. L., Northrop, D. B., and Royer, C. A., eds.), Oxford University Press, New York, NY, pp. 298–313.
16. Huang, J. and Feigenson, G. W. (1999) A microscopic interaction model of maximum solubility of cholesterol in lipid bilayers. *Biophys. J.* **76,** 2142–2157.
17. Parasassi, T. G. and Gratton, E. (1995) Membrane lipid domains and dynamics as detected by laurdan fluorescence. *J. Fluoresc.* **5,** 59–69.

18. Cannon, B., Heath, G., Huang, J., Somerharju, P., Virtanen, J. A., and Cheng, K. H. (2003) Time-resolved fluorescence and Fourier transform infrared spectroscopic investigations of lateral packing defects and superlattice domains in compositionally uniform phosphatidylcholine bilayers. *Biophys. J.* **84,** 3777–3791.
19. Bartlett, G. R. (1959) Phosphorus assay in column chromatography. *J. Biol. Chem.* **234,** 466–468.
20. Chong, P. L. -G. and Wong, P. T. T. (1993) Interactions of laurdan with phosphatidylcholine liposomes: a high pressure FTIR study. *Biochim. Biophys. Acta* **1149,** 260–266.
21. Chong, P. L. -G. and Thompson, T. E. (1986) Depolarization of dehydroergosterol in phospholipid bilayers. *Biochim. Biophys. Acta* **863,** 53–62.
22. Lin, H. and Huang, C. (1988) Eutectic phase behavior of 1-stearoyl-2-caprylphosphatidylcholine and dimyristoylphosphatidylcholine mixtures. *Biochim. Biophys. Acta* **946,** 178–184.
23. Chong, P. L. -G., Tang, D., and Sugar, I. P. (1994) Exploration of physical principles underlying lipid regular distribution: effects of pressure, temperature, and radius of curvature on E/M dips in pyrene-labeled PC/DMPC binary mixtures. *Biophys. J.* **66,** 2029–2038.
24. Venegas, B., Sugar, I., P., and Chong, P. L. -G. (2007) Critical factors for detection of biphasic changes in membrane properties at specific sterol mole fractions for maximal superlattice formation. *J. Phys. Chem. B.* **111,** 5180–5192.

12

Differential Scanning Calorimetry in the Study of Lipid Phase Transitions in Model and Biological Membranes

Practical Considerations

Ruthven N. A. H. Lewis, David A. Mannock, and Ronald N. McElhaney

Summary

Differential scanning calorimetry (DSC) is a relatively rapid, straightforward, and nonperturbing technique for studying the thermotropic phase behavior of hydrated lipid dispersions and of reconstituted lipid model or biological membranes. However, because of the diversity of lipid thermotropic phase behavior, data-acquisition and data-analysis protocols must be modified according to the nature of the phase transition under investigation. In this chapter, the theoretical basis of the DSC experiment is examined and, with the aid of specific examples, also how the information content of DSC thermograms is affected by the nature of the lipid phase transition examined. The overall goal is to provide practical guidelines for the development of data-acquisition and data-analysis protocols, which are compatible with the instrumentation available and the nature of the lipid phase transition under investigation.

Key Words: Biological membranes; differential scanning calorimetry (DSC); lipid bilayers; lipid model membranes; lipid thermotropic phase behavior; lipid phase transitions; transition cooperativity; transition enthalpy; transition temperature.

1. Introduction

Membrane lipids can exist in many polymorphic forms and temperature-induced interconversions between these forms (phase transitions) readily occur. Differential scanning calorimetry (DSC) is a simple yet powerful nonperturbing physical technique, which is well suited for detection and for the thermodynamic characterization of thermotropic phase transitions in both lipid model and biological membranes. This technique involves the simultaneous heating (or cooling) of a sample and an inert reference, a material of comparable thermal mass and heat capacity, which exhibits no thermotropic events within the temperature range of interest, at a predetermined constant rate in a calorimeter designed to measure the differential rate of heat flow into the sample relative to that of the inert reference. The temperatures of the sample and reference may either be actively varied by independently controlled units (power-compensation calorimetry) or passively changed through contact with a common heat sink, which has a thermal mass that greatly exceeds the combined thermal masses of the sample and reference (heat-conduction calorimetry).

At temperatures far from any thermotropic events, the temperatures of the sample and reference cells change linearly with time and the temperature difference between them remains constant. Under such conditions, the DSC instrument records a constant difference between the rates of heat flow into the sample and reference cells, which ideally, is reflected by a straight, horizontal baseline. Whenever a thermotropic event occurs within the sample, a temperature differential between the sample and reference occurs and the instrument either actively changes the power input to the sample cell to negate the temperature differential (power-compensation calorimetry), or passively records the resulting changes in the rate of heat flow into the sample cell until the temperature differential eventually dissipates (heat-conduction calorimetry).

In both instances, the differential rates of heat flow into the sample and reference cells change and either an exothermic or endothermic deviation from the baseline occurs. On completion of the thermal event, the instrument either re-establishes its original baseline or establishes a new one if a change in the specific heat of the sample has occurred, thereby generating a plot of differential heat flow as a function of temperature. DSC can thus accurately measure the temperature, enthalpy, and cooperativity of lipid phase transitions, as well as their kinetics under certain conditions.

In this chapter, the practical aspects of using DSC to study the thermotropic phase transitions of synthetic membrane lipids dispersed in excess water will be examined. However, there is so much variation in the thermotropic characteristics of the various types of phase transitions which one can encounter in studying membrane lipids that no single experimental protocol can be devised that will be optimal for the detection and accurate thermodynamic characterization of each type of thermotropic phase transition that can be encountered. Therefore, also how these experimental protocols can be modified to optimize the detection and characterization of the various types of lipid phase transitions most commonly observed will be examined.

2. Theoretical Considerations

As illustrated in **Table 1**, the thermodynamic and kinetic properties of various lipid thermotropic phase transitions vary considerably, so that the thermotropic events observed by DSC may not always be occurring under equilibrium conditions. Therefore, some of the theoretical aspects of applying DSC in studying thermotropic phase transitions under both equilibrium and nonequilibrium conditions shall be examined first, in order to provide a basis for the proper interpretation of the experimental data obtained and for the design of experimental protocols which will optimize the usefulness of this data.

2.1. Thermotropic Processes Proceeding at Thermodynamic Equilibrium

This aspect of the application of DSC can be illustrated by examining a simple, two-state, temperature-induced conversion between A and B, for which one can define an equilibrium constant (K_e):

$$A \underset{}{\overset{K_c}{\rightleftharpoons}} B \text{ and } K_e = \frac{[B]}{[A]} \tag{1}$$

Table 1
Commonly Observed Thermotropic Phase Transitions of Hydrated Lipids

Transition type	Kinetics	Thermotropic properties		Example lipids
		Enthalpy	Cooperativity	
$L_c \Leftrightarrow L_\beta$	Slow	High	Moderate	Medium to long-chain diacyl PCs, PGs, PSs, and glucosyl diacylglycerols
$L_c \Leftrightarrow L_{\beta'}$	Slow	High	Moderate	Some highly asymmetric mixed-chain diacyl PCs
$L_c \Leftrightarrow L_\alpha$	Slow	High	Moderate	Short-chain saturated diacyl PCs, saturated diacyl PEs, short to medium-chain glucosyl diacylglycerols
$L_c \Leftrightarrow Q_{II}$	Slow	High	Moderate	Short-chain glycosyl diacylglycerols
$L_c \Leftrightarrow H_{II}$	Slow	High	Moderate	Long-chain glycosyl diacylglycerols
$L_\beta \Leftrightarrow L_\alpha$	Fast	High	High	Saturated diacyl phospholipids
$L_{\beta'} \Leftrightarrow L_\alpha$	Fast	High	High	Saturated dialkyl PCs
$L_\beta \Leftrightarrow Q_{II}$	Fast	High	High	Long-chain monoglycosyl glycerolipids
$L_\beta \Leftrightarrow H_{II}$	Fast	High	High	Long-chain monoglycosyl glycerolipids
$L_{\beta'} \Leftrightarrow L_\alpha$	Fast	High	High	Platelet-activating factor, lysophosphatidylcholines
$L_\alpha \Leftrightarrow H_{II}$	Moderate	Weak	Moderate	Long-chain diacyl PEs, medium and long-chain monoglycosyl diacylglycerols
$L_\alpha \Leftrightarrow Q_{II}$	Moderate	Weak	Moderate	Short-chain monoglycosyl glycerolipids
$Q_{II} \Leftrightarrow H_{II}$	Moderate	Weak	Moderate	Short-chain monoglycosyl glycerolipids

where [A] and [B] represent the concentrations of the initial and final states, respectively. Assuming that the volume changes accompanying this thermotropic process are small relative to the total volume of the system, K_e can be defined in terms of the fractional conversion (α)

$$K_e = \frac{\alpha}{1-\alpha} \text{ where } \alpha = \frac{[B]}{[A]+[B]} = \frac{K_e}{1+K_e} \qquad (2)$$

This thermotropic process can then be defined in terms of $\frac{d\alpha}{dT}$, the temperature dependence of the fractional change, as:

$$\frac{d\alpha}{dT} = \frac{\Delta H_0}{RT^2} \cdot \frac{K_e}{(1+K_e)^2} \qquad (3)$$

where ΔH_0 is the van't Hoff enthalpy change and R is the gas constant. Finally, this thermotropic process can be described in terms of the response of the calorimeter, $\frac{dE}{dt}$, by calibrating equation (**Eq. 3**) in terms of the total energy change:

$$\frac{dE}{dT} = \eta \cdot \Delta H_{cal} \cdot \frac{d\alpha}{dT} = \eta \cdot \Delta H_{cal} \cdot \frac{\Delta H_0}{RT^2} \cdot \frac{K_e}{(1+K_e)^2} \quad (4)$$

where η is the number of moles of sample, ΔH_{cal} is the molar calorimetric enthalpy of the process, and $\eta \cdot \Delta H_{cal}$ is the sample thermal load, and then correcting for the temperature scan rate:

$$\frac{dE}{dt} = \frac{dE}{dT} \cdot \frac{dT}{dt} = S \cdot \eta \cdot \Delta H_{cal} \cdot \frac{\Delta H_0}{RT^2} \cdot \frac{K_e}{(1+K_e)^2} \quad (5)$$

where S is the temperature scan rate $= \frac{dT}{dt}$.

The idealized expression in **Eq. 5** reveals the following points. First, aside from the size of the sample (η), the scan rate (S), and the temperature (T), all of the terms involved in the definition of the instrument response are either fundamental constants or intrinsic properties of the thermotropic process under investigation. Thus, in any DSC measurement, the flexibility of the experimenter will be limited to the choice of a suitable scanning range, to adjustments of sample size, and/or to alterations in heating (or cooling) rate, in order to obtain signals of acceptable intensity. Second, **Eq. 5** also shows that the calorimetric response is dependent on two enthalpy terms, ΔH_{cal} and ΔH_0. The first of these (ΔH_{cal}) is the molar calorimetric enthalpy of the process, which in combination with the size of the sample, defines the total enthalpy change associated with the thermotropic process under investigation. From the expressions shown in **Eqs. 4** and **5**, one can demonstrate that

$$\frac{1}{S} \cdot \frac{dE}{dt} = \frac{dE}{dT} = \eta \cdot \Delta H_{cal} \cdot \frac{d\alpha}{dT} \quad (6)$$

where $\frac{1}{S} \cdot \frac{dE}{dt}$ is the scan rate-normalized DSC signal. By integrating **Eq. 6** between T_a and T_b, the starting and ending temperatures of the experiment, one finds that:

$$\int_{T_a}^{T_b} \frac{dE}{dt} = \eta \cdot \Delta H_{cal} \cdot \int_{T_a}^{T_b} \frac{d\alpha}{dT} \quad (7)$$

If the temperature range of the experiment ($T_b - T_a$) is large enough to span the entire range of the thermotropic process under investigation, the integral term, $\int_{T_a}^{T_b} \frac{d\alpha}{dt}$, will become unity, and the value of the integral term, $\int_{T_a}^{T_b} \frac{dE}{dT}$, will be completely defined by $\eta \cdot \Delta H_{cal}$, the thermal load of the sample. In practical terms, this means that as long as the temperature range scanned in the DSC experiment exceeds that of the thermotropic process under investigation, the calorimetric enthalpy of that process can be directly accessed by integrating the area of the scan rate-normalized DSC thermogram relative to the experimental baseline. However, more importantly, this measurement is independent of all other parameters, and in principle, the only limits to the accuracy with which this measurement can be made are the sensitivity of the instrument used and the intensity of the calorimetric signal relative to the noise level of the measurement.

The second enthalpy term described in **Eq. 5** is the van't Hoff enthalpy change (ΔH_0). This term is derived from the van't Hoff isochore, $\frac{d}{dT} \ln K_e = \frac{\Delta H_0}{RT^2}$, and is thus a measure of the sharpness or cooperativity of the thermotropic transition under investigation. Estimates of the ΔH_0 can be obtained thus:

The value of the $\frac{K_e}{(1+K_e)^2}$ term in **Eq. 5** is subject to the following limits:

$$\text{for } K_e < 1, \lim_{K_e \to 0} \frac{K_e}{(1+K_e)^2} = K_e$$

$$\text{for } K_e = 1, \frac{K_e}{(1+K_e)^2} = 0.25$$

$$\text{for } K_e > 1, \lim_{K_e \to \infty} \frac{K_e}{(1+K_e)^2} = 0.$$

Thus, $\frac{K_e}{(1+K_e)^2}$ describes a symmetrical biphasic curve (**Fig. 1**) which attains its maximal value (0.25) when the value of the equilibrium constant K_e is unity and approaches zero when K_e is either very small or very large. The shape of the DSC thermogram is described by the value of the $\frac{K_e}{(1+K_e)^2}$ term in **Eq. 5** and the curve attains its maximal value at a temperature (T_m) where the process is 50% complete. Moreover, if the temperature range spanned by the thermotropic process is relatively small, the intensity of the DSC signal relative to the baseline is approximately proportional to $\frac{K_e}{(1+K_e)^2}$, and at temperatures where this intensity is half maximal, $\frac{K_e}{(1+K_e)^2}$ should approach values near 0.125 thus:

$$\frac{K_e}{(1+K_e)^2} = 0.125 \quad \text{i.e., } \ln K_e = \pm 1.76275 \tag{8}$$

The roots of the quadratic **Eq. 8** can then be substituted into the van't Hoff isochore to determine the ΔH_0 for the process. From the symmetry properties of the DSC trace, one can demonstrate that, as long as the width of the transition is small, the temperatures at which DSC signal intensity are half maximal can be approximated by the expression $T_m \pm \frac{\Delta T}{2}$, where ΔT is the transition width at half height. By taking definite integrals to the van't Hoff isochore, one can demonstrate that:

$$\left[(\ln K_e)_{T_2} - (\ln K_e)_{T_1} \right] = \frac{\Delta H_0}{R} \left[\frac{1}{T_2} - \frac{1}{T_1} \right] \tag{9}$$

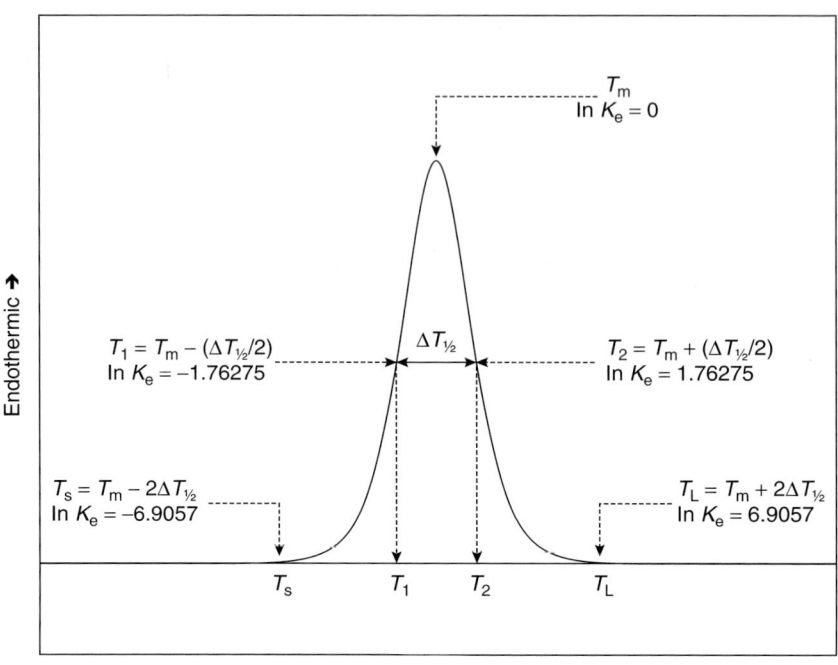

Fig. 1. Idealized heating thermogram exhibited by a two-state thermotropic process at equilibrium. Values for the natural logarithm of the equilibrium constant are listed relative to the transition midpoint temperature (T_m) and the transition width at half-height ($\Delta T_{1/2}$) thus: T_S: Effective starting temperature of transition. T_L: Effective completion temperature of the transition. T_1 and T_2: Temperatures at half maximal peak intensity.

where $(\ln K_e)_{T_1}$ and $(\ln K_e)_{T_2}$ are the roots of the quadratic **Eq. 8**.

$$T_1 = T_m - \frac{T_{1/2}}{2} \qquad T_2 = T_m + \frac{T_{1/2}}{2}$$

From these expressions, ΔH_0 can be expressed in terms of the transition width at half height thus:

$$\Delta H_0 = 3.5255 \cdot R = 3.5255 \cdot R \cdot \frac{T_m^2 \cdot (\Delta T_{1/2})^2}{\Delta T_{1/2}} \tag{10}$$

Thus, if ΔT is small relative to T_m, ΔH_0 can be obtained by the following approximation:

$$\Delta H_0 \approx 3.5255 \cdot R \frac{T_m^2}{\Delta T_{1/2}} \tag{11}$$

As illustrated by **Eq. 11**, the value of ΔH_0 term is inversely proportional to width of the transition. With most lipid phase transitions, the ΔH_0 term is considerably greater than the calorimetric enthalpy (ΔH_{cal}), and the ratio $\dfrac{\Delta H_0}{\Delta H_{cal}}$ is interpreted in terms of a cooperativity parameter or cooperative unit size (CUS), which in effect is an estimate of the number of lipid molecules in the cooperative unit from which the thermotropic phase change is nucleated. The CUS is thus a measure of the degree of intermolecular cooperation between lipid molecules. For a completely cooperative, first-order phase transition of an absolutely pure substance, this ratio should approach infinity, whereas for a completely noncooperative process, this ratio should approach unity. However, one should note that the absolute values of the CUS calculated from a DSC measurement should be regarded as tentative, because they are markedly sensitive to the presence of impurities and may also be limited by heat transfer capacities of the instrument (*see* below). Nevertheless, carefully determined CUS values can be useful in assessing the purity of synthetic lipids and in quantitating the degree of cooperativity of lipid phase transitions.

2.2. Thermotropic Processes Proceeding Under Nonequilibrium Conditions

Thermotropic interconversions between lipid phases do not always proceed under equilibrium conditions. In fact most of the thermotropic interconversions between lamellar gel and crystalline lipid phases are inherently slow processes for which kinetic considerations either determine or strongly influence the temperatures at which these processes are actually observed. In such cases, the thermodynamic characteristics of the thermotropic process (apart from ΔH_{cal}) are essentially inaccessible by temperature scanning techniques such as DSC and the data obtained must be interpreted accordingly. However, as illustrated by the example below, DSC can be used to characterize the kinetic characteristics of the thermotropic process *(1)*.

For a kinetically driven first-order thermotropic conversion of state A to state B, the rate of reaction $\left(\dfrac{dA}{dt}\right)$ is determined by the concentration of the starting state A, and one can define a rate equation thus:

$$A \xrightarrow{K_r} B$$

$$\frac{dA}{dt} = -K_r [A] \tag{12}$$

where K_r is the first-order rate constant and [A] is the relative concentration of state A. Within the context of a DSC experiment, the differential rate of heat flow into the sample will be directly proportional to the rate of loss of state A, and **Eq. 12** can thus be expressed in terms of the relative intensity of the DSC signal $\left(\dfrac{dE}{dt}\right)$ thus:

$$\frac{dE}{dt} \propto \frac{dA}{dt} = -K_r [A] \tag{13}$$

If the volume changes involved are small relative to the total volume of the system and the progress of the reaction is proportional to the cumulative differential heat flow into the sample, the relative intensity of the DSC signal at any given temperature (T) is then:

$$\left[\frac{dE}{dt}\right]_T = -[K_r]_T \cdot \eta \cdot \Delta H_{cal} \cdot [\beta]_T \tag{14}$$

where β is the residual fraction of the starting species A, which can be expressed in terms of the partial area of the DSC thermogram thus:

$$[\beta]_T = \frac{\int_T^{T_\infty} \frac{dE}{dT}}{\int_{T_0}^{T_\infty} \frac{dE}{dT}} \quad (15)$$

where T_0 and T_∞ are the starting and ending temperatures of the DSC experiment. At any temperature, the rate constant of the process will be given by the expression:

$$\frac{1}{\eta \cdot \Delta H_{cal}} \cdot [\phi]_T = -[K_r]_T \quad (16)$$

where

$$[\phi]_T = \frac{\left[\frac{dE}{dt}\right]_T}{[\beta]_T}$$

and, as the rate constants are describable by the Arrhenius equation, **Eq. 16** can be expressed as a function of temperature thus:

$$\frac{1}{\eta \cdot \Delta H_{cal}} \cdot [\phi]_T = -[K_r]_T = -A e^{\frac{-U}{RT}} \quad (17)$$

$$-\ln(\eta \cdot \Delta H_{cal}) + \ln(\phi) = -\ln(A) + \frac{U}{RT} \quad (18)$$

where U is the Arrhenius activation energy of the process.

The relationships expressed in the preceding equations reveal a number of points pertinent to the application of DSC to slow thermotropic phase transitions. First, the characteristics of the DSC thermogram obtained are determined exclusively by the thermal load of the sample and the kinetics of the process under investigation. Thus, although a phase transition may be reproducibly observed over a specific temperature range, there is no thermodynamic significance to the transition peak temperature midpoints and transition widths of the thermograms observed. Second, the temperature scanning rates do not necessarily factor in the determination of the kinetic parameters. Thus, as long as the sample thermal load ($\eta \cdot \Delta H_{cal}$) is large enough to generate a signal of reasonable intensity and the scanning rates are slower than the kinetics of the process under investigation, the data obtained should be essentially independent of any finite scanning rate used. Finally, the intensity of the signals generated is potentially interpretable in terms of the rate constant of the process at all temperatures over which there is significant signal intensity relative to the baseline. Moreover, as the temperature dependences of the rate constants are usually describable by the Arrhenius equation, the data can also used to determine the Arrhenius activation energy of the process. Thus, as illustrated in **Eq. 18**, the Arrhenius activation energy (U) of processes driven by first-order kinetics can be obtained from the slope of a plot of $\ln(\phi)$ against the reciprocal of the absolute temperature.

3. Instrumentation

Under ideal conditions, the thermograms recorded by any DSC instrument would reflect the temperature dependence of the thermotropic changes occurring in the sample without significant distortion from the recording instrument. However, such conditions are extremely difficult to achieve in practice because the finite conductivity of the thermal path between the heat source of the instrument and the contents of its sample capsule, along with the thermal mass of the sample itself, form a *de facto* kinetic barrier between the heat source of the instrument and the processes occurring within the sample. These kinetic limitations are reflected by an instrument time constant, which essentially defines the capacity of the instrument to transfer heat between its heat source and sample capsule. Because of these kinetic considerations, DSC thermograms acquired by any instrument will be distorted from the ideal whenever the rate of heat transfer required to achieve thermal equilibrium significantly exceeds the thermal transfer capacity of the instrument as embodied in its time constant. (The ways in which some of these distortions are manifest will be examined later in this section.) However, these considerations also imply that the acquisition of accurate DSC thermograms may be possible if instrument-related distortions can be minimized. **Equation 5** suggests that this should be feasible, in principle, if protocols based on slow temperature scanning rates and relatively small sample thermal loads are used. However, this is only practical if the configuration of the instrument is such that the combined thermal mass of the sample and reference forms a large fraction of the thermal mass. Currently, the only instruments which approach these conditions are the high-sensitivity power-compensation calorimeters (*see* below). However, one should note that the experimental protocols which facilitate the recording of accurate DSC thermograms are inherently incompatible with the acquisition of DSC thermograms with high signal-to-noise ratios (*see* **Eq. 5**), especially in cases where the thermotropic processes under investigation are fairly broad (when CUS values are low and/or ΔH_0 values are small). Thus, in such cases one must sacrifice some signal fidelity in order to obtain an acceptable level of signal intensity.

Currently, commercially available DSCs are designed to operate as either heat-conduction or power-compensation instruments. With heat-conduction instruments, both sample and reference are heated (or cooled) indirectly through contacts with a common heat sink, which typically has a thermal mass that greatly exceeds their combined thermal masses. Heat-conduction instruments are thus configured to passively measure the differential heat flow into the sample as the heat sink is heated (or cooled) at a constant predetermined rate. Because the thermal mass of these types of calorimeters is large, the time constants of heat-conduction instruments tend to be fairly long (≥ 120 s) and it is difficult to maintain thermal equilibrium throughout the system except when operating at very slow scan rates. Thus, even at moderate temperature scanning rates, DSC data acquired with these types of instruments can be so severely distorted (through broadening, end-tailing, and up/down temperature-shifting of the heating/cooling thermograms) that estimation of all thermodynamic parameters save ΔH_{cal} can be severely compromised (*see* **Fig. 2A**).

However, these distortions are markedly diminished at lower scan rates (*see* **Fig. 2B**), and in principle, can be effectively eliminated if the scanning rates are slow relative to the instrument time constant. However, as noted earlier, the latter approach will incur a proportional loss of signal intensity (**Eq. 5**). Ironically, however, the fact that the thermal mass of the sample is a very small fraction of the thermal mass under instrumental control makes it feasible for the user to markedly increase sample size to compensate for the decline of signal intensity at very slow scanning rates, and for all practical purposes, this capacity is probably limited only by the

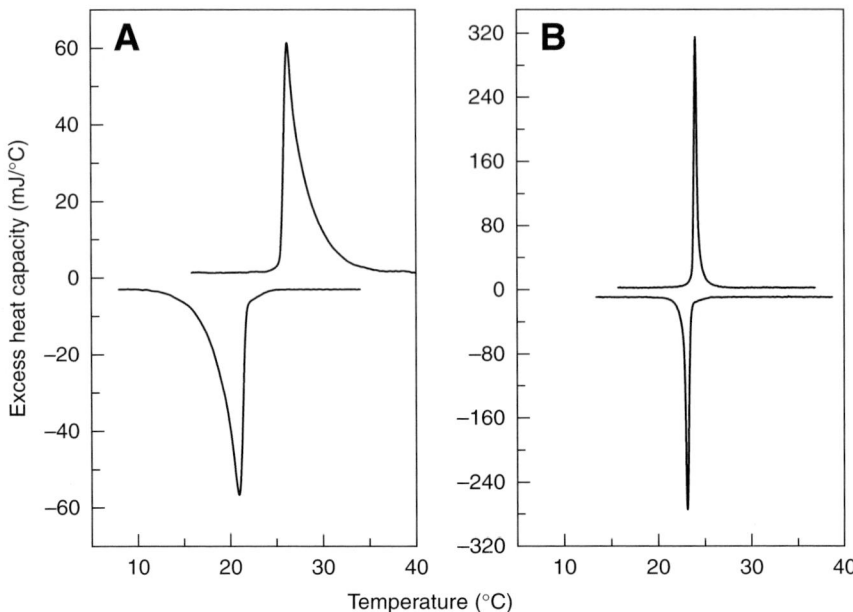

Fig. 2. Effect of scan rate on the heating endotherms and cooling exotherms of the gel/liquid-crystalline phase transition of DMPC as observed with a high-sensitivity heat-conduction calorimeter. The thermograms shown were recorded with the same sample at scan rates of 48°C/h (**panel A**) and 10°C/h (**panel B**). The sample consisted of ~6.5 µM of lipid dispersed in 500 µL of aqueous buffer.

capacity of the sample capsule of the instrument and the dynamic range of its control and measurement electronics. However, this approach is also inherently costly of both sample and instrument time and may thus be unsuitable for routine application. Nevertheless, the fact that one does have the flexibility to maintain signal intensity by a proportional increase in sample size also enables an alternative option, which does not require extraordinarily slow scanning rates and very large sample sizes. In principle, if the time constant of the instrument is known and the heat transfer properties of the instrument are properly calibrated over its effective operational temperature range, one can acquire data at a reasonable scanning rate and use the Tian equation to correct for instrument-related distortions (*see* **ref. 2** and references cited therein). However, it is cautioned that this approach is possible only when the scanning rates are compatible with or slower than the instrument time constant (i.e., when the distortions are relatively small). Nevertheless, as long as the proper experimental protocols are implemented, it is feasible to use heat-conduction instruments to obtain high-quality DSC thermograms, which approach the fidelity attainable with modern high-sensitivity power-compensation instruments (e.g., *see* **Fig. 3**).

Unlike heat-conduction calorimeters, power-compensation instruments are designed with independent heat sources of the sample and reference capsules which can each be heated (and cooled) independently of the adiabatic shielding between them and the environment at large. Moreover, because the thermal masses controlled by power-compensation instruments are considerably smaller, their instrument time constants are also considerably shorter (~5–30 s) and it is thus feasible to operate such instruments at considerably faster temperature scanning rates than is practical with heat-conduction instruments. Thus, instrument-related distortions

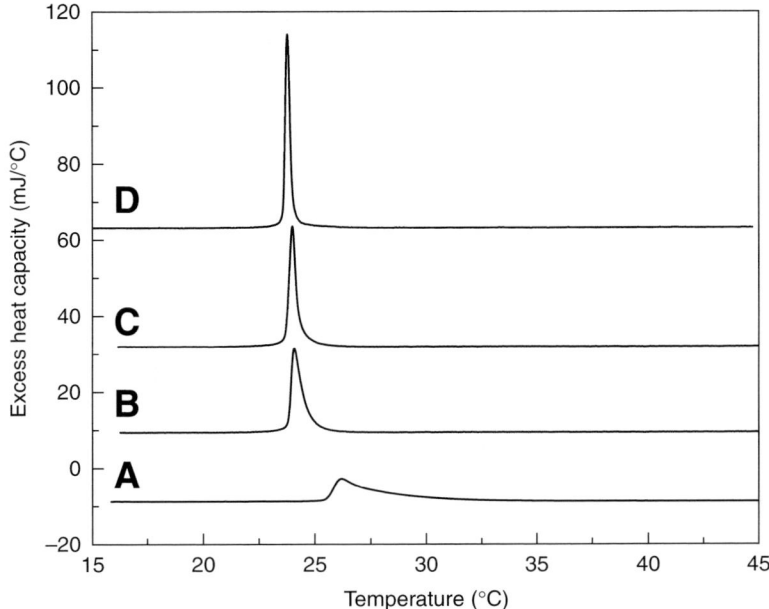

Fig. 3. Illustration of the Tian correction of DSC data acquired by a heat-conduction calorimeter. **Curves A,B** show uncorrected heating endotherms of the sample acquired at temperature scanning rates 48°C and 10°C/h, respectively. **Curve C** shows the heating endotherm acquired at 10°C/h after application of the Tian correction. **Curve D** shows the heating endotherm exhibited by the same sample when recorded with a high-sensitivity power-compensation instrument operating at 10°C/h. The sample used to obtain **curves A–C** was the same sample described in **Fig. 1**. The data illustrated by **curve D** was obtained from a 323-μL aliquot of a sample prepared by diluting 100 μL of sample used in **Fig. 1** with 1.5 mL of buffer. For comparative purposes, the heating endotherm obtained was normalized to the same sample mass as **curves A–C**.

of DSC thermograms of the type and magnitude illustrated in **Fig. 2A** are rarely observed with power-compensation calorimeters unless the instruments are operated at extraordinarily high scanning rates (*see* **Fig. 4**).

However, because these types of instruments are configured to operate in the power-compensation mode, it is also possible for the system to "over compensate" if the response time characteristics of the instrument and the kinetics of the thermotropic process are not balanced. This phenomenon is most likely to occur when recording the heating endotherms of fast, highly energetic and highly cooperative thermotropic processes such as the gel/liquid-crystalline phase transitions of synthetic phospholipids. When this phenomenon occurs, the instrument supplies more energy than required for the thermotropic process under investigation and the temperature of the sample actually rises above that of the reference. This effect is manifest by an asymmetric sharpening of the endothermic peak, by a slight down-shifting of the peak maximum to lower temperatures, and by an exothermic baseline excursion at temperatures near the normal completion of the process (*see* **Fig. 5A**).

Although these types of distortions are usually not as severe as those illustrated in **Fig. 2A**, they are serious enough to compromise the accuracy with which thermodynamic (or kinetic) parameters can be extracted from the data. However, these distortions can be easily avoided

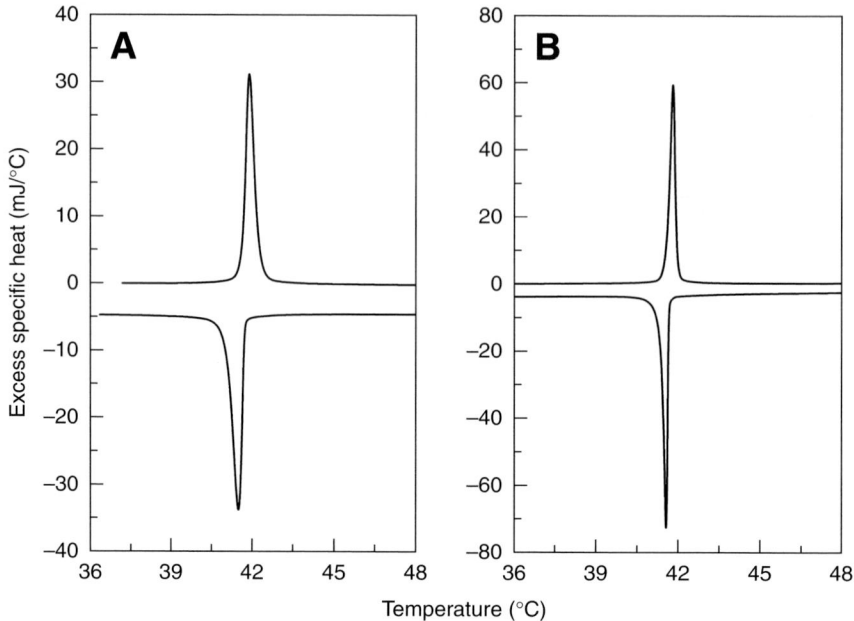

Fig. 4. Effect of scan rate on the heating endotherms and cooling exotherms of the gel/liquid-crystalline phase transition of DPPC as observed with a high-sensitivity power-compensation calorimeter. The thermograms shown were recorded with the same sample at scan rates near 30°C/h (**panel A**) and 10°C/h (**panel B**). The sample consisted of approx 1 μM of lipid dispersed in 323 μL of aqueous buffer.

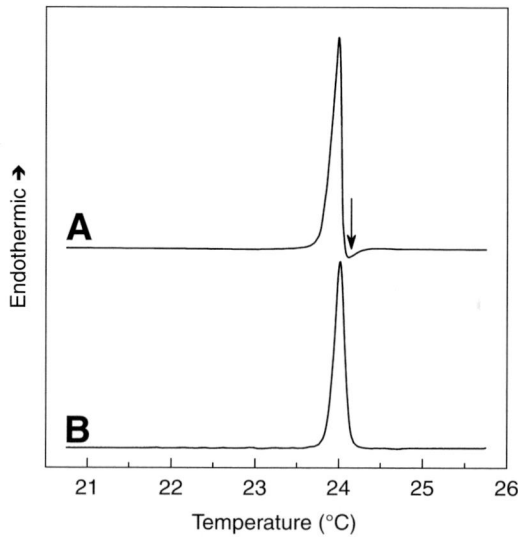

Fig. 5. Comparison of overcompensation-distorted (**A**) and undistorted (**B**) DSC thermograms recorded by a high-sensitivity power-compensation DSC. The heating endotherms shown represent the gel/liquid-crystalline phase transitions exhibited by a 876-μL aliquot of a 2 mM dispersion of DMPC in water. The thermograms were recorded with the same sample at temperature scanning rates of approx 12°C/h (**A**) and approx 6°C/h (**B**). The exothermic baseline excursion caused by overcompensation is indicated by the arrow.

by reducing the effective heat transfer rates of the calorimeter by reducing the scanning rate (*see* **Fig. 5B**) and/or the size of the sample being analyzed. These points, and the others described earlier, emphasize quite vividly the importance of understanding both the capabilities and limitations of the available instruments, in order to ensure the application of suitable experimental protocols.

4. Sample Preparation

Adequate sample preparation is obviously an important aspect of the application of DSC to the study of thermotropic processes in general and to the study of lipid thermotropic phase behavior in particular. However, most of the practical aspects of sample preparation for DSC are actually related to the assurance of sample purity and homogeneity, to the complete hydration and uniform dispersal of the lipid samples, and to the preparation of appropriately stable liposomes. These issues are not specific to DSC *per se* and will not be discussed here. However, from the narrower perspective of DSC, the practical aspects of preparing lipid samples for DSC can be addressed through considerations of sample size and sample equilibration before data acquisition. As illustrated below, these issues are ultimately related to the nature of the phase transition under investigation.

4.1. Sample Size

Although the preparation of an adequately sized sample is obviously an important aspect of any sound DSC protocol, the issue of what constitutes an adequate sample size for the effective DSC detection and characterization of any thermotropic phase transition has received little attention in the literature. However, from the practical perspective, the important aspect of sample size is actually the sample thermal load ($\eta \cdot \Delta H_{cal}$), which defines the total amount of energy that will either be absorbed or dissipated during the thermotropic transition under investigation. However, regardless of the type of instrument used, any thermotropic phase transition can be adequately detected and characterized if the mean rate of heat transfer in the temperature range defined by $T_{max} \pm \dfrac{\Delta T_{1/2}}{2}$ is at least 25–30 times the noise level of the instrument. Thus, for example, with thermotropic processes observed at or near to thermodynamic equilibrium, the minimal thermal sample load required for effective detection at any given scan rate (S) can be derived from the expression:

$$\eta \cdot \Delta H_{cal} \approx 20 \cdot N^{*} \cdot \frac{\Delta T_{1/2}}{S} \qquad (19)$$

where T_{max} is the transition peak temperature, $\Delta T_{1/2}$ is the transition width at half height, N^{*} is the instrument noise level, and S is the temperature scanning rate.

Thus, whenever practical, the size of the sample should be adjusted such that the thermal load of the least energetic thermotropic process under investigation would meet these guidelines. With a modern, high-sensitivity, power-compensation instrument operating at a temperature scanning rate of 10°C/h, these conditions are easily met for characterizing the pretransition of dimyristoylphosphatidylcholine (DMPC) with a 160-μg lipid sample for which the pretransition thermal load is some 250 μcal.

4.2. Sample Equilibration

Hydrated lipids exhibit phase transitions, which vary considerably owing to their kinetic characteristics. Indeed, as illustrated in **Table 1**, some form of kinetic limitation can be expected with many types of phase transitions, which occur in hydrated lipid assemblies, with the exception of simple hydrocarbon chain-melting phase transitions. Within the context of a DSC experiment, these kinetic limitations are usually manifest, in part, by a scan rate dependent up-shifting of heating endotherms and more pronounced down-shifting of cooling exotherms, relative to the temperature range over which the processes should occur under conditions of thermodynamic equilibrium. Because of these kinetic problems, lipid thermotropic behavior as observed by temperature scanning techniques such as DSC can be strongly influenced by the thermal history of the sample before the initiation of the DSC scan. The effects of such kinetic problems on the likely outcome of a DSC experiment, and the nature of the protocols required to circumvent or otherwise address these problems, are illustrated by the three types of lipid phases transitions discussed below.

4.2.1. The Pretransitions of Linear Saturated 1,2-Diacyphosphatidylcholines

Figure 6A shows some DSC thermograms illustrating the thermotropic phase behavior of an aqueous dispersion of dipalmitoylphosphatidylcholine (DPPC) as detected by a high-sensitivity, power-compensation calorimeter operating at temperature scanning rates of 10°C/h. Under these conditions, the pretransition ($L_{\beta'}/P_{\beta'}$ phase transition) of this lipid is observed as a fairly sharp ($\Delta T_{1/2} \approx 1.5°C$), weakly energetic peak centered near 35.4°C on heating, and as a broader ($\Delta T_{1/2} \approx 2.5°C$) peak of reduced area centered near 31.4°C on cooling. Also, when examined as a function of heating rate, the peak temperatures of the observed endotherms describe a sigmoid curve with low- and high-range asymptotes near 35.5° and 38.5°C, respectively (*see* **Fig. 6B**). Finally, the shapes of the heating thermogram can also be affected by the rate at which the sample was cooled before the initiation of the heating scan. Thus, for example, when cooled very rapidly through the range of the pretransition, the succeeding heating thermogram can actually be distorted by exothermic excursions before the onset of the heating endotherm (*see* **Fig. 6C**).

The pattern of behavior exhibited by the pretransitions of the linear saturated 1,2-diacyl phosphatidylcholines (PCs) typifies the behavior observed when the rate at which a process approaches thermodynamic equilibrium is somewhat slower than the temperature scanning rate at temperatures near to the equilibrium phase transition temperature (i.e., when $T = T_m \pm 5 \cdot \Delta T_{1/2}$). These effects will also be exacerbated when these types of thermotropic processes are examined in the cooling mode, because the rates of approach to thermodynamic equilibrium decrease exponentially with decreasing temperature. Thus, accurate reporting of the characteristics of these types of lipid phase transitions by temperature scanning techniques such as DSC will only be feasible with experiments performed at slow heating rates, using samples that have been equilibrated at temperatures well above the phase transition of interest and slowly cooled to low temperatures. Data quality can be further improved if such samples are subsequently slowly heated to temperatures just below the normal onset temperature of the phase transition, and then recooled before recording the DSC heating scan.

4.2.2. Thermotropic Interconversions Involving Lamellar-Crystalline Lipid Phases

Lamellar-crystalline (L_c) lipid phases are a diverse family of highly ordered, largely dehydrated, crystal-like structures, which have been observed in both model and biological membranes. Typically, the lipid molecules forming these bilayer structures are characterized by tightly-packed all-*trans* hydrocarbon chains, partially (or possibly fully) dehydrated headgroups and polar/apolar interfaces, and specific lateral close-contact interactions between

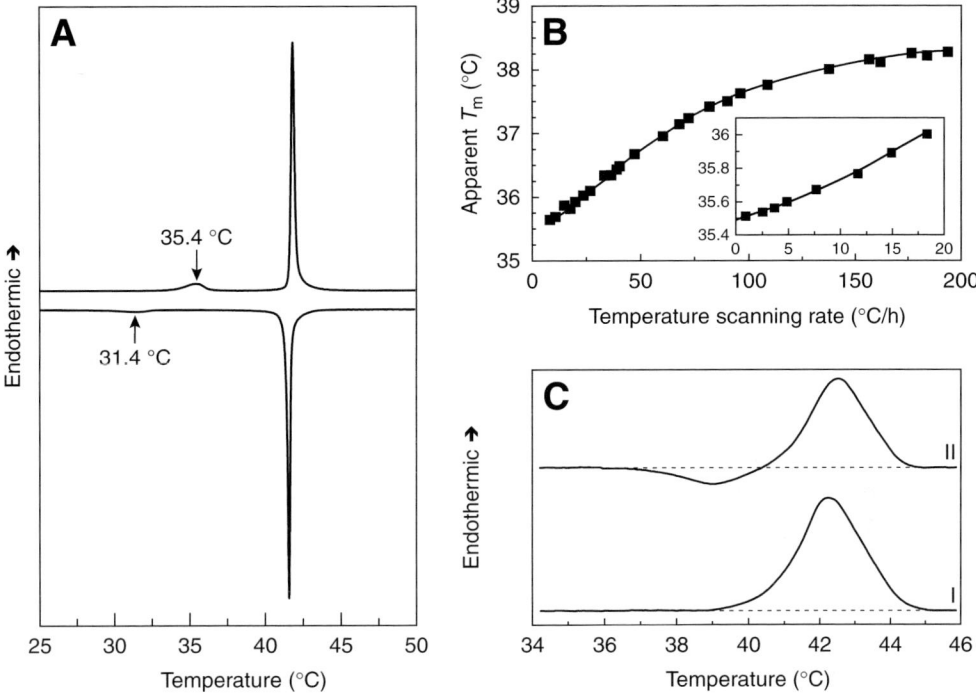

Fig. 6. Kinetic distortion of the properties of the pretransition of linear saturated 1,2-diacyl PCs as observed by DSC. **(A)** Heating/cooling hysteresis of the pretransition of DPPC. **(B)** Effect of heating scan rate on the apparent transition temperature of DPPC. The inset shows the low range of temperature scanning rates on an expanded scale. **(C)** Effect of cooling rate on the properties of the heating endotherm of the pretransition of diheptadecanoyl PC. The heating endotherms shown were obtained at a temperature scanning rate of 10°C/h after: **(I)** slow cooling (~5°C/h) through the pretransition and **(II)** fast cooling (~5°C/min) through the pretransition. Note the exothermic excursion (arrowed) before the onset of the pretransition endotherm.

moieties on the polar headgroup. On heating, these structures may convert to either the lamellar gel (L_β) or to one or more of the chain-melted lipid phases (i.e., L_α, Q_{II}, or H_{II}) through fairly energetic processes, which are often somewhat slower than the temperature scanning rates commonly used in DSC measurements (5–30°C/h). However, on cooling, the process of reforming the L_c phase is typically very slow and rarely occurs at temperatures comparable with those of the phase transitions observed in the heating mode.

Consequently, the characterization of the thermodynamic properties of phase transitions involving lipid L_c phases by DSC can only be reliably achieved with heating mode experiments performed on samples that have been carefully equilibrated under conditions suitable for the nucleation and growth of the L_c phase. For most lipids, this requires the formation of its L_β phase and subsequent periods of incubation at low temperatures, which are both highly specific to the individual lipid species (for examples of the conditions required for the formation of lipid L_c phases, *see* **refs. 3–5** and references cited therein). Consequently, there are no generic sample handling protocols for optimizing the detection and characterization of lipid L_c phases by DSC (or indeed any other physical technique), and appropriate protocols will need to be developed on a case-by-case basis.

However, the following guidelines are useful in the development of such protocols.

1. L_c phase formation can be facilitated by ensuring that the L_β phase from which it is nucleated is "pure." Because of the considerable variability of the kinetic characteristics of the various types of lipid phases which may be formed, small amounts of other lipid phases can often be kinetically trapped within the L_β phase of the lipid. This can be avoided if the samples are slowly cooled through the temperature range at which the lipid L_β phase forms.
2. The kinetics of L_c phase formation tends to decrease exponentially as hydrocarbon chain length increases (*3–6*). Thus, the time and temperature conditions suitable for short-chain lipids may not be suitable for longer-chain homologs.
3. Superimposed on the hydrocarbon chain length effects described immediately above the kinetics of lipid L_c phase formation may vary significantly depending on whether the chains contain an odd- or an even-number of carbon chains (*3–6*). Thus, the kinetics of L_c phase formation with lipid homologs with hydrocarbon chains containing an odd-number of carbon atoms are often significantly slower than for their neighboring even-numbered counterparts. Thus, different protocols may have to be developed for the odd- and even-numbered members of a homologous series of lipids.
4. The conditions required for the nucleation and growth of lipid L_c phases are not always the same. Therefore, the L_c phase of a lipid can be nucleated by incubation at temperatures well below the onset of its gel/liquid-crystalline phase transition temperature, but the growth of the L_c phase may only occur at higher temperatures, or even at temperatures closer to the onset of the L_c phase transition. For such samples, the process of assuring proper sample equilibration can be quite tedious, especially if the incubation time and temperature conditions required for the nucleation and growth of the L_c phase are unknown.

An effective approach to addressing this problem is to cool the sample slowly to a predefined low temperature, incubate the sample at that temperature for about 12–24 h, and then heat the sample slowly to temperatures either below the onset of the L_β/L_α phase transition or the expected L_c/L_α phase transition, whichever is lower. If significant nucleation of the L_c phase occurs under the incubation conditions used, the growth of the L_c phase usually occurs quite rapidly and is manifest by either exothermic baseline excursions and/or considerable baseline instability at temperatures whereby significant growth of the L_c phase occurs. For example, when cooled to temperatures near 0°C, aqueous DMPE forms an L_β phase, which melts at temperatures near 50°C (**Fig. 7**, **curve A**), and when incubated at temperatures near 0°C, the L_β phase slowly transforms to the L_c phase, which melts at temperatures near 58°C (**Fig. 7**, **curve C**). On heating, this transformation is manifest by the gradual disappearance of the endotherm near 50°C and a concomitant growth of a considerably larger endotherm near 58°C (**Fig. 7**).

However, this process is usually accompanied by noisy exothermic baseline excursions at temperatures near 15–25°C (**inset, curve B**), especially when the samples have been incubated at 0°C for relatively short periods of time. From this one can conclude that the process of transforming the L_β phase of dimyristoylphosphatidylethanolamine (DMPE) to its L_c phase can be accelerated by first nucleating the sample at lower temperatures and then warming to temperatures between 15 and 25°C, wherein the growth of the L_c phase is quite rapid. Indeed, formation of the L_c phase of DMPE can be fairly rapidly induced by three cycles of approx 6–12 h incubation at 0°C and subsequent warming to temperatures near 15–25°C.

Thus, through the application of procedures such as these, and insightful variation of the incubation time and incubation temperature, one usually can identify the conditions favorable for the nucleation and growth of the L_c phase of interest and develop incubation protocols accordingly. Finally, once appropriate protocols for sample equilibration have been developed and applied, data quality can be further improved if the sample is slowly heated to temperatures

Study of Lipid Thermotropic Phase Behavior by DSC

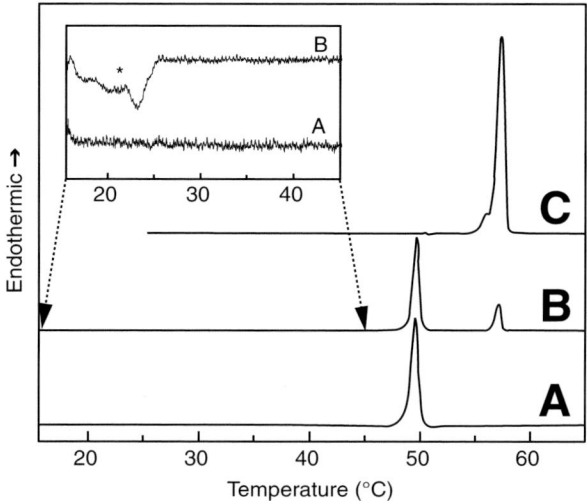

Fig. 7. Nucleation and growth of the L_c phase of DMPE from its L_β phase. The thermograms shown were obtained after incubating the L_β phase of DMPE at 0°C thus: **curve A**: no incubation; **curve B**: incubated approx 12 h at 0°C; **curve C**: incubated 5–7 d at 0°C. The inset highlights the low-temperature regions **curves A,C** on an expanded scale and the exothermic baseline excursions (*), which indicate the temperature range wherein rapid growth of the L_c phase of this lipid occurs.

below the normal onset temperature of the L_c phase transition and recooled before recording the DSC heating scan (*see* **Subheading 4.2.1.**).

4.2.3. Thermotropic Interconversions Involving Cubic Phases

The term cubic phase actually refers to a family of three dimensionally ordered, nonlamellar, liquid-crystalline lipid phases with cubic symmetry, which are typically formed when hydrated samples of so-called nonlamellar phase forming lipids are heated to temperatures above the hydrocarbon chain-melting phase transition temperature (*7*). Inverted cubic (Q_{II}) phases may be formed directly from any of the lamellar lipid phases either as end products or as intermediates *en route* to other nonlamellar phases such as the inverted hexagonal (H_{II}) phase (*8–10*). When formed from either the L_α phase or another cubic phase, the process is usually manifest by fairly broad ($\Delta T_{1/2} \approx 2-6°C$), weakly energetic (~100–1000 cal/mol) thermotropic transitions. However, because of the considerable variations in the three-dimensional geometries of the different types of Q_{II} phases (*7*), interconversion between the Q_{II} phases of any given lipid are often subject to significant kinetic limitations and as a result, the nature of Q_{II} phase observed often varies with both temperature and the scanning rate. Moreover, once some types of Q_{II} phases are formed, they can remain kinetically trapped at all temperatures above the lipid hydrocarbon chain-melting phase transition temperature and the system will "reset" only after cooling the sample to temperatures below the lipid gel/liquid-crystalline phase transition temperature (*8,10*). For these reasons, it is often very difficult to accurately measure the thermodynamic characteristics of L_α/Q_{II} and Q_{II}/Q_{II} phase transitions by DSC under conditions of thermodynamic equilibrium, without using sample sizes with hydrocarbon chain-melting thermal loads large enough to overload most modern high-sensitivity instruments. Moreover, for these types of phase transitions, useable data is usually only obtainable with heating mode experiments, using samples that are initially equilibrated to the lamellar gel or lamellar crystalline phases. It is often necessary to sacrifice some signal fidelity to attain an acceptable level of signal intensity. Inevitably, this decreases the accuracy of the information, which can be derived from the data obtained (*see* **Section 3.**).

5. The Acquisition, Analysis, and Interpretation of DSC Thermograms

It is clear from the issues discussed above that, aside from instances whereby DSC experiments are intended to simply detect or verify the occurrence of a lipid phase transition within a given temperature range, any meaningful analysis of DSC thermograms will depend on their properties being determined primarily by either the kinetic or thermodynamic characteristics of the thermotropic process under investigation. Thus, a requirement for meaningful analysis and interpretation of DSC data is the application of experimental protocols suitable for the acquisition of either accurate kinetic or thermodynamic data. As noted in **Section 3** above, the acquisition of such data is best achieved when data are acquired at slow temperature scanning rates using suitably equilibrated, appropriately sized samples that are matched to the sensitivity of the instrument. Given this, the remaining aspects of DSC data acquisition analysis and interpretation can be addressed by adopting data acquisition protocols, such that, stable baseline conditions can be established for the acquisition of accurate DSC thermograms, and the application of analytical procedures appropriate to the nature of the lipid phase transition being investigated. The following are some general guidelines in this regard.

5.1. The Establishment of Stable Operational Baseline Conditions

It is evident that establishing a proper baseline requires that the temperature range scanned during a DSC experiment be broad enough for the instrument to establish stable and identifiable baselines before the onset and after the completion of the phase transition of interest. In the authors' experience, the minimal condition for meeting this requirement is for the temperature range scanned to be wide enough for the instrument to establish stable baselines, which span at least five transition half widths before the onset and after the completion of the phase transition of interest. However, following these guidelines may not always be practical. For example, because of the necessity of maintaining ample sample hydration during the experiment, there will certainly be problems when the initiation or completion of the phase transitions of interest occur at temperatures closer to either the boiling point or freezing point of water. With many commercially available DSC instruments, the limitations of their design may make the boiling point problem virtually insurmountable. However, there are a few instruments, which are equipped with pressure-resistant sample cells that offer some capacity for operating at temperatures above the boiling point of water. However, considerable caution must always be exercised when operating under such conditions, because of the risk of personal injury and/or instrument damage if the pressure within the sample cell becomes excessive.

At the other end of the scale, problems arising from proximity to the freezing point of water are actually more tractable, mainly because considerable supercooling (~5–7°C) of water usually occurs before freezing actually begins. However, it is cautioned that supercooling of aqueous samples in calorimeters designed with nonremovable sample cells is not recommended, because costly and irreparable instrument damage may occur if water actually freezes within the sample cells of those types of instruments. However, supercooling of aqueous samples can be safely performed with calorimeters equipped with removable sample capsules and even dilute aqueous lipid dispersions can be held at temperatures near −6°C for several hours

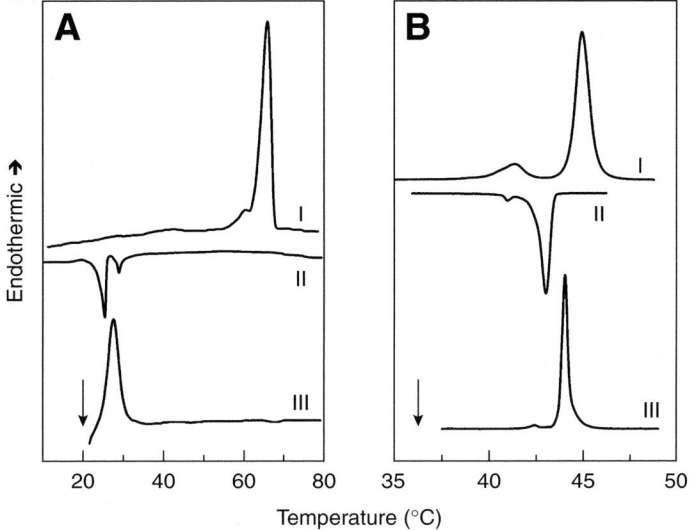

Fig. 8. DSC of lipids with metastable gel phases. The data shown were obtained with samples of 1,2-di-*O*-dodecanoyl-3-*O*-(β-D-galactopyranosyl)-*sn*-glycerol (**A**) and 1,4-di-hexadecyl, glutamic acid-*N*-succinate (**B**). (**I**) Heating thermograms obtained at 1°C/min (**panel A**) and 10°C/h (**panel B**) after low-temperature equilibration. (**II**) Cooling thermogams obtained at 1°C/min (**panel A**) and 10°C/h (**panel B**). (**III**) Heating thermograms obtained immediately after cooling samples to the temperatures indicated by the arrows. These thermograms were acquired at 5°C/min (**panel A**) and 10°C/h (**panel B**).

before ice formation actually begins. Indeed, with such instruments, it is also feasible to study lipid phase transitions at temperatures where the bulk aqueous phase is actually frozen, because dispersions of lipids in excess water contain a tightly bound population of so-called "unfreezable water," which is sufficient to assure full hydration *(11)*.

Thus, with the appropriate instrument, many of the problems arising from proximity to the freezing point of water can be circumvented, as long as a temperature range suitable for observing the phase transition can be chosen to avoid the freezing or melting of the bulk aqueous phase during the experiment. The capacity to study lipid phase transitions when the bulk aqueous phase is frozen should also obviate the need for using antifreeze agents such as ethylene glycol. In fact, the use of such reagents while studying lipid phase transitions is best avoided, as it is well known that such reagents are prone to alter the thermotropic phase behavior of lipids (e.g., *see* **refs. *12,13***).

Other problems, which might compromise the capacity to establish good baseline conditions before the onset of a lipid phase transition, usually stem from the fact that many lipid phases are metastable and are thus likely to convert to other forms under conditions wherein one would normally attempt to establish baseline conditions for studying the metastable phase. The galacto- (**panel A**) and peptido- (**panel B**) lipids illustrated in **Fig. 8** are examples of lipids with lamellar gel (L_β) phases that are metastable relative to their L_c phases. On heating, their L_c phases convert to lamellar liquid-crystalline (L_α) phases at relatively high temperatures (**curve I**), and on cooling, the L_α phases convert to the L_β phases (**curve II**). However, the L_β phases of these lipids are very unstable and their L_c phases tend to nucleate and grow rapidly if the samples are cooled to temperatures below 20° or 36°C, respectively (*see* arrows in **panels A,B**).

Thus, in order to characterize the L_β/L_α phase transitions of these lipids, the samples must not be cooled to temperatures less than 20°C (**panel A**) and 36°C (**panel B**) and must be quickly reheated after cooling even to these temperatures. However, with the peptide–lipid sample, it is feasible to establish a good baseline and obtain an accurate DSC thermogram of the L_β/L_α phase transition provided that the sample is not cooled to temperatures much below 36°C (**panel B, curve III**). However, with the galactolipid sample, it is evident that conditions are not conducive to establishing a stable baseline before the onset of the L_β/L_α phase transition (**panel A, curve III**). The information provided by DSC thermograms obtained under such circumstances will obviously be less accurate and its interpretation will be more difficult.

In closing this section, it is noted that sample baselines should be obtained by recording DSC thermograms in which the sample and reference cells are both filled with buffer and these baseline datasets should be subtracted from the sample data before any data analysis. However, this may be unnecessary when examining lipids, which exhibit sharp, highly energetic phase transitions (e.g., highly pure synthetic lipids). In such cases, one can usually reliably interpolate a baseline under the heating endotherm (or cooling exotherm), even when there is a significant coincident heat capacity change, and most modern data processing computer software is well equipped for doing this. However, constructing sample baselines is more difficult when one is investigating broad, poorly cooperative thermotropic phase transitions such as those exhibited by cholesterol-rich lipid bilayers or the compositionally heterogenous mixture of lipids found in natural cell membranes. For such systems, interpolation routines are highly prone to error, and the acquisition of so-called buffer–buffer baseline data is essential. Regarding this it is noted that considerable care must be expended to ensure that such baseline data are acquired under the same general environmental conditions as the sample data. Moreover, it is also advisable that broad, low-resolution smoothing routines be applied to the baseline data sets before subtraction to minimize degradation of the quality of the sample data.

5.2. Data Analysis and Interpretation

Given the considerable variability in the thermodynamic and kinetic characteristics of lipid phase transitions, it is evident that there will be considerable variation in the quality of the data encoded in DSC thermograms, even those recorded under near ideal conditions. The analysis and interpretation of such data thus requires the assumption of some knowledge of the nature of the information encoded therein, so that appropriate procedures can be applied. Nevertheless, despite the variability in the nature of the lipid phase transition that one is likely to encounter, there are in practice only three general approaches to data analysis and interpretation of DSC data. As illustrated in **Fig. 9**, these approaches are determined primarily by the kinetics of the phase transitions over the equilibrium temperature range of the thermotropic process under investigation and will be briefly examined below.

At one extreme are systems wherein the progress of the phase transition is fast throughout the equilibrium temperature range of the process under investigation. Under such circumstances, kinetic issues essentially become irrelevant and DSC thermograms should approach the ideal form expected of thermotropic phase changes proceeding at thermodynamic equilibrium (*see* **Fig. 9, curves B,C**). These types of phase transitions should also be equally observable in both heating and cooling mode experiments and, under ideal conditions, heating and cooling thermograms should exhibit mirror image symmetry relative to the baseline. The symmetry properties of these types of phase transitions can thus be used as a gauge of

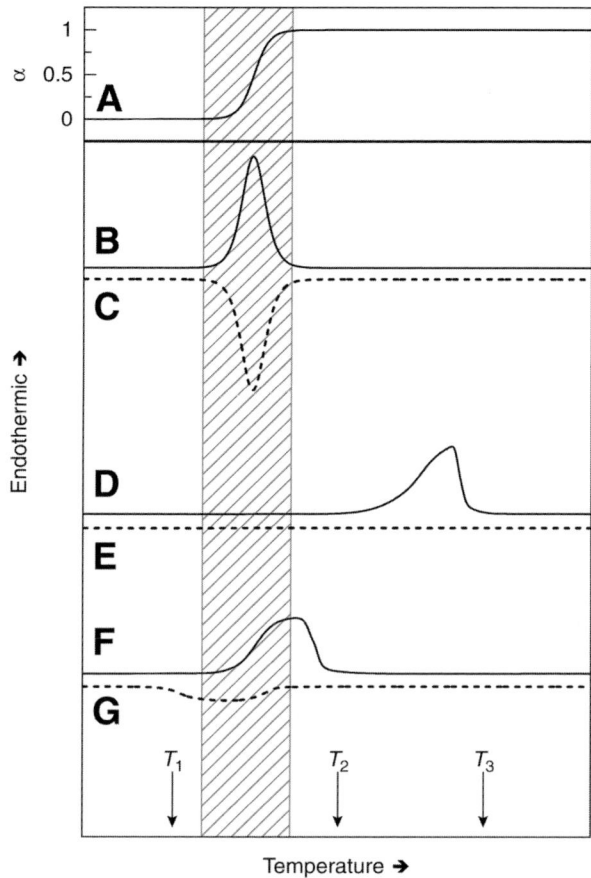

Fig. 9. Effects of reaction kinetics on the nature of thermotropic phase transitions as observed by DSC. **Curve A** shows the temperature dependence of the fractional conversion (α) of a thermotropic process at thermodynamic equilibrium. The shapes of the heating (—) and cooling (--) thermograms expected when phase transition kinetics are as follows: **Curves B,C**: fast kinetics at temperatures $\geq T_1$. **Curves D,E**: slow kinetics at temperatures $\leq T_2$ and fast kinetics at temperatures $\geq T_3$. **Curves F,G**: slow kinetics at temperatures $\leq T_1$ and fast kinetics temperatures $\geq T_2$. The hatched rectangle marks the effective equilibrium temperature range of the phase transition.

the fidelity of the data-acquisition process. With hydrated lipids, such behavior is commonly observed with simple hydrocarbon chain-melting phase transitions such as the L_β/L_α type of gel/liquid-crystalline phase transitions exhibited by most membrane lipids. DSC heating and cooling thermograms of L_β/L_α phase transitions can thus be analyzed on the basis of equilibrium thermodynamics using the approaches described earlier, or indeed, any other equilibrium thermodynamic models that are appropriate for the system under investigation (e.g., *see* **refs. *2,14***).

At the other extreme, are phase transitions for which reaction kinetics are slow throughout the equilibrium temperature range of the process. These types of processes are usually only observed in DSC heating experiments as highly asymmetric endotherms when kinetic conditions are favorable, and generally only at temperatures well above the equilibrium temperature range of the process (*see* **Fig. 9, curves D,E**). With hydrated lipids, this pattern of

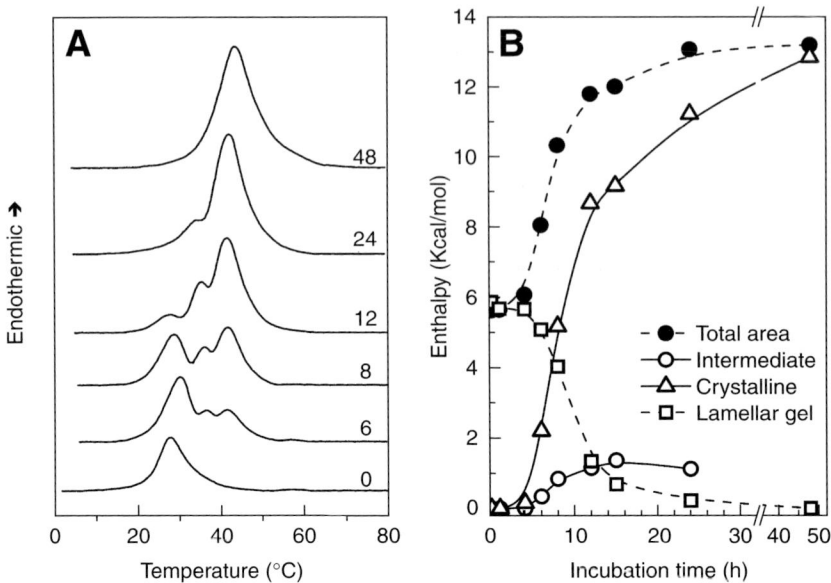

Fig. 10. Time-resolved DSC studies of the kinetics of formation of the lamellar crystalline phase of 1,2-di-*O*-dodecanoyl-3-*O*-(β-D-galactopyranosyl)-*sn*-glycerol. **Panel A** shows the DSC thermograms acquired by fast temperature scanning (~5°C/min) of a sample that was incubated at −3°C for the time (hours) indicated. **Panel B** shows some time-resolved partial analysis of the DSC thermograms.

thermotropic phase behavior is often (although not always) observed with L_c/L_β phase transitions or when lipid lamellar crystalline phases convert to the lamellar-gel state (e.g., the subtransitions exhibited by medium and long-chain PCs and phosphatidylglycerols (PGs) *[3,6]*).

For these types of phase transitions, there is no thermodynamic significance to the "transition temperature," "transition width," or indeed any of the properties of the DSC thermogram except the calorimetric enthalpy, which is encoded by the area of the thermogram relative to the baseline. Moreover, the calorimetric enthalpy so determined will be accurate only if the system is fully equilibrated before data acquisition. However, because the DSC thermograms obtained under these conditions are determined exclusively by the kinetic characteristics of the process under investigation, it is feasible to interpret these high-fidelity DSC thermograms in terms of the kinetic characteristic of the phase transition using approaches similar to those described above, or indeed any other kinetic model which may be appropriate for the process under investigation.

DSC can also be used to study the kinetics of lipid phase transition with the aid of classical time-resolved rapid sampling techniques. As illustrated in **Fig. 10**, such techniques are well suited for studying the kinetics of slow thermotropic process such as the thermotropic transformations involving the lamellar gel and crystalline phases of lipids. In the example shown, a hydrated sample of the glycolipid, 1,2-di-*O*-dodecanoyl-3-*O*-(β-D-glucopyranosyl)-*sn*-glycerol was incubated at −3°C, and DSC was used for time-resolved monitoring of the changes in the thermotropic phase behavior of the sample as its L_β phase converts to L_c phase.

Because the application of DSC to this type of study requires that the temperature scanning rate used to record the DSC thermograms be considerably faster than the kinetics of the process(es) under investigation, this type of work cannot be done with high-sensitivity instruments, which are usually constrained to operating at relatively slow scan rates. However, fast scanning, low-sensitivity instruments are well suited for this kind of work, especially because the only calorimetric parameter required for this analysis is the calorimetric enthalpy, as encoded in the area of the thermograms relative to the baseline. In the example shown, this application of DSC was quite effective at characterizing the kinetics of the transformations taking place (see **panel B**) and also detected the presence of an intermediate, which was subsequently confirmed by Fourier-transform infrared (FTIR) spectroscopy *(15)*.

In between these extremes of the thermodynamically controlled and kinetically determined types of phase transitions described above, are phase transitions which approach thermodynamic equilibrium at moderate rates over the equilibrium temperature range of the phase transition. These types of transitions exhibit heating thermograms which tend to be slightly distorted at the low temperature end and are upward shifted from the equilibrium temperature range of the process (**Fig. 9**, **curve F**). Also, the extent of the upshifting of the observed DSC thermogram is scan rate dependent and typically describes a sigmoid curve with low and high temperature asymptotes near the thermodynamic equilibrium temperature, and near the lower temperature boundary of the fast kinetic region (temperature T_3 in **Fig. 9**).

When examined in the cooling mode, these types of phase transitions exhibit pronounced scan rate-dependent cooling hysteresis, the observed DSC thermogram is usually considerably broader, and its resolvable area relative to the baseline is also significantly diminished (*see* **Fig. 10**, **curve F**). With hydrated lipids, this pattern of thermotropic phase behavior is commonly observed at thermotropic interconversions between different types of lamellar gel phases (e.g., the pretransition of DPPC), and between different types of inverted nonlamellar phases (e.g., Q_{II}/Q_{II} and Q_{II}/H_{II} transitions). For these types of lipid phase transitions, meaningful analysis and interpretation of the data is only feasible with thermograms recorded in the heating mode, and even then the process of accessing the kinetic or thermodynamic characteristics of these phase transitions is extremely difficult.

Thus, these DSC thermograms should always be reported alongside the temperature scanning rates used during data acquisition, and one should not attempt to attach any thermodynamic significance to these properties. However, because the properties of the observed thermograms should approach the form expected of thermodynamic equilibrium at infinitely slow scan rates, one might be able to estimate the thermodynamic characteristics of these types of phase transitions by extrapolation to zero scan rate, using a series of high-fidelity heating thermograms recorded at very slow temperature scanning rates (e.g., *see* **Fig. 6B**). Obviously, there is considerable potential for error in this latter approach and the parameters obtained should be interpreted cautiously.

6. Conclusions

From the aforementioned, it is apparent that most of the problems which one might encounter in the application of DSC to study of lipid phases stem from the fact that lipid thermotropic phase behavior is very diverse. Thus, effective application of DSC in this field essentially requires that the user have some general knowledge of the nature and properties of the lipid phase transitions which may be encountered, so that appropriate protocols for the acquisition, analysis, and interpretation of the data can be used. It is thus recommended that the user be fairly well acquainted with the general field of lipid thermotropic phase behavior, and with the ways in which lipid phase behavior can be affected by the presence of proteins, peptides, and other guest molecules (for reviews in this area, *see* **refs.** *16–18*). The user should also be fully acquainted with the capabilities and limitations of the instruments available to enable the application of data acquisition protocols, which maximize the information content of the data acquired. Finally, the user should also recognize that there is considerably more potential for the effective application of DSC in the field lipid and biomembrane research beyond what has been commonly used so far.

Acknowledgments

Research performed in the authors' laboratory was supported by operating and major equipment grants from the Canadian Institutes of Health Research and by major equipment and personnel support grants from the Albert Heritage Foundation for Medical Research.

References

1. Borchardt, H. J. and Daniels, F. (1957) The application of differential thermal analysis to the study of reaction kinetics. *J. Am. Chem. Soc.* **79,** 41–46.
2. Hinz, H. -J. and Schwarz, F. W. (2001) Measurement and analysis of results obtained on biological substances with differential scanning calorimetry. *Pure Appl. Chem.* **73,** 745–759.
3. Lewis, R. N. A. H., Mak, N., and McElhaney, R. N. (1987) A differential scanning calorimetric study of the thermotropic phase behavior of model membranes composed of phosphatidylcholines containing linear saturated fatty acyl chains. *Biochemistry* **26,** 6118–6126.
4. Mannock, D. A., Lewis, R. N. A. H., Sen, A., and McElhaney, R. N. (1988) The physical properties of glycosyl diacylglycerols. Calorimetric studies of a homologous series of 1,2-di-O-acyl-3-O-(β-D-Glucopyranosy)-*sn*-glycerols. *Biochemistry* **27,** 6852–6859.
5. Lewis, R. N. A. H. and McElhaney, R. N. (1993) Calorimetric and spectroscopic studies of the polymorphic phase behavior of the homologous series of N-saturated 1,2-diacyl phosphatidylethanolamines. *Biophys. J.* **64,** 1081–1096.
6. Zhang, Y. -P., Lewis, R. N. A. H., and McElhaney, R. N. (1997) Calorimetric and spectroscopic studies of the thermotropic phase behavior of the n-saturated 1,2-diacylphosphatidylglycerols. *Biophys. J.* **72,** 779–793.
7. Luzzati, V., Delacroix, H., Gulik, A., Gulik-Krzywixki, T., Mariani, P., and Vargas, R. (1997) The cubic phases of lipid, in *Lipid Polymorphism and Membrane Properties*, (Epand, R. M., ed.), Academic Press, New York, pp. 3–24.
8. Shyamsunder, E., Gruner, S. M., Tate, M. W., Turner, D. C., So, P. T. C., and Tilcock, C. P. S. (1988) Observation of inverted cubic phase in hydrated dioleoylphosphatidylethanolamine membranes. *Biochemistry* **27,** 2332–2336.
9. Veiro, J. A., Khalifah, R. G., and Rowe, E. S. (1990) P-31 Nuclear magnetic resonance studies of the appearance of an isotropic component in dioleoylphosphatidyletanoamine. *Biophys. J.* **57,** 637–641.
10. Lewis, R. N. A. H., McElhaney, R. N., Harper, P. E., Turner, D. C., and Gruner, S. M. (1994) Studies of the thermotropic phase behavior of phosphatidylcholines containing 2-alkyl substituted fatty acyl chains. a new class of phosphatidylcholines forming inverted nonlamellar phases. *Biophys. J.* **66,** 1088–1103.
11. Kodama, M., Kuwabara, M., and Seki, S. (1982) Successive phase-transition phenomena and phase diagram of the phoshatidylcholine-water system as revealed by differential scanning calorimetry. *Biochim. Biophys. Acta* **689,** 567–570.
12. Lewis, R. N. A. H., Sykes, B. D., and McElhaney, R. N. (1987) Thermotropic phase behavior of model membranes composed of phosphatidylcholines containing *dl*-methyl anteisobranched fatty acids. 1. Differential scanning calorimetric and ^{31}P-NMR spectroscopic studies. *Biochemistry* **26,** 4036–4044.
13. Lewis, R. N. A. H., Sykes, B. D., and McElhaney, R. N. (1988) Thermotropic phase behavior of model membranes composed of phosphatidylcholines containing *cis*-monounsaturated acyl chain homologues of oleic acid. Differential scanning calorimetric and ^{31}P-NMR spectroscopic studies. *Biochemistry* **27,** 880–887.

14. Ladbury, J. E. and Chowdhry, B. Z. (1998) Biocalorimetry. Applications of calorimetry in the biological sciences. Wiley & sons, New York.
15. Lewis, R. N. A. H., Mannock, D. A., McElhaney, R. N., Mantsch, H. H., and Wong, P. T. T. (1990) The physical properties of glycosyl diacylglycerols. An infrared spectroscopic study of the gel phase polymorphism of the 1,2-di-O-acyl-3-O-(β-D-Glucopyranosy)-*sn*-glycerols. *Biochemistry* **29,** 8933–8943.
16. McElhaney, R. N. (1982) The use of differential scanning calorimetry and differential thermal analysis in studies of model and biological membranes. *Chem. Phys. Lipids* **30,** 229–259.
17. McElhaney, R. N. (1986) Differential scanning calorimetric studies of lipid-protein interactions in model membrane systems. *Biochim. Biophys. Acta* **864,** 361–421.
18. Lewis, R. N. A. H. and McElhaney, R. N. (2004) The mesophilic phase behavior of lipid bilayers, in *The Structure of Biological Membranes, 2nd ed.*, (Yeagle, P. L., ed.), CRC Press, Boca Raton, FL, pp. 53–120.

13

Pressure Perturbation Calorimetry

P. D. Heiko Heerklotz

Summary

Pressure perturbation calorimetry is a rather new technique which serves to measure the temperature-dependent thermal volume expansion of a solute or particle in aqueous dispersion. It can be used to detect thermotropic transitions in lipid systems and to characterize their accompanying volume changes and kinetics. The results are of highest precision and obtained in a very convenient, fully automated experiment, requiring relatively little material. The strategy of the technique is to measure the heat response to a very little, isothermal pressure perturbation in a high-sensitivity isothermal calorimeter. On the basis of such data, thermodynamic laws and considerations yield the thermal expansion of the partial volume of the solute or colloidal particle.

Key Words: Expansivity; melting; phase transition; PPC; pressure dependence; volume change.

1. Introduction

Pressure perturbation calorimetry (PPC) was introduced rather recently as a method to study thermotropic changes in expansivity and partial volume of proteins, polymers, and other solutes or colloidal particles in aqueous solution *(1,2)*. A recent criticism on the use of PPC for measuring volumetric properties of solutes *(3)* was shown to be not relevant *(4,5)*. Only a few studies have applied PPC to characterize lipid membranes *(6–10)*. The method, however, has been used to derive significant information, as follows:

1. Integration of a "PPC peak" from the baseline yields the *volume change of lipid melting* and other thermotropic phase transitions *(6,9)*, provided the transitions are fast and reversible, and the lipid remains microdispersed (*see* **Note 11**). In turn, the pressure dependence of the melting point is obtained if the melting heat is known.
2. PPC provides an independent means to *validate the existence* and peak shape of a thermotropic transition. This may be of particular value for weak, broad transitions in complex lipid mixtures that show ambiguities regarding the correct baseline and interpretation of, for example, a differential scanning calorimetry (DSC) curve *(7,10,11)*. The peak shape is related to the cooperativity of the transition.
3. Independent of a transition, PPC measures the temperature-dependent *coefficient of thermal expansion* of the lipid aggregate, which depends mainly on the molecular packing in the membrane, and the volume effects of hydration of polar and apolar surfaces.
4. The *kinetics* of the processes induced by a small pressure jump can be quantified within a time window of about 1 s, which requires deconvolution of data with instrument response-function for a few minutes (limited by the long-term baseline stability of the instrument) *(8)*. Metastable states can be identified *(9)*.

The instrument measures the heat response of a sample to a small pressure change under isothermal conditions: $\partial Q/\partial p|_T$. This information serves to quantify the thermal volume expansion at constant pressure $(\partial V/\partial T|_p)$, according to the Maxwell relation to the isothermal, pressure-dependent change in entropy $(\partial S/\partial p|_T)$:

$$\left.\frac{\partial S}{\partial p}\right|_T = \left.\frac{\partial V}{\partial T}\right|_p \qquad (1)$$

where, in a reversible process, dS is given by:

$$dS = \frac{dQ}{T} \qquad (2)$$

After deducing the change in the partial volume of the solute (absolute [V_S] and molar [v_S]), the results can be expressed as the coefficient of thermal expansion of the solute (α_S):

$$\alpha_S = \frac{\partial V_S}{V_S \partial T} \qquad (3)$$

In certain cases, for example, when contributions from different components or moieties to the total expansivity are discussed, it is preferable to derive the thermal change of the partial molar volume ($\partial v_S/\partial T$) in mL/mol K (see **Note 6**). Whereas α_S averages over different components or moieties (weighted by their volume), contributions to $\partial v_S/\partial T$ are additive if the different parts do not perturb each other. Alternatively, the relative volume expansion of the whole solution can be computed (see **Note 12**).

2. Materials

2.1. Instrument

The PPC experiment is based on the measurement of the heat response of a sample to an isothermal pressure change $(\partial Q/\partial p|_T)$. Thus, it is related to experimental techniques that have been termed piezothermal analysis, scanning transitiometry, and so on (see **ref. 4** and, references therein for more details). The term PPC denotes a new approach utilizing very small, sudden pressure changes in the order of 5 bar, which induce only a differential change in the state of the system (see **Note 1**). The resulting heats are extremely small, and thus, require a calorimeter with state-of-the-art sensitivity. The calorimeter must work in isothermal mode, but should allow for a fast change in experimental temperature between two individual measurements; thus, power-compensation calorimeters seem superior for this purpose. All published data were obtained on a commercially available PPC accessory and calorimeter by MicroCal (Northampton, MA) *(1,2,6,9,12–14)*, or a home-built setup utilizing an insert cell *(8,15)*.

2.2. Samples

The PPC sample should be dispersed in buffer or water; concentrations used in lipid measurements range from 4 *(9)* to 100 mM *(8)*. It is important to note that salt solutions show a strong PPC signal (see **Note 2**). This makes it essential to perform the measurement with the reference cell being filled with buffer identical to that in the sample. Thus, biological samples should be extensively dialyzed to make sure that the aqueous phase matches the content of the reference cell. If possible, low-concentration buffers or pure water are preferable to keep blanks and corrections small and thus yield most accurate results.

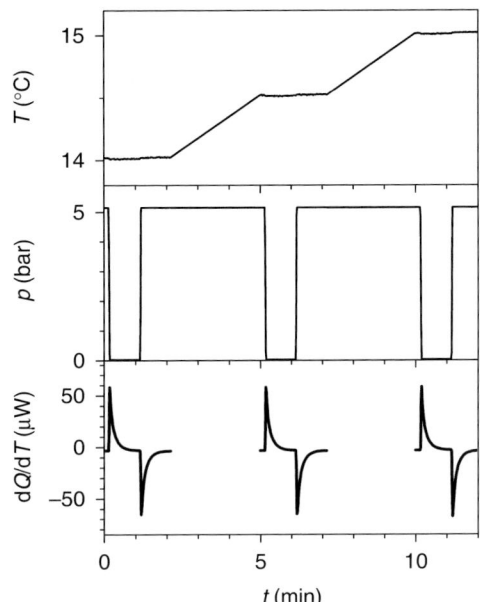

Fig. 1. PPC curve (open spheres, right ordinate) and DSC curve (solid line, left ordinate) of the chain melting transition of egg sphingomyelin. The area under the PPC peak (from the base line) yields the partial molar volume change of melting, $\Delta v_S = 21$ mL/mol (corresponds to $\Delta v_S/v_S = 3$ vol%). The area under the DSC curve corresponds to an enthalpy change of $\Delta H = 7.3$ kcal/mol; the shape of both peaks (with baselines subtracted or being virtually constant as in the example) is the same (*see also* **ref. 8**). (Reproduced with permission from **ref. 7**, ©2002 Biophysical Society.)

3. Method

3.1. Experiment

The experiment starts with filling the sample and buffer into sample and reference cells, respectively (e.g., 0.5 mL, each). It is crucial to avoid air bubbles in the cells (*see* **Note 7**). The experimental schedule lists all temperatures at which measurements (one down- and one upjump in pressure, each) will be performed. Repetitions at the same temperature are possible, and sometimes desirable, to improve the statistics and to further check the reversibility of the measurement. In transition ranges, the density of temperature-points should be increased when compared with regions without transitions, to make sure that the peak is well resolved (*see* different spacing of points in **Fig. 1**). For pure saturated lipids, which show melting peaks with a width of the order of 0.1 K, the temperature resolution poses a problem and the minimum step width between two subsequent experimental temperatures must be markedly diminished compared with the default value optimized for protein samples (*see* **Note 13**).

A commercial system performs the whole series of down- and upjumps in pressure at, for example, 90 selected temperatures automatically. **Figure 2** shows details of such an experiment. The system is equilibrated at a selected temperature and a slightly enhanced pressure of approx 5 bar above ambient (applied from a nitrogen tank). After equilibration, a valve disconnects the nitrogen tank from the sample and lets the pressure drop to ambient. This should induce a reduction in the temperature of the system, but this temperature change is actively compensated by a feedback heater. Hence, the heat response of the sample is equal in value

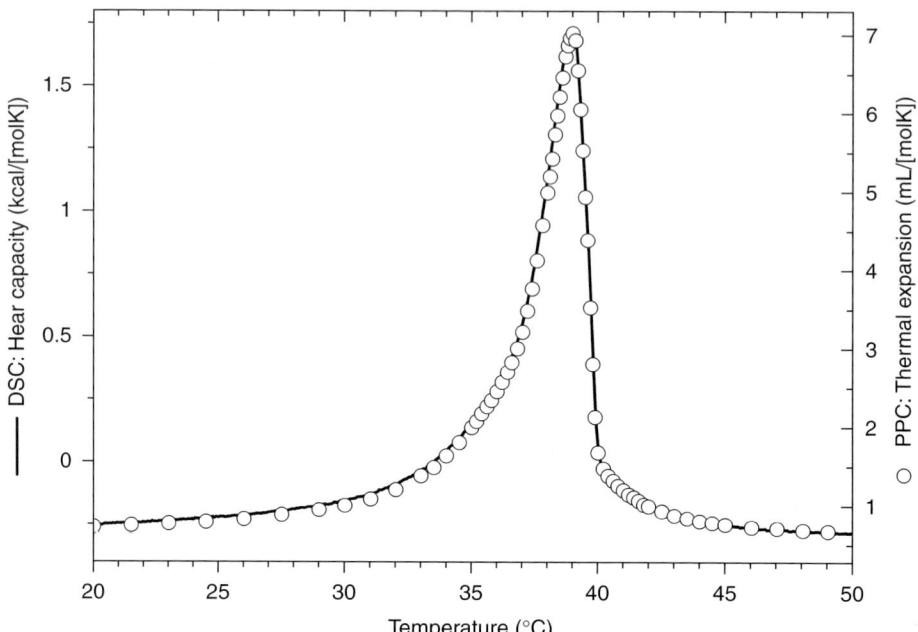

Fig. 2. Detail of a PPC experiment. The temperature T (**top**) is kept constant at a desired value, the pressure (**middle**) drops, and the resulting heat is compensated by a power peak of the feedback heater (**bottom**). After re-equilibration, the pressure is switched back and the heat response is reversed. Then, the detection (**bottom**) is switched off, and the instrument scans automatically to the next desired temperature. (Reproduced with permission from **ref. 6**, ©2002 Biophysical Society.)

and opposite in sign to the (electrical) power peak of the heater, which is the experimental output. Following the return of the signal to baseline (which is automatically detected) after approx 1 min, the pressure is switched back up to 5 bar, yielding an experimentally independent, second measurement of $\partial Q/\partial p|T$ (which should agree with that after the downjump; otherwise, check **Notes 7–10**). After re-equilibration at the high pressure, the system is brought automatically to the next desired temperature and equilibrated. The procedure continues as explained until the whole schedule of experimental temperatures is completed.

3.2. Data Evaluation to Yield the "PPC Curve," $\alpha_S(T)$

After the run, all power peaks are automatically integrated over time from an automatically adjusted baseline to yield the heat response (in µcal) as a function of temperature. This run yields the difference in heat responses between the cell (with the sample S dispersed in buffer) and the reference cell (with buffer, B): ΔQ_{S-B}. Very small heats obtained in a blank experiment with buffer in both cells, ΔQ_{B-B}, are subtracted to correct for minimal differences between the volumes and other properties of the two cells, yielding a corrected value of ΔQ_{S-B} (ΔQ_{S-B}^{corr}). This blank run can be performed with a low temperature resolution, as it shows no peaks or sudden changes. The resulting curve is fitted by a second order polynomial in order to interpolate ΔQ_{B-B} to all temperatures where ΔQ_{S-B} was measured.

In order to derive the thermal expansion of the partial volume of the lipid or protein *per se*, two more blank runs are needed for each buffer. A run with buffer vs water yields ΔQ_{B-W}, and another run with both cells filled with water detects ΔQ_{W-W}, yielding a technically corrected value of $\Delta Q_{B-W}^{corr} = \Delta Q_{B-W} - \Delta Q_{W-W}$, analogously to the ΔQ_{B-W}^{corr} explained previously. The evaluation software calculates the coefficient of thermal expansion of the sample, α_S, automatically using **Eq. 4** (*see* derivation in **Note 5**):

$$\alpha_S = \frac{Q_S}{TV_S \Delta p} = -\frac{\Delta Q_{S-B}^{corr} + C_S \tilde{V}_S \left(\Delta Q_{B-W}^{corr} + Q_W \right)}{TV_{cell} C_S \tilde{V}_S \Delta p} \quad (4)$$

To this end, one needs to open the data set of ΔQ_{S-B} to provide data or polynomials for ΔQ_{B-W}, ΔQ_{B-B}, and ΔQ_{W-W}, and specify the sample (e.g., lipid) concentration (c_S), its specific volume ($\tilde{v}_S$), and the cell volume (V_{cell}). The absolute heat response of a cell containing water (Q_W) cannot be measured, but is considered in the evaluation software using published, high-precision data on the thermal expansion of pure water. Specific volumes of monounsaturated lipids such as palmitoyloleoylphosphatidylcholine are almost exactly 1 mL/g; with saturated lipids showing somewhat smaller, and lipids with higher unsaturation showing somewhat larger values of $\tilde{v}_S$. Such data are given, for example, by Nagle et al. (*16*), and can be measured by a densitometry experiment (such as vibrating tube densimetry, isopycnic centrifugation in H_2O/D_2O mixtures, and so on) at a single temperature.

If the results obtained show pressure decreases and increases, and if performed at subsequent pressure perturbations at the same temperature show no systematic deviation, they may be averaged. Systematic differences may be because of different problems, which must be identified (*see* **Notes 7–10**) before it can be decided whether it is warranted to average data (for how to do that in Origin™) (*see* **Note 14**) or whether they indicate the failure of the experiment.

3.3. Deriving Additional Information From PPC Data

If the PPC curve shows a peak representing a thermotropic transition the accompanying relative volume change ($\Delta V_S/V_S$) may be derived as follows:

1. Arbitrary baselines for the low- and high-temperature phases, $\alpha_{below}(T)$ and $\alpha_{above}(T)$, are assumed on the basis of the PPC curve in the temperature ranges lower and higher than the transition region.
2. The baseline, $\alpha_{base}(T)$, starts as α_{below} and shows a smooth transition to α_{above} in the transition region. For a two-state transition, it is thermodynamically justified and reasonable to compute a progress baseline (which is possible with a routine included in the instrument software). In some other cases, a cubic function can, for example, be used to derive a smooth α_{base}. The difference between the extrapolated, one-phase baselines at the midpoint temperature (T_m) is the change in the coefficient of thermal expansion accompanying the transition ($\Delta \alpha_S$):

$$\Delta \alpha_S = \alpha_{above}(T_m) - \alpha_{below}(T_m) \quad (5)$$

for interpretation *see* **Note 3**.

3. Integration of the PPC peak from the baseline yields the relative volume change ($\Delta V_S/V_S$):

$$\frac{\Delta V_S}{V_S} = \int_{below}^{above} \alpha_S - \alpha_{base} \, dT \quad (6)$$

4. Integration from below the transition to a selected temperature in the transition range (T^*) yields the progress variable of the transition, $\xi_V(T^*)$, which runs from 0 (below) to 1 (above):

$$\xi_V(T^*) = \int_{below}^{T^*} \alpha_S - \alpha_{base} dT \bigg/ \int_{below}^{above} \alpha_S - \alpha_{base} dT \qquad (7)$$

thus, ξ_V is related to the progress of the volume change.

5. For a two-state transition, ξ_V agrees (analogously to the progress of a heat capacity peak measured by DSC, *see* **ref. 17**) with the fraction of molecules in the high-temperature state. Then, the PPC curve can also be used to determine the van't Hoff enthalpy of the transition (ΔH_{vH}):

$$\alpha_S(T) - \alpha_{base}(T) = \frac{\Delta V_S}{V_S} \frac{K(T) \Delta H_{vH}}{[1+K(T)]^2 RT^2} \qquad (8)$$

with

$$K(T) = \frac{\xi_V(T)}{1-\xi_V(T)} = \exp\left[\frac{-\Delta H_{vH}}{RT}\left(1-\frac{T}{T_m}\right)\right] \qquad (9)$$

As ΔH_{vH} denotes the enthalpy per elementary process (melting of a mole of cooperative units) and the calorimetric ΔH (measured by DSC) is the enthalpy per mole of lipid, the quotient $\Delta H_{vH}/\Delta H$ represents the number of molecules per cooperative unit (*see* **ref. 12** for an analogous approach applied to a micellar transition).

6. If ΔH is known from a DSC experiment, the results for Δv_S (in mL/mol) obtained by PPC can be used to compute the pressure dependence of the transition temperature (dT_m/dp):

$$\frac{dT_m}{dp} = T_m \frac{\Delta V_S}{\Delta V} \qquad (10)$$

derived from the Clausius Clapeyron equation (*see* **Note 4** on results). Heerklotz and Tsamaloukas *(11)* describes the case that an additive (in this example cholesterol) changes the partial molar volume of a fraction ξ of the lipid by ΔV, with both parameters depending on temperature (T), and the mole fraction of the additive (X). The composition-dependent expansivity, $V' = dV/dT$, becomes:

$$V'(X) = X \cdot V'(1) + (1-X)[V'(0) + \xi \Delta V' + \xi' \Delta V] \qquad (11)$$

The study derives expressions for $\xi(X)$ and $\xi(X)'$ for the cases of random mixing and phase transitions, and applies them to test the validity of phase approximation for cholesterol-rich domains in membranes.

4. Notes

4.1. Physical Background

1. A pressure change of +5 bar causes a T_m-shift of only about 2.5×10^{-4} K (for $dT_m/dp = 20$ K/kbar) (*see* **Note 4**). Hence, it "freezes" far less than 1% of the previously fluid lipid. Whereas parameters measured in the kbar range must be extrapolated to low pressure when properties under physiological conditions are concerned, PPC data do not need to be corrected. PPC serves

primarily to measure thermal expansion at ambient pressure rather than focusing on pressure effects.
2. Salts and other polar solutes exhibit a temperature dependent "binding" of water molecules. Water molecules in a hydration shell of a polar moiety are more tightly packed than in bulk water, wherein a space-consuming water–water hydrogen bond network is established (particularly at not too high temperature). The thermally activated release of condensed water molecules gives rise to a strong positive contribution to the thermal expansion *(18)*.
3. $\Delta\alpha_S$ values of protein denaturation can also be interpreted in terms of the accompanying change in water-exposed surface area. For lipids, $\Delta\alpha_S$ appears to be small, although lipid melting is also accompanied by an increased hydration. A detailed understanding of lipid $\Delta\alpha_S$ remains to be established.
4. It is interesting to observe that many different lipids with saturated chains share a pressure dependence in the melting point of dT_m/dp about 20 K/kbar (reviewed in **ref. 4**). This seems to be governed by an intrinsic property of the *trans–gauche* isomerization of the chains. Somewhat smaller values were found for unsaturated compounds with the chain packing being perturbed by one or more *cis*-double bonds.

4.2. Derivation and Background of Eq. 4

5. The goal of the procedure is to derive the absolute heat response of the sample (Q_S) from the difference between the overall signals from sample and reference cell (*see also* ref. 2). Denoting the partial volume occupied by the sample as V_S, it is stated that the sample cell contains V_S of sample and $V_{cell} - V_S$ of buffer, whereas the reference cell contains V_{cell} of buffer. The contributions of the volume $V_{cell} - V_S$ cancel each other because this volume is filled with buffer in both cells. Contributions to the differential heat arise only from the sample (Q_S) and from the fraction of the buffer in the reference cell that is not matched in the sample cell (V_S/V_{cell}). The corrected, differential signal, ΔQ^{corr}_{S-B}, reads:

$$\Delta Q^{corr}_{S-B} = Q_S - \frac{V_S}{V_{cell}} \cdot Q_B \qquad (12)$$

where Q_B is the total signal from a cell filled with buffer and the negative sign applies to the heat in the reference cell. In order to determine Q_B for a selected buffer, two more blanks are required with buffer (sample cell) vs water (reference), and with water in both cells as a blank, yielding $\Delta Q^{corr}_{B-W} = \Delta Q_{B-W} - \Delta Q_{W-W}$. Based on literature values of the heat response of pure water on a given pressure change (Q_W), which are saved in the evaluation program, one may replace Q_B in **Eq. 12** as:

$$Q_B = \Delta Q^{corr}_{B-W} + Q_W \qquad (13)$$

The absolute volume of the sample (V_S) can be expressed as:

$$V_S = m_S \tilde{V}_S = c_S \tilde{V}_S V_{cell} \qquad (14)$$

with the mass (m_S), mass-per-volume concentration (c_S), and specific volume ($\tilde{v}_S$), of the, for example, lipid. Based on **Eqs. 1** and **2**, the coefficient of thermal expansion of the sample (α_S) is given by **Eq. 4**.

6. In some cases, it is advisable to convert the results for the expansion coefficients (α_S) into molar expansivities, $\partial v_S/\partial T|_p$, i.e., in mL/mol K. To this end, the values for α_S must be multiplied by the molar volume (v_S) and the molecular weight.

4.3. (Dis)agreement of Results of Pressure Increase and Decrease

7. Usually, a not significant yet systematic deviation between the *results obtained on pressure decrease and increase* should be observed. For runs with thermally similar solutions in the cells (same buffer with and without a small amount of lipid or protein), typical heats are less than about 100 µcal, and those measured by pressure decrease and increase at a given temperature agree within about 2%. Several effects can lead to systematic deviations between the two values, so that the resulting curve, $\Delta Q(T)$, exhibits a saw-tooth behavior.
8. One reason is a poor filling leaving a small *air bubble* in one of the cells. The resulting data are wrong and must not be averaged; the correct value may be far out the range between the data points. The cells must be carefully refilled.
9. If the solutions in the two cells differ strongly from each other, for example, in buffer vs water runs with buffers containing *high salt or denaturant concentrations or cosolvents* (glycerol and so on), the heats become much larger than 100 µcal and the deviations may reach to 5% and more. These results are fully reproducible in an independent measurement. In this case, the correct value is approximately the average of the values obtained by pressure decrease and increase.
10. If the heats measured secondly (typically the one induced by pressure increase) are smaller in absolute value than the first, and if subsequent pressure perturbations at the same temperature yield even smaller heats, the system shows an *irreversible transition*. Again, it is not justified to average such data and interpret them in terms of equilibrium properties *(9)*.
11. Even a very good *constancy in the temperature* during an isothermal pressure perturbation of the order of 0.05 K may not be fully sufficient to deal with very sharp and strong transition peaks, as obtained for the melting of pure saturated lipids (e.g., width approx 0.1 K, ΔV ~3%). Because of the very high heat capacity in the transition range, even a slight temperature increase causes a significant endothermic heat effect (like a DSC peak) and a temperature decrease causes an exothermic one. For example, the heat accompanying a temperature shift by +0.02 K in the transition range of a lipid (20 mM, 0.5 mL cell, and 20 kcal/[mol K]) is 4 mcal; PPC heats for DMPC show a maximum of about 20 mcal *(6)*. Such effects may contribute to the fact that for DMPC, pressure-increase data alone suggest a somewhat larger $\Delta V_S/V_S$ of melting (3%) than pressure decrease data (2.8%) *(6)*.

4.4. Technical Solutions for the MicroCal VP PPC

As all available literature (apart from a home-built setup) has been derived using the same instrument, it seems appropriate to provide some specific information that may be useful to perform PPC experiments with lipids using this calorimeter.

12. Flotation and sedimentation of the sample, for instance, on formation of a macroscopically separating (e.g., inverse hexagonal) phase may change the amount of sample within the calorimetrically active cell. This is because of the fact that the access capillary is filled with the same material. The effect of sedimentation of multilamellar vesicles of saturated (relatively high density) lipids can be minimized by reducing the amount of the sample in the access tube to a well-defined, minimum amount using a labelled syringe needle provided by the manufacturer.
13. Deriving the overall expansivity of the solution in the cell: in specific cases, it may be of interest to consider the overall expansivity of a solution in the sample cell (measured vs water). In a data evaluation based on **Eq. 4**, this can be achieved by reading the water–water blank data, ΔQ_{W-W}, into the channels for ΔQ_{W-W}, ΔQ_{B-B}, and ΔQ_{B-W} (thus eliminating the buffer correction), and by specifying apparent concentrations of $c_S = 1$ g/mL and a specific volume of $v_S = 1$ mL/g; thus, the effective reference volume becomes V_{cell} rather than V_S (*see* **Eq. 4**).
14. For melting experiments with lipids, the default tolerance for the experimental temperature (0.1 K) is usually too large, given that a resolution of approx 0.01 K is needed to determine the transition peak. If the next desired temperature is within the tolerance of the previous (current) temperature, the system does not move to the new desired temperature but conducts the next pressure

jumps immediately. To resolve this problem for the Microcal VP DSC&PPC system, one has to edit the file vpviewer.ini, section (PPC), and change the entry "PPCTemperatureTolerance" to a smaller value (check manual for more information).

15. Averaging in origin: to average pairs of data in a column denoted A (such as α and T values of pressure increase and decrease), "set column values" of another column to $[col(A)(2*i) + col(A)(2*i-1)]/2$.

Acknowledgments

The author thanks Alekos Tsamaloukas for valuable comments on the manuscript. Financial support from the Swiss National Science Foundation (grant 31-67216.01) is gratefully acknowledged.

References

1. Kujawa, P. and Winnik, F. M. (2001) Volumetric studies of aqueous polymer solutions using pressure perturbation calorimetry: A new look at the temperature-induced phase transition of poly(N-isopropylacrylamide) in water and D2O. *Macromolecules* **34**, 4130–4135.
2. Lin, L. N., Brandts, J. F., Brandts, J. M., and Plotnikov, V. (2002) Determination of the volumetric properties of proteins and other solutes using pressure perturbation calorimetry. *Anal. Biochem.* **302**, 144–160.
3. Randzio, S. L. (2001, 2003) Comments on "volumetric studies of aqueous polymer solutions using pressure perturbation calorimetry" *Macromolecules* **34**, 4130; *Thermochim. Acta* **398**, 75–80.
4. Heerklotz, H. (2004) Microcalorimetry of lipid membranes *J. Phys. Cond. Matter* **16**, R441–R467.
5. Brandts, J. and Lin, L. (2003, 2004) Rebuttal to communication critical of the use of pressure perturbation calorimetry for measuring volumetric properties of solutes. *Thermochim. Acta* 75–80; *Thermochim. Acta* **414**, 95–100.
6. Heerklotz, H. and Seelig, J. (2002) Application of pressure perturbation calorimetry to lipid bilayers. *Biophys. J.* **82**, 1445–1452.
7. Heerklotz, H. (2002) Triton promotes domain formation in lipid raft mixtures. *Biophys. J.* **83**, 2693–2701.
8. Grabitz, P., Ivanova, V. P., and Heimburg, T. (2002) Relaxation kinetics of lipid membranes and its relation to the heat capacity. *Biophys. J.* **82**, 299–309.
9. Wang, S. L. and Epand, R. M. (2004) Factors determining pressure perturbation calorimetry measurements: evidence for the formation of metastable states at lipid phase transitions. *Chem. Phys. Lipids* **129**, 21–30.
10. Chong, P. L. G., Ravindra, R., Khurana, M., English, V., and Winter, R. (2005) Pressure perturbation and differential scanning calorimetric studies of bipolar tetraether liposomes derived from the thermoacidophilic archaeon Sulfolobus acidocaldarius. *Biophys. J.* **89**, 1841–1849.
11. Heerklotz, H. and Tsamaloukas, A. (2006) Gradual change or phase transition—characterization of fluid lipid-cholesterol membranes on the basis of thermal volume changes. *Biophys. J.* **91**, (in press).
12. Heerklotz, H., Tsamaloukas, A., Kita-Tokarczyk, K., Strunz, P., and Gutberlet, T. (2004) Structural, volumetric and thermodynamic characterization of a micellar sphere-to-rod transition. *J. Am. Chem. Soc.* **126**, 16,544–16,552.
13. Ravindra, R. and Winter, R. (2004) Pressure perturbation calorimetry: A new technique provides surprising results on the effects of co-solvents on protein solvation and unfolding behaviour. *Chemphyschemistry* **5**, 566–571.
14. Ravindra, R., Royer, C., and Winter, R. (2004) Pressure perturbation calorimetic studies of the solvation properties and the thermal unfolding of staphylococcal nuclease. *Phys. Chem. Chem. Phys.* **6**, 1952–1961.
15. Heimburg, T., Grabitz, P., and Ivanova, V. P. (2001) The coupling of relaxation times to the heat capacity of lipid vesicles studied by pressure jump calorimetry. *Biophys. J.* **80**, 2148.

16. Nagle, J. F. and Tristram-Nagle, S. (2000) Structure of lipid bilayers. *Biochim. Biophys. Acta* **1469,** 159–195.
17. Leharne, S. A. and Chowdhry, B. Z. (1998) Thermodynamic background to differential scanning calorimetry, in *Biocalorimetry*, (Chowdhry, J. L. A. B. Z., ed.), J. Wiley and Sons, Chichester, pp. 157–182.
18. Chalikian, T. V. (2003) Volumetric properties of proteins *Ann. Rev. Biophys. Biomol. Struct.* **32,** 207–235.

14

Fourier Transform Infrared Spectroscopy in the Study of Lipid Phase Transitions in Model and Biological Membranes

Practical Considerations

Ruthven N. A. H. Lewis and Ronald N. McElhaney

Summary

Fourier transform infrared (FTIR) spectroscopy is a powerful, nonperturbing technique that has been used to good effect for the detection and characterization of lipid phase transitions in model and natural membranes. The technique is also quite versatile, covering a wide range of sophisticated applications, from which fairly detailed information about the structure and organization of membranes and other lipid assemblies can be obtained. In this chapter, an introduction to this particular application of FTIR spectroscopy is presented. Special emphasis is put on how the technique can be used to study lipid phase transitions under biologically relevant conditions. The chapter is intended to give an overview of the capabilities of FTIR spectroscopy in the field of lipid and biomembrane research, and provide the reader with some practical guidelines for the design and execution of simple FTIR spectroscopic experiments suitable for the detection and characterization of lipid phase transitions in hydrated lipid bilayers.

Key Words: Biological membranes; Fourier-transform infrared spectroscopy (FTIR); lipid bilayers; lipid model membranes; lipid phase transitions; lipid thermotropic phase behavior; transition temperature.

1. Introduction

Fourier transform infrared (FTIR) spectroscopy has now become a powerful yet relatively rapid, inexpensive, and nonperturbing technique for the detection and characterization of phase transitions in both model and natural membranes (for reviews, *see* **refs.** *1–4*), mainly because of the fact that accurate and reproducible high-resolution, low-noise spectra can be obtained even with nonideal samples, such as aqueous lipid dispersions. Also, because modern computer technology is now an integral part of the instrumentation, most aspects of the data-acquisition process can be automated, thus freeing the worker from many of the tedious and repetitive tasks involved. Furthermore, the data obtained are automatically stored in digitally encoded formats, thereby facilitating spectral analysis with the aid of postacquisition data manipulation algorithms. Thus, even small changes in the contours of weak absorption bands in the IR spectra of molecules of interest can be accurately and reproducibly measured, which provides incentive for the user to extend the applications of FTIR spectroscopy to increasingly more technically demanding situations (for discussions of such applications, *see* **refs.** *5–9*). However, from the perspective of the detection and characterization of lipid phase transitions, which is the focus here, many of these applications are rather specialized and will not be examined in this chapter. Instead, the primary focus will be on the practical aspects of using FTIR spectroscopy to study the lipid phase behavior, with special emphasis on transmission FTIR spectroscopy of hydrated lipids. However, the reader will be referred to other FTIR applications that may be relevant to this task, especially where the application of such techniques may enable the user to improve or otherwise extend the information content of the data obtained.

2. General Considerations

The simplest and most widely used approach is transmittance FTIR spectroscopy, in which the sample is probed by an unpolarized beam of IR radiation directed perpendicular to the windows of the sample cell. However, when considering hydrated lipid assemblies, the presence of strong absorption bands from the dispersal buffer poses special challenges in both the preparation of suitable samples and the analysis and interpretation of the data obtained (*see* **Subheadings 4–7** below). In principle, spectral subtraction methods can be used to address this problem. However, such procedures fail when the absorption bands are very strong, because little or no IR radiation reaches the detector in spectral regions encompassing these bands. Thus, to minimize the problem of strong solvent absorption, the use of short pathlength samples is essentially mandatory. However, this requires the use of fairly concentrated lipid samples, which brings a number of disadvantages (*see* below). Thus, for example, it may be impractical to record the IR spectra of some micelle-forming lipids at concentrations below their critical micellar concentrations. Moreover, significantly increasing sample concentration in order to accurately observe the more weakly absorbing bands in the IR spectrum of the sample may not always be feasible because absorption from other more strongly absorbing sample bands may be too high. However, such experimental circumstances are the exception rather than the rule, and for the most part, high sample concentrations *per se* pose no major problems in routine transmission IR spectroscopic studies of hydrated lipid samples.

FTIR spectroscopic studies of hydrated lipid assemblies have also been approached through the use of reflectance techniques, of which attenuated total reflectance (ATR) applications are the most widespread and potentially useful. In such studies, the sample is placed on the surface of an inorganic crystal, and probed by IR radiation through the edge of the crystal at an incident angle higher than the critical angle of total internal reflection. Under such conditions, the sample in contact with the surface of the crystal is probed by an evanescent wave through which information about the absorption characteristics of the sample are eventually carried to the detector (for details, *see* **refs. *10,11***). This application is particularly useful for hydrated lipids because the effective sample path length is considerably smaller than that normally used in transmittance experiments, thereby minimizing potential problems caused by the presence of highly absorptive solvent bands. Nevertheless, because this also requires sample concentrations in contact with the surface of the ATR crystal to be appropriately high, the procedure incurs in penalties similar to those of transmittance procedures (*see* above paragraph).

However, a potentially serious problem with all ATR techniques is that many of the materials suitable for use as ATR crystals are highly polar, and it is therefore possible that the behavior of lipids in contact with the surfaces of such materials may be perturbed. Because in all ATR applications the evanescent wave only probes subpopulations of the sample in contact with the crystal, the observed IR spectrum may reflect only those lipids whose behavior may be affected by interactions with the polarized surface of the crystal. This problem may not be serious for most zwitterionic and nonionic lipids because the contact of such lipids with a polarized surface does not normally have significant effects on their phase behavior or organization. However, this limitation in the observed IR spectrum may be significant with charged lipids, whose behavior is known to be significantly affected by ionic strength *(12)*. However, although the materials that are used as windows for transmittance experiments may well be the same as those used for ATR crystals, similar problems are rarely encountered in transmittance applications, as under most circumstances only a small fraction of the sample will be in contact with the polarized surface of the window.

Reflectance IR techniques can also be used to study lipid monolayers at the air–water interface (*see* **refs. 6,8,13**, and references therein). In this procedure, monomolecular lipid films layered on the surface of an aqueous substrate are probed by an IR beam at an angle of incidence small enough for the beam to be reflected off the surface of the medium. Similar procedures (i.e., grazing- or glancing-angle reflection techniques) have also been used in IR spectroscopic studies of amphiphiles adsorbed onto the surfaces of solid supports *(8,14)*. The application of such techniques to the study of lipid monomolecular films at the air–water interface offers a number of advantages. First, because the procedure offers the possibility of simultaneously making IR spectroscopic and traditional force/area measurements on the same sample, direct correlations between IR spectroscopic observations and phase transitions and/or molecular area estimates can be made. The usefulness and effectiveness of such procedures have been demonstrated in studies of phospholipid monolayers at the air–water interface *(13,15–18)*. Second, lipid monolayers are macroscopically oriented at the air–water interface at all biologically relevant lateral pressures. Consequently, polarized-IR techniques can also be used to study the orientation of the transition moments of the IR-active groups of lipid molecules relative to their axis of reorientation. Moreover, because lipid monolayers should be macroscopically very well aligned, the accuracy of any orientational information that is obtained should be very high. Also, the same sample can be probed at different angles of incidence, and thus, good estimates of the orientational order parameters of various band vectors can be obtained *(15–18)*. However, these fairly specialized applications will not be examined here.

3. Lipid Phase Transitions

Hydrated lipids can exist in one or more of numerous polymorphic forms, and depending on the conditions, interconversions between these forms (i.e., phase transitions) readily occur. The detection, assignment, and characterization of lipid phase transitions is one of the simplest and more widespread applications of FTIR spectroscopy, an application for which the technique has proven to be very effective and extremely useful. The process involves an examination of the IR spectroscopic changes that accompany lipid phase transitions (however induced) and a correlation of observed spectroscopic changes with the lipid conformational and organizational changes known to occur during those processes. The details of the spectroscopic changes occurring at these phase transitions will vary with the structure of lipid being studied, as well as with the nature of the phase transition under observation. The IR bands commonly used to observe and characterize lipid phase transitions are summarized in **Table 1** (for a more detailed listing of the IR absorption bands of lipids, *see* **ref. 2** and references therein). As illustrated in **Table 1** of chapter 12 in this volume, the types of phase transitions known to occur in hydrated lipid assemblies varies considerably, often encompassing variable combinations of related and/or unrelated conformational and organizational events. Thus, although a single IR spectroscopic marker may detect the occurrence of a lipid phase transition, no single IR spectroscopic marker can uniquely identify the type of phase transition it detects. Consequently, the use of FTIR spectroscopy for the assignment of specific types of lipid phase transitions often requires the collation of information obtained from several IR spectroscopic markers. Furthermore, it may even require in some instances the application of techniques other than

Table 1
IR-Active Marker Groups for Studying Lipid Phase Transitions

Functional group	Vibrational mode	Frequency range (cm^{-1})	Comments
C–CH$_3$	Methyl asymmetric stretch ($_{as}\nu_{CH_3}$)	2962	–
C–CH$_3$	Methyl symmetric stretch ($_s\nu_{CH_3}$)	2872	–
C–CH$_3$	Methyl asymmetric bend ($_{as}\gamma_{CH_3}$)	1450	–
C–CH$_3$	Methyl symmetric bend ($_s\gamma_{CH_3}$)	1370–1380	Methyl umbrella band. Sensitive to hydrocarbon chain lateral packing.
C–CD$_3$	CD$_3$ asymmetric stretch ($_{as}\nu_{CD_3}$)	2212	–
C–CD$_3$	CD$_3$ symmetric stretch ($_{as}\nu_{CD_3}$)	2169	–
–(CH$_2$)$_n$–	Methylene asymmetric stretch ($_{as}\nu_{CH_2}$)	2916–2936	Sensitive to hydrocarbon chain conformation
–(CH$_2$)$_n$–	$_s\nu_{CH_2}$	2843–2863	Sensitive to hydrocarbon chain conformation
–(CD$_2$)$_n$–	$_{as}\nu_{CD_2}$	2190–2200	Sensitive to hydrocarbon chain conformation
–(CD$_2$)$_n$–	$_s\nu_{CD_2}$	2085–2095	Sensitive to hydrocarbon chain conformation
–(CH$_2$)$_n$–	Methylene bend (γ_{CH_2})	1466–1473	Sensitive to hydrocarbon chain lateral packing; CH$_2$ scissoring, CH$_2$ deformation vibrations
–(CH$_2$)$_n$–	γ_{CH_2}	1468	Rotationally disordered, hexagonally packed, hydrocarbon chains
–(CH$_2$)$_n$–	γ_{CH_2}	1471–1473	Hydrocarbon chains in triclinic parallel (T$_{//}$) subcell
–(CH$_2$)$_n$–	γ_{CH_2}	1476, 1473	Correlation field splitting; hydrocarbon chains in orthorhombic perpendicular (O$_\perp$) subcell
–(CD$_2$)$_n$–	γ_{CD_2}	1088	Rotationally disordered, hexagonally packed, perdeuterated hydrocarbon chains
–(CD$_2$)$_n$–	γ_{CD_2}	1093	Perdeuterated hydrocarbon chains in triclinic parallel (T$_{//}$) subcell
–(CD$_2$)$_n$–	γ_{CD_2}	1088, 1093	Correlation field splitting; perdeuterated hydrocarbon chains in orthorhombic perpendicular (O$_\perp$) subcell

(Continued)

Table 1 (Continued)

Functional group	Vibrational mode	Frequency range (cm^{-1})	Comments
$-CH_2-COOR$	α-Methylene bend (γ_{CH_2})	1414–1422	Sensitive to the conformation of the linkage to the glycerol backbone
$-(CH_2)_n-$	CH_2 wagging	1300–1400	Sensitive to hydrocarbon chain conformation
$-(CH_2)_n-$	CH_2 wagging	1180–1330	All-*trans* polymethylene chains exhibit band progressions in this region
$-(CH_2)_n-$	Methylene rocking (δ_{CH_2})	718–731	Sensitive to hydrocarbon chain lateral packing
$-(CH_2)_n-$	δ_{CH_2}	720	Rotationally disordered, hexagonally packed, hydrocarbon chains
$-(CH_2)_n-$	δ_{CH_2}	718	Hydrocarbon chains in triclinic parallel ($T_{//}$) subcell
$-(CH_2)_n-$	δ_{CH_2}	718, 731	Correlation field splitting; hydrocarbon chains in orthorhombic perpendicular ($O_\perp$) subcell
$-(CD_2)_n-$	δ_{CD_2}	517	Rotationally disordered, hexagonally packed, perdeuterated hydrocarbon chains
$-(CD_2)_n-$	δ_{CD_2}	515	Perdeuterated hydrocarbon chains in triclinic parallel ($T_{//}$) subcell
$-(CD_2)_n-$	δ_{CD_2}	515, 521	Correlation field splitting; perdeuterated hydrocarbon chains in orthorhombic perpendicular ($O_\perp$) subcell
$-CH_2-COOR$	C=O stretch ($\nu_{C=O}$)	1720–1750	Sensitive to hydration and hydrogen-bonding; frequencies of hydrated or H-bonded groups occur toward the lower end of the frequency range
$-CH_2-{}^{13}COOR$	^{13}C=O stretch ($\nu_{C=O}$)	1675–1705	Sensitive to hydration and hydrogen-bonding; frequencies of hydrated or H-bonded groups occur toward the lower end of the frequency range

(Continued)

Table 1 *(Continued)*

Functional group	Vibrational mode	Frequency range (cm^{-1})	Comments
R–CONHR	Amide I predominantly ($\nu_{C=O}$)	1610–1680	Sensitive to hydration and hydrogen-bonding; frequencies of hydrated or H-bonded groups occur toward the lower end of the frequency range
R–CONHR	Amide II predominantly N–H bend	~1550	Sensitive to hydration and hydrogen-bonding; H–D exchange causes a shift to ~1465/cm^{-1}
RO–P(OR′)OO–	PO$_2$ asymmetric stretch ($_{as}\nu_{PO_2}$)	1200–1260	Sensitive to hydration and hydrogen-bonding; frequencies of hydrated or H-bonded groups occur toward the lower end of the frequency range

IR spectroscopy. For simplicity, the application of FTIR spectroscopy to the detection and assignment of lipid phase transitions will be examined from the perspectives of three general (although not mutually exclusive) groups of lipid phase transitions outlined immediately below.

3.1. Hydrocarbon Chain-Melting Phase Transitions

In principle, the IR absorption bands of virtually all fundamental vibrations of the CH$_2$ group can be used to detect lipid hydrocarbon chain-melting phase transitions. However, the CH$_2$-asymmetric and -symmetric stretching bands are the most widely used; these bands occur near 2920 and 2850cm^{-1}, respectively. Regardless of the structural details of the process (or how it is induced), all hydrocarbon chain-melting phase transitions are accompanied by discontinuous increases in both the frequency of the absorption maxima and the bandwidth of these two vibrational modes. These changes reflect the increase in hydrocarbon chain conformational disorder and mobility that occur during hydrocarbon chain-melting phase transitions, and are diagnostic of the onset of *gauche* rotamer formation in all-*trans* polymethylene chains *(19)*.

Although both 2920 and 2850 cm^{-1} absorption bands can be used to monitor hydrocarbon chain-melting phase transitions, the 2920cm^{-1} band is rarely used in practice because it is not a pure CH$_2$ stretching band. The CH$_2$-asymmetric stretching vibrational mode is a major constituent of IR absorption near 2920cm^{-1}; however, the 2920cm^{-1} absorption band contains significant overlapping contributions from methyl groups, and depending on the nature of the lipid gel phase, the CH$_2$ asymmetric stretching vibrational mode can itself be perturbed by Fermi resonance interactions with the first overtones of the methylene scissoring vibration. In contrast, the absorption band near 2850cm^{-1} is considerably freer of such effects, and under most circumstances, arises predominantly from the symmetric stretching vibrations of methylene groups. Thus, as long as lipid hydrocarbon chains are the predominant source of methylene groups in the sample under observation, the absorption band near 2850cm^{-1} can be reliably used to detect lipid hydrocarbon chain-melting phase transitions, and to some extent, semiquantitatively characterize concomitant changes in hydrocarbon chain conformational disorder. For these reasons, the CH_2 symmetric stretching band is widely used as the main marker band for the detection of lipid hydrocarbon chain-melting phase transitions in virtually every model and natural membrane system (for reviews, see **refs. 1–3,20–22**).

The melting of hydrocarbon chains is usually accompanied by a 1.5–3cm^{-1} increase in the frequency of CH_2 symmetric stretching band, regardless of the type of hydrocarbon chain-melting transition under study. However, the magnitude of this change can vary with the length and chemical structure of the lipid hydrocarbon chains, the structure of the lipid polar head group, the nature of any additives that might be present, as well as the nature of the chain-melting phase transition itself (e.g., see **refs. 23–25**). Consequently, the aforementioned changes in the frequency and bandwidth of the CH_2 symmetric stretching band are only useful for the detection of lipid hydrocarbon chain-melting events, and cannot be used to assign the type of chain-melting phase transition under observation.

Lipid hydrocarbon chain-melting phase transitions can also be detected by examining the CH_2 scissoring and rocking bands (~1470cm^{-1} and ~720cm^{-1}, respectively). With lipid polymethylene chains, these vibrational modes each give rise to sharp absorption bands of moderate intensity when the chains are in an all-*trans* conformation and the widths of these bands increase significantly and their integrated intensities decrease substantially on melting. These changes are also diagnostic of hydrocarbon chain-melting phase transitions, as they reflect concomitant increases in hydrocarbon chain mobility and *gauche* rotamer content. However, CH_2 scissoring and rocking bands are particularly useful because their contours in the gel state are also very sensitive to lateral packing interactions between the all-*trans* polymethylene chains *(26–27)*. Consequently, an examination of the contours of these bands at the onset of the hydrocarbon chain-melting process can provide useful information about the mode of hydrocarbon chain packing before chain melting; this information can then be used in the assignment of the type of chain-melting phase transition under observation (*see* **Subheadings 3.2. and 3.3.** below).

The wagging vibrations of the hydrocarbon chain methylene groups can also be used to detect lipid hydrocarbon chain-melting phase transitions (e.g., see **refs. 28–31**). This approach is based on the fact that the CH_2 wagging vibrations of all-*trans* polymethylene chains couple to produce band progressions, the intensities of which depend on the length of the all-*trans* polymethylene segments involved. Because the introduction of *gauche* conformers into all-*trans* polymethylene chains results in a reduction of the mean lengths of the residual all-*trans* polymethylene segments, and significantly weakens the coupling of their CH_2 wagging vibrations, the band progressions usually disappear when lipid hydrocarbon chains melt. However, although the disappearance (or marked diminution in the intensities) of CH_2 wagging band progressions can be diagnostic of hydrocarbon chain-melting phase transitions, it should be noted that such changes are not always indicative of a chain-melting phase transition. This is because CH_2 wagging vibrations may also couple to one of the fundamental vibrations of an attached polar moiety, a fact that makes the intensities of these progression bands very dependent on the structure and conformation of the polar group to which the all-*trans* segment is attached. The sensitivity of the CH_2 wagging band progressions to these parameters is amply illustrated by the observation that CH_2 wagging band progressions of dipalmitoylphosphatidylcholine (DPPC) are considerably more intense than those of 1,2-dihexadecyl-*sn*-glycero-3-phosphorylcholine, and the observation is that CH_2 wagging band progressions arising from the *sn*-1 fatty acyl chain of DPPC are stronger than those arising from the *sn*-2 fatty acyl chain *(32)*.

Clearly, the aforementioned considerations place limits on the overall usefulness of this particular method for detecting lipid hydrocarbon chain-melting phase transitions. Moreover, the implementation of this procedure is not as straightforward as those previously described (*see* immediately above). Unlike the methylene stretching, scissoring, and rocking vibrations bands described earlier, CH_2 wagging band progressions are intrinsically weak; they are usually located in the midst of considerably stronger absorptions arising from the solvent and/or other IR-active groups located on the lipid polar-head group. Consequently, the use of CH_2 wagging band progressions to detect lipid hydrocarbon phase transitions usually requires the subtraction of appropriately weighted solvent and other reference spectra. It also requires further postacquisition data-processing procedures, which, although facilitated by modern computer-assisted procedures, may well make the process too cumbersome for routine applications. Nevertheless, there are some instances where the special properties of the CH_2 wagging band progressions offers some advantages over other IR marker bands that are typically used to detect and characterize lipid hydrocarbon chain-melting phase transitions. For example, the disappearance of the CH_2 wagging band progressions was used to accurately define the upper boundaries of the hydrocarbon chain-melting phase transitions of *Acholeplasma laidlawii* B membranes *(30)* and sterol-containing linear saturated phosphatidylcholine (PC) bilayers *(33,34)*. In these particular cases, the hydrocarbon chain-melting phase transitions were very broad and their upper boundaries were too diffused and ill-defined to be accurately mapped using other IR spectroscopic markers.

3.2. Phase Transitions Involving Crystalline and/or Quasi-Crystalline Lipid Structures

Transitions involving crystalline or *quasi*-crystalline (L_c) phases of lipids can be detected by an examination of spectroscopic markers that are sensitive to intermolecular close-contact interactions between lipid molecules. In crystalline and *quasi*-crystalline lipid phases, the rates and amplitudes of molecular motions are generally low, resulting in sharp bands that reflect short-range interactions between adjacent lipid molecules. Such interactions include specific modes of hydrocarbon chain packing, specific patterns of interfacial hydration, as well as distinctive types of headgroup–headgroup and headgroup–solvent interactions. Because of this, crystalline and *quasi*-crystalline phases tend to exhibit distinctive patterns of IR absorption that are generally specific to the structure of the crystalline or *quasi*-crystalline lipid phase concerned. Because the distinctive features of the IR spectra of these various lipid phases are also very sensitive to conformational changes, the detection and assignment of all phase transitions involving such lipid phases are usually fairly facile. For example, with short-chain homologs of the *n*-saturated diacyl PCs and *n*-saturated β-D-glucosyl diacylglycerols, the conversion of metastable *quasi*-crystalline phase to the stable forms are accompanied by a collapse of the correlation field splitting of the CH_2 scissoring bands of these lipids and fairly drastic alterations of the fine structure of their ester carbonyl stretching bands *(35–37)*.

However, throughout these transformations the general spectroscopic characteristics of crystalline lipids are maintained, which is consistent with the assignment of the process as the conversion of one crystalline form to another. Although the interconversions between different crystalline or *quasi*-crystalline polymorphs of most lipids may not always involve such drastic spectroscopic changes, the spectroscopic manifestations of such transformations are usually sufficiently distinct to enable detection and correct assignment. Similarly, IR spectroscopic detection and assignment of phase transitions between the *quasi*-crystalline and L_β-type lipid gel phases, is usually a straightforward process characterized by the appearance/disappearance of the spectroscopic features characteristic of crystalline lipid phases, while retaining the spectroscopic features characteristic of all-*trans* hydrocarbon chains. For example, at the L_c/L_β phase transitions of longer chain iso-branched diacyl PCs, a 4–5 cm^{-1} decrease in the frequency of the CH$_2$ scissoring band (from 1473 to 1468cm^{-1}) and the loss of the fine structure of the ester carbonyl stretching bands are generally observed *(38)*. However, the frequencies of the symmetric and asymmetric CH$_2$ stretching bands remain unchanged during that process, and all IR-absorption bands associated with the hydrocarbon chains remain very sharp, indicating that the hydrocarbon chains are not melted. Because these spectroscopic features indicate that the *quasi*-crystalline structure of the L_c phase breaks down before the melting of the lipid hydrocarbon chains, unambiguous detection and assignment of those L_c/L_β lipid phase transitions is possible. However, there are circumstances where some lipids may exhibit a comparable array of spectroscopic change in the absence of any discernible phase transition. For example, when the L_β phases of long-chain *n*-saturated diacyl PCs are cooled to temperatures well below the gel/liquid-crystalline (L_β/L_α) phase-transition temperature, a pronounced correlation field splitting of the CH$_2$ scissoring band (a feature usually associated with crystalline or *quasi*-crystalline lipid phases) does occur in the absence of any discernible lipid phase transition. Although in this particular case the absence of significant changes in the carbonyl stretching bands provides a clear indication that a crystal-like lipid phase is not formed, the situation may not always be so clear. Consequently, some caution should always be exercised when IR spectroscopy is the sole means available to detect and assign these types of lipid phase transitions.

Finally, the generally distinctive features of crystalline lipid phases also enable facile detection of phase transitions between crystalline and melted lipid phases. In such cases, the disappearance of the spectroscopic features of the crystalline phase as described previously is also accompanied by spectroscopic features characteristic of the hydrocarbon chain-melting process (*see* **3.1.** above). However, as stated previously, the IR spectroscopic signatures of the various melted forms of lipids are not sufficiently distinct to enable reliable differentiation of these phases. Consequently, IR spectroscopy will enable effective detection of crystalline to liquid-crystalline phase transitions, but usually cannot single out the type of liquid-crystalline phase that is being formed.

3.3. Phase Transitions Involving Nonlamellar Lipid Phases

For our purposes, transitions between lamellar and nonlamellar lipid phases will be broadly grouped into those involving interconversions between lipid lamellae and normal (type I) micellar structures, and those involving interconversions between lipid lamellae and inverted (type II) micellar, cubic, or hexagonal phases. At present, the use of IR spectroscopy to detect and characterize both types of lipid phase transitions is probably the least developed of the applications discussed here. In part, this can be attributed to the fact that lipid hydrocarbon chain-melting phase transitions often occur simultaneously with lamellar/nonlamellar-phase transitions. In addition, spectroscopic changes specific to the conversion from a lamellar to a nonlamellar phase (and vice versa), are often obscured by the larger and more distinctive spectroscopic features of the hydrocarbon chain-melting phase transition. Under these circumstances, FTIR spectroscopy would detect the melting of the hydrocarbon chains yet would give little or no indication of whether the melted form of the lipid is lamellar or not. For example, when aqueous dispersions of 1-octadecyl 2-acetyl PC (platelet-activating factor) are heated to temperatures near 22°C, a transition from a lamellar phase with fully extended hydrocarbon chains to a type I micellar phase with conformationally disordered (melted) hydrocarbon chains is observed *(39)*. Among the FTIR spectroscopic changes observed at this phase transition are discontinuous increases in the frequency and bandwidth of the CH_2 symmetric stretching band, and a discontinuous decrease in the frequency of the ester carbonyl stretching band *(40)*. The changes in the methylene symmetric stretching band are diagnostic of an increase in hydrocarbon chain conformational disorder, whereas the change in the ester carbonyl stretching band is consistent with an increase in the polarity and/or hydration of the polar–apolar interfacial region of the lipid assembly (*see* above). However, such structural changes are also typical of the L_β/L_α phase transition that is commonly observed in most fully hydrated-lipid bilayers, and the magnitudes of the spectroscopic changes observed at L_β/L_α phase transitions are usually only slightly smaller than those observed at the lamellar–micellar-phase transition of platelet-activating factor. Consequently, it is difficult to identify a set of spectroscopic changes that are uniquely characteristic of the lamellar–micellar-phase transition of this material. The aforementioned observations underscore the general problem of using IR spectroscopy to detect different types of lamellar/nonlamellar phase transitions that have been examined; they are probably the critical reason for the slow pace at which FTIR spectroscopy is being applied to this particular aspect of lipid and biomembrane research.

Despite the difficulties exemplified previously, FTIR spectroscopy can be used to detect and characterize interconversions between the L_α and H_{II} phases of hydrated nonlamellar phase-forming lipids *(41–43)*. It has been demonstrated that the L_α/H_{II} phase transition of hydrated phosphatidylethanolamines is accompanied by a small discontinuous increase (~1 cm^{-1}) in the frequency of the CH_2 symmetric stretching band, a small increase (~4 cm^{-1}) in the frequency of the ester carbonyl stretching band, and a small increase (~3 cm^{-1}) in the frequency of the phosphate O–P–O asymmetric stretching band *(41)*. These observations suggest that L_α/H_{II} phase transition of the lipid is accompanied by an increase in hydrocarbon chain conformational disorder as well as a small decrease in the hydration of the polar–apolar interfacial and polar head group regions of the lipid assembly. These conclusions are consistent with current ideas about the structural and mechanistic requirements for the conversion of a lamellar phase to an inverted nonlamellar structure (for a recent review, *see* **ref. 44**).

However, the L_α/H_{II} phase transitions observed in the articles cited earlier are fairly cooperative processes (i.e., they occur over a relatively narrow temperature range), and are uncoupled from the melting of the lipid hydrocarbon chains and/or the reorganization of polymorphic crystalline forms of the lipid. Thus, although the observed spectroscopic changes are fairly small and not necessarily unique to the lamellar/nonlamellar phase transitions, a straightforward correlation between the observed spectroscopic changes and structural events known or presumed to occur at L_α/H_{II} phase transitions is still feasible. However, it is unlikely that comparable correlations can be confidently established under conditions wherein the lamellar/nonlamellar phase transition is very broad, or wherein coupled to a process involving the melting of the lipid hydrocarbon chains (e.g., L_β/H_{II} phase transitions) or the reorganization of crystalline polymorphic forms of the lipid (e.g., L_c/H_{II} phase transitions). In principle, it should be feasible to use similar FTIR spectroscopic approaches to detect and characterize phase transitions involving other inverted nonlamellar structures (e.g., inverted micellar and inverted cubic phases), provided that the phase transitions are also fairly cooperative and uncoupled from major structural changes.

4. Hardware Considerations

The basic hardware required for using transmission FTIR spectroscopy to study phase transitions in lipids and lipid membranes is a midrange (~400–4000cm^{-1}) interferometer with an optical resolution of 2cm^{-1} or better. With this basic hardware, interferograms recorded with two levels of zero-filling will enable the acquisition of FTIR spectra with a digital encoding of approx 0.5cm^{-1}, which is more than adequate for the majority of the most demanding applications in this particular field. The instrument should also be equipped with the accessories needed for inducing phase transitions in lipids and lipid membranes. In most applications, FTIR spectroscopy is used to examine thermotropically induced lipid phase transitions, which require the capacity for controlled manipulation of sample temperature. However, FTIR spectroscopy can also be applied to the study of barotropically and lyotropically induced lipid phase transitions, for which one would require hardware accessories for controlled manipulation of sample pressure and water content, respectively (for examples of such applications and the hardware, *see* **refs. *45–50***). Ideally, these accessories would be under computer control, so that the user would have the flexibility to automate the process of data acquisition.

Finally, given the nature of the samples being examined (concentrated aqueous lipid dispersions/pastes), it is prudent that the system be equipped with demountable sample cells to facilitate proper cleaning. These cells should be equipped with appropriate spacers (thickness ≤25 µm), and windows made up of water-insoluble IR-transparent materials with an effective transmission range between 400 and 4000cm^{-1}. Although windows made up of materials such as zinc selenide, silver chloride, and silver bromide easily meet these criteria, their routine use in transmission IR applications is impractical, mainly because they are extremely expensive and quite prone to deformation and breakage through thermal and mechanical shock. On the other hand, windows made up of calcium or barium fluoride are less costly and considerably more resistant to thermal and mechanical shock, and thus better suited for routine use, despite their having a narrower effective transmission range (frequency cutoff: ~1000–1100cm^{-1}). However, this range limitation is not particularly significant because most of the IR absorption bands routinely used to monitor lipid phase transitions occur at frequencies between 1000 and 3000cm^{-1} (*see* **Table 1**). Thus, windows made up of inexpensive materials such as BaF$_2$ and CaF$_2$ can be used for routine work, whereas those made up of the more expensive and fragile materials can be reserved for those specialized applications requiring the extended transmission ranges of such materials (e.g., *see* **refs. *31,33,34***).

5. Sample Preparation

Typically, hydrated lipid samples for FTIR spectroscopy are prepared by forming a homogenous dispersion of the lipid in an appropriate aqueous buffer, usually by vigorous mixing at temperatures well above the gel/liquid-crystalline phase transition of the lipid. Here, the dominant issues involved in preparing such samples all stem from the requirement that the sample be fully hydrated throughout the experiment, the result of which is that the IR spectra obtained are usually dominated by strong absorption bands from the dispersal buffer. Although the intrinsic molar extinction coefficients of the major absorption bands of water are not very large (~10^3–10^4) (see **ref. 51**), the sheer quantity of water required to ensure full hydration is such that the IR spectra of a fully hydrated lipid sample will be dominated by very strong solvent-derived absorption bands, which make the resolution of the weaker bands emanating from the sample fairly difficult. Moreover, the latter may be further complicated because some of the strong solvent-derived bands actually overlap many structurally significant sample absorptions. These issues are usually addressed by preparing samples with very short transmission path lengths (≤25 μm) and by the use of D_2O-based media as dispersal buffers (see two paragraph below). However, because of the short transmission path lengths required, it is necessary that sample concentrations be quite high (~100 mM) to ensure that the intensities of sample-derived absorption bands are high enough to be usable. Typically, for a 25-μm path length sample, this is achieved by dispersing 2–3 mg of lipid in 50 μL of buffer; the resulting lipid paste is used to form a thin homogenous film squeezed between the windows of a sample cell equipped with a spacer of appropriate thickness. With zwitterionic and nonionic lipids, such high concentrations usually do not cause any problems, mainly because the buffering capacity of the dispersal medium is not normally exceeded by these lipids. However, this problem should be considered when preparing samples of anionic or cationic lipids, and it may be necessary to increase the buffering capacity of the dispersal medium. This usually requires a *de facto* increase in the ionic strength of the dispersal medium and, as the thermotropic phase behavior of both cationic and anionic lipids are strongly influenced by ionic strength *(12)*, it is often the case with charged lipid membranes that phase behavior (in general) and the phase-transition temperature (in particular), as resolved by FTIR spectroscopy, may differ somewhat from those resolved by techniques requiring more dilute sample concentrations. There being no practical way of avoiding this particular problem, the latter possibility should thus be factored into the interpretation of the data obtained.

Notwithstanding the necessity of preparing relatively high sample concentrations, it is also important that sample concentrations remain below the threshold at which significant distortion of important IR marker bands will occur. Preserving the fidelity of the fine structure of IR marker bands is especially important in cases wherein the intent is the extraction of structural information from the IR spectra. Given the considerable variations in the extinction coefficients of the IR marker bands commonly examined, achieving the proper balance between signal intensity and signal fidelity is sometimes quite difficult, and in the extreme cases this can only be done by trial and error. However, with most D_2O-based dispersions of the common diacylglycerolipids, it is generally found that this can be achieved if sample concentrations are such that the total sample absorbance (including background solvent contributions) in the C=O stretching region between 1500 and 1800cm^{-1} is between 0.5 and 0.6 absorbance units. Typically, this is achieved with lipid concentrations of 80–100 mM lipid. However, lipid samples of such high concentrations are often fairly surface active, and one must be quite careful to avoid the incorporation of air bubbles in the sample film and the problems that may occur because of this (see **Subheading 6.**).

As noted earlier, most of the lipid and membrane samples prepared for FTIR spectroscopy are dispersed in D_2O-based buffers because the broad O–H stretching and O–H bending bands of water occur in the same regions of the IR spectrum as the absorption bands from the C–H stretching vibrations of lipid hydrocarbon chains (2800–3000cm^{-1}) and the C=O stretching vibrations of ester carbonyl groups (1500–1800cm^{-1}), respectively, which are found in the lipid polar/apolar interface. The substitution of D_2O for H_2O circumvents this particular problem by shifting the solvent absorptions to lower frequencies, thereby enabling observation of the underlying sample bands. However, this approach is not entirely free of potential problems because substituting D_2O for H_2O shifts the solvent absorption bands to other regions of the IR spectrum, wherein they may compromise the observation of other absorption bands of interest. For example, the dispersal of diacyl phospholipids in D_2O (instead of H_2O) results in the replacement of the strongly absorbing H–O–H bending band near 1645cm^{-1} by the comparably absorptive D–O–D bending band near 1215cm^{-1}. Thus, whereas substituting D_2O for H_2O facilitates the observation of lipid ester carbonyl stretching bands near 1735cm^{-1}, this is achieved at the expense of decreasing the capacity to observe the head group O–P–O$^-$ asymmetric stretching band near 1230cm^{-1}.

Given the problems mentioned in the previous paragraph, observation of all bands in the IR spectrum of a hydrated lipid sample can only be achieved by an examination and comparison of spectra acquired from samples dispersed in both H_2O and D_2O. However, substituting H_2O for D_2O can pose an even potentially more serious problem, because H/D exchange of exchangeable sample protons will almost certainly occur under such circumstances. Inevitably, H/D exchange will result in loss of the absorption band of the protonated species and its replacement with the absorption band of the deuterated species. The latter is observed at a lower frequency than the former and, depending on the band concerned, spectroscopic observation may not be feasible or convenient. In sphingomyelin bilayers, for example, H/D exchange results in the loss of the amide II band near 1550cm^{-1} and its replacement with the amide II band near 1465cm^{-1}. However, it may well be inconvenient to study the latter, because it will occur in the midst of the CH_2 scissoring band and other headgroup absorption bands. Furthermore, if the rates of H/D exchange are fairly slow, the disappearance of the amide I band can itself be exploited to monitor the rates and extent of H/D exchange. From such a study, information on the exposure of the exchangeable proton site to the bulk solvent phase can be obtained. It should also be noted that H/D exchange might also alter the IR absorption characteristics of potential hydrogen-bonding acceptor groups. Thus, H/D exchange may shift the frequencies of, for example, hydrogen-bonded ester carbonyl and amide I absorption bands of phospho- and sphingolipids, and depending on the extent of H/D exchange, fairly complex spectra can be obtained. However, such occurrences may not always pose major problems, as long as they are appropriately considered when the data are interpreted.

6. Data Acquisition

The practical issues of acquiring transmission FTIR spectra suitable for the detection and characterization of lipid phase transitions are fairly simple. Essentially, the process involves the equilibration of the sample under a defined set of experiment conditions (i.e., temperature, pressure, relative humidity, and so on), followed by the acquisition and coaddition of a number of interferograms suitable for the generation of spectra of acceptable signal-to-noise quality. This process is repeated while experimental conditions are systematically varied over a range that is calculated to induce a phase transition in the lipid under study. In most modern laboratories, this entire process is under computer control and thus can be easily automated, with the user having the flexibility of designing quite intricate protocols for manipulating sample conditions before and during the process of recording spectra.

However, in designing such protocols, the user should be aware of the following potential pitfalls, which might degrade the usefulness to use the data acquired.

1. Many of the phase transitions that occur in hydrated lipid assemblies are inherently slow processes. Consequently, data acquisition protocols should allow an appropriate period of time for sample equilibration.
2. Second, with most transmission IR experiments, the instrument sample compartment is continually purged with dry gas; hydrated lipid samples will thus be prone to slow moisture loss and possible desiccation over the time course of an experiment. This is a potentially serious problem that can markedly affect the quality and usability of the data obtained. For example, moisture loss may be severe enough to prevent full sample hydration, in which case data would obviously be unusable. Also, slow uneven leakage of moisture can also result in the formation of gas bubbles within the sample film, in which case the data obtained will be severely distorted because of the combined effect of differences in effective sample thickness and refractive indices of the liquid paste and the gas bubbles. Such distortions are usually manifest by the appearance of an undulating sinusoidal baseline.
3. The slow progressive loss of sample moisture over the course of an experiment, also causes the relative background contributions from the dispersal solvent and water vapor to the observed spectra to change over time, further complicating the process whereby the sample spectra can be corrected for such effects (*see* below). Given the nature of a hydrated lipid sample, such problems cannot be avoided completely. In effect, there is a practical limit to the usable lifetime of such samples, and this fact should be factored in the design of all protocols for data acquisition. Moreover, the data acquired should always be inspected for evidence of such distortion before analysis and interpretation.

7. Data Analysis and Interpretation

All protocols for the analysis and interpretation of hydrated lipid IR spectra should initially include spectral correction for background absorptions from the dispersal buffer and residual water vapor. This is normally achieved through subtraction of appropriately weighted reference spectra. However, the process of detecting and characterizing lipid phase transitions by FTIR spectroscopy usually requires the acquisition of numerous (usually 20–50) spectra as a function of temperature (or pressure, relative humidity, and so on), and as the contours of the IR bands of the dispersal buffer and residual water vapor are influenced by the prevailing conditions, such protocols also require the accumulation of vast numbers of solvent and water vapor reference spectra acquired under the same conditions as those used for the acquisition of sample spectra. Moreover, because of the considerable variations in the relative contributions of both solvent and water vapor bands to the observed sample spectra (*see* above), the process of spectral subtraction cannot be easily automated and must therefore be performed manually. For these reasons, strict adherence to such protocols can be a very tedious exercise that may well be too costly in both instrument and personnel time for routine applications. Thus, for routine purposes, a more practical approach is the use of uncorrected spectra to determine the range of temperature (pressure, relative humidity, and so on) over which lipid phase transitions occur and, subsequently, select a few spectra representative of the various lipid phases that have been detected. These spectra can then be appropriately corrected for solvent and water vapor interference before further analysis.

After correcting spectra for background contributions from the solvent and water vapor, postacquisition processing of FTIR spectra may also involve spectral enhancement procedures designed to facilitate the identification of key features of the substructure of structurally significant absorption bands. Absorption band envelopes observed in the IR spectra of semisolid and liquid-crystalline materials, such as aqueous lipid dispersions, are usually summations of subcomponents reflecting variations in the structure, conformation, and interactions between various subpopulations of vibrating groups whose parameters are usually encoded in the frequencies, widths, and relative intensities of these components. However, the intrinsic widths of these component bands are often quite broad when compared with the differences between their peak frequencies, and as a result, it is usually quite difficult to resolve the observed band envelope into its various subcomponents. This problem can be addressed with the aid of band-narrowing procedures such as Fourier self-deconvolution and Fourier derivatization *(52–55)*, in which information about the intrinsic widths of the component bands is sacrificed in order to obtain more accurate information about the component frequencies and relative intensities.

These procedures usually require the user to provide an estimate of the mean bandwidth of the components, and under ideal conditions, they can generate spectra of enhanced band resolution without significant distortion or compromise of the information content of the spectra *(52–55)*. However, given the nature of the hydrated lipid samples used for FTIR spectroscopy, the finite noise component embedded in all real spectra and the problems inherent in the correcting of such spectra for background contributions from the dispersal solvent and water vapor (*see* above), the input spectra for such procedures are far from ideal. This fact, along with the inaccuracies inherent in using a single mean bandwidth (i.e., assuming that all component bands are of comparable bandwidth), means that, the application of Fourier self-deconvolution and other band-narrowing procedures to such spectra usually result in significant baseline shift and other distortions that may make the band-narrowed spectra unsuitable for further manipulative processing. Thus, it is better to apply the data processing procedures to the original input spectra (i.e., those corrected for solvent and water background contributions), and the spectra obtained by Fourier self-deconvolution or other band-narrowing procedures should only be used to obtain estimates of the frequencies and the number of resolvable components in the original absorption band. The frequencies, intensities, and widths of the component bands should thus be estimated by reconstructing the contours of the original band envelope, using the results of the band-narrowing analysis as input for peak-fitting procedures.

Further analysis and interpretation of FTIR spectra of lipids may be driven by many objectives, the simplest of which is the detection and identification of lipid phase transitions. Lipid phase transitions are normally manifest by a discontinuous change in the contours of structurally sensitive IR spectroscopic markers, as a function of temperature, pressure, and so on (e.g., *see* **Fig. 1**). Thus, depending on the lipid and the nature of the phase transition under investigation, discontinuities are usually observed in the C–H stretching (~2800–3000cm^{-1}), C=O stretching (1500–1800cm^{-1}), and/or CH$_2$ deformation (1400–1500cm^{-1}), but may also be observed in other regions of the IR spectrum (for examples of the latter, *see* **ref. 2**). In addition, these phase change-induced discontinuities in the properties of the IR absorption bands may occur in some or all regions of the IR spectrum, depending on the nature of the phase transition under investigation. Thus, for example, the L$_c$/L$_\beta$ phase transition of DPPC is usually accompanied by fairly drastic changes in the C=O stretching and CH$_2$ deformation regions of the IR spectrum with little or no discernable changes in the C–H stretching region. In contrast, the L$_\beta$/L$_\alpha$ phase transition is accompanied by discontinuous increases in the frequency and widths of all bands in the C–H stretching region, and discontinuous broadening of all bands in the CH$_2$ deformation region, but little discernable change in the contours of the C=O stretching bands (*see* **refs. 2,4**). Given this, and the fact that some lipids may exhibit a comparable array of spectroscopic changes in the absence of any discernible phase transition (*see* **Subheading 3.3.** above), due caution must always be exercised when using FTIR

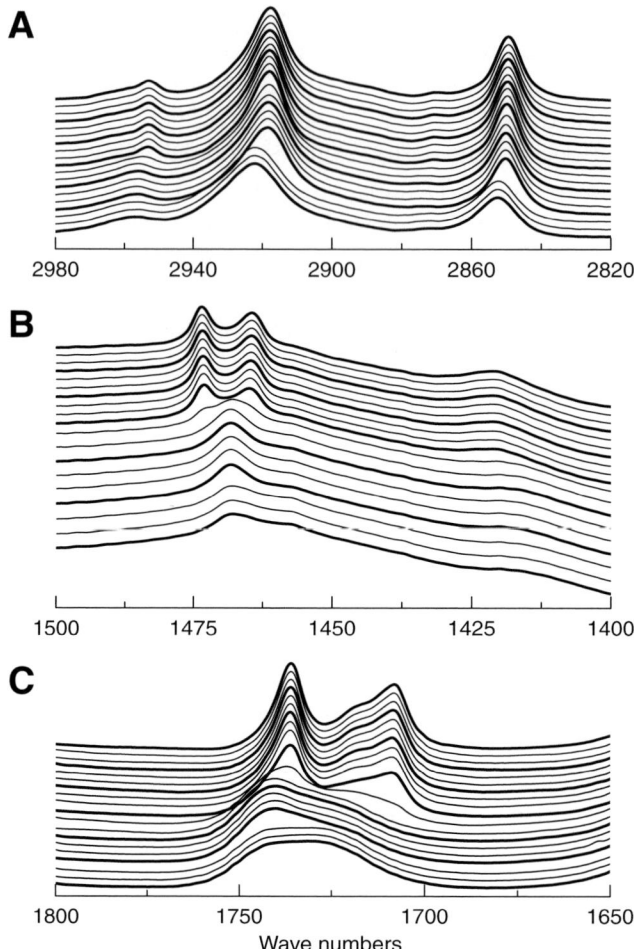

Fig. 1. Stacked plots illustrating FTIR spectroscopic detection of the thermotropic phase transitions exhibited by aqueous (D$_2$O) dispersions of dimyristoylphosphatidylserine. The spectra shown in each panel were acquired between 6 (top spectrum) and 42°C (bottom spectrum). The panels show the spectroscopic changes occurring in: **(A)** the C–H stretching region, **(B)** the CH$_2$ deformation region, and **(C)** the C=O stretching region of the IR spectrum.

spectroscopy to assign the nature of lipid phase changes. General guidelines for assigning the nature of the main IR-detectable lipid phase transitions are summarized above in the **Subheading 3.**

The data obtained can also be used to characterize the structural changes occurring at IR detectable lipid phase changes. Phase transitions involving crystalline or *quasi*-crystalline lipid phases are better suited for such studies because the intermolecular distances and relative orientations between molecules in these phases exhibit patterns of IR absorption that are largely determined by close-contact, intra- and intermolecular interactions between neighboring vibrating groups. Moreover, because the overall rates and amplitudes of molecular motions of lipid molecules within crystal-like lipid phases are relatively low, the environments around vibrating groups tend to be less heterogeneous than those found in more mobile lipid phases.

As a result, relatively sharp absorption bands are usually observed. These properties are highly conducive to the identification of specific patterns of IR absorption from which very detailed information on the molecular organization of the lipid phase is potentially available if these patterns are correctly identified and interpreted.

Although detailed structural interpretation of the FTIR spectroscopic data may be too specialized for routine application (for discussions on applying FTIR spectroscopy to structure determination studies of lipids and membranes, *see* **refs. 2** and **4** and references therein), sufficient structural information can be obtained from even these simple applications to provide valuable insight into the structure of the lipid phases examined, and the nature of the structural changes occurring on conversion to other lipid phases. In the case of diacylglycerolipids, for example, information about the conformation and hydration about the lipid polar/apolar interfacial region is largely encoded in the fine structure of the C=O stretching bands, whereas information about lateral packing interactions between hydrocarbon chains can be deduced from both the CH_2 deformation and rocking absorption bands. The general guidelines for structural interpretation of the fine structure of these IR bands are summarized in **Table 1** and in the **Subheading 3**.

8. Concluding Remarks

It is clear that FTIR spectroscopy is a very powerful, nonperturbing tool for studying phase transitions in lipids and biological membranes. Indeed, even when fairly basic equipment is used, one can easily detect, identify, and perform some structural characterization of lipid phase transitions in both model and biological membranes. In addition, provided judicious and careful sample preparation is conducted, one can also examine how lipid phase transitions are affected by the presence of sterols, proteins, peptides, and other lipophilic inclusions (e.g., *see* **ref. 20**). Moreover, with appropriate investment in isotopically labeled lipid samples, these simple techniques can also be used to study the behavior of the components of lipid mixtures, and how the behavior of each component of a mixture is affected by the presence of lipophilic inclusions. However, this article has covered only a small aspect of the applications of FTIR spectroscopy in the field of lipid and biomembrane research, and the simple techniques discussed here are intended to be introductory in nature. Thus, the reader is encouraged to use these simple techniques as a foundation for more in depth studies, and to work toward extending such studies to more specialized applications, which can provide more detailed information about the structure and organization of lipids and biological membranes.

Acknowledgments

Research performed in the authors' laboratory was supported by operating and major equipment grants from the Canadian Institutes of Health Research and by major equipment and personnel support grants from the Albert Heritage Foundation for Medical Research.

References

1. Mantsch, H. H. and McElhaney, R. N. (1991) Phospholipid phase transitions in model and biological membranes as studied by infrared spectroscopy. *Chem. Phys. Lipids* **57,** 213–226.
2. Lewis, R. N. A. H. and McElhaney, R. N. (1996) FTIR spectroscopy in the study of hydrated lipids and lipid bilayer membranes, in *Infrared Spectroscopy of Biomolecules*, (Mantsch, H. H. and Chapman, D., eds.), John Wiley & Sons, New York, pp. 159–202.
3. Lewis, R. N. A. H. and McElhaney, R. N. (1998) The structure and organization of phospholipid bilayers as revealed by infrared spectroscopy. *Chem. Phys. Lipids* **96,** 9–21.

4. Lewis, R. N. A. H. and McElhaney, R. N. (2002) Vibrational Spectroscopy of Lipids, in *Handbook of Vibrational Spectroscopy*, vol. 5, (Chalmers J. M. and Griffith, eds.), John Wiley and Sons, Chichester, England, pp. 3447–3464.
5. Scheuing, D. R. (1991) Fourier transform infrared spectroscopy in colloid and interface science. *ACS symposium series* 447. American Chemical Soc. Washington, DC.
6. Mendelsohn, R., Brauner, J. W., and Gericke, A. (1995) External infrared reflection-absorption spectrometry. Monolayer films at the air-water interface. *Ann. Rev. Phys. Chem.* **46,** 305–334.
7. Picard, F., Buffeteau, T., Desbat, B., Auger, M., and Pezolet, M. (1999) Quantitative orientation measurements in thin lipid films by attenuated total reflection infrared spectroscopy. *Biophys. J.* **76,** 539–551.
8. Dicko, A., Bourque, H., and Pezolet, M. (1998) Study by infrared spectroscopy of the conformation of dipalmitoylphosphatidylglycerol monolayers at the air-water interface and transferred on solid substrates. *Chem. Phys. Lipids* **96,** 125–139.
9. Nabet, A., Auger, M., and Pezolet, M. (2000) Investigation of the temperature behavior of the bands due to the methylene stretching vibrations of phospholipid acyl chains by two-dimensional infrared correlation spectroscopy. *Appl. Spectrosc.* **54,** 948–955.
10. Frengcli, U. P. (1977) The structure of lipids and proteins studied by attenuated total reflectance (ATR) infrared spectroscopy. *Z. Naturforsch.* **32B,** 20–45.
11. Fringeli, U. P. and Günthard, Hs. H. (1981) Infrared membrane spectroscopy, in *Membrane Spectroscopy*, (Grell, E., ed.), Springer-Verlag, New York, pp. 270–332.
12. Cevc, G. C. and Marsh, D. (1987) Phospholipid Bilayers, Physical Principles and Models. Wiley, New York, pp. 246–257.
13. Dluhy, R. A. and Cornell, D. G. (1991) Monolayer structure at gas-liquid and gas-solid interfaces, in *Fourier Transform Infrared Spectroscopy in Colloid and Interface Science*, (Scheiung D. R., ed.) ACS Symposium Series 447. American Chemical Society, Washington, DC, pp, 24–43.
14. Ulman, A. (1991) Fourier transform infrared spectroscopy of Langmuir-Blodgett and self assembled films. An overview, in *Fourier Transform Infrared Spectroscopy in Colloid and Interface Science*, (Scheiung D. R., ed.) ACS Symposium Series 447. American Chemical Society, Washington, DC, pp. 144–159.
15. Dluhy, R. A. (1986) Quantitative external reflection infrared spectroscopic analysis of insoluble monolayers spread at the air-water interface. *J. Phys. Chem.* **90,** 1373–1379.
16. Dluhy, R. A., Mitchell, M. L., Pettenski, T., and Beers, J. (1988) Design and interfacing of an automated Langmuir-type film balance to an FTIR spectrometer. *Appl. Spectrosc.* **42,** 1289–1293.
17. Mitchell, M. L. and Dluhy, R. A. (1988) In situ FT-IR investigation of phospholipid monolayer phase transitions at the air-water interface. *J. Am. Chem. Soc.* **110,** 712–718.
18. Dluhy, R. A., Keilly, K. E., Hunt, R. D., Mitchell, M. L., Mautone, A. J., and Mendelsohn, R. (1989) Infrared spectroscopic investigations of pulmonary surfactant-surface film transitions at the air–water interface and bulk phase thermotropism. *Biophys. J.* **56,** 1173–1181.
19. Snyder, R. G. (1967) Vibrational study of chain conformation in the liquid n-paraffins and molten polyethylene. *J. Chem. Phys.* **47,** 1316–1360.
20. Mendelsohn, R. and Mantsch, H. H. (1986) Fourier transform infrared studies of lipid-protein interactions, in *Progress in Lipid Protein Interactions*, vol. 2, (Watts A. and De Pont, J. J. H. H. M., eds.), Elsevier, New York, pp. 103–146.
21. Lee, D. C. and Chapman, D. (1986) Infrared spectroscopic studies of biomembranes and model membranes. *Biosci. Rep.* **6,** 235–256.
22. Jackson, M. and Mantsch, H. H. (1993) Biomembrane structure from FT-IR spectroscopy. *Spectrochim. Acta* **15,** 53–69.
23. Mantsch, H. H., Madec, C., Lewis, R. N. A. H., and McElhaney, R. N. (1985) The thermotropic phase behaviour of model membranes composed of phosphatidylcholines containing isobranched fatty acids. II. Infrared and ^{31}P-NMR spectroscopic studies. *Biochemistry* **24,** 2440–2446.

24. Casal, H. L. and Mantsch, H. H. (1983) The thermotropic phase behavior of n-methylated dipalmitoyl phosphatidylethanolamines. *Biochim. Biophys. Acta* **735**, 387–396.
25. Casal, H. L. and Mantsch, H. H. (1984) Polymorphic phase behavior of phospholipid membranes studied by infrared spectroscopy. *Biochim. Biophys. Acta* **779**, 381–402.
26. Snyder, R. G. (1961) Vibrational spectra of n-paraffins II. Inter molecular effects. *J. Mol. Struct.* **7**, 116–144.
27. Snyder, R. G. (1979) Vibrational correlation splitting and chain packing for the crystalline alkanes. *J. Chem. Phys.* **71**, 3229–3235.
28. Casal, H. L. and McElhaney, R. N. (1990) Quantitative determination of hydrocarbon chain order in bilayers of saturated phosphatidylcholines of various chain lengths by Fourier transform infrared spectroscopy. *Biochemistry* **29**, 5423–5427.
29. Senak, L., Davies, M. A., and Mendelsohn, R. (1991) A quantitative IR study of hydrocarbon chain conformation in alkanes and phospholipids: CH_2 wagging modes in disordered bilayer and H_{II} phases. *J. Phys. Chem.* **95**, 2565–2571.
30. Moore, D. J., Wyrwa, M., Reboulleau, C. P., and Mendelsohn, R. (1993) Quantitative IR studies of acyl chain conformational order in fatty acid homogenous membranes of live cells of *Acholeplasma laidlawii* B. *Biochemistry* **32**, 6281–6287.
31. Chia, N. -C. and Mendelsohn, R. (1992) CH_2 wagging modes of unsaturated acyl chains as IR probes of conformational order in methyl alkenoates and phospholipid bilayers. *J. Phys. Chem.* **96**, 10,543–10,547.
32. Lewis, R. N. A. H., Pohle, W., McElhaney, R. N. (1996) The interfacial structure and the phospholipid bilayers: Differential scanning calorimetry and Fourier transform infrared spectroscopic of 1,2-dipalmitoyl-sn-glycero-3-phosphocholine and is dialkyl and acyl-alkyl analogs. *Biophys. J.* **70**, 2736–2746.
33. Senak, L., Moore, D., and Mendelsohn, R. (1992) CH_2 wagging progressions as IR probes of slightly disordered phospholipid acyl chain states. *J. Phys. Chem.* **96**, 2749–2754.
34. Chia, N. -C., Vilcheze, C., Bittman, R., and Mendelsohn, R. (1993) Interactions of cholesterol and synthetic sterols with phosphatidylcholines as deduced from infrared CH wagging progression intensities. *J. Am. Chem. Soc.* **115**, 12,050–12,055.
35. Lewis, R. N. A. H., Mantsch, H. H., and McElhaney, R. N. (1989) The thermotropic phase behavior of phosphatidylcholines with ω-tertiary-butyl fatty acyl chains. *Biophys. J.* **56**, 183–193.
36. Lewis, R. N. A. H. and McElhaney, R. N. (1990) The subgel phases of *n*-saturated diacyl phosphatidylcholines. A Fourier transform infrared spectroscopic study. *Biochemistry* **29**, 7946–7953.
37. Lewis, R. N. A. H. and McElhaney, R. N. (1992) Structures of the subgel phases of n-saturated diacyl phosphatidylcholine bilayers: FTIR spectroscopic studies of $^{13}C=O$ and ^2H labeled lipids. *Biophys. J.* **61**, 63–77.
38. Mantsch, H. H., Madec, C., Lewis, R. N. A. H., and McElhaney, R. N. (1985) The thermotropic phase behaviour of model membranes composed of phosphatidylcholines containing isobranched fatty acids. II. Infrared and ^{31}P-NMR spectroscopic studies. *Biochemistry* **24**, 2440–2446.
39. Huang, C., Mason, J. T., Stephenson, F. A., and Levin, I. W. (1986) Polymorphic phase behavior of platelet-activating factor. *Biophys. J.* **49**, 587–595.
40. Mushayakarara, E. C. and Mantsch, H. H. (1985) Thermotropic phase behavior of platelet activating factor. An infrared spectroscopy. *Can. J. Biochem. Cell Biol.* **63**, 1071–1076.
41. Mantsch, H. H., Martin, H. A., and Cameron, D. G. (1981) Characterization by infrared spectroscopy of the bilayer to nonbilayer phase transition of phosphatidylethanolamines. *Biochemistry* **20**, 3138–3145.
42. Cheng, K. H. (1991) Infrared study of the polymorphic phase behavior of dioleoylphosphatidylethanolamine and dioleoylphosphatidylcholine mixtures. *Chem. Phys. Lipids* **60**, 119–125.
43. Castresana, J., Nieva, J.-L., Rivas, E., and Alonso, A. (1992) Partial dehydration of phosphatidylethanolamine headgroups during hexagonal phase formation, as seen by I.R. spectroscopy. *Biochem. J.* **282**, 467–470.

44. Lewis, R. N. A. H., Mannock, D. A., and McElhaney, R. N. (1997) Membrane lipid molecular structure and polymorphism, in *Current Topics in Membranes*, vol. 44, (Epand, R. M., ed.), Publ Academic Press, NY. pp. 25–102.
45. Wong, P. T. T. and Moffatt, D. G. (1983) Glass anvil cell for high pressure Raman and infrared spectroscopic studies up to 12.6 Kbar. *Appl. Spectrosc.* **37,** 85–87.
46. Wong, P. T. T. and Klug, D. D. (1983) Reevaluation of Type I diamonds for infrared and Raman spectroscopy in high-pressure diamond anvil cells. *Appl. Spectrosc.* **37,** 284–286.
47. Wong, P. T. T. (1985) Temperature variable high-pressure cell and assembly for Raman spectroscopic studies of aqueous systems. *Rev. Sci. Instrum.* **56,** 1417–1419.
48. Wong, P .T. T. and Mantsch, H. H. (1985) Pressure induces phase transitions and structural changes in aqueous dipalmitoylphosphatidylcholine and distearoylphosphatidylcholine. *J. Phys. Chem.* **89,** 883–886.
49. Pohle, W., Selle, C., Fritzsche, H., and Binder, H. (1998) Fourier transform infrared spectroscopy as a probe for the study of the hydration of lipid self-assemblies. I. Methodology and general phenomena. *Biospectroscopy* **4,** 267–280.
50. Pohle, W. and Selle, C. (1996) Fourier-transform infrared spectroscopic evidence for a novel lyotropic phase transition occurring in dioleoylphosphatidylethanolamine. *Chem. Phys. Lipids* **82,** 191–198.
51. Bertie, J. E. and Lan, Z. D. (1996) Infrared intensities of liquids 20. The intensity of the OH stretching band of liquid water revisited, and the best current values of the optical constants of $H_2O(1)$ at 25 degrees C between 15,000 and 1 cm^{-1}. *Appl. Spectrosc.* **50,** 1047–1057.
52. Kauppinen, J. K., Moffatt, D. J., Mantsch, H. H., and Cameron, D. G. (1981) Fourier self deconvolution. A method for resolving intrinsically overlapped bands. *Appl. Spectrosc.* **35,** 271–276.
53. Kauppinen, J. K., Moffatt, D. J., Hollberg, M. R., and Mantsch, H. H. (1991) Fourier self deconvolution. A new line narrowing procedure based on Fourier self deconvolution, maximum entropy, and linear prediction. *Appl. Spectrosc.* **45,** 411–416.
54. Moffatt, D. G. and Mantsch, H. H. (1992) Fourier resolution enhancement of infrared spectral data. *Methods Enzymol.* **210,** 192–200.
55. Mantsch, H. H., Moffatt, D. J., and Casal, H. L. (1988) Fourier transform methods for spectral resolution enhancement. *J. Mol. Struct.* **173,** 285–298.

15

Optical Dynamometry to Study Phase Transitions in Lipid Membranes

Rumiana Dimova and Bernard Pouligny

Summary

The fluidity of the lipid matrix of cell membranes is crucial for the mobility of various inclusions like proteins. When the lipid bilayer undergoes phase transition from fluid-to-gel phase, the shear surface viscosity of the membrane diverges, thus hindering the motion of the membrane inclusions. On the other hand, the membrane bending stiffness drops down, and below the main phase transition, drastically increases with lowering the temperature. A tool to study the membrane properties when the lipid bilayer crosses the phase transition is provided by optical trapping and manipulation of microspheres attached to the membrane. Giant unilamellar vesicles are used, which allow for direct visualization of the membrane response, as model membranes. Following the motion of one or two particles attached to a vesicle, the microscope can provide evidence for the membrane elasticity and state of fluidity. As forces acting on the spheres, one can use gravity, thermal noise, or radiation pressure force.

Key Words: Bending stiffness; giant vesicles; model membranes; optical trapping; phase transition; shear surface viscosity.

1. Introduction

Optical trapping has been extensively exploited for biophysical applications, especially in recent years, when single molecule manipulation has become of intense interest. The great advantage of the technique of optical trapping consists of the possibility to manipulate objects without having to establish a direct mechanical contact with them. Instead, the momentum of light from a laser comes into play to form tiny tweezers for handling microscopic objects. Nowadays, two main versions of optical trapping are broadly used. One is the conventional (also commercially available), single-beam tweezers formed by focusing a laser beam in a sample through a high numerical aperture objective. The latter limits the studies with such a setup to short working distances. This version of the optical trap is currently applied to trapping of particles up to a few microns in diameter. The second version (which, as a matter of fact, was historically first) (*see* **ref.** *1*) is based on two counter-propagating beams focused by two long-focal-distance objectives. From the viewpoint of optical architecture and tuning, the single-beam trap is simple; whereas the double-beam trap is more complex. However the latter offers the advantage of a much larger working distance, and is better suited to manipulation of large (several micrometers) objects and structures. In addition, because the beams are only weakly focused, optical damage (essentially heating) is in general weak. Basically, the choice of one or the other type of optical tweezers depends on the application and the studied system. This chapter reports on experiments involving large particles (up to 10 µm in radius) inside a thick chamber, for which

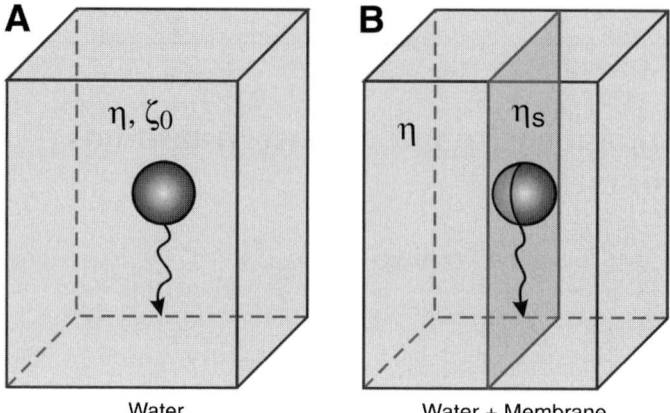

Fig. 1. **(A)** When a particle moves freely in bulk solution of viscosity η the drag coefficient (ζ_0) is given by the Stokes friction. **(B)** A particle attached to a membrane of shear surface viscosity η_S experiences an additional friction resulting in $\zeta = \zeta_0 + \zeta_m$.

the double-beam version of optical trapping was most appropriate. The setup is described in details elsewhere (*2*) (*see* also **ref. 3** for additional characteristics).

Latex particles of size between 1 and 20 μm were used to explore the mechanical and rheological properties of model lipid membranes. These were in the form of giant vesicles (*4*), which are closed sacs made of a lipid bilayer in a water environment. Their relatively large size (several tens of micrometers), compared with conventional vesicles (few hundreds of nanometer in size), allows for direct microscopy observation.

1.1. Shear Surface Viscosity Measurements

When a latex microsphere is attached to a membrane, it can be used as a probe of the membrane state (*see* **Fig. 1**). In spite of the fact that the particle is several orders of magnitude larger than the lipid membrane (~5 nm in thickness), the particle motion can be affected by the displacement of the surrounding lipids. Motion of the lipids and of the water that is confined inside the vesicle increase the particle hydrodynamic drag coefficient (ζ). When the vesicle is much larger than the particle, ζ can be approximately decomposed according to:

$$\zeta = \zeta_0 + \zeta_m \tag{1}$$

where ζ_0 is the background or Stokes friction of the particle in the bulk solution, $\zeta_0 = 6\pi\eta a$, and ζ_m is the excess friction because of shearing the membrane. η is the viscosity of the surrounding fluid (the water solution), and a is the particle radius. The problem of a sphere straddling a membrane has already been theoretically treated in detail (*5,6*) and the results experimentally applied (*7*). The membrane contribution to the friction coefficient was found to depend almost linearly on the membrane shear surface viscosity (η_S) and only weakly on the particle radius:

$$\zeta_m \cong 2.93\eta_S \left(\frac{\eta a}{\eta_S}\right)^{0.1} \tag{2}$$

Optical Dynamometry

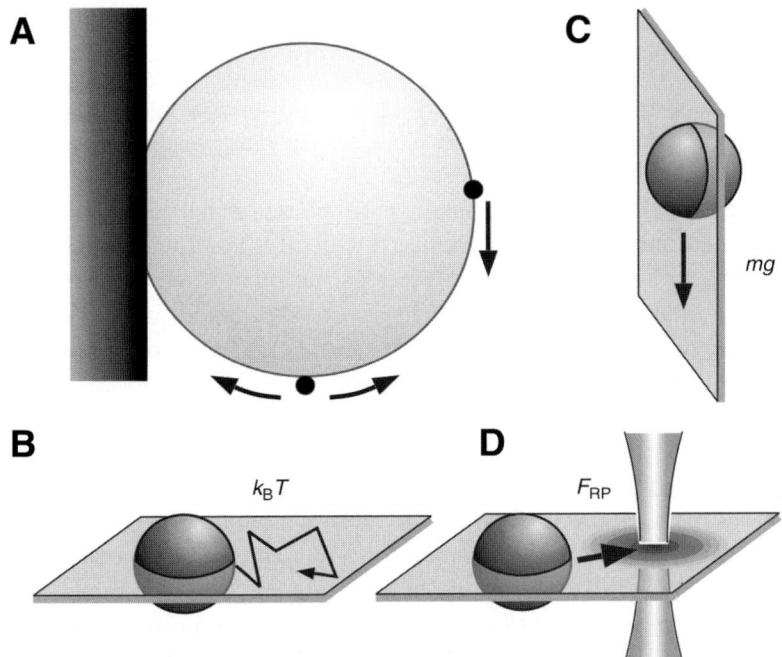

Fig. 2. Three approaches to measure the friction coefficient of a particle moving on a membrane. (**A**) Vesicle fixed to a surface with two possible positions of a particle adhered to the vesicle membrane. (**B**) A Brownian particle experiences a random walk at the bottom (or top) of the vesicle. (**C**) A heavy particle sediments toward the vesicle bottom due to gravity. (**D**) A particle is driven toward the axis of the trap beam because of the radiation pressure force (F_{RP}). The sketch is exaggerated in terms of particle-trap distance.

Note that the surface viscosity has units of (bulk viscosity × length), i.e., dyn.s/cm or surface poises (sp). Using the relation in **Eq. 2** one can estimate the membrane viscosity by measuring the friction coefficient of a small particle attached to the membrane.

Three possible approaches can be used to measure ζ_m (*see* **Fig. 2**). The first is applicable to small Brownian particles of diameter up to about 2 μm (**Fig. 2B**). Analyzing the trajectory in time of such a particle, one can extract the diffusion coefficient (D), which is inversely proportional to the friction coefficient, according to the Einstein relation. The second approach can be used with larger (heavier) particles attached to the membrane (**Fig. 2C**). By means of the optical trap the particle is brought to the upper part of the vesicle and released from the trap. Under the influence of its own weight, the particle then starts to sediment following the surface of the vesicle.

Analysis of the sedimentation trajectory gives the friction experienced by the particle. In the third method, the particle motion is driven by the radiation pressure force of the trap (F_{RP}) (*see* **Fig. 2D**). The particle is brought to either the top or the bottom of the vesicle, wherein the membrane is essentially flat and horizontal and the trap is switched on in the direct vicinity of the particle. This results in a sudden motion of the particle center toward the beam axis. Knowing the stiffness of the trap, analysis of the particle trajectory provides the value of the friction coefficient. Finally, the membrane viscosity can be extracted from the friction coefficient using **Eqs. 1** and **2** in all three cases. Performing any of these three procedures at

1.2. Measuring the Membrane Bending Stiffness Close to the Fluid–Gel Phase Transition

When the temperature is lowered below the main phase-transition temperature of the lipids in the bilayer, the membrane properties change drastically. The fluid-to-gel transition leads to divergence in the shear surface viscosity, whereas the membrane acquires a nonzero shear modulus (in the fluid-state, lipid membranes have zero shear modulus). The membrane elastic properties also change. Lipid membranes in the fluid state are characterized by a bending rigidity in the order of 10 $k_B T$ (exact values for different lipid bilayers can be found in **refs. 8** and **9**), where k_B is the Boltzmann constant and T is temperature. When the lipid bilayer undergoes the fluid-to-gel transition, the bending rigidity drops down, and in the gel phase, it increases by a few orders of magnitudes *(10,11)* when temperature is decreased. Measuring the bending stiffness of membranes in the fluid state is a handled task, and a large number of methods have already been developed. To mention a few, fluctuation spectroscopy (*see* **ref. 12**) is based on the analysis of membrane undulation of giant vesicle membranes; the micropipet aspiration technique *(13)* in the low-tension or entropic regime is based on pulling out membrane fluctuations; analysis of the degree of deformation of giant vesicles subjected to alternating electric fields can also provide the bending stiffness modulus *(14)*. All these techniques have been applied to lipid membranes in the fluid state. However, in the gel phase the bending stiffness increases dramatically and most of the classical methods cannot be applied. A few new techniques have been recently developed for measuring the bending stiffness of membranes in the gel phase *(10,11,15)* and one of them is presented in this article.

When particles are attached to the membrane and the bilayer is brought to the gel phase, the particle motion is hindered (i.e., their motion along the membrane becomes frozen). In a typical experiment, the penetration depth of the particles is usually small, i.e., the contact line with the membrane is far from the equator of the particle, and the particles are located more to the outside of the vesicle (*see* **Fig. 3A**). Two particles (spaced by a few particle diameters) can be manipulated simultaneously using two optical tweezers. The latter are characterized by the trap stiffness, or the trapping constant (k_{RP}). A force (F_{RP}) is applied to displace one of these particles in the membrane plane by a mobile trap while holding the other in the potential well of an immobile or fixed trap. For particle configurations as the one shown in **Fig. 3A**, the main membrane deformation caused by the particle's displacement is bending (*see* **Fig. 3C**). The experiment leads to measuring an apparent membrane spring constant, k_M *(10)*:

$$k_M = k_{RP}\left[(x_m + x_f)/(l_1 - l_0) - 1\right] \qquad (3)$$

where x_m and x_f are the displacements of the beads in the mobile and the fixed trap, respectively, and l_0 and l_1 are the distances between the particles before and after applying the forces. The relation between k_M and the membrane bending modulus (κ) is given by an empirical formula *(10)*:

$$k_M \cong 60\,\kappa/a^2 \qquad (4)$$

Optical Dynamometry

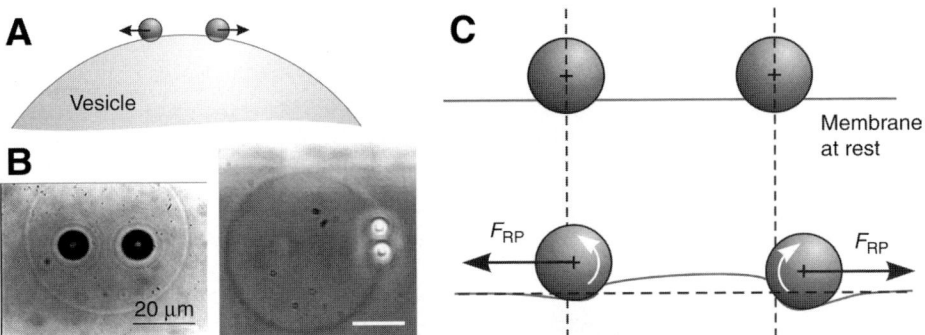

Fig. 3. **(A)** Sketch of two particles adhering to the membrane of a vesicle. **(B)** **(Left)** Top view (in classical transmission microscopy) of two large particles located on top of a vesicle. The particles are approximately in focus. The image of the vesicle equator, well below the focal plane of the microscope, appears as a clear faint ring. **(Right)** Two smaller particles are located at the vesicle equator. The scene is observed in phase contrast, which provides a dark image of the membrane and bright images of the particles. **(C)** Sketch of the out-of-plane deformation of the membrane (side view) induced by pulling on one of the two particles and keeping the other in the potential well of the fixed trap. The apparent spring constant of the membrane can be calculated by measuring the relative particle displacement.

2. Materials
2.1. Lipids for Vesicle Preparation

Any synthetic phosphatidylcholine can be used as long as the main phase-transition temperature is easily accessible in the laboratory environment. 1,2-dimyristoyl-*sn*-glycero-3-phosphocholine (DMPC) (Avanti Polar Lipids, Alabaster, AL) (no additional purification of the lipid is needed) was used. The DMPC membrane undergoes fluid-to-gel transition around 24°C *(16)*. The lipid is dissolved in chloroform solution and stored at –20°C. At this temperature, the lipid is stable for up to 3 months.

2.2. Particles for Optical Dynamometry

Both latex (polystyrene) particles and glass beads can be used. In this work mainly latex spheres (Polyscience, Warrington, PA) will be referred to, with diameters ranging from 2 to 12 µm. The particles are stored at 4°C and are stable for a few years. Right before using, a droplet of the particle stock solution is diluted in 2 mL pure water (*see* **Note 1**).

3. Methods
3.1. Vesicle Preparation, Particle Size Determination, and Trap Calibration

The vesicle formation and the particle dynamometry experiments are performed in the same chamber. Giant unilamellar vesicles are prepared following the electroformation method *(17–19)* in a chamber with the geometry sketched in **Fig. 4**. During vesicle formation, the temperature was set to 30°C, which is well above the main phase transition of DMPC. The preparation steps are the following:

1. A few droplets of the solution of the lipid dissolved in chloroform at concentration approx 2 mg/mL are deposited on the electrodes (platinum wires of ~1-mm diameter) (*see* **Fig. 4**). The electrodes are left under a nitrogen stream or under vacuum for 2 h, for complete evaporation of the organic solvent.

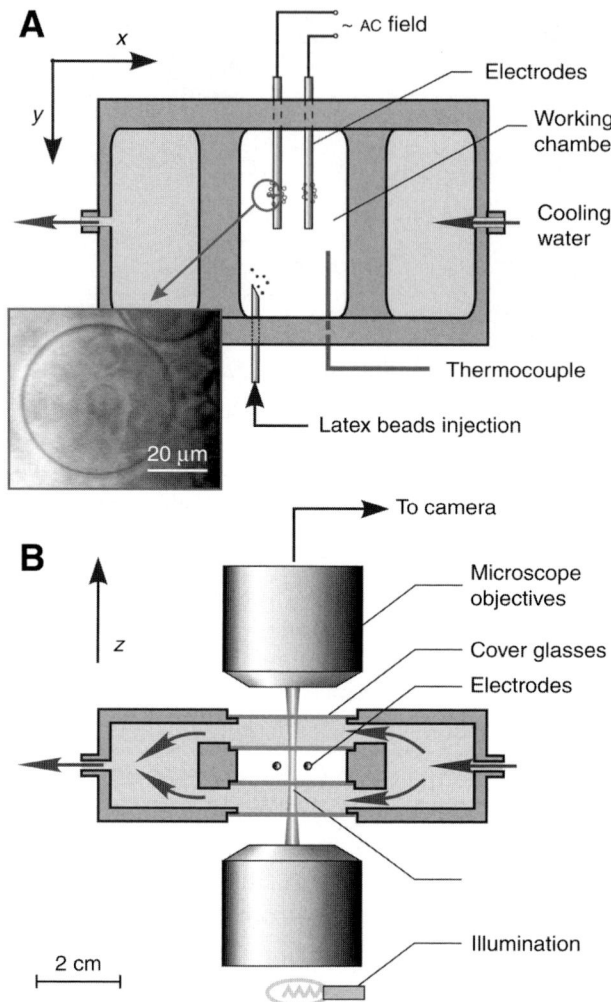

Fig. 4. Sketch of the experimental chamber. (**A**) Top view. (**B**) Side view. Vesicles are grown on both electrodes. The working chamber is surrounded by a cooling water jacket assuring temperature control. A thermocouple measures the current temperature in the working chamber. The top microscope objective (×40, numerical aperture [N.A.] = 0.6) is operated in the phase-contrast mode. The latex particles are injected through a needle (*see* **panel A**). The chamber is mounted on a motorized *x-y-z* stage.

2. The electrodes are inserted in the chamber. The external compartment (water jacket) is filled with water and a constant flow is assured from a thermostat. The temperature in the chamber is adjusted to 30°C. The electrodes are connected to an AC-field generator and a voltage at 10 Hz and 0.3 V/mm is supplied for 10 min. The working chamber is slowly filled with water (*see* **Note 2**), which is introduced through a needle (the same needle is later used for introducing the latex particles).
3. After the first 20 min the voltage in the chamber is gradually increased to about 1.2 V/mm in 0.3 V/mm steps every 20 min and the sample is left at these field conditions for another couple of hours. The vesicles can be found along the electrodes. The procedure is terminated by slowly decreasing the voltage to 0.5 V/mm and the frequency changed to 5 Hz for about 5 min, which helps detaching the vesicles from the electrode. The vesicles near the outer boundary of the clusters formed at the electrodes (*see* snapshot in **Fig. 4A**) are most convenient for working (*see* **Note 3**).

4. A few tens of microliter of the latex bead solution are introduced in the working chamber (a microliter syringe is used). A particle is captured "in flight" by the laser beam at the exit of the syringe needle (*see* **Note 4**).
5. Particle size (a) measurement: if the particle is small, a is determined from the diffusion path of the particle in water (D_{free}). Conversely, the size of a heavy particle is deduced from its sedimentation velocity (v_{sed}). In the former case, the bead radius is given by the Stokes–Einstein relation $a = k_B T / 6\pi \eta D_{free}$ (*see* **Note 5**). In the latter case, the application of Stokes' law gives $a = (9\eta v_{sed} / 2\Delta\rho g)^{1/2}$, where $\Delta\rho$ is the density difference between water and polystyrene ($\approx 0.05 g/cm^3$) and g is the gravity acceleration (*see* **Note 6**).
6. To calibrate the radiation pressure force (F_{RP}), the trapped sphere is submitted to a constant counter-flow of known velocity (this is done by moving the stage on which the chamber is mounted with a fixed velocity). The "escape" velocity (v_{esc}) at which the particle leaves the trap is determined. The corresponding trapping force, which is the maximum of F_{RP}, balances the viscous drag force (Stokes' law): $F_{RP} \cong 6\pi a \eta v_{esc}$. The trap spring constant is given by $K_{RP} \cong 6\pi \eta v_{esc}$.
7. Finally, the particle is brought in contact to a selected vesicle. In this procedure, the particle is held fixed in space, while the whole chamber is moved by the motorized stage. When in contact with the vesicle, the particle spontaneously adheres to the membrane. Once completed, adhesion is irreversible: it is not possible to detach the particle applying radiation pressure forces (several tens of pN).

3.2. Shear Surface Viscosity Measurements

The procedure involves measuring ζ at different temperatures, and deducing η_S at each step. It starts at $T = 30°C$, and proceeds through successive temperature steps. After each step, the sample is left at rest until thermal equilibrium is reached (this takes about 15 min). At temperatures $\leq 22°C$, the shear viscosity of DMPC membranes is so high that motion of the latex beads is no longer optically detectable. An example of data obtained from the three procedures applied to the same vesicle-particle system is given in **Fig. 5**.

3.2.1. Brownian Motion Dynamics

1. A small particle (radius <2 µm) adhering to the vesicle membrane is brought to the bottom or top of the vesicle and the trap is switched off (*see* **Fig. 2B**).
2. The particle trajectory is recorded for a period of time enough to collect about 400 data points (when the lipid bilayer is in the fluid phase, 100 s of recording at acquisition frequency of approx 6 Hz suffices; however, at temperatures approaching the gel phase, longer acquisition times are necessary).
3. The mean square displacement of the particle is determined and the diffusion coefficient extracted:

$$\left\langle (\Delta x)^2 + (\Delta y)^2 \right\rangle = 4D\Delta t \qquad (5)$$

where t is time.

4. The friction coefficient is then determined from the Stokes–Einstein relation, $\zeta = k_B T / D$.

3.2.2. Sedimentation Velocity Measurements

1. A large or heavy particle (for latex particles the size should be above 3 µm in radius) adhered to the vesicle membrane is brought to the top of the vesicle and the trap is switched off (*see* **Fig. 2C**).
2. The trajectory over time of the particle sedimenting toward the vesicle bottom is recorded. As the observation is from above the moving particle quickly gets out of focus, leading to frequent

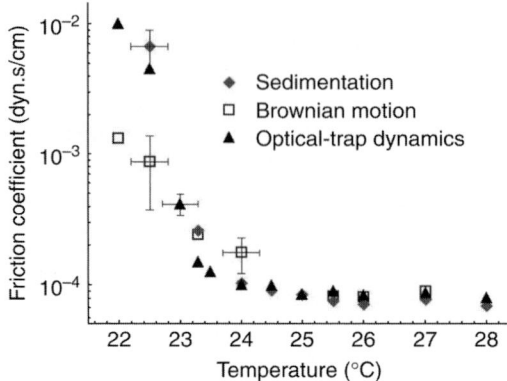

Fig. 5. Examples of data collected using the three described procedures with the same vesicle-particle system. The particle size is $a = 3.2$ µm, and the vesicle size is 21 µm.

refocusing. The coordinates of the vesicle center are determined from a snapshot taken in the equatorial plane.

3. The distance between the particle and vesicle centers, $r(t)$, is extracted from the recorded particle trajectory. It is described by the equation (7):

$$r(t) = r_{max} \sin\{2\arctan[\exp(4\pi a^3 \Delta\rho g t/3\zeta r_{max})]\tan(\theta_0/2)\} \quad (6)$$

where the angle θ_0 is defined by $\sin\theta_0 = r(t=0)/r_{max}$.

This equation is used to fit the experimental curve and to extract the bead friction coefficient.

3.2.3. Optical Trapping Dynamics

1. A particle of several micrometers in radius is brought to the top or the bottom of the vesicle and the trap is switched off. The experimental chamber is displaced in the horizontal plane by a distance approx 0.6a. The trap is then switched on, and the trajectory of the particle toward the beam axis is recorded over time (*see* **Fig. 2D**).
2. The time dependence of the distance $d(t)$ between the particle center and the trap axis is extracted from the particle trajectory. The data is fitted to by $d(t) = d(t=0)\exp(-tk_{RP}/\zeta)$, providing the particle friction coefficient.

3.3. Measuring the Membrane Bending Stiffness Close to the Fluid–Gel Phase Transition

1. Two particles of similar size and several microns in radius are brought to the top or the bottom of the vesicle by means of the double optical trap. It is a requirement that both penetration depths of the particles are small (*see* **Fig. 3C**). This condition is in general fulfilled when more than one particle are brought to adhere to the vesicle membrane.
2. The temperature of the chamber is lowered to 15°C, and the sample is let to equilibrate.
3. The initial distance between the traps (l_0) is measured, and the location of the particles centers is determined. The mobile trap is shifted by approx 0.6a in direction away from its initial position, away from the fixed trap. The new intertrap distance (l_1) is measured, as well as the relative displacements of the particle in the mobile trap (x_m) and the particle in the fixed trap (x_f).
4. The membrane spring constant (k_M) is determined using **Eq. 3**. The bending stiffness of the membrane is deduced from **Eq. 4**.

5. The temperature is increased by a few degrees, and **steps 3** and **4** are repeated. Thus, the temperature dependence of the membrane bending stiffness is obtained when the lipid bilayer approaches its gel-to-fluid phase transition. The work is completed when $T = 24°C$, and the membrane is in the fluid state.

4. Notes

1. Unless stated otherwise, all solutions should be prepared in water that has a resistivity of 18.2 MΩ.cm. This standard is referred to as "water" in this text.
2. The water used to fill the working chamber should be degassed before introducing it in the chamber. This prevents the formation of bubbles during work.
3. At the end of the electroformation procedure, the vesicles are usually interconnected and clustered. Target vesicles are selected at the outer rim of such clusters for experiments wherein vesicles that are unilamellar (as far as one can determine from phase contrast views) and without obvious internal structures are easily found.
4. It is important to trap a particle "in flight" at the exit of the syringe needle. One might be tempted to simply catch one of the many particles that have been released and lie on the floor of the chamber. Experience shows that such particles often do not adhere on the vesicle membranes. The reason for this is unclear; it is supposed that the surfaces of the particles, when hitting the walls of the chamber, get contaminated by traces of lipids. Apparently (and fortunately) the bulk water inside the chamber is free from lipids.
5. When the temperature in the chamber is changed, the bulk water viscosity changes; this has to be taken into account when performing the calculations for the particle radius and the friction coefficient; *see* **Eq. 2**.
6. The density difference between latex (or particle material) and water can be measured independently in the following way: the particles are diluted in solutions of glycerol of various densities (i.e., various concentrations of glycerol in water). The solutions are centrifuged. The density at which the particles neither sediment to the bottom of the centrifuge tube nor cream to the surface of the solution is corresponding to the particle material density.

References

1. Ashkin, A. (1970) Acceleration and trapping of particles by radiation pressure. *Phys. Rev. Lett.* **24,** 156–159.
2. Angelova, M. I. and Pouligny, B. (1993) Trapping and levitation of a dielectric sphere with off-centered Gaussian beams. I. Experimental. *Pure Appl. Opt. A* **2,** 261–276.
3. Dietrich, C., Angelova, M. I., and Pouligny, B. (1997) Adhesion of latex spheres to giant phospholipid vesicles: statics and dynamics. *J. Phys. II France* **7,** 1651–1682.
4. Dimova, R., Dietrich, C. and Pouligny, B. Motion of particles attached to giant vesicles: falling ball viscosimetry and elasticity measurements on lipid membranes, in Giant Vesicles, ed. Luisi, P. L., and Walde, P. (1999) John Wiley and Sons, New York, pp. 221–230.
5. Danov, K. D., Dimova, R., and Pouligny, B. (2000) Viscous drag of a solid sphere straddling a spherical or flat surface. *Phys. Fluids* **12,** 2711–2722.
6. Dimova, R., Danov, K., Pouligny, B., and Ivanov, I. B. (2000) Lateral motion of a large solid particle trapped in a thin liquid film. *J. Coll. Interface Sci.* **226,** 35–43.
7. Dimova, R., Dietrich, C., Hadjiisky, A., Danov K., and Pouligny, B. (1999) Falling ball viscosimetry of giant vesicle membranes: finite-size effects. *Eur. Phys. J. B* **12,** 589–598.
8. Seifert, U. and Lipowsky, R. (1995) Morphology of vesicles. in *Structure and Dynamics of Membranes*, (Lipowsky, R. and Sackmann, E., eds.), Elsevier, Amsterdam, pp. 403–463.
9. Rawicz, W., Olbrich, K. C., McIntosh, T., Needham, D., and Evans, E. (2000) Effect of chain length and unsaturation on elasticity of lipid bilayers. *Biophys. J.* **79,** 328–339.
10. Dimova, R., Pouligny, B., and Dietrich, C. (2000) Pretransitional effects in dimyristoylphosphatidylcholine vesicle membranes: Optical dynamometry study. *Biophys. J.* **79,** 340–356.

11. Mecke, K. R., Charitat, T., and Graner, F. (2003) Fluctuating lipid bilayer in an arbitrary potential: theory and experimental determination of bending rigidity. *Langmuir* **19,** 2080–2087.
12. Méléard, P., Gerbaud, C., Pott, T., et al. (1997) Bending elasticities of model membranes—influences of temperature and sterol content. *Biophys. J.* **72,** 2616–2629.
13. Evans, E. and Rawicz, W. (1990) Entropy-driven tension and bending elasticity in condensed-fluid membranes. *Phys. Rev. Lett.* **17,** 2094–2097.
14. Kummrow, M. and Helfrich, W. (1991) Deformation of giant lipid vesicles by electric fields. *Phys. Rev. A* **44,** 8356–8360.
15. Lee, C. -H., Lin, W. -C., and Wang, J. (2001) All-optical measurements of the bending rigidity of lipid-vesicle membranes across structural phase transitions. *Phys. Rev. E* **64,** 020901.
16. Koynova, R. and Caffrey, M. (1998) Phases and phase transitions of the phosphatidylcholines. *Biochim. Biophys. Acta* **1376,** 91–145.
17. Angelova, M. I. and Dimitrov, D. S. (1986) Liposome electroformation. *Faraday Discuss. Chem. Soc.* **81,** 303–311.
18. Angelova, M. I., Soléau, S., Méléard, P., Faucon, J. F., and Bothorel, P. (1992) Preparation of giant vesicles by external AC electric fields. Kinetics and applications. *Prog. Colloid Polym.* **89,** 127–131.
19. Dimova, R., Aranda, S., Bezlyepkina, N., Nikolov, V., Riske, K. A. and Lipowsky, R. (2006) A practical guide to giant vesicles. Probing the membrane nanoregime via optical microscopy. *J. Phys. Condens. Matter* **18,** S1151–S1176.

16

Fluorescence Assays for Measuring Fatty Acid Binding and Transport Through Membranes

Kellen Brunaldi, Jeffrey R. Simard, Frits Kamp, Charu Rewal, Tanong Asawakarn, Paul O'Shea, and James A. Hamilton

Summary

The authors' laboratory has applied a series of different fluorescence assays for monitoring the binding and transport of fatty acids (FA) in model and biological membranes. The authors recently expanded their fluorescent assays for monitoring the adsorption of FA to membranes to a total of three probes that measure different aspects of FA binding: (1) an acrylodan-labeled FA-binding protein, which measures the partitioning of FA between membranes and the aqueous buffer; (2) the naturally occurring fluorescent *cis*-parinaric acid, which specifically measures the insertion of the FA acyl chain into the hydrophobic core of the phospholipid bilayer, and (3) a fluorescein-labeled phospholipid (N-fluorescein-5-thiocarbomoyl-1,2,dihexadecanoyl-*sn*-glycero-3-phosphoethanolamine), which specifically measures the arrival of the FA carboxyl at the outer leaflet of the membrane. None of these probes allow the transmembrane movement of FA to the inner leaflet to be measured. FA translocation (flip–flop) is typically measured directly, using a pH-sensitive fluorophore such as 8-hydroxypyrene-1.3.6-trisulfonic acid or 2′,7′-*bis*-(2-carboxyethyl)-5-(and-6)- carboxyfluorescein. These probes detect the release of protons from unionized FA that have diffused through the membrane to the inner leaflet. Because adsorption of FA to the outer leaflet must occur before flip–flop, these probes measure the effects of the combined steps of adsorption and translocation.

In this chapter, detailed methods are provided on how to monitor the transport of FA through protein-free model membranes, and some of the fluorescent artifacts that may arise with the use of these probes are addressed. Also, experiments designed to investigate such artifacts, and improve the reliability and interpretation of the data are described.

Key Words: ADIFAB; BCECF; cis-parinaric acid; DHPE fatty acid transport; fatty acids; flip–flop; fluorescence; FPE; membranes; pyranine.

1. Introduction

Long-chain fatty acids (FA) are one of the major nutrients in human physiology, serving as an immediate energy source through β-oxidation or as stored energy in the form of intracellular triglycerides, which can be hydrolyzed to release FA. In contrast to other major nutrients, such as glucose and amino acids, FA interact strongly with both the polar interface and hydrophobic interior of lipid membranes. In fact, the lipid phase of cell membranes is a better solvent for FA than the surrounding aqueous phase. The amphipathic structure of the FA molecule, with its polar carboxyl end and its hydrocarbon chain, presents unique opportunities for following its transport in membranes.

Figure 1 illustrates fluorescence approaches described in this chapter and also in Chapter 3 of this volume to follow the transport of FA in membranes, beginning with the unbound FA molecule in the external medium. The overall mechanism involves a minimum of three steps

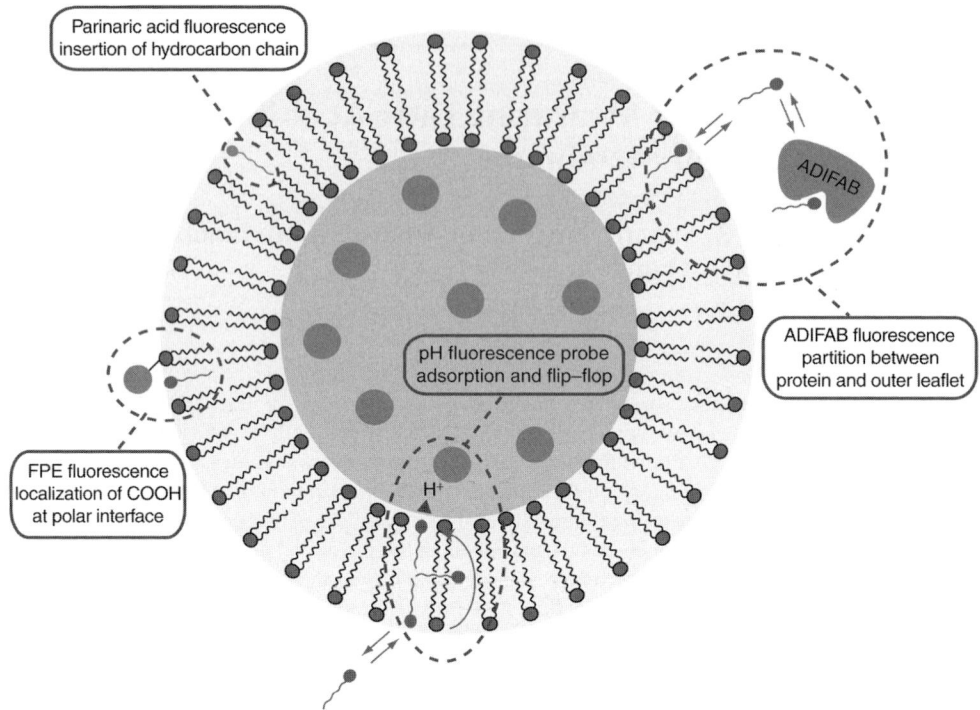

Fig. 1. Fluorescence approaches developed to study the mechanism of FA transport in membranes. Trapping a pH dye inside a lipid vesicle measures the release of protons as unionized FA arrive at the inner leaflet. As this cannot occur without FA first binding to the outer leaflet, the pH dye reports the combined kinetic steps of adsorption and flip–flop. ADIFAB, the fluorescent-labeled FABP, is placed in the external buffer and is used to monitor the binding of FA to the outer leaflet. ADIFAB provides quantitative data for calculating the partitioning of FA between aqueous and membrane phases. However, ADIFAB does not provide information about the transmembrane movement of FA. The naturally fluorescent FA, *cis*-PA, measures the insertion of the acyl chain into the hydrophobic membrane core. A FPE molecule inserts into the outer leaflet of membranes when added to the external buffer. FPE is sensitive to the presence of the charged FA carboxyl at the water–lipid interface.

that can and should be distinguished: adsorption, transmembrane movement, and desorption *(1–3)*. Rigorous and quantitative characterization of each step is important for evaluating the contribution of the lipid phase of membranes in the membrane transport of FA. Such studies provide a foundation for evaluating potential roles of proteins in transporting FA in membranes or modulating the uptake of FA into cells.

The adsorption step can be measured by several methods. As discussed in Chapter 3 of this volume, the soluble fluorescent acrylodan-labeled FA-binding protein (ADIFAB) can be added to the external medium to acquire kinetic and thermodynamic data (partitioning between membrane-bound FA and unbound FA). This approach does not provide information about the translocation step. The naturally fluorescent FA *cis*-parinaric acid (PA) provides kinetic data reflecting insertion of the FA tail into the phospholipid bilayer as the fluorescence of this part of the FA changes drastically when it moves from the aqueous to the lipid phase. Because the tail is buried in the same hydrocarbon region in both leaflets of the bilayer, PA fluorescence does not provide information about transmembrane movement of the FA molecule **(Fig. 2)**.

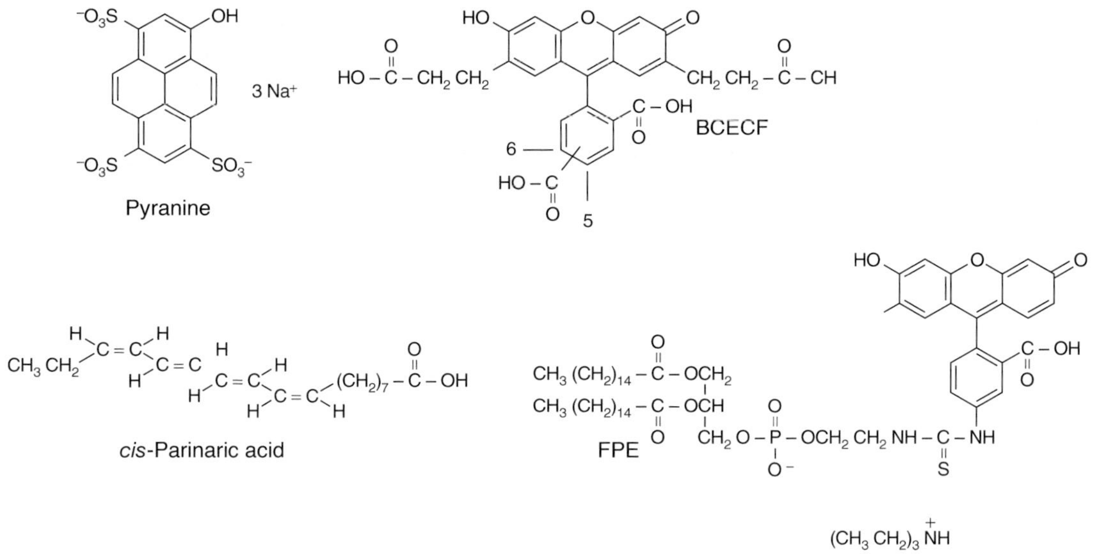

Fig. 2. Structures of fluorescence probes used to study FA transport in membranes. Pyranine and BCECF acid are water-soluble pH probes that are trapped inside lipid vesicles. FPE and cis-PA are membrane probes used to detect the arrival of the FA carboxyl and the insertion of hydrocarbon chain into the membrane, respectively. All probes are available from Molecular Probes (Eugene, OR).

PA is very slowly metabolized, which can eliminate complications of metabolism in studies of FA transport into cells *(5)*.

Incorporation of the polar end of the FA molecule into the membrane can be detected by interaction of the charged head group of the FA with fluorescein-labeled phosphatidylethanolamine (*N*-fluorescein-5-thiocarbomoyl 1,2,dihexadecanoyl-*sn*-glycero-3-phosphoethanolamine [FPE or DHPE]) (**Fig. 2**). On addition to a suspension of vesicles, FPE inserts into the outer leaflet of membranes such that the fluorescent group protrudes into the aqueous solution. In a well-buffered system, it is sensitive to local charge in the vicinity of phospholipids. The model of FA flip–flop in membranes is based on the ionization properties of FA in the membrane, which gives rise to both unionized and ionized FA in each membrane leaflet. This probe will detect the arrival of the FA carboxyl anion at the outer leaflet, thereby monitoring insertion of the FA head group into the membrane–water interface.

The mechanism of the transmembrane step has been the most difficult to measure and the most debated *(4–7)*. After the FA becomes incorporated into the lipid bilayer, the ionization pK_a of the carboxyl group increases to approx 7.5 *(8)*. As the energy barrier for transfer of the uncharged FA across the membrane is expected to be much lower than that of the ionized species, the hypothesis was tested that unfacilitated FA "flip–flop" across the lipid bilayer of model membrane vesicles causes measurable changes in internal pH *(2)*. Intravesicular pH can be measured with a fluorescent pH probe, 8-hydroxypyrene-1.3.6-trisulfonic acid (pyranine), or 2′,7′-*bis*-(2-carboxyethyl)-5-(and-6)-carboxyfluorescein (BCECF) (**Fig. 2**). Either dye responds to the arrival of protons in the aqueous compartment inside the vesicle that have been transported by unionized FA. Because FA arriving at the inner leaflet must reach ionization equilibrium, approx 50% of the FA that flip release a proton. When FA is added to the external buffer, pH probes measure the combined steps of adsorption and translocation;

it is predicted that for every four FA molecules that bind to the external leaflet of a vesicle, one H^+ is delivered into the inner volume of the vesicle. It is important to note that only the proton and not the FA molecule must desorb from the membrane in order to give this pH change.

An advantage of all the fluorescence methods considered above is that, real time measurements can be made without separation procedures. Fluorescent probes that detect adsorption can be combined with the pH assay to monitor binding and transmembrane movement simultaneously, as shown in initial studies with cells *(9,10)*.

2. Materials

2.1. Stock Solution of FA—K⁺ Salt

1. 99% pure FA in the form of powder (saturated) or oil (mono- and polyunsaturated) (Sigma, St. Louis, MO). Store at room temperature (saturated) or at −20°C (mono- and polyunsaturated).
2. Reagent grade chloroform.
3. 0.1 N potassium hydroxide solution.
4. Briefly, approx 250 mg of pure FA (*see* **Note 1**) is placed inside a glass of scintillation vial and dissolved to a concentration of 20–30 mg/mL with chloroform. Determine the dry weight of the dissolved FA in order to calculate the exact concentration of FA **(Chapter 3, section 3.1)**.
5. The FA dissolved in chloroform is dried completely under a stream of N_2 gas and lyophilized for 1 hour to remove all chloroform. The dried FA is dissolved to a concentration of 10 mM with 0.1 mM KOH. The pH of the FA stock is then adjusted to more than 10 to ensure complete dissolution of the FA into micelles (*see* **Note 2**). FA stock solutions can also be prepared to a concentration of 200 mM by dissolving FA in either ethanol or dimethyl sulfoxide (DMSO).

2.2. Stock Solution of FA Complexed to Bovine Serum Albumin

1. Bovine serum albumin (BSA). Essentially, FA-free, lyophilized power (Sigma). Store at 4°C.
2. 99% pure FA in the form of powder (saturated) or oil (mono- and polyunsaturated) (Sigma). Store at room temperature (saturated) or at −20°C (mono- and polyunsaturated).
3. Supor® Acrodisc® 0.2 µm (Gelman Sciences, Ann Arbor, MI).
4. Briefly, BSA is dissolved to a concentration of approx 3 mM in water (*see* **Note 3**) and filtered using a 0.2-µm Acrodisk. A 20-µL aliquot of BSA stock is then diluted to a total volume of 2 mL using deionized water (1:100 dilution) and absorbance is measured at 279 and 350 nm to determine the exact concentration of the BSA stock solution (*see* **Note 4**).
5. FA–BSA complex is made by diluting 50 µL of BSA stock solution to a total volume of 500 µL with water (1:10 dilution) and then adding a desired volume of 200 mM FA stock solution (in ethanol or DMSO) to the diluted BSA solution (*see* **Note 5**).

2.3. Stock Solution of cis-PA

1. *cis*-PA (Molecular Probes, Eugene, OR) is dissolved to 40 mM in DMSO **(Fig. 2)**. Store at −20°C.

2.4. Preparation of Small Unilamellar Vesicles Containing a pH Probe

1. Phosphatidylcholine isolated from chicken egg (eggPC) (Avanti Polar Lipids, Alabaster, AL) and dissolved in chloroform to 20 mg/mL. Tightly seal the cover of the bottle and store at −20°C (*see* **Note 6**).
2. 50-mL glass round-bottom flasks and 3-mm glass beads.
3. HEPES/KOH buffer: 50 mM HEPES titrated with KOH, pH 7.4 (*see* **Note 7**). Buffer is stable at 4°C for up to 3 wk. Warm to working temperature of 25°C before fluorescence experiments.
4. Pyranine (trisodium salt) **(Fig. 2)** (Molecular Probes) is dissolved in water to 10 mM. Store at −20°C (*see* **Note 8**).
5. Sephadex G-25 (Amersham Biosciences/GE Healthcare, Piscataway, NJ).

Table 1
Fluorescence Probes Used to Study FA Transport in Membranes

Fluorescence probe	Location in the vesicle	Steps monitored in a system where free FA is added to vesicles
Pyranine/BCECF acid	Internal buffer	Adsorption and flip–flop
FPE	Outer leaflet	Adsorption
cis-PA	Outer and inner leaflet	Adsorption
ADIFAB	External buffer	Adsorption
	Internal buffer	Adsorption, flip–flop, and desorption

6. 50-mL size-exclusion column with a stopcock to regulate the flow of eluate.
7. Open-ended UltraClear™ centrifuge tubes ($\frac{1}{2} \times 2\frac{1}{2}$ in^2. or 13×64 mm^2) (Beckman, Fullerton, CA).
8. Sonicator (Sonifier® Cell Disruptor 350, Branson Sonic Power Co., Danbury, CT).

2.5. Preparation of Small Unilamellar Vesicles Containing the Surface pH Probe FPE

1. Materials as described in **steps 1–3**, *see* **Subheading 2.4.**
2. Fluorescein FPE (thiethylammonium salt) (Molecular Probes) is dissolved to 2.143 mg/mL in a 5:1 (v/v) chloroform:methanol solution **(Fig. 2)**. Store at –20°C.

2.6. Nigericin pH Calibration of Small Unilamellar Vesicles

1. Materials: as described in **Subheading 2.4.**, for preparing small unilamellar vesicles (SUV) containing a pH probe.
2. Nigericin (Sigma) is dissolved to 1.4 mg/mL in absolute ethanol. Store at 4°C.
3. 8 *M* Potassium hydroxide solution.
4. 8 *M* Sulfuric acid solution.

2.7. Fluorescence Measurements

1. Polystyrene cuvets with four clear sides and spinbars (spectral cell stir bars) (Fisher Scientific, Pittsburgh, PA).
2. Gel-loading pipet tips (1–200 µL) (VWR Scientific, Bridgeport, NJ).
3. Stopped-flow apparatus equipped with pressure-driven pneumatic drive unit and two glass syringes (2 mL volume capacity; total sample mixing volume of 400 µL) (Hi-Tech Scientific, Salisbury, UK).

3. Methods

The laboratory has applied a series of novel fluorescence assays to monitor directly the movement of FA through model and biological membranes **(Table 1)**. These assays use a variety of fluorescent probes to discriminate between the adsorption and transbilayer movement of FA in membranes *(1,2,4–7,11,12)*. The structures of these probes are shown in **Fig. 2** and in Chapter 3 of this volume, **Fig. 1** (ADIFAB). ADIFAB, PA, and FPE can be used to measure partitioning or binding of FA to the outer membrane leaflet (adsorption). The pH-sensitive fluorophores, pyranine, and BCECF acid, both detect protons released by unionized FA as they diffuse through the membrane (transmembrane movement) *(2)*. Studies performed with this assay suggest that FA can diffuse rapidly across model lipid membranes *(1,2,5–7,12,13)* and across the plasma membranes of isolated rat adipocytes *(11,14)*, undifferentiated and differentiated 3T3-L1 adipocytes (*unpublished*; J.A. Hamilton), and HepG2 cells *(15,16)*. These studies also reveal that FA diffusion into isolated rat adipocytes is rapid enough to supply the intracellular metabolism of FA into triacylglycerol *(14)*.

However, in order to accurately interpret the data from these studies, it is important to find the caveats and define the limitations of each fluorescence approach. The pH assay has been widely used in model systems and in cells to establish the kinetics of FA flip–flop across membranes. Some potential artifacts have been described, but these did not alter the overall conclusions *(17,18)*. The use of PA fluorescence to monitor FA binding to membranes predates the pH assay. Recently, it was reported that the PA assay, when used in conjunction with other molecules, can be subject to artifacts that do alter conclusions *(16)*. The more recent FPE assay for FA binding has not yet been described, as to its usefulness or its limitations. Here, typical experiments involving each of these probes are presented. Also investigated are (1) the effect of the proposed inhibitor of FA transport, phloretin, on the efficacy of PA for FA binding to protein-free model membranes and (2) the report that external untrapped pyranine produces artifacts in the pH assay of FA flip–flop *(17)*.

3.1. Preparation of SUV Containing a pH Probe

1. Place 50 mg of eggPC in chloroform (20 mg/ml) into a 50-mL round-bottom flask and dry under a vacuum in a rotary evaporator (*see* **Note 9**).
2. Place the dried lipid film in a lyophilizer for 1 h to remove all traces of chloroform.
3. Prepare a 0.1 m*M* pyranine solution by diluting 20 µL of the 10 m*M* pyranine stock solution to a total volume of 2 mL with HEPES/KOH buffer (*see* **Note 10**).
4. Remove the lipids from the lyophilizer and place 5–10 glass beads and 2 mL of HEPES/KOH buffer containing 0.1 m*M* pyranine into the round-bottom flask. Gently swirl the mixture by hand until all lipids have been removed from the sides of the flask. The lipid mixture will appear green and opaque. Purge the flask with N_2 and seal it with parafilm (*see* **Note 6**).
5. Allow the lipids to hydrate either overnight at 4°C or for 1 h at room temperature (*see* **Note 11**).
6. Weigh out 4.3 g of Sephadex 25 and place it inside a 50-mL conical tube with 40–50 mL of HEPES/KOH buffer. Cover the tube and shake it gently until all of the gel is mixed with the buffer. Allow the gel to hydrate using conditions similar to those used to hydrate the lipids in **step 5** (*see* **Note 12**).
7. Transfer the hydrated lipids to an open-ended polypropylene centrifuge tube. Place the tip of the sonicator into the lipid mixture and center it approx 1 mm from the bottom of the tube (*see* **Note 13**). Deliver a light flow of N_2 gas into the centrifuge tube through a small rubber tube connected to a pressurized gas tank (*see* **Note 6**). The rubber tubing is held in place by a piece of Teflon tape that is stretched around the sonicator tip and covers the tube (*see* **Note 14**).
8. With the sample tube partially submerged in an ice-water bath, pulse-sonicate for 1 h (output control 3, 30% duty cycle), or until the suspension changes from highly turbid to nearly clear in appearance (*see* **Note 15**). Following sonication, SUV are formed when the lipid mixture has become completely translucent.
9. Centrifuge the SUV suspension for 30 min at approx 5000*g* (*see* **Note 16**).
10. Add the hydrated Sephadex gel to a 50-mL size-exclusion column and allow the gel to settle into the column. Allow the buffer to elute from the column until it reaches the top of the packed gel.
11. Add the centrifuged SUV suspension to the column and allow it to sink completely into the gel. Add approx 1 mm of pure sand to the column (*see* **Note 17**). Fill the remainder of the column with HEPES/KOH buffer.
12. Collect the eluting SUV suspension into a 15-mL conical tube (*see* **Note 18**). Generally, 1.5 mL of SUV suspension is obtained for each 1 mL of SUV suspension added to the column.
13. Estimate the final eggPC concentration in SUV based on the quantity of eggPC dried and the volume of SUV collected (*see* **Note 19**).

3.2. Preparation of SUV Containing the Surface Probe FPE on the Outer Leaflet of the Vesicle

1. Dry the lipids by performing **steps 1** and **2**, *see* **Subheading 3.1.**
2. Remove the lipids from the lyophilizer and place 5–10 glass beads and 2 mL of HEPES/KOH buffer (pyranine-free) into the round-bottom flask (*see* **Note 20**). Gently swirl the mixture by hand until all lipids have been removed from the sides of the flask. The lipid mixture will appear white and opaque. Purge the flask with N_2 and seal it with parafilm (*see* **Note 6**).
3. Allow the lipids to hydrate either overnight at 4°C or for 1 h at room temperature (*see* **Note 11**).
4. Sonicate and centrifuge the lipids as described in **steps 7–9**, *see* **Subheading 3.1.** (*see* **Note 21**).
5. Dry an aliquot of the FPE stock solution under a steady stream of N_2 gas (*see* **Note 22**).
6. Resuspend the dried FPE in an equivalent volume of 100% ethanol or DMSO, add it to the SUV suspension (i.e., if 70 µL was dried, resuspend in 70 µL ethanol or DMSO), gently vortex and incubate for 1 h in the dark at room temperature (*see* **Note 23**).
7. Estimate the final eggPC concentration in SUV as described in **step 13**, *see* **Subheading 3.1.**

3.3. Nigericin pH Calibration of SUV

1. Prepare SUV entrapped with a pH probe as described in **Subheading 3.1.**
2. Add a desired volume of the SUV suspension to a Fisher polystyrene cuvet containing a spinbar and bring to a total volume of 3 mL with HEPES/KOH buffer. Turn on the magnetic stir plate of the fluorimeter to rapidly stir the sample.
3. Add 1 µL of nigericin stock solution to the cuvet (*see* **Note 24**).
4. Open the fluorescence software program Datamax for Windows (Jobin Yvon, Edison, NJ) and select the fluorescence protocol for measuring the emission of the pH probe.
5. Measure the fluorescence until a stable baseline is established. Add 1–2 µL aliquots of 8 M sulfuric acid solution to the cuvet to reduce the pH in a step-wise fashion. Measure the fluorescence after the change in pH until a new stable baseline is achieved.
6. Using a fresh suspension of SUV and nigericin, repeat measurements using the step-wise addition of 8 M potassium hydroxide stock solution. After the change in pH, measure the fluorescence until a new stable baseline is achieved.
7. Plot a pH calibration curve (pH vs fluorescence) as shown in **Fig. 3**.

3.4. Measurement of FA Binding and Diffusion Through Membranes by Online Fluorescence Spectroscopy—Single Probe

3.4.1. pH Probe—Pyranine

1. Prepare SUV containing entrapped pyranine as described in **Subheading 3.1.**
2. Add the desired amount of SUV suspension to a polystyrene cuvet containing a spinbar and 3 mL of HEPES/KOH buffer.
3. Turn on the magnetic stir plate of the fluorimeter to rapidly stir the sample and open the excitation and emission slits to 3 nm.
4. Place a cuvet into the fluorimeter, open a fluorescence software program and select the desired fluorescence protocol for measuring pyranine fluorescence (excitation: 455 nm and emission: 509 nm).
5. When a stable fluorescence baseline is established, add a desired volume of a K^+-oleate stock solution using a micropipet equipped with a gel-loading tip (or a Hamilton syringe [Hamilton Company, Reno, NV]). After the pH change, measure the fluorescence until a new stable baseline is achieved (*see* **Note 25**). **Figure 4A** illustrates typical data from this assay.
6. Repeat **step 4** with various concentrations of FA to demonstrate the sensitivity and dose-dependant response of pyranine to FA (**Fig. 4B**).

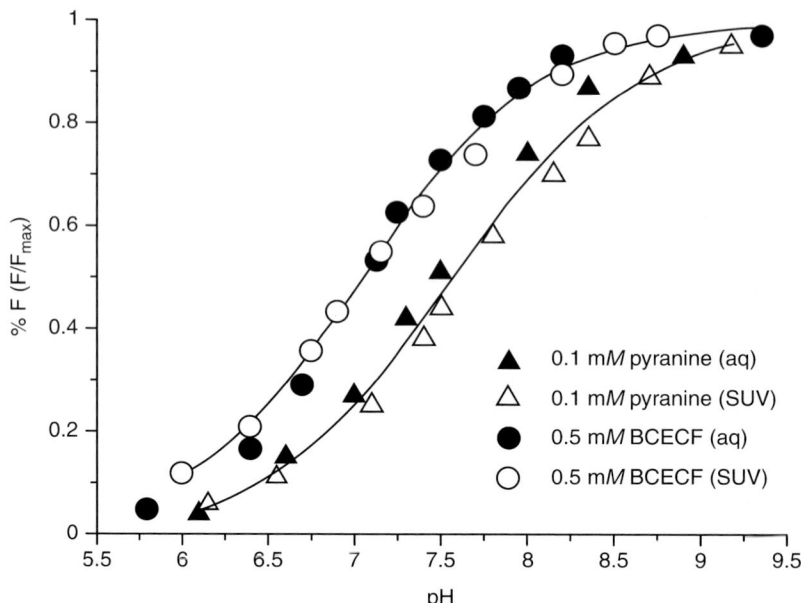

Fig. 3. Calibration of BCECF and pyranine fluorescence with pH. Both probes titrate similarly when placed in 50 m*M* HEPES/KOH buffer (*closed symbols*) or trapped within SUV that have been permeablized to protons by nigericin (*open symbols*). The pK_a of pyranine (*triangles*) and BCECF (*circles*) are 7.5 and 7.2, respectively. These curves demonstrate that each probe is not affected by the environment in which the probe is placed during the experiment.

3.4.2. Surface Membrane Potential Probe for Detecting the Binding of the FA Carboxyl at the Membrane–Water Interface—FPE

1. Prepare SUV labeled with FPE as described in **Subheading 3.2.**
2. Add the desired amount of SUV suspension to a polystyrene cuvet containing a spinbar and 3 mL of HEPES/KOH buffer.
3. Turn on the magnetic stir plate of the fluorimeter to rapidly stir the sample and open the excitation and emission slits to 3 nm.
4. Place a cuvet into the fluorimeter, open a fluorescence software program and select the desired fluorescence protocol for measuring FPE fluorescence (excitation: 490 nm and emission: 520 nm).
5. When a stable fluorescence baseline is established, deliver different concentrations of K^+-oleate to the cuvet, as described in **step 5**, *see* **Subheading 3.4.1.** (*see* **Note 26**). Typical data from this assay are shown in **Fig. 5**.

3.4.3. Probe for Measuring Insertion of the FA Hydrocarbon Chain—cis-PA

1. Prepare SUV as described in **Subheading 3.1.**, with or without the entrapped pH probe.
2. Add the desired amount of SUV suspension to a polystyrene cuvet containing a spinbar and 3 mL of HEPES/KOH buffer.
3. Turn on the magnetic stir plate of the fluorimeter to rapidly stir the sample and open the excitation and emission slits to 3 nm.
4. Place a cuvet into the fluorimeter, open a fluorescence software program, and select the desired fluorescence protocol for measuring *cis*-PA fluorescence (excitation 303 nm and emission 416 nm) (*see* **Note 27**).
5. When a stable fluorescence baseline is established, deliver the desired concentrations of *cis*-PA to the cuvet as described in **step 5**, *see* **Subheading 3.4.1.** (*see* **Note 28**). Typical data from this assay are shown in **Fig. 6**.

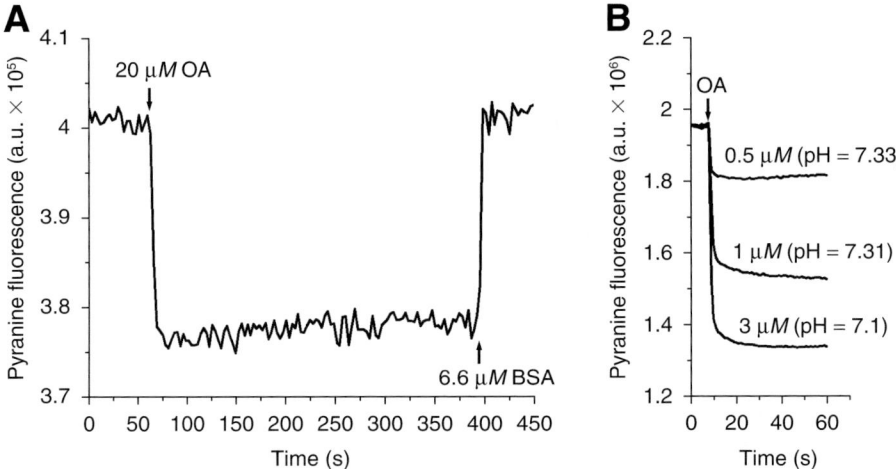

Fig. 4. The pyranine assay detects the arrival of FA at the inner membrane leaflet. (**A**) a classical result is presented; addition of K$^+$-oleate (20 µM) to a suspension of SUV (~700 µM eggPC) containing pyranine (0.1 mM) produces a rapid decrease in pyranine fluorescence (<2 s), or pH drop. This drop is followed by a gradual increase in pyranine fluorescence as protons slowly leak out of SUV down their concentration gradient and the pH inside the SUV slowly increases. Addition of BSA (6.6 µM) extracts all of the oleate from the SUV (3 high affinity FA sites per albumin molecule) and rapidly (<2 s) restores the internal pH and pyranine fluorescence to their original levels. (**B**) Changes in pyranine fluorescence are dose-dependant with respect to the amount of FA added (0.5, 1, and 3 µM of K$^+$-oleate) to the SUV suspension (100 µM eggPC; 0.5 mM entrapped pyranine). Using this assay, it has been possible to detect flip–flop of oleate at concentrations well below its solubility at pH 7.4 (as low as ~0.5 µM), and show that the pH drops seen at higher oleic acid concentrations are not artifacts. A pH calibration with nigericin was used to relate changes in pyranine fluorescence to the intravesicular acidification that occurs on addition of FA to a vesicle suspension.

3.4.4. Probe for Measuring the Binding and Partitioning of FA Into Membranes—ADIFAB

1. Refer to Chapter 3 for a complete overview of the uses of ADIFAB for measuring FA binding and partitioning into lipid membranes.

3.5. Online Fluorescence Spectroscopy–Multiple Probes

1. As described in **Subheading 3.1.**, prepare two SUV suspensions: one containing pyranine and other containing BCECF.
2. Add FPE to the SUV suspension with pyranine as described in **steps 5** and **6**, see **Subheading 3.2.**
3. Add the desired amount of the suspension of SUV containing FPE and pyranine to a polystyrene cuvette containing a spinbar and 3.0 ml of Hepes/KOH Buffer.
4. Turn on the magnetic stir plate of the fluorimeter to rapidly stir the sample and open the excitation and emission slits to 3 nm. Select the desired fluorescence protocol for measuring dual-fluorescence emissions of FPE and pyranine.
5. When a stable fluorescence baseline is established, add a single (or multiple) dose of K$^+$-oleate to the cuvette (**Fig. 7A-B**).
6. Repeat **steps 3** and **4** using the suspension of SUV containing BCECF. Select the desired fluorescence protocol for measuring dual-fluorescence emissions of BCECF and *cis*-parinaric acid.
7. When a stable fluorescence baseline is established, add various concentrations of *cis*-PA to the cuvette (**Fig. 7C-D**).

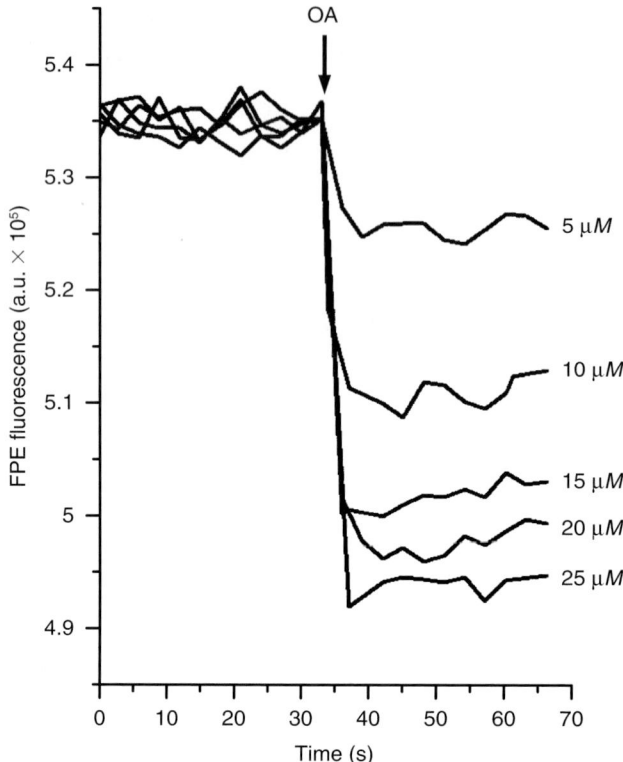

Fig. 5. The FPE assay detects the binding of FA to membranes, specifically the arrival of the FA carboxyl at the aqueous interface. Addition of K^+-oleate (5, 10, 15, 20, and 25 μM) to a suspension of FPE-labeled SUV (~700 μM eggPC, 0.2 mol% FPE) produces dose-dependant and rapid decreases in FPE fluorescence (<2 s). In this assay, changes in FPE fluorescence reflect the event of FA adsorption as this probe is situated in the outer membrane leaflet. However, the equilibrium fluorescence of FPE, reached within the time resolution of the experiments (1–2 s), reports the net-negative charge in the outer leaflet, which is expected to change with flip–flop, according to the model. Although the probe does not reveal direct information about the transmembrane movement of FA, the results indicate that flip–flop is fast; otherwise, a slow component would be observed following the initial binding step.

3.6. Stopped-Flow Fluorescence Spectroscopy

3.6.1. Measurement of the Transmembrane Movement of FA Delivered Unbound or Complexed to a Donor (Vesicles or BSA)

1. As described in **Subheading 3.1.**, prepare two SUV suspensions: one containing entrapped pyranine, the other without pyranine.
2. Add a desired amount of K^+-oleate to the suspension of SUV without pyranine and incubate overnight at 4°C (*see* **Note 30**).
3. Prepare a solution of K^+-oleate complexed to BSA, as described in **Subheading 2.2.**
4. Prepare diluted suspensions of SUV (complexed with oleate) and SUV-containing pyranine (containing no FA) by combining the desired amount of vesicle stock and HEPES/KOH buffer in a 50-mL conical tube (*see* **Note 31**).
5. Prepare a diluted suspension of BSA (complexed with oleate) by combining the desired amount of the BSA complex and HEPES/KOH buffer in a 50-mL conical tube.

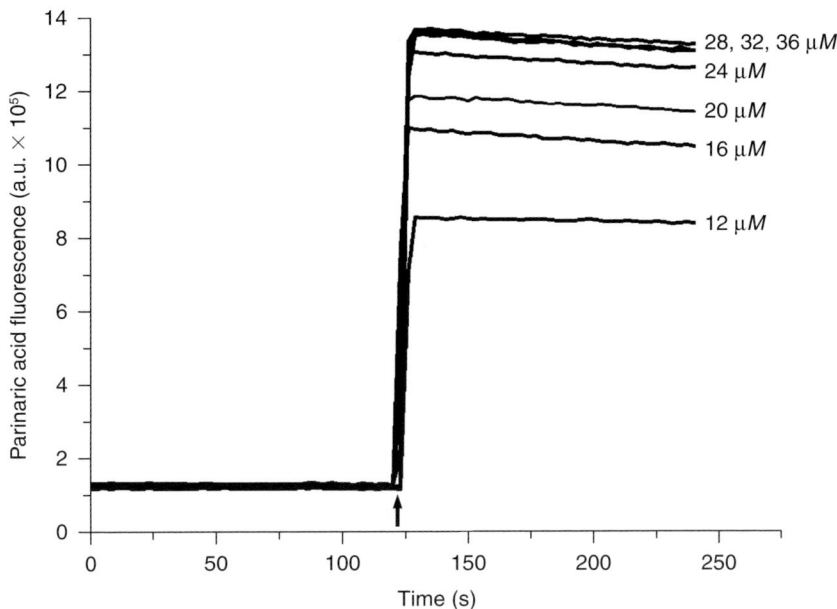

Fig. 6. The *cis*-PA assay detects the binding and insertion of the FA hydrocarbon tail into the lipid bilayer. Addition of PA (12, 16, 20, 24, 28, 32, and 36 μM) to a suspension of SUV (~700 μM eggPC) produces dose-dependant and rapid increases in PA fluorescence. Equilibrium fluorescence values are reached within 2 s. In this assay, changes in PA fluorescence reflect only the event of FA adsorption. This probe does not reveal any information about the transmembrane movement of FA. Note that fluorescence level reaches an apparent saturation; this effect is not owing to limited binding capacity of the egg PC for the PA, but to self-quenching of the PA in the membrane *(16)*.

6. Prepare a diluted solution of oleate (uncomplexed) by combining the desired amount of K^+-oleate stock and HEPES/KOH buffer in a 15-mL conical tube (*see* **Note 32**).
7. Load the suspension of SUV containing pyranine into one of the two stopped-flow syringes and the uncomplexed FA solution into the other syringe.
8. Open a fluorescence software program and select the desired fluorescence protocol for measuring pyranine fluorescence by stopped-flow (*see* **Note 33**). Open the excitation slits to at least 5 nm and emission slits to at least 10 nm (*see* **Note 34**).
9. Rapidly mix the two suspensions together under high pressure five to six times to wash the sample through the stopped-flow apparatus and into the closed cuvet inside the fluorimeter (*see* **Note 35**).
10. Mix the suspensions from each syringe while monitoring pyranine fluorescence. Repeat this measurement at least five times and average the spectra to further improve signal-to-noise.
11. Repeat **steps 7–10** for oleate complexed to SUV and BSA, washing before measurements are made for FA bound to different donor vehicles. Typical pyranine data obtained by stopped-flow spectroscopy for oleate delivered to SUV complexed to different vehicles is shown in **Fig. 8**.

3.6.2. Measurement of FA Binding and Transmembrane Movement with FPE and Pyranine in the Same Vesicle

1. Prepare two SUV suspensions: one labeled with FPE and containing entrapped pyranine (both probes in same preparation), and other without any fluorescence probe, as described in **Subheadings 3.1.** and **3.2.**
2. Add a desired amount of K^+-oleate to the suspension of SUV without any fluorescence probe.

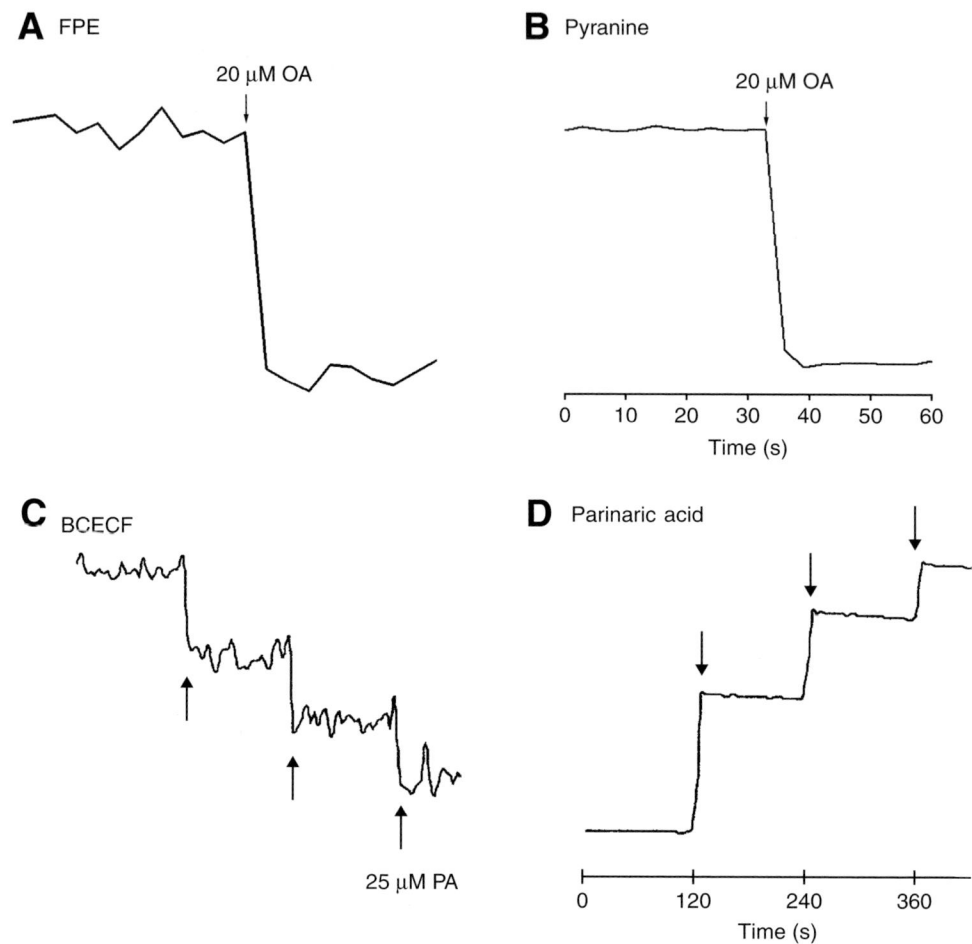

Fig. 7. Dual-fluoresJ254cence assays utilize two probes to monitor simultaneously the distinct processes of FA adsorption to, and diffusion across, a phospholipid bilayer. Addition of K$^+$-oleate (20 µM) to a suspension of FPE-labeled SUV (~700 µM eggPC; 1 mol% FPE) containing entrapped pyranine (0.1 mM) produces simultaneous and rapid decreases (<2 s) in the fluorescence of FPE (**A**) and pyranine (**B**). The simultaneous fluorescence change for both external and internal probes suggests that FA rapidly diffuse through membranes after binding to the outer leaflet. Similar results are obtained using *cis*-parinaric and BCECF. Addition of PA (3 × 25 µM doses) to a suspension of SUV (~700 µM eggPC) containing entrapped BCECF (0.5 mM) produces a dose-dependant, simultaneous, and rapid change (<2 s) in the fluorescence of BCECF (**C**) and PA (**D**). Again, this assay suggests that FA diffusion through membranes is extremely fast. The *y*-axis show changes in fluorescence (units not shown). Pyranine and BCECF can both be used as a pH probe, as the figure illustrates the sensitivity of pyranine is higher.

3. Prepare diluted suspensions of SUV with FPE and pyranine, as described in **step 4**, *see* **Subheading 3.6.1.**
4. Load the diluted suspension of SUV with FPE and pyranine in one of the two syringes and the other SUV suspension witout any probe into the other syringe of the stopped-flow apparatus.
5. Open a fluorescence software program to measure pyranine fluorescence by stopped-flow, wash the sample through the stopped-flow apparatus, and then mix suspensions form each syringe while monitoring pyranine fluorescence as described in **steps 8, 9** and **10**, *see* **Subheading 3.6.1.**
6. Repeat **step 5** while measuring FPE fluorescence (*see* **Notes 36** and **37**). Typical data obtained by stopped-flow spectroscopy for the binding and diffusion of oleate into SUV are shown in **Fig. 9A,B**.

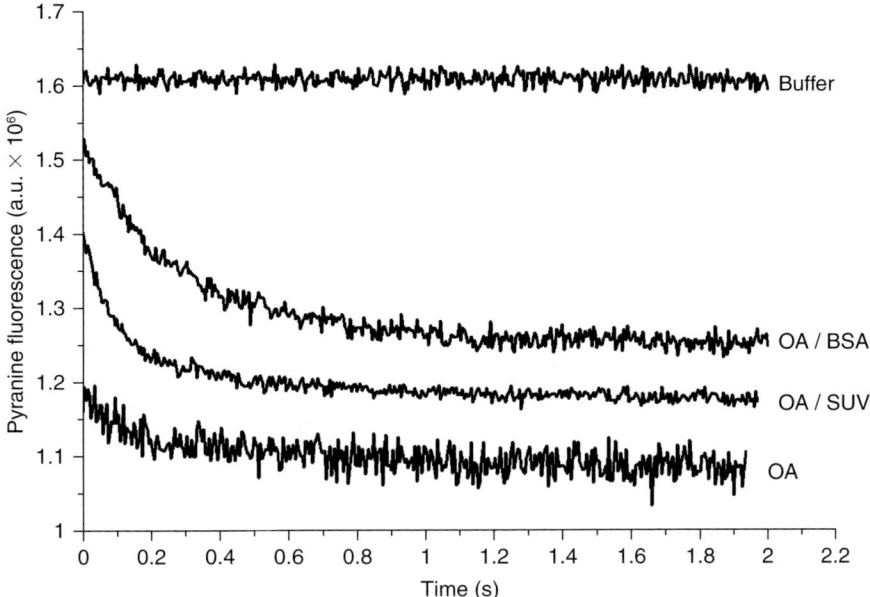

Fig. 8. Stopped-flow fluorescence assay for measuring the transfer of K^+-oleate to SUV from different donors. Solutions containing uncomplexed oleate (6 μM) **(bottom trace)**, oleate complexed to donor SUV (6 mol% FA; 100 μM eggPC) **(middle trace)**, and oleate complexed to BSA (8:1 FA/BSA ratio; 25 μM BSA) **(upper trace)** were rapidly mixed with a suspension of acceptor vesicles (100 μM eggPC) containing pyranine (0.1 mM). In these assays, changes in pyranine fluorescence reflect the steps of FA desorption from the donor (SUV or BSA), binding to the membrane, and diffusion to the inner leaflet. In each case, oleate produces rapid decreases in pyranine fluorescence. The $t_{1/2}$ of pyranine fluorescence change was <10, ~100, and ~205 ms for uncomplexed K^+-oleate, oleate/SUV complexes, and oleate/BSA complexes, respectively. In the case of uncomplexed oleic acid, most of the fluorescence decrease occurred within the dead time/mixing time of the instrument. The kinetics of oleate transfer depends on the donor used to deliver the oleate to the acceptor SUV. These results suggest that FA flip–flop is extremely fast and that desorption is the slowest step of the transfer process. Mixing dead time was 10 ms. All fluorescence traces are the average of four to eight measurements.

4. Notes

1. Density must be used to calculate the amount of pure mono- and polyunsaturated FA because these FA are liquid at room temperature.
2. To produce a 10 mM FA stock solution in ethanol or DMSO, simply dissolve the lyophilized FA in either solvent.
3. To prevent the generation of foam, do not vortex the BSA solution.
4. The exact BSA concentration (in milligram per milliliter) is calculated according to the following equation:

$$BSA = (A_{279} - A_{350}/0.667 \text{ mg/mL}) \times 100$$

A_{279} and A_{350} are the absorbance of BSA measured at 279 and 350 nm, respectively; 0.667 mg/mL is the extinction coefficient of BSA; and 100 is the correction for dilution.

5. Do not add more than 10 μL of FA dissolved in either ethanol or DMSO to the BSA solution to avoid any unknown effects that these solvents may have on BSA. More advanced protocols for creating FA/albumin complexes have also been reported elsewhere *(16)*.
6. To help prevent oxidation of unsaturated lipid chains, trap either N_2 or Ar gas inside the bottle of eggPC and store at −20°C.

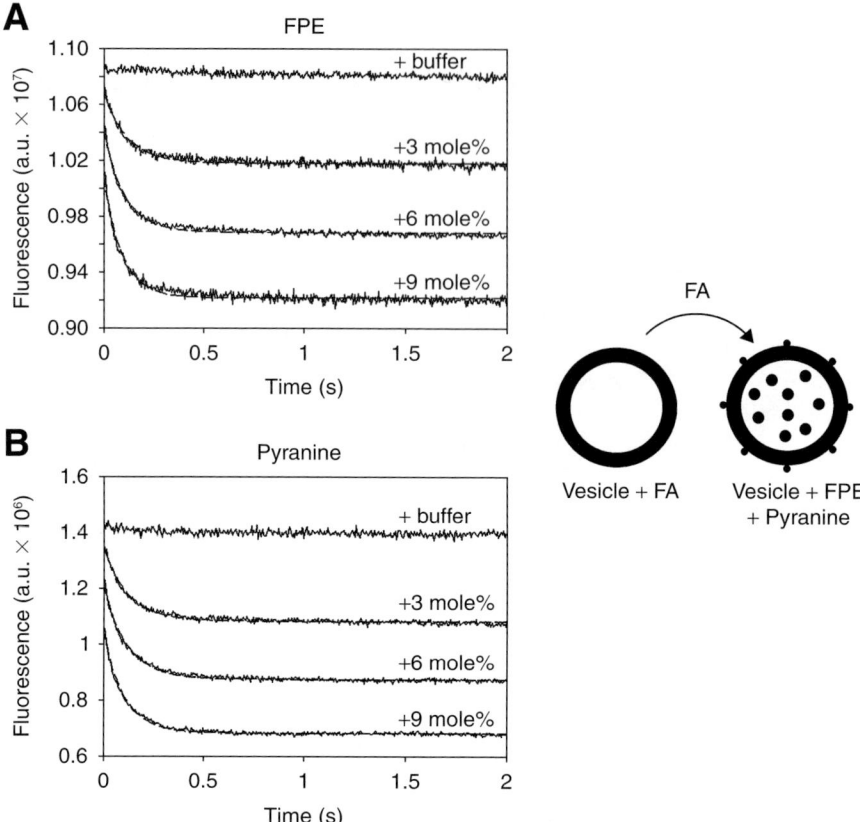

Fig. 9. Stopped-flow fluorescence assay for measuring the transfer of oleate between SUV using acceptor SUV with both FPE and pyranine. The transfer of oleate from donor SUV to acceptor SUV can be monitored by rapidly mixing SUV (100 μM eggPC) loaded with variable amounts of oleate (6, 12, and 18 mol%) and FPE-labeled SUV (100 μM eggPC; 1 mol% FPE), as well as trapped pyranine (0.1 mM). A dose-dependent and rapid decrease in FPE (**A**) and pyranine (**B**) fluorescence is observed with similar kinetics ($t_{1/2}$ ~ 100 msec), The change in the FPE reflects arrival of FA that have desorbed from the donor SUV; the change in pH requires the additional step of transmembrane movement of the FA. Thus, flip–flop occurs as soon as the FA binds to the outer leaflet of the SUV (with this time resolution), and FA desorption is slower than the combined steps of adsorption and flip–flop. HEPES/KOH buffer (free of FA/SUV complex) was used as a mixing control. Mixing dead time was 10 ms. All fluorescence traces are the average of four to eight measurements.

7. HEPES concentration can range between 20 and 100 mM. Modify the buffer concentration to optimize signal-to-noise when measuring changes in pH. Avoid using buffer with Cl⁻ anions, which may make lipid vesicles more permeable to protons. However, Cl⁻ anions are required in experiments involving ADIFAB (*see* Chapter 3).
8. If the pH probe BCECF (Molecular Probes) (**Fig. 2**) is used in these experiments, prepare a stock solution of 1 mM BCECF acid in HEPES/KOH buffer (use same buffer concentration that is used for the SUV preparation). Store at −20°C.
9. During this time, the round-bottom flask containing the lipids is spun over a heated water bath to facilitate evaporation of the chloroform and create a thin film of dried eggPC.
10. Pyranine concentration should range from 0.1 to 0.5 mM. Better signal-to-noise is observed with higher pyranine concentrations, which is particularly useful for insensitive techniques such as

stopped-flow fluorescence (*see* **Subheading 3.6.**). In typical online fluorescence experiments, high pyranine concentrations tend to increase fluorescence drift. If the pH probe BCECF acid is used, a 0.5 mM BCECF solution is typically prepared by mixing 1 mL of 1 mM BCECF acid stock solution with 1 mL of HEPES/KOH buffer. The signal-to-noise is typically much poorer with BCECF. However, BCECF can be used in cellular systems by incubating cells with BCECF-acetoxymethyl ester (Molecular Probes), whereas pyranine cannot be introduced into intact cells.

11. The lipid hydration has to be carried out above the gel–liquid crystal transition temperature (Tm) of the lipid. The Tm of eggPC is approx 0°C.
12. To save time, this step should be carried out simultaneously with the lipid hydration step.
13. It has been found that this placement of the sonicator tip maximizes the dispersal of energy throughout the entire sample, whereas minimizing the loss of sample through splashing.
14. Care must be taken not to wrap the Teflon tape around the sonicator tip too tightly so that the N_2 gas can escape.
15. The sonication has to be carried out above the Tm of the lipid. The ice-water bath prevents overheating the sample, and consequently, the breakdown of lipid acyl chains with generation of free FA.
16. Centrifugation removes any metal particulates released from the sonicator tip during the sonication step.
17. The layer of sand is used to protect the top of the gel from becoming perturbed when HEPES/KOH buffer is added to the column.
18. The gel filtration eliminates nearly all of the pyranine remaining in the external buffer. It has been proposed that some pyranine may remain tightly adsorbed to the outer leaflet of the vesicles, which may lead to fluorescent artifacts when monitoring changes in the pH inside vesicles *(17)*. The data shown in **Fig. 10** do not support this idea. Small volumes of FA stock (pH >10.0) added to SUV suspended in 20–100 mM HEPES/KOH buffer do not change the pH outside of the vesicle. In fact, the presence of only 50 µM pyranine in the external buffer can mask the detection of pH changes inside SUV containing 500 µM pyranine. Although in this example, the concentration of pyranine outside the SUV is 10-fold lower, the volume in which pyranine is dissolved is quite large compared with the small internal volume of SUV. Thus, there are far more molecules of pyranine outside the SUV than inside, and it is the signal from external pyranine that is detected by the fluorimeter. The authors contend that if pyranine were in fact present outside of the SUV, it would reduce the sensitivity of the measurements rather than enhance or create fluorescence changes that are artifacts.
19. The exact eggPC concentration is determined by assaying total phosphorus content *(20)*.
20. For dual fluorescence measurements using FPE and a pH probe in the same vesicle, add the pH probe in this step as described in **steps 3** and **4**, *see* **Subheading 3.1.**
21. For dual fluorescence measurements using FPE and a pH probe, perform the gel filtration as described in **steps 10–12**, *see* **Subheading 3.1.**
22. Typically, 73 µL of the FPE stock is dried per 50 mg of eggPC used to make SUV. The final concentration of FPE in the SUV suspension is 0.2 mol% of total lipid, but the authors have successfully used up to 1 mol% FPE.
23. The FPE incorporation to vesicles has to be carried out above the gel–liquid crystal transition temperature (Tm) of the lipid. There is no need to perform gel filtration as FPE has a high solubility in lipids, and consequently, all FPE added binds to the external leaflet of the vesicles. FPE is located only in the outer membrane leaflet. Diffusion of FPE to the internal leaflet is extremely slow (≥1 wk) *(3)*.
24. To completely permeabilize vesicles to protons, typically 1 µg of nigericin/mg of eggPC is used.
25. The signal-to-noise in pyranine fluorescence traces can be increased by: (1) reducing the buffer concentration to allow for more robust pH changes inside the vesicle and (2) opening the emission and excitation slits.
26. The signal-to-noise in FPE fluorescence traces can be increased by: (1) increasing the mol% of FPE in the outer membrane leaflet and (2) opening the emission and excitation slits.

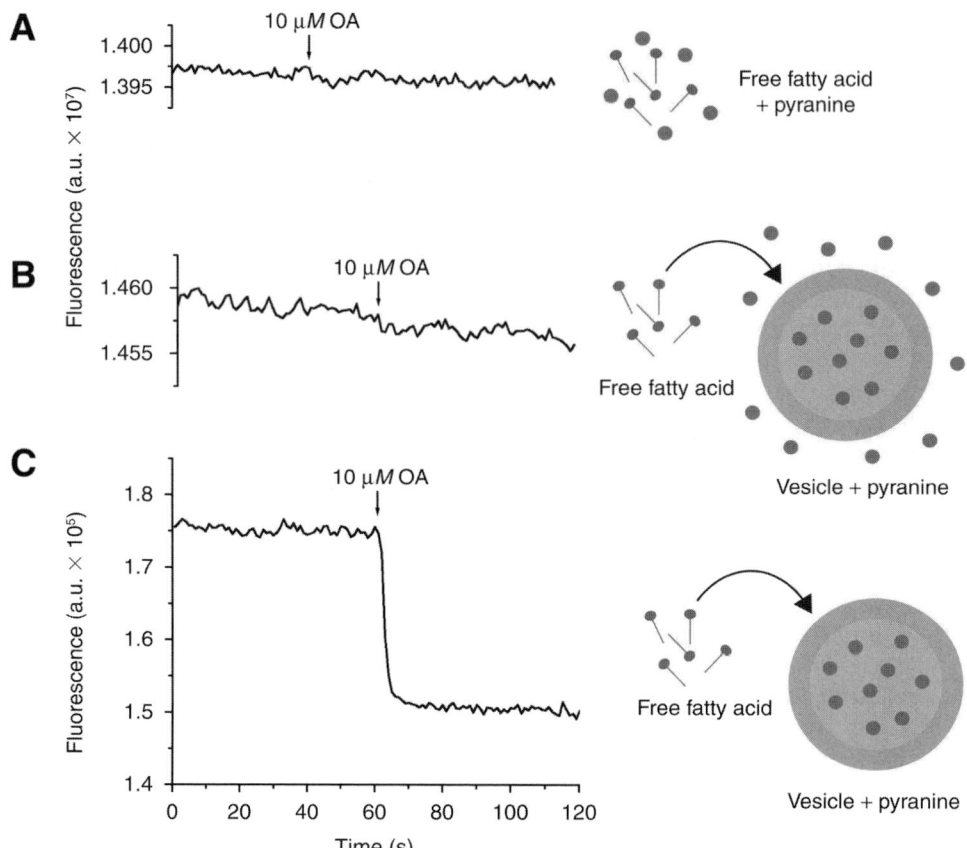

Fig. 10. The effect of external pyranine on measurements of FA diffusion through membranes. **(A)** The addition of K$^+$-oleate (10 μM) to a solution of 50 μM pyranine in 100 mM HEPES/KOH buffer (pH 7.4) without vesicles does not produce a change in pyranine fluorescence. This demonstrates the strong buffering capacity of the buffer against small volumes of 0.1 mM KOH that are typically added when delivering uncomplexed oleate to a suspension of SUV. **(B)** The addition of K$^+$-oleate (10 μM) to a suspension of SUV (500 μM eggPC) containing 0.5 mM internally trapped pyranine in 100 mM HEPES/KOH buffer together with 50 μM external pyranine also fails to cause a significant change in pyranine fluorescence. **(C)** Alternatively, the addition of K$^+$-oleate (10 μM) to a suspension of SUV (500 μM eggPC) containing 0.5 mM internally trapped pyranine in 100 mM HEPES/KOH buffer in the absence of external pyranine resulted in large rapid decrease in pyranine fluorescence (<2 s). Note that the 100-fold lower pyranine emission intensity in this sample is due to the absence of external pyranine. The data in panels **B,C** suggest that an excess of external pyranine molecules could mask the detection of intravesicular pH changes produced by FA diffusion through membranes, and that the decrease in pyranine fluorescence that is normally observed in our an assays cannot be ascribed to the presence of contaminating pyranine in the external buffer. In addition, should the FA result in disruption of the bilayer and increased permeability of the membrane to pyranine, pyranine would diffuse into the external buffer, and a decrease in fluorescence would not be seen.

27. The signal-to-noise in *cis*-PA fluorescence traces can also be increased by opening the emission and excitation slits. Increasing the quantity of *cis*- PA that is added to vesicles can also increase the signal-to-noise. However, it is important to note that *cis*-PA self-quenches at high concentrations when it is in a membrane *(16)*. Evidence has also been found that *cis*-PA should not be used in conjunction with phloretin, a proposed inhibitor of FA transport into cells, which

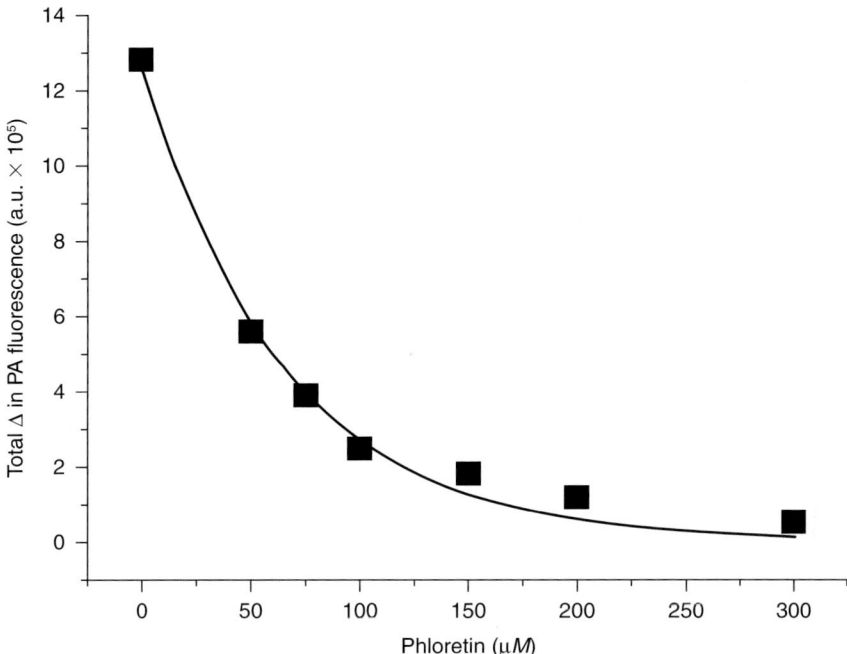

Fig. 11. Phloretin quenches the fluorescence of *cis*-PA in membranes. *Cis*-PA (40 µ*M*) was added to a suspension of SUV (~700 µ*M* eggPC) in the presence of increasing concentrations of phloretin (0, 50, 75, 100, 150, 200, and 300 µ*M*). The fluorescence of *cis*-PA normally increases rapidly on binding to membranes (**Fig. 6**). For **Fig. 11**, the magnitude of this fluorescence increase was measured in the presence of phloretin and plotted against the concentration of phloretin. As observed with HepG2 cells *(13)*, the fluorescence of *cis*-PA in the presence of 300 µ*M* phloretin, was completely quenched. In this example, there are no proteins present that could serve as transporters; thus, the effects of phloretin occur in the lipid bilayer, and are independent of proteins. The data were fitted with a firs order exponential function.

quenches cis-PA. One study *(20)* suggested that *cis*-PA is taken up into cells by a protein-mediated transport mechanism, which is inhibited by phloretin (i.e., decreased *cis*-PA fluorescence in the presence of phloretin). However, it has been shown that phloretin quenches *cis*-PA fluorescence in model membranes (**Fig. 11**). Phloretin has been shown to permeate the lipid bilayer and intercalate into the membrane lipids *(5)*. Therefore, in the study with cells *(20)* the decrease in *cis*-PA fluorescence in the presence of phloretin is most likely because of fluorescence quenching rather than inhibition of the protein-mediated uptake of *cis*-PA into cells.

28. Because of its higher solubility, higher amounts of *cis*-PA are typically added when compared with those of oleate.
29. Some combinations of probes should be avoided because fluorescent artifacts will disturb the accuracy of the measurements (FPE and BCECF; *cis*-PA and ADIFAB).
30. Addition of free FA to a suspension of vesicles has been shown to decrease the pH inside the vesicle *(2)*. Large pH gradients can reduce the net diffusion of FA across the membrane *(12)*. Incubation overnight will allow this pH gradient to dissipate.
31. As stopped-flow spectroscopy involves the rapid mixing of two solutions that are equal in volume, the concentration of SUV in suspension should be doubled to achieve the desired concentration after mixing.
32. At pH 7.4, the solubility limit of oleate has been estimated at approx 5–6 µ*M* by ADIFAB measurements *(22)*. Higher concentrations of oleate will lead to the aggregation of the FA in aqueous

solution and could potentially complicate the results. Using an upper limit of 6 µM oleate, only a maximum of 3 µM oleate can be achieved on mixing.

33. The total time of these measurements is 2 s. The dead time, or time required to measure the first data point, is 10 ms.
34. The slits should be increased in stopped-flow experiments in order to increase the fluorescence signal detected by the fluorimeter from the small volumes of sample used for each measurement. If the slits are not opened sufficiently, the experiment will suffer from poor signal-to-noise.
35. This must be done to ensure that enough volume of the current sample is used to flush the previous sample out of the tubing running between the syringes and the cuvet wherein fluorescence measurements are recorded.
36. The time-resolution of stopped-flow fluorescence does not allow the emission of two probes to be monitored simultaneously, requiring separate experiments with each probe. In order to record the fluorescence of two probes using stopped-flow, each probe must be monitored separately while rapidly mixing the same suspensions. Because the concentration and lipid composition of donor and acceptor SUV are equal, the FA will equilibrate, such that approx 50% will transfer to the acceptor.
37. In performing the similar experiments of in which FA are transferred from FPE-labeled "donor" SUV to "acceptor" SUV containing pyranine; some problems have been experienced resulting from a 10–20% overlap in the excitation spectra of each probe. As FA desorb from the FPE-labeled SUV, FPE fluorescence will increase. As FA bind to SUV containing pyranine and flip–flop across the bilayer, pyranine fluorescence decreases. Thus, the fluorescence emission of each probe changes in different directions. For this particular experiment the authors sometimes observe a slight "canceling" effect of one probe on the sensitivity of the other, which hampers kinetic analysis of the fluorescence changes. This effect was minimized by adjusting the slit widths of the fluorescence instrumentation or by making slight modifications to the excitation wavelengths used to monitor each probe or by using filters.

References

1. Kamp, F., Zakim, D., Zhang, F., Noy, N., and Hamilton, J. A. (1995) Fatty acid flip-flop in phospholipid bilayers is extremely fast. *Biochemistry* **34,** 11,928–11,937.
2. Kamp, F. and Hamilton, J. A. (1992) pH gradients across phospholipid membranes caused by fast flip-flop of unionized fatty acids. *Proc. Natl. Acad. Sci. USA* **89,** 11,367–11,370.
3. Cooper, R. B., Noy, N., and Zakim, D. (1989) Mechanism for binding of fatty acids to hepatocyte plasma membranes. *J. Lipid Res.* **30,** 1719–1726.
4. Hamilton, J. A. (1998) Fatty acid transport: Difficult or easy? *J. Lipid Res.* **39,** 467–481.
5. Hamilton, J. A., Johnson, R. A., Corkey, B., and Kamp, F. (2001) Fatty acid transport: The diffusion mechanism in model and biological membranes. *J. Mol. Neurosci.* **16,** 99–107.
6. Hamilton, J. A. and Kamp, F. (1999) How are free fatty acids transported in membranes? Is it by proteins or by free diffusion through the lipids? *Diabetes* **48,** 2255–2269.
7. Hamilton, J. A., Guo, W., and Kamp, F. (2002) Mechanism of uptake of long-chain fatty acids: Do we need cellular proteins? *Mol. Cell Biochem.* **239,** 17–23.
8. Hamilton, J. A. (1994) ^{13}C-NMR studies of the interactions of fatty acids with phospholipid bilayers, plasma lipoproteins, and proteins. In (*Carbon-13 NMR Spectroscopy*) of Biological systems (Beckman, N.,), Academic Press, San Diego, pp. 117–157.
9. Hamilton, J. A., Johnson, R. A., Corkey, B., and Kamp, F. (2001). Fatty acid transport: the diffusion mechanism in model and biological membranes. *J Mol Neurosci* **16,** 99–108.
10. Kamp, F., Guo, W., Souto, R., Pilch, P., Corkey, B. E., and Hamilton, J. A. (2003) Rapid flip-flop of oleic acid across the plasma membrane of adipocytes. *J. Biol. Chem.* **278,** 7988–7995.
11. Civelek, V. N., Hamilton, J. A., Tornheim, K., Kelly, K. L., and Corkey, B. E. (1996) Intracellular pH in adipocytes: Effects of free fatty acid diffusion across the plasma membrane, lipolytic agonists, and insulin. *Proc. Natl. Acad. Sci. USA* **93,** 10,139–10,144.

12. Kamp, F. and Hamilton, J. A. (1993) Movement of fatty acids, fatty acid analogues, and bile acids across phospholipid bilayers. *Biochemistry* **32,** 11,074–11,085.
13. Hamilton, J. A. (1998) Fatty acid transport: difficult or easy? *J. Lipid Res.* **39,** 467–481.
14. Kamp, F., Guo, W., Souto, R., Pilch, P., Corkey, B. E., and Hamilton, J. A. (2003). Rapid flip-flop of oleic acid across the plasma membrane of adipocytes. *J. Biol. Chem.* **278,** 7988–7995.
15. Hamilton, J. A. and Guo, W. (2001) Fatty acid uptake and metabolism in HepG2 cells. *Biophys. J. (Annual Meeting Abstract)* **365a** 45th Annual Biophysical Society Meeting, Baltimore, MD.
16. Guo, W., Huang, N., Cai, J., Xie, W., and Hamilton, J. A. (2006) Fatty acid transport and metabolism in HepG2 cells. *Am. J. Physiol. Gastrointest. Liver Physiol.* **290,** G528–G534.
17. Cupp, D., Kampf, J. P., and Kleinfeld, A. M. (2004) Fatty acid-albumin complexes and the determination of the transport of long chain free fatty acids across membranes. *Biochemistry* **43,** 4473–4481.
18. Thomas, R. M., Baici, A., Werder, M., Schulthess, G., and Hauser, H. (2002) Kinetics and mechanism of long-chain fatty acid transport into phosphatidylcholine vesicles from various donor systems. *Biochemistry* **41,** 1591–1601.
19. Bojessen, I. N. (2002) Studies of membrane transport mechanism of long-chain fatty acids in human erythrocytes. *Recent Res. Dev. Membr. Biol.* **1,** 33–50.
20. Bartlett, G. R. (1959) Phosphorous assay in column chromatography. *J. Biol. Chem.* **234,** 466–468.
21. Fraser, H., Coles, S. M., Woodford, J. K., et al. (1997) Fatty acid uptake in diabetic rat adipocytes. *Mol. Cell Biochem.* **117,** 56–60.
22. Richieri, G. V., Ogata, R. T., and Kleinfeld, A. M. (1999) The measurement of free fatty acid concentration with the fluorescent probe ADIFAB: A practical guide for the use of the ADIFAB probe. *Mol. Cell Biochem.* **192,** 87–94.

17

Measurement of Lateral Diffusion Rates in Membranes by Pulsed Magnetic Field Gradient, Magic Angle Spinning-Proton Nuclear Magnetic Resonance

Klaus Gawrisch and Holly C. Gaede

Summary

Membrane organization, including the presence of domains, can be characterized by measuring lateral diffusion rates of lipids and membrane-bound substances. Magic angle spinning (MAS) yields well-resolved proton nuclear magnetic resonance (NMR) of lipids in biomembranes. When combined with pulsed-field gradient NMR (rendering what is called "pulsed magnetic field gradients-MAS-NMR"), it permits precise diffusion measurements on the micrometer lengths scale for any substance with reasonably well-resolved proton MAS-NMR resonances, without the need of preparing oriented samples. Sample preparation procedures, the technical requirements for the NMR equipment, and spectrometer settings are described. Additionally, equations for analysis of diffusion data obtained from unoriented samples, and a method for correcting the data for liposome curvature are provided.

Key Words: Solid state NMR; pulsed magnetic field gradient; magic angle spinning; lateral diffusion; membranes; lipids.

1. Introduction

Magic angle spinning (MAS)-nuclear magnetic resonance (NMR) is a desirable approach for obtaining solution-like proton spectra of biomembranes. MAS-NMR reduces the linewidth of lipid resonances to the range of 10–30 Hz *(1)*. A resolution of 3 Hz has been achieved for signals from polyunsaturated hydrocarbon chains like docosahexaenoic acid *(2)*. Linewidth is equivalent to or better than the resolution of resonances of very small unilamellar liposomes that tumble rapidly enough to eliminate anisotropic interactions. MAS frequencies up to 10 kHz have negligible influence on lipid packing in bilayers. Experiments can be conducted at any level of hydration, provided that the lipids are in the biologically relevant liquid-crystalline lamellar phase. The investigation of natural biological membranes that contain lipids and proteins is feasible as well. Spectra on milligram quantities of lipid can be obtained within seconds. With a longer acquisition time, experiments on micrograms of membranes are feasible.

If MAS-NMR is combined with the application of pulsed magnetic field gradients (PFG), diffusion rates of all substances with resolved proton resonances can be measured. The approach does not require preparation of oriented samples. However, the technical requirements of the NMR equipment are rather stringent, and the useable range of gradient strength may be limited, which prevents the study of very low lateral diffusion rates. In recent years, several studies reported the use of this approach to successfully study rates of lateral diffusion of lipids *(3–7)*, membrane bound drugs *(3,4,8)*, peptides *(9)*, water *(3,6)*, and even the formation of domains in liposomes *(10)*.

2. Materials

1. Lipids, cell suspensions, or tissue samples.
2. Ideally, water for ^{1}H PFG-MAS-NMR experiments should be 98% deuterated.
3. For cell suspensions or tissue samples, ultracentrifuge capable of spinning samples with a volume of 1–5 mL at 100,000g or higher.
4. Desktop centrifuge with swinging buckets capable of spinning samples with a volume of 15 mL at 2000g (optional).
5. Solid state NMR spectrometer equipped with a high-resolution ^{1}H MAS probe, a coil for application of PFG oriented along the magic angle (54.7°) to the magnetic field, and a gradient amplifier. The achievable-pulsed field gradient strength should be at least 0.3 T/m. The proton resonance frequency of the instrument is of secondary importance. However, higher resonance frequencies yield better-resolved spectra and higher sensitivity. The probe must have provisions to control sample temperature to ±1°C or better. The instrument must permit the shimming of the proton resonances to a width of 0.1 ppm or better at half height and 0.5 ppm or better at 5% intensity. Most commercial ^{1}H MAS-NMR probes easily exceed these specifications.
6. MAS rotors with Kel-F inserts for work with liquid samples (e.g., Bruker Biospin Inc., Billerica, MA; Doty Scientific, Columbia, SC). Inserts with a small (10–20 µL) spherical sample volume improve* resolution of resonances.

3. Methods

3.1. Preparation of Samples

Preparation procedures must ensure that the radii of curvature of membranes are larger than 1 µm. Samples with smaller radii of curvature require application of PFG that are much stronger than 0.3 T/m.

1. Dry lipids are hydrated by direct addition of D_2O or of buffer solutions prepared with D_2O (*see* **Note 1**). Lipids dissolved in organic solvent are dried in a stream of nitrogen gas or argon. It is advantageous to spin the vials during drying to generate a thin layer of lipid on the glass surface. Residual quantities of solvent are removed by application of vacuum. Sample preparation with polyunsaturated lipids must be conducted in a glove box filled with an oxygen-free atmosphere. Addition of an antioxidant like butylated hydroxytoluene at a lipid/butylated hydroxytoluene molar ratio of 200/1 is recommended. The complete removal of solvent is easily tested by recording ^{1}H MAS-NMR spectra of the sample. An appropriate amount of buffer (e.g., 50 wt%) is added to the dry lipid and the sample is homogenized by centrifugation, annealing, and/or freeze–thaw cycles. Addition of smaller quantities of water favors formation of larger liposomes. Most biophysical properties of bilayers are independent of water concentration down to 20 water molecules per lipid or even less.
2. Cell membranes are concentrated by ultracentrifugation. It is desirable to replace most of the H_2O with D_2O. This is conveniently achieved by discarding the supernatant after centrifugation and subsequent addition of a buffer prepared with D_2O.
3. The membrane pellet is transferred to a MAS rotor equipped with an insert for liquid samples. The efficient transfer of small samples is conveniently achieved by centrifugation at low g-force using a swinging bucket rotor. For the transfer, the container with the sample is inserted into the wider end of a plastic pipet tip of appropriate size. The thin end of the tip is cut such that it fits snugly into the end of an MAS rotor. This assembly is placed into a tube that fits into a swinging bucket, and the membrane pellet is transferred to the MAS rotor by centrifugation. Please note that the sample must be in a liquid crystalline phase for efficient transfer.

3.2. NMR Acquisition and Processing Parameters

1. The NMR instrument is set up for acquisition of a ^{1}H NMR signal. The spectral bandwidth must cover all resonances of the MAS spinning centerband. The backfolding of spinning sidebands

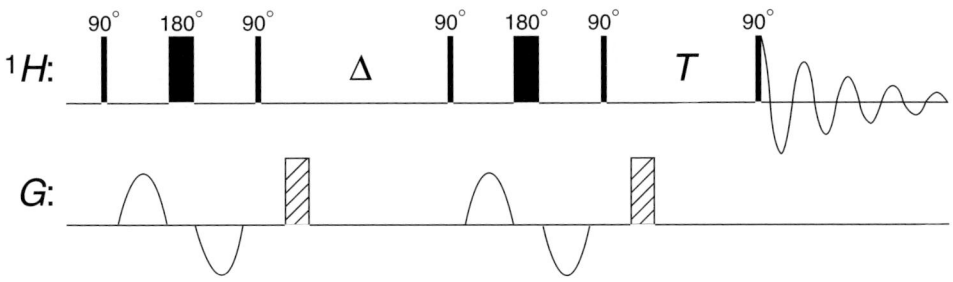

Fig. 1. Schematic presentation of the stimulated echo sequence using shaped bipolar gradient pulses. The magnetization of protons is spatially encoded by the first pair of bipolar gradients. At the end of this period, magnetization is oriented parallel to the magnetic field B_0. Residual magnetization in the xy-plane is dephased by a spoil pulse. Molecules change location by diffusive motions during the diffusion time (Δ). The second pair of bipolar gradients has opposite polarity and reverses spatial encoding. Magnetization is turned back to being parallel to B_0 and xy-magnetization dephased with a second spoil pulse. The purpose of the period T is to give the main magnetic field time to recover from perturbations resulting from application of gradient pulses. At the end of this period, magnetization is turned to the xy-plane for detection of the free induction decay.

into the spectral centerband is suppressed by selecting a proper filter width. Insert the MAS rotor into the probe and spin the sample. Most lipid samples yield good resolution at spinning frequencies as low as 2 kHz. However, the resonances of hydrocarbon chains and the glycerol protons in particular will have significant intensity in the spinning sidebands at such low frequencies. Also, resolution of resonances from cholesterol-containing samples may benefit greatly from spinning at higher frequencies (*see* **Note 2**). It is very important that the sample spins without movement along the MAS axis (*see* **Note 3**).

2. The free induction decay is acquired with a stimulated echo sequence using shaped bipolar gradient pulses *(11)* (*see* **Fig. 1**) (*see* **Note 4**). Bipolar pulses enable registration of resonance signals with minimal baseline distortions. Typical acquisition parameters or guidance in determining them are listed next.

   - Spectral width 10 ppm or higher.
   - Delay time between data acquisitions equal to three to five times the longest spin-lattice relaxation time of any resonance that will be used for data analysis.
   - Determine the length of the π/2- and π-pulses on the proton channel.
   - Adjust maximal gradient strength, gradient length, and diffusion time to achieve a significant reduction of signal intensity at highest gradient strength (*see* **Note 5**).
   - Adjust the length of the delay time *T* (**Fig. 1**) to minimize spectral perturbations (phase shifts, baseline roll) from the PFG. Field recovery times on the order of a few milliseconds are typical.

3. The spectra are acquired as a function of gradient strength (e.g., in 16 increments that are equally spaced from 2 to 95% of maximal strength) (*see* **Note 6**). It is strongly recommended to acquire the spectra by interleaving the high and low gradient strengths; i.e., the highest gradient strength first, followed by the lowest gradient strength, the second highest gradient strength, the second lowest gradient strength, and so on. Following this method it is possible to establish, whether the results are tainted by a continuous drift of spectrometer parameters during the time-course of data acquisition.
4. Experiments are conducted at three to five different diffusion times to determine liposome size (*see* **Subheading 3.3.2.**).
5. Amplitudes of resonances are measured after careful phase- and base plane correction.

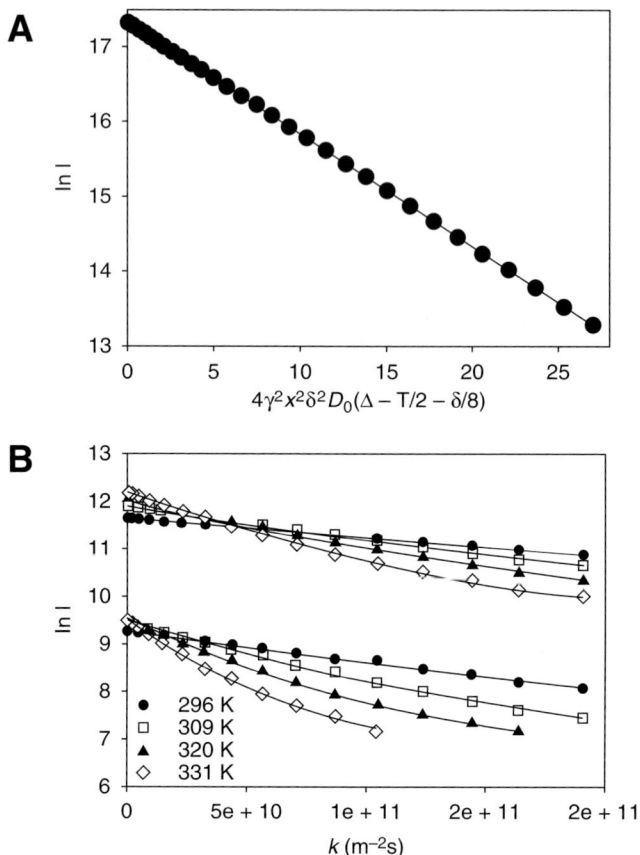

Fig. 2. **(A)** Logarithm of signal attenuation of water in a spinning MAS rotor as a function of the square of gradient strength (variable k). Please note that water diffusion is isotropic and yields a linear decay. The effective gradient strength was 0.3875 T/m achieved at a gradient amplifier current of 10 A. **(B)** Logarithmic plot of signal attenuation of 1-palmitoyl-2-oleoyl-sn-glycero-3-phosphocholine (POPC) (upper) and ibuprofen (lower) as a function of increasing k for different temperatures. Please note that the decay is nonexponential, as expected for lipid bilayers that are oriented at random. The data were fit to **Eq. 3** to obtain diffusion constants. Only points for which $kD < 5$ were included in the fit.

3.3. Interpretation of Results

3.3.1. Signal Attenuation in Powder Samples

The diffusional motions of lipids and interlamellar water in multilamellar liposomes are confined to layers that are randomly oriented with respect to the applied PFG. For such randomly oriented bilayers, a logarithmic plot of signal attenuation as a function of the square of gradient strength yields a nonlinear dependence (*see* **Fig. 2**). The squared gradient strength is typically expressed as a variable k. A sine-shaped gradient pulse yields:

$$k = 4\gamma^2 g^2 \delta^2 \left(\Delta - \frac{d_2}{2} - \frac{\delta}{8} \right) \quad (1)$$

where γ is the gyromagnetic ratio of protons, g is the gradient strength, δ is the gradient pulse length, and d_2 is the time between the gradient pulses sandwiching the π-pulses *(12)*. The formula for square-shaped pulses is slightly different:

$$k = 4\gamma^2 g^2 \delta^2 \left(\Delta - \frac{d_2}{16} - \frac{\delta}{3} \right) \qquad (2)$$

It was reported earlier *(3,4)* that for $kD \leq 5$, where k is the diffusion constant, the signal attenuation, I/I_0, can be approximated as:

$$\ln\left(\frac{I}{I_0}\right) = -\frac{2}{3}kD + \frac{2}{45}(kD)^2 \qquad (3)$$

This formula is valid down to a signal intensity of 10%, which is sufficient for most applications. The first term of **Eq. 3** contains the familiar factor of 2/3 that has been cited in previous work as a first order correction for powder samples *(13,14)*. An example of signal attenuation that was fitted by **Eq. 3** is shown in **Fig. 2B**. Although the data for the lipid POPC appear to be linear, the second term of **Eq. 3** is required for an adequate fit to the data. A linear fit underestimates the slope systematically with increasing diffusion time because of the increase in barely* perceptible curvature of the data.

3.3.2. Correction for Bilayer Curvature

Owing to the finite curvature of liposomes, diffusion rates are underestimated. It has been shown previously that true diffusion constants of substances in bilayers are obtained by analyzing the apparent diffusion rate (D_{app}) as a function of diffusion time (Δ) according to *(3,4)*:

$$D_{app} = \frac{r^2}{2\Delta} \sin^2\left[\frac{(2D\Delta)^{\frac{1}{2}}}{r} \right] \qquad (4)$$

where D is the true diffusion constant, and r the effective radius of liposome curvature. Fitting the measured diffusion values D_{app} to **Eq. 4** gives a radius of curvature and the true diffusion constant D (*see* **Fig. 3**). For sufficiently small values of Δ, D_{app} is equal to the value of D in good approximation. If liposomes have diameters of a few micrometers, this is achieved at diffusion times on the order of 25 ms. Therefore, at sufficiently short diffusion times the correction for finite liposome size may not be required.

4. Notes
1. There is a systematic difference between pH readings taken with a glass electrode in H_2O and pD readings in D_2O. The approximate conversion is pD ≈ pH + 0.4. Small quantities of deuterated buffers are conveniently prepared by lyophylization of the proper quantity of protonated buffer with subsequent addition of D_2O to the lyophilized residue.
2. The control of sample temperature at high MAS spinning frequencies is more difficult. For a rotor with an outer diameter of 4 mm, spinning at frequencies up to 5 kHz yields a sample temperature that is a few degrees lower than the temperature of the bearing gas, owing to the Joule–Thompson effect of the expanding gas. At higher spinning frequencies sample temperature

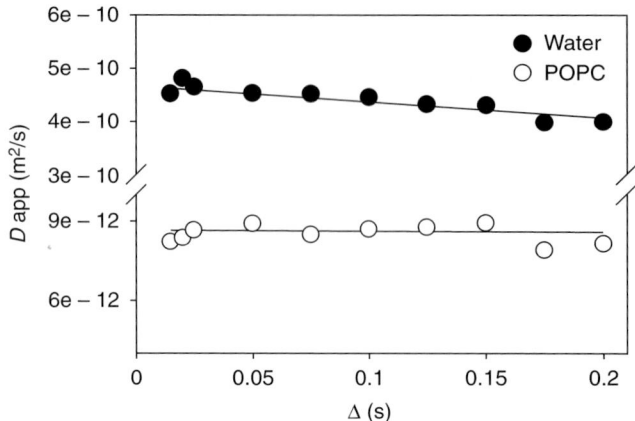

Fig. 3. Plot of apparent diffusion constants of water and POPC as a function of diffusion time (Δ) at 322 K. The sample contained 15 water molecules per lipid. Please note that the dependence of D_{app} on diffusion time is stronger for higher diffusion rates, as in the case of water. The fit to **Eq. 4** yields an average radius of curvature of 21 µm, and the true lateral diffusion rates of water, $D_{water} = (4.7 \pm 0.1) \times 10^{-10}$ m²/s, and the lipid POPC, $D_{POPC} = (8.6 \pm 0.2) \times 10^{-12}$ m²/s.

is increased because of frictional heating at the bearings. Highly conductive samples may also heat because of induction of eddy currents in the sample. However, the latter is rarely noticed at physiological concentrations of saline. At a spinning frequency of 10 kHz, frictional heating typically raises sample temperature by a few degrees above the set temperature. The magnitude of temperature increase depends on the type of MAS stator and the rotor material. At spinning frequencies of 15 kHz, sample temperature is on the order of 20°C higher than the temperature of the bearing gas.

Furthermore, significant temperature gradients over the sample are to be expected. Changes of sample temperature are conveniently followed by recording the chemical shift of the water resonance. Please note that the chemical shift of the water depends also on lipid hydration. Lower water concentrations shift the resonance to a higher magnetic field strength. Absolute values of temperature inside the spinning rotor are determined by recording the spectra of lipid bilayers with known gel–fluid phase transition temperatures. The transition to the gel phase is easily detected by the broadening of hydrocarbon chain resonances.

Sample spinning results in moderate dehydration of lipids because of the centrifugal force in the fast spinning rotor. The level of sample dehydration depends on the rotor diameter, the spinning frequency, and the density difference between membranes and water *(15)*. Dehydration of membranes is prevented by matching the water density to the membrane density, for example, by mixing H_2O and D_2O at a proper ratio.

3. The requirements for stable spinning of the MAS rotor are very stringent. Any movement of the rotor along the MAS axis (which is also the orientation of the axis of PFG) during the diffusion time will result in signal attenuation. Requirements for spinning stability increase with increasing gradient strength. If the gradient coils are attached to the MAS stator, the application of gradient pulses also exerts a force on the stator that may result in mechanical vibrations that disturb NMR signal acquisition. Therefore, not all MAS equipment with gradient coils may be suitable for diffusion measurements. Indications for spinning instability and mechanical vibrations are: (1) modulations of the probe resonance curve that are visible while tuning and matching the resonance circuits, (2) imperfections in signal reproducibility between consecutively acquired spectra, and (3) instability of spinning frequency.

4. Sine-shaped gradient pulses yield low-field perturbation but the effective gradient strength is significantly reduced. Application of square gradient pulses yields highest effective gradient

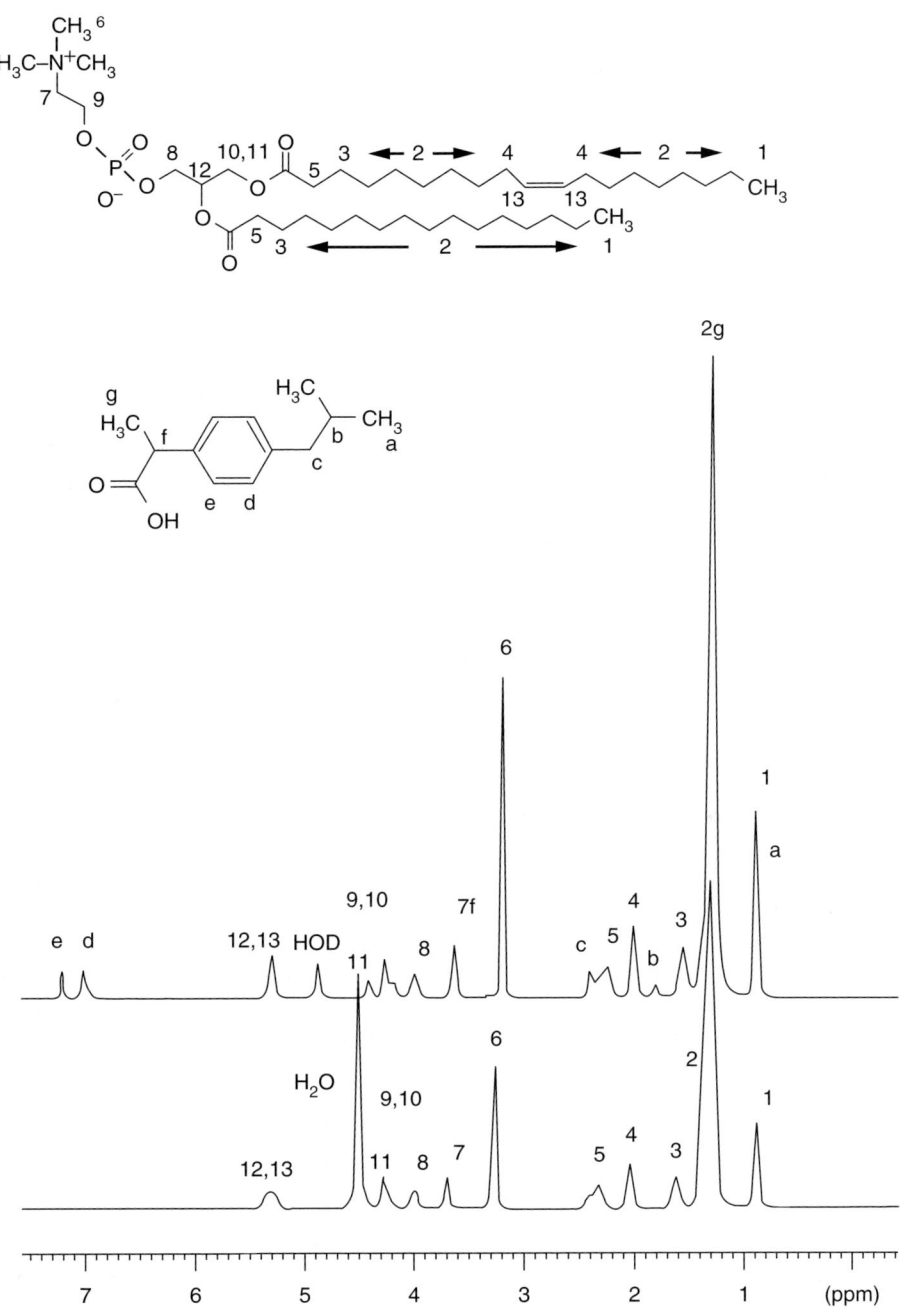

Fig. 4. ^{1}H NMR spectra of POPC:H$_2$O 1:8.2 (mol/mol) (lower) and POPC:ibuprofen:D$_2$O 1:0.56:15 (mol/mol/mol) (upper) with resonance assignments as indicated in the POPC and ibuprofen structures shown above the spectra. Please note the excellent resolution that can be achieved in experiments on fluid lipid membranes. Most resonances are 10–30 Hz wide.

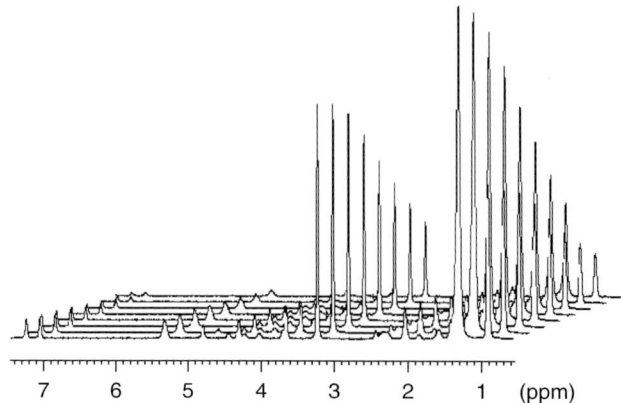

Fig. 5. Signal attenuation in spectra of POPC:ibuprofen:D_2O (1:0.56:15, mol/mol/mol) at 309 K as a function of increasing gradient strength in a stimulated echo diffusion sequence using sine-shaped bipolar gradient pulses of 5 ms duration. Diffusion measurements were conducted at sixteen different values of gradient strength varying from 0.01 to 0.37 T/m. A longitudinal eddy current delay of $T = 5$ ms was used and the diffusion time $\Delta = 200$ ms. At every gradient strength eight scans were acquired with a recycle delay of 4 s.

strength, but also strong field perturbations. Trapezoidal-shaped gradient pulses with short raise and decay times that yielded almost the same effective gradient strength as square pulses were successfully used, but without significant signal perturbation. The optimal gradient shape is equipment-dependent.

5. Ideally, signal intensity at highest gradient strength should be in the range from 1 to 10% compared with signal intensity at very low gradient strength. However, in practice the available maximum gradient strength may be insufficient to achieve such strong attenuation. The amplitude and length of the bipolar gradient pulses is limited by the equipment and cannot be set higher than the values recommended by the manufacturer without risking damage to the NMR probe. A gradient pulse length on the order of few milliseconds is a typical limit. The length of the diffusion time is limited by several factors: (1) average liposome size (*see* **Subheading 3.3.2.**), (2) cross-relaxation between protons, and (3) spin-lattice relaxation times. The diffusion time must be selected such that lipids cover, on average, distances that are a fraction of liposome size. In most cases, this limits diffusion times to 200 ms. Magnetization is flipped back to the B_0 field axis during the diffusion time, which may result in magnetization exchange between dipolar coupled protons through the nuclear overhauser effect *(16)*.

As a result, the apparent diffusion rates between membrane components may equalize with increasing diffusion time. For most lipid membranes, diffusion times of 200 ms or less yield insignificant magnetization transfer through nuclear overhauser effect cross-relaxation. If the water is not deuterated, cross-relaxation with water resonances may contribute as well. The latter is particularly critical for resonances from the lipid water interface, like choline. Spin-lattice relaxation is usually the least stringent restriction to the length of diffusion time. Some of the resonances may have relaxation times of up to several seconds. However, relaxation times in the 500 ms range are most typical. Because of differences in spin-lattice relaxation times, the relative signal intensities at longer diffusion times may be very different compared with spectra that are recorded with a single $\pi/2$ hard pulse (compare signal intensities in **Fig. 4** vs **Fig. 5**).

6. Gradient strength and linearity are determined by measurement of water diffusion using the same NMR equipment, MAS rotors, and rotor inserts for liquid samples as for diffusion measurements on membranes. The diffusion rate of water in H_2O and of residual protonated water in D_2O as a function of temperature were reported by R. Mills *(17)* (*see also* **Fig. 2**).

Acknowledgments

This work was supported by the Intramural Research Program of the National Instituted on Alcohol Abuse and Alocoholism and National Institute of Health.

References

1. Holte, L. L. and Gawrisch, K. (1997) Determining ethanol distribution in phospholipid multilayers with MAS-NOESY spectra. *Biochemistry* **36**, 4669–4674.
2. Eldho, N. V., Feller, S. E., Tristram-Nagle, S., Polozov, I. V., and Gawrisch, K. (2003) Polyunsaturated docosahexaenoic vs docosapentaenoic acid—Differences in lipid matrix properties from the loss of one double bond. *J. Am. Chem. Soc.* **125**, 6409–6421.
3. Gaede, H. C. and Gawrisch, K. (2003) Lateral diffusion rates of lipid, water, and a hydrophobic drug in a multilamellar liposome. *Biophys. J.* **85**, 1734–1740.
4. Gaede, H. C. and Gawrisch, K. (2004) Multidimensional PFG-MAS-NMR experiments on membranes. *Magn. Reson. Chem.* **42**, 115–122.
5. Scheidt, H. A., Huster, D., and Gawrisch, K. (2005) Diffusion of cholesterol and its precursors in lipid membranes studied by ^{1}H pulsed field gradient magic angle spinning NMR. *Biophys. J.* **89**, 2504–2512.
6. Wattraint, O. and Sarazin, C. (2005) Diffusion measurements of water, ubiquinone and lipid bilayer inside a cylindrical nanoporous support: A stimulated echo pulsed-field gradient MAS-NMR investigation. *Biochim. Biophys. Acta* **1713**, 65–72.
7. Pampel, A., Zick, K., Glauner, H., and Engelke, F. (2004) Studying lateral diffusion in lipid bilayers by combining a magic angle spinning NMR probe with a microimaging gradient system. *J. Am. Chem. Soc.* **126**, 9534–9535.
8. Pampel, A., Michel, D., and Reszka, R. (2002) Pulsed field gradient MAS-NMR studies of the mobility of carboplatin in cubic liquid-crystalline phases. *Chem. Phys. Lett.* **357**, 131–136.
9. Pampel, A., Kärger, J., and Michel, D. (2003) Lateral diffusion of a transmembrane peptide in lipid bilayers studied by pulsed field gradient NMR in combination with magic angle sample spinning. *Chem. Phys. Lett.* **379**, 555–561.
10. Polozov, I. V. and Gawrisch, K. (2004) Domains in binary SOPC/POPE lipid mixtures studied by pulsed field gradient H-1 MAS NMR. *Biophys. J.* **87**, 1741–1751.
11. Cotts, R. M., Hoch, M. J. R., sun, T., and markert, J. T. (1989) pulsed field gradient stimulated echo methods for improved NMR diffusion measurements in heterogeneous systems. *J. Magn. Reson.* **83**, 252–266.
12. Fordham, E. J., Mitra, P. P., and Latour, L. L. (1996) Effective diffusion times in multiple-pulse PFG diffusion measurements in porous media. *J. Magn. Reson. A.* **121**, 187–192.
13. Callaghan, P. T. and söderman, O. (1983) examination of the lamellar phase of aerosol OT-water using pulsed field gradient nuclear magnetic-resonance. *J. Phys. Chem.* **87**, 1737–1744.
14. Lindblom, G., Wennerström, H., and Arvidson, G. (1977) translational diffusion in model membranes studied by nuclear magnetic resonance. *int. J. quant. chem.* **12**, 153–158.
15. Nagle, J. F., Liu, Y. F., Tristram-Nagle, S., Epand, R. M., and Stark, R. E. (1999) Re-analysis of magic angle spinning nuclear magnetic resonance determination of interlamellar waters in lipid bilayer dispersions. *Biophys. J.* **77**, 2062–2065.
16. Huster, D., Arnold, K., and Gawrisch, K. (1999) Investigation of lipid organization in biological membranes by two-dimensional nuclear Overhauser enhancement spectroscopy. *J. Phys. Chem. B.* **103**, 243–251.
17. Mills, R. (1973) self-diffusion in normal and heavy-water in range 1–45 degrees. *J. phys. chem.* **77**, 685–688.

18

Using Fluorescence Recovery After Photobleaching to Measure Lipid Diffusion in Membranes

Conrad W. Mullineaux and Helmut Kirchhoff

Summary

The lateral diffusion of lipids is crucial to the biogenesis and function of biological membranes. In this chapter, approaches for observing the lateral diffusion of lipids using fluorescence recovery after photobleaching are described. The procedures described can be carried out with a standard laser-scanning confocal microscope. The membrane of interest is stained with a lipophilic fluorophore or fluorescent lipid analog. The confocal laser spot is then used to photobleach fluorescence in a small region of the sample. Subsequent spread and recovery of the bleach, reports on the diffusion of the fluorophore. The results provide a measure of membrane fluidity, and the extent to which lipid diffusion in the membrane might be constrained. Fluorescence recovery after photobleaching measurements may be carried out in vivo, in systems ranging from bacteria to mammalian cells, or in vitro in isolated membrane fragments. Procedures for preparing the biological sample, performing the measurement, and quantitative data analysis are explained.

Key Words: Biological membranes; confocal microscopy; fluorescence recovery after photobleaching; lipid diffusion; membrane fluidity; lipophilic fluorophore.

1. Introduction

Fluorescence recovery after photobleaching (FRAP) provides a relatively simple and direct method for measuring large-scale lateral diffusion of lipids in membranes *(1,2)*. FRAP reports only on the averaged behavior of many thousands of molecules. More sophisticated methods are needed to track the diffusion of individual lipid molecules *(3,4)*. Because of the limited spatial resolution of FRAP, measurements will miss some important aspects of biological membranes such as microdomains *(5)*. However, advantages of FRAP are: (1) it is rather straightforward and noninvasive and (2) measurements can readily be carried out in vivo. FRAP measurements of lipid diffusion give an overview of the fluidity of the membrane; for example, they can be used to show changes in fluidity owing to phase transition and lipid composition *(1,2)*. Comparison of the diffusion coefficient of a lipophilic fluorophore in different biological membranes reveals considerable variability in lipid diffusion rates, probably depending mainly on the membrane lipid composition and lipid–protein interaction. The information has direct biological relevance, because long-range lipid diffusion is vital to many membrane processes, including membrane biogenesis, signal transduction, and electron transport mediated by quinones.

The approaches described in this chapter can all be carried out using a standard laser-scanning confocal microscope. The principle of the method is that the membrane of choice is stained with a lipophilic fluorophore or fluorescent lipid analog *(1,2)*. The confocal microscope is used to image the membrane, visualizing fluorescence from the fluorophore. The laser power is increased and the confocal spot is scanned over a small, localized region of the membrane, bleaching the fluorophore in this area. The laser power is then decreased to prevent further bleaching, and the sample is scanned to acquire a series of images at different time-points after the bleach *(6,7)*. Diffusion of the fluorophore results in a characteristic "blurring" of the bleached area *(6,7)*. The bleach is initially deep and very localized: diffusion causes spread of the bleach and recovery of fluorescence at the center of the bleached area, so the bleach becomes broader and shallower. Quantitative data can be extracted from the images and analyzed to obtain the diffusion coefficient of the fluorophore *(6,7)*.

Traditional FRAP measurements were carried out without laser scanning, simply using a stationary laser spot to bleach and then record fluorescence recovery in one place. Imaging the sample by laser scanning reduces time resolution but makes the measurement very much more robust *(7)*. Imaging the sample allows the visualization of diffusion barriers, and makes it easy to spot experimental problems such as movement of the sample, loss of focus, or the initial bleach being broader than expected. This chapter describes the use of laser-scanning confocal microscopy for FRAP measurements of lipid diffusion. Approaches will be described to measuring lipid diffusion in vivo, and in isolated membranes in vitro. In the case of smaller samples, such as bacterial cells and small membrane fragments, a significant problem is preventing any movement of the sample during measurements. This chapter describes approaches for solving this problem.

2. Materials
2.1. The Fluorescent Probe

The best choice of lipophilic fluorescent probe depends on the wavelengths available in the confocal microscope. The favorite is 4,4-difluoro-5,7-dimethyl-4-bora-3a,4a-diaza-*s*-indacene-3-dodecanoic acid (BODIPY® FL C_{12}), which combines a green fluorophore with a 12-carbon saturated hydrocarbon tail. It can be efficiently excited with the 488-nm line of an argon laser (Spectra-Physics, Mountain View, CA), and has a peak fluorescence emission at about 512 nm. It is supplied by the Molecular Probes division of Invitrogen (http://probes.invitrogen.com/). Molecular Probes also supply the same fluorophore linked to some specific biologically important lipids. Prepare a 10 m*M* stock solution of BODIPY FL C_{12} in dimethylsulfoxide, and store at –20°C.

2.2. Immobilizing the Sample—In Vivo Measurements on Bacteria and so on

1. A suitable growth medium for the bacteria being studied.
2. Difco bacto-agar (BD Biosciences, Franklin Lakes, NJ).
3. Glass cover slips (0.1 mm thickness) (VWR International Ltd., West Chester, PA).
4. Cell division inhibitor. For *Escherichia coli* use cephalexin (Sigma-Aldrich, St. Louis, MO): prepare a 10 mg/mL stock solution in water and filter sterilize.
5. Temperature-controlled microscope sample holder, with well to accommodate a small block of agar, with a cover slip on top. The author's is laboratory-built. VWR supply disposable slides with wells, but these do not have temperature control.
6. Silicone vacuum grease (VWR International Ltd.).

2.3. Immobilizing the Sample—In Vivo Measurements on Cultured Mammalian Cells

1. Suitable growth medium (e.g., Dulbecco's Modified Eagles Medium) (VWR International Ltd.).
2. Glass cover slips (0.1 mm thickness).

3. Presterilized culture vessel with chambers for cover slips (VWR International Ltd.).
4. Temperature-controlled microscope sample holder as in **Subheading 2.2.**
5. Silicone vacuum grease (VWR International Ltd.).

2.4. Immobilizing the Sample—In Vitro Measurements on Isolated Membrane Fragments or Vesicles

1. Egg yolk phosphatidyl choline (Sigma-Aldrich) made into a 10-mM solution in 2,2,2-trifluoroethanol (TFE) (Sigma-Aldrich).
2. Glucose (Sigma-Aldrich).
3. Glucose oxidase (Sigma-Aldrich).
4. Catalase (Sigma-Aldrich).
5. Glass microscope slides (VWR International Ltd.).
6. Glass cover slips (0.1 mm thickness) (VWR International Ltd.).

2.5. Performing the FRAP Measurement

This requires a laser-scanning confocal microscope. The authors use a Nikon PCM2000 (Nikon Instech Co., Kanagawa, Japan) equipped with a 100-mW argon laser (Spectra-Physics). For BODIPY FL C_{12} they use the 488-nm line from the argon laser, selected by an interference band-pass filter. Fluorescence is selected by a 505-nm dichroic mirror and a band-pass filter transmitting at approx 500–527 nm. The measurements require suitable acquisition software for switching between XY and X-scanning modes, and recording a series of images at defined time intervals. The authors' use the EZ2000 software (Nikon) supplied with the PCM2000 microscope. It must be possible to rapidly change the light output from the laser (e.g., with neutral density filters).

2.6. Data Analysis

1. Suitable software for extracting quantitative data from images, for example, Image-Pro Plus 5.1 (MediaCybernetics, Silver Spring, MD).
2. Suitable software for data analysis and curve fitting. SigmaPlot 9.0 (Systat, Point Richmond, CA) is suitable for most of the applications described here.

3. Methods
3.1. Staining the Sample With a Lipophilic Fluorophore

For BODIPY FL C_{12}, add enough stock solution to the incubation medium to give a final concentration of 1–10 µM. Incubate with the cells or membranes for about 30 min and then, if possible, wash in dye-free growth medium or buffer to remove unincorporated dye *(2)*.

3.2. Immobilizing the Sample—In Vivo Measurements on Bacteria and so on

1. Grow a cell culture in liquid medium. For example, for *E. coli* grow a culture overnight at 37°C in Luria–Bertani medium.
2. FRAP measurements are very much easier to perform on bacterial cells that have been elongated by growth in the presence of cell division inhibitors. In *E. coli* this can be achieved by growth for 3 h in the presence of 30 µg/mL cephalexin before the measurement *(8)*.
3. Take the cell suspension and check the cell concentration by measuring apparent absorbance at 750 nm in a spectrophotometer. The optimal cell concentration must be determined by trial and error. For *E. coli* an overnight culture diluted 10-fold is about right.
4. Stain with the lipophilic fluorophore as described in **Subheading 3.1.**

5. Take a Petri dish with 1.5% bacto-agar made up in growth medium. Put drops of the diluted cell culture onto the agar surface, and allow to dry down. This can be achieved in about 15 min by placing the open Petri dish in a laminar flow hood.
6. When all the liquid has been adsorbed, use a scalpel to cut out a small block of agar with cells adsorbed to the surface. Place in the well of the sample holder.
7. Smear a thin layer of vacuum grease around the rim of the well, and use this to stick down a glass cover slip, so that the cover slip is gently pressing onto the agar surface.
8. Place on the microscope stage.

3.3. Immobilizing the Sample—In Vivo Measurements on Cultured Mammalian Cells and so on

1. Culture the cells in wells containing cover slips.
2. Rinse cover slip in growth medium to remove unattached cells.
3. Examine in a phase-contrast microscope to check for adhesion of cells to the cover slip.
4. Stain with the lipophilic fluorophore as described in **Subheading 3.1**.
5. Invert the cover slip and use vacuum grease to seal it over a microscope sample holder, with a well containing growth medium.
6. Place on the microscope stage.

3.4. Immobilizing the Sample—In Vitro Measurements on Isolated Membrane Fragments or Vesicles

A significant problem with isolated membrane fragments is keeping the membrane fragment immobile while simultaneously minimizing interactions between the membrane and a solid support, which might perturb the mobility of membrane components. Following procedure is recommended in which the membrane fragments are spread on a glass slide coated with an artificial phosphatidyl choline bilayer (**Fig. 1**). The lipids within the phosphatidyl choline bilayer are fully mobile (as judged from FRAP on lipophilic fluorophores), and thus, are unlikely to prevent the mobility of the isolated membrane fragment components that rest on them. Ideally, the phosphatidyl choline support should also be stained with a lipophilic fluorophore to allow it to be visualized separately from the biological membrane of interest. This makes it possible to check whether the membrane fragment of interest is sitting on the lipid support (**Fig. 1**). The choice of fluorophore will depend on the excitation and emission wavelengths available in the confocal microscope.

1. Take 100 µL of phosphatidyl choline (10 mM solution in TFE). Add a lipophilic fluorophore if desired. Spot 20 µL onto a glass slide. TFE has a very-low surface tension, which allows it to spread efficiently on glass.
2. Evaporate the TFE and dry the lipid film under a stream of nitrogen for 1 h.
3. Warm the dried lipid film to 40°C for 1 min (the temperature must be more than the phase-transition temperature of the lipid).
4. Keeping the slide at 40°C, add 30 µL of buffer (the same buffer used to suspend the membrane fragments). Incubate for 2 min at 40°C. Stacks of phosphatidyl choline bilayer are formed during this time (**Fig. 1**).
5. Take membrane fragments in a suitable buffer, stained with the lipophilic fluorophore as described in **Subheading 3.1**.
6. It may help to use anaerobic conditions. Under anaerobic conditions photobleaching is generally impeded. This makes it easier to image the sample, although it also means that more laser power must be used to do the bleach. Anaerobic conditions can be achieved by adding to the membrane fragment suspension glucose (20 mM), glucose oxidase (48 U/mL), and catalase (800 U/mL).
7. Add 30 µL of the membrane suspension to the bilayer stack.

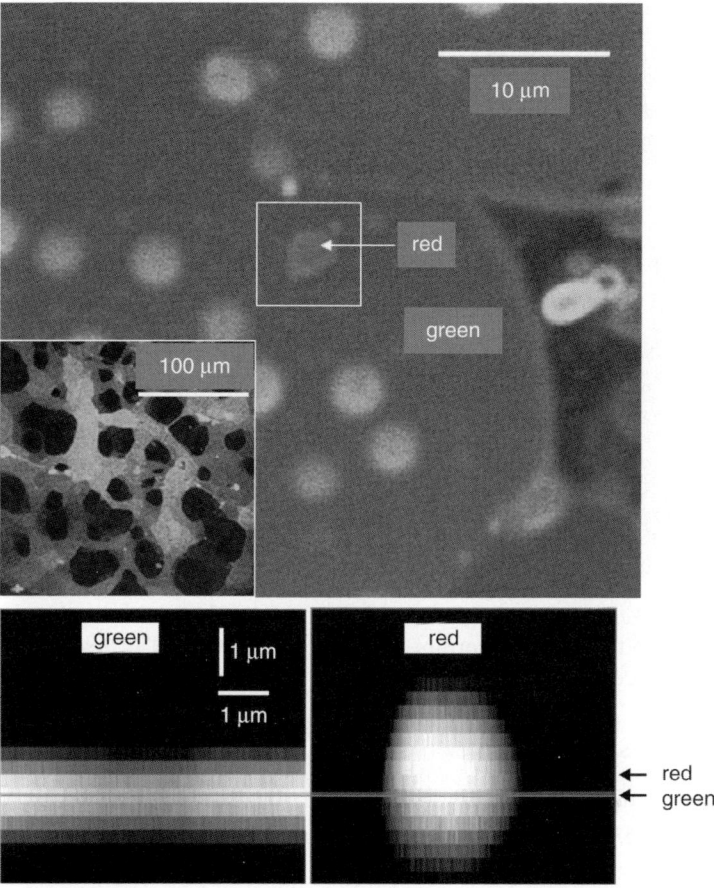

Fig. 1. Use of a phosphatidyl choline bilayer to support and immobilize a small membrane fragment for FRAP. In this case the membrane fragments were diluted granal membranes from spinach chloroplasts. The phosphatidyl choline bilayer was stained with BODIPY and the granal membrane was visualized using chlorophyll autofluorescence. **(Top)** XY-scan showing a granal membrane fragment (highlighted by the square) on the phosphatidyl choline support. The inset gives a smaller-scale view of the phosphatidyl choline bilayer stacks **(bottom)** Z-projections showing green fluorescence from the phosphatidyl choline support, and red fluorescence from the membrane fragment. The green and red images have been separated for clarity in a monochrome image. The membranes appear very much thicker than they really are, because of the limited Z-resolution. However, the higher "center of gravity" of the red fluorescence indicates that the granal membrane fragment is on top of the phosphatidyl choline support.

8. Incubate for 15–30 min.
9. Seal the sample with a cover slip.
10. Place on the microscope stage.

3.5. Performing the FRAP Measurement

1. Focus the microscope on the sample. This is usually most easily done in simple transmission or epifluorescence mode. The authors use a ×60 oil-immersion objective lens (numerical aperture 1.4).
2. Switch to laser-scanning confocal mode and image fluorescence from the lipophilic fluorophore. Trial-and-error must be used to establish conditions under which the fluorescence can be imaged without significantly bleaching the fluorophore (*see* **Notes 1** and **2**). Record a series of successive images, and see whether the fluorescence intensity decreases from image to image. If it

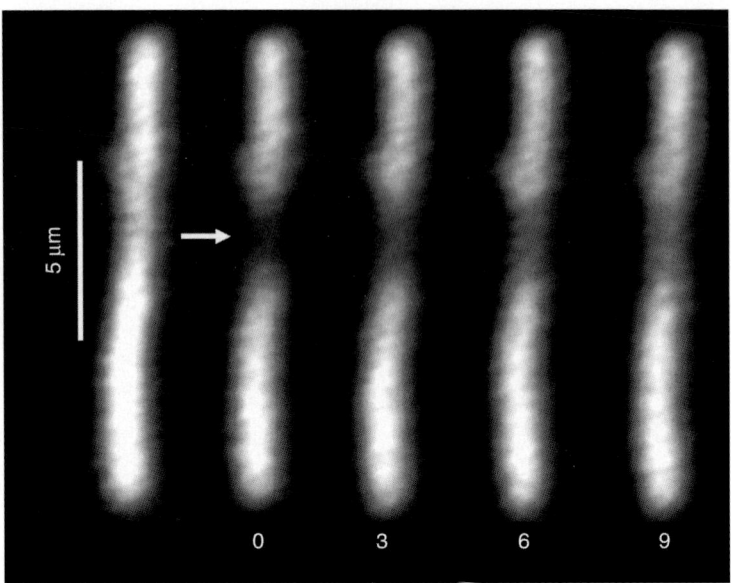

Fig. 2. FRAP image sequence showing diffusion of BODIPY in the membranes of an elongated bacterial cell (in this case the cyanobacterium *Synechococcus* 7942). The arrow indicates where a line was bleached across the cell. (Data were reproduced with permission from **ref. 2**.)

    does, it will be necessary to decrease the laser intensity. Routinely a 100-mW argon laser with intensity decreased by a factor of 32 with neutral density filters is used.
3. Select a suitable cell or membrane fragment of interest (*see* **Note 3**). In the case of elongated bacterial cells *(2,8–10)*, select a long cell aligned roughly in the Y-direction (**Fig. 2**). In the case of isolated membrane fragments, select a good-sized fragment that is securely superimposed on the phosphatidyl choline support (**Fig. 1**).
4. Zoom in so that the cell or membrane fragment chosen fills most of the field of view. Record a prebleach image of the sample. Check that the sample can still be imaged without bleaching, because scanning the laser over a smaller area increases the sample exposure.
5. Bleach fluorescence from a small area of the sample. The way the bleach is done depends on the geometry of the sample, but when possible, the bleach should always be done away from the edges of the sample, in the center of a homogeneous area of membrane (*see* **Note 4**). Usually, it is much more convenient to bleach a line rather than a spot (*see* **Note 5**). To do this, switch the confocal microscope from XY-scanning mode to X-scanning mode, so that the line of the X-scan passes through the center of the sample. If possible, start the X-scan increasing the laser power at the same time. Generally, the laser power by a factor of 32 is increased by briefly removing neutral density filters. After about a second, the fluorescence signal should be significantly decreased (*see* **Note 6**). It is not necessary or desirable to bleach away all the fluorescence in the center of the bleached zone. Generally, something between 50 and 90% bleaching is ideal.
6. Quickly replace the neutral density filters, switch back to XY scanning mode, and record a series of images at set time intervals. The best time intervals depend on the lipid-diffusion rate and the size of the sample. In general, it is sufficient to record a series of about 10 images at intervals of 1–3 s. One longer time-point is useful to check whether recovery is complete (recovery will be incomplete if there is an immobile population of the fluorophore). In a successful FRAP measurement the initial bleach will appear as a clearly defined dark zone in the fluorescence image (*see* **Note 7**). With time, the dark zone will spread and fill in, as the bleached fluorophore diffuses away and unbleached fluorophore diffuses into the bleached area (*see* **Note 8**).

FRAP and Lipid Diffusion in Membranes

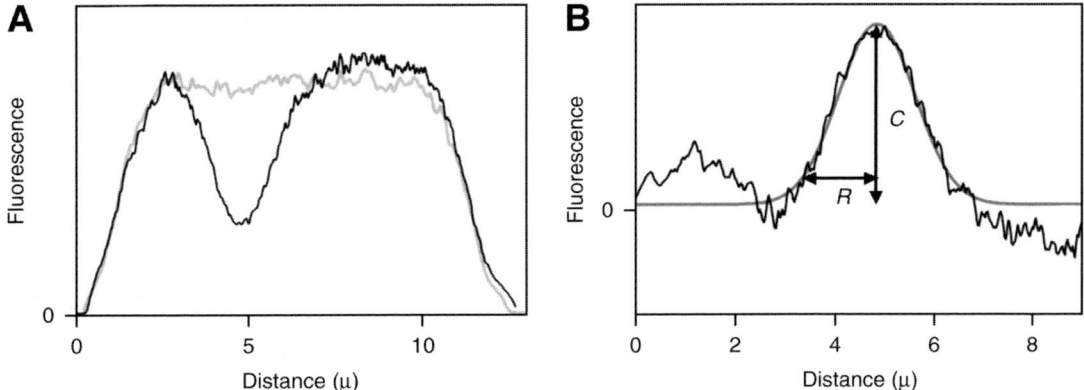

Fig. 3. 1D fluorescence profiles extracted from images of an elongated bacterial cell before and after bleaching a line. **(A)** Prebleach profile (gray line) and postbleach profile (black line). **(B)** Difference profile obtained by subtracting the postbleach profile from the prebleach profile. The fitted Gaussian curve and the parameters C (depth of the bleach) and R ($1/e^2$ radius of the bleach), which are required for calculation of the diffusion coefficient, are also shown. (Reproduced with permission from **ref. 12**.)

3.6. Data Analysis

A prerequisite for quantitative data analysis is that the topography of the membrane must be simple and predictable. Examples include elongated bacterial cells, wherein the plasma membrane can be approximated as a long cylinder *(2,8,10)*; many cultured mammalian cells, wherein the plasma membrane can be approximated as an extended flat sheet *(7)*; and isolated flat membrane fragments. Quantitative data analysis is invariably based on the assumption that the membrane environment is uniform over the measured area. If the membrane geometry is simple and the sample size is large relative to the bleach, it is usually straightforward to use an analytical solution relating the evolution in the bleaching profile to the diffusion coefficient (*see* **Note 9**). The following procedure relates to the estimation of the diffusion coefficient from an elongated sample, such as the plasma membrane of an elongated bacterial cell. In this case, a line is bleached across the cell **(Fig. 2)** and diffusion is observed in one dimension (1D) along the long axis of the cell *(10)* (*see* **Note 10**).

1. Import the images into Image ProPlus or similar software.
2. Use the software to extract a 1D fluorescence profile along the long axis of the cell, summing the fluorescence values across the width of the cell. This is done by setting a sampling line width greater than the width of the cell, then drawing the line down the long axis of the cell, and extracting the data **(Fig. 3)**.
3. Repeat for each time-point in the sequence, including the prebleach image.
4. Import the fluorescence profiles into SigmaPlot (Systat Software Inc., Richmond, CA) or similar software. Subtract each postbleach profile from the prebleach profile to generate a series of difference profiles, which should be approximately Gaussian in shape.
5. Use SigmaPlot to fit the profiles to Gaussian curves, noting the parameters C (the peak height of the Gaussian curve) and R (the $1/e^2$ radius of the Gaussian curve) **(Fig. 3)**. If the changes observed are because of diffusion, then C should decrease with time and R should increase with time, but the area under the Gaussian curve should remain constant.
6. In this 1D case, C is related to the diffusion coefficient D as follows: $C = C_0 R_0 (R_0^2 + 8 Dt)^{-0.5}$, where t is time and C_0 and R_0 are the values for C and R extracted from the first postbleach image *(10)*.

D can be obtained by plotting $1/C^2$ vs time. The plot should give a straight line. The gradient can be obtained by linear regression and is equal to $8D/C_0^2 R_0^2$ (*see* **Note 11**). C_0 and R_0 are known, so D can be calculated (*see* **Note 12**). If t is in seconds and R_0 is in micrometers, then D is in $\mu m^2/s$.

4. Notes

1. Photobleaching is increased by high laser power, low scanning speeds, and a small scanning area (field of view). The use of the highest possible scanning speeds is recommended, because this also increases the time-resolution of the measurement. In general, high Z-resolution is not required for FRAP measurements, so it is best to use a larger confocal pinhole, which gives a higher signal-to-noise ratio. This in turn allows images to be obtained at lower laser intensity, which minimizes the problem of photobleaching during imaging.
2. The use of anaerobic conditions, as described in **Subheading 3.4.**, generally minimizes the problem of photobleaching during imaging. However, anaerobiosis makes it harder to perform the FRAP bleach.
3. The larger the sample, the easier the measurement will be. Data analysis is easier and more robust if you have a tightly localized bleach in the middle of a large sample (it gets difficult when the bleach removes a large proportion of the total fluorescence). A larger sample makes it much easier to follow rapid diffusion. With a small sample and rapid diffusion, the bleach might re-equilibrate over the whole sample before you have time to record any images (*see also* **Note 7**).
4. It is not always possible to the bleach in the center of the sample. When a small sample is used, it may be necessary to bleach at one edge to prevent the whole sample from being bleached away. This is feasible, but it makes the data analysis more difficult *(11)*.
5. Some confocal microscopes are not well equipped to bleach spots, because when the laser spot is stationary it is automatically diverted away from the sample. In any case, a line bleach makes it easier to hit the sample.
6. The filters are lifted by hand. This can be done very quickly with practice. Shutters would allow shorter exposure times, but in the authors' experience very short exposure times do not give sufficient bleaching, in any case. With manual control you can watch the fluorescence signal decreasing and drop the filters when the bleach is deep enough. It is not necessary to time the bleach accurately, as long as the postbleach image sequence is timed accurately. The first postbleach image provides the first time-point for the data analysis.
7. The minimum diameter of the bleach could theoretically be about half the wavelength of the laser light. In practice, the width of the bleach is generally increased by light scattering by the sample and diffusion occurring during the bleach and before the first image is recorded. A common problem with small samples and rapid diffusion is that the bleach spreads over the whole sample before the first image can be recorded. Then the first postbleach image shows decreased fluorescence overall, but a similar fluorescence distribution to the prebleach image. This problem can be tackled in various ways:
    a. Do the bleach more quickly. This may require more laser power.
    b. Record the images quicker. This can often be achieved by reducing the pixel resolution of the images.
    c. Try to find a sample with a larger membrane area.
8. It is best to use a simple control to check that fluorescence recovery is because of diffusion, rather than reversible fluorescence quenching of the fluorophore. Try bleaching out an entire small cell or membrane fragment, and check that fluorescence does not recover.
9. Analytical solutions depend on simplifying assumptions, and the easiest assumption to make is that the membrane area is large compared with the bleach. If it is not, then you have to factor in the exact dimensions of the membrane, and the width, depth, and position of the bleach. Under these conditions, analytical solutions often are not worthwhile. As a more practical approach, measuring the fluorescence profile after the bleach is recommended; then, use an iterative computer routine

to predict how it will evolve with time, assuming simple random diffusion. Comparison of the simulation with the experimental data allows estimation of the diffusion coefficient.
10. In the case of a spot bleach in the center of a flat membrane, diffusion is observed in 2D. Data extraction and analysis are somewhat more complicated than in the 1D case described. *See* **ref. 7** for an example.
11. Take care to weight the linear regression accurately (weight each point according to the reciprocal of the estimated variance). In a plot of $1/C^2$ vs time the variance increases sharply as C decreases.
12. Individual measurements of D can be very accurate if obtained with care. But real biological samples often show variations: every cell or membrane fragment is different. Always do enough replicates to check that the result is representative.

Acknowledgments

Dr. Mark Tobin and Dr. Mary Sarcina contributed to the development of the methods described. Work in Conrad Mullineaux laboratory is supported by the Biotechnology and Biological Sciences Research Council and the Wellcome Trust. Helmut Kirchhoff is currently supported by a research fellowship from the Deutsche Forschungsgemeinschaft.

References

1. Fulbright, R. M., Axelrod, D., Dunham, W. R., and Marcelo, C. L. (1997) Fatty acid alteration and the lateral diffusion of lipids in the plasma membrane of keratinocytes. *Exp. Cell Res.* **233,** 128–134.
2. Sarcina, M., Murata, N., Tobin, M. J., and Mullineaux, C. W. (2003) Lipid diffusion in the thylakoid membranes of the cyanobacterium *Synechococcus* sp.: effect of fatty acid desaturation. *FEBS Lett.* **533,** 295–298.
3. Lee, G. M., Ishihara, A., and Jacobson, K. A. (1991) Direct observation of Brownian motion of lipids in a membrane. *Proc. Natl. Acad. Sci. USA* **88,** 6274–6278.
4. Fujiwara, T., Ritchie, K., Murakoshi, H., Jacobson, K., and Kusumi, A. (2002) Phospholipids undergo hop diffusion in compartmentalized cell membrane. *J. Cell Biol.* **157,** 1071–1081.
5. Douglass, A. D. and Vale, R. D. (2005) Single-molecule microscopy reveals plasma membrane microdomains created by protein-protein networks that exclude or trap signaling molecules in T cells. *Cell* **121,** 937–950.
6. Blonk, J. C. G., van Aalst, D. H., and Birmingham, J. J. (1993) Fluorescence photobleaching recovery in the confocal scanning light microscope. *J. Microsc.* **169,** 363–374.
7. Kubitscheck, U., Wedekind, P., and Peters, R. (1994) Lateral diffusion measurements at high spatial resolution by scanning microphotolysis in a confocal microscope. *Biophys. J.* **67,** 948–956.
8. Ray, N., Nenninger, A., Mullineaux, C. W., and Robinson, C. (2005) Location and mobility of twin-arginine translocase subunits in the *Escherichia coli* plasma membrane. *J. Biol. Chem.* **280,** 17,961–17,968.
9. Elowitz, M. B., Surette, M. G., Wolf, P.-E., Stock, J. B., and Leibler, S. (1999) Protein mobility in the cytoplasm of *Escherichia coli*. *J. Bacteriol.* **181,** 197–203.
10. Mullineaux, C. W., Tobin, M. J., and Jones, G. R. (1997) Mobility of photosynthetic complexes in thylakoid membranes. *Nature* **390,** 421–424.
11. Cowan, A. E., Koppel, D. E., Setlow, B., and Setlow, P. (2003) A soluble protein is immobile in dormant spores of *Bacillus subtilis* but is mobile in germinated spores: implications for spore dormancy. *Proc. Natl. Acad. Sci. USA* **100,** 4209–4214.
12. Mullineaux, C. W. (2004) FRAP analysis of photosynthetic membranes. *J. Exp. Bot.* **55,** 1207–1211.

19

Single-Molecule Fluorescence Microscopy to Determine Phospholipid Lateral Diffusion

Michael J. Murcia, Sumit Garg, and Christoph A. Naumann

Summary

The single-molecule detection (SMD) of individual fluorophores represents a powerful experimental technique that allows for the observation of individual membrane molecules in their different dynamic states without having to average over a large number of molecules. Spatial resolution in ensemble-averaging techniques such as fluorescence recovery after photobleaching, is limited by the diffraction limit of light (~250 nm). In contrast, SMD (as well as single-molecule tracking of gold-labeled biomolecules through Nanovid microscopy) provides a tracking accuracy of approx 10–30 nm (dependent on experimental conditions). This level of accuracy makes single-molecule tracking techniques much better suited to detect nanometer-size heterogeneous structures in membranes. SMD can be easily applied to model and cellular membranes using a variety of fluorescent labels including organic dyes, quantum dots, and dye/quantum dots-doped nanoparticles.

The main focus of this chapter is to outline the SMD methodology to study lateral diffusion of lipids in model membranes. **Subheading 1.** provides an overview about the development of single-molecule tracking techniques, and explains the basic concept of single-molecule tracking. **Subheading 2.** lists all relevant chemicals necessary to successfully conduct lipid lateral diffusion studies on model membranes. **Subheading 3.** describes a typical experimental setup for SMD using wide-field illumination; thus, this setup can be utilized to track single-lipid tracers in solid-supported phospholipid bilayers and phospholipid monolayers at the air–water interface. Furthermore, some general considerations are included about different fluorescent labels for lipid-tracking studies. In addition a description of sample preparation procedures for the design of solid-supported phospholipid bilayers and Langmuir monolayers of phospholipids are described. Finally, **Subheading 4.** lists additional relevant information helpful for conducting SMD experiments on lipid membranes.

Key Words: Fluorescence; lateral diffusion; optical microscopy; phospholipid monolayer; single-molecule tracking; solid-supported phospholipid bilayer.

1. Introduction

1.1. Background

Many important biological functions of biomembranes are dependent on the fluidity of the membrane. The membrane fluidity, which facilitates the translational and rotational movements of lipid and membrane proteins along the sheet of the bilayer, is caused by very weak intermolecular forces among membrane constituents. Because of its enormous biological relevance, lateral and rotational diffusion processes of lipids and proteins have been studied in cellular and model membranes for quite some time (**ref. 1**, and references therein). One of the hallmarks in modern cell biology and biophysics is that biomembranes are not just homogeneous mixtures of proteins and lipids, but instead are made up of rapidly changing arrangements of widely differing patches called domains *(2–5)*. Furthermore, the cytoskeleton linkages to the membrane were found to compartmentalize the plasma membrane, thus affecting the lateral diffusion of lipids and membrane proteins *(6,7)*. This current perspective is largely the result of experimental techniques, such as nanovid microscopy and single-molecule fluorescence microscopy, which are able to: (1) detect the movement of individual tracer molecules, (2) probe the heterogeneity of membranes with a spatial resolution better than the diffraction limit of light, and (3) detect individual biomolecules in their different dynamic states (**ref. 8**, and references therein).

The first successful single-molecule detection (SMD) experiments on individual fluorophores were performed using low-temperature spectroscopy on solids *(9,10)*. There are multiple experimental strategies that allow the imaging of individual fluorescent probes, including near-field scanning optical microscopy *(11–15)*, confocal-scanning microscopy *(16–19)*, and wide-field single-molecule fluorescence microscopy *(20–28)*. A very important goal in single-molecule fluorescence imaging experiments is to maximize the signal-to-background and signal-to-noise ratios. This goal is best achieved by using a confocal imaging setup. However, mobile probes, such as diffusing lipids, are much better investigated using a wide-field imaging configuration wherein multiple probes are monitored at the same time over an area of about 10×10 μm^2. Importantly, SMD-based diffusion studies can be conducted on model membranes and cellular membranes (*see* **Note 1**).

Tracking experiments on individual lipid molecules were first reported using 30-nm size colloidal gold probes conjugated to lipids imaged through Nanovid microscopy *(29)*. Lipid lateral diffusion experiments based on SMD were first successfully applied by tracking individual tetra-methylrhodamine (TMR)-labeled phospholipids in a solid-supported phospholipid bilayer *(30)*. Using SMD, lipid-tracking studies have also been reported on free-standing planar lipid bilayers and cell surfaces *(31,32)*. The authors first applied this technique to track lipids in a lipid monolayer at the air–water interface *(33)*. More recently, they also reported on the obstructed diffusion of phospholipids in a polymer-tethered phospholipid bilayer using SMD *(34)*.

1.2. Concept of Single-Molecule Tracking

The goal of single-molecule tracking techniques is to monitor the trajectories of individual probe molecules over time and to analyze these trajectories on the basis of diffusion theory *(8,35)*. Trajectories can be made visible using specific labels covalently linked to the diffusing biomolecules of interest. The lateral diffusion of such tracer molecules along a flat membrane sheet is of a two-dimensional (2D) nature. As illustrated in **Fig. 1**, the position of a tracer molecule at a given time (time lag) is measured in terms of characteristic x- and y-coordinates. Trajectory data can be obtained by successively determining the displacement r_n for different time lags $\tau_n = n\,\tau_0$ over n jumps (τ_0: time lag between two successive observations and n: number of tracer jumps at r_n). Herein, r_n is simply related to the x- and y-data by:

$$r_n(\tau_n) = \sqrt{(x_n - x_0)^2 + (y_n - y_0)^2} \qquad (1)$$

with x_0, y_0 being the coordinates of the original position of the tracer, and x_n, y_n being the coordinates after n jumps. With respect to lipid diffusion studies, two different cases need to be distinguished: (1) tracer diffusion in a homogeneous membrane environment and (2) tracer diffusion in a heterogeneous membrane environment. In the following, the basic concepts of single-molecule tracking are explained for these two cases.

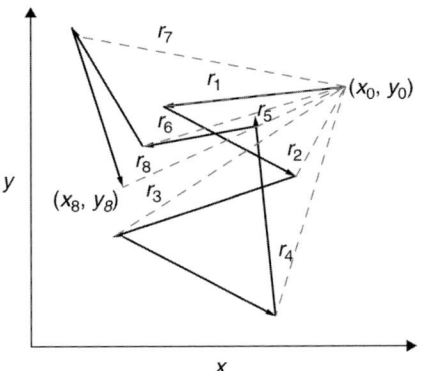

Fig. 1. Schematic of a 2D-track consisting of eight individual steps. The progression of the track, which starts at the coordinates x_0, y_0, is described by the coordinates x_n, y_n and the displacements relative to the origin of the track r_n.

1.2.1. Tracer Diffusion in a Homogeneous Membrane Environment

Individual lipid tracer molecules in a homogeneous environment of a planar fluid lipid bilayer (e.g., one-component lipid bilayer) show 2D-Brownian motion. As aforementioned, this process is well described by a 2D-random walk. The stochastic nature of the Brownian motion is depicted in **Fig. 2** in the form of a representative r^2-histogram (**Fig. 2A**) and a cumulative distribution function (CDF) (**Fig. 2B**), which show the distribution of r^2 of a lipid tracer molecule in a planar solid-supported phospholipid bilayer at a constant time lag τ_c. Here the CDF can be expressed by: (*see* **ref. 36**)

$$P(r^2, \tau_c) = 1 - \exp\left[-\frac{r^2(\tau_c)}{\langle r^2(\tau_c)\rangle}\right] \quad (2)$$

where $r^2(\tau_c)$ and $\langle r^2(\tau_c)\rangle$ are the square displacements and the mean-square displacement at a characteristic τ_n, which indicates Brownian motion. In the limits of long random walks or large number of tracer molecules, the laws of thermodynamics tell us that the information of microscopic random walks of individual molecules can be translated into a long-range lateral diffusion coefficient (D), which represents a macroscopic thermodynamic parameter. Thus, to determine D using a single-molecule tracking technique, the motion of several tracers is typically monitored over time. In the case of Brownian motion, D can be determined from individual tracks through:

$$\langle r^2(\tau_n)\rangle = 4D\tau_n \quad (3)$$

where D is the lateral diffusion coefficient. Alternatively, D can be obtained by analyzing the $r^2(\tau_c)$ distribution from multiple tracks based on one time lag τ_c. If the $r^2(\tau_c)$ distribution can be expressed by **Eq. 2**, D can be calculated through $D = \langle r^2(\tau_c)\rangle/4\tau_c$. The linear relationship between mean-square displacement and time lag in **Eq. 3** implies that the diffusion coefficient D is independent of the observation time. If the tracer diffusion occurs in a membrane environment characterized by flow (e.g., lipid diffusion in a monolayer at the air–water interface), $\langle r^2(\tau_c)\rangle$ is defined as the superposition of Brownian diffusion and membrane flow

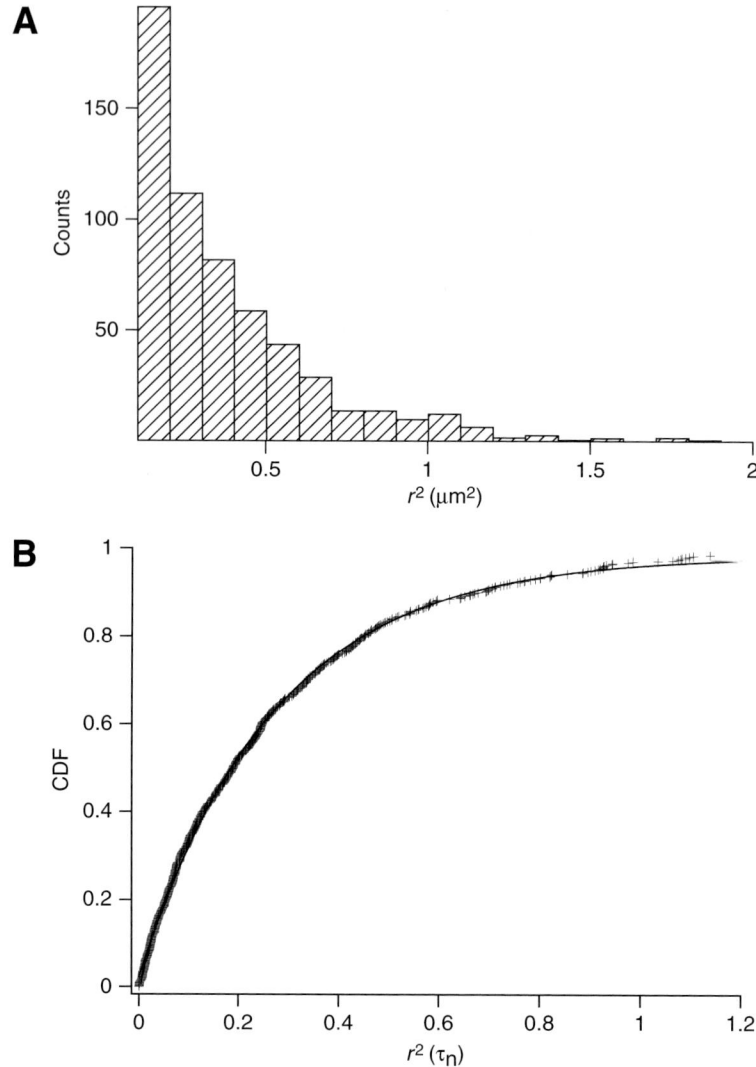

Fig. 2. Representative square-displacement r^2 histogram (**A**) and CDF (**B**) of the r^2 from a lipid tracer tracking experiment in a planar supported bilayer. The data show the distribution of r^2 after a specific time lag τ_c. The CDF illustrates that the tracer diffusion shows Brownian diffusion at τ_c because the tracking data are well described by the fitting curve based on **Eq. 2**.

$$\langle r^2(\tau_n)\rangle = 4D\tau_n + (V\tau_n)^2 \qquad (4)$$

where V is the velocity. This case does not show the linear relationship between $\langle r^2(\tau_n)\rangle$ and τ_n. To determine the self-diffusion of molecules in a membrane system characterized by flow, the relative movement of multiple tracers (multiple particle tracking) should be analyzed.

1.2.2. Tracer Diffusion in a Heterogeneous Membrane Environment

In heterogeneous membrane environments, such as domain-forming mixtures of lipids, diffusion processes can show non-Brownian behavior. Common non-Brownian processes

Single-Molecule Fluorescence and Phospholipid Diffusion

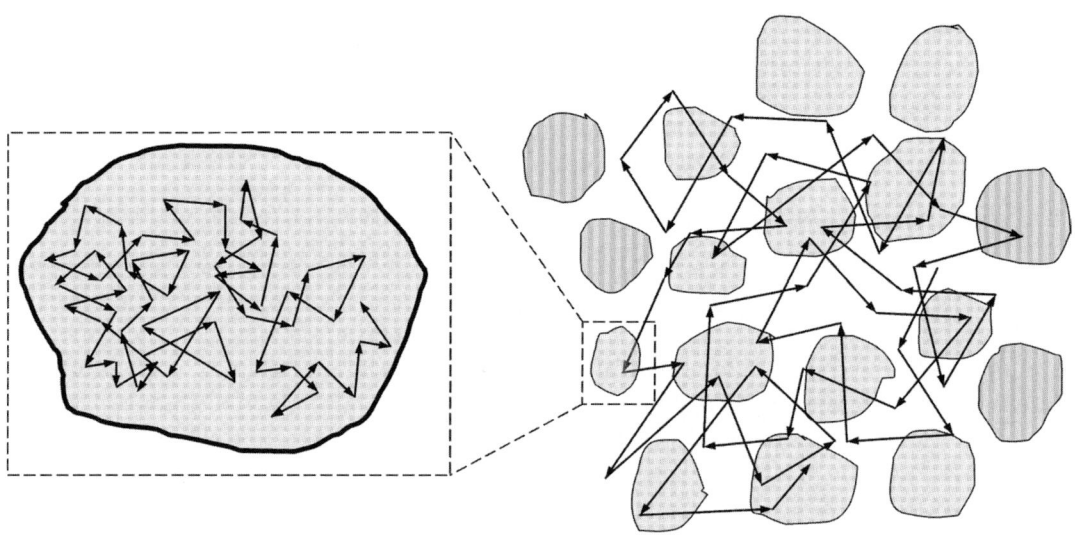

Fig. 3. Schematic of two types of tracking experiments on the same heterogeneous membrane sample. In the first case (**inset** on the left side), the time resolution is high enough so that the tracer does not experience the boundary region of the domain; thus, probing a homogeneous membrane environment. In the second case (right side), the jump distance is of similar length as the average domain size, because time resolution is much lower. Here the tracer probes the heterogeneous membrane environment.

observed include anomalous and corralled diffusion where $\langle r^2(\tau_n)\rangle$ is described by the following relationships (8)

$$\text{anomalous:} \quad \langle r^2(\tau_n)\rangle = 4D\tau_n^\alpha \qquad (\alpha < 1) \tag{5}$$

$$\text{corralled:} \quad \langle r^2(\tau_n)\rangle = \langle r_c^2\rangle\left[1 - A\exp\left(-4BDt/\langle r_c^2\rangle\right)\right]. \tag{6}$$

Here, α is the anomalous diffusion exponent, $\langle r_c^2\rangle$ is the average size of the coral, and A and B are geometry parameters. Anomalous diffusion, which results in the distributional broadening of jump times or jump lengths (in comparison with Brownian diffusion), can be observed in membranes if the diffusion is affected by obstacles and traps influencing the distribution of binding energies and escape times *(1,8)*. For example, SMD lipid-tracking studies in polymer-tethered phospholipid bilayers showed that Brownian diffusion can be observed at low tethering concentration, whereas anomalous diffusion was detected at higher tethering *(34)*.

Corralled diffusion can be found if tracer molecules diffuse within a domain without being able to leave the domain. This situation occurs when tracer molecules have a very high affinity for one particular domain type. An important consequence of the nonlinear relationship between $\langle r^2(\tau_n)\rangle$ and τ_n in heterogeneous membrane environments is that the tracking results are typically dependent on the length scale of the experiment relative to the average size of membrane heterogeneities. The concept of this length scale dependence is depicted in **Fig. 3**. The problem can be treated as a Brownian diffusion inside the domain if the average jump distance of tracers between successive observations is small in comparison with the average size of the heterogeneities. If this is not the case, the tracer molecule may diffuse over an area bigger than an individual domain, thus probing the domain boundary and the heterogeneous membrane environment, respectively. Note that different diffusion coefficients between both cases in **Fig. 3** are only measurable if the domain boundary acts as a diffusion barrier or if each domain region exhibits different tracer diffusion.

1.3. Lipid Lateral Diffusion: SMD vs Fluorescence Recovery After Photobleaching

There is still some confusion within the scientific community about the main differences between an ensemble-averaging technique like fluorescence recovery after photobleaching (FRAP) and a single-molecule tracking technique like SMD if membrane diffusion processes are studied. SMD and FRAP not only differ because one technique detects single molecules, whereas the other probes large ensembles of molecules; they also show different spatial resolutions. Whereas the spatial resolution of FRAP is given by the size of the diffraction spot of approx 250 nm, the position of individual tracer molecules can be measured with an accuracy of up to 10 nm *(37)*.

Based on these differences, the cases of lipid diffusion in homogeneous and heterogeneous membrane environments are considered. The case of lipid diffusion in a homogeneous membrane region is described by **Eq. 3**. This equation implies that there is a linear relationship between mean-square displacement and time, and that the diffusion coefficient is independent of the length of individual tracks and the spatial resolution of the technique used. Under this condition, FRAP and SMD should provide identical values of D. In contrast, single-molecule-based techniques outperform ensemble-averaging techniques when diffusion studies are conducted on heterogeneous membrane systems, in which anomalous and corralled diffusion may occur. Here, the diffusion properties depend on the length scale of the diffusion process. In particular, SMD is advantageous if the average size of the membrane heterogeneities is smaller than the spatial resolution of FRAP, but bigger than that of SMD (which is often the case).

2. Materials

2.1. Preparation Through Vesicle Fusion

Chloroform (high-performance liquid chromatography [HPLC] grade) is utilized to prepare lipid mixtures of labeled and nonlabeled lipids. Milli-Q water (pH = 5.5, 18 MΩ resistivity) or buffered salt solutions (e.g., 5 mM Tris, pH 8.0; 50 mM NaCl) can be used. The choice of the solution depends on the lipid composition. A large variety of purified phospholipids is commercially available (e.g., from Avanti Polar Lipids, Alabaster, AL). Dye-labeled phospholipids, such as N-(6-tetramethylrhodaminethiocarbamoyl)-1,2-dihexadecanoyl-sn-glycerophos TMR-DHPE (Invitrogen/Molecular Probes, Eugene, OR) can be purchased as well (e.g., www.invitrogen.com).

2.2. Preparation Through Film Deposition Techniques

Butyl acetate and ethanol are used to clean the Teflon trough. Chloroform (Fisher Scientific, Pittsburgh, PA) (HPLC grade) is used to prepare lipid-containing spreading solutions. Milli-Q (Millipore, Billerica, MA) water (pH = 5.5, 18 MΩ resistivity) or buffer is used as a subphase material. Again, the choice of the subphase material depends on the lipid system studied. The following glass cleaning solutions are utilized to prepare the glass substrates before formation of solid-supported bilayers: (1) 1% sodium dodecyl sulfate (SDS), (2) methanol saturated with NaOH, and (3) 1% HCl.

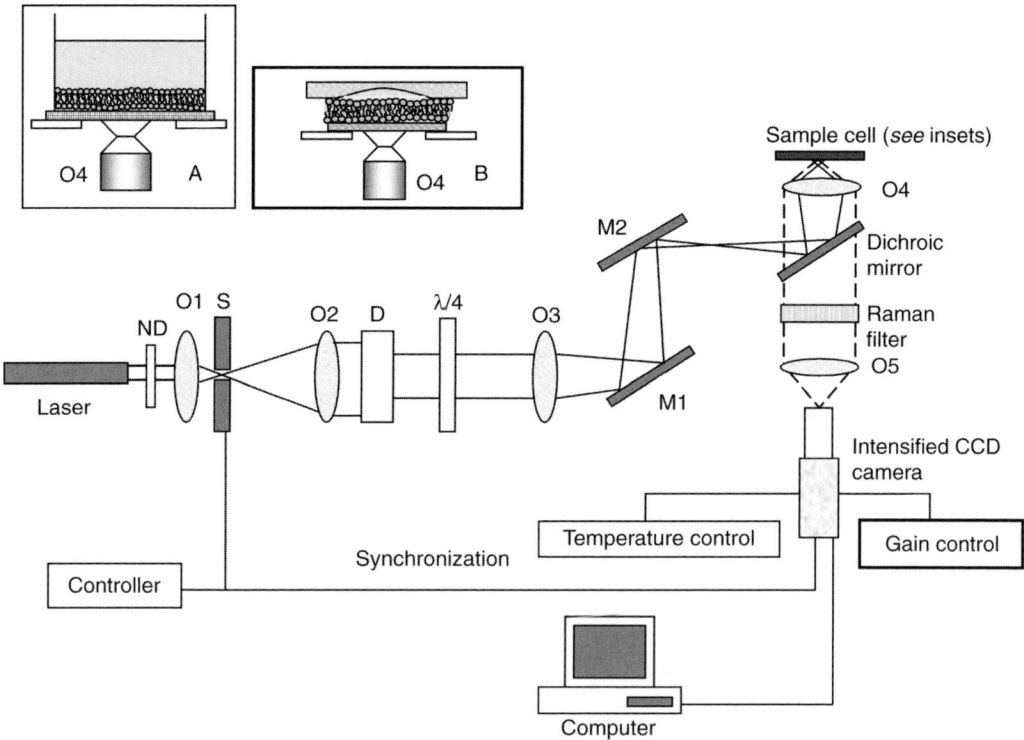

Fig. 4. Schematic of a wide-field single-molecule fluorescence microscopy setup as utilized to track the lateral diffusion of individual fluorophore-labeled lipid molecules in model and cell membranes. The insets depict two common sample cells for the imaging of solid-supported phospholipid bilayer. One sample cell shows a solid-supported bilayer in an open cuvet (**A**). The other shows a bilayer sandwiched between a cover slip and a welled microscopy slide (**B**).

2.3. Monolayer at the Air–Water Interface

Butyl acetate and ethanol are used to clean the Teflon trough. Chloroform (HPLC grade) is utilized to prepare lipid-containing spreading solutions. Milli-Q water (pH = 5.5, 18 MΩ resistivity) or buffer is used as a subphase material. Again, 1% SDS, methanol saturated with NaOH, and 1% HCl are used as glass cleaning solutions.

3. Methods
3.1. Wide-Field Single-Molecule Fluorescence Microscopy: Experimental Setup

To explore the lateral diffusion of membrane constituents in model membranes and cellular membranes, SMD is usually used using either regular wide-field or total internal reflection illuminations (*see* **Note 2**). Next, a wide-field single-molecule fluorescence microscopy setup is described, as it is used in the laboratory to study the lateral diffusion of phospholipids in model membranes, including solid-supported phospholipid bilayers and phospholipid monolayers at the air–water interface *(33,34)*.

Figure 4 shows the schematic of the experimental setup. A 100-mW frequency-doubled Neodymium-doped yttrium aluminium garnet (Nd:YAG) laser at a wavelength of 532 nm

(Crystalaser, Reno, Nevada) is used as a continuous excitation light source. The laser should be mounted appropriately to ensure the proper alignment of the laser beam (parallel to the surface of the optical bench). The laser intensity is regulated by a circular, continuously variable neutral density filter ND. The laser beam is expanded and collimated using lenses O1 and O2 and spatially filtered using a diaphragm D1 to obtain a clean Gaussian intensity profile. A quarter wave plate ($\lambda/4$) is inserted into the beam path to convert the linear polarized laser light into a circular polarized one. The use of circular polarized light prevents elongation artifacts induced by the interaction of the electric field of the laser light with transition dipoles of individual fluorophores, thus enhancing the tracking accuracy. The beam is then delivered to the epifluorescence microscopy (EPI) port of an inverted optical microscope (Zeiss Axiovert S100, Carl Zeiss Optical, Chester, VA) (*see* **Note 3**) by mirrors M1 and M2 and the dichroic mirror DM (Omega XF1051, Omega Optical Inc., Brattleboro, VE), and focused onto the sample by lens O3 and a high magnification (40 × 100), high numerical aperture (1.1–1.45) objective O4 to achieve a spot size of about 20 Êm in diameter (*see* **Note 4**). Here, the position of O3 is adjustable to vary the spot size at the sample.

To guide the laser beam correctly into the microscope, mirrors M1 and M2 need to be mounted onto adjustable mirror mounts, thus allowing for small adjustments of the beam alignment. The fluorescent light from the sample is guided through the dichroic (Omega XF1051) and a Raman edge filter (Omega XR3002 540AELP), and is focused through lens O5 to an intensified CCD camera (iPentaMAX 512EFT, Princeton Instruments, Roper Scientific, Trenton, NJ), which acts as a position-sensitive detector with ultrahigh light sensitivity (*see* **Note 5**). This filter combination ensures the passage of fluorescent light from individual fluorescent probes centered at 566-nm (optimized for TMR-labeled lipids) with optically high signal-to-noise ratio.

The exposure time and the time lag are controlled by one external, mechanical shutter of 3-mm opening aperture (Uniblitz, VMM-D1, Uniblitz, Rochester, NY) and the electronic camera shutter, which both are operated in a synchronized manner. Single, fluorescently labeled lipids are usually imaged with exposure times of 8–10 ms, and time lags between successive exposures of $\tau_0 \geq 25$ ms. Shutter control, image recording, and single-molecule tracking are achieved using ISee imaging software (ISee Imaging Systems, Raleigh, NC) operated on a Dell PC Dell, Round Rock, Texas (2.4 Ghz/512 KB cache, 1 GB RAM, Pentium 4) running on a Linux platform. To achieve high tracking accuracy, the optical setup is mounted on a regular vibration isolation table. It has been previously reported that a similar setup equipped without quarter wave plate but with a Wollaston prism (Zeta Optics, NJ) polarizer positioned between the dichroic and CCD detector can be used to measure the rotational diffusion of dye-labeled lipids in a phospholipid bilayer *(38)*.

The insets of **Fig. 4** show two typical sample geometries for the study of the lateral diffusion of phospholipids in a phospholipid bilayer supported on a glass substrate. One geometry wherein the bilayer is located on top of a glass slide (thickness of a cover slip) at the bottom of a cuvet **(Fig. 4A)**, is well suited to study solid-supported bilayers created through vesicle fusion (VF) (*see also* **Subheading 3.3.**). The other geometry, wherein the bilayer is sandwiched between a cover slip and a microscopy slide with a well, is used to investigate bilayers formed layer by layer using Langmuir–Blodgett (LB) and Schaefer film deposition techniques. Solid-supported phospholipid bilayers are typically imaged using a Zeiss 100× Plan Neofluar (Carl Zeiss, Inc., Thornwood, NY) oil-immersion objective (NA 1.3, WD 0.07 mm). **Figure 5** shows a schematic (side and top views) of a custom-built Langmuir trough for

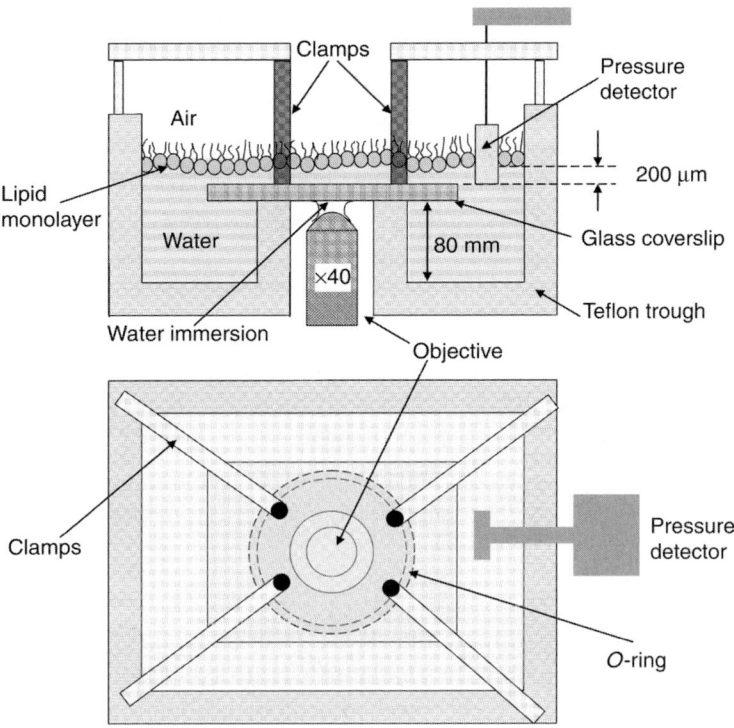

Fig. 5. Schematic of a custom-built mini Langmuir trough for SMD showing side (**above**) and top (**below**) views.

the study of lipid lateral diffusion in Langmuir monolayers at the air–water interface, as used in the laboratory. The trough is built from Teflon. To allow imaging using an inverted microscope, the bottom of the trough has an opening at the center, which is sealed by an optically transparent microscopy cover slip (dimension: 45 × 50 mm^2). To suppress the surface flow of the monolayer and to match the 0.25-mm working distance of the 40× water-immersion objective (Olympus UAPO 40× (Olympus, Center Valley, PA) water immersion, NA 1.15) that is used, the thickness of the water layer above the cover glass has to be maintained at approx 200 µm. The center of the trough is surrounded by a notably deeper well region to allow for the immersion of a stand-alone pressure detector NIMA Technology, Coventry, England (NIMA) and enable the thin water layer in the central area despite of the contact angle between water and Teflon. To avoid water leakage, the cover glass is slightly pressed against an *O*-ring using four to six clamps (*see* **Note 6**).

3.2. Fluorescent Labels

The choice of the right label represents a very important aspect in single-molecule fluorescence imaging applications. Most importantly, the label should not compromise the properties (e.g., lateral diffusion) of the molecule labeled. The choice of the best label is usually determined by both location to be studied and information being sought after. The perfect fluorescence-based single-molecule tracking label should show: (1) small size in comparison with the labeled molecule, (2) high-quantum yield, (3) large extinction coefficients, (4) excellent photostability, and (5) no on–off blinking (typically linked to long residency times in the triplet state).

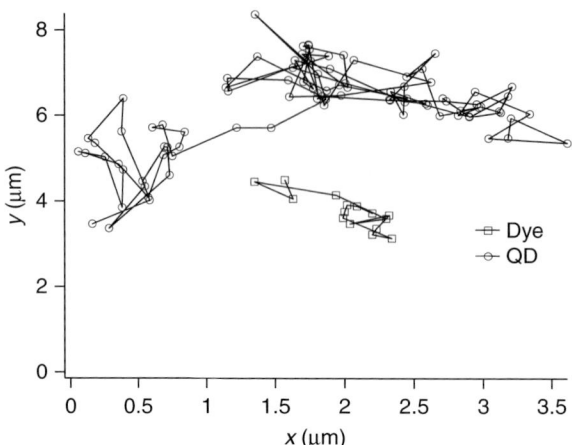

Fig. 6. Representative tracks of dye- and QD-labeled tracer molecules in the same type of planar solid-supported phospholipid bilayer. The longer tracks observed from the QD-conjugated lipid tracers illustrate the enhanced photostability of QDs.

Currently, dye-labeled phospholipids like TMR-DHPE represent the only lipid tracers for SMD studies, which are commercially available (Invitrogen). SMD lipid-tracking experiments are mostly conducted using lipid tracer molecules labeled through covalently attached fluorescent dyes like the cyanine dye Cy3 and TMR *(21,31,34)*. Fluorescent dyes have the advantage of being very small (~1 nm). In addition, imaging of single-lipid molecules is guaranteed because each dye molecule is only bound to one lipid molecule.

The major drawback of dye-labeled lipids in single-molecule imaging applications is their poor photostability and the occurrence of on–off blinking, thus preventing the observation of long tracks from individual tracer molecules. Traditionally, gold-conjugated phospholipids have been used to obtain long tracks *(29)*. However, in this case, the intensity of the scattered signal is dependent on the particle size, which makes gold probes of less than 50 nm size difficult to detect. Quantum dots (QDs) are very promising imaging probes because they combine a relatively small size of less than 10 nm with high photostability and enhanced brightness. Our group has been successful in developing QD-conjugated phospholipids, which show identical diffusion properties as TMR-labeled ones *(39)*. To illustrate the different photostability of TMR-labeled lipids vs QD-labeled lipids, **Fig. 6** shows representative tracks of both lipid probes in planar solid-supported bilayer systems of identical lipid composition. However, it should be noted that individual QDs (like individual dyes), exhibit on–off blinking, thus complicating the tracking of molecules. To overcome on–off blinking, dye- or QD-doped nanoparticles represent an interesting alternative. Although being of larger size (≥20 nm) than dyes and QDs, these probes are attractive because their brightness can be enhanced by increasing the average number of fluorophores per nanoparticle. Here, the maximum loading concentration is only limited by self-quenching. However, extreme care should be taken if nanoparticles are conjugated to lipids to ensure that each nanoparticle is not linked to more than one lipid and that no nanoparticle cross-linking occurs. In addition, these nanoparticles should have bioinert surface properties, thus preventing any nonspecific interaction with their environment *(40)*.

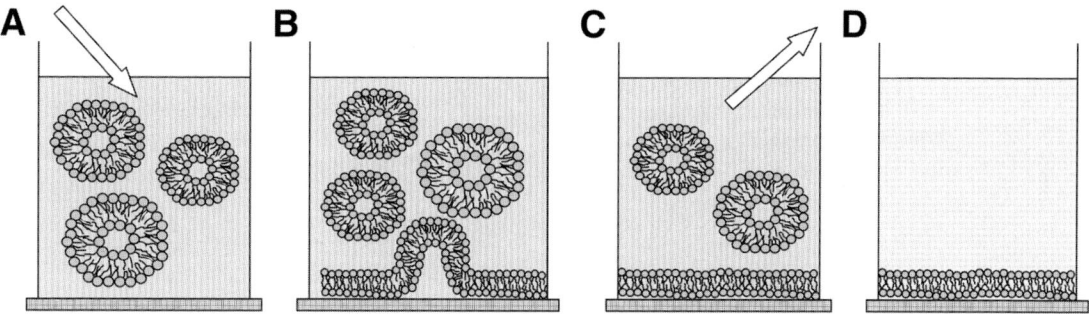

Fig. 7. Scheme of the preparation of a solid-supported phospholipid bilayer through fusion of SUVs. The scheme shows the following steps: (**A**) injection of SUVs, (**B**) VF, (**C**) removal of excess SUVs, and (**D**) solid-supported bilayer ready for imaging experiments. The procedure leads to symmetric compositions of nonlabeled and labeled lipids.

3.3. Sample Preparation

3.3.1. Solid-Supported Phospholipid Bilayer

Most commonly, planar solid-supported phospholipid bilayers are prepared using established procedures based on VF on glass substrates or lipid monolayers formed through LB film deposition, and successive LB and Schaefer (S) film deposition steps *(41–44)*. Whereas VF on a glass slide represents the simplest method to create bilayers of symmetric nonlabeled and labeled lipid compositions, LB/VF and LB/S are used to build asymmetric bilayers with respect to lipids and tracer molecules, respectively. For example, the latter two approaches have been utilized to build polymer-supported phospholipid bilayers *(34,45,46)*. Representative approaches of these two bilayer preparation procedures are described in **sections 3.3.1.1.** and **3.3.1.2.**

3.3.1.1. Preparation Through VF

First, separate chloroform stock solutions of labeled and nonlabeled phospholipids are prepared. Then, a lipid mixture of appropriate molar ratio (~10^{-8} mol% labeled lipid) from these stock solutions is formed. To evaporate chloroform, place the sample solutions under nitrogen gas. Once the sample is completely dried, Milli-Q (pH = 5.5, 18 MΩ/cm resistivity) or buffer should be added to obtain a lipid concentration of approx 8–10 mg/mL (*see* **Note 7**). To form multilamellar vesicles, the sample needs to be vortexed using a table-top vortexer. Because multilamellar vesicles are not fusogenic, they need to be transferred into small unilamellar vesicles (SUVs) of ≤100 nm size using a rod sonifier (e.g., Branson Model 450 (Branson Ultrasonics, Danbury, CT) or a hand extruder (*see* **Note 8**). If sonication is used, the titanium dust needs to be removed from the vesicle solution by centrifugation at approx 925g for 5 min using a table-top centrifuge. In addition, it is beneficial to separate smaller SUVs from bigger ones by adding a higher-speed centrifugation step *(42)*. The assembly of the solid-supported bilayer is illustrated in **Fig. 7**. If the cuvet geometry is utilized, the solid-supported bilayer is formed by simply adding the SUV solution into the cuvet (**Fig. 7A**). SUVs are allowed to fuse for about 30–45 min (**Fig. 7B**). Finally, excess (nonfused) vesicles are removed by exchanging the solution several times (**Fig. 7C**). After all SUVs are removed, the solid-supported bilayer is ready for imaging (**Fig. 7D**).

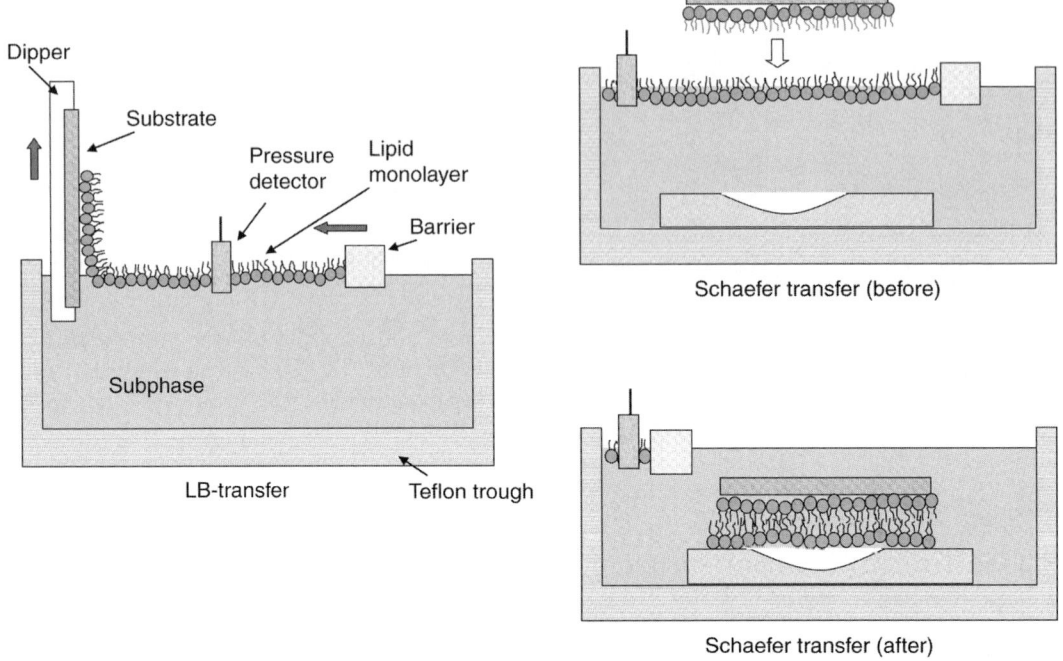

Fig. 8. Preparation of the solid-supported bilayer using Langmuir (**left**) and Schaefer (**right**) film transfer steps. This approach is used to achieve asymmetric bilayer composition or label only one leaflet of the bilayer.

3.3.1.2. Preparation Through Film Deposition Techniques

To prepare phospholipid bilayers of asymmetric lipid composition or to only label one leaflet of the bilayer, LB and Schaefer film deposition procedures should be applied. During the LB film deposition procedure, lipids are transferred from the air–water interface to a glass substrate, thus forming a solid-supported monolayer (**Fig. 8**, **left side**). Here, the glass substrate is mounted onto a computer-controlled dipper of a film-balance system (Labcon, Labcon, Darlington, UK), which positions the substrate vertical to the air–water interface. Before use, the Teflon trough needs to be cleaned using butyl acetate- or ethanol-soaked kimwipes followed by extensive rinsing with Milli-Q water. After the trough is filled with Milli-Q or buffer, the lipids are dissolved in a spreading solvent such as chloroform and spread onto the subphase using a microsyringe. Some phospholipids do not dissolve very well in pure chloroform. In these cases, add a small amount of methanol. After the spreading solvent is evaporated, the phospholipids self-assemble into a monolayer at the air–water interface. The computer-controlled moveable barrier of the trough compresses the film to the desired surface pressure of 30–35 mN/m (area per molecule) and maintains the pressure during dipping. The surface tension is measured using a partially submerged Wilhelmy plate (part of the Labcon trough system). A dipper with the solid substrate (glass cover slide) attached is lowered into the water before the film is spread onto the subphase. Once the film is spread and compressed to the desired surface pressure, the dipper can be pulled at a set speed out of the water transferring the monolayer from the air–water interface to the substrate.

Note that best film transfer results are achieved if the glass cover slides show very hydrophilic surface properties (*see* **Note 9**). Constant pressure can be maintained during film transfer by a computer-controlled feedback system between the electrobalance measuring the surface pressure and the barrier-moving mechanism. The surface pressure should remain constant to maintain a constant area per molecule A_{lipid} ~65 Å² because this area corresponds to typical values found for lipids in bilayers. A constant area per molecule is important because the lateral diffusion of phospholipids is described by a free area model, which predicts a change in diffusion coefficient (*D*) for different values of A_{lipid} *(47,48)*.

There are several parameters that affect what type of LB film is produced: the nature of the spread film, the subphase composition and temperature, the surface pressure during the deposition and the deposition speed, the type and nature of the solid substrate, and the time the solid substrate is stored in air or in the subphase between the deposition cycles. The quality of the transfer can be monitored by the transfer ratio, which is the ratio of the decrease in monolayer area during deposition and the area of the substrate (ideally the ratio would be one). Once the monolayer is transferred onto a solid substrate the lipid molecules are essentially immobile. Importantly, the LB film retains the structural aspects of the monolayer before deposition. The solid-supported bilayer is completed by transferring a lipid monolayer at an average area per lipid of approx 65 Å² from the air–water interface to the LB-monolayer using the Schaefer dipping technique (**Fig. 8, right side**). The glass substrate should be kept slightly tilted during the dipping process to avoid the entrapment of air bubbles. The completed solid-supported bilayer needs to be stored in an aqueous environment.

3.3.2. Phospholipid Monolayer (Air–Water Interface)

Before use, the Teflon trough is cleaned using butyl acetate and extensively rinsed with Milli-Q. As mentioned, the cover glasses used for sealing the trough (**Fig. 5**) need to be very hydrophilic (*see* **Note 9**). A cleaned cover glass is mounted onto the trough using four to six clamps (*see* **Note 6**). Herein, proper sealing can be achieved by an *O*-ring. Before spreading, the trough needs to be filled to obtain a water layer of approx 5 mm above the cover glass (*see* **Note 10**). Then, the monolayer is spread onto the subphase. After spreading, the thickness of the water layer is reduced to approx 200 μm to match the working distance of the ×40 objective used. To reduce the perturbation of the monolayer by airflow before imaging, the trough needs to be covered and sufficient equilibration time should be considered before imaging can start.

3.4. Tracking Analysis

As illustrated in **Fig. 1**, the goal of single-molecule tracking techniques is to determine the position of tracer molecules over time. **Figure 9** shows a typical particle tracking software approach, as it is used in the laboratory. Two frames are defined here: tracking and search frame. The former, based on *n* pixels, contains the signal from the tracer molecule of interest and its direct surrounding. The search frame defines the distance from the tracer molecule whereby the tracer molecule is searched in the following image. Herein, the screening is executed by correlating the intensity pattern of the pixels with those of the tracking frame (*see* **Note 11**).

Importantly, this tracking method allows the identification of the proper tracer molecule out of an ensemble of tracers characterized by different shapes and intensity levels. Using this tracking procedure, the x_n, y_n positions of tracer molecules can be determined as a function of τ_n. If the individual tracks are long enough (e.g., in the case of QD-conjugated lipids), $\langle r^2(\tau_n) \rangle$ the value can be simply determined by analyzing the $r^2(\tau_n)$ of multiple tracks (~150–200) as a function of τ_n and by comparing the data with the different diffusion models shown in **Eqs. 3–6**. When photobleaching limits the length of the tracks, SMD

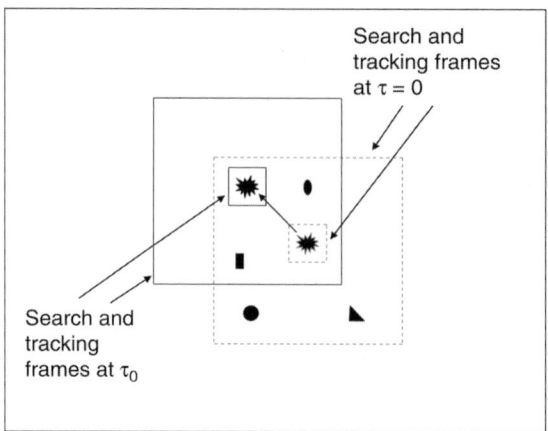

Fig. 9. Process of single-molecule tracking based on tracking and search frames. The tracking frame defines the intensity profile of the tracer and its surrounding area. The search frame defines the area in the next image where to search for the intensity profile defined by the tracking frame. If the tracer is found, its position becomes the new center position of the search frame for the next following image. This process is repeated until the tracer molecule cannot be identified within the search frame.

data are best analyzed in terms of CDFs. In this case, the information about particular diffusion processes can be derived by comparing the CDF data with theoretical models describing Brownian diffusion (**Eq. 2**), two-component diffusion (**Eq. 7**) *(36)*, and anomalous diffusion (**Eq. 8**) *(34)*:

$$P(r^2, \tau_0) = 1 - \left\{ \alpha \exp\left[-\frac{r^2(\tau_0)}{\langle r_I^2(\tau_0) \rangle} \right] + (1-\alpha)\left[-\frac{r^2(\tau_0)}{\langle r_{II}^2(\tau_0) \rangle} \right] \right\} \quad (7)$$

$$P(r^2, \tau_0) = \gamma(a, br^c) / \Gamma(a) \quad (8)$$

with $\langle r_I^2(\tau_0) \rangle$ and $\langle r_{II}^2(\tau_0) \rangle$ being the mean-square displacements of the two populations and $\gamma(a, br^c)$ and $\Gamma(a)$ being incomplete and complete Γ-functions.

The data can also be analyzed in terms of the probability density function $P(r, \tau)$ *(49)*. However, CDF are advantageous because there is no need to impose bin sizes, and the summation integrates out much of the noise *(34)*. The aforementioned tracking procedures are well suited to analyze diffusion data on solid-supported lipid bilayers. When single lipids are tracked in a fluid lipid monolayer at the air–water interface, single-molecule tracking as described earlier becomes very challenging. This is because the surface flow and vibrations of the setup amplified by the approx 200-μm thick water layer between cover slip and monolayer are difficult to suppress. In this situation, it is often necessary to analyze the relative motion of multiple tracers to obtain information about their lateral diffusion properties. For example, multiple-particle tracking analysis has been applied to study the microrheology of inhomogeneous soft materials, and also to correlate microstructure and rheology of a bundled and cross-linked F-actin network *(50–52)*.

To track dye-labeled lipids in a phospholipid monolayer, the best approach is to determine the relative square displacements $(\delta r_n)^2$ of an ensemble of tracers at one time lag. The resulting distribution of $(\delta r_n)^2$ can be analyzed in terms of the CDF, wherein the data can be fitted with respect to the different diffusion models (**Eqs. 2,7,8**). In addition, $\langle(\Delta r_m)^2\rangle$ and D can be determined as well.

4. Notes

1. If cells are studied, care should be taken that the lateral diffusion of tracer molecules is of 2D nature. Cell surface tracking experiments are best performed using cells that can be easily immobilized on a cover glass and show flat membrane areas. If curved membranes are studied, out-of-plane diffusion will occur, which is detected as an anomalous diffusion process by the imaging system.
2. Wide-field single-molecule fluorescence microscopy applied to diffusion studies on individual fluorescently labeled lipids can be operated through EPI illumination or total internal reflection fluorescence (TIRF). SMD through EPI illumination (illustrated in **Fig. 4**) represents the least demanding strategy. However, the optical background in EPI illumination is relatively high in comparison with TIRF, where a thin layer of approx 100-nm thickness is excited. The excitation through an evanescent field that is required in TIRF can be generated either through an objective of high magnification and very-high numerical aperture (>1.3) or through a prism *(53,54)*. TIRF is particularly valuable for tracking experiments on cells wherein autofluorescence is a serious issue. However, EPI- and TIRF-based SMD are of similar quality when applied to model membranes, as EPI can bring more light to the sample than TIRF, thus maintaining a comparable signal-to-background ratio with respect to the evanescent field approach.
3. Because SMD is a very low light-level application, it is important to keep the optical beam path as short as possible. For that reason, an inverted optical microscope with a TV port is often used, whereby the CCD camera is mounted underneath the microscope.
4. Solid-supported bilayers are typically imaged using a Zeiss 100× oil-immersion Plan Neofluar objective (NA: 1.3, WD: 0.07 mm), whereas Langmuir monolayers are studied using an Olympus ×40 water-immersion objective (NA: 1.15, WD: 0.25 mm). In both cases, an additional magnification lens can be used (e.g., Optovar ×1.6 or 2.5, Carl Zeiss Inc., Thornwood, NY).
5. A back-illuminated CCD camera of very high quantum yield represents an alternative that is less expensive yet also provides lower frame rates.
6. Care should be taken by tightening the clamps because bending or even breakage of the cover glass might occur.
7. The choice of the p*I* and pH depends on the composition of the bilayer. For example, it was found that SUV fusion depends on the net charge (charge density) of the bilayer. SUV fusion will improve with decreasing pH and increasing p*I* if the bilayer is partially negatively charged, but will not be negatively affected if the bilayer is neutral or positively charged *(44)*.
8. SUVs are only stable if the lipids remain in the fluid-phase state.
9. To obtain a very hydrophilic-glass surface, bake the glass cover slides (2×3.5 cm^2) at 515°C in a kiln for 3 h. Sonicate the glass slides in 1% SDS, followed by methanol saturated with NaOH and 1% HCl for 30 min each. After each sonication step, rinse the slides extensively with Milli-Q. Store the glass slides in Milli-Q, to be used within a week *(34)*.
10. The spreading of the lipid-containing chloroform solution onto the air–water interface may disrupt the water film above the glass substrate if the water layer is too thin.
11. Care should be taken to assure that the tracer molecule remains in the tracking frame throughout the whole track. Otherwise, noise will be tracked. In the latter case, the measured lateral diffusion increases with increasing size of the search frame. Also, make sure that the concentration of tracer molecules is not too high because this could cause the jumping of the tracking frame from one tracer to another within each track.

References

1. Saxton, M. J. (1999) Lateral diffusion of lipids and proteins. *Curr. Top. Membr.* **48,** 229–282.
2. Sheetz, M. P. (1995) Cellular plasma membrane domains. *Mol. Membr. Biol.* **12,** 89–91.
3. Edidin, M. (1997) Lipid microdomains in cell surface membranes. *Curr. Opin. Struct. Biol.* **7,** 528–532.
4. Edidin, M. (2001) Shrinking patches and slippery rafts: scales of domains in the plasma membrane. *Trends Cell Biol.* **11,** 492–496.
5. Simons, K. and Ikonen, E. (1997) Functional rafts in cell membranes. *Nature* **387,** 569–572.

6. Sako, Y. and Kusumi, A. (1994) Compartmentalized structure of the plasma membrane for receptor movements as revealed by nanometer-level motion analysis. *J. Cell Biol.* **125,** 1251–1264.
7. Sako, Y. and Kusumi, A. (1995) Barriers for lateral diffusion of transferrin receptor in the plasma membrane as characterized by receptor dragging by laser tweezers: fence versus tether. *J. Cell Biol.* **129,** 1559–1574.
8. Saxton, M. J. and Jacobson, K. (1997) Single particle tracking: Applications to membrane dynamics. *Annu. Rev. Biophys. Biomol. Struct.* **26,** 373–399.
9. Moerner, M. E. and Kador, L. (1989) Optical-detection and spectroscopy of single molecules in a solid. *Phys. Rev. Lett.* **62,** 2535–2538.
10. Orrit, M. and Bernard, J. (1990) Single pentacene molecules detected by fluorescence excitation in a para-terphenyl crystal. *Phys. Rev. Lett.* **65,** 2716–2719.
11. Betzig., E. and Chichester, R. J. (1993) Single molecules observed by near-field scanning optical microscopy. *Science* **262,** 1422–1425.
12. Ambrose, W. P., Goodwin, P. M., Martin, J. C., and Keller, R. A. (1994) Single-molecule detection and photochemistry of a surface using near-field optical excitation. *Phys. Rev. Lett.* **72,** 160–163.
13. Ambrose, W. P., Goodwin, P. M., Martin, J. C., and Keller, R. A. (1994) Alterations of single molecule fluorescence lifetimes in near-field optical microscopy. *Science* **265,** 364–367.
14. Xie, X. S. and Dunn, R. C. (1994) Probing single-molecule dynamics. *Science* **265,** 361–364.
15. Trautmann, J. K., Macklin, J. J., Brus, L. E., and Betzig, E. (1994) Near-field spectroscopy of single molecules at room temperature. *Nature* **369,** 40–42.
16. Nie, S., Chiu, D. T., and Zare, R. N. (1994) Probing individual molecules with confocal fluorescence microscopy *Science* **266,** 1018–1021.
17. Macklin, J. J., Trautmann, J. K., Harris, T. D., and Brus, L. E. (1994) Imaging and time-resolved spectroscopy of single molecules at an interface. *Science* **272,** 255–258.
18. Ha, T., Enderle, T., Chemla, D. S., Selvin, P. R., and Weiss, S. (1996) Single molecule dynamics studied by polarization modulation. *Phys. Rev. Lett.* **77,** 3979–3982.
19. Lu, H. P. and Xie, X. S. (1997) Single-molecule spectral fluctuations at room temperature. *Nature* **385,** 143–146.
20. Funatsu, T., Harada, Y., Tokunaga, M., Saito, K., and Yanagida, T. (1995) Imaging of single fluorescent molecules and andividual ATP turnovers by single myosin molecules in aqueous solution. *Nature* **374,** 555–559.
21. Schmidt, Th., Schuetz, G. J., Baumgartner, W., Gruber, H. J., and Schindler, H. (1995) Characterization of photophysics and mobility of single molecules in a fluid lipid membrane. *J. Phys. Chem.* **99,** 17,662–17,668.
22. Guettler, F., Irngartinger, T., Plakhotnik, T., Renn, A., and Wild, U. P. (1994) Fluorescence microscopy of single molecules. *Chem. Phys. Lett.* **217,** 393–397.
23. Sase, I., Miyata, H., Corrie, J. E. T., Craik, J. S., and Kinosita, K. (1995) Real-time imaging of single fluorophores on moving actin with an epifluorescence microscope. *Biophys. J.* **69,** 323–328.
24. Schmidt, Th., Schuetz, G. J., Baumgartner, W., Gruber, H. J., and Schindler, H. (1996) Imaging of single molecule diffusion. *Proc. Natl. Acad. Sci. USA* **93,** 2926–2929.
25. Dickson, R. M., Cubitt, A. B., Tsien, R. Y., and Moerner, W. E. (1996) 3-dimensional imaging of single molecules solvated in pores of poly(acrylamide) gels. *Science* **274,** 966–969.
26. Dickson, R. M., Cubitt, A. B., Tsien, R. Y., and Moerner, W. E. (1997) On/off blinking and switching behavior of single molecules of green fluorescent proteins. *Nature* **388,** 355–358.
27. Xu, X. H. and Yeung, E. S. (1997) Direct measurement of single- molecule diffusion and photodecomposition in free solution. *Science* **275,** 1106–1109.
28. Tokunaga, M., Kitamura, K., Saito, K., Iwane, A. H., and Yanagida, T. (1997) Single molecule imaging of fluorophores and enzymatic reactions achieved by objective-type total internal-reflection fluorescence microscopy. *Biochem. Biophys. Res. Commun.* **235,** 47–53.

29. Lee, G. M., Ishihara, A., and Jacobson, K. (1991) Direct Brownian motion of lipids in a membrane. *Proc. Natl. Acad. Sci.* **88,** 6274–6278.
30. Sonnleitner, A., Schuetz, G. J., and Schmidt, Th. (1999) Free Brownian motion of individual lipid molecules in biomembranes. *Biophys. J.* **77,** 2638–2642.
31. Fujiwara, T., Ritchie, K., Murakoshi, H., Jacobson, K., and Kusumi, A. (2002) Phospholipids undergo hop diffusion in compartmentalized cell membrane. *J. Cell Biol.* **157,** 1071–1081.
32. Ke, P. C. and Naumann, C. A. (2001) Single molecule fluorescence imaging of phospholipid monolayers at the air-water interface. *Langmuir* **17,** 3727–3733.
33. Ke, P. C. and Naumann, C. A. (2001) Hindered diffusion in polymer-tethered phospholipid monolayers at the air-water interface: A single molecule fluorescence imaging study. *Langmuir* **17,** 5076–5081.
34. Deverall, M. A., Gindl, E., Sinner, E. -K., et al. (2005) Membrane lateral mobility obstructed by polymer-tethered lipids studied at the single molecule level. *Biophys. J.* **88,** 1875–1886.
35. Qian, H., Sheetz, M. P., and Elson, E. L. (1991) Single particle tracking. Analysis of diffusion and flow in two-dimensional systems. *Biophys. J.* **60,** 910–921.
36. Schuetz, G. J., Schindler, H., and Schmidt, Th. (1997) Single-molecule microscopy on model membranes reveals anomalous diffusion. *Biophys. J.* **73,** 1073–1080.
37. Bobroff, N. (1986) Position measurement with a resolution and noise-limited instrument. *Rev. Sci. Instrum.* **57,** 1152–1157.
38. Harms, G. S., Sonnleitner, M., Schuetz, G. J., Gruber, H. J., and Schmidt, Th. (1999) Single-molecule anisotropy imaging. *Biophys. J.* **77,** 2864–2870.
39. Murcia, M. J., Minner, D. E., Ritchie, K., Naumann, C. Design of Monovalently-Labeled Quantum Dot-Conjugated Lipids for High-Speed Tracking Experiments on Cell Surfaces (Submitted).
40. Murcia, M. J. and Naumann, C. A. (2005) Biofunctionalization of fluorescent nanoparticles, in *Biofunctionalization of Nanomaterials*, (Kumar, C., ed.) Wiley-VCH Weinheim, Germany pp. 1–40.
41. Tamm, L. K. and McConnell, H. M. (1985) Supported phospholipid bilayers. *Biophys. J.* **47,** 105–113.
42. Kalb, E., Frey, S., and Tamm, L. K. (1992) Formation of supported planar bilayers by fusion of vesicles. *Biophys. J.* **47,** 105–113.
43. Kalb, E., Frey, S., and Tamm, L. K. (1992) Formation of supported planar bilayers by vesicle fusion to supported monolayers. *Biochim. Biophys. Acta* **1103,** 763–765.
44. Cremer, P. S. and Boxer, S. G. (1999) Formation and spreading of lipid bilayers on planar glass supports. *J. Phys. Chem. B* **103,** 2554–2559.
45. Wagner, M. L. and Tamm, L. K. (2000) Tethered polymer-supported planar bilayers for reconstitution of integral membrane proteins: Silane-polyethyleneglycol-lipid as a cushion and covalent linker. *Biophys. J.* **79,** 1400–1414.
46. Naumann, C. A., Prucker, O., Lehmann, T., Ruehe, J., Knoll, W., and Frank, C. W. (2002) The polymer-supported phospholipid bilayer: tetherering as a new approach toward substrate-membrane stabilization. *Biomacromolecules* **3,** 667–678.
47. Traeuble, H. and Sackmann, E. (1972) Studies of the crystalline-liquid phase transition of lipid model membranes, III. Structure of a steroid-lecithin system below and above the lipid phase transition. *J. Am. Chem. Soc.* **94,** 4499–4510.
48. Galla, H. J., Hartmann, W., Theilen, U., and Sackmann, E. (1979) On two-dimensional passive random walk in lipid bilayers and fluid pathways in biomembranes. *J. Membr. Biol.* **48,** 215–236.
49. Anderson, C. M., Georgiou, G. N., Morrison, I. E., Stevenson, G. V., and Cherry, R. J. (1992) Tracking of cell surface receptors by fluorescence digital imaging microscopy using a charge-coupled device camera. Low-density lipoprotein and influenca virus receptor mobility at 4°C. *J. Cell Sci.* **101,** 415–425.
50. Crocker, J. C., Valentine, M. T., Weeks, E. R., et al. (2000) Two-point microrheology of inhomogeneous soft materials. *Phys. Rev. Lett.* **85,** 888–891.

51. Valentine, M. T., Kaplan, P. D., Thota, D., et al. (2001) Investigating the microenvironments of inhomogeneous soft materials with multiple particle tracking. *Phys. Rev. E* **64,** 061,506–061,515.
52. Shin, J. H., Gardel, M. L., Mahadevan, L., Matsudaira, P., and Weitz, D. A. (2000) Relating microstructure to rheology of a bundled and cross-linked F-actin network in vitro. *Proc. Natl. Acad. Sci. USA* **101,** 9636–9641.
53. Axelrod, D. (1989) Total internal-reflection fluorescence microscopy. *Methods Cell Biol.* **30,** 245–270.
54. Ambrose, W. P., Goodwin, P. M., and Nolan, J. P. (1999) Single molecule detection with total internal reflection excitation: Comparing signal-to-background and total signals in different geometries. *Cytometry* **36,** 224–231.

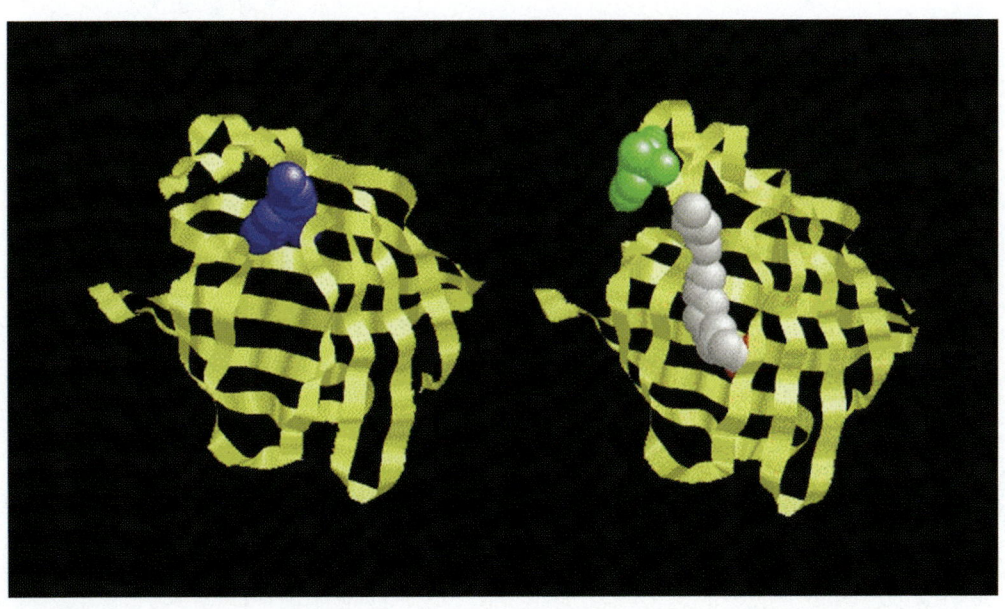

Color Plate 1. Modeled structure of ADIFAB with **(left)** and without **(right)** FA bound *(26)*. (Fig. 1, Chapter 3; *see* complete caption on p. 28.)

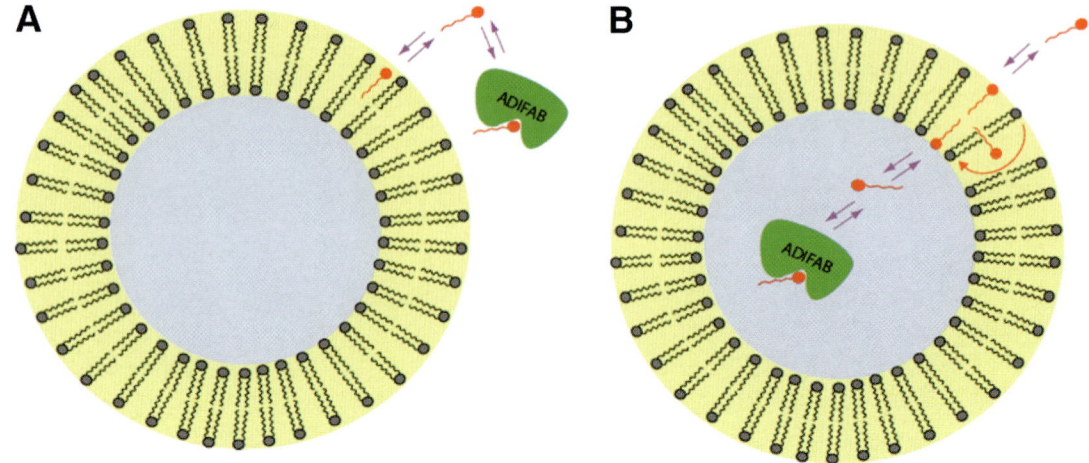

Color Plate 2. Fluorescence assays utilizing the ADIFAB probe to detect unbound concentrations of FA. (Fig. 2, Chapter 3; *see* complete caption on p. 31.)

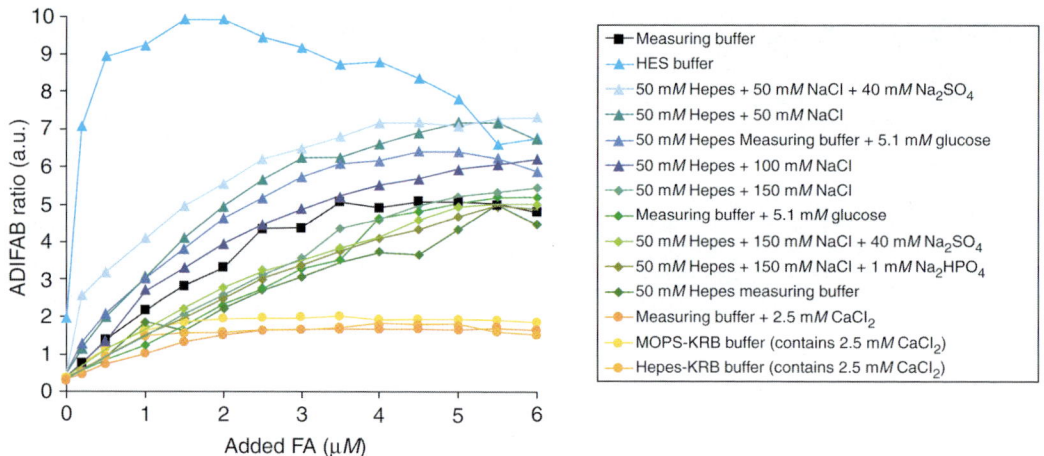

Color Plate 3. Effects of various buffers on ADIFAB titration curves. (Fig. 6, Chapter 3; *see* complete caption on p. 37.)

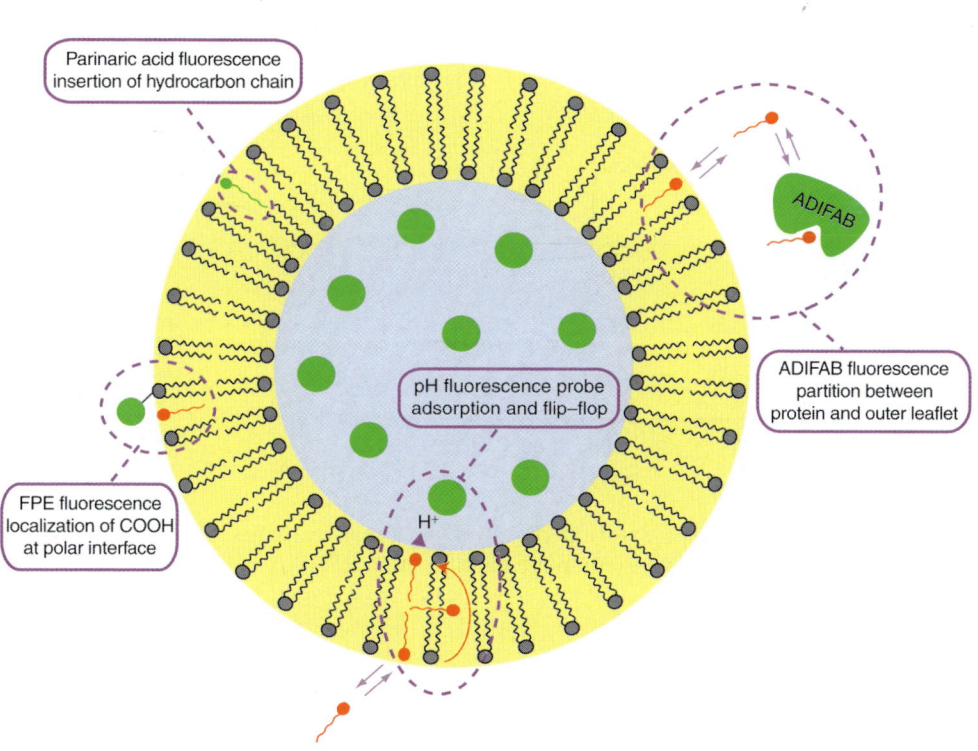

Color Plate 4. Fluorescence approaches developed to study the mechanism of FA transport in membranes. (Fig. 1, Chapter 16; *see* complete caption on p. 238.)

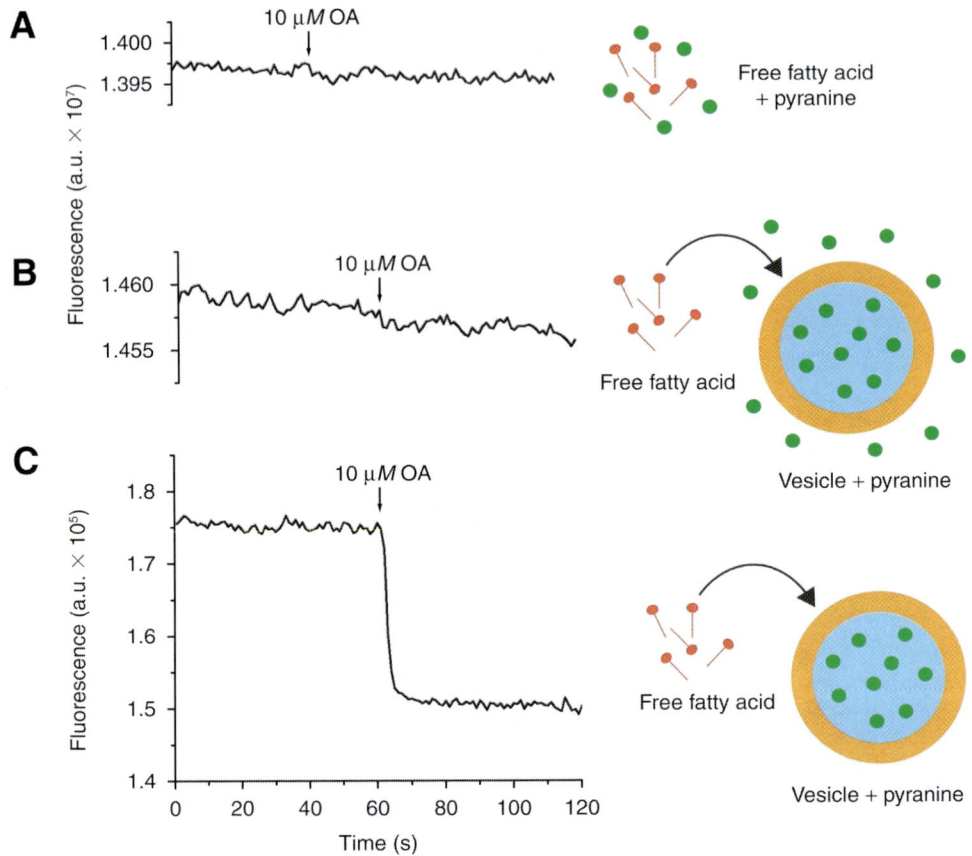

Color Plate 5. The effect of external pyranine on measurements of FA diffusion through membranes. (Fig. 10, Chapter 16; *see* complete caption on p. 252.)

20

Modeling 2D and 3D Diffusion

Michael J. Saxton

Summary

Modeling obstructed diffusion is essential to the understanding of diffusion-mediated processes in the crowded cellular environment. Simple Monte Carlo techniques for modeling obstructed random walks are explained and related to Brownian dynamics and more complicated Monte Carlo methods. Random number generation is reviewed in the context of random walk simulations. Programming techniques and event-driven algorithms are discussed as ways to speed simulations.

Key Words: Brownian dynamics; diffusion; event-driven algorithm; Monte Carlo; n-fold way; random number generation; random walk.

1. Introduction

Diffusion in a cell—whether two-dimensional (2D) diffusion in a membrane or 3D diffusion in the nucleus or the cytoplasm—occurs at a high concentration of obstacles. To model a diffusion-mediated process, it is necessary to account for the effects of mobile and immobile obstacles. The biophysics of lateral diffusion in membranes is reviewed elsewhere *(1)*. High concentrations of nonreactants also have a major effect on reaction equilibria and rates, as studied extensively by Minton and colleagues *(2)*. One major application is in the modeling of kinetics. Another is in the analysis of single-particle tracking experiments, in which the motion of individual lipids, proteins, or vesicles is followed *(3)*. Simulations are essential here. First, a pure random walk is the null hypothesis for any claim of anomalous, directed, or confined motion, so one must compare the observed trajectory to random ones. Second, any treatment applied to the experimental data must also be applied to simulated data with a similar number of time points and noise, as a control. Third, simulations can be used to fill in gaps in the experimental argument. This chapter discusses how to model diffusion in 2D and 3D, mostly from the standpoint of writing a program. Much of this is essential to understand even when using a packaged program. The chapter is an introduction; a full treatment would be of book length *(4–6)*. Before writing a diffusion program, one should also read the methods sections from a variety of publications in the area of interest.

2. Principles of Diffusion

The basic quantity characterizing motion in diffusion simulations is the mean-square displacement (MSD) $\langle r^2 \rangle$ as a function of time t. **Table 1** lists the main types of motion. In normal diffusion, the MSD is proportional to time and the diffusion coefficient is constant. In hindered normal diffusion, the diffusion coefficient is reduced but the time dependence of the MSD is unchanged. In anomalous subdiffusion, however, the form of the time dependence is

From: *Methods in Molecular Biology, vol. 400: Methods in Membrane Lipids*
Edited by: A. M. Dopico © Humana Press Inc., Totowa, NJ

Table 1
Types of Motion in *d* Dimensions

Type of motion	MSD	Examples
Normal	$\langle r^2 \rangle = 2dD_0 t$	Dilute membrane or solution
Hindered normal	$\langle r^2 \rangle = 2dDt, D < D_0$	Mobile obstacles: rod outer segment, mitochondria, chloroplasts
Anomalous	$\langle r^2 \rangle = \Gamma t^\alpha, \alpha < 1$	CTRW, traps, percolation cluster
Directed	$\langle r^2 \rangle = 2dD_0 t + v^2 t^2$	Diffusion in flowing membrane or cytoplasm
Confined	$\langle r^2 \rangle \approx \langle r^2 \rangle_0 [1 - \exp(-t/\tau)]$	Trapping in domains or cytoskeletal corral

changed and the MSD is proportional to some power of time less than one. Diffusion is hindered and the diffusion coefficient is time-dependent,

$$D(t) = \Gamma/t^{1-\alpha}, \qquad (1)$$

decreasing to zero at large times *(7)*. Anomalous subdiffusion is the result of deviations from the central limit theorem, such as strong correlations in diffusive motion or a wide distribution of trapping times *(8)*. Here a wide distribution is one for which no mean value exists, such as the continuous-time-random walk (CTRW) of **Eq. 11**. Anomalous subdiffusion due to obstacles *(9)* and traps *(10)* is discussed elsewhere. Confined diffusion is normal at short times but at long times the MSD approaches a constant proportional to the size of the domain confining the particle. For simplicity the time dependence is shown as a single exponential in **Table 1**. The full functions are infinite series of exponentials, although they are often well approximated by the constant plus first exponential terms. The full functions are available for square *(11)* and circular *(12)* corrals. Programs to calculate the Bessel functions needed for the circular case are in *(13)* and in Mathematica (Wolfram Research, Champaign, IL, www.wolfram.com). **Figure 1** shows the MSD as a function of time in linear and log–log plots.

The basic mathematical description of diffusion is the propagator or Green's function $P(r,t|r_0,t_0)dV$, giving the probability that if a particle is at position r_0 at time t_0, it is in an element of volume dV at r at time t. For simplicity, take $r_0 = 0$, $t_0 = 0$ and write the propagator as $P(r,t)$. For pure diffusion in a d-dimensional medium,

$$P(\vec{r},t)dV = \frac{1}{(4\pi Dt)^{d/2}} \exp\left[\frac{-r^2}{4Dt}\right] dV \qquad (2)$$

where D is the diffusion coefficient. The MSD is

$$\langle r^2 \rangle = 2dDt \qquad (3)$$

and the standard deviation σ of the MSD is proportional to the MSD

$$\sigma = \sqrt{\frac{2}{d}} \langle r^2 \rangle. \qquad (4)$$

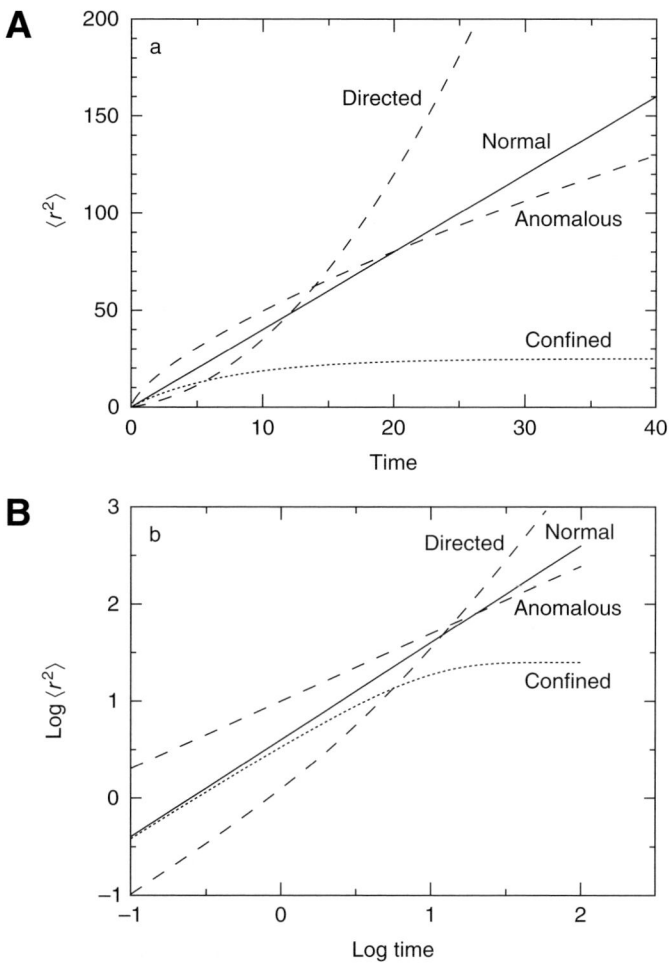

Fig. 1. Mean-square displacement $\langle r^2 \rangle$ as a function of time in (a) linear and (b) log–log plots for normal diffusion, anomalous subdiffusion, directed motion, and confined motion. The anomalous subdiffusion exponent $\alpha = 0.695$ is the value for diffusion on a percolation cluster.

For details *see* the Appendix. In 2D, the propagator can be rewritten as

$$P(r,t)dV = \left[2r\,dr/\langle r^2(t)\rangle\right]\exp\left[-r^2/\langle r^2(t)\rangle\right] \quad (5)$$

but the corresponding formulas in 1D and 3D are less convenient. In the simplest approximation *(7)*, a propagator for anomalous subdiffusion can be obtained from **Eq. 2** by replacing D with $D(t)$ from **Eq. 1**. This approximation was sufficient to model anomalous fluorescence photobleaching recovery *(14)*. The next simplest approximation is discussed elsewhere *(15)*; Metzler and Klafter *(16)* provide an excellent review of anomalous diffusion. The propagator for confined diffusion is given in *(12)* for an impermeable circular corral; forms for other geometries are given by Carslaw and Jaeger *(17)*; and the propagator for a grid of penetrable squares is given in *(18)*, as used by the Kusumi group *(19)* to analyze single-particle tracking data in terms of their membrane skeleton fence model.

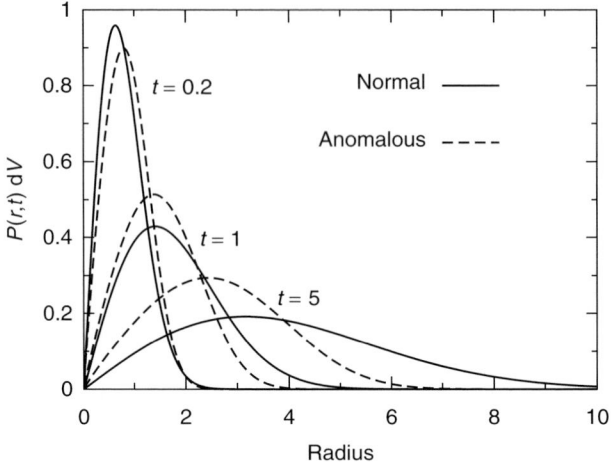

Fig. 2. Propagators $P(r,t)$ dV for normal diffusion in 2D (solid line) and for anomalous subdiffusion (dashed line) for the indicated times. The normal propagator is from **Eq. 2** and the anomalous propagator is the approximation used in **ref. 15**, with the exponents given there for a random walk on a percolation cluster. As time increases, the hindrance to diffusion becomes more apparent.

The propagator (**Fig. 2**) is the probability distribution function (PDF) for displacements, and single-particle tracking experiments can be analyzed by fitting the PDF to histograms of measured square displacements r^2 for prescribed times *(12,20)*. Alternatively, one can integrate the PDF to give a cumulative distribution function (CDF) and use this in the analysis *(15,21)*. PDFs have the advantage that they are easy to visualize; everyone knows what a Gaussian PDF looks like and can recognize deviations. CDFs are harder to visualize in detail but they integrate out much of the experimental noise, and one can apply the Kolmogorov–Smirnov test *(22)* to test whether two CDFs are from the same distribution.

3. Simulation of Diffusion

The programming of diffusion simulations for a dense fluid on the continuum is beyond the scope of this chapter and is well discussed elsewhere *(4–6)*. The complication is in the force calculations. For a small number of particles N, one can find the $N(N-1)/2$ distances between all the pairs of particles and calculate the forces. But this becomes prohibitively expensive for large N, so one must make a list of the neighbors of each particle and update the list appropriately. These neighbor or Verlet lists are discussed in *(4–6)*.

The fluid and colloid literature includes much work on 2D and 3D diffusion simulations, including diffusion of particles interacting by various potentials and particles at high concentrations. The results, the comparisons with experiment, and the computational techniques are all of interest. Single-particle tracking has become a major experimental tool here *(23–26)*, based on the idea that a colloidal suspension is analogous to a liquid but at a scale observable by light microscopy *(27,28)*. Charge-stabilized microspheres interact by a screened Coulomb (Yukawa) potential and sterically stabilized microspheres interact by a hard sphere potential. Diffusion coefficients have been obtained from simulations for hard-particle interactions *(29–31)* and a screened Coulomb potential *(32)*. Modeling related to protein crystallization is an important intersection of biophysics with this area of colloid science *(33,34)*.

3.1. Approaches to Calculation

Several approaches may be taken to diffusion calculations: molecular dynamics, Brownian dynamics, Monte Carlo calculations, and dissipative particle dynamics. The problem of evaluating forces or energies is similar in all; the difference is in the type of potential and how the move is generated.

In the simplest sort of Monte Carlo calculations on a lattice, a diffusing particle is moved randomly to an adjacent unblocked lattice site. On the continuum the particle is moved randomly to an unblocked position according to the propagator **Eq. 2**.

In molecular dynamics, the diffusing species are represented as an atomic-level structure or nearly so, and the motion of the atoms obtained from Newtonian mechanics. This is a practical mean of calculating diffusion of simple fluids, as shown in Rahman's *(35)* pioneering paper on 864 atoms of liquid argon. Note the difference in time scales between argon and phospholipid dynamics. The time for an atom in liquid argon to diffuse its Lennard–Jones radius is of the order of 10 ps. For a phospholipid, trans-gauche isomerization in the hydrocarbon chains takes 50–100 ps; molecular rotation, 1–2 ns; and lateral diffusion, >100 ns *(36)*. Molecular dynamics calculations can also be done for proteins or colloidal particles represented as structureless particles interacting by a prescribed potential.

In Brownian dynamics *(37–39)*, each move includes a deterministic displacement based on the total force on a particle, plus a random displacement representing Brownian motion. For the jth particle,

$$\vec{r}_j(t + \Delta t) = \vec{r}_j(t) + (D/kT)\vec{F}_j(r)\Delta t + R \tag{6}$$

where $\vec{r}_j$ is the position of the jth particle, $D = D_{trans}$ is the translational diffusion coefficient, kT is the thermal energy,

$$\vec{F}_j(r) = -\sum_i \nabla V_{ij}(r) \tag{7}$$

is the force on the jth particle, V_{ij} is the potential energy, and R is the random displacement, defined by $\langle R \rangle = 0$, $\langle R^2 \rangle = 2D\Delta t \delta_{ij}$, where δ_{ij} is the Kronecker delta, so that $\langle x^2 \rangle = \langle y^2 \rangle = \langle z^2 \rangle = 2D\Delta t$ and the cross-terms are uncorrelated, $\langle xy \rangle = \langle xz \rangle = \langle yz \rangle = 0$. A similar equation is used for rotational diffusion,

$$\vec{\Omega}_j(t + \Delta t) = \vec{\Omega}_j(t) + (D_{rot}/kT)\vec{T}_j(r)\Delta t + \vec{W}, \tag{8}$$

where $\vec{\Omega}_j$ is a vector of orientational angles of the jth particle, $\vec{T}_j$ is the torque on the jth particle, D_{rot} is the rotational diffusion coefficient, and $\vec{W}$ is the random angular displacement, defined by $\langle W \rangle = 0$, $\langle W^2 \rangle = 2D_{rot}\Delta t\, \delta_{ij}$. These equations neglect hydrodynamic interactions; in the full form the diffusion coefficient is a tensor and the random displacement includes correlations *(37,40)*. If $V(r)=0$, Brownian dynamics reduces to the simple Monte Carlo scheme first discussed.

In the basic Brownian dynamics algorithm, the force $\vec{F}(r)$ is calculated at the current position and used to generate the move. Higher-order algorithms similar to the Runge–Kutta algorithm can be used to permit a larger time step *(41)*. Here the first-order position $\vec{r}'$ is found from $\vec{F}(\vec{r})$, $\vec{F}(\vec{r}')$ is calculated, and the final position is found from the average force $\frac{1}{2}[\vec{F}(\vec{r}) + \vec{F}(\vec{r}')]$.

In Brownian dynamics the time step must be much greater than the momentum relaxation time, $\Delta t \gg mD/kT$, where m is mass. It is often said that the time step must be small enough that the forces are effectively constant over a time step, but in the author's view this is unduly restrictive. Consider a particle near the minimum in a Lennard–Jones potential. The force is close to zero, and almost any move will increase the force by a large factor. But a tripling of the force, say, is insignificant here. A better requirement might be that the absolute value of the change in the force be much less than the mean absolute value of the random force. Brownian dynamics simulations ought to include a test of the sensitivity of the results to Δt.

The standard Metropolis Monte Carlo algorithm allows moves as follows. Calculate the energy change ΔE for a move. If $\Delta E \leq 0$, accept the move. If $\Delta E > 0$, generate a random number $RAND$ uniformly distributed on $(0,1)$; and move according to a Boltzmann factor: accept the move if $RAND \leq \exp(-\Delta E/kT)$ and reject it otherwise. The simple Monte Carlo scheme is a special case of this for hard particles. If the particles overlap, ΔE is infinite and the move is always rejected; if they do not, ΔE is zero and the move is always accepted. The Metropolis algorithm satisfies the condition of detailed balance, so the algorithm leads to the Boltzmann equilibrium distribution. For details of Monte Carlo techniques *see* **refs. 4,5,42,43**.

A general Monte Carlo algorithm may allow complicated or nonphysical moves, *see* pp. 38–52 of **ref. 5**. In a simulation involving phase separation, each Monte Carlo step could attempt to translate each particle locally, rotate each particle, exchange positions of a prescribed number of pairs of random particles, and adjust the size of the system to maintain constant tension *(44)*. The advantage of such a complicated set of moves is that the system equilibrates quickly, and the disadvantage is that there is no physical time scale associated with such a set of moves. If the moves were just translation and rotation, the probabilities could be chosen according to the known ratio of D_{rot} to D_{trans}, but for more complicated moves there is no time scale. As a result one can calculate equilibrium structures but not dynamics. Of course one could use this sort of algorithm to equilibrate the system and then switch to purely physical moves for a diffusion measurement. The Weinberg group *(45,46)* has done key work on time dependence in Monte Carlo calculations. These references discuss surface rate processes such as adsorption, desorption, and diffusion, and include the effect of activation barriers.

Of particular interest to membrane researchers is a discussion of Monte Carlo calculations of a minimal model of lipid–sterol mixtures *(47)*. This was an off-lattice calculation with three species: ordered lipid chains, disordered lipid chains, and a sterol, either cholesterol or lanosterol. Highly simplified pair interaction potentials were used. A variety of moves were used but were combined randomly; the author argued that their diffusion simulations give meaningful relative diffusion coefficients.

Dissipative particle dynamics is often used in simulations in which long physical times must be simulated and hydrodynamic interactions included. This technique is a molecular dynamics simulation of particle motion in which the particles represent collections of atoms, say a few water molecules or a few methylene groups of a lipid hydrocarbon chain. The particles are assumed to be soft spheres, with a potential proportional to $(1 - r/r_c)^2$, where r is the radius, r_c is the cutoff radius, and the potential is zero for $r > r_c$. The softness of the potential allows long time steps. Also included are a random force and a dissipative force that reduces the velocity difference between two particles. These forces provide a thermostat for the system. Particles may be joined by springs or rigid rods to form amphiphiles or polymers *(48–50)*.

3.2. Geometry

Three cases are considered, 2D diffusion "in the plane of the membrane," 3D diffusion as in the cytoplasm, and diffusion in a 2D surface with a 3D shape, such as diffusion in an organelle membrane, or the reorientation of a spin label or transition moment of a lipid analog diffusing in a spherical vesicle.

Diffusion calculations can be done on a lattice or the continuum. From the computational point of view, the advantage of the lattice is in treating obstructions. One can trivially test whether the destination point is occupied or not, and accept or reject the move accordingly. For continuum diffusion, one has to keep track of all the neighboring particles of each mobile particle. The disadvantage of the lattice is its lattice structure, especially at short distances. The distribution of parameters describing the random walk—such as the asymmetry parameter obtained from the moment of inertia tensor *(12)*—show spikes, which are real for a lattice and do not disappear on averaging. Another disadvantage is that the percolation thresholds of a lattice and a continuum are different. The physical limitations of approximating a continuum by a lattice are summarized in **ref. 51** and discussed in more detail in **refs. 52** and **53**.

3.3. Periodic Boundary Conditions

Periodic boundary conditions are used to reduce edge effects. Consider a random walk in a square. The original square is surrounded by an infinite array of copies, each with the same configuration of obstacles or traps as the original. Each tracer then has two x-coordinates, one, X, representing its position in the original square and the other, integer IX, representing which square of the array it is in. If the tracer moves out of the right side of the square, it reenters the square at the corresponding point on the left side, but IX is changed to record the crossing. For a square with boundaries [0,L], the new position X is

```
IF (X < 0) THEN            IF (X > L) THEN
    X = X + L                  X = X − L
    IX = IX − 1                IX = IX + 1
END IF                     END IF
```

The absolute position of the particle $X(abs) = X + L * IX$ is used to find $\langle r^2 \rangle$.

3.4. Diffusive Steps for Normal Diffusion

For the square lattice, moves are simply steps of one lattice constant in the $+x$, $-x$, $+y$, or $-y$ direction, each with probability 1/4, and similarly for the cubic lattice. After each step, periodic boundary conditions are applied. For speed one can apply only those tests required for the particular move made. For an $L \times L$ lattice with lattice constant ℓ,

$$r = \ell^2 \left[X(abs)^2 + Y(abs)^2 \right]. \tag{9}$$

For the corresponding triangular lattice, unit vectors $\hat{a}$ and $\hat{b}$ (**Fig. 3**), the moves are $\pm a$, $\pm b$, and $\pm c = \pm(a - b)$, so in Cartesian coordinates (unit vectors $\hat{i}$, $\hat{j}$), $\hat{a} = \hat{i}$, $\hat{b} = \frac{1}{2}\hat{i} + \frac{\sqrt{3}}{2}\hat{j}$, and $\hat{c} = \frac{1}{2}\hat{i} - \frac{\sqrt{3}}{2}\hat{j}$. If the absolute position is $A\hat{a} + B\hat{b}$, then

$$r^2 = \ell^2[A^2 + B^2 + AB]. \tag{10}$$

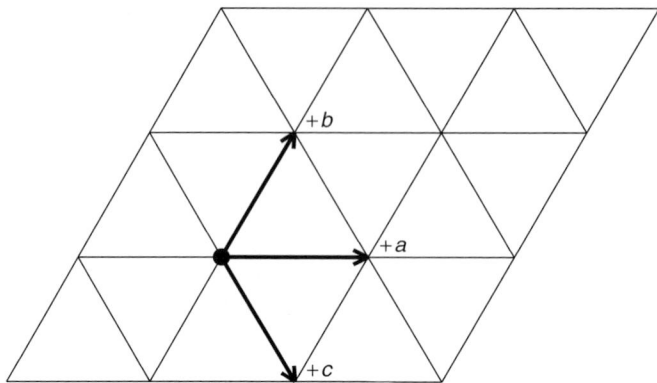

Fig. 3. Moves on the triangular lattice.

For the continuum, the simplest choice of moves is to take Δx, Δy, and Δz from a Gaussian distribution with mean zero and variance $2D\Delta t$.

For diffusion on a curved 2D surface, the moves are more complicated. If the surface is defined mathematically—a sphere, an ellipsoid, or a 3D periodic surface—one can move the particle randomly in the tangent plane to the surface and then project the end point back onto the surface *(54)*, but *see also* **ref. 55**. If the surface is irregular—say a mitochondrial membrane—one could represent it by a triangulation and carry out a random walk on the lattice defined by the triangulation. Sbalzarini et al. *(56)* discuss how to model 3D diffusion in a region with highly complex boundaries, emphasizing methods faster than random walks.

3.5. Simulating the Types of Diffusion

Normal diffusion can be simulated easily on a lattice or the continuum, and diffusion with directed motion is equally easy on the continuum. Obstructed diffusion is trivial to simulate on a lattice but more difficult to program on the continuum. In the simplest simulation of confined diffusion, the boundary is simply a line, and a move across the boundary succeeds with a prescribed probability p. The corresponding analytical solution to the diffusion equation in a circular corral uses the boundary condition $\partial C/\partial r = hC$ on the corral wall. Here C is concentration, and the parameter h is the logarithmic derivative of the concentration at the boundary. If the wall is perfectly impermeable, $h = 0$ and if the wall is perfectly permeable, $h \to \infty$. A simple relation between p and h is discussed elsewhere *(57)*. A better approach is to do a Brownian dynamics calculation in which confinement is by a potential energy barrier of some assumed width, height, and shape.

Some types of anomalous subdiffusion are easy to simulate. In the CTRW, the tracer carries out an ordinary lattice or continuum random walk, but at each time step the waiting time is a random number generated from the distribution

$$P(t) = \beta/(1+t)^{1+\beta}, \tag{11}$$

$\beta > 0$, which gives $\langle r^2 \rangle \sim t^\beta$ for large t. The waiting time at a given point in space varies randomly from visit to visit *(58)*. Diffusion on an infinite percolation cluster is anomalous on all time scales *(9)* with $\langle r^2 \rangle \sim t^{0.695}$. But implementing fractional Brownian motion is more of a research problem than a recipe. The problem is reviewed briefly in the methods section of **ref. 14**. The key point is that published algorithms vary enormously in quality, and one must read the review articles before choosing an algorithm. In addition to the review articles cited there, *see* **ref. 59**. Different algorithms may give the same exponent for the time dependence of $\langle r^2 \rangle$ at large times but the behavior at short times and the behavior of higher moments may differ considerably *(14)*.

3.6. Particles

To introduce obstructions or tracers, a loop is used: Choose a random position for the particle; test for occupancy or overlap; continue until the required number of obstacles has been added. If nonpoint tracers are symmetric enough, say hexagons on the triangular lattice or circles on the continuum, the finite size of the tracers can be taken into account by enlarging the obstacles by the tracer radius and reducing the tracer to a point *(51)*.

If one tries to fill a finite area with a high area fraction of random nonoverlapping nonpoint particles, one will rediscover the so-called random parking limit, well studied in the physics literature on random sequential adsorption *(60,61)*. There is a well-defined average maximum area fraction of random particles that can be placed in a finite area. The limit depends on the particle shape and size, and is well short of complete filling. For circles on the continuum *(62)* the parking limit is 0.547 and for hexagons of unit radius on the triangular lattice (Saxton, unpublished), 0.670. To fill above the parking limit, fill the area as uniformly as possible with nonoverlapping particles, randomly choose and remove the excess particles, and randomize the configuration by running a random walk in which all particles are mobile. Randomization is measured by the radial distribution function. After randomization one can begin to record data.

In a system with multiple mobile particles, the unit of time is one Monte Carlo step per particle, that is, one attempt to move each mobile particle. It is bad practice to number the particles in lattice order and then move them in numerical order; this introduces unnecessary correlations. Minimally, one can introduce the particles randomly as above, number them in order of filling, and then move them in numerical order. Better practice is to move them in random order. Then for N particles, one Monte Carlo step is N moves of randomly chosen particles, and in any particular Monte Carlo step some particles might not be moved and some might be moved more than once. Another reasonable approach is to generate a random permutation of N at every Monte Carlo step and move each particle once according to that permutation. For a discussion of the same principle in lattice updates in the Ising model *see* **ref. 63**.

For a random walk on an obstructed lattice, two definitions are possible, the "blind ant" and "myopic ant" algorithms *(64)*. ("Ant" is from de Gennes' image of diffusion on a percolation cluster as an ant walking in a labyrinth.) At each time step, the blind ant can move to each nearest neighbor point with equal probability. If the new point is blocked, the ant remains at the old point and the clock is incremented. The myopic ant can move to each unblocked nearest neighbor point with equal probability, and necessarily moves at every time step. As $t \to \infty$, the blind ant is uniformly distributed but the myopic ant is not. The position of the myopic ant can oscillate in a way that the position of the blind ant cannot, *see* p. 92 in **ref. 65**. The two algorithms yield the same asymptotic behavior but the blind ant converges faster, *see* p. 264 in **ref. 66**. Tobochnik et al. *(67)* argue that the myopic ant algorithm is in fact incorrect.

3.7. Potentials and Traps

In simulations of hard particles on the continuum, any overlap leads to an infinite energy and an infinite force. In Monte Carlo equilibration, this is no problem because the move is summarily rejected. But in Brownian dynamics and molecular dynamics, precautions must be taken. The problem in molecular dynamics is discussed in section 3.61 in **ref. 4** and Chapter 14 in **ref. 6**. In Brownian dynamics, the overlapping particles may be backed off *(30)*, or a move giving overlap may be rejected as in Monte Carlo calculations but then the results must be extrapolated to zero time step *(68)*. Alternatively, in either molecular or Brownian dynamics the infinite force may be replaced by a large but finite force from a steep repulsive potential *(68)* such as $1/r^{36}$.

A random configuration of interacting particles is a highly nonequilibrium state, and two methods of generating a thermally equilibrated state are useful. First, one can start with a random or an ordered state, and do an annealing run before beginning to collect data, monitoring the energy and radial distribution function during the annealing run to test for equilibration. Second, for a system of traps one can produce an equilibrium state directly from the Boltzmann distribution *(10)*.

Certain interactions require specialized approaches: for electrostatic interactions in general, *see* section 5.5. in **ref. 4** and Chapter 13 in **ref. 6**; for protein electrostatics in electrolyte solution, *(70–73)*; for hydrodynamic interactions in membranes, *(74)*; and for 3D hydrodynamic interactions in solution, *(37,38,40,69,75)*. For membrane proteins, one must also take into account lipid-mediated protein–protein interactions *(76–80)*.

3.8. Data Collection and Analysis

For runs with many time points, it is convenient to record data at varying intervals, as shown in **Table 2**. With this scheme, for a run of 128K time points, 3584 points are recorded and for 128M time points, 7424 points are recorded. The amount of averaging changes from interval to interval, so the noise level in $\langle r^2 \rangle$ changes abruptly as well, a useful test of the alertness of reviewers. This sampling scheme makes it difficult to calculate an autocorrelation function using the entire data set, although one could calculate the autocorrelation for lag 1 using the first interval, lag 4 using the second interval, and so on, in the spirit of a multitau correlator.

Diffusion simulations have the advantage that noise can be reduced by averaging over many repetitions. Single-particle tracking experiments do not provide that luxury. The usual practice is to calculate the MSD using all pairs of points with the appropriate lags. For N time points, the nth average is

$$\langle r^2(n) \rangle = \frac{1}{N} \sum_{i=0}^{N-1} [\vec{r}(i+n) - \vec{r}(i)]^2. \tag{12}$$

In this way all points are weighted as equally as possible, and all the information is extracted. Note, however, that the terms in the average are highly dependent. For example, in $\langle r^2(4) \rangle$ the averages are for times 0,1,2,3, and 1,2,3,4, and 2,3,4,5, and so on. As a result the averaging smooths the data only slightly more than averaging over independent segments 0,1,2,3 and 4,5,6,7, and so on would. Calculation of the MSD is discussed in detail in the appendix of **ref. 81**. Given that the standard deviation of $\langle r^2 \rangle$ is proportional to $\langle r^2 \rangle$, in single-particle tracking it is necessary to cut off the data at, say, one-quarter of the total number of time steps.

For the practical aspects of evaluating percolation thresholds, *see* the methods section of **ref. 51** and references to the physics literature there. As shown there, if the tracer has a nonzero radius, percolation depends on the excluded area, not simply the area of the obstacles. The threshold is therefore very sensitive to tracer size, and this sensitivity is a useful experimental test to identify percolation.

Table 2
Time Sampling

Time range	Interval	Number of points
1–1024	1	1024
1028–4096	4	768
4112–16384	16	768
16448–65536	64	768
65792–131072	256	256

4. Random Number Generators

A true random number generator (RNG) must be based on a random physical process such as radioactive decay, so pseudo-RNGs are normally used instead. Given a specified seed number as input, these algorithms generate a deterministic sequence that looks enough like a true random sequence for the purposes of the simulation. More precisely, the pseudorandom sequence cannot have correlations that distort the simulation. The general requirements for an RNG are that it must have a long period; it must be efficient (high speed and low memory use); it must be repeatable and portable; and the random numbers (RNs) must be uniform and independent *(82)*. Random number generation is an active area of research and space does not permit a full review. This section will provide an introduction to the use of RNGs and their problems, with references to some recent authoritative reviews. Knuth's book *(83)* has been a fundamental reference in the field since its first edition.

4.1. Types of RNGs

For purposes of a brief review, a useful way to classify RNGs is by the number of points in the recursion:

1. One-point: linear congruential: $x_{n+1} = a * x_n + c \pmod{m}$.
2. Two-point: linear recursion: $x_{n+1} = x_{n-p} \otimes x_{n-q} \pmod{m}$.
3. Multipoint: $x_{n+1} = x_{n-p} \otimes x_{n-q1} \otimes x_{n-q2} \ldots \pmod{m}$.
4. Combined: typically, two or three of the recursions above.

Here x_n are the RNs; x_0 is the seed; a and c are constants; $\otimes$ represents addition, subtraction, multiplication, or exclusive or (XOR); and $x \pmod{m}$ is the remainder when x is divided by m. The actual algorithms are often complicated because they must break up arithmetic operations involving large integers so that the operations can be done without overflow. XOR does not have this problem.

The linear congruential generator simply cycles through the integers without repetition until it reaches its original value and begins to repeat. A different seed just starts the sequence somewhere else in the loop. For a 32-bit computer, the linear congruential generator yields a sequence of $2^{31} - 1 \sim 2.2 \times 10^9$ random integers, which could easily be used up in a single simulation. Another limitation is the lattice structure of the RNs. In a famous article entitled "Random numbers fall mainly in the planes," Marsaglia *(84)* showed that the RANDU RNG then standard on IBM computers produced RNs that taken in pairs, triplets, and so on were localized to a few planes. If one generates RNs uniform on the interval (0,1) and plots successive pairs on an expanded scale, say the square $[0.4995, 0.5005]^2$, one will see lattice structure. Linear recursion generators remove some of these shortcomings. Multipoint and combined generators eliminate some of the correlations inherent in linear recursion generators. For example, in one multipoint generator, three XORs shuffle the bits more effectively than a single XOR would *(85)*. Multipoint and combined generators may have extremely long periods.

4.2. Which RNG to Use?

Using the system RNG is not acceptable unless the algorithm is stated in detail and has been tested in the literature. Personal computer RNGs may be suitable for computer games but not for research *(82,86,87)*. For several default RNGs in Pentium software, plots of very long random walks show dramatic periodicity and symmetries, and a plot of $\log\sqrt{\langle r^2 \rangle}$ vs log t is linear at first but shows artifactual structure after only 10^5 time steps *(88)*. Periodicity is unavoidable but one can choose an algorithm with a periodicity much greater than the run length. The default RNG problem is not limited to personal computers. The IBM RANDU RNG is infamous *(13)* and one test of RNGs in a physics simulation led the author to conclude that "one lesson from these results is not to trust random number generators provided by computer vendors" *(89)*. Furthermore, *do not (90)* use the RNGs from the first edition of *Numerical Recipes (91)*. They have been greatly improved in later versions *(13,92)*. Whatever RNG is used, in any Monte Carlo or Brownian dynamics publication one must specify the algorithm used. A reviewer would be justified in rejecting a paper if authors could not specify the RNG, on grounds that the work could not be reproduced independently. Merely stating that one used the system RNG on a particular computer is not sufficient; others should not have to buy a new computer to test a result.

Among the RNGs that seem appropriate for further consideration are the Ziff 4-tap *(85,93)*, the Mersenne twister *(94)*, the combined generator mzran *(95)*, RANLUX at a luxury level of 2 or above *(96)*, and ran2 *(13)*. The so-called "minimal standard" 32-bit linear congruential generator *(97)* does well for its type but has major limitations on account of the short period. It would be worth testing a 64-bit version. In Ising model tests, a 32-bit linear congruential generator failed but two 48-bit ones passed *(89)*. The consensus negative control seems to be R250

$$x_{n+1} = x_{n-250} \text{ XOR } x_{n-103} \pmod{m}. \tag{13}$$

Recommended reviews and test results are *(63,82,96,98–103)*. Code for many RNGs is available on the Web. NETLIB has many RNGs and test programs in Fortran and in C (www.netlib.org). The GNU Scientific Library has C versions of many RNGs (www.gnu.org/software/gsl), and Matpack has C++ versions (http://www.matpack.de/). Other highly useful starting points are the websites of Luc Devroye (http://cgm.cs.mcgill.ca/~luc/rng.html) and Peter Hellekalek (http://random.mat.sbg.ac.at).

4.3. Tests

Systematic testing of RNGs is essential. The tests probe for "some property that a truly random sequence should not possess" *(100)*. As L'Ecuyer puts it, "*bad* RNGs are those that fail simple tests, whereas *good* RNGs fail only complicated tests that are hard to find and run" *(103)*. Before an RNG is used in applications, it ought to be tested by its author and in an independent comparative study. Standard general tests for good statistics include Diehard (George Marsaglia, stat.fsu.edu/pub/diehard), the NIST Statistical Test Suite *(104)*, SPRNG = Scalable Parallel RN Generators *(105)*, and TestU0l (L'Ecuyer and Simard. www.iro.umontreal.ca/~simardr/testu01/tu01.html).

An RNG that passes general statistical tests still needs to be tested in applications *(98–100, 106)*. Ideally, it would be tested in applications in which exact answers are known, but comparisons with other RNGs, good and bad, are also useful. The comparisons ought to include RNGs of totally different types *(82)*. Tests on simulations of phase transitions, often some form of the Ising model, showed the limitations of RNGs that had been previously considered satisfactory *(63,89,98,107,108)*. Other physical simulations that have proved useful in testing include the structure of crystal lattices *(109,110)*, bond percolation *(85)*, and self-avoiding walks *(93,111)*.

Of particular interest here are tests based on random walks. Vattulainen et al. *(98)* ran n-step random walks starting at the origin, recorded the final position by quadrant, and applied a χ^2 test for uniformity. Some tests used a fixed value of $n = 1000$ and others varied n to show that the range of correlations in the RNG was approximately the maximum offset p in the RNG. Another approach was to examine the average of blocks of n successive numbers, again varying n to identify the range of correlations. The PDF of the final position of 2D random walks was also tested *(100)*. Two of the RNGs gave very similar PDFs as shown in a difference plot, but another RNG showed clear systematic error, although comparison with the analytical expression might be better. Tretiakov et al. *(110)* tested the smoothness of the distribution of RNs by generating 10^{10} RNs between 0 and 1 and binning them into 10^6 bins. A histogram of the counts per bin was expected to be a narrow Gaussian centered on 10^4. Two generators gave this result but the Press ran2 *(13)* gave two peaks, both displaced from 10^4. Gammel *(112)* tested RNGs using rescaled range analysis, a method used to characterize fractional Brownian motion. Vattulainen *(106)* examined the intersections of independent random walks as a test of RNGs for parallel computation.

4.4. Correlations in RNGs

RNGs fail these tests on account of correlations among RNs. All pseudo-RNGs have correlations. The recursion relation is one correlation; substituting the recursion relation into itself implies another correlation, and so forth *(113)*. Errors arise in simulations when there is, so to speak, "constructive interference" between the correlations in the simulated system and the correlations in the RNG *(114)*. Errors can arise from correlations between nearby RNs in the sequence and from correlations having a range close to a dimension of the system, particularly in phase transition simulations on a lattice. A completely avoidable source of correlations is using an RNG with a short period. The RNG ought to have a period much greater than is needed in the run (*see* section 4.2.5 in **ref. *102***). A common recommendation is that the number of RNs used be no greater than the square root of the period of the RNG *(101)*.

In a given simulation, a key test for errors resulting from correlations is to see whether the results are changed significantly when a different RNG based on a distinct algorithm is used. In addition, correlations can be broken by decimating the RN sequence, that is, using say only every third RN in the sequence, or by periodically (or randomly) generating and discarding a (random) number of RNs *(115)*. Decimation has been shown to fix RNGs in physics simulations *(107,116)*. The RANLUX RNG has various "luxury levels" setting the fraction of RNs discarded *(96)*. As the fraction discarded increased, this RNG shifted from failing certain tests to barely passing them to passing them *(106)*.

One could argue by analogy with molecular dynamics calculations *(109)* that obstruction by random obstacles decorrelates a random walk after a few collisions, and so the choice of RNG is not critical here. But Holian et al. *(109)* present molecular dynamics results in which the initial state is not forgotten quickly. Furthermore, an application-specific test of an RNG requires a comparison of the results of various RNGs with exact results, and few exact results are available for obstructed diffusion. Many results are known for percolation problems involving diffusion, but they involve finite-size corrections so it is hard to tell whether an observed discrepancy is the result of problems in the RNG or in the scaling correction. The best test might use a 2D continuum random walk in regular arrays of cylinders or spheres, for which high-quality approximations are available, *see* p. 414 in **ref. 117**.

4.5. Transformations of RNs

Typically, RNs uniform on (0,1) are generated and transformed to other distributions if necessary. Commonly used transformations are discussed in several sources *(13,118–120)* and Devroye treats the question comprehensively *(121)*. The most important transformation for diffusion modeling is to produce Gaussian RNs. One method that has been used is to sum N uniform RNs on (0,1) and subtract $N/2$, often with $N = 12$. By the central limit theorem this is close to Gaussian. But the method is slow, it is not exact, and it has a systematic error in the tails because the largest step size possible is $N/2$. Bratley et al. *(119)* describe the method as "not very quick and unnecessarily dirty." A much better approach is the Box–Muller transformation, which converts a pair of uniform RNs to a pair of Gaussian RNs. An efficient program to do this without trigonometric functions is in **ref. 13**. A warning: Unless the RNG producing the uniform RNs is of high quality, lattice structure in the pairs of uniform RNs produces beautiful but unwanted spiral structure in the pairs of Gaussian RNs, *see* pp. 161–162 and pp. 223–224 in **ref. 119**, Chapter 3 in **ref. 118**, and pp. 88–89 in **ref. 120**. Another high-quality approach is the so-called ziggurat method *(122)* but *see also* **ref. 123**.

The transformation or inverse CDF method can be useful for some distributions (*see* section 7.2. in **ref. 13** and section 2.1. in **ref. 120**). For a 2D continuum random walk, one can use this method to generate r from **Eq. 2**, and then find $\Delta x = r \cos \theta$ and $\Delta y = r \sin \theta$, where θ is a random number uniformly distributed on $[0,2\pi)$. This method is slower than the Box–Muller transformation; both require two RNs, a logarithm and a square root, and this method also requires two trigonometric functions. The transformation method can also be used with **Eq. 11** to generate waiting times for a CTRW.

4.6. Seeds

What is the best way to generate the seed? The system clock is too regular to use unless one has access at the nanosecond level. Better practice is to generate a list of random seeds from a different RNG with an arbitrary seed. It may be convenient to use the Mathematica RNG, which by default uses as its seed the system time when the program is first started on the machine. One must save the seed to be able to repeat a sequence of moves for debugging. A major advantage of a pseudo-RNG over a true RNG is that the sequence from the pseudo-RNG is reproducible and the sequence from the true RNG is not.

5. Programming and Computational Speed

The first principle in programming is cleanliness of structure, cleanliness of notation, and uniformity of notation. This will help both correctness and speed. Like cockroaches, computer bugs hide in clutter and thrive in messes.

One should think about speed in terms of the inner loop of the program, the part of the program that is executed the most often. In the pseudocode obstructed diffusion program of **Fig. 4**, the inner loop is the loop over time steps. This is where improvements in programming will

```
Initialize
LOOP OVER OBSTACLE CONFIGURATIONS
    Insert obstacles
    LOOP OVER TRACERS
        Insert tracers
        LOOP OVER TIME INCREMENTS
            LOOP OVER TIME STEPS
                Generate trial move
                Test for obstruction
                Accept or reject move
            END LOOP
            Record motion of tracer
        END LOOP
    END LOOP
    Write intermediate data
END LOOP
Data analysis
```

Fig. 4. Pseudocode for obstructed diffusion simulation.

have the most effect. There are likely to be many parts of the program, such as data analysis or insertion of obstacles, where reducing the run times to zero would have little effect. The use of a code profiler is recommended to find out where the program is in fact spending its time.

Some useful principles are discussed in the literature of high-performance computing, though some of the optimizations given there are better left to compilers *(124,125)*. The discussion of the programming strategy for the GROMACS 3.0 molecular simulation package is instructive *(126)*. One important point is that CPU speeds have increased much more than memory chip speeds, so that the limiting factor in a program can be memory access, not CPU speed. There is a hierarchy of memory and access times: access times for registers and level 1 cache, nanoseconds; level 2 cache, tens of nanoseconds; memory, hundreds of nanoseconds; and disk, milliseconds, *see* p. 35 in **ref. 124** and p. 49 in **ref. 125**.

Another consideration is that instructions may be pipelined in the CPU, that is, the CPU may at the same time execute instruction 1, fetch the operands for instruction 2, decode instruction 3, and fetch instruction 4 from memory. Branching may force the pipeline to be cleared and reloaded. Some branching is unavoidable, such as in choosing moves on a lattice, but unnecessary branching slows execution.

Locality of reference is important to speed. Accessing adjacent locations in memory is faster than accessing disjointed locations. So when operating on an array $A(I,J)$ in a pair of loops one should make the inner loop over adjacent memory locations (I in Fortran and J in C). For the same reason, beware of linked lists. These are elegant structures, excellent for saving memory *(6)*, but with poor locality of reference.

For speed, programs may need to be organized differently for nonparallel and parallel processors. For nonparallel processors, one should do expensive computations only if necessary. In a force calculation for interacting particles on the continuum, one divides the system into cells and tests only the particles in the same cell as the particle in question, or in an adjacent cell. One then calculates the square distance between pairs of particles and compares that with the square of the cutoff distance for the potential. (In practice this is done periodically as a neighbor list, as mentioned at the beginning of **Section 3.**) One takes the square root to find the distance only if it is needed. On the other hand, in a parallel program one tries to maximize the amount of computation done in parallel so it may be faster to do necessary and unnecessary calculations uniformly.

The RNG is used at every move, so improving its efficiency may be important. First, although a few RNGs inherently produce real numbers, most first produce random integers. If the RNG produces random integers NRAND in (0,N) and then normalizes them as RAND = FLOAT(NRAND)/FLOAT(N), one can sometimes avoid the multiplication by 1/FLOAT(N) by using the unnormalized form. Here FLOAT represents conversion from integer to floating point. For a square lattice, the move

```
IF(RAND < 3/4)THEN
    MOVE = +X
ELSE IF (RAND < 1/2)THEN
    MOVE = - X...
```

can be replaced by

```
IF(NRAND < 3N/4)THEN
    MOVE = +X
ELSE IF (NRAND < N/2)THEN
    MOVE = -X...
```

where the constants 3N/4, N/2... are precomputed. Second, RNG subroutines sometimes have different procedures to initialize the RNG and to generate the RNs. Some subroutines *(13)* do this by a switch. This is inefficient, because every move requires an unnecessary IF statement and a branch. Better practice is to separate the operations and isolate the RNG initialization in the initialization subroutine of the program. Third, if the RNG is a subroutine, it introduces the overhead of a subroutine call into the inner loop. This ought to be avoided by inline coding, either by copying the code into every place it is used, or setting the compiler optimization to do this automatically.

6. Special Methods

Some diffusion simulations can be speeded considerably by changing the fundamental algorithm instead of tweaking details of the programming.

6.1. The n-*Fold Way*

The approach known as the *n*-fold way or an event-driven algorithm has appeared independently in various parts of the physics, chemistry, and simulation literature *(127)*. The essentials of this approach can be seen in the problem of escape of a tracer from a binding site. Let the escape probability from the binding site per time step be p. There are three ways to set up the escape algorithm. First, at each time step one can generate a random number *RAND* uniformly distributed in (0,1). If *RAND* < p, the particle escapes. This method is simple because all events—escape, random walk, and so on—are on the same clock. It correctly takes into account the statistical fluctuations in the escape time, but it is highly inefficient. Second, one can use a mean-field approach in which the particle stays in the trap for a time of exactly $1/p$ and then escapes. This approach is fast and simple but it eliminates fluctuations in the escape time (which can be considered an advantage or a disadvantage). And one has to carry forward the position of the particle for $1/p$ time steps. Third, one can generate the escape time as a random number from the probability distribution of escape times $p(n) = (1 - p)^{n-1} p$. This approach is fast and includes the statistical fluctuations correctly, although again one has to carry forward the position. The second and third methods are examples of the *n*-fold way.

The *n*-fold way originated as an improved Monte Carlo algorithm for the Ising model. The algorithm "accounts for the a priori probability of changing spins before, rather than after, choosing the spin or spins to change" so it can efficiently handle situations in which the probability of a spin flip is small *(128)*. The term "*n*-fold" refers to a system in which there is a finite number *n* of possible events, so that the required probabilities and time increments can be calculated. A lucid discussion of the *n*-fold way and kinetic Monte Carlo calculations is in **ref. 129**. Event-driven algorithms were essential in the work on random sequential adsorption described in **Section 3.6**. Here the difficulty is that when the coverage is nearly complete, the number of random attempts needed to find an available region is prohibitively large. The event-driven algorithm attempts adsorption only in available regions and increments the clock by an amount that accounts for the elimination of unavailable regions *(62,130)*.

6.2. First Passage Time Method

The first passage time method is a special case of the *n*-fold way. Consider continuum diffusion of a point tracer in the presence of circular obstacles. In the usual algorithm, the tracer makes small steps Δr with a time increment Δt, and the tracer spends most of its time moving between obstacles. In the improved algorithm, at each time step one constructs the largest circle that is centered on the tracer but does not intersect any obstacles. One then chooses a random angle in $[0,2\pi)$, moves the tracer to the corresponding point on the circumference of the circle, and increments the clock by a random number chosen from the first passage time distribution for a circle of that radius. This approach was applied to diffusion-controlled reaction rates *(131)* and to the conductivity and the diffusion coefficient *(67,132)*. In a modification of this method, the time increment was chosen to be the mean first passage time *(133)*. For further developments of the first passage time method by the Torquato group, *see* **ref. 134**. Another variant of the method, the "walk on spheres" algorithm and its extension the Green's function first passage method, has given highly efficient calculations of reaction rates of a diffusing particle with spherical traps, and conductivity in composites (equivalent to obstructed diffusion) *(135)*. The Green's function reaction dynamics method uses an event-driven algorithm to simulate chemical reactions using the propagator for the Smoluchowski problem of reactive collisions of spheres *(136)*.

A complication in the event-driven method is that in a system with many particles one must determine which pair of particles will be the next to collide or react. For an approach to this problem in a well-mixed reacting system, *see* **ref. 137**. Methods for collisions in a hard-sphere fluid are discussed on pp. 391–417 in **ref. 6**. A disadvantage of the event-driven algorithm is the irregularity of the time coordinate.

7. Packaged Programs

This chapter emphasizes how to write a diffusion program, but diffusion can also be simulated using high-quality packaged programs for modeling molecular interactions and cellular processes. Programs for cellular modeling are reviewed by Slepchenko et al. *(138)* and programs for molecular modeling are reviewed by Baker *(73)* and in a special issue of the Journal of Computational Chemistry *(139)*. An excellent review by Takahashi et al. *(140)* compares a variety of algorithms and programs in the context of modeling intracellular crowding. The molecular modeling programs use molecular dynamics, Brownian dynamics, or both. The cellular modeling programs may be able to include realistic cellular geometry in the simulations. For example, Dix et al. *(141)* used GROMACS in a comprehensive model of biophysical fluorescence correlation spectroscopy experiments, taking into account photophysics as well as diffusion. Effects of crowding and system geometry can also be included. Coggan et al. *(142)* used MCell to model neurotransmission at a synapse. Atomic-level modeling of protein–protein diffusion and association is reviewed by Gabdoulline and Wade *(75)*.

8. References on the Physics of Diffusion

In biophysics, the usual first reference on diffusion is Berg's book *(143)*. The standard references on diffusion and equivalently heat conduction are Carslaw and Jaeger *(17)* and Crank *(144)*. Hughes *(145)* provides an encyclopedic treatment of random walks from a mathematician's point of view but written at the level of abstraction of the physics literature. Torquato's book *(117)* on random heterogeneous materials is highly helpful. Some useful books discuss random walks *(66,146,147)*, percolation *(148)*, and fractals *(149–151)* from the standpoint of contemporary physics. For general numerical methods, an essential reference is Numerical Recipes *(13)*, available in Fortran, C, and C++. For criticisms *see* http://www.lysator.liu.se/c/ num-recipes-in-c.html and for replies *see* http://www.nr.com/. For inspiration on how to approach nonstandard problems in diffusion simulation, it is useful to go outside the literature of Monte Carlo simulations. *See* textbooks on algorithms *(152,153)*, computational geometry *(154)*, and computer graphics, where it is necessary to calculate many intersections of lines and surfaces quickly *(155)*. For references to formulas on the diffusion coefficient as a function of obstacle concentration, and the percolation threshold of ellipses *see* **Section III-B** in **ref.** *1* and for percolation thresholds for ellipsoids, *see* **ref.** *156*.

9. General Comments

Modeling an entire organelle or cell is becoming popular. How useful is this? It depends on the information available. For example, Tremmel et al. *(157)* modeled the diffusion of plastoquinol in thylakoid membranes. The simulation used realistic concentrations and protein shapes. The effect of the shapes was found to be small but this was not known *a priori*. Later work *(158)* examined the effect of short-range interactions on protein–protein association. Another model *(80)* of the thylakoid membrane assumed cylindrical proteins but treated the interactions in more detail, including a screened Coulomb repulsion, a van der Waals attraction, and lipid-mediated interactions. This work focused on the interactions required for segregation of photosystems I and II. Elcock *(159)* modeled removal of a protein from a chaperonin cage as a function of the concentration of inert polysaccharides. Coggan et al. *(142)* modeled neurotransmission at a synapse using 3D geometry from serial-section electron tomography, and experimental values of reactant distributions and rate constants.

The first problem in such modeling is that one needs the concentrations and shapes of the major components. These are known for chloroplasts and synapses, but not for an entire cell. Even finding the total area fraction of membrane proteins in the plasma membrane is problematic *(1)*. For each component, one needs the translational and rotational diffusion coefficient in a dilute solution. This could be estimated from size and asymmetry using the Saffman–Delbrück equations for a 2D membrane *(160)*, or the Stokes–Einstein equations for spheres, or numerically for more complex 3D shapes *(161)*. A serious problem is characterizing the interactions, both protein–protein and protein–organelle. The interaction of mobile species with immobile species is particularly important because it can lead to transient anomalous subdiffusion *(10)*.

Given these difficulties, whole-cytoplasm simulations are good ways to capture the imagination of audiences but are hard to prove or disprove. For example, Bicout and Field *(162)* modeled *Escherichia coli* cytoplasm—the best-characterized example—as a mixture of spherical particles representing proteins, ribosomes, and tRNA interacting by Lennard-Jones and electrostatic potentials. Testability is the key factor. One useful approach would be the simulation of model systems, such as the actin/Ficoll mixtures used as a model of cytoplasm in diffusion experiments *(163,164)*. Entangled F-actin filaments represented the cytoskeletal network, and Ficoll represented the high concentration of mobile species in the cytoplasm. For a more abstract approach, one could model diffusion of interacting particles with various distributions of interactions, comparing uniform interactions with normal, lognormal, and singular distributions of interaction parameters.

Acknowledgments

I thank Adrian Elcock, Joseph Gall, Klaus Schulten, Nancy Thompson, and Matthias Weiss for helpful discussions. This work was supported by NIH grant GM-038133.

Appendix: Properties of the Diffusion Propagator in d Dimensions

The simplest approach to verifying the normalization of the propagator and finding its moments is to solve the general case for even moments in d dimensions. The formulas are not conveniently accessible in the literature. For free diffusion in d dimensions, the propagator is *(165)*

$$P(\vec{r},t) = \frac{1}{(4\pi Dt)^{d/2}} \exp\left[\frac{-r^2}{4Dt}\right]. \tag{A1}$$

The volume of a d-dimensional sphere is *(166)*

$$V_d = \frac{2\pi^{d/2} r^d}{d\Gamma(d/2)} \tag{A2}$$

so that the volume element is

$$dV_d = \frac{2\pi^{d/2} r^{d-1}}{\Gamma(d/2)} dr. \tag{A3}$$

Here Γ is the gamma function, defined by *(167)*

$$\Gamma(z) = \int_0^\infty t^{z-1} e^{-t} dt. \tag{A4}$$

The $2n$th moment of r is defined as

$$\langle r^{2n} \rangle = \int_0^\infty P(\vec{r},t) r^{2n} dV_d \tag{A5}$$

Substitute **Eqs. A1** and **A3** into **A5**, and use the definition of the gamma function to obtain

$$\langle r^{2n} \rangle = (4Dt)^n \frac{\Gamma\left(n + \frac{d}{2}\right)}{\Gamma\left(\frac{d}{2}\right)}. \tag{A6}$$

Specifically,

$$\langle r^0 \rangle = 1 \tag{A7}$$

verifying the normalization of $P(r,t)$. The MSD is

$$\langle r^2 \rangle = 2dDt \tag{A8}$$

and

$$\langle r^4 \rangle = (d/2)(d/2 + 1)(4Dt)^2 \tag{A9}$$

so that the variance in the MSD is

$$\text{Var}\langle r^2 \rangle = \langle r^4 \rangle - \langle r^2 \rangle^2 = \frac{2}{d}\langle r^2 \rangle^2 \tag{A10}$$

and the standard deviation of the MSD is proportional to the MSD in all dimensions

$$\sigma = \sqrt{\frac{2}{d}}\,\langle r^2 \rangle. \tag{A11}$$

References

1. Saxton, M. J. (1999) Lateral diffusion of lipids and proteins. *Curr. Topics Membr.* **48,** 229–282.
2. Hall, D. and Minton, A. P. (2003) Macromolecular crowding: Qualitative and semiquantitative successes, quantitative challenges. *Biochim. Biophys. Acta* **1649,** 127–139.
3. Saxton, M. J. and Jacobson, K. (1997) Single-particle tracking: Applications to membrane dynamics. *Annu. Rev. Biophys. Biomol. Struct.* **26,** 373–399.
4. Allen, M. P. and Tildesley, D. J. (1989) *Computer Simulation of Liquids*. Oxford University Press, Oxford. Fortran programs from the appendix are available at www.ccp5.ac.uk/librar.shtml.
5. Frenkel, D. and Smit, B. (1996) *Understanding Molecular Simulation: From Algorithms to Applications.* Academic Press, San Diego.
6. Rapaport, D. C. (2004) *The Art of Molecular Dynamics Simulation*, 2nd ed., Cambridge University Press, Cambridge.
7. Feder, T. J., Brust-Mascher, I., Slattery, J. P., Baird, B., and Webb, W. W. (1996) Constrained diffusion or immobile fraction on cell surfaces: A new interpretation. *Biophys. J.* **70,** 2767–2773.
8. Bouchaud, J.-P. and Georges, A. (1990) Anomalous diffusion in disordered media: Statistical mechanisms, models and physical applications. *Phys. Reports* **195,** 127–293.
9. Saxton, M. J. (1994) Anomalous diffusion due to obstacles: A Monte Carlo study. *Biophys. J.* **66,** 394–401.
10. Saxton, M. J. (1996) Anomalous diffusion due to binding: A Monte Carlo study. *Biophys. J.* **70,** 1250–1262.
11. Kusumi, A., Sako, Y., and Yamamoto, M. (1993) Confined lateral diffusion of membrane receptors as studied by single particle tracking (nanovid microscopy). Effects of calcium-induced differentiation in cultured epithelial cells. *Biophys. J.* **65,** 2021–2040.
12. Saxton, M. J. (1993) Lateral diffusion in an archipelago: Single-particle diffusion. *Biophys. J.* **64,** 1766–1780.
13. Press, W. H., Teukolsky, S. A., Vetterling, W. T., and Flannery, B. P. (1992) *Numerical Recipes in FORTRAN: The Art of Scientific Computing*, 2nd ed., Cambridge University Press, Cambridge. Also available in C and C++ editions.
14. Saxton, M. J. (2001) Anomalous subdiffusion in fluorescence photobleaching recovery: A Monte Carlo study. *Biophys. J.* **81,** 2226–2240.
15. Deverall, M. A., Gindl, E., Sinner, E.-K., et al. (2005) Membrane lateral mobility obstructed by polymer-tethered lipids studied at the single molecule level. *Biophys. J.* **88,** 1875–1886.

16. Metzler, R. and Klafter, J. (2004) The restaurant at the end of the random walk: Recent developments in the description of anomalous transport by fractional dynamics. *J. Phys. A* **37,** R161–R208.
17. Carslaw, H. S. and Jaeger, J. C. (1959) *Conduction of Heat in Solids,* 2nd ed., Oxford University Press, Oxford.
18. Powles, J. G., Mallett, M. J. D., Rickayzen, G., and Evans, W. A. B. (1992) Exact analytic solutions for diffusion impeded by an infinite array of partially permeable barriers. *Proc. R. Soc. Lond. A* **436,** 391–403.
19. Kusumi, A., Nakada, C., Ritchie, K., et al. (2005) Paradigm shift of the plasma membrane concept from the two-dimensional continuum fluid to the partitioned fluid: High-speed single-molecule tracking of membrane molecules. *Annu. Rev. Biophys. Biomol. Struct.* **34,** 351–378.
20. Anderson, C. M., Georgiou, G. N., Morrison, I. E. G., Stevenson, G. V. W., and Cherry, R. J. (1992) Tracking of cell surface receptors by fluorescence digital imaging microscopy using a charge-coupled device camera. Low-density lipoprotein and influenza virus receptor mobility at 4 C. *J. Cell Sci.* **101,** 415–425.
21. Schütz, G. J., Schindler, H., and Schmidt, T. (1997) Single-molecule microscopy on model membranes reveals anomalous diffusion. *Biophys. J.* **73,** 1073–1080.
22. Barlow, R. J. (1993) *Statistics: A Guide to the Use of Statistical Methods in the Physical Sciences.* Wiley, Chichester, pp. 155–156.
23. Schaertl, W. and Sillescu, H. (1993) Dynamics of colloidal hard spheres in thin aqueous suspension layers—Particle tracking by digital image processing and Brownian dynamics computer simulations. *J. Colloid Interface Sci.* **155,** 313–318.
24. Crocker, J. C. and Grier, D. G. (1996) Methods of digital video microscopy for colloidal studies. *J. Colloid Interface Sci.* **179,** 298–310.
25. Kegel, W. K. and van Blaaderen, A. (2000) Direct observation of dynamical heterogeneities in colloidal hard-sphere suspensions. *Science* **287,** 290–293.
26. Weeks, E. R. and Weitz, D. A. (2002) Properties of cage rearrangements observed near the colloidal glass transition. *Phys. Rev. Lett.* **89,** 095704.
27. Pusey, P. N., Lekkerkerker, H. N. W., Cohen, E. G. D., and de Schepper, I. M. (1990) Analogies between the dynamics of concentrated charged colloidal suspensions and dense atomic liquids. *Physica A* **164,** 12–27.
28. Cohen, E. G. D. and de Schepper, I. M. (1991) Note on transport processes in dense colloidal suspensions. *J. Stat. Phys.* **63,** 241–248.
29. Heyes, D. M. (1994) Molecular simulations of colloidal liquids. *Adv. Colloid Interface Sci.* **51,** 247–268.
30. Schaertl, W. and Sillescu, H. (1994) Brownian dynamics simulations of colloidal hard spheres. Effects of sample dimensionality on self-diffusion. *J. Stat. Phys.* **74,** 687–703.
31. Lahtinen, J. M., Hjelt, T., Ala-Nissila, T., and Chvoj, Z. (2001) Diffusion of hard disks and rod-like molecules on surfaces. *Phys. Rev. E* **64,** 021204.
32. Löwen, H. and Szamel, G. (1993) Long-time self-diffusion coefficient in colloidal suspensions: theory versus simulation. *J. Phys. Cond. Matter* **5,** 2295–2306.
33. Piazza, R. (2000) Interactions and phase transitions in protein solutions. *Curr. Opin. Colloid Interface Sci.* **5,** 38–43.
34. Stradner, A., Sedgwick, H., Cardinaux, F., Poon, W. C. K., Egelhaaf, S. U., and Schurtenberger, P. (2004) Equilibrium cluster formation in concentrated protein solutions and colloids. *Nature* **432,** 492–495.
35. Rahman, A. (1964) Correlations in the motion of atoms in liquid argon. *Phys. Rev.* **136,** A405–A411.
36. Feller, S. E., Huster, D., and Gawrisch, K. (1999) Interpretation of NOESY cross-relaxation rates from molecular dynamics simulation of a lipid bilayer. *J. Am. Chem. Soc.* **121,** 8963–8964.
37. Ermak, D. L. and McCammon, J. A. (1978) Brownian dynamics with hydrodynamic interactions. *J. Chem. Phys.* **69,** 1352–1360.

38. Madura, J. D., Briggs, J. M., Wade, R. C., et al. (1995) Electrostatics and diffusion of molecules in solution: Simulations with the University of Houston Brownian dynamics program. *Comput. Phys. Comm.* **91,** 57–95.
39. Elcock, A. H. (2004) Molecular simulations of diffusion and association in multimacromolecular systems. *Methods Enzymol.* **383,** 166–198.
40. Schlick, T., Beard, D. A., Huang, J., Strahs, D. A., and Qian, X. (2000) Computational challenges in simulating large DNA over long times. *Comput. Sci. Eng.* **2,** 38–51.
41. Iniesta, A. and Garcia de la Torre, J. (1990) A second-order algorithm for the simulation of the Brownian dynamics of macromolecular models. *J. Chem. Phys.* **92,** 2015–2018.
42. Binder, K. and Heermann, D. W. (1992) *Monte Carlo Simulation in Statistical Physics: An Introduction*, 2nd ed., Springer, Berlin.
43. Landau, D. P. and Binder, K. (2000) *A Guide to Monte Carlo Simulations in Statistical Physics.* Cambridge University Press, Cambridge.
44. Brannigan, G. and Brown, F. L. H. (2005) Composition dependence of bilayer elasticity. *J. Chem. Phys.* **122,** 074905.
45. Fichthorn, K. A. and Weinberg, W. H. (1991) Theoretical foundations of dynamic Monte Carlo simulations. *J. Chem. Phys.* **95,** 1090–1096.
46. Kang, H. C. and Weinberg, W. H. (1992) Dynamic Monte Carlo simulations of surface-rate processes. *Acc. Chem. Res.* **25,** 253–259.
47. Zuckermann, M. J., Ipsen, J. H., Miao, L., et al. (2004) Modeling lipid-sterol bilayers: Applications to structural evolution, lateral diffusion and rafts. *Methods Enzymol.* **383,** 198–229.
48. Groot, R. D. and Warren, P. B. (1997) Dissipative particle dynamics: Bridging the gap between atomistic and mesoscopic simulation. *J. Chem. Phys.* **107,** 4423–4435.
49. Shillcock, J. C. and Lipowsky, R. (2002) Equilibrium structure and lateral stress distribution of amphiphilic bilayers from dissipative particle dynamics simulations. *J. Chem. Phys.* **117,** 5048–5061.
50. Symeonidis, V., Karniadakis, G. E., and Caswell, B. (2005) A seamless approach to multiscale complex fluid simulation. *Comput. Sci. Eng.* **7,** 39–46.
51. Saxton, M. J. (1993) Lateral diffusion in an archipelago: Dependence on tracer size. *Biophys. J.* **64,** 1053–1062.
52. Abney, J. R. and Scalettar, B. A. (1992) Molecular crowding and protein organization in biological membranes, in *Thermodynamics of Membrane Receptors and Channels.* (Jackson, M. B., ed.), CRC Press, Boca Raton, FL, pp. 183–226.
53. Almeida, P. F. F. and Vaz, W. L. C. (1995) Lateral diffusion in membranes, in *Structure and Dynamics of Membranes*, vol. 1, (Lipowsky, R. and Sackmann, E., eds.) Elsevier, Amsterdam, pp. 305–357.
54. Holyst, R., Plewczyński, D., Aksimentiev, A., and Burdzy, K. (1999) Diffusion on curved, periodic surfaces. *Phys. Rev. E* **60,** 302–307.
55. Christensen, M. (2004) How to simulate anisotropic diffusion processes on curved surfaces. *J. Comput. Phys.* **201,** 421–438.
56. Sbalzarini, I. F., Mezzacasa, A., Helenius, A., and Koumoutsakos, P. (2005) Effects of organelle shape on fluorescence recovery after photobleaching. *Biophys. J.* **89,** 1482–1492.
57. Saxton, M. J. (1995) Single-particle tracking: Effect of corrals. *Biophys. J.* **69,** 389–398.
58. Nagle, J. F. (1992) Long tail kinetics in biophysics? *Biophys. J.* **63,** 366–370.
59. Coeurjolly, J. -F. (2000) Simulation and identification of the fractional Brownian motion: a bibliographical and comparative study. *J. Stat. Software* 5, Issue 7 (www.jstatsoft.org).
60. Evans, J. W. (1993) Random and cooperative sequential adsorption. *Revs. Mod. Phys.* **65,** 1281–1329.
61. Talbot, J., Tarjus, G., Van Tassel, P. R., and Viot, P. (2000) From car parking to protein adsorption: an overview of sequential adsorption processes. *Colloids Surf. A* **165,** 287–324.

62. Wang, J. S. (1994) A fast algorithm for random sequential adsorption of disks. *Int. J. Mod. Phys.* **C5,** 707–715.
63. Ossola, G. and Sokal, A. D. (2004) Systematic errors due to linear congruential random-number generators with the Swendsen–Wang algorithm: A warning. *Phys. Rev. E* **70,** 027701.
64. Havlin, S. and Ben-Avraham, D. (1987) Diffusion in disordered media. *Adv. Phys.* **36,** 695–798.
65. Stauffer, D. (1985) *Introduction to Percolation Theory.* Taylor & Francis, London.
66. ben-Avraham, D. and Havlin, S. (2000) *Diffusion and Reactions in Fractals and Disordered Systems.* Cambridge University Press, Cambridge.
67. Tobochnik, J., Laing, D., and Wilson, G. (1990) Random-walk calculation of conductivity in continuum percolation. *Phys. Rev. A* **41,** 3052–3058.
68. Cichocki, B. and Hinsen, K. (1990) Dynamic computer simulation of concentrated hard sphere suspensions. I. Simulation technique and mean square displacement data. *Physica A* **166,** 473–491.
69. Heyes, D. M. and Branka, A. C. (1994) Molecular and Brownian dynamics simulations of self-diffusion in inverse power fluids. *Phys. Chem. Liq.* **28,** 95–115.
70. Gabdoulline, R. R. and Wade, R. C. (1998) Brownian dynamics simulation of protein-protein diffusional encounter. *Methods* **14,** 329–341.
71. Elcock, A. H. (2002) Modeling supramolecular assemblages. *Curr. Opin. Struct. Biol.* **12,** 154–160.
72. Baker, N. A. and McCammon, J. A. (2003) Electrostatic interactions. *Meth. Biochem. Anal.* **44,** 427–440.
73. Baker, N. A. (2004) Poisson-Boltzmann methods for biomolecular electrostatics. *Methods Enzymol.* **383,** 94–118.
74. Dodd, T. L., Hammer, D. A., Sangani, A. S., and Koch, D. L. (1995) Numerical simulations of the effect of hydrodynamic interactions on diffusivities of integral membrane proteins. *J. Fluid Mech.* **293,** 147–180.
75. Gabdoulline, R. R. and Wade, R. C. (2002) Biomolecular diffusional association. *Curr. Opin. Struct. Biol.* **12,** 204–213.
76. Marcelja, S. (1999) Toward a realistic theory of the interaction of membrane inclusions. *Biophys. J.* **76,** 593–594.
77. Kim, K. S., Neu, J., and Oster, G. (2000) Effect of protein shape on multibody interactions between membrane inclusions. *Phys. Rev. E* **61,** 4281–4285.
78. Lagüe, P., Zuckermann, M. J., and Roux, B. (2001) Lipid-mediated interactions between intrinsic membrane proteins: Dependence on protein size and lipid composition. *Biophys. J.* **81,** 276–284.
79. Bohinc, K., Kralj-Iglic, V., and May, S. (2003) Interaction between two cylindrical inclusions in a symmetric lipid bilayer. *J. Chem. Phys.* **119,** 7435–7444.
80. Borodich, A., Rojdestvenski, L., and Cottam, M. (2003) Lateral heterogeneity of photosystems in thylakoid membranes studied by Brownian dynamics simulations. *Biophys. J.* **85,** 774–789.
81. Saxton, M. J. (1997) Single-particle tracking: The distribution of diffusion coefficients. *Biophys. J.* **72,** 1744–1753.
82. L'Ecuyer, P. (2001) Software for uniform random number generation: distinguishing the good and the bad. *Proceedings of the 2001 Winter Simulation Conference* (Peters, B. A., Smith, J. S., Medeiros, D. J., and Rohrer, M. W., eds.) Association for Computing Machinery, New York, Vol. 1, pp. 95–105.
83. Knuth, D. E. (1997) *The Art of Computer Programming. Seminumerical Algorithms,* vol. 2, 3rd ed., Addison-Wesley, Reading, Mass.
84. Marsaglia, G. (1968) Random numbers fall mainly in the planes. *Proc. Natl. Acad. Sci. USA* **61,** 25–28.
85. Ziff, R. M. (1996) Effective boundary extrapolation length to account for finite-size effects in the percolation crossing function. *Phys. Rev. E.* **54,** 2547–2554.
86. Knüsel, L. (2005) On the accuracy of statistical distributions in Microsoft Excel 2003. *Comput. Statist. Data Anal.* **48,** 445–449.

87. McCullough, B. D. and Wilson, B. (2005) On the accuracy of statistical procedures in Microsoft Excel. *Comput. Statist. Data Anal.* **49**, 1244–1252.
88. Nogués, J., Costa-Krämer, J. L., and Rao, K. V. (1998) Are random walks random? *Physica A* **250**, 327–334.
89. Coddington, P. D. (1996) Tests of random number generators using Ising model simulations. *Int. J. Mod. Phys. C* **7**, 295–303.
90. Press, W. H. and Teukolsky, S. A. (1992) Portable random number generators. *Comput. Phys.* **6**, 522–524.
91. Press, W. H., Flannery, B. P., Teukolsky, S. A., and Vetterling, W. T. (1986) *Numerical Recipes: The Art of Scientific Computing.* Cambridge University Press, Cambridge.
92. Press, W. H., Teukolsky, S. A., Vetterling, W. T., and Flannery, B. P. (1996) *Numerical Recipes in Fortran 90: The Art of Parallel Scientific Computing.* Cambridge University Press, Cambridge.
93. Ziff, R. M. (1998) Four-tap shift-register-sequence random-number generators. *Comput. Phys.* **12**, 385–392.
94. Matsumoto, M. and Nishimura, T. (1998) Mersenne twister: a 623-dimensionally equidistributed uniform pseudo-random number generator. *ACM Trans. Modeling Comput. Simul.* **8**, 3–30. Available in various languages at www.math.sci.hiroshima-u.ac.jp/~m-mat/MT/eindex.html.
95. Marsaglia, G. and Zaman, A. (1994) Some portable very-long-period random number generators. *Comput. Phys.* **8**, 117–121.
96. Hamilton, K. G. and James, F. (1997) Acceleration of RANLUX. *Comput. Phys. Comm.* **101**, 241–248. Has a Fortran version.
97. Park, S. K. and Miller, K. W. (1988) Random number generators—good ones are hard to find. *Commun. ACM* **31**, 1192–1201.
98. Vattulainen, I., Ala-Nissila, T., and Kankaala, K. (1995) Physical models as tests of randomness. *Phys. Rev. E.* **52**, 3205–3214.
99. Vattulainen, I., Kankaala, K., Saarinen, J., and Ala-Nissila, T. (1995) A comparative study of some pseudorandom number generators. *Comput. Phys. Comm.* **86**, 209–226.
100. Vattulainen, I. and Ala-Nissila, T. (1995) Mission impossible: find a random pseudorandom number generator. *Comput. Phys.* **9**, 500–504.
101. Hellekalek, P. (1998) Good random number generators are (not so) easy to find. *Math. Comput. Simulation* **46**, 485–505.
102. L'Ecuyer, P. (1998) Random number generation, in *Handbook on Simulation: Principles, Methodology, Advances, Applications, and Practice*. Wiley, New York, pp. 93–137.
103. L'Ecuyer, P. (2006) Random number generation, in *Simulation* (Henderson, S. G. and Nelson, B. L., eds.), Elsevier, Amsterdam, pp. 55–81.
104. Rukhin, A., Soto, J., Nechvatal, J., et al. (2001) A statistical test suite for random and pseudorandom number generators for cryptographic applications. National Institute of Standards and Technology (NIST) Special Publication 800-22 (with revisions May 15, 2001). Available at csrc.nist.gov/publications/nistpubs/ along with Unix and PC forms of the NIST Statistical Test Suite.
105. Mascagni, M. and Srinivasan, A. (2000) SPRNG: a scalable library for pseudorandom number generation. *ACM Trans. Math. Software* **26**, 436–461.
106. Vattulainen, I. (1999) Framework for testing random numbers in parallel calculations. *Phys. Rev. E* **59**, 7200–7204.
107. Ferrenberg, A. M., Landau, D. P., and Wong, Y. J. (1992) Monte Carlo simulations: Hidden errors from "good" random number generators. *Phys. Rev. Lett.* **69**, 3382–3384.
108. Schmid, F. and Wilding, N. B. (1995) Errors in Monte Carlo simulations using shift register random number generators. *Int. J. Mod. Phys. C* **6**, 781–787.
109. Holian, B. L., Percus, O. E., Warnock, T. T., and Whitlock, P. A. (1994) Pseudorandom number generator for massively parallel molecular-dynamics simulations. *Phys. Rev. E* **50**, 1607–1615.

110. Tretiakov, K. V. and Wojciechowski, K. W. (1999) Efficient Monte Carlo simulations using a shuffled nested Weyl sequence random number generator. *Phys. Rev. E* **60,** 7626–7628.
111. Grassberger, P. (1993) On correlations in "good" random number generators. *Phys. Lett. A* **181,** 43–46.
112. Gammel, B. M. (1998) Hurst's rescaled range statistical analysis for pseudorandom number generators used in physical simulations. *Phys. Rev. E* **58,** 2586–2597.
113. Shchur, L. N., Heringa, J. R., and Blöte, H. W. J. (1997) Simulation of a directed random-walk model. The effect of pseudo-random-number correlations. *Physica A* **241,** 579–592.
114. Compagner, A. (1995) Operational conditions for random-number generation. *Phys. Rev. E* **52,** 5634–5645.
115. Shchur, L. N. (1999) On the quality of random number generators with taps. *Comput. Phys. Comm.* **121–122,** 83–85.
116. Filk, T., Marcu, M., and Fredenhagen, K. (1985) Long range correlations in random number generators and their influence on Monte Carlo simulations. *Phys. Lett. B* **165,** 125–130.
117. Torquato, S. (2002) *Random Heterogeneous Materials: Microstructure and Macroscopic Properties*. Springer, New York.
118. Ripley, B. D. (1987) *Stochastic Simulation*. Wiley, New York. Has Fortran listings.
119. Bratley, P., Fox, B. L., and Schrag, L. E. (1987) *A Guide to Simulation,* 2nd ed., Springer-Verlag, New York.
120. Gentle, J. E. (1998) *Random Number Generation and Monte Carlo Methods*. Springer-Verlag, New York.
121. Devroye, L. (1986) *Non-Uniform Random Variate Generation*. Springer-Verlag, New York.
122. Marsaglia, G. and Tsang, W. W. (2000) The ziggurat method for generating random variables. *J. Stat. Software* 5, Issue 8 (www.jstatsoft.org). Has a C listing.
123. Leong, P. H. W., Zhang, G., Lee, D. -U., Luk, W., and Villasenor, J. D. (2005) A comment on the implementation of the ziggurat method. *J. Stat. Software* 12, Issue 7 (www.jstatsoft.org).
124. Dowd, K. and Severance, C. R. (1998) *High Performance Computing,* 2nd ed., O'Reilly & Associates, Sebastopol, CA.
125. Wadleigh, K. R. and Crawford, I. L. (2000) *Software Optimization for High Performance Computing*. Prentice Hall PTR, Upper Saddle River, NJ.
126. Lindahl, E., Hess, B., and van der Spoel, D. (2001) GROMACS 3.0: a package for molecular simulation and trajectory analysis. *J. Mol. Model.* **7,** 306–317.
127. Novotny, M. A. (1995) A new approach to an old algorithm for the simulation of Ising-like systems. *Comput. Phys.* **9,** 46–52.
128. Bortz, A. B., Kalos, M. H., and Lebowitz, J. L. (1975) New algorithm for Monte Carlo simulation of Ising spin systems. *J. Comput. Phys.* **17,** 10–18.
129. Battaile, C. C., Srolovitz, D. J., and Butler, J. E. (1997) A kinetic Monte Carlo method for the atomic-scale simulation of chemical vapor deposition: Application to diamond. *J. Appl. Phys.* **82,** 6293–6300.
130. Brosilow, B. J., Ziff, R. M., and Vigil, R. D. (1991) Random sequential adsorption of parallel squares. *Phys. Rev. E* **43,** 631–638.
131. Zheng, L. H. and Chiew, Y. C. (1989) Computer simulation of diffusion-controlled reactions in dispersions of spherical sinks. *J. Chem. Phys.* **90,** 322–327.
132. Tobochnik, J. (1990) Efficient random walk algorithm for computing conductivity in continuum percolation systems. *Comput. Phys.* **4,** 181–184.
133. Torquato, S. and Kim, I. C. (1989) Efficient simulation technique to compute effective properties of heterogeneous media. *Appl. Phys. Lett.* **55,** 1847–1849.
134. Torquato, S., Kim, I. C., and Cule, D. (1999) Effective conductivity, dielectric constant, and diffusion coefficient of digitized composite media via first-passage-time equations. *J. Appl. Phys.* **85,** 1560–1571.

135. Hwang, C. O., Given, J. A., and Mascagni, M. (2001) The simulation-tabulation method for classical diffusion Monte Carlo. *J. Comput. Phys.* **174,** 925–946.
136. van Zon, J. S. and ten Wolde, P. R. (2005) Simulating biochemical networks at the particle level and in time and space: Green's function reaction dynamics. *Phys. Rev. Lett.* **94,** 128103.
137. Gillespie, D. T. (1977) Exact stochastic simulation of coupled chemical reactions. *J. Phys. Chem.* **81,** 2340–2361.
138. Slepchenko, B. M., Schaff, J. C., Carson, J. H., and Loew, L. M. (2002) Computational cell biology: Spatiotemporal simulation of cellular events. *Annu. Rev. Biophys. Biomol. Struct.* **31,** 423–441.
139. Brooks, C. L. (2005) Special issue: Modern tools for macromolecular simulation and modeling. *J. Comput. Chem.* **26,** 1667.
140. Takahashi, K., Arjunan, S. N. V., and Tomita, M. (2005) Space in systems biology of signaling pathways—towards intracellular molecular crowding in silico. *FEBS Lett.* **579,** 1783–1788.
141. Dix, J. A., Hom, E. F. Y., and Verkman, A. S. (2006) Fluorescence correlation spectroscopy simulation of photophysical phenomena and molecular interactions: A molecular dynamics/Monte Carlo approach. *J. Phys. Chem. B* **110,** 1896–1906.
142. Coggan, J. S., Bartol, T. M., Esquenazi, E., et al. (2005) Evidence for ectopic neurotransmission at a neuronal synapse. *Science* **309,** 446–451.
143. Berg, H. C. (1993) *Random Walks in Biology,* 2nd ed., Princeton University Press, Princeton, NJ.
144. Crank, J. (1975) *The Mathematics of Diffusion,* 2nd ed., Oxford University Press, Oxford.
145. Hughes, B. D. (1995, 1996) *Random Walks and Random Environments,* 2 vols. Oxford University Press, Oxford.
146. Redner, S. (2001) *A Guide to First-Passage Processes.* Cambridge University Press, Cambridge.
147. Rudnick, J. and Gaspari, G. (2004) *Elements of the Random Walk: An Introduction for Advanced Students and Researchers.* Cambridge University Press, Cambridge.
148. Stauffer, D. and Aharony, A. (1992) *Introduction to Percolation Theory,* 2nd ed., Taylor & Francis, London.
149. Bassingthwaighte, J. B., Liebovitch, L. S., and West, B. J. (1994) *Fractal Physiology.* Oxford University Press, Oxford.
150. Dewey, T. G. (1997) *Fractals in Molecular Biophysics.* Oxford University Press, Oxford.
151. Liebovitch, L. S. (1998) *Fractals and Chaos Simplified for the Life Sciences.* Oxford University Press, New York.
152. Sedgewick, R. (1988) *Algorithms.* Addison-Wesley, Reading, MA.
153. Skiena, S. S. (1998) *The Algorithm Design Manual.* Springer TELOS-The Electronic Library of Science, Santa Clara, CA.
154. O'Rourke, J. (1994) *Computational Geometry in C.* Cambridge University Press, Cambridge.
155. Glassner, A. S. (1990) *Graphics Gems.* Academic Press, Boston. There are now 5 volumes in the series, with various editors.
156. Yi, Y. -B. and Sastry, A. M. (2004) Analytical approximation of the percolation threshold for overlapping ellipsoids of revolution. *Proc. R. Soc. Lond. A* **460,** 2353–2380.
157. Tremmel, I. G., Kirchhoff, H., Weis, E., and Farquhar, G. D. (2003) Dependence of plastoquinol diffusion on the shape, size, and density of integral thylakoid proteins. *Biochim. Biophys. Acta* **1607,** 97–109.
158. Tremmel, I. G., Weis, E., and Farquhar, G. D. (2005) The influence of protein-protein interactions on the organization of proteins within thylakoid membranes. *Biophys. J.* **88,** 2650–2660.
159. Elcock, A. H. (2003) Atomic-level observation of macromolecular crowding effects: Escape of a protein from the GroEL cage. *Proc. Natl. Acad. Sci. USA* **100,** 2340–2344.
160. Saffman, P. G. and Delbrück, M. (1975) Brownian motion in biological membranes. *Proc. Natl. Acad. Sci. USA* **72,** 3111–3113.
161. Garcia de la Torre, J., Huertas, M. L., and Carrasco, B. (2000) Calculation of hydrodynamic properties of globular proteins from their atomic-level structure. *Biophys. J.* **78,** 719–730.

162. Bicout, D. J. and Field, M. J. (1996) Stochastic dynamics simulations of macromolecular diffusion in a model of the cytoplasm of *Escherichia coli*. *J. Phys. Chem.* **100,** 2489–2497.
163. Hou, L., Luby-Phelps, K., and Lanni, F. (1990) Brownian motion of inert tracer macromolecules in polymerized and spontaneously bundled mixtures of actin and filamin. *J. Cell Biol.* **110,** 1645–1654.
164. Hou, L., Lanni, F., and Luby-Phelps, K. (1990) Tracer diffusion in F actin and Ficoll mixtures. Toward a model for cytoplasm. *Biophys. J.* **58,** 31–44.
165. Weiss, G. H. and Rubin, R. J. (1983) Random walks: Theory and selected applications. *Adv. Chem. Phys.* **52,** 363–505, Eq. 2.156.
166. Gradshteyn, I. S. and Ryzhik, I. M. (2000) *Table of Integrals, Series, and Products,* 6th ed., Academic Press, San Diego, p. 609, Eq. 4.632.2.
167. Abramowitz, M. and Stegun, I. A. (1972) *Handbook of Mathematical Functions.* U. S. Government Printing Office, Washington, DC, p. 255, Eq. 6.1.1.

21

Measurement of Water and Solute Permeability by Stopped-Flow Fluorimetry

John C. Mathai and Mark L. Zeidel

Summary

Osmotic water permeability and solute permeability coefficient are measured using stopped-flow fluorimetry. In a vesicle that behaves as a perfect osmometer, water flux is directly proportional to imposed osmotic pressure, and solute flux is proportional to the chemical gradient across the vesicle. The flux of water and solute leads to a change in vesicle volume. This change in volume is measured by fluorescence quenching of entrapped carboxyfluorescein in the vesicle. Equations relating volume change of the vesicle to flux of water or solute from the vesicle are given to enable computation of water and solute permeability coefficients.

Key Words: Diffusion; fluorescence quenching; membrane; liposomes; osmotic water permeability; solute permeability.

1. Introduction

Small molecules (e.g., water, urea, glycerol, and so on) permeate most lipid membranes at an appreciable rate through a nonspecific mechanism in the absence of transporters. The mechanism of water and solute permeability across lipid membranes is not entirely defined. Finkelstein's book *(1)* provides an excellent introduction to water permeation, as well as access to early work in this field. Water permeability is measured most often by osmotic methods whereby an osmotic gradient is applied to a cell or vesicle and the resulting volume change owing to water flux, is measured. The osmotic method is widely used to measure permeabilities of liposomes, black lipid membranes, erythrocytes, oocytes, and cells cultures *(2–8)*. Passive or diffusional water permeability is traditionally measured by using tritiated water *(9)*. Recently, diffusion weighted nuclear magnetic resonance methods have been used to measure permeability in erythrocytes and oocytes *(10,11)*. Fluorescence methods based on enhanced quantum yield of fluorophore amino napthaline sulphonic acid (ANTS) in D_2O have been used in liposomes and red-cell ghosts to measure diffusional D_2O permeability *(12,13)*.

Diffusional water permeability is difficult to measure because the diffusion of water across membranes is so rapid that diffusion across the unstirred layers adjacent to the lipid bilayer might occur more slowly than diffusion across the lipid bilayer, which leads to underestimation of water permeability. By contrast, solute permeabilities across bilayers are much slower than those of water, so that unstirred layers rarely present a problem. On this basis, solute permeabilities are usually measured as fluxes of radio-labeled solute molecules. These methods are technically straightforward and because solute permeabilities are relatively low, the lack of time resolution of tracer methods does not affect the measurement. However, in some

From: *Methods in Molecular Biology, vol. 400: Methods in Membrane Lipids*
Edited by: A. M. Dopico © Humana Press Inc., Totowa, NJ

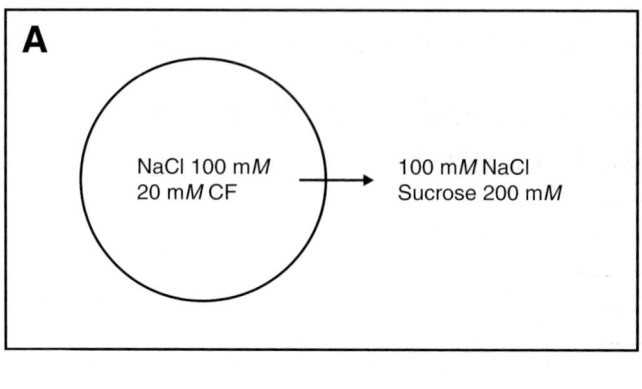

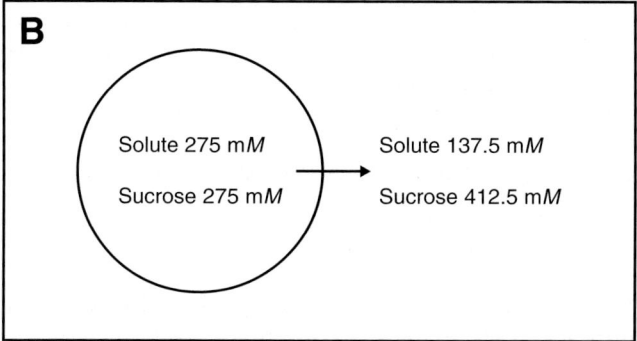

Fig. 1. Schematic of water and solute flux from vesicle: **(A)** water efflux from the vesicle because of a higher external osmotic gradient created after a 1:1 mixing of HB and isotonic solutions (IB). **(B)** Solute permeability measurement in absence of osmotic gradient. The total osmolarity of internal and external solutions remain constant, whereas a gradient for permeate solute is formed on 1:1 mixing of solutions S1 and NP. This leads to exit of permeate solute from the vesicle, and thus, to a reduction in vesicle volume, which is measured in the stopped-flow device.

cases, stopped-flow methods for measuring solute permeability have been used in liposomes and erythrocytes *(14,15)*. This chapter will focus on measurement of osmotic water permeability (P_f) and solute permeability across membrane vesicles using stopped-flow methods.

1.1. Osmotic Water Permeability

Abrupt exposure of vesicles to an inwardly directed osmotic gradient will result in water efflux from the vesicles and a concomitant reduction of vesicle volume until a new osmotic equilibrium is reached (**Fig. 1A**). The rate of volume change of the vesicle is directly related to the rate at which water leaves the vesicle. The volume change can be reliably measured by entrapping self-quenching concentrations of a fluorescent dye, 5,6-carboxyfluorescein (CF), in the vesicle. Light scattering can also be used to measure volume changes. However, caution should be exercised because of potential problems arising from mixing artifacts, vesicle size heterogeneity, and shape changes, leading to anomalous scattering signals. The rate of volume change is related to osmotic water permeability by the following equation:

$$dV_{(t)}/dt = (P_f)(SAV)(MVW)(C_{in}/V_t - C_{out}) \qquad (1)$$

where, V_t, is the relative volume of the vesicle at time t, (i.e., volume at time t, divided by the initial volume), P_f (cm/s) is the osmotic water permeability coefficient (P_s), and SAV is the surface area to volume ratio, MVW is the molar volume of water (18 cm^3/mol), C_{in} and C_{out}

are initial solute concentrations inside and outside the vesicle. As the external solute volume is very large compared with internal vesicle volume, the C_{out} is assumed to be constant. **Equation 1** can be simplified to a first-order process, and as $t \to 0$, the data can be fitted to a single exponential curve *(16)*. Using mathematical modeling software such as MathCad (Parametric technology corp., Needham, MA), **Eq. 1** can be simulated and P_f can be calculated as shown in **Note 1**.

1.2. Solute Permeability Measurements

Solute permeability measurements are performed under iso-osmotic conditions to avoid any volume changes resulting from osmotic pressure differences. Vesicles are loaded with the solute whose permeability is to be determined, and the external solute conditions are adjusted such that there is no osmotic gradient but only a gradient of the permeating solute between vesicle and external solution at the time of mixing. This results in solute efflux, followed by water, and a concomitant change in volume of the vesicle. Because the vesicles are loaded with quenching concentrations of the fluorophore (CF), the vesicle volume can be monitored as a function of time. The rate of volume change is directly proportional to the rate of solute efflux from the vesicle. **Figure 1B** shows an example of experimental conditions used for urea permeability. Solute flux from the vesicle is governed by the general equation *(17)*:

$$J_s = ds/dt = (P_s)(SA)(\Delta C) \qquad (2)$$

where J_s is the flux of the permeanting solute s, P_s (cm/s) the permeability of the solute, SA is the surface area of the vesicle, and ΔC is the concentration difference of the solute(s) between inside and outside of the vesicle. As the measured parameter is volume, it has to be related to solute flux. The equations necessary for calculation of solute permeability are given in **Note 2**. The parameters of the single exponential that fit to the experimental data can be used to compute the solute permeability (*see* **Note 3**).

2. Materials

CF (cat: C1904 Molecular Probes Invitrogen, Carlsbad, CA), antifluorescein antibody (Molecular probes, rabbit IgG fraction; cat. no. A889), *Escherichia coli* lipids (Avanti Polar Lipids, Alabaster, AL), Avanti mini-extruder device for making liposomes, 100–200 nm nucleopore membrane, PD-10 Sephadex column (Amersham Biosciences, Piscataway, NJ), fluorimeter Aminco Bowman Series 2, Urabana, IL, osmometer Osmette A, Precision systems, Natick, MA, stopped-flow spectrophotometer (SF 17. MV Applied photophysics, Leatherhead, UK), particle sizer DynaPro LSR, Wyatt technology corp., Santa Barbara, CA, and MathCad software.

2.1. Solutions for Water Permeability Determination

1. Entrapment buffer (EB): 100 m*M* NaCl, 10 m*M* HEPES buffer, pH 7.4, 20 m*M* CF. Make a 50 m*M* stock solution of CF in water. Addition of small aliquots of 1 *M* NaOH will be required to completely dissolve CF. Then, CF is dissolved and the pH is adjusted back to 7.0 with HCl.
2. Isotonic buffer (IB): measure the osmolarity of the EB and make the above buffer (EB) without CF. Adjust the osmolarity of the solution to that of the EB by adding additional NaCl.
3. Hypertonic buffer (HB): weight sucrose equivalent to twice the osmolarity of the isotonic solution and add it directly to IB (osmolarities of several solutions are given in the CRC Hand-Book of Chemistry and Physics 64th Edition, Ed R. C. Weast, CRC Press Inc., Boca Raton, FL.). The idea is to get a doubling of external osmolarity of the solution on 1:1 mixing of IB and HB solution in the stopped-flow device (**Fig. 2**).

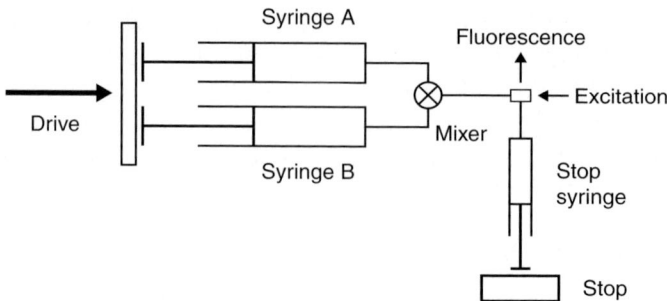

Fig. 2. Schematic of the stopped-flow device. The syringes are filled with appropriate solution, and on trigger, the syringe pistons are pushed. Thus, the solutions are mixed in the mixer within a millisecond. The solution then flows into the cuvet and finally into the stop syringe whereby flow stops; fluorescence is monitored from that point.

2.2. Solutions for Solute Permeability Measurements

1. Solute EB (S1): 275 mM sucrose, 10 mM HEPES, pH 7.4, 20 mM CF.
2. Permeate buffer (PB): 275 mM of test solute (urea) in the S1 buffer.
3. Non-PB (NP): 550 mM sucrose with 10 mM HEPES, pH 7.4.
4. Wash buffer (WB): measure the osmolarity of S1 buffer and make S1 buffer without CF, but adjust the osmolarity of the solution to that of S1 using additional sucrose.

3. Methods

3.1. Preparation of Liposomes or Lipid Vesicles

1. Dissolve 10 mg of *E. coli* lipids in 2:1 chloroform:methanol mixture and evaporate the organic solvent under a stream of nitrogen at room temperature.
2. Place the dried lipids under vacuum for 4–5 h to remove any residual organic solvent.
3. Rehydrate the lipids using 1 mL of EB (for water permeability) or S1 (for solute permeability), and vortex the preparation until all the lipids in solution get multilamellar liposomes.
4. To obtain unilamellar vesicles, extrude the multilamellar solution 25 times through a 100 or 200-nm nucleopore membrane using an extrusion device such as Avanti mini-extruder. This is performed at room temperature or above the phase-transition temperature of the lipid. Depending on the lipids a brief bath sonication of 20–40 s will help in initial dispersion of the lipids making them easier to extrude.
5. Monitor the size of the vesicles during the procedure to obtain consistent vesicle size distribution from one experiment to another.
6. To remove unentrapped CF from the vesicles, equilibrate a PD-10 Sephadex column with five column volumes of IB or WB for solute and pass the vesicle solution over the column. Vesicles will come out in the void volume and can be used directly for transport studies. Alternatively, the vesicles can be spun at 100,000g for 45 min, and the pellet washed two to three times in WB until all the external CF is removed.

3.2. To Check Integrity and Osmotic Response of the Vesicle

1. Aliquot 20–30 µL of the vesicles in 2 mL of IB into a fluorescence cuvet. Use WB if the vesicles were made in S1 buffer.
2. Set excitation and emission wavelengths of the fluorimeter to 490 and 520 nm, respectively.
3. Add antifluorescence antibody in 2 µL increments to the cuvet, until addition of antibody does not result in decrease of fluorescence. This ratio of vesicles to antibody is used for all further studies

using these vesicles. If more than 20–30 µL of antifluorescence antibody is required for quenching all external fluorescence, then either there is an excess of external CF or the vesicles are leaky.

4. Next, add 200 µL of 2 M sucrose to the cuvet to make the solution hypertonic and note the decrease in fluorescence.
5. To check whether the decrease in fluorescence is not owing to dilution of vesicles, add 200 µL of IB and note the decrease in fluorescence. If the fluorescence decrease on addition of sucrose is much more than that caused by addition of 200 µL of IB or WB, then the vesicles are osmotically responsive, and thus, can be used for further transport studies.

3.3. Water Permeability Measurements

1. Set the excitation wavelength of the stopped-flow device to 490 nm, and use a 515 nm long pass cutoff filter for emission.
2. Aliquot 20–30 µL of vesicles into 2 mL of IB with appropriate amount of antibody determined from antibody titration step (*see* **Subheading 3.2.**).
3. Load the vesicle solution into one of the drive syringes of the stopped-flow device and load the other syringe loaded with HB.
4. Trigger the stopped-flow device and collect the data.
5. To enhance the signal-to-noise ratio make 8–16 measurements and average them.
6. Fit the averaged time trace to a single exponential.
7. The rate obtained from the fit is used to calculate the osmotic permeability coefficient by using **Eq. 1**, which can be solved in MathCad software as shown in **Note 1**.

3.4. Solute Permeability Measurements

1. Load the vesicles with the solute of interest by incubating the vesicles (urea was used here) in 2 mL of PB buffer (*see* **Subheading 2.1.**) for 30 min.
2. Load the above vesicle solution into one of the drive syringes and load the other syringe with NP solution (*see* **Subheading 2.1.**).
3. Trigger the stopped-flow device to fire and collect the data. On mixing (**Fig. 2**), an iso-osmotic urea gradient is formed, which causes urea efflux, and thus, volume change.
4. Average 8–10 traces and fit the averaged data to a single exponential curve.
5. Input the rate obtained from the fit into solute permeability **Eq. n15** (**Note 2**), and solve it using MathCad (*see* **Note 3**) to get the permeability coefficient (**Fig. 3**).

4. Notes

1. *Calculation of osmotic water permeability using MathCad:* a doubling of external osmolarity will result in decrease of relative volume from 1 to 0.5, the equations have been set for this condition and hence input only rate, diameter of the vesicle, and internal and external osmolarity.

   Amplitude a $a = 2.17$ End point $c = -0.858$
   Rate constant $b = 1C$
   Diameter of vesicle (nm) $vd = 100$
   Radius of the vesicle (cm)

   $$r = \left(\frac{vd}{2}\right) \cdot 10^{-7}$$

   Surface area to volume ratio

   $$SAV = \frac{(4 \cdot \pi \cdot r^2)}{\left(\frac{4}{3}\right) \cdot \pi \cdot r^3} \quad\quad SAV = 6 \times 10^5$$

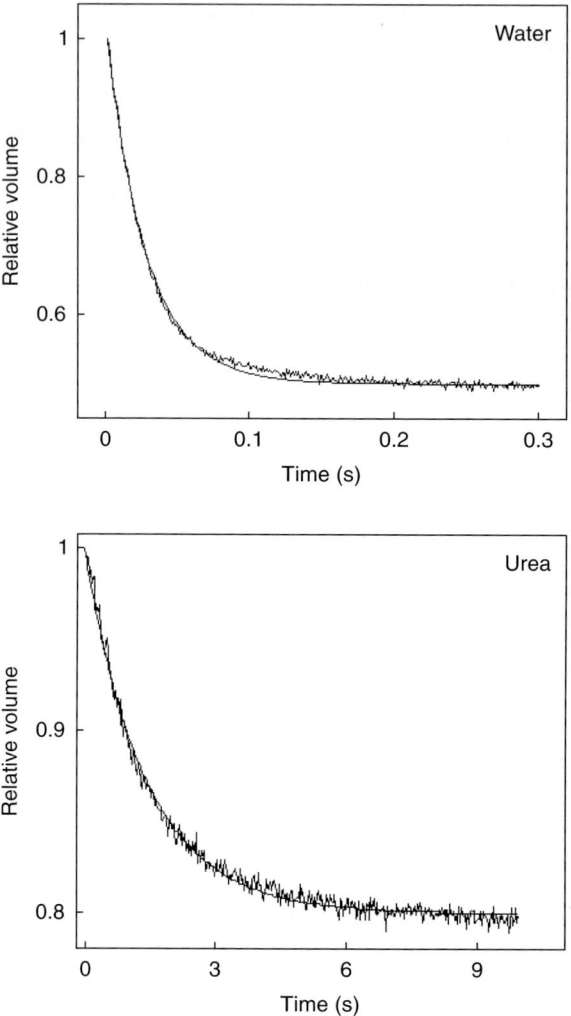

Fig. 3. Time trace of water and solute efflux from the vesicles. Top figure shows changes in relative volume as a function of time as water moves out of the vesicle. The solid line is the single exponential fit to the data. The bottom figure shows the efflux of urea down its concentration gradient as a function of time. Urea takes longer than water to cross the membrane.

Molar volume of water MVW = 18
Initial solute concentration $C_{in} = 292 \times 10^{-6}$
Solute concentration outside vesicle $C_{out} = 821 \times 10^{-6}$
Time range of exp $t = 0, 0.01, \ldots 0.5$
Equation
Determination of relative fluorescence as a function of time, t

$$F(t) = (a) \cdot e^{-b \cdot t} + c$$

Normalization of fluorescence and calculation of relative volume from 1 to 0.5.

$$k1 = (a + c) - c$$

$$k1 = 2.17$$

$$N = \frac{0.5}{k1}$$

$$N = 0.23$$

$$xx = 0.5 - N \cdot c$$

$$xx = 0.698$$

$$V(t) = N \cdot F(t) + xx$$

$$V(0.1) = 0.684$$

$$F(0.1) = -0.06$$

Initial guess of z $\qquad z = 0.000$
Given

$$\frac{d}{dt}V(t) = (z \cdot SAV \cdot MVW) \cdot \left(\frac{C_{in}}{V(t)} - C_{out} \right)$$

$p(t) = \text{Find}(z)$ $\qquad p(0) = 8.752 \times 10^{-4}$
P_f of water (cm/s) $\qquad p(0) = 8.752 \times 10^{-4}$

2. *Derivation of equation for calculation of solute permeability*: the general solute flux equation is

$$J_{urea} = ds/dt = P_{urea}(SAV)(\Delta C)$$

where SAV is the surface area to volume ratio and ΔC is the difference in concentration of urea, inside to that of outside.

Because volume is the measured parameter, all variables (J_{urea} and ΔC) need to be expressed in terms of volume.

Initial conditions and assumptions: V_0 is the initial volume, X is the osmoles of urea inside, X_2 is the osmoles of sucrose inside, and s is the osmoles of urea that effluxes (*see* **Fig. 1B**).

$$V_{rel} = V/V_0 \qquad \text{n1}$$

$$(X + X_2)/V = 550 \qquad \text{n2}$$

Substituting **Eq. n1** into **Eq. n2** results in

$$(X + X_2) = 550 V_0 V_{rel} \qquad \text{n3}$$

$$(X + S) = 275 V_0 \qquad \text{n4}$$

$$X_2 = 275 V_0 \qquad \text{n5}$$

To define J_s or ds/dt in terms of relative volume as a function of time, subtract **Eq. n3** from **Eq. n4**

$$X + X_2 - (X + S) = 550 V_0 V_{rel} - 275 V_0 \qquad \text{n6}$$

Rearranging equation **Eq. n6** and substituting **Eq. n5**

$$S = 550 V_0 - 550 V_0 \qquad \text{n7}$$

and $ds/dt = -550 V_0 (dV_{rel}/dt)$ $\qquad$ n8

ΔC needs to be expressed in terms of measured quantity (i.e., volume) with the assumption that external urea concentration does not change, as the external volume is large compared with the internal volume.

$$\Delta C = C_{urea}\text{ in} - C_{urea}\text{ out} \qquad \text{n9}$$

$$\Delta C = [(X - S)/V] - 137.5 \qquad \text{n10}$$

$$\Delta C = [(X - S)/V_0 V_{rel}] - 137.5 \qquad \text{n11}$$

Express X and S in terms of V_0 and V_{rel}
Subtracting **Eq. n3** from **Eq. n5** results in

$$X = 550 V_0 V_{rel} - 275 V_0 \qquad \text{n12}$$

Substituting **Eq. n7** and **Eq. n12** into **Eq. n11** results in

$$\Delta C = ([(550 V_0 V_{rel} - 275 V_0 - 550 V_0 + 550 V_0 V_{rel})/V_0 V_{rel}] - 137.5 \qquad \text{n13}$$

$$\Delta C = 962.5 - (825/V_{rel}) \qquad \text{n14}$$

Substituting **Eq. n8** and **Eq. n14** into general equation

$$J_s = ds/dt = P_s \, (SA) \, (\Delta C)$$

$$- 550 V_0 \, d \, V_{rel}/dt = P_s \, (SA) \, (\Delta C) \qquad \text{n15}$$

$$dV_{rel}/dt = P_s \, (SA/V_0) \, (1/550) \, [(825/V_{rel}) - 962.5]$$

As this equation can be simplified as a first-order process and as $t \to 0$, a single exponential equation can be used to fit the data. This equation can be solved in MathCad to get P_s.

Using these guidelines permeability equations can be derived to compute the permeability of any solute under variant conditions.

3. *Solving solute permeability equation using MathCad:* this equation only holds when the conditions of the solute flux experiment are as follows: Concentration of permeable solute inside vesicles is 275 mM and outside the vesicle after mixing is 137.5 mM. For other conditions use **Note 2** to derive the permeability equation and solve it as shown below.

$F(t)$ defines the relationship between relative fluorescence (fluorescence at time t divided by fluorescence at time zero) and t or time. $V(t)$ defines the relationship between relative volume (volume at time t divided by volume at time zero) and $F(t)$. This relationship is determined empirically and may be linear or quadratic. $d/dt \, V(t)$ defines the relationship between relative volume and the permeability coefficient (s), as derived in **Note 2**. Parameters of single exponential fit from stopped-flow data.

a. Input only rate b obtained from fitting of the experimental curve

Amplitude	$a = 2.376$
Rate constant	$b = 0.15$
End point	$c = 2.44$
Diameter of vesicle (nm)	$vd = 250$
Radius of the vesicle (cm)	

$$r = \left(\frac{vd}{2}\right) \cdot 10^{-7}$$

Surface area to volume

$$SAV = \frac{(4 \cdot \pi \cdot r^2)}{\left(\frac{4}{3}\right) \cdot \pi \cdot r^3}$$

$$SAV = 6 \times 10^5$$

Time range of experiment (s) $\qquad t = 0, 1, \ldots 7.$
Initial guess for solute P_f, permeability coefficient $\qquad s = 0.00005.$

b. Equations
Single exponential of fluorescence as a function of time, t:

$$F(t) = (a) \cdot e^{-b \cdot t} + c$$

Converting relative volume as a function of relative fluorescence:
Under conditions used, the initial relative volume is 1, and the maximal relative volume change expected is 0.25:

$$k1 = a = 2.376$$

$$N = \frac{0.25}{k1} \qquad\qquad N = 0.105$$

$$h = 0.75 - N \cdot c \qquad\qquad h = 0.493$$

$$V(t) := N \cdot F(t) + h$$

$$V(0.1) = 0.996 \qquad\qquad F(0.00) = 4.816$$

$$V(7) = 0.837 \qquad\qquad F(7) = 3.271$$

Solving the solute permeability **Eq. n15** derived in **Note 2**.
Given

$$\frac{d}{dt}V(t) = s \cdot SAV \cdot \left(\frac{825}{V(t)} - 962.5 \right)$$

$$p(t) = \text{Find}(s)$$

$p(0.0) = 6.25 \times 10^{-7}$ cm/s is the solute permeability coefficient.

References

1. Finkelstein, A. (1987) *Water Movement Through Lipid Bilayers, Pores, and Plasma Membranes: Theory and Reality*. Ed. Finkelstein Alau. John Wiley & Sons, New York, NY.
2. Zeidel, M. L., Ambudkar, S. V., Smith, B. L., and Agre, P. (1992) Reconstitution of functional water channels in liposomes containing purified red cell CHIP28 protein. *Biochemistry* **31,** 7436–7440.
3. Tsunoda, S. P., Wiesner, B., Lorenz, D., Rosenthal, W., and Pohl, P. (2004) Aquaporin-1, nothing but a water channel *J. Biol. Chem.* **279,** 11,364–11,367.
4. Preston, G. M., Carroll, T. P., Guggino, W. B., and Agre, P. (1992) Appearance of water channels in *Xenopus* oocytes expressing red cell CHIP28 protein. *Science* **256,** 385–387.
5. Mlekoday, H. J., Moore, R., and Levitt, D. G. (1983) Osmotic water permeability of the human red cell. Dependence on direction of water flow and cell volume. *J. Gen. Physiol.* **81,** 213–220.
6. Mathai, J. C., Sprott, G. D., and Zeidel, M. L. (2001) Molecular mechanisms of water and solute transport across archaebacterial lipid membranes. *J. Biol. Chem.* **276,** 27,266–27,271.
7. Saparov, S. M., Antonenko, Y. N., Koeppe, R. E., 2nd., and Pohl, P. (2000) Desformylgramicidin: a model channel with an extremely high water permeability. *Biophys. J.* **79,** 2526–2534.

8. Verkman, A. S. (2000) Water permeability measurement in living cells and complex tissues. *J. Membr. Biol.* **173,** 73–87.
9. Brahm, J. (1982) Diffusional water permeability of human erythrocytes and their ghosts. *J. Gen. Physiol.* **79,** 791–819.
10. Mathai, J. C., Mori, S., Smith, B. L., et al. (1996) Functional analysis of aquaporin-1 deficient red cells. The Colton-null Phenotype. *J. Biol. Chem.* **271,** 1309–1313.
11. Pfeuffer, J., Broer, S., Broer, A., Lechte, M., Flogel, U., and Leibfritz, D. (1998) Expression of aquaporins in Xenopus laevis oocytes and glial cells as detected by diffusion-weighted 1H NMR spectroscopy, and photometric swelling assay. *Biochim. Biophys. Acta* **1448,** 27–36.
12. Ye, R. G. and Verkman, A. S. (1989) Simultaneous optical measurement of osmotic and diffusional water permeability in cells and liposomes. *Biochemistry* **28,** 824–829.
13. Karan, D. M. and Macey, R. I. (1980) The permeability of the human red cell to deuterium oxide (heavy water). *J. Cell Physiol.* **104,** 209–214.
14. Lande, M. B., Priver, N. A., and Zeidel, M. L. (1994) Determinants of apical membrane permeabilities of barrier epithelia. *Am. J. Physiol.* **267,** C367–C374.
15. Levitt, D. G. and Mlekoday, H. J. (1983) Reflection coefficient and permeability of urea, and ethylene glycol in the human red cell membrane. *J. Gen. Physiol.* **81,** 239–253.
16. Harris, H. W., Jr., Handler, J. S., and Blumenthal, R. (1990) Apical membrane vesicles of ADH-stimulated toad bladder are highly water permeable. *Am. J. Physiol.* **258,** F237–F243.
17. Priver, N. A., Rabon, E. C., and Zeidel, M. L. (1993) Apical membrane of the gastric parietal cell: water, proton, and nonelectrolyte permeabilities. *Biochemistry* **32,** 2459–2468.

22

Fluorescence Microscopy to Study Pressure Between Lipids in Giant Unilamellar Vesicles

Anna Celli, Claudia Y. C. Lee, and Enrico Gratton

Summary

The authors developed a technique to apply high hydrostatic pressure to giant unilamellar vesicles and to directly observe the consequent structural changes with two-photon fluorescence microscopy imaging using high numerical aperture oil-immersion objectives. The data demonstrate that high pressure has a dramatic effect on the shape of the vesicles, and both fluidity and homogeneity of the membrane.

Key Words: Microscopy; pressure; vesicles.

1. Introduction

Lipid membranes are the envelopes that separate the interior of the cell and its organelles from external media. Extensive research in the past two decades has shown that the lipids forming the matrix of the cell membranes do not merely play a structural role but have a crucial function in regulating many vital processes *(1,2)*. Recently, much attention has been paid to the dynamics of lipid interaction and organization in the membrane. In particular, the lipid phase determines the fluidity of the membrane, and thus, regulates the diffusion of membrane proteins. Many studies have pointed out that microdomains of different fluidity and lipid organization (rafts) exist under physiological conditions in biomembranes. For this reason, the direct visualization of membrane heterogeneity has received particular attention in the past few years. Fluorescence microscopy has proved to be a powerful tool to directly study biological membranes noninvasively under physiological conditions.

Among different artificial membranes, giant unilamellar vesicles (GUV) are particularly good models of cell membranes because of their similarity to cells in terms of their dimensions (10–100 μm), radius of curvature, and lamellarity. Moreover, GUV can be visualized individually under the microscope, allowing researchers to study the local properties of the membrane. The study of the physics involved in lipid–lipid interactions requires perturbation of the system and the eventual observation of how the system reacts to the perturbation. In this chapter, a technique is presented that allows to visualize the local changes in GUV membranes caused by high-pressure perturbation.

The response of biomembranes to pressure has been studied for a variety of reasons: from the need to understand the adaptation of deep-sea organisms to high pressures (up to a 1 Kbar) to the effects of pressure on anesthesia *(3–5)*. Moreover, pressure offers a very powerful means to perturb the biophysical conditions of the membranes, and therefore, can be used to investigate the physics of the interactions of biomolecules contained in the bilayer. Many studies have been performed to study the response of lipid vesicles to pressure *(6–8)*. However, to the knowledge of the authors no high-pressure study has been conducted on individual vesicles.

As a general observation, membranes in the liquid-crystalline state transform to the gel phase under pressure. Chong et al. put forward the idea of some compensation existing between temperature and pressure effects. As temperature increases, lipid–lipid interactions tend to decrease. On the other hand, high hydrostatic pressure tends to put the lipids closer together, which results in compensation between increasing the temperature and increasing the pressure. In the gel phase, the total volume and surface area occupied by lipids in the membrane is reduced, and water is expelled from the membrane. However, in the case of vesicles a change in the surface area will cause a change in the volume of the entire vesicle. However, unless the internal vesicle volume can adapt to the reduced surface area by expelling the internal water, the membrane will be under stress. Although water will slowly permeate across the membrane, rapid changes in volume cannot be rapidly compensated, and changes in shape are needed to account for the difference in volume-surface area caused by the change in pressure.

An interesting observation is that if the initial pressure is high and the pressure is decreased up to values in which the gel-like membrane becomes fluid, the surface area increases. If the internal volume remains the same, then the shape of the vesicle must change, breaking the spherical symmetry of the vesicle. This departure from the spherical shape will produce regions of different curvature. These regions would have particular properties and determine the local phase state of the membrane. However, these phenomena are transient, because water will eventually leak through the membrane. Therefore, a cycle of increasing–decreasing pressure will show different behavior depending on whether the pressure is increased or decreased. In this chapter a technique is presented that couples high pressure with two-photon fluorescence microscopy. In the setup, hydrostatic pressure is applied to GUVs in aqueous solution contained in a thin-fused silica capillary. The capillary is filled with the GUVs in buffer, sealed and connected to a high-pressure pump. The vesicles inside the capillary can be visualized with high numerical aperture immersion objectives under the microscope. This microscopy method allows the study of local membrane heterogeneities caused by pressure.

2. Materials

1. The experiments reported in this contribution were performed using palmitoyl-oleoyl-3-glycerol-phosphoatidilcholine GUVs, labeled with LAURDAN.
2. Palmitoyl-oleoyl-3-glycerol-phosphatidilcholine (both in powder and in chloroform, 99% pure) was purchased from Avanti Polar Lipids (Alabaster, AL).
3. LAURDAN was purchased from Molecular Probes (Eugene, OR).
4. The pressure is applied by means of a high-pressure pump connected to a cylindrical fused silica capillary with an outer diameter of 360 µm and an inner diameter of 50 µm, as described in **ref. 9**. The thickness of the wall of the capillary (155 µm) enables the use of high numerical aperture oil-immersion objectives. The capillary was purchased from Polymicro technologies (Phoenix, AX).

3. Methods

1. The spectral behavior of LAURDAN at lipid interfaces has been extensively described (*see* **ref. 10** for a review). LAURDAN emission spectrum is sensitive to the degree of water penetration in the membrane, and therefore, to the degree of lipid packing in the membrane. In the gel phase, the spectrum emission maximum is at 440 nm. In the liquid-crystalline phase, the spectrum shifts to approx 490–500 nm.

2. To quantify the spectral shift of LAURDAN, and thus, the lipid phase, the GP function has been introduced *(11)*, which is defined as:

$$GP = \frac{I_b - I_r}{I_b + I_r}$$

where I_b and I_r are the intensities at 440 and 490 nm, respectively. High values of GP correspond to highly ordered lipids (gel phase), whereas low values correspond to the disordered, liquid-crystalline phase.

3. GUVs are grown at room temperature following the electroformation method previously described by Angelova et al. *(12)*. The chamber used for the GUV formation was previously described by Bagatolli et al. LAURDAN in dimethyl sulfoxide is added to the lipids dissolved in chloroform in a ratio of 4:100 molar dye:lipid. The mixture is then dried to remove the solvent, and dissolved again in chloroform to a concentration of 0.2 mg/mL. On each of two (2-cm long) platinum electrodes 6 µL of lipid solution is spread. To the chamber 300 µL of nanopure water is added. An AC field with a frequency of 10 Hz and amplitude of 3 V field is applied to the electrodes for 40 min.

4. The vesicles are then detached from the electrodes by gradually decreasing the frequency and increasing the amplitude of the applied field to 0.1 Hz and 8 V. The vesicle solution is then transferred to an Eppendorf tube using a pipet. The capillary is filled up with the vesicle solution applying suction with a syringe. One end of the capillary is sealed using a blow torch. The capillary is then connected to the pressure pump, and secured to a custom-made stage on the microscope. The experimental setup is schematically shown in **Fig. 1**.

5. A home-built, two-photon, two-channel fluorescence microscope is used to image the GUVs in the capillary. Two-photon microscopy allows high-spatial resolution, whereas minimizing out-of-focus photo damage. The design was described in **ref. 13**. A Ti:Sapphire (Mira 900, Coherent Palo, Alto, CA) optically pumped by a Nd:Vanadate laser (Verdi; Coherent) is used as the light source. An excitation wavelength of 780 nm is used. The laser is driven into the microscope by a couple of galvanometric mirrors (Cambridge Technology, Watertown, MA), which allow the scanning of the beam in the x and y directions. A 63X Zeiss objective (PLAN-apochromat) is used to image the vesicles in the capillary.

6. The fluorescence light is detected by two photomultipliers (Hamamatsu R-5600, Hamamatsu city, Japan). A dichroic filter in the emission path reflects the light with a wavelength smaller than 470 nm and transmits the rest. Two additional interference filters centered about 440 and 490 nm, before the two detectors minimize cross talk and detection of scatter.

7. Intensity images are acquired for both the blue and the green channels. Then, they are processed point by point during data acquisition to yield the GP image. The obtained images are analyzed with the same acquisition software. The background is subtracted, and the average GP for each vesicle in the image is calculated together with the standard deviation.

8. Once a GUV is localized in the capillary, the pressure is increased and images of the GUV at constant pressure intervals are acquired. With this setup, the same vesicle can be imaged as a function of pressure throughout the whole experiment. The GUV does not move significantly during the acquisition of each image; however, in a sequence of images changes in shape and dimension of the GUV can be seen clearly.

4. Notes

1. The data shown were acquired every 200 bars. **Figure 2** shows the series of images taken from a GUV at an atmospheric pressure up to 2500 bars. The top part of the image shows the intensity images in one channel, whereas the bottom part shows the changes in GP. As the pressure is increased, an increase in the average GP value and a decrease in the diameter of the vesicle can be clearly observed, whereas the vesicle maintains its spherical shape. A volume reduction of about 40% was estimated.

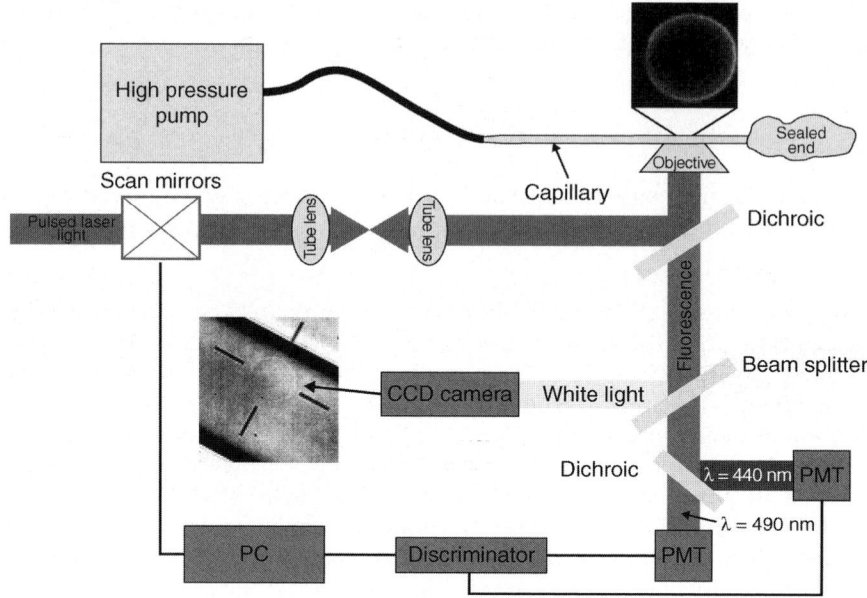

Fig. 1. Experimental setup.

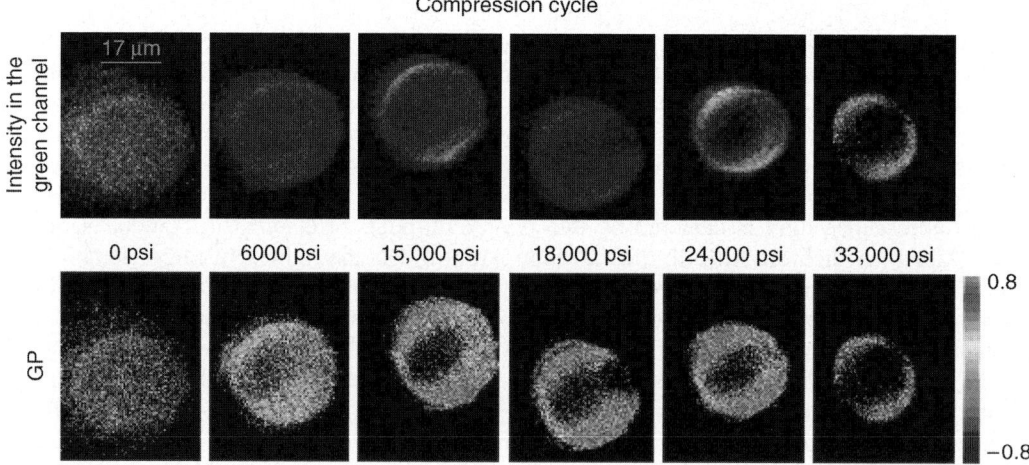

Fig. 2. Effects of compression on the GUV. The top series of images show the intensity in the green channel. It is evident that the vesicle shrinks drastically when pressure increases. On the bottom part the GP images are shown. Notice how the GP average value increases with pressure.

2. **Figure 3** shows the effects of decompression on the vesicle. It was noticed that GUV looses its spherical shape around 1500 bars. This is consistent with the observation of shape changes in temperature-driven phase transitions reported in **ref. 14**. As the pressure is further released, it was observed that the vesicle abruptly looses its tension and becomes flabby at a pressure of about 1500 bars.
3. The GP is not homogeneous throughout the whole vesicle. In particular, it can be seen in **Fig. 4** that parts of the membrane with high curvature radius tend to have a higher value of GP.

Fluorescence Microscopy on Lipid Pressure in GUVs

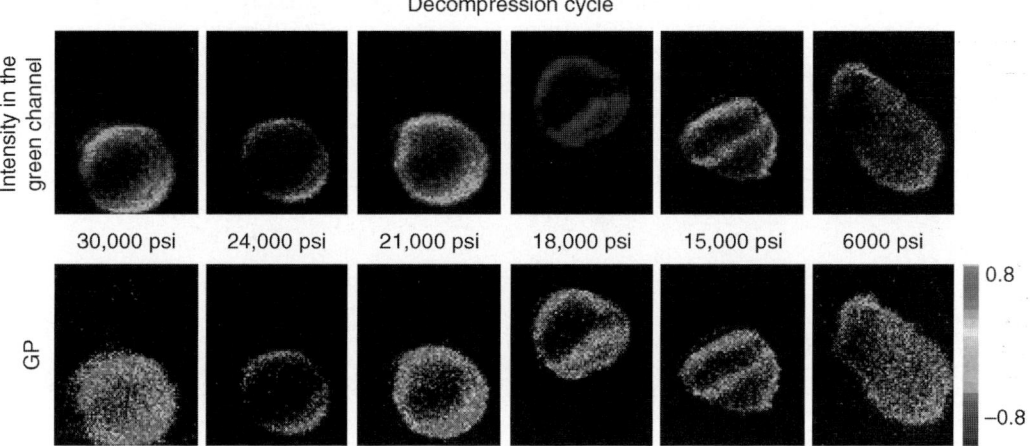

Fig. 3. Effects of decompression on the GUV. The most striking effect of decompression is the loss of membrane tension and spherical shape of the vesicle. As the pressure is released, the GUV starts wobbling and becomes very dim. Another interesting observation is that the GP is not homogeneous throughout the vesicle.

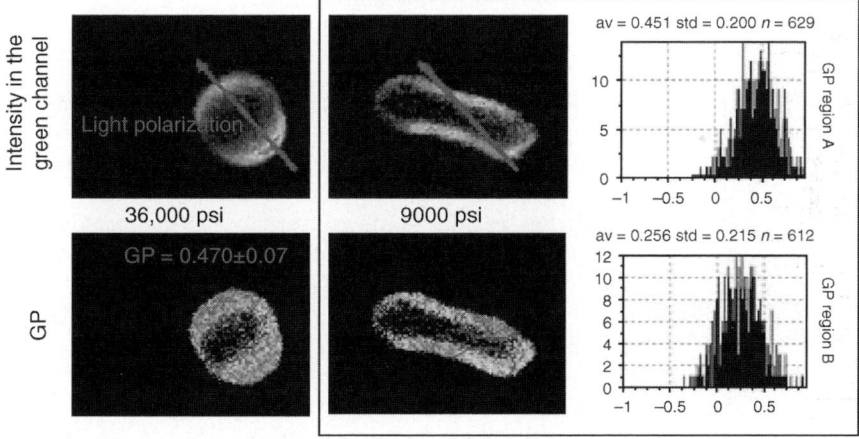

Fig. 4. Inhomogeneity of the GP. At high pressure, the vesicle GP appears to be homogeneous; the variations in intensity are because of the nonperfect circular polarization of the excitation light. The arrow in the intensity images indicates the direction of the polarization of the light. Notice that the high GP regions are not in the direction of the polarization of light for the vesicle at 9000 psi during the decompression cycle. Moreover, it is interesting that the high GP regions appear to be flat. This shows that the technique can be useful to study the correlation between the morphology of the membrane and the degree of lipid hydration and fluidity.

4. A graph of changes in the vesicle GP as a function of pressure clearly shows a transition at a pressure of about 1300 bars (*see* **Fig. 5**). The curve relative to the decompression of the vesicle is very noisy, which reflects the shape hysteresis that the vesicle undergoes and the inhomogeneity of GP values across the membrane.
5. In the experiments, a drastic reduction in the volume of the vesicle was observed. The change in volume of the vesicle is more than the change in volume of water because of the compression of

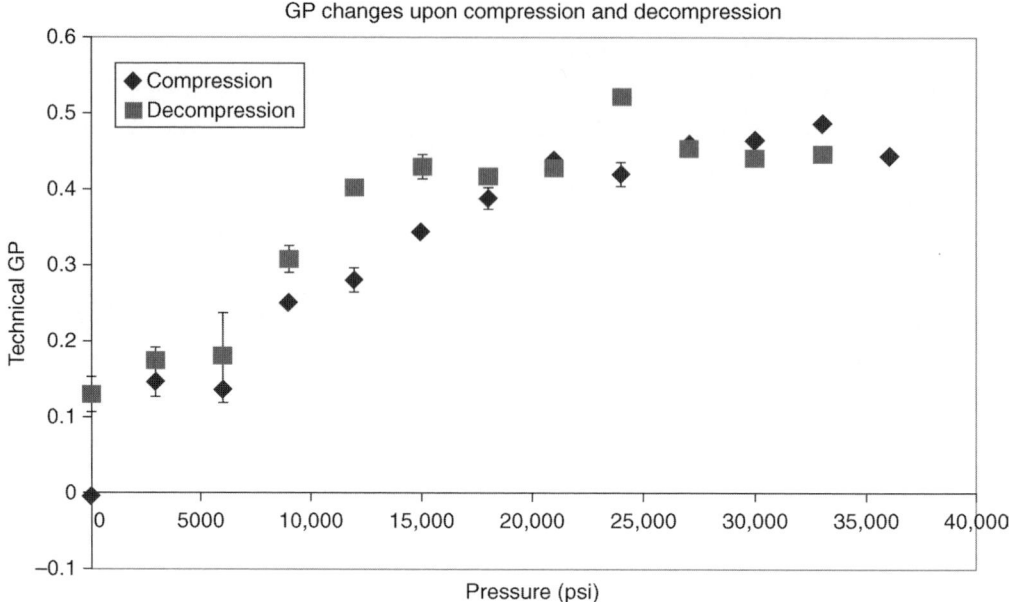

Fig. 5. GP graph. In this graph, the GP value calculated for each pressure is shown. The hysteresis in the curve indicates that the process is not reversible.

water (the change of volume for water is of ~5% at 1000 bar). From this, it is concluded that water is expelled from the GUV during the compression.
6. The increase in the GP value clearly indicates a reduction in the degree of water penetration in the membrane, and therefore, a tighter packing membrane lipids.
7. When the pressure is released, the vesicle starts to "crack" as the packing of the lipids becomes less tight. As the pressure is further released, the lipid surface area increases, and the membrane regain its fluidity. However, the vesicle cannot absorb the amount of water that is ejected during the compression cycle, and therefore, the membrane of the GUV appears to have lost its tension.

5. Conclusions

1. The microscope setup here described allows researchers to directly observe the effect of high pressure on the structure and fluidity of model membranes. The ability to visualize individual vesicles during the compression and decompression cycles enabled the study of the changes in structure and fluidity of the membrane on a local scale. This type of information is lost in steady-state bulk measurements, which is the result of averaging over many vesicles.
2. By observing the GP value locally, the curvature of the membrane can be related with its fluidity. Moreover, the setup here presented allows also the use of other techniques, such as fluorescence correlation spectroscopy, which can give the information about the diffusion of particles in the membrane as a function of pressure.
3. Using the setup here described the authors intend to study the dynamics of phase transition on single lipid membranes as well as on membranes made of lipids mixtures displaying phase separation at given temperatures. This method can also be used to study the behavior of membrane proteins under pressure, and finally, that of live cells.

Acknowledgment

The work presented was founded by the NIH grant: NIH, PHS 5 P41 RR03155.

References

1. Brown, D. and London, E. (1998) Structure and origin of ordered lipid domains in biological membranes. *J. Membr. Biol.* **164,** 104–114.
2. Mouritsen, O. and Jorgensen, K. (1994) Dynamical order and disorder in lipid bilayers. *Chem. Phys. Lipids* **73,** 3–25.
3. Beham, M. K., Macdonald, A. G., Jones, G. R., and Cossin, A. R. (1992) Homeviscus adaptation under the pressure: the pressure dependence of membrane order in brain myelin membranes of deep-sea fish. *Biochim. Biophys. Acta* **1103,** 317–323.
4. Lakowicz, J. R. and Thompson, R. B. (1983) Differential polarized phase fluorimetric studies of phospholipid bilayers under high hydrostatic pressure. *Biochim. Biophys. Acta* **732,** 359–371.
5. Tauc, P. C., Mateo, R., and Brochon, J. -C. (2002) Investigation of the effect of high hydrostatic pressure on proteins and lipidic membranes by dynamic fluorescence spectroscopy. *Biochim. Biophys. Acta* **1595,** 103–115.
6. Chong, P. L. G. and Cossing, A. R. (1984) Interacting effects of temperature, pressure and cholesterol content upon the molecular order of dioleoylphosphatidylcholine vesicles. *Biochim. Biophys. Acta* **772,** 197–201.
7. So, P. T. C., Gruner, S. M., and Erramilli, S. (1993) Pressure-Induced Topological Phase Transition in Membranes. *Phys. Rev. Lett.* **70,** 3455–3458.
8. Winter, R. (2002) Synchroton X-ray and neutron small-angle scattering of lyotropic lipid mesophases, model biomembranes and protein in solution at high pressure. *Biochim. Biophys. Acta* **1595,** 160–184.
9. Muller, J. D. and Gratton, E. (2003) High-Pressure Fluorescence Correlation Spectroscopy. *Biophys. J.* **85,** 2711–2791.
10. Parasassi, T. and Gratton, E. (1995) Membrane lipid domains and dynamics detected by LAURDAN. *J. Fluoresc.* **5,** 59–70.
11. Parasassi, T., De Stasio, G., d'Ubaldo, A., and Gratton, E. (1991) Quantization of lipids phases in phospholipids vesicles by the generalized polarization of LAURDAN fluorescence. *Biophys. J.* **60,** 179–189.
12. Angelova, M. I. and D. D. S. (1986) Liposome electroformation. *Faraday Discuss. Chem. Soc.* **81,** 303–311.
13. So, P. T. C., French, T., Yu, W. M., Berland, K. M., Dong, C. Y., and Gratton, E. (1995) Time resolved fluorescence microscopy using two-photon excitation. *Bioimaging* **3,** 49–63.
14. Bagatolli, L. A. and Gratton, E. (1999) Two-Photon Fluorescence Microscopy Observation of Shape Changes at the Phase transition in Phospholipid Giant Unilamellar Vesicles. *Biophys. J.* **77,** 2090–2101.

23

X-Ray Scattering and Solid-State Deuterium Nuclear Magnetic Resonance Probes of Structural Fluctuations in Lipid Membranes

Horia I. Petrache and Michael F. Brown

Summary

Molecular fluctuations are a dominant feature of biomembranes. Cellular functions might rely on these properties in ways yet to be determined. This expectation is suggested by the fact that membrane deformation and rigidity, which govern molecular fluctuations, have been implicated in a number of cellular functions. However, fluctuations are more challenging to measure than average structures, which partially explain the small number of dedicated studies. Here, it is shown that two accessible laboratory methods, small-angle X-ray scattering and solid-state deuterium nuclear magnetic resonance (NMR), can be used as complementary probes of structural fluctuations in lipid membranes. In the case of X-ray scattering, membrane undulations give rise to logarithmically varying *positional* correlations that generate scattering peaks with long (power-law) tails. In the case of ^{2}H NMR spectroscopy, fluctuations in the magnetic-coupling energies resulting from molecular motions cause relaxation among the various spin energy levels, and yield a powerful probe of *orientational* fluctuations of the lipid molecules. A unified interpretation of the combined scattering and ^{2}H NMR data is provided by a liquid-crystalline membrane deformation model. The importance of this approach is that it is possible to utilize a microscopic model for positional and orientational observables to calculate bulk material properties of liquid-crystalline systems.

Key Words: Bending rigidity; liquid crystals; membranes; molecular dynamics; order parameters; relaxation.

1. Introduction
1.1. Molecular Fluctuations in Biology

Under physiological conditions, biological processes necessarily involve molecular rearrangement, fluctuations, and disorder. Most biological processes occur in solution and must obey laws of either equilibrium or nonequilibrium thermodynamics. For example, the activation of receptor molecules by chemical or mechanical stimulants occurs against the inherent tendency of fluid solutions to maximize their entropy (disorder). Whereas the average molecular structure, be it that of a protein or of a lipid bilayer, is of immediate and obvious relevance, fluctuations around these averages can be at least equally important. Structural changes undergone by membrane receptors are typically measured on the order of thermal (k_BT) units.

By definition of thermal units, energies associated with spontaneous molecular fluctuations are on the same energy scale. How biological systems use fluctuations is still an outstanding question, but there is only one way to find out: measure them. In this chapter, an overview of two complementary experimental methods is provided for studying the kinds of molecular fluctuations that are accessible through measurement of either positional or orientational order: small-angle X-ray scattering *(1)* and solid-state ^{2}H nuclear magnetic resonance (NMR) spectroscopy *(2)*.

From: *Methods in Molecular Biology, vol. 400: Methods in Membrane Lipids*
Edited by: A. M. Dopico © Humana Press Inc., Totowa, NJ

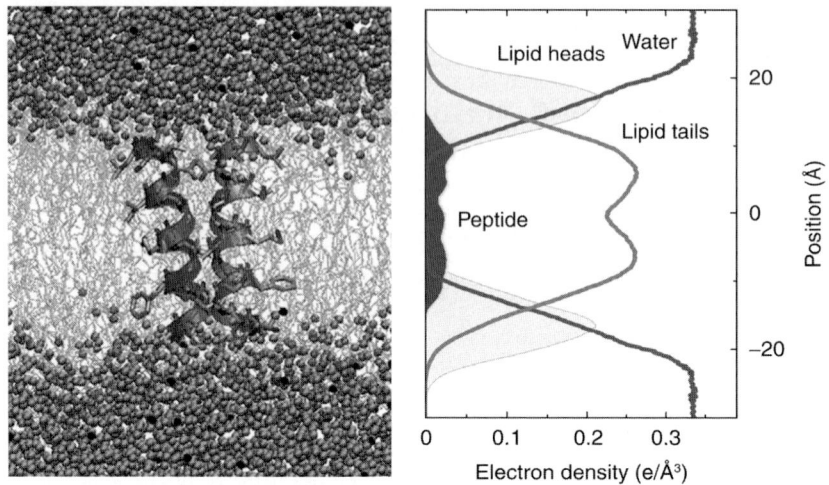

Fig. 1. Molecular fluctuations can be simulated by computer modeling. (**Left**) A snapshot is shown of a DMPC bilayer containing a glycophorin A helix dimer. (**Right**) Spatial distributions are indicated for various molecular components sampled over a 5-ns trajectory (*see* **ref. 6**).

1.2. Fluctuation–Dissipation Theorem and Ergodicity

Scattering and spectroscopic methods are typically used to measure structural parameters, and also reveal that fluctuations around these averages are significant at physiological temperatures. How important is measuring these fluctuations? According to the fluctuation–dissipation theorem, a direct relationship exists between spontaneous thermal fluctuations and the macroscopic response of the system (here the membrane) to external perturbation. Typically, the mechanical (elastic) properties of the membrane material are measured by the structural deformations upon external stress. This external stress can be imposed by osmotic pressure or mechanical pressure *(3)*. As an attractive alternative, material properties can be obtained from measurements of spontaneous thermal fluctuations, a direct consequence of the fluctuation–dissipation theorem and ergodicity *(4)*.

According to the ergodic hypothesis of statistical mechanics, the configurations sampled by a single molecule over a sufficiently long time correspond to those of the entire system at any given instant. This correspondence provides a link between ^{2}H NMR and X-ray observables. As mentioned above, it is expected that fluctuations and membrane deformation *per se* are involved in biological function *(5)*, further motivating the development of techniques capable of measuring thermal fluctuations of lipid membranes.

1.3. Molecular Motions in Biomembranes

As shown in **Fig. 1**, large amplitudes of molecular motions within biomembranes give rise to broad spatial distributions *(6,7)*. A hierarchy of molecular motions exists, from local uncorrelated fluctuations to collective concerted motions *(1,8)*. The collective molecular motions are related to material (elastic) parameters describing physical quantities, such as compressibilities and bending rigidities *(9)*. Because lipid bilayers combine properties of solids and fluids, the appropriate theoretical framework is the physics of liquid crystals *(10)*. A key issue is the length scale at which the material properties begin to emerge (the so-called mesoscopic regime). Is it more fruitful to consider an all-atom molecular dynamics simulation, as in **Fig. 1**,

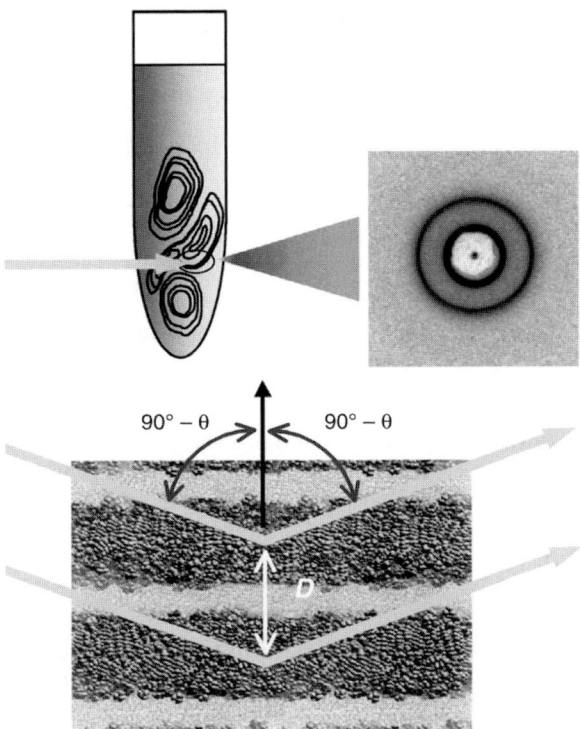

Fig. 2. Self-assembled lipid multilayers diffract incident X-rays to generate characteristic Bragg rings. Diffraction angles θ satisfy the relationship $2D\sin\theta = h\lambda$, where D is the interlamellar repeat spacing, λ is the X-ray wavelength, and h is an integer index. (In X-ray scattering θ is defined relative to the detector and is the glancing angle to the lamellar surface.) Unlike a typical crystalline material, fully hydrated DMPC at 35°C produces only two Bragg rings. Higher order rings ($h > 2$) vanish because of membrane thermal fluctuations.

and focus on extension to larger length scales? Or rather is it more appropriate to begin with the macroscopic bilayer, and extrapolate down to the mesoscale approaching the bilayer and even molecular dimensions? This question will be addressed later.

2. Fluctuations and Correlations

2.1. Liquid-Crystalline Membranes are Described by Characteristic Intermolecular Correlation Functions

In multilamellar lipid suspensions, molecular correlations extend beyond individual membranes. Because of the van der Waals attraction between membranes, lipids form multilamellar vesicles. The average spacing between lamellae is measurable by X-ray scattering (**Fig. 2**). The scattering rings shown in **Fig. 2** are characteristic of unoriented multilayers, for which scattering domains are uniformly distributed relative to an incoming X-ray beam. Scattering rings appear at angles satisfying the Bragg law:

$$2D\sin\theta = h\lambda \qquad (1)$$

where D is the interlamellar repeat spacing, λ is the X-ray wavelength, and h is an integer index. Although multilamellar vesicles are aqueous suspensions, this scattering is called "powder scattering" by analogy with the random orientation obtained from powder crystallites (as in

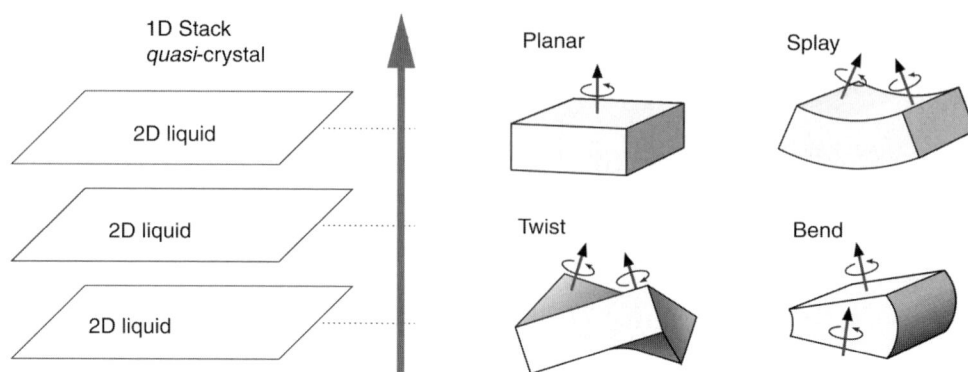

Fig. 3. Lipids exhibit anisotropic properties. **(Left)** A stack of lipid membranes in the fluid state can be described as a 1D array of 2D fluids. Thermal undulations lead to loss of positional correlations of the stacked membranes. **(Right)** At shorter length scales, the finite thickness of each membrane involves additional splay, twist, and bend deformations. The order/disorder properties are described by positional and/or orientational correlation functions. (Figure on right by T. Huber.)

the Debye–Scherrer method). However, the lipid multilayers are imperfect crystals. As shown in **Fig. 3** lipid multilayers can be regarded as a one-dimensional (1D) array of fluid membranes *(8,9)*. However, because of thermally driven deformations **(Fig. 3)**, correlations along the membrane stack decay rather quickly *(11–13)*.

The Caillé-de Gennes model considers the local displacement of membrane patches as shown in **Fig. 4A** to calculate how intermembrane correlations behave along the membrane stack (i.e., the variation of pairwise relative displacement from nearest neighbors to the middle of the stack). Because periodic boundaries are used in the calculation, the system is symmetric about the middle of the stack ($k = N/2$). By letting N go to infinity, the asymptotic limit of interneighbor displacement fluctuations shown in **Fig. 4B** can be determined. Membranes further apart are progressively less correlated; however, this decoupling diverges only logarithmically.

2.2. Intermembrane Correlations are Measured by X-Ray Diffraction

The kind of collective motions considered previously generate X-ray scattering peaks with long tails, as shown in **Fig. 5**. The numerical exponent of the power-law decay (η_1) is in fact a (dis)order parameter that describes the collective (smectic) disorder of the membrane stack. The larger the value the more flexible (more disordered) the membrane. A simple relationship exists between this measurable parameter and two material parameters that describe membrane elasticity, namely, the bending rigidity (K_C) and the stacking compression parameter (B). K_C measures the energy needed to bend membranes locally, whereas B measures the harmonic term of interbilayer interactions (for mathematical relationships, *see* **Subheading 4**). It is possible then to use X-ray diffraction to determine the effect of lipid composition on membrane elasticity *(11–14)*.

2.3. Lipid Dynamics are Probed by Deuterium NMR Spectroscopy

Correlated lipid motions are also measured by solid-state ^{2}H NMR spectroscopy of deuterium-labeled acyl chains *(15–17)*. However, as opposed to X-ray scattering, which measures *positional* correlations between lipid domains, NMR measures *orientational* correlations relative to the

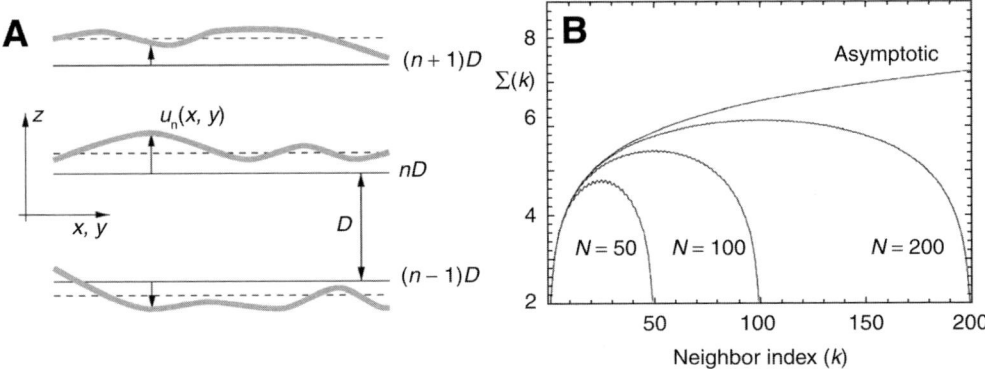

Fig. 4. Caillé-de Gennes smectic liquid crystal model used to analyze experimental X-ray data. (**A**) Local deformation of membrane n at point (x,y) is described by the field $u_n(x,y)$. (**B**) Mean-square fluctuations between membranes n and $n + k$ increase logarithmically with k, as shown by the (normalized) relative displacement function $\Sigma(k)$ on the right. Collective undulatory motions in liquid crystals generate X-ray scattering peaks with long tails, as compared with Gaussians in the case of crystals.

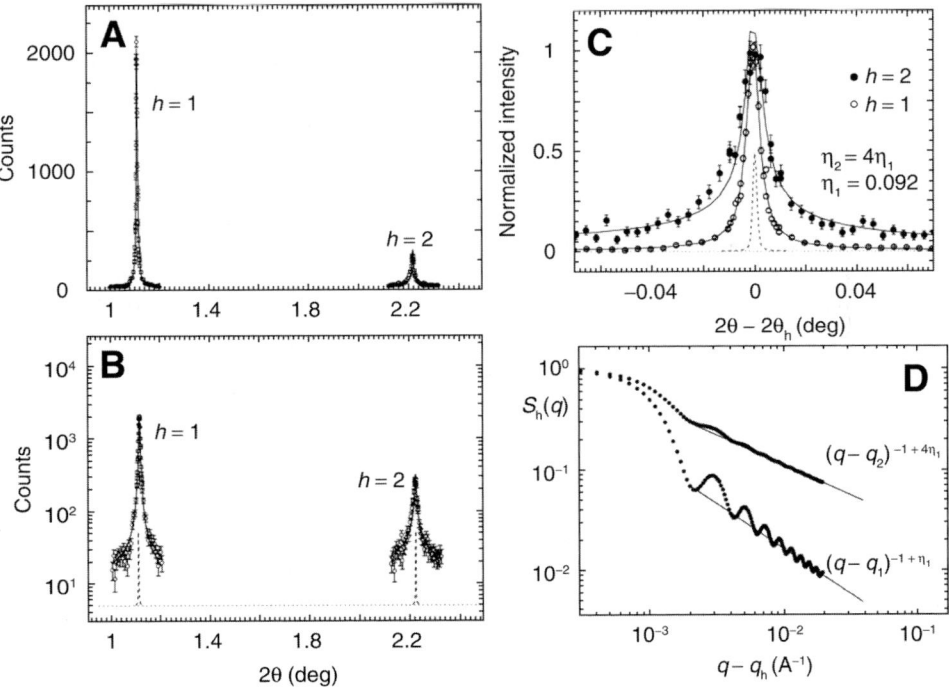

Fig. 5. (**A,B**) Multilamellar DMPC samples give sharp scattering peaks. (**C**) High-resolution X-ray scattering reveals long, power-law tails characteristic of smectic liquid crystals (*1*). The power-law exponent depends on the (Caillé) order parameter η_1, with a harmonic scaling for high-order scattering peaks, i.e., $\eta_2 = 4\eta_1$. (**D**) Theoretical scattering profiles in log–log plots show this scaling (*1*).

externally applied magnetic field (*2,18*). Moreover, one can expect the fluctuations measured with ^{2}H NMR to correspond with the *quasi*-3D limit over short wavelengths on the order of the bilayer thickness (*19,20*), as shown in **Fig. 6**. Coupling of the ^{2}H-nuclear electric quadrupole moment to the electric field gradient of the C–^{2}H bond gives an orientation-dependent

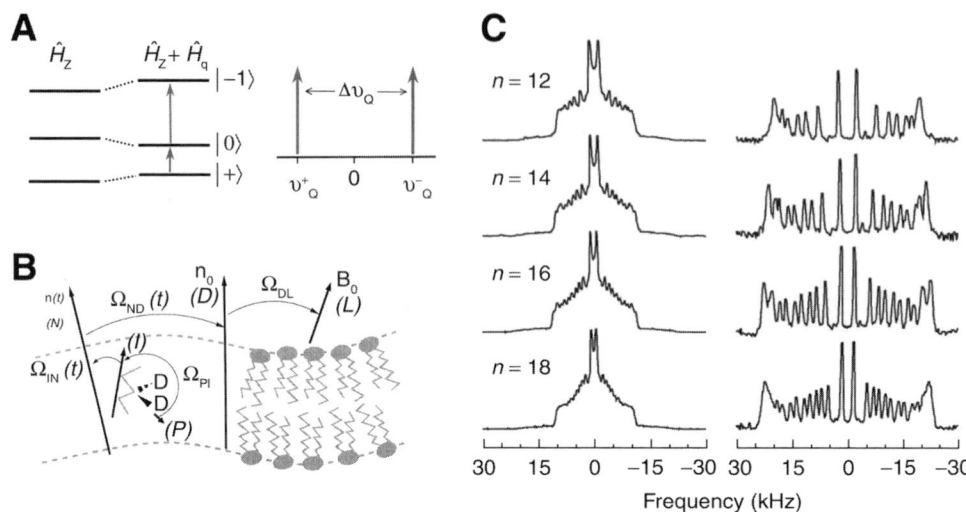

Fig. 6. **(A)** Energy levels and resonance lines showing perturbation of ^{2}H-nuclear Zeeman energy levels by quadrupolar interaction in solid-state ^{2}H NMR spectroscopy. The Zeeman Hamiltonian $\hat{H}_Z$ is perturbed by the quadrupolar Hamiltonian $\hat{H}_Q$ giving an unequal spacing of the nuclear spin energy levels, indicated by $|m\rangle$ where $m = 0, \pm 1$. The residual quadrupolar coupling (RQC) is designated by Δv_Q; it represents the difference in frequencies ($v_Q^\pm$) of the single quantum transitions owing to interaction of the ^{2}H-nuclear quadrupole moment with the electric field gradient of the C–^{2}H bond. **(B)** Illustration of the types of fluctuations in the lipids that give rise to NMR relaxation (*see* **ref. 15** for an explanation of coordinate systems). **(C)** Representative ^{2}H NMR spectra of unoriented powder samples of polyunsaturated DHA phospholipid bilayers having perdeuterated *sn*-1 acyl chains. Powder-type spectra of randomly oriented mutilamellar dispersions **(right)** were numerically inverted (de-Paked) **(left)** to yield spectra corresponding to the θ = 0° orientation. (In NMR θ is defined with respect to the normal to the lamellar surface, which differs from X-ray scattering.) Note that a distribution of RQCs is evident, corresponding to contributions from the various C^{2}H$_2$ and C^{2}H$_3$ groups. The quadrupolar splittings differ for C–^{2}H bonds along the lipid acyl chain because of variations in orientational potential for the different C^{2}H$_2$ groups and C^{2}H$_3$ groups *(21)*.

perturbation ($\hat{H}_Q$) of the magnetic Zeeman Hamiltonian ($\hat{H}_Z$). The two energy gaps shown in **Fig. 6A** then differ by a certain amount (Δv_Q), which is measurable by ^{2}H NMR. Because the dynamics of C–^{2}H bonds vary along the acyl chain (owing to the bilayer depth), the measured Δv_Q values differ. A profile of Δv_Q values as a function of segment index is thus generated *(21)*.

Molecular motions affect the ^{2}H NMR measurement in two ways. This is because there are two kinds of observables: *order* parameters, indicative of equilibrium properties, and *relaxation* parameters, which characterize the molecular dynamics. As shown below **(Subheading 2.4)**, the relationship between these quantities tells us about the types of molecular motions within the bilayer. For fluid bilayers, the dynamics of the lipids may encompass significant contributions from collective motions of the bilayer aggregate, in addition to the local motions. The observed NMR relaxation rates are then determined by the composite or resultant (effective) motions of the lipids. Relaxation spectra for 1,2-dimyristoyl-*sn*-glycero-3-phosphocholine (DMPC) are shown in **Fig. 7**. Within a continuum description of lipid bilayers, the contribution of collective motions to relaxation depends on material properties of lipid bilayers; for example, bilayer softness or membrane deformability over relatively

X-Ray and ^{2}H NMR of Lipid Fluctuations

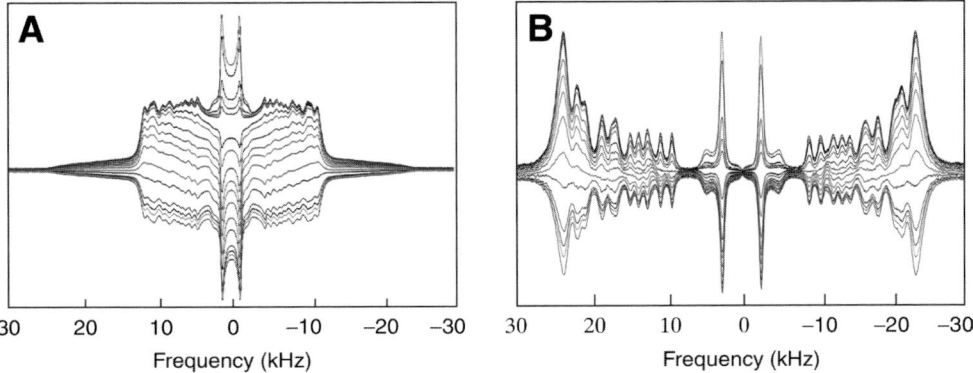

Fig. 7. Partially relaxed ^{2}H NMR spectra of randomly oriented DMPC-d_{54} mutilamellar dispersion in the liquid-crystalline (L_α) state at $T = 44$°C. **(A)** Experimental ^{2}H NMR spectra. **(B)** Numerically inverted (de-Paked) ^{2}H NMR spectra ($\theta = 0°$). Data were obtained at 76.8 MHz (magnetic field strength of $11.7\,T$) using an inversion-recovery pulse sequence, with the variable delay τ ranging from 5 ms to 3 s *(22)*.

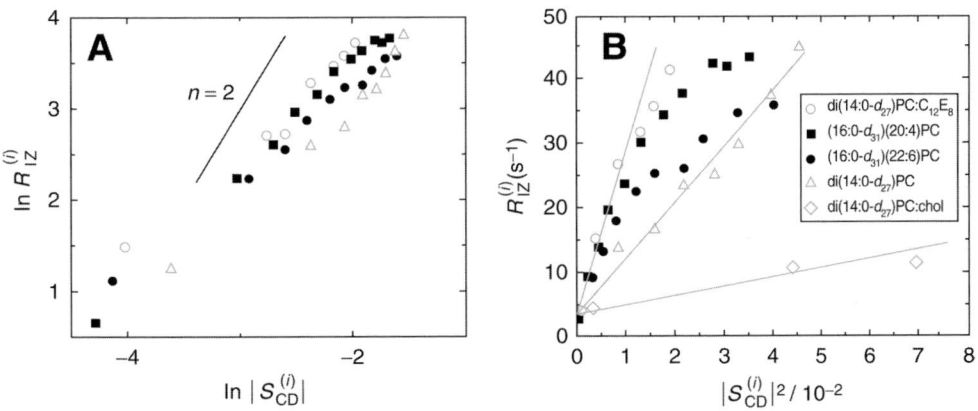

Fig. 8. Dependence of relaxation rate $R_{1Z}^{(i)}$ and order parameter $|S_{CD}^{(i)}|$ profiles for various phospholipids in the liquid-crystalline (L_α) state. Data for randomly oriented (powder-type) samples were acquired at 76.8 MHz (magnetic field strength of 11.7 T) and numerically inverted (de-Paked). **(A)** Logarithmic plots are shown, wherein data are compared with the slope of $n = 2$ predicted for order fluctuations. **(B)** Square-law functional dependence of corresponding relaxation and order profiles. The square-law plots of NMR observables indicate differences in membrane rigidities, wherein a steeper slope reflects a softer, more deformable bilayer *(23)*.

short length scales *(8,22)*. According to this view, bulk bilayer properties may be emergent on length scales approaching the bilayer thickness, and even less.

Square-law plots of relaxation rates and order parameters constitute a model-free relation between the structural and dynamical ^{2}H NMR observables. The slope of this relationship is predicted theoretically and found experimentally to be inversely related to the softness of the membrane bilayer. When evaluated at the same amplitude of motion as in **Fig. 8**, ^{2}H NMR relaxation parameters provide a naturally calibrated scale for comparison of lipid bilayers. Reference data are provided by benchmark DMPC and its mixtures with cholesterol and a nonionic detergent, whose viscoelastic properties are known. When a detergent is present as cosurfactant, an increase in slope is found *(20)*. In contrast, a bilayer-stiffening agent such as cholesterol or lanosterol leads to a dramatic reduction in the square-law slope *(22)*.

By this method, as an interesting example, significant differences have been found in the viscoelasticity of the ω-3 and ω-6 bilayers, a possibly biologically relevant property that distinguishes between the two fatty acid types *(23)*. One interesting possibility is that an increased deformability of bilayers containing arachidonic (22:4, ω-6) lipids might be related to the fatty acid selectivity of phospholipase A_2 enzymes. It would be interesting to see whether X-ray scattering measurements confirm these results.

2.4. Motional Models Describe Collective Excitations of Lipid Bilayers

By considering combinations of deformation modes as shown in **Fig. 3**, one can in principle derive a number of motional models, ranging from molecular models with local correlations to full 3D models that consider both membrane displacement and thickness deformation *(16)*. Continuum models represent bulk bilayer properties, where the length scale of the emergence of bulk properties may be more transparent than in molecular dynamics simulations. The weakness is that molecular detail is relinquished. One can consider a membrane deformation model, which assumes two types of motions for the relaxation mechanism. The segmental motions are fast when compared with the resonance frequency, and the ordering represents a small correction. In contrast, the slow motions are owing to collective fluctuations of the lipids, which modulate the residual quadrupolar coupling (RQC) left over from the fast motions, and depend on the square of the local-order parameter. If relatively slow motions are the major contribution to relaxation, an approximately linear dependence of the spin-lattice relaxation rate $R_{1Z}^{(i)}$ vs $\left|S_{CD}^{(i)}\right|^2$ along the chain (index i) is expected *(18)*, as observed for disaturated lipids (**Fig. 8**). See **Subheading 4**, for further details of the mathematical description of NMR relaxation in terms of membrane elasticity.

Further insight into the types of bilayer fluctuations is obtained from the frequency dependence of the relaxation. Motional models can be distinguished based on the frequency dependence of the nuclear spin relaxation, which corresponds to the fluctuating part of the electric quadrupolar coupling *(2)*. **Figure 9** presents spin-lattice relaxation rates for DMPC obtained at a series of different magnetic field strengths, each corresponding to a different resonance frequency. The relaxation is distinctly nonexponential, with very long relaxation tails being observed in the time domain. In this regard, the data are quite analogous to the case of X-ray scattering. It is plausible that the origin of long relaxation tails in both cases is similar, namely that they originate from membrane motions owing to collective interactions among the bilayer lipids with each other and with water.

Basically, a model involving only molecular motions is doubtful as an explanation of the relaxation data. A 2D flexible surface model, which is shown by X-ray diffraction to be applicable to longer wavelength fluctuations, predicts a ω^{-1} frequency-dependence and fails to account for the ^2H R_{1Z} relaxation dispersion. However, such a frequency-dependence explains the relaxation dispersion obtained at appreciably lower frequencies in Carr–Purcell–Meiboom–Gill experiments *(24)*, which may correspond to longer wavelength undulations as seen in X-ray scattering experiments (*see* **Subheading 2.2**). For the higher frequency end of the relaxation dispersion (in the MHz regime), a 3D membrane deformation model predicts a $\omega^{-1/2}$ frequency law for the R_{1Z} measurements, and explains the experimental data in **Fig. 9**.

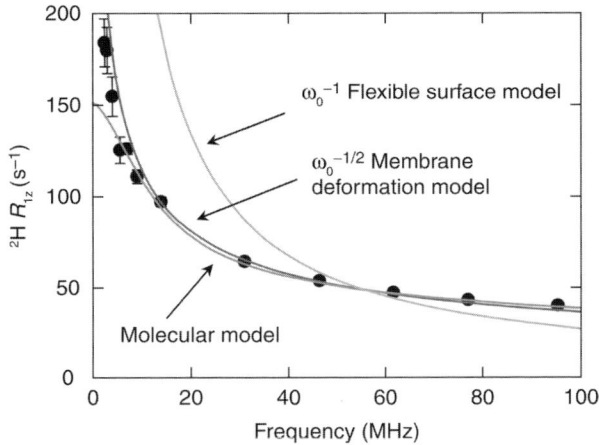

Fig. 9. Frequency-dependence of relaxation for different motional models. Experimental ^{2}H R_{1Z} relaxation rates are plotted as a function of frequency for vesicles of DMPC deuterated at the C3 acyl segment in the liquid crystalline state at $T = 30°C$. Data at 12 different resonance frequencies (magnetic field strengths) are shown, together with theoretical fits to various motional models that are described in the text. *Quasi*-elastic lipid motions give long relaxation tails, and govern relaxation mechanisms measurable by solid-state ^{2}H NMR spectroscopy. The results allow one to distinguish among various limiting relaxation laws *(16)*.

3. Conclusions

As complementary methods, X-ray scattering and ^{2}H NMR of multilamellar lipid vesicles reveal a unified view of the hierarchy of molecular motions within biomembranes. Spontaneous thermal fluctuations as well as intermolecular correlation functions can be measured and analyzed in terms of material parameters relevant to membrane function. For example, an antagonistic effect of cholesterol and polyunsaturated lipids on membrane bending rigidity is found by both X-ray scattering and NMR spectroscopy. A dual analysis by X-ray and NMR methods explains the molecular origin of these effects. Continuum analysis of X-ray and ^{2}H NMR data is complementary to all-atom molecular dynamics simulations and both approaches have much to offer. Although line shape analysis of either X-ray or NMR measurements are not yet common, considerations of motional models (possibly inspired by computer simulations) are promising approaches for a full description of biomembrane function and interactions.

4. Notes

4.1. Membrane Deformation and Elasticity–Positional Fluctuations

Scattering methods provide information about molecular positions as in the case of X-ray *(1,25)* and neutron *(14,26,27)* diffraction studies of membrane bilayers. To calculate positional correlations owing to spontaneous membrane fluctuations, the Caillé-de Gennes model considers the energies associated with 2D membrane deformations. In the fluid state, lipid membranes can bend locally with an energy cost proportional to the square of the local membrane curvature and K_C *(28)*. Interactions between nearest-neighbor membranes are accounted for by a harmonic compressibility term depending on the material parameter B and the relative membrane displacement. The interaction per unit area between neighboring membranes of size $L \times L$ is represented as an integral of the deformation terms plus a bare interaction term, $V(D_W)$, where D_W represents the interlamellar water spacing. This bare interaction term represents the intermembrane interaction in the absence of fluctuations *(1)*:

$$H = \frac{1}{NL^2} \int_0^L dx \int_0^L dy \sum_{n=1}^{N} \left[\frac{1}{2} K_C \left(\frac{\partial^2 u_n}{\partial x^2} + \frac{\partial^2 u_n}{\partial y^2} \right)^2 + \frac{1}{2} B (u_{n+1} - u_n)^2 \right] + V(D_W) \quad (2)$$

Using statistical mechanics, the mean-square fluctuation of the interbilayer separation can be calculated in relation to the material parameters K_C and B:

$$\sigma^2 = \left\langle (u_n - u_{n+1})^2 \right\rangle = \frac{k_B T}{2\pi} \frac{1}{\sqrt{K_C B}} \quad (3)$$

Moreover, correlations between neighboring membranes n and $n + k$, for $k = 1, 2, 3, \ldots N$ can be calculated to determine the scattering correlation function which is given by:

$$G_h(k) = \left\langle \exp[iq(u_{n+k} - u_n)] \right\rangle \cong \exp\left[-\frac{1}{2} q_h^2 \Delta^2(k) \right] = \exp\left[-\eta_1 h^2 \Sigma(k) \right] \quad (4)$$

Here, $\Sigma(k)$ (shown in **Fig. 4B**) represents the normalized mean-square fluctuations of the relative displacement between membranes n and $n + k$, and the Caillé order parameter η_1 is a function of K_C and B:

$$\eta_1 = \pi^2 \frac{\sigma^2}{D^2} = \frac{\pi}{2} \frac{k_B T}{D^2} \frac{1}{\sqrt{K_C B}} \quad (5)$$

The Caillé parameter η_1 containing information on K_c and B is obtained experimentally by fitting X-ray data using the correlation function in Eq. (4).

4.2. Membrane Deformation and Elasticity–Angular Fluctuations

Solid-state ^{2}H NMR spectroscopy detects angular fluctuations of the ^{2}H-labeled molecules and is complementary to scattering methods. ^{2}H NMR data are obtained at an atomically specific level and correspond to a membrane ensemble. Temporal fluctuations of the molecules are described by a liquid-crystalline membrane-deformation model. The bilayer is approximated as a continuous material, and the relaxation is assumed to be owing to a distribution of overdamped modes.

In ^{2}H NMR, membrane fluctuations are described by spectral densities, which are the Fourier transform of correlation functions that depend on changes in the orientation of the C–^{2}H bonds of the labeled lipid molecules. The symbols $J_m(\omega)$ denote the spectral densities of motion ($m = 0, \pm 1$, and ± 2), which correspond to the different irreducible components of the second-rank quadrupolar-coupling tensor in the laboratory frame. For liquid-crystalline lipids, a composite model for membrane fluctuations is considered *(16)*. Collective bilayer excitations are superimposed with collective axial rotations of the flexible phospholipid molecules, giving for the spectral densities:

$$J_m(\omega) = J_m^{mol}(\omega) + J_m^{col}(\omega) + J_m^{mol-col}(\omega) \quad (6)$$

Spectral densities $J_m^{col}(\omega)$ and $J_m^{mol}(\omega)$ correspond to collective fluctuations and molecular rotations, respectively, and $J_m^{mol-col}(\omega)$ is a geometrical cross-term.

For collective fluctuations, the spectral densities $J_m^{col}(\omega)$ are given by:

$$J_m^{col}(\omega) = \frac{5}{2}\left|\left\langle D_{00}^{(2)}(\Omega_{PD})\right\rangle\right|^2 D_{col}\omega^{-(2-d/2)}\left[\left|D_{-1m}^{(2)}(\Omega_{DL})\right|^2 + \left|D_{1m}^{(2)}(\Omega_{DL})\right|^2\right] \qquad (7)$$

Here, $D_{col} = (3/5)(k_B T \eta^{1/2}/2^{1/2}\pi K^{3/2} S_s^{(2)^2})$ is a viscoelastic constant that includes the elastic constant (K), viscosity (η), temperature (T), and the order parameter ($S_s^{(2)}$) for collective slow motions; d is the dimensionality of the bilayer thermal excitations. Euler angles Ω_{PD} transform from the frame of the C–^{2}H bond to the director frame (D); they are time-dependent and correspond to the fast order parameter ($S_f^{(2)}$). Angles Ω_{DL} are fixed and transform from the bilayer director to the laboratory axis (L). Formulas for the spectral densities for molecular motions $J_m^{mol}(\omega)$ and the cross-term $J_m^{mol-col}(\omega)$ are given elsewhere *(16)*.

Moreover, for distances approaching the bilayer thickness, the dimensionality of the fluctuations comes into play, for example, for 3D vs 2D membranes. According to the above formula, collective order fluctuations in 2D are described by a *flexible-surface model,* wherein the finite thickness of the membrane bilayer is neglected ($d = 2$). This gives an ω^{-1} frequency-dependence, and effectively corresponds to the Caillé-de Gennes model. Alternatively, 3D order fluctuations are described by a collective *membrane-deformation model* ($d = 3$), which predicts an $\omega^{-1/2}$ frequency-dependence of the spectral densities $J_m^{col}(\omega)$, in analogy with a Pincus-de Gennes liquid-crystal model. K_C may also differ in the two limits because of a possible length scale dependence. In either case, $J_m^{col}(\omega)$ depends on the square of the observed order parameter, $S_{CD} = S_f^{(2)}S_s^{(2)}$. The slope of a square-law plot of $R_{1Z}^{(i)}$ against $|S_{CD}^{(i)}|^2$ (where i is the chain index) is inversely related to the softness of the membrane.

The C–^{2}H bond segmental order parameters, $S_{CD}^{(i)}$, are obtained as experimental observables in ^{2}H NMR spectroscopy. They are evaluated from the RQCs ($\Delta\nu_Q$) using the relation:

$$\Delta\nu_Q = \frac{3}{2}\chi_Q|S_{CD}|P_2(\cos\theta) \qquad (8)$$

Here, $\chi_Q = e^2qQ/h$ is the static quadrupolar coupling constant (≈ 167 kHz), P_2 is the second Legendre polynomial, and θ is the angle between the bilayer normal and the static magnetic field. ^{2}H NMR spectra in the L_α phase are numerically deconvoluted with the de-Pakeing algorithm *(29–31)* to obtain subspectra corresponding to the $\theta = 0°$ orientation of the bilayer normal to the main magnetic field.

Finally, information on membrane motions is accessible through measurement of various ^{2}H NMR relaxation rates. To compute the relaxation curve, spin-lattice (Zeeman) relaxation times (T_{1Z}) corresponding to each of the resolved splittings are obtained by fitting the data to a three-parameter decaying exponential function:

$$S(\tau) = S(\infty) - [S(\infty) - S(0)]\exp(-R_{1Z}\tau) \qquad (9)$$

where $R_{1Z} = 1/T_{1Z}$ is the spin-lattice relaxation rate and τ is a variable delay. R_{1Z} is related to the spectral densities $J_m(\omega)$ of the membrane motions by:

$$R_{1Z} = \frac{3}{4}\pi^2\chi_Q^2[J_1(\omega_0) + 4J_2(2\omega_0)] \qquad (10)$$

where ω_0 is the deuteron resonance (Larmor) frequency. For membrane lipids, the relaxation data are described by a simple law of the form:

$$R_{1Z}^{(i)} = a\tau_f + b\omega_0^{-1/2}\left|S_{CD}^{(i)}\right|^2 \qquad (11)$$

where τ_f is the correlation time for segmental fast motions of the lipids, and a and b are held constant. A model for collective bilayer fluctuations explains the combined angular and frequency dependencies of the NMR relaxation for lipid bilayers in the fluid phase *(16)*.

References

1. Petrache, H. I., Gouliaev, N., Tristram-Nagle, S., Zhang, R. T., Suter, R. M., and Nagle, J. F. (1998) Interbilayer interactions from high-resolution x-ray scattering. *Phys. Rev. E.* **57,** 7014–7024.
2. Brown, M. F. (1996) Membrane structure and dynamics studied with NMR spectroscopy, in *Biological Membranes. A Molecular Perspective from Computation and Experiment*, (Merz, K., Jr. and Roux, B., eds.), Birkhäuser, Basel, pp. 175–252.
3. Rand, R. P., Fuller, N. L., Gruner, S. M., and Parsegian, V. A. (1990) Membrane curvature, lipid segregation, and structural transitions for phospholipids under dual-solvent stress. *Biochemistry* **29,** 76–87.
4. Landau, L. D. and Lifshitz, E. M. (1986) *Theory of Elasticity, 3rd ed.*, Pergamon, Oxford.
5. Brown, M. F. (1994) Modulation of rhodopsin function by properties of the membrane bilayer. *Chem. Phys. Lipids* **73,** 159–180.
6. Petrache, H. I., Grossfield, A., MacKenzie, K. R., Engelman, D. M., and Woolf, T. B. (2000) Modulation of glycophorin A transmembrane helix interactions by lipid bilayers: molecular dynamics calculations. *J. Mol. Biol.* **302,** 727–746.
7. Huber, T., Botelho, A. V., Beyer, K., and Brown, M. F. (2004) Membrane model for the GPCR prototype rhodopsin: hydrophobic interface and dynamical structure. *Biophys. J.* **86,** 2078–2100.
8. Brown, M. F., Thurmond, R. L., Dodd, S. W., Otten, D., and Beyer, K. (2002) Elastic deformation of membrane bilayers probed by deuterium NMR relaxation. *J. Am. Chem. Soc.* **124,** 8471–8484.
9. Bloom, M., Evans, E., and Mouritsen, O. G. (1991) Physical properties of the fluid lipid-bilayer component of cell membranes: a perspective. *Q. Rev. Biophys.* **24,** 293–397.
10. De Gennes, P. G. (1974) *The Physics of Liquid Crystals,* Clarendon, Oxford.
11. Nagle, J. F. and Tristram-Nagle, S. (2000) Structure of lipid bilayers. *Biochim. Biophys. Acta* **1469,** 159–195.
12. Als-Nielsen, J., Litster, J. D., Birgeneau, R. J., Kaplan, M., Safinya, C. R., Lindegaard-Andersen, A., and Mathiesen, S. (1980) Observation of algebraic decay of positional order in a smectic liquid crystal. *Phys. Rev. B.* **22,** 312–320.
13. Wack, D. C. and Webb, W. W. (1989) Measurement by x-ray diffraction methods of the layer compressional elastic constant B in the lyotropic smectic-A (L_α) phase of the lecithin-water system. *Phys. Rev. A.* **40,** 1627–1636.
14. Salditt, T. and Brotons, G. (2004) Biomolecular and amphiphilic films probed by surface sensitive X-ray and neutron scattering. *Anal. Bioanal. Chem.* **379,** 960–973.
15. Nevzorov, A. A. and Brown, M. F. (1997) Dynamics of lipid bilayers from comparative analysis of ^{13}C and ^{2}H NMR relaxation data as a function of frequency and temperature. *J. Chem. Phys.* **107,** 10288–10310.
16. Nevzorov, A. A., Trouard, T. P., and Brown, M. F. (1998) Lipid bilayer dynamics from simultaneous analysis of orientation and frequency dependence of deuterium spin-lattice and quadrupolar order relaxation. *Phys. Rev. E* **58,** 2259–2281.

17. Martinez, G. V., Dykstra, E. M., Lope-Piedrafita, S., Job, C., and Brown, M. F. (2002) NMR elastometry of fluid membranes in the mesoscopic regime. *Phys. Rev. E* **66,** 050902-1–050902-4.
18. Brown, M. F. (1982) Theory of spin-lattice relaxation in lipid bilayers and biological membranes. ^{2}H and ^{14}N quadrupolar relaxation. *J. Chem. Phys.* **77,** 1576–1599.
19. Brown, M. F., Thurmond, R. L., Dodd, S. W., Otten, D., and Beyer, K. (2001) Composite membrane deformation on the mesoscopic length scale. *Phys. Rev. E* **64,** 010901-1–010901-4.
20. Otten, D., Brown, M. F., and Beyer, K. (2000) Softening of membrane bilayers by detergents elucidated by deuterium NMR spectroscopy. *J. Phys. Chem. B* **104,** 12119–12129.
21. Petrache, H. I., Dodd, S. W., and Brown, M. F. (2000) Area per lipid and acyl length distributions in fluid phosphatidylcholines determined by ^{2}H NMR spectroscopy. *Biophys. J.* **79,** 3172–3192.
22. Martinez, G. V., Dykstra, E. M., Lope-Piedrafita, S., and Brown, M. F. (2004) Lanosterol and cholesterol-induced variations in bilayer elasticity probed by ^{2}H NMR relaxation. *Langmuir* **20,** 1043–1046.
23. Rajamoorthi, K., Petrache, H. I., McIntosh, T. J., and Brown, M. F. (2005) Packing and viscoelasticity of polyunsaturated ω-3 and ω-6 lipid bilayers as seen by ^{2}H NMR and X-ray diffraction. *J. Am. Chem. Soc.* **127,** 1576–1588.
24. Prosser, R. S., Heaton, N. J., and Kothe, G. (1996) Application of the quadrupolar Carr-Purcell-Meiboom-Gill pulse train for sensitivity enhancement in deuteron NMR of liquid crystals. *J. Magn. Reson. B.* **112,** 51–57.
25. Liu, Y. and Nagle, J. F. (2004) Diffuse scattering provides material parameters and electron density profiles of biomembranes. *Phys. Rev. E* **69,** 040901-1–040901-4.
26. Yang, L., Harroun, T. A., Heller, W. T., Weiss, T. M., and Huang, H. W. (1998) Neutron off-plane scattering of aligned membranes. I. Method of measurement. *Biophys. J.* **75,** 641–645.
27. Endress, E., Heller, H., Casalta, H., Brown, M. F., and Bayerl, T. (2002) Anisotropic motion and molecular dynamics of cholesterol, lanosterol, and ergosterol in lecithin bilayers studied by quasielastic neutron scattering. *Biochemistry* **41,** 13078–13086.
28. Helfrich, W. (1973) Elastic properties of lipid bilayers. Theory and possible experiments. *Z. Naturforsch.* **28c,** 693–703.
29. Lafleur, M., Fine, B., Sternin, E., Cullis, P. R., and Bloom, M. (1989) Smoothed orientational order profile of lipid bilayers by ^{2}H-nuclear magnetic resonance. *Biophys. J.* **56,** 1037–1041.
30. McCabe, M. A. and Wassall, S. R. (1997) Rapid deconvolution of NMR powder spectra by weighted fast Fourier transformation. *Solid State Nucl. Magn. Reson.* **10,** 53–61.
31. Sternin, E., Schäfer, H., Polozov, I. V., and Gawrisch, K. (2001) Simultaneous determination of orientational and order parameter distributions from NMR spectra of partially oriented model membranes. *J. Magn. Reson.* **149,** 110–113.

24

Determination of Lipid Spontaneous Curvature From X-Ray Examinations of Inverted Hexagonal Phases

Michael M. Kozlov

Summary

Structure of a lipid monolayer can be characterized by its spontaneous shape corresponding to a stress-free state. In the chapter, the rigorous description of monolayer shape in terms of curvature of a Gibbs dividing surface is reviewed, and the notions of the neutral and pivotal surfaces are discussed. Further, the effective and exact method of finding the position of the pivotal surface and the corresponding monolayer spontaneous curvature of cylindrically curved monolayers based on the results of the X-ray scattering experiments are presented. Briefly, some limitations of this method are discussed.

Key Words: Elasticity; inverted hexagonal phase; lipid monolayer; neutral surface; spontaneous curvature; X-ray scattering.

1. Introduction

Amphiphilic molecules (or amphiphiles) are characterized by dual hydrophobic–hydrophilic properties. Phospholipids are biological amphiphiles, whose molecules consist of hydrophilic polar heads, and in most cases, two hydrophobic hydrocarbon tails *(1)*. Phospholipids (referred in the following as lipids, for simplicity) of different types differ in the chemical structure of their polar heads as well as in length and degree of saturation of their hydrocarbon chains. In spite of that, all lipids exhibit a similar behavior in aqueous surrounding. Driven by the hydrophobic effect *(2)*, lipids self-organize into assemblies referred to as mesophases, which are characterized by a common feature—the hydrophobic moieties of the lipid molecules are shielded from water by layers of polar heads. If the amount of water is limited or in special stabilizing conditions, the lipid mesophases have properties of lyotropic liquid crystal phases (*see* **refs. *3–7***). In excess water, the mesophases can be seen as isotropic solutions of amphiphilic aggregates.

Building blocks of mesophases are lipid monolayers, which are monomolecular layers wherein all the lipid molecules are oriented similarly: the polar heads toward the one and the hydrocarbon chains toward the other side of the layer. As a result, one monolayer side is covered by polar heads, whereas the opposite side is hydrophobic. The monolayer thickness (δ), determined by the lipid molecular length, equals approx 2 nm. The structures of lipid molecules and interactions between them determine the preferable shapes of monolayers. The monolayer shape serves for characterization and classification of lipid mesophases. In turn, lipids can be classified according to shapes of monolayers and the corresponding types of mesophases they are forming in aqueous solutions (for review, *see* **refs. *8,9***). Whereas a large

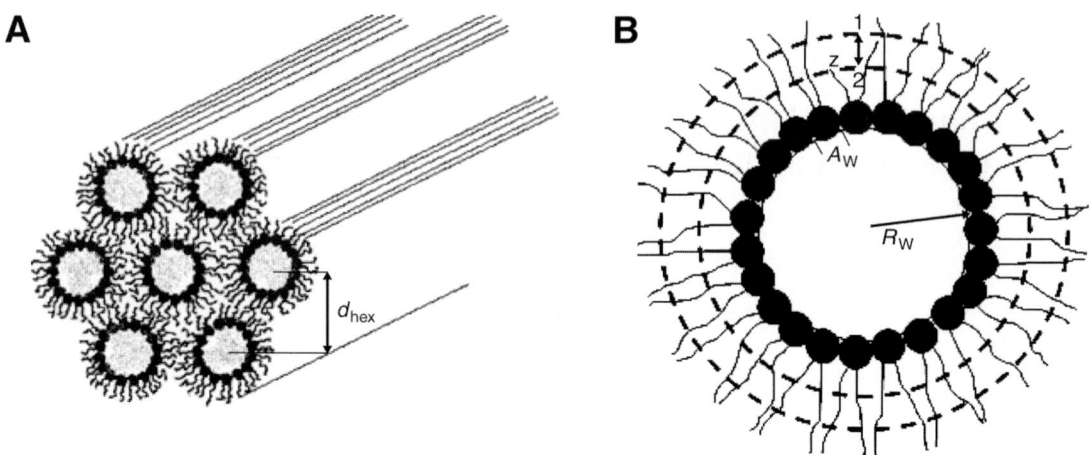

Fig. 1. Inverted hexagonal (H_{II}) phase: schematic illustration and notations. (**A**) The H_{II} phase consists of infinite prisms packed parallel to each other. The prism cross-sections form a 2D hexagonal lattice. The first-order Bragg parameter (d_{hex}) is measured by X-ray scattering; (**B**) cross-section of one prism of the H_{II} phase. The Luzzati dividing surface is located in the polar head region. It is characterized by radius R_W and the area per lipid molecule A_W. Two arbitrary dividing surfaces separated by distance z are denoted by indices 1 and 2.

number of lipid mesophases has been described (for review, see **ref. 10**), just some characterized by the simplest and most common shapes of their constituting monolayers will be addressed herein.

The most familiar type of lipid assembly is a planar bilayer consisting of two monolayers, contacting each other along the hydrophobic planes and whose hydrophilic surfaces cover the inner and outer bilayer planes. The propensity of lipids to self-assemble into planar bilayers underlies formation of cell membranes (the universal biological wrappers), which form boundaries of cells and intracellular organelles. A biological membrane is, basically, a multicomponent lipid bilayer with numerous proteins inserted into the lipid matrix or bound to the bilayer surface (*11*). Because of the powerful hydrophobic interactions, the bilayer lipid matrix provides the cell boundary with mechanical stability. The lyotropic liquid crystal phase formed by planar lipid bilayers is called the lamellar phase (*4*). It consists of a stack of bilayers separated by a few nanometer-thick water layers. Lipids such as lecithins forming the lamellar phases are often referred to as "lamellar" lipids.

Other type of lipid assembly consists of strongly curved lipid monolayers, whose shape can be seen within a good approximation as a narrow cylinder with few nanometer radius (**Fig. 1A**). The lipid polar heads of such monolayer are oriented toward the internal space of the cylinder, which is filled with water, whereas the external surface of the cylindrical monolayer is hydrophobic. Within the mesophase, multiple lipid cylinders are oriented in parallel and contact each other along the hydrophobic surfaces. This leads to a most compact packing of the monolayers so that the mesophase cross-section perpendicular to the cylinder axes reveals a hexagonal lattice of cross-sections of the individual cylinders (**Fig. 1A**). This mesophase is called the inverted hexagonal (H_{II}) phase, and the corresponding lipids are referred to as the "hexagonal" ones. A typical example of a "hexagonal" lipid is dioleoyl-phosphatidyl-ethanolamine (*see* **ref. 7**).

The H_{II} phase will be the focus of the present review. At the same time, few other mesophases should be briefly mentioned in order to illustrate the wide spectrum of possible shapes of lipid monolayers and the corresponding diversity of the lipid-phase behavior (for a review *see* **ref. 10** and references therein). Molecules of lysolipids having only one hydrocarbon tail tend to form monolayers, which are strongly curved in the direction opposite to that of the "hexagonal" lipids. The resulting assemblies are represented by cylindrical or spherical micelles with internal volumes filled by the hydrocarbon chains. In a limited amount of water the cylindrical micelles pack into hexagonal phases, which in contrast to the inverted hexagonal phases, are denoted as H_I phases. Finally, there are lipids and lipid mixtures forming bicontinuous cubic phases whose monolayers have shapes of saddles organized into periodic three-dimensional surfaces.

This short phenomenological consideration aims to illustrate the diversity of shapes adopted spontaneously by lipid monolayers and the degree of monolayer spontaneous curving, which varies from zero in the case of "lamellar" lipids to high values for the "hexagonal," "micellar," and "cubic" lipids. In the following, we will discuss how the spontaneous shapes of lipid monolayers can be addressed based on X-ray studies of H_{II} phases. First, we introduce a rigorous description of monolayer shapes in terms of Gibbs dividing surface, and define the notion of monolayer spontaneous curvature. Further, we introduce the notion of monolayer elasticity apply it to the cylindrical monolayers of H_{II} phases, and define the neutral surface. Based on these definitions, we present protocol for measuring the position of neutral surface and its spontaneous curvature using the X-ray data.

2. Description of Monolayer Shape—Spontaneous Curvature
2.1. Dividing Surface

A few nanometer thickness of a lipid monolayer is, for most membrane structures, few orders of magnitude smaller than the dimension measured along the monolayer plane. Therefore, it seems straightforward to describe lipid monolayers as thickness-less surfaces. Whereas such approach did prove to be productive, there is a delicate issue of definition of the surface with which the monolayer is identified. This issue becomes especially important for strongly curved monolayers, such as those forming the inverted hexagonal or micellar phases, whose radii of curvature are of the same order of magnitude as the monolayer thickness.

The problem of dealing with layers of small but finite thickness is not specific for lipid monolayers but rather originated from attempts to treat transition regions between immiscible liquids. Herein the Gibbs method will be used *(12)*, which has been developed for description of the liquid–liquid interfaces, but has a general character and can be applied to any type of thin layers including lipid monolayers. According to the Gibbs approach, one has to choose within the lipid monolayer a geometrical surface (called the *dividing surface*) that lies parallel to the plane of the polar heads. The monolayer shape is identified with that of the dividing surface and the monolayer thermodynamic properties are assigned to the dividing surface. To this end, a reference system is considered where bulk phases are extended with their unchanged properties up to the dividing surface. The thermodynamic values of the dividing surface are, by definition, the excess values (e.g., excess entropy $[S^s]$, energy $[U^s]$, and masses of components $[m^s]$) determined as differences between the thermodynamic values of the real system and those of the reference system.

The exact position of the dividing surface along the monolayer thickness can be chosen arbitrarily. However, it is important to realize that both the geometrical and most of the thermodynamic characteristics (except the total free energy) of the system depend on the choice of the dividing surface, so that once the dividing surface is chosen within the monolayer, the following description has to be performed consequently for this specific dividing surface (*see* **ref. *13***). Whereas the Gibbs method places no limitations on the choice of the dividing surface, there are physical reasons for selection of particular dividing surfaces such as Gibbs surface of tension *(12)* and neutral surface, which simplify the treatment of monolayer properties and thermodynamic behavior. The neutral surface is a center of the following discussion.

2.2. Spontaneous Cuvature

Once the dividing surface is chosen the monolayer shape can be characterized at each point by the curvatures of the dividing surface. According to the differential geometry of surfaces (*see* **refs. *14,15***), the local shape is determined by two principal curvatures, c_1 and c_2, or equivalently, by the total curvature, $J = c_1 + c_2$, and the Gaussian curvature, $K = c_1 \cdot c_2$. The two latter values represent the two independent invariants of the curvature tensor (*14,15*) and are convenient for description of lipid monolayers having properties of two-dimensional (2D) fluids *(16)*. (Note that in the mathematical literature the notion of the mean curvature $H = J/2$ rather than the total curvature J is commonly used.) To determine the signs of the curvatures, the orientation of the normal vector of the dividing surface has to be chosen. It is convenient and common to define the normal vector to be oriented from the hydrocarbon chains toward the polar heads. The principal curvatures are defined as positive if the monolayer bulges in the direction of polar heads and negative in the case of bulging toward the hydrophobic tails.

Herein the monolayer spontaneous curvature (J_s) is defined, as the total curvature the monolayer adopts on spontaneous self-assembly. Note that this definition differs from the original one, which has been given by Helfrich *(16)* to quantify the tendency of a flat monolayer to bend spontaneously. The Helfrich spontaneous curvature represents, practically, a bending stress existing within a monolayer when the latter has a flat shape (indeed, the bending stress equals the Helfrich spontaneous curvature multiplied by the bending modulus). According to the present definition, the spontaneous curvature is a geometrical characteristic of a stress-free state of a monolayer.

Whereas the spontaneous curvature is an essentially macroscopic, rather than molecular, characteristic of a lipid monolayer, the notion of *spontaneous curvature of individual lipid molecules* is frequently used in the literature. The molecular spontaneous curvature has to be understood as the spontaneous curvature of a monolayer consisting of molecules of a given kind. In addition to the spontaneous curvature, the stress-free state of a monolayer can be characterized by a spontaneous area per given macroscopic number of lipid molecules. This area has to be determined at the dividing surface, and it's meaning is the area of the monolayer cross-section by the dividing surface. Commonly, one considers a spontaneous molecular area (A_s), which is an effective value equal to the total spontaneous area divided by the number of lipid molecules.

3. Monolayer Elasticity—Neutral Surface
3.1. Elastic Stresses and Energy

To advance toward consideration of the experimental procedure used to determine spontaneous curvatures of individual lipids, a step beyond the spontaneous state of the monolayer has to made and the monolayer deformations have to be considered. The considerations will be limited to cylindrical monolayers of H_{II} phases because this is the system in which the most productive X-ray measurements have been performed *(7,17–25)*. Geometrical description of a cylindrical surface is simplified because its Gaussian curvature equals zero, $K = 0$, and the absolute value of the mean curvature equals the cylinder radius, $|J| = 1/R$. There are two types of deformations (strains) of a cylindrical monolayer: deformation of stretching leading to deviation of the molecular area (A) from its spontaneous value (A_s), and bending deformation resulting in difference between the total curvature J and the spontaneous curvature J_s.

In the following, for simplicity, positive value $1/R$ and $1/R_s$ will be used to characterize the deformed and spontaneous states of the H_{II} phase monolayers. But it has to be kept in mind that, according to the earlier convention on curvature sign, the corresponding curvatures are negative, $J = -1/R$ and $J_s = -1/R_s$. Deformations give rise to elastic stresses and the elastic energy. A complete consideration of membrane strains, stresses, and relationships among them is given in **ref. 26**. Here, instead, a simplified case of cylindrical monolayers is sketched *(27,28)*. The elastic stresses corresponding to the stretching and bending strains are the lateral tension (γ) and bending moment (τ), respectively. In the limit of small deformations,

$$\left|\frac{A-A_s}{A_s}\right| \ll 1 \text{ and } \left|\frac{R-R_s}{R_s}\right| \ll 1 \tag{1}$$

the stresses depend linearly on the strains according to:

$$\gamma = \Gamma \cdot \frac{A-A_s}{A_s} + E \cdot \left(\frac{1}{R} - \frac{1}{R_s}\right) \tag{2}$$

and

$$\tau = \kappa \cdot \left(\frac{1}{R} - \frac{1}{R_s}\right) + E \cdot \left(\frac{A-A_s}{A_s}\right) \tag{3}$$

The coefficients Γ and κ are the monolayer stretching modulus and bending modulus, respectively. The coefficient E determines coupling between the deformations of stretching and bending.

The elastic energy of deformation per unit area of the dividing surface is

$$F = \frac{1}{2} \cdot \Gamma \cdot \left(\frac{A-A_s}{A_s}\right)^2 + \frac{1}{2} \cdot \kappa \cdot \left(\frac{1}{R} - \frac{1}{R_s}\right)^2 + E \cdot \left(\frac{A-A_s}{A_s}\right) \cdot \left(\frac{1}{R} - \frac{1}{R_s}\right) \tag{4}$$

Note that in case of vanishing spontaneous curvature: $J_s = 0$ or, equivalently, $\frac{1}{R_s} \to 0$; validation of the expression (**Eq. 3**) for the bending moment and of the curvature-dependent contributions to the energy (**Eq. 4**) requires, instead of the condition (**Eq. 1**), fulfillment of inequality $|J \cdot \delta| \ll 1$, where $\delta = 2$ nm is the monolayer thickness.

3.2. Elastic Moduli for Different Dividing Surfaces

All elastic and structural characteristics of a monolayer (κ, Γ, E, A_s, and R_s) depend on the position of the dividing surface. Consider an arbitrary dividing surface whose characteristics will be indicated by subscript 1, and another dividing surface, indicated by superscript 2, which is shifted by z with respect to the first toward the external surface of the cylindrical monolayer (**Fig. 1B**). The relationships between the radii of spontaneous curvatures and spontaneous molecular areas corresponding to the two dividing surfaces are given by

$$R_{s2} = R_{s1} + z \tag{5}$$

$$A_{s2} = A_{s1} \cdot \left(1 + \frac{z}{R_{s2}}\right) \tag{6}$$

A thorough analysis of relationships between the elastic moduli related to different dividing surfaces is given in **refs. 24,27,28**. Here, only simplified equations are presented. For the case of vanishing spontaneous curvature ($J_s = 0$), the set of elastic moduli (κ_1, Γ_1, and E_1) is related to the corresponding set (κ_2, Γ_2, and E_2), by

$$\Gamma_2 = \Gamma_1 \tag{7}$$

$$\kappa_2 = \kappa_1 + z^2 \cdot \Gamma_1 - 2 \cdot z \cdot E_1 \tag{8}$$

$$E_2 = E_1 - z \cdot \Gamma_1 \tag{9}$$

3.3. Neutral Surface

Analysis of the elastic moduli dependences on the position of the dividing surface shows that the bending and stretching moduli adopt only positive values. Along with the spontaneous geometrical characteristics A_s and R_s, they describe macroscopically the monolayer structure and elasticity, and represent material properties of membrane monolayers, which within the approximation (**Eq. 1**), do not depend on deformations. At the same time, the coupling modulus E can adopt positive or negative values depending on the position of the dividing surface. For a special dividing surface, called the *neutral surface*, this coefficient vanishes: $E = 0$ *(26–28)*, and the whole description of the membrane elastic properties *(2–4)* simplifies. Specifically, for the neutral surface the lateral tension γ_N does not depend on the bending deformation $J-J_s$, and the bending moment τ_N is independent of the deformation of stretching $A-A_s$.

It is instructive to present (**Eqs. 2 and 3**) in the form that determines the membrane deformations $\frac{1}{R} - \frac{1}{R_s}$ and $A - A_s$, which result from stresses γ and τ applied to the membrane by external sources:

$$\frac{A - A_s}{A_s} = \frac{1}{\left(1 - \frac{E^2}{\kappa \cdot \Gamma}\right)} \left(\frac{\gamma}{\Gamma} - \frac{E \cdot \tau}{\Gamma \cdot \kappa}\right) \tag{10}$$

$$\frac{1}{R} - \frac{1}{R_s} = \frac{1}{\left(1 - \frac{E^2}{\kappa \cdot \Gamma}\right)} \left(\frac{\tau}{\Gamma} - \frac{E \cdot \gamma}{\Gamma \cdot \kappa}\right) \tag{11}$$

Following from **Eqs. 10 and 11**, a special property of the neutral surface ($E = 0$) is that its molecular area (A_N) does not change and remains equal to the spontaneous value (A_{sN}) if only a bending moment τ is applied to the monolayer. Deviation of A_N from A_{sN} requires application of tension γ. Analogously, the curvature of the neutral surface ($J_N = 1/R_N$) is independent of γ and changes only on application of τ.

The neutral surface is the most convenient to treat the elastic behavior of lipid monolayers. On the other hand, the position of the neutral surface within a monolayer depends on the monolayer structure and is usually unknown. The **Eq. 9** provides a key for finding the neutral surface. Assume that there is a way to measure the whole set of the elastic moduli (κ_1, Γ_1, E_1) for a certain dividing surface with a known position. The distance z_N between this dividing surface and the neutral surface is given, according to (**Eq. 9**) by $z_N = \dfrac{E_1}{\Gamma_1}$. Hence, measurement of the elastic moduli for one of the dividing surfaces gives enough information for determination of the location of the neutral surface, and calculation of the corresponding elastic moduli according to the relationships (**Eqs. 7–9**) or the more general ones presented in **refs. 26–28**. In the following, a specific protocol for finding the dividing surface and the related elastic and structural characteristics of a monolayer based on X-ray studies of the inverted hexagonal (H_{II}) phases is presented.

4. Structural Studies of H_{II} Phases

4.1. Directly Measurable Characteristics

An H_{II} phase can be seen as a 2D hexagonal lattice formed by axes of infinitely long and parallel regular prisms (**Fig. 1A**). The water core of each prism is lined with the lipid polar groups, and the rest of the prism volume is filled with the hydrocarbon chains. The essence of the experimental approach that is being described in this review is variation of the water content of H_{II} phase at fixed and known lipid content, and measurement of the related changes in the structural parameters of the H_{II} prisms. The measurements are performed by X-ray diffraction.

There are two methods used to change the water content (*see* **refs. 7,24,29–31** and references therein for detailed descriptions). According to the first, referred to as the gravimetric method, dry lipid samples are hydrated in water vapor atmosphere to different degrees quantified by the weight fraction of water within the samples. The volume fraction of water in the sample (ϕ_w) is then calculated using specific molecular volumes of water, lipids, and lipid polar groups. In the second approach, called the osmotic pressure method, a lipid sample is placed in an excess amount of aqueous polyethylene glycol solution of known osmotic pressure (Π). The amount of water within H_{II} phase is then controlled by the value of Π.

The structural dimension of the H_{II} phase that is measured in X-ray diffraction experiments, is the first order Bragg-spacing d_{hex} illustrated in **Fig. 1A**. Usually, an H_{II} phase is characterized by at least three X-ray spacings bearing ratios to d_{hex}, of 1, $1/\sqrt{3}, 1/\sqrt{4}, 1/\sqrt{7}$, and so on. These measurements yield d_{hex} with an accuracy of ±0.01 nm. The direct outcome of the gravimetric studies is the dependence of d_{hex} on the volume fraction of water in the H_{II}-phase. The osmotic pressure studies give d_{hex} as a function of Π. The two experimental functions, $d_{hex}(\phi_w)$ and $d_{hex}(\Pi)$, contain sufficient information for determining the structural-elastic parameters κ, Γ, E, a_s, and R_s of the H_{II} phase monolayers.

4.2. Luzzati Plane

The cross-section of each prism of H_{II} phase is hexagonal at the hydrocarbon chain boundary. The cross-section of the prism-water core is represented by a hexagon with smoothened vertexes *(7,32–34)*. Analysis shows that, as far as determination of spontaneous curvature, bending and stretching moduli is concerned, the water core can be approximated by a circular cylinder. According to the method introduced by Luzzati *(35)*, a prism can be subdivided into lipid and water compartments. The boundary between the two compartments is represented by the Gibbs dividing surface of zero excess of water, which means that it encloses a volume equal to the total volume of water within the prism. This dividing surface (referred to as the Luzzati surface) lies within the region of polar heads, and by convention, has a cylindrical shape.

The characteristics of the cylindrical Luzzati plane are its radius (R_W) and the area (A_W) per lipid molecule (**Fig. 1B**). Both of them can be determined from the measured values of d_{hex} and ϕ_W using the relationships (*see* **ref. 24**)

$$R_W = d_{hex} \cdot \sqrt{\frac{2 \cdot \phi_w}{\pi\sqrt{3}}} \tag{12}$$

$$A_W = \frac{2 \cdot \phi_w \cdot V_1}{(1-\phi_w) \cdot R_W} \tag{13}$$

where V_1 is the volume of a lipid molecule. It is assumed that the value of d_{hex} determines unambiguously the values of R_W and a_W independently of whether the gravimetric or osmotic pressure method is applied. This allows one to determine, based on combination of the gravimetric and osmotic data, the characteristics of the Luzzati plane as functions of the osmotic pressure, $R_W(\Pi)$ and $A_W(\Pi)$ *(29,30)*. The determination is performed by analyzing the monolayer mechanical equilibrium on pressure Π, and using the relationships between the elastic and structural characteristics of different dividing surfaces (*see* above). Description of this procedure in full detail and its application to specific experimental data are described in **refs. 27,28**. Although such treatment is feasible in principle, the attempts to perform it have demonstrated that the accuracy of experimental determination of variations in A_W is not sufficiently high to guarantee reliable estimation of the whole set of parameters characterizing the neutral surface. Below, determination of the elastic parameters of a so-called pivotal plane is described, which only approximately describes the neutral surface but allows for a more exact characterization based on the H_{II} phase dehydration experiments.

4.3. Pivotal Plane as Approximation for the Neutral Surface—Determination of Spontaneous Geometrical Characteristics

The pivotal dividing surface is defined as the dividing surface for which the molecular area A does not change in the course of deformation (**ref. 24** and references therein). As aforementioned, when the monolayer deformation is solely generated by application of τ, the neutral surface keeps its molecular area, and therefore, represents the pivotal plane. In the dehydration experiments, both a bending moment and a compressing lateral stress are applied to the cylindrical monolayers. Hence, the neutral surface must decrease its area and deviate from the pivotal plane. The pivotal plane is a dividing surface for which the area changes related to bending and compression mutually compensate. However, a thorough analysis shows that deviation of the pivotal plane from the neutral surface is relatively small *(24)*, and differences between the structural characteristics related to the neutral and the pivotal surfaces are of the order of parameter $\gamma = \frac{\kappa}{\Gamma \cdot R_s^2}$. The values of this parameter are: $\gamma \approx 0.03-0.1$, and hence, the pivotal plane describes the neutral surface with an accuracy higher than 10%.

Verification of the existence of the pivotal plane and determination of its position with respect to the Luzzati plane can be done based on the pure geometrical relationships. The molecular A and the radius of curvature R at any cylindrical dividing surface, which is separated from the Luzzati plane by a volume V per lipid molecule, are given by

$$A^2 = A_w^2 + 2 \cdot V \cdot \frac{A_w}{R_w} \tag{14}$$

$$R = R_w \cdot \sqrt{1 + \frac{1-\phi_w}{\phi_w} \cdot \frac{V}{V_l}} \tag{15}$$

The relationship (**Eq. 14**) can be rewritten in the form

$$A_w^2 = A^2 + 2 \cdot V \cdot \frac{A_w}{R_w} \tag{16}$$

A plot A_w^2 vs $\frac{A_w}{R_w}$ can serve to make a "diagnosis," i.e., to find out whether the pivotal plane exists. Indeed, this plot is linear if A does not change with dehydration, which means that the corresponding dividing surface is a pivotal plane. The intercept of this linear plot gives the square value of the molecular area at the pivotal surface (A_p), whereas the slope determines the molecular volume (V_p) between the Luzzati and the pivotal planes.

Finding the values of R_W and A_W for the Luzzati plane at full hydration (according to **Eqs. 12** and **13**), determining the position of the pivotal plane by means of the diagnostic plot *(16)*, and finally, calculating R_p and A_p for the pivotal plane using expressions (**Eqs. 15 and 16**), is the way to determine the corresponding spontaneous curvature $J_{sp} = -\frac{1}{R_p}$, and molecular area $A_{sp} = A_p$.

4.4. Determination of the Bending Modulus

Once the pivotal plane is found, osmotic pressure measurements allow for determination of the monolayer bending modulus κ. Based on the expression (**Eq. 2**), taking into account constancy of the area of the pivotal plane, and considering the mechanical equilibrium of a monolayer under pressure, the relationship between the osmotic pressure Π and the radius of the pivotal plane R_p is found *(30,31)*:

$$\Pi \cdot R_p^2 = 2 \cdot \kappa_p \cdot \left(\frac{1}{R_p} - \frac{1}{R_{ps}} \right) \tag{17}$$

From the intercept and the slope of the plot $\Pi \cdot R_p$ vs $\frac{1}{R_p}$ one obtains the spontaneous curvature $-\frac{1}{R_{ps}}$ and the κ_p related to the pivotal plane.

4.5. Discussions and Conclusions

The described protocol for analysis of the X-ray diffraction data has been used during the last decade in studies of H_{II} phases formed by mixtures of dioleoyl-phosphatidyl-ethanolamine with numerous lipids *(5,7,17–22,24,25,28–31,34)*. The major goal was to determine the spontaneous curvatures of physiologically relevant lipids. This could be done because of the fact that the spontaneous curvatures of different lipid species are additive. It has been shown experimentally *(17,20–22,24,25)* and predicted theoretically *(36)* that, in many cases, the total spontaneous curvature of a mixed monolayer is a weighted average of the spontaneous curvatures of the individual lipid components. A table collecting the values of the lipid spontaneous curvature measured to date can be found in **ref. 8**.

Spontaneous curvatures of different lipids vary in a broad range depending on the lipid molecular structure. At the same time, positions of the pivotal planes, and hence, of the neutral surfaces of lipid monolayers proved to be fairly similar for different lipid species and their mixtures. The pivotal surfaces have been always found close to the plane of glycerol backbones of lipid molecules (*see* **refs. 17,20,21,24**). This result can be explained by considering the internal monolayer factors that determine the position of the neutral and pivotal planes. The two planes have to be located in the most rigid region within the monolayer. The narrower this region is, the closer the two surfaces are to each other. Most probably, the local rigidities are distributed nonhomogeneously through the monolayer thickness, and the region of the glycerol backbones is the most rigid one, independently of the monolayer composition. This would determine the locations of the neutral and pivotal surfaces next to each other and to the glycerol backbone.

In conclusion, it has to be mentioned that whereas the described protocol proves to be productive in determining the radii of spontaneous curvatures of lipids, the results obtained for the bending modulus are, in many cases, not accurate enough to find the contributions of individual lipids. Moreover, attempts to determine the values of the stretching modulus give only order of magnitude estimations rather than exact values. Reaching a better quality of the elastic moduli determination requires improved accuracy in the X-ray diffraction measurements.

Acknowledgments

The work of MMK is supported by the Israel Science Foundation (ISF), The Binational USA-Israel Science Foundation (BSF), and Marie Curie Network "Flippases."

References

1. Alberts, B., et al. (2002) *Molecular Biology of the Cell*. Garland, New York.
2. Tanford, C. (1973) *The hydrophobic effect:Formation of Micelles and Biological Membranes*. Wiley & Sons, New York, 200 p.
3. Koynova, R. and Caffrey, M. (1994) *Phases and phase transitions of the hydrated phosphatidylethanolamines*. Chem. Phys. Lipids **69(1),** 1–34.
4. Luzzati, V. (1968) X-ray Diffraction Studies of Lipid-Water Systems, in *Biological Membranes*, (Chapman, D., ed.), Academic Press, New York, pp. 71–123.
5. Gruner, S. M., (1989) *Stability of lyotropic phases with curved interfaces*. J. Phys. Chem. **93,** 7562–7570.
6. Seddon, J. and Templer, R. (1993) Cubic phases of self-assembled amphiphilic aggregates. *Philos. Trans. R. Soc. London A* **344,** 377–401.
7. Rand, R. P. and Fuller, N. L. (1994) Structural dimensions and their changes in a reentrant hexagonal-lamellar transition of phospholipids. *Biophys. J.* **66(6),** 2127–2138.
8. Zimmerberg, J. and Kozlov, M. M. (2006) How proteins produce cellular membrane curvature. *Nat. Rev. Mol. Cell Biol.* advance online publication; published online 15 November 2005 | doi:10.1038/nrm1784.

9. Israelachvili, J. N. (1985) *Intermolecular and surface forces*. Academic Press, London.
10. Seddon, J. M. and Templer, R. H. (1995) Polymorphism of lipid-water systems, in *Structure and Dynamics of Membranes*, (Lipowsky, R. and Sackmann, E., eds.), Elsevier, Amsterdam, pp. 97–160.
11. Engelman, D. M. (2005) Membranes are more mosaic than fluid. *Nature* **438(7068),** 578–580.
12. Gibbs, J. W. (1961) *The scientific papers*. Dover, New York.
13. Markin, V. S. and Kozlov, M. M. (1989) Is it really impermissible to shift the Gibbs dividing surface in he classical theory of capillarity. *Langmuir* **5(4),** 1130–1132.
14. Spivak, M. (1970) A comprehensive introduction to differential geometry. Brandeis University.
15. Vekua, I. N. (1978) *Basics of tensor analysis and theory of covariants*. Nauka, Moscow.
16. Helfrich, W. (1973) *Elastic Properties of Lipid Bilayers: Theory and Possible Experiments*, (Naturforsch, Z., ed.), 28c: pp. 693–703.
17. Chen, Z. and Rand, R. P. (1997) The influence of cholesterol on phospholipid membrane curvature and bending elasticity. *Biophys. J.* **73(1),** 267–276.
18. Chen, Z. and Rand, R. P. (1998) *Comparative study of the effects of several n-alkanes on phospholipid hexagonal phases*. Biophys. J. **74(2 Pt 1),** 944–952.
19. Epand, R. M., Fuller, N., and Rand, R. P. (1996) Role of the position of unsaturation on the phase behavior and intrinsic curvature of phosphatidylethanolamines. *Biophys. J.* **71(4),** 1806–1810.
20. Fuller, N. and Rand, R. P. (2001) The influence of lysolipids on the spontaneous curvature and bending elasticity of phospholipid membranes. *Biophys. J.* **81(1),** 243–254.
21. Fuller, N., Benatti, C. R., and Rand, R. P. (2003) Curvature and bending constants for phosphatidylserine-containing membranes. *Biophys. J.* **85(3),** 1667–1674.
22. Kooijman, E. E., et al. (2005) Spontaneous Curvature of Phosphatidic Acid and Lysophosphatidic Acid. *Biochemistry* **44(6),** 2097–2102.
23. Kozlov, M. M., Leikin, S., and Rand, R. P. (1994) Bending, hydration and void energies quantitatively account for the hexagonal-lamellar-hexagonal reentrant phase transition in dioleoylphosphatidylethanolamine. *Biophys. J.* **67,** 1603–1611.
24. Leikin, S., et al. (1996) Measured effects of diacylglycerol on structural and elastic properties of phospholipid membranes. *Biophys. J.* **71(5),** 2623–2632.
25. Szule, J. A., Fuller, N. L., and Rand, R. P. (2002) The Effects of Acyl Chain Length and Saturation of Diacylglycerols and Phosphatidylcholines on Membrane Monolayer Curvature. *Biophys. J.* **83(2),** 977–984.
26. Kozlov, M. M., Leikin, S. L., and Markin, V. S. (1989) Elastic properties of interfaces. Elasticity moduli and spontaneous geometrical characteristics. *J. Chem. Soc. Faraday Trans.* **2(85),** 277–292.
27. Kozlov, M. M. and Winterhalter, M. (1991) Elastic moduli and neutral surface for strongly curved monolayers. Analysis of experimental results. *J. Phys. II France* **1,** 1085–1100.
28. Kozlov, M. M. and Winterhalter, M. (1991) Elastic moduli for strongly curved monolayers. Position of the neutral surface. *J. Phys. II France* **1,** 1077–1084.
29. Gawrisch, K., et al. (1992) Energetics of a hexagonal-lamellar-hexagonal-phase transition sequence in dioleoylphosphatidylethanolamine membranes. *Biochemistry* **31(11),** 2856–2864.
30. Rand, R. P., et al. (1990) Membrane curvature, lipid segregation, and structural transitions for phospholipids under dual-solvent stress. *Biochemistry* **29(1),** 76–87.
31. Gruner, S. M., Parsegian, V. A., and Rand, R. P. (1986) Directly measured deformation energy of phospholipid HII hexagonal phases. *Faraday Discuss. Chem. Soc.* **81,** 29–37.
32. Turner, D. C. and Gruner, S. M. (1992) X-ray reconstitution of the inverted hexagonal (HII) phase in lipid-water system. *Biochemistry* **31,** 1340–1355.
33. Hamm, M. and Kozlov, M. (1998) Tilt model of inverted amphiphilic mesophases. *Eur. Phys. J. B* **6(4),** 519–528.

34. Ding, L., et al. (2005) Distorted hexagonal phase studied by neutron diffraction: lipid components demixed in a bent monolayer. *Langmuir* **21(1),** 203–210.
35. Luzzati, V. and Husson, F. (1962) X-ray diffraction studies of lipid-water systems. *J. Cell Biol.* **12,** 207–219.
36. Kozlov, M. M. and Helfrich, W. (1992) Effects of a Cosurfactant on the Stretching and Bending Elasticities of a Surfactant Monolayer. *Langmuir* **8,** 2792–2797.

25

Shape Analysis of Giant Vesicles With Fluid Phase Coexistence by Laser Scanning Microscopy to Determine Curvature, Bending Elasticity, and Line Tension

Samuel T. Hess, Manasa V. Gudheti, Michael Mlodzianoski, and Tobias Baumgart

Summary

Membrane shape parameters such as curvature, bending elasticity, and lateral tension, are relevant to the lateral organization and function of biomembranes, and may critically influence the formation of lateral clustering patterns observed in living cells. Fluorescence laser-scanning microscopy can be used to image vesicles and cell membranes, and from shape analysis of these images mechanical membrane parameters can be quantified. Methods to analyze images of equatorial sections obtained by confocal or multiphoton microscopy are detailed, in order to estimate curvature, lateral tension, line tension, relative differences in mean curvature and Gaussian curvature bending moduli, and fluorescence dye intensity profiles, typically within coexisting liquid-ordered and liquid-disordered membrane domains. A variety of shape tracing and shape fitting methods are compared.

Key Words: Bending modulus; cholesterol; curvature; domains; energy minimization; fluorescence imaging; force balance; line tension; lipid phases; rafts; sphingomyelin.

1. Introduction

The mechanisms of lateral and spatial membrane organization and their interdependence remain poorly understood *(1)*. A number of models *(1–6)* attempt to describe the interactions between proteins and lipids to form functional biological structures (rafts) in cell membranes *(7)*, yet consensus is not found on the size, shape, lifetime, and composition of these structures. Despite the tremendous importance and relevance of biological membranes to health and disease *(8–11)*, quantitative descriptions of the mechanics and physical chemistry that govern domain formation in model membranes and living cells demand further attention.

Herein, methods to study these phenomena are described. The effects of pressure differences, membrane curvature, bending elasticity, lateral tension, and phase boundary line tension can be measured, compared with theoretical estimates *(12–14)* and related to membrane lipid composition, phase behavior, and lateral organization. In particular, the interest is in fluid-lipid phases that are proposed to exist in living cell membranes: the liquid-ordered (L_o) and liquid-disordered (L_d) phases *(15)*. In the L_o phase, acyl chains are relatively densely packed, and thus, conformationally constrained. This results in reduced area per head group, increased bilayer thickness *(12,16)*, and increased bending rigidity *(17,18)* but still allows considerable lateral and rotational mobility *(19–21)*. On the other hand, in the L_d phase, acyl chains are less constrained, and thus, have higher area per head group, resulting in a thinner bilayer, less orientational order, decreased bending rigidity, and high lateral mobility. The lateral patterns and out-of-plane morphologies that form in the coexistence of these two phases show a wide range of conformations *(17–19,22–25)*, which reveal mechanics of lipid–lipid (and in more complex systems, lipid–protein and protein–protein) interactions. Thus, studies of such structures and determination of physical shape parameters can be of considerable relevance to the understanding of biological membranes.

2. Biomembrane Models: Giant Unilamellar Vesicles

Giant unilamellar vesicles (GUVs) offer a means to study the forces that control the shapes of membrane bilayers as well as their lateral organization *(17,18,22–24,26–29)*. GUVs do not suffer from the complications introduced by interfacial interactions in solid-supported bilayers and monolayers at the air–water interface, are conveniently prepared *(30)*, and can be easily visualized by light microscopy because of their significantly large size (up to ~100 µm) compared with large and small unilamellar vesicles.

Unilamellar vesicles, in comparison with multilamellar vesicles, are advantageous because the interaction between membrane domains of different phases can be studied. In particular, three-component (ternary) mixtures of a saturated lipid, an unsaturated lipid, and cholesterol (Chol) often exhibit extended regions within the phase diagram of fluid–fluid (L_o–L_d) phase coexistence under a variety of compositions and temperatures. Although this increased number of constituents leads to increased complexity (e.g., compared with two-component mixtures), such three component mixtures are arguably more realistic models of cellular membranes than binary mixtures, and exhibit a surprising number of biologically relevant phenomena and structures even in the absence of protein *(17,18,22)*. **Figure 1** shows an example of a confocal image of a GUV made with a composition (molar ratios) of 1:1:1 sphingomyelin (Spm):*cis*-dioleoylphosphatidylcholine (DOPC):Chol, labeled with fluorescent lipid analogs Bodipy-FL-C_{12}-sphingomyelin and lissamine–rhodamine-B-dioleoylphosphatidylethanolamine (LR-DOPE).

3. Visualization by Laser-Scanning Fluorescence Microscopy

Qualitative characterization of membrane shapes can provide insights into phase behavior and membrane structure, but quantitative analysis is necessary for testing and rejection of some of the contending membrane models. Because of their sensitivity, specificity, and access to a wide range of time-scales (~10^{-11} to >10^3 s) with multiple simultaneous labels, fluorescence methods are well suited to studies of membrane structure. Furthermore, confocal and multiphoton microscopy provide an image, which (under well-chosen conditions) is linearly proportional to the concentration of fluorophore within the sample. Therefore, probe partitioning may also be studied using such linear methods wherein ratios of intensities, corrected for difference in probe brightness, correspond to ratios of concentrations to ratios of concentrations.

Fluorescence microscopy methods, of course, depend on the use of fluorescent probes. Key criteria for such probes include: (1) differential partitioning into specific membrane phases, (2) high brightness (photons per fluorophore per second), (3) minimal perturbation of membrane phase behavior, (4) resistance to photobleaching, (5) excitation spectra easily accessible by common lasers, and (6) nonoverlapping emission spectra to minimize channel overlap (bleed-through). Fluorescence microscopy methods are also usually limited by the optical resolution (equal to a fraction of a wavelength, typically approx 250 nm).

Fluorescence microscopy permits acquisition of images over a wide range of time-scales (from ~10^{-3} s to days). Multiple fluorophores can be used simultaneously (two to four is

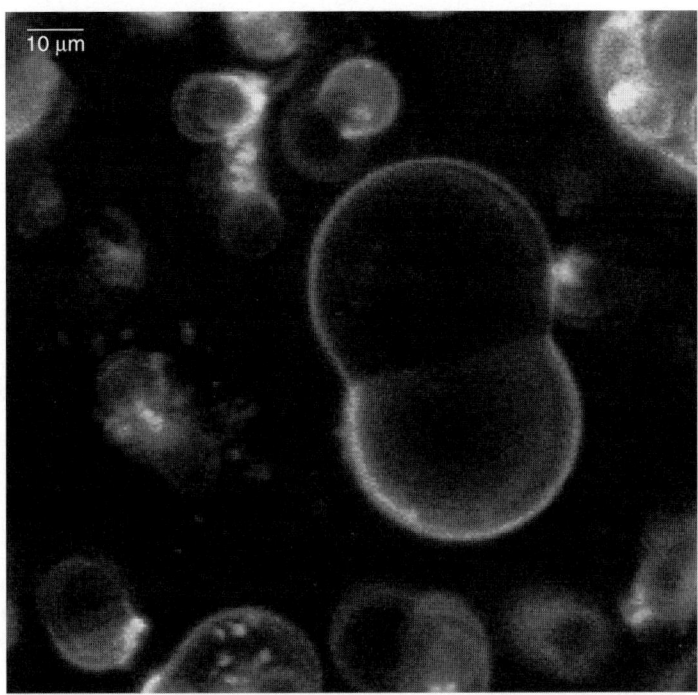

Fig. 1. Confocal z-stack projection of GUVs made with a mixture of egg sphingomyelin, DOPC, and cholesterol, with fluid–fluid (L_o–L_d) phase coexistence at room temperature, labeled with lissamine-rhodamine B-DOPE (LR-DOPE) and Bodipy-FL-C_{12}-sphingomyelin (BFL-Spm). These GUVs were made with a ratio of 1:1:1 Spm: DOPC: Chol by electroformation in 100 mM sucrose and imaged with a Leica TCS-4D confocal microscope using a 60 × 1.35 NA oil objective, fluorescin emission (515–545 nm) and rhodamine emission (590–660 nm) filters, and excitation with 488 and 568 nm lines from a Krypton-Argon laser.

common), allowing colocalization of lipids and proteins to be quantified. However, fluorescence probes must be carefully chosen. The partitioning of fluorescently labeled lipid analogs often differs significantly from that of the unlabeled original lipid. Spm labeled with a large fluorophore attached to the head group, for example, does not partition in the same way as unlabeled Spm, and changes in the head group strongly affect partitioning. For example, Bodipy-FL-C_{12}-sphingomyelin has been shown to partition into both L_o and L_d phases in GUVs made up of 1:1:1 Chol:DOPC:Egg Spm, whereas Bodipy-TR-C_{12}-sphingomyelin under identical conditions partitions almost exclusively into the L_d phase. Furthermore, probe-partitioning behavior appears to be strongly environment dependent. Perylene, for example, in some regions of the Chol:DOPC:Spm phase diagram partitions preferentially into the L_o phase, but in other regions (e.g., wherein the mole fraction of Spm is lower), does not show a strong preference for either phase. Thus, experiments using perylene can be useful in conjunction with another probe, such as LR-DOPE, which partitions strongly into the L_d phase.

The environment-dependence of probe partitioning is a strong indication that lipids and membrane proteins themselves must have strong environment-dependent partitioning. This dependence is very likely to have significant biological consequences. For example, the introduction of a population of signaling lipid molecules such as phosphatidylinositol 4, 5-*bis*-phosphate into the membrane, could result in change in the local partitioning of proteins with PH

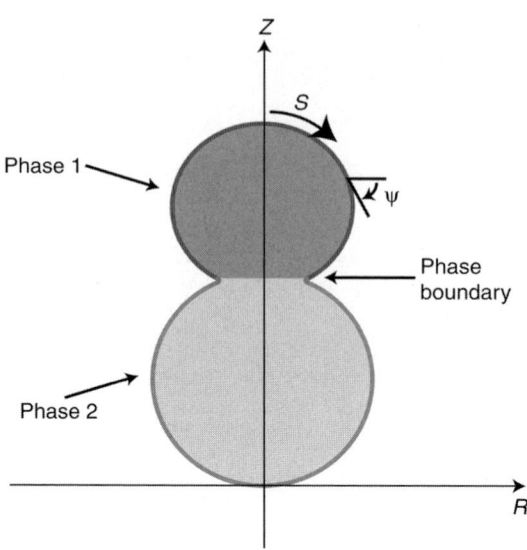

Fig. 2. Equatorial section of an axially symmetric phase-separated GUV. The quantity S is the arc length and ψ is the angle between a tangent to the arc length and the horizontal. Note that the phase boundary and vesicle neck do not necessarily coincide. Differences in Gaussian bending modulus between L_o and L_d phases, or other conditions, can cause the phase boundary to be shifted toward one phase relative to the neck position *(17)*.

domains near the site of phosphatidylinositol 4, 5-*bis*-phosphate localization. This change in local partitioning could then increase the local concentration to a level wherein the rate of cross-phosphorylation of key proteins becomes significant, leading to a downstream cascade. Evidence for environment-dependent partitioning also suggests that cells could produce and benefit from a variety of local membrane environments regulated to have specific lipid compositions. The following sections discuss experimental methods to obtain membrane traces that can be analyzed in terms of probe-partitioning behavior and membrane geometry.

4. Shape Tracing and Analysis of Equatorial Sections

Equatorial membrane sections may be traced using a variety of algorithms to determine mechanical membrane properties as well as extract information on phase coexistence. For the analysis of a confocal or multiphoton fluorescence image of the equatorial plane of a GUV, two such algorithms (the maximum brightness [MB], and the least squares [LSQ] methods) will be discussed.

Shape analysis of membrane equatorial sections relies fundamentally on determination of the position and angle (orientation) of the membrane obtained by analysis of images, typically from a confocal equatorial slice through the vesicle. From this slice, which is rotated to orient the symmetry (Z) axis in the vertical direction, the R-Z coordinates of the membrane as a function of arc length are determined (*see* **Fig. 2**). The angle of the membrane path ψ (measured from the horizontal) and its derivative ψ' as a function of arc length S are then used for numerical analysis of physical shape parameters.

4.1. Tracing in the Direction of Maximum Brightness (MB Method)

The MB algorithm involves following the direction of maximum brightness along the curve of the membrane shape. The algorithm is initiated by manually positioning an marker at some

Fluorescence Microscopy and Membrane Shape Parameters

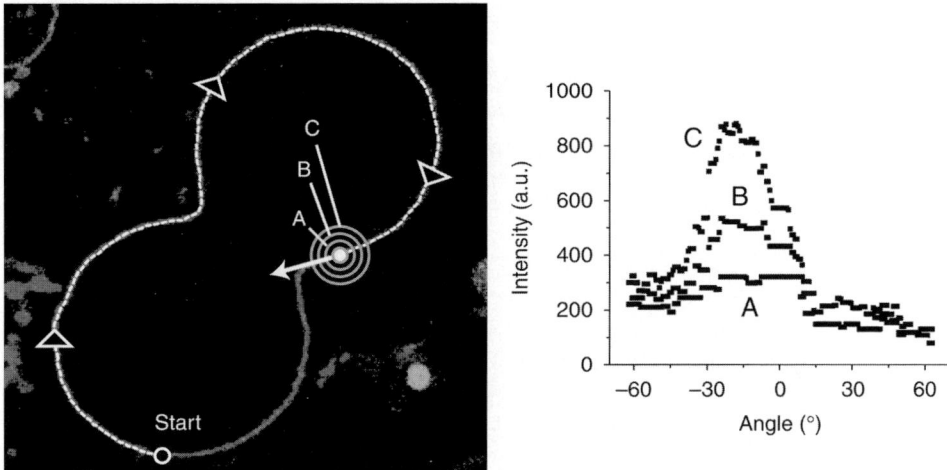

Fig. 3. MB Angular Tracing Algorithm. Shape tracing is accomplished by an automated MATLAB script. A marker (*gray circle with white center*) is initialized manually to a location on the shape (start). Then, the intensity as a function of angle with increasing radius (**curves A,B**) is determined from the pixel values in the image summed up for several radii up to the maximum radius at a given angle (**curve C**). The direction with the maximum intensity (*white arrow*) is then chosen for a small stepwise motion of the marker (and the location and direction recorded). From the new marker location, the process is repeated to trace the entire shape.

point along the membrane shape. Then the intensity profile as a function of angle is determined at several fixed distances from the marker, as shown in **Fig. 3**. The shape is traced by fitting the sum of the intensity profiles at the various radii as a function of angle and determining the direction of maximum intensity turning toward that direction, and then taking a small step. The process is repeated, recording the *x*- and *y*-coordinates, the pixel intensity, the arc length, and the direction angle measured from the *x*-axis. The following section discusses the accuracy of this tracing algorithm.

4.1.1. Estimation of the Uncertainty in Membrane Angles and Curvature Determined by Image Tracing of Equatorial Sections; Determination of Uncertainty by Tracing Ideal Images

Figure 4 shows an example of an ideal image generated from a rasterized circle (representing a membrane 8 pixels wide to avoid pixelization artifacts) blurred by convolution with a Gaussian of width 8 pixels, to which a constant background of pixel value 30 is added. Shot noise is then added to the image assuming a Poisson distribution with the brightest pixel value (i.e., 255) corresponding to $n = \phi \cdot 255$ photons, where ϕ is equal to the number of photons per pixel intensity value. The total number of photons N is then equal to

$$\sum_i n_i = \sum_i \phi p_i$$

where n_i is the number of photons in the *i*-th pixel and p_i is the pixel value in the *i*-th pixel. Note the decrease in noise of $\psi(S)$ and $d\psi/dS$ in **Fig. 5** as the value of ϕ is increased. Thompson

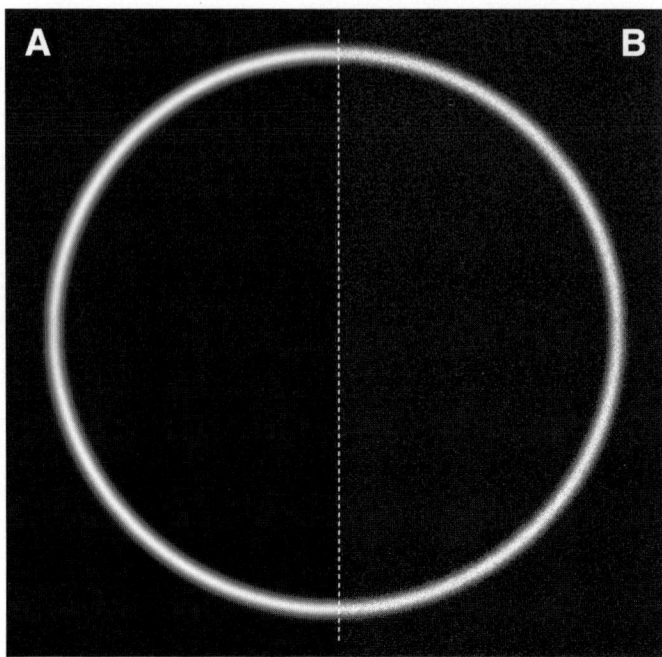

Fig. 4. Theoretical image of a GUV. This image was created using a Matlab program to simulate an image of a spherical GUV with the same visual properties as an image recorded with a confocal microscope. The initial circle has been blurred by convolution with a Gaussian of $1/e^2$ width R_0 ~8 pixels, had background added, and then shot noise applied. (**A**) The image has a photon per pixel intensity value of ϕ ~4.7, or approx 25-fold more total photons compared with (**B**), which has ϕ ~0.19.

et al. *(31)* calculate the maximum localization precision σ_x for objects smaller than the diffraction-limited resolution, localized in one dimension by fluorescence microscopy:

$$\sigma_x^2 = \frac{r_0^2 + q^2/12}{N} + \frac{4\sqrt{\pi}r_0^3 b^2}{qN^2} \qquad (1)$$

where r_0 is the standard deviation of the point spread function, N is the number of photons collected, q is the width of the pixel in the image, and b is the background noise. Using σ_x, the error value associated with the angle (σ_ψ) is estimated from the error in the positions of two points (σ_{x1}, σ_{x2}) if $\bar{x} = \bar{x}_1 - \bar{x}_2$ and the arc length ΔS between the two points in the limit of small $\bar{x}$ is $|\bar{x}| \approx \Delta S$.

$$\sigma_x^2 = \sigma_{x1}^2 + \sigma_{x2}^2 = 2\left[\frac{r_0^2 + q^2/12}{N} + \frac{4\sqrt{\pi}r_0^3 b^2}{qN^2}\right] = 2\rho^2 \qquad (2)$$

where ρ is defined as $\rho = \sqrt{\frac{r_0^2 + q^2/12}{N} + \frac{4\sqrt{\pi}r_0^3 b^2}{qN^2}}$. From the uncertainty σ_x in the position x, the uncertainty σ_ψ in the angle ψ is estimated (for small angles) as

$$\sigma_\psi = \frac{\sigma_x}{\Delta S} = \frac{\rho}{\Delta S}\sqrt{2} \qquad (3)$$

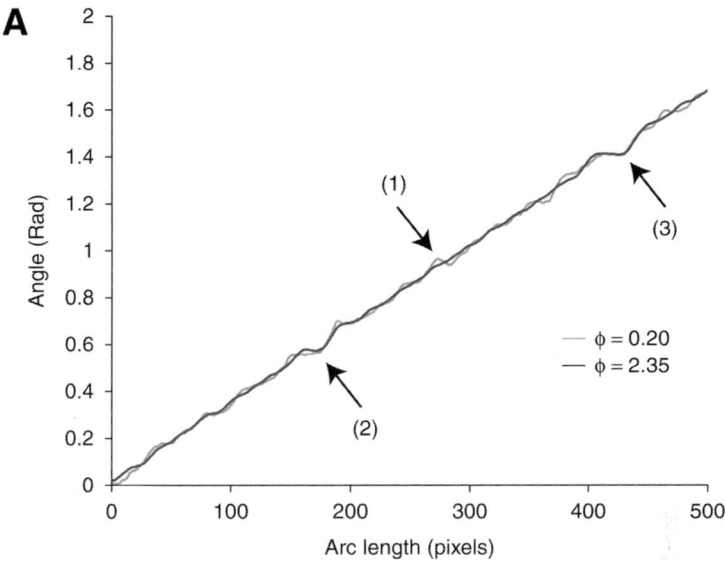

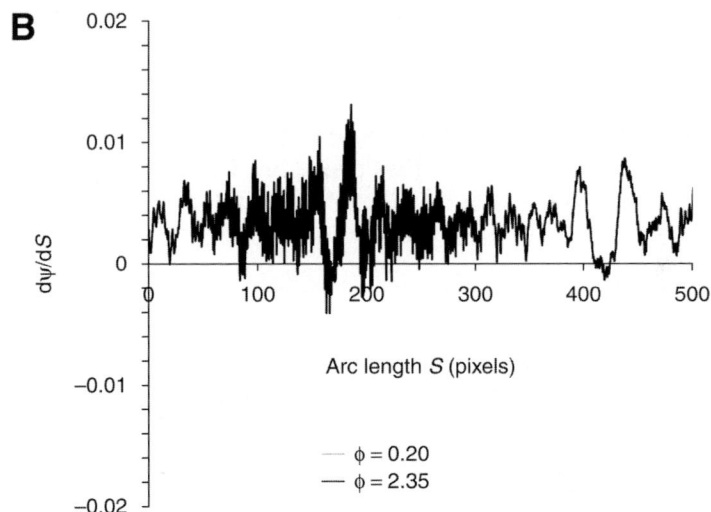

Fig. 5. Effect of image noise on the uncertainty in angles from vesicle traces. (**A**) The resulting angle ψ is shown as a function of arc length S, from a trace of a simulated confocal image of a spherical GUV, for different values of photons per pixel value ϕ = 0.20 and ϕ = 2.35. Because the simulated vesicle in this case is circular, the expected result for ψ vs arc length is a straight line. Deviations from this expected behavior are shown by arrows designating *(1)* an example of noise, which decreased when the ϕ increased, as expected for shot noise, and *(2,3)* noise owing to the finite pixel size, which does not decrease with increasing ϕ. (**B**) The resulting curvature (ψ' = dψ/dS) for both traces, corresponding to the same ϕ as in **A**.

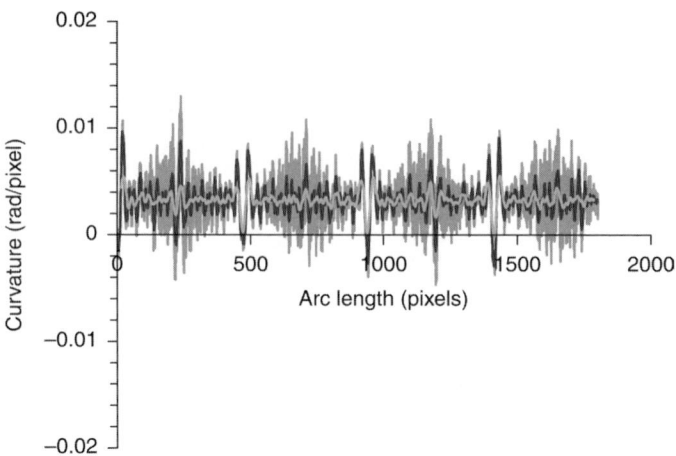

Fig. 6. The effect of smoothing on ψ' as a function of arc length is to reduce noise, in particular noise owing to finite pixel size, which is hard to eliminate for a finite-sized image even with large numbers of line or frame averages. Shown are four complete successive traces of the same circular simulated vesicle whose trace is also shown in **Fig. 5**. Repeated tracing of the same shape provides a measure of reproducibility (uncertainty) of the tracing algorithm. However, smoothing also reduces the information content of $\psi'(S)$, particularly at high spatial frequencies such as occur at the neck of a sharply pinched shape. Unsmoothed $\psi'(S)$ (medium gray), and smoothed curves with box size 20 (black) and 100 (light gray).

From the values of ψ one can calculate numerically the derivative $d\psi/dS$, or ψ', which is approximately equal to $\Delta\psi/\Delta S$ for small step size ΔS. **Figures 5** and **6** show the ψ and ψ' determined from the trace of a vesicle identical to that shown in **Fig. 4** except with $\phi = 0.20$ and $\phi = 2.35$. For a given value of ΔS, this will result in an estimated uncertainty in $\psi' = \dfrac{d\psi}{dS}$ of

$$\sigma_{\psi'} = \frac{\sigma_\psi}{\Delta S} = \frac{\sqrt{2}}{(\Delta S)^2}\sqrt{\frac{r_0^2 + q^2/12}{N} + \frac{4\sqrt{\pi}r_0^3 b^2}{qN^2}} \qquad (4)$$

Note that this expression is expected to decrease monotonically as a function of N for all finite N and fixed b. However, in the situation where $\beta = b/N$ is constant (the ratio of background b to total photons N), the expression will approach a limiting value as N becomes large. Such a situation could occur, for example, when the number of emitted signal and background photons is constant but the detection efficiency is changed.

Figure 7 shows the mean squared uncertainty in the curvature $\sigma_{\psi'}$ as determined from traces of a simulated circular vesicle with varying (inverse) numbers of photons. The value of $\sigma_{\psi'}$ does not appear to continue to decrease indefinitely with increasing N, but instead approaches a limiting value of $\sigma_{\psi'} \approx \dfrac{\beta}{(\Delta S)^2}\sqrt{\dfrac{8\sqrt{\pi}r_0^3}{q}}$ at large N (i.e., large ϕ), which is consistent with predictions by the model of Thompson et al., but may also include other sources of uncertainties such as from the algorithm itself and differences owing to the one-dimension geometry of the membrane (compared with 2D localization of a spot). Clearly, localization precision will improve as the number of collected signal photons increase, but will only help

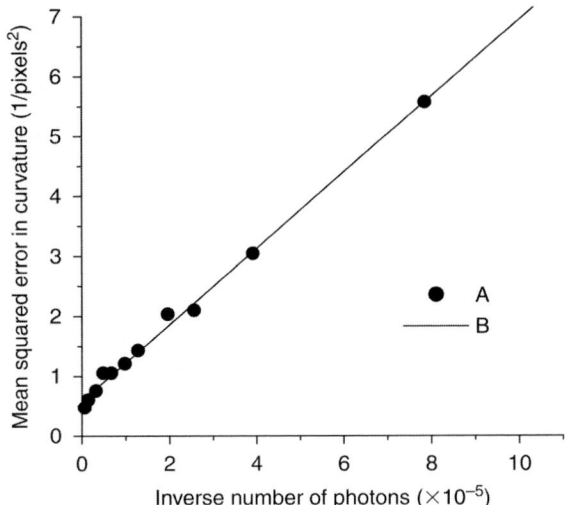

Fig. 7. Effect of number of photons on image tracing accuracy. (**A**) The value of $\sigma_{\psi'}^2$ equal to the $1/e^2$ width squared of the distribution of curvature values obtained from simulated images, is plotted vs the inverse of N, where N is total number of photons within the region used to estimate that curvature. (**B**) The $\sigma_{\psi'}^2$ obtained from simulated images was well described by the model of Thompson et al. (*see* **ref. *31***) for 2D localization extended here to membrane tracing. The straight line was generated using the actual ratio of background pixel value ($b = 10$) to total photons N ($\beta = b/N = 4.5 \times 10^{-4}$), the pixel size ($a = 1$), the effective arc length per step ($\Delta S = 4$), the total number of photons N, and a value for the resolution ($r_0 = 6.4$ pixels) close to but different from the ideal resolution (8 pixels). For increasing values of N, there is significant improvement in the precision of curvature values determined by the MB algorithm, up to a limiting value (given by the nonzero intercept as $1/N$ goes to zero). For higher values of N (smaller values of $1/N$), $\sigma_{\psi'}^2$ decreases (and the precision in the curvature increases) less rapidly with N, because of the increased relative contribution of background in these images, which have a constant signal-to-background ratio.

improve the precision in curvature determination (using this algorithm) up to a point that is ultimately limited by signal-to-background ratio. A decrease in the number of microns per pixel (q) will also be necessary to achieve optimal tracing, but this will have the counterproductive effect of spreading the acquired photons out over a larger number of pixels, increasing the fractional uncertainty in the intensity in each pixel. Additionally, smoothing of the measured ψ' can improve the signal-to-noise ratio, whereas adversely affecting the upper limit of resolvable curvature. **Figure 6** shows various boxcar smoothing of the profile of ψ' obtained from **Fig. 4**.

4.2. LSQ Algorithm

Comparison of the MB algorithm and the LSQ algorithm described below, shows that the MB algorithm is particularly sensitive to photon noise and background noise in the image. Thus, acquisition of images that have been averaged over a significant number of frames is necessary to obtain as smooth an image as possible. **Figure 8** explains the LSQ method and shows an example trace for the same simulated spherical vesicle image traced using the MB algorithm discussed previously. The basis of the LSQ algorithm is to use each pixel in the image as a data point, weighted by its uncertainty (estimated as the square root of the number of photons in that pixel), and then fitted by a polynomial (typically of second order) through the data points.

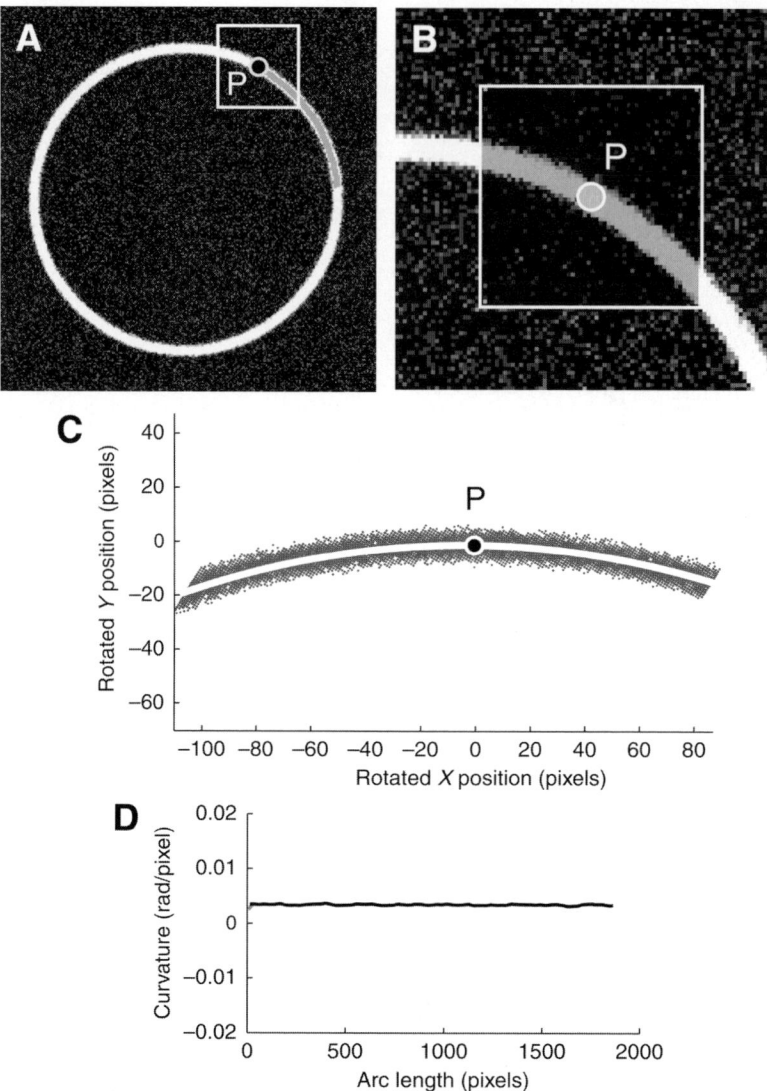

Fig. 8. LSQ vesicle tracing algorithm. The image of an equatorial section of the vesicle is traced by treating the image points as data points, which can be fitted locally with a function $y(x)$. **(A)** Example of a simulated confocal equatorial section of a spherical vesicle with radius 300 pixels, approx 50 photons detected in the brightest pixels, realistic background, and shot noise applied appropriate to the number of photons. The trace (gray line) shown up to the current point (*P*, black spot with white edge) is calculated by fitting the pixels in the image locally around the point (those within the gray box), each pixel weighted according to the square root of its intensity; **(B)** zoom of image region being fitted. Next, the image within the box is rotated such that the direction along the membrane at P is horizontal (*X*) and the direction perpendicular to the membrane at P is vertical (*Y*). The rotated *X* and *Y* coordinates for the image shown in the box in **A** and **B** are shown (dark gray points), with LSQ second-order polynomial fit (white line) $y = \alpha x^2 + \beta x + \gamma$ in **C**. Note that the size of each point is equal to the number of photons at the corresponding pixel in the image. The fit is then used to calculate the direction of a step along the membrane, and the value of -2α is equal to the meridional curvature of the membrane at P, shown in **D**. Results from several repeated traces around the vesicle are shown in **D** to demonstrate the reproducibility of the method.

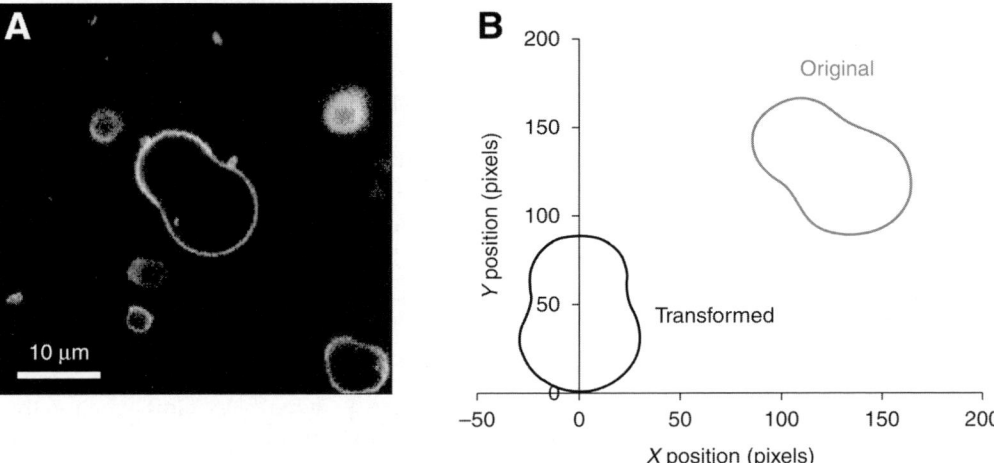

Fig. 9. **(A)** Example of a GUV made up of egg Spm, DOPC, and cholesterol in the mixture 1:(0.5, 0.5):1 Spm:(*trans*-DOPC: *cis*-DOPC):Chol, created by incubation at 60°C in aqueous 100 mM sucrose, and the fluorescent labels LR-DOPE and bodipy-FL-Spm. This 256 × 256 pixel image is the line-average of 32 successive frames imaged on a Leica TCS-4D confocal microscope a using 60 × 1.35 NA oil objective, fluorescin emission (515–545 nm) and rhodamine emission (590–660 nm) filters, and excitation with 488 and 568 nm lines from a Krypton-Argon laser. The total image size was 50.3 × 50.3 µm^2 and imaged at 25°C (stage temperature). **(B)** Rotation and transformation of vesicle trace to center and align the equatorial section of the vesicle with the longitudinal axis of symmetry (in this case the *y*-axis). Intersection of the trace with the *y*-axis should be at a 90° angle.

From the fit, the position and curvature of the membrane can be estimated (for further detail, *see* **Fig. 8**). **Figure 8** also shows that the LSQ algorithm is able to extract the local curvature with improved certainty compared with the MB algorithm: the uncertainties in curvatures in **Fig. 8** (LSQ algorithm, unsmoothed) are much smaller for the same image than those shown in **Fig. 5** (MB algorithm), even those which were boxcar smoothed shown in **Fig. 6** (black and lightgray). Thus, because of the sensitivity of the MB algorithm to typical levels of image noise, particularly on short-length scales, the LSQ algorithm is better suited to portions of the vesicle shape with high curvature, or wherein greater precision in curvature is needed.

4.3. Determination of Area Fractions of Lo and Ld Phases for an Equatorial Section of a Phase-Separated GUV

Area fractions of L_o and L_d phases can be determined directly from an equatorial section of an axially symmetric phase-separated GUV. First, the RGB image (shown in **Fig. 9A**) is converted to a grayscale image. This grayscale image is traced by the LSQ method using a program written in MATLAB (TheMathworks, Inc. Natick, MA). The starting pixel values for the tracing routine are chosen manually by selecting a coordinate on the vesicle wherein the membrane path is horizontal ($\psi = 0$). Some of the parameters that can be controlled in MATLAB to optimize the trace (**Fig. 9B**) are the threshold value, the step size, and the maximum turn rate. The output of the shape tracing routine provides *x*- and *y*-coordinates, membrane path angles, and pixel intensities (for each channel). The output from the MATLAB tracing routine is compiled in Microsoft Excel and the transformations below are performed to align the shape to be symmetric with the *y*-axis:

$$x_1 = \alpha x \cos\theta - \beta y \sin\theta + x_0 \quad (5A)$$

$$y_1 = \alpha x \sin\theta + \beta y \cos\theta + y_0 \quad (5B)$$

where x_1 and y_1 are the transformed *x*- and *y*-coordinates, α and β are *x* and *y* stretch factors, x_0 and y_0 are *x* and *y* offsets, and θ is the rotation angle. The transformation results in a rotated translated shape, shown in **Fig. 9B**. After these transformations, the equatorial section should resemble **Fig. 2**. Each side of the symmetric section is analyzed separately. The equations listed below are used for computing the area fractions. The differential arc length (Δs_i) for the *i*-th step is determined using $\Delta s_i = \sqrt{(\Delta x_i)^2 + (\Delta y_i)^2}$ where Δx_i and Δy_i are the displacements in *x*- and *y*-coordinates, respectively. Thus, the differential area (ΔA_i) for the *i*-th step in the trace is computed using $\Delta A_i = 2\pi x_i \Delta s_i$, and summed over the entire surface to yield the total area $A_t = \sum \Delta A_i$. Then, the location of the phase boundary is determined by analysis of the intensity as a function of arc length (*see* above), or by manual estimation using image processing software. The area of phase 1 (A_1) is calculated by summing up to the phase boundary and the area of phase 2 (A_2) is obtained from $A_2 = A_t - A_1$, namely, the difference between the total area A_t and the area of the first phase A_1, where $A_1 = \sum_0^b \Delta A_i$, where *b* is the location of the phase boundary.

4.4. Analysis of Dye Partitioning in GUVs by Quantification of Fluorescence Intensity

As equatorial sections are traced, the fluorescence intensity can be quantified either at each pixel of the trace or averaged for all pixels within a region centered on the given pixel of the trace. The intensity in both channels is quantified separately and plotted as a function of arc length. **Figure 10** shows the results of an intensity trace of the same GUV image from **Fig. 9**.

The spatial profile of the two dye intensities across the phase boundary can be analyzed for several pieces of information. First, the ratio well inside the phases (after background subtraction) is related to the partition coefficient of the two dyes in the two phases: the LR-DOPE showed 43 ± 27-fold higher fluorescence in the L_d phase, whereas Bodipy-FL-Spm showed 1.9 ± 0.6-fold ratio of L_d/L_o fluorescence. Knowing the relative brightness of the probes in these environments, the partition coefficient can be determined. Fluorescence correlation spectroscopy (FCS) is one method, which can in principle be used to determine such concentration and brightness ratios simultaneously. Second, one can extract the position S_{50}, the value of the arc length wherein the dye intensity is halfway between its value well inside the L_o and its value well inside the L_d phase. Differences in S_{50} for different probes may result from an extended phase boundary, differences in dye partitioning behavior, and proximity to a critical point. Third, resolution permitting, in some cases it is possible to determine the profile of the fluorescence intensity across the phase boundary. In many cases the phase boundary will be too sharp to resolve such a profile by light microscopy, because the (spatially narrow) profile will be convolved with a much broader optical point spread function.

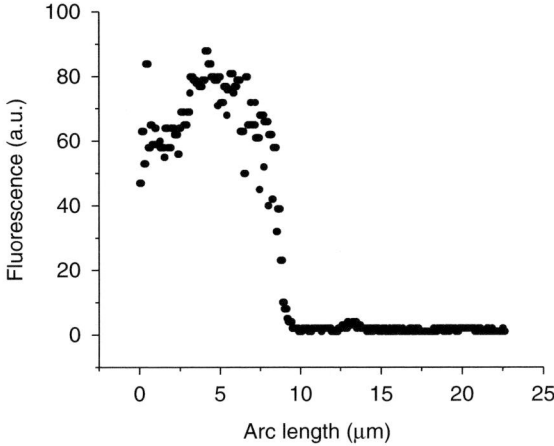

Fig. 10. Plot of fluorescence intensity as a function of arc length S for the GUV shown in **Fig. 9**. The red channel (LR-DOPE, a L_d probe) is shown as black circles. The position (as a function of arc length S) at which the fluorescence intensity drops to halfway between the mean in the one phase and the mean in the other phase is defined as S_{50}, in this case $S = 8.43 \pm 0.04$ μm.

5. Shape Analysis of Vesicle Traces
5.1. Shape Fitting Using First Integral of Shape Equations

The analysis is based on determination of the properties that minimize the free energy of the membrane shape. The total free energy of the membrane can be expressed as a sum of contributions *(32)* from bending energy, inner–outer pressure difference (P), lateral tension (Σ), and phase boundary line tension (σ):

$$F = PV + \sum_{i=1}^{n} F_b^{(i)} + \Sigma^{(i)} A^{(i)} + \sum_{j=1}^{m} \sigma^{(j)} \ell^{(j)} \tag{6}$$

summed over each domain *i* and each phase boundary *j*, where V is the total vesicle volume, $A^{(i)}$ is the area of the *i*-th domain, and $\ell^{(j)}$ is the perimeter of the *j*-th phase boundary. The bending energy terms can be formulated as an integral over the vesicle area:

$$F_b^{(i)} = \frac{\kappa^{(i)}}{2} \int_{A^{(i)}} (C_1 + C_2 - C_0)^2 \, dA + \kappa_G^{(i)} \int_{A^{(i)}} C_1 C_2 \, dA \tag{7}$$

and combined with the other terms to yield

$$F = PV + \sum_{i=1}^{n} \int_{A^{(i)}} \left[\frac{\kappa^{(i)}}{2} (C_1 + C_2 - C_0^{(i)})^2 + \kappa_G^{(i)} C_1 C_2 + \Sigma^{(i)} \right] dA^{(i)} + \sum_{j=1}^{m} \sigma^{(j)} \ell^{(j)} \tag{8}$$

where C_1 and C_2 are the local radii of curvature of the vesicle around the principal and meridional axes, respectively, $\kappa^{(i)}$ is the mean curvature bending modulus, $\kappa_G^{(i)}$ is the Gaussian curvature bending modulus, and $C_0^{(i)}$ is the spontaneous curvature, for the *i*-th domain, respectively.

Now shapes with cylindrical (axial) symmetry are considered, which allow to parameterize the vesicle shape in its equatorial plane, along the coordinates of the cylindrical axis Z and radial axis R, the arc length S along the shape in the equatorial plane, and the angle ψ of the path of the shape as measured from the R axis (*see* **Fig. 2**). Within this parameterization, Jülicher and Lipowsky *(32)* show that the total energy can be written for a vesicle with cylindrical symmetry and two-phase domains as

$$F = F_{\text{boundary}} + \int_{A^{(i)}} L dt = \sigma R_{12} + (\kappa_G^{(1)} - \kappa_G^{(2)})\cos\psi_{12}$$
$$+ \sum_{i=1}^{2} 2\pi \int_{A^{(i)}} \left[\frac{\kappa^{(i)}}{2} R \frac{dS}{dt} \left(\frac{d\psi/dt}{dS/dt} + \frac{\sin\psi}{R} - C_0^{(i)} \right)^2 + \Sigma^{(i)} R \frac{dS}{dt} + \frac{1}{2} PR^2 \frac{dS}{dt} \sin\psi + \gamma \left(\frac{dR}{dt} - \frac{dS}{dt}\cos\psi \right) \right] dt \quad (9)$$

where the value of F (other than the boundary energy terms F_{boundary}) is calculated by integrating the Lagrangian-like quantity L as a function of t, where t is the independent variable with respect to which R, S, and ψ have been parameterized. At the boundary between phases 1 and 2, R_{12} and ψ_{12} are the radius and angle of the path of the shape, respectively. The parameter γ is a Lagrange multiplier introduced to force the constraint that $\frac{dR}{dt} = \frac{dS}{dt}\cos\psi$, which is a direct consequence of geometry.

The problem can now be solved by variational methods. The free energy has been written as an integral over the shape plus two terms evaluated at the phase boundary, which must be minimized. In classical mechanics, the action or the integral of the Lagrangian (L) over time, is minimized, and typically, a number of relationships between derivatives of L with respect to time and spatial coordinates are obtained. Here, derivatives of F with respect to the coordinates directly follow as a result of $\frac{\partial L}{\partial X} - \frac{d}{dt}\left(\frac{\partial L}{\partial \dot{X}}\right) = 0$, where X is any one of the three coordinates R, ψ, or S, and $\dot{X}$ denotes the derivative of X with respect to t. These derivatives lead to a set of coupled ordinary differential equations, which can be solved simultaneously to calculate membrane shapes given a known set of shape parameters (i.e., given known values for the bending moduli, the line tension, the pressure difference, and lateral tension). However, in order to extract these same parameters from a vesicle of known shape, one can exploit another important consequence of the minimization of the action-like quantity F. As in classical mechanics the Hamiltonian (kinetic plus potential energy) of a closed system in an inertial frame is a conserved quantity, in this system the analogous quantity

$$H = \sum_{i=1}^{2} H^{(i)} = \sum_{i=1}^{2} \frac{\kappa^{(i)} R}{2}\left[(\psi')^2 - \left(\frac{\sin\psi}{R} - C_0^{(i)}\right)^2\right] - \Sigma^{(i)} R - \frac{P}{2} R^2 \sin\psi + \gamma\cos\psi \quad (10)$$

is conserved. In fact, because H is constant at every point along the equatorial vesicle path, it can be evaluated at the poles together with γ to show that $H(t) = H^{(1)} = H^{(2)} = 0$ and that $\gamma = 0$ at both the north and south poles of the shape. Elsewhere, resulting from derivatives of F with respect to γ, Jülicher and Lipowsky find that

$$\gamma' = \frac{\kappa^{(i)}}{2}(\psi' - C_0^{(i)})^2 - \frac{\kappa^{(i)} \sin^2\psi}{2R^2} + \Sigma^{(i)} + PR\sin\psi \quad (11)$$

Additionally, the following phase boundary jump condition

$$\gamma(S+\varepsilon) = \gamma(S-\varepsilon) + \sigma \quad (12)$$

can be used directly to determine the line tension, given $\gamma(S)$. These results allow to determine the membrane parameters that give the minimum shape energy. Specifically, the set of parameters that results in $H = 0$ (or as near to zero as possible) at all points along the shape are the parameters that yield the lowest shape energy.

Vesicles are imaged by laser-scanning microscopy, and then the image of the equatorial section of the vesicle is traced to find R, ψ, and S. Because H contains only the quantity γ that must be obtained by integration of γ', and because γ' depends only on the shape and the shape parameters, one can iteratively guess a set of parameters, calculate γ', and integrate with respect to S to get γ. Then, determine H, and test whether that set of parameters gives the smallest possible value (value closest to zero) for H everywhere along the path. The parameters are thus adjusted, so that the sum of $[H(S)]^2$ summed over each point in the vesicle is minimized. The parameters, which can in principle be considered are $\kappa^{(i)}, \kappa_G^{(i)}, \Sigma^{(i)}, P, C_0^{(i)}$, and σ, but the overall energy is arbitrary within a multiplicative constant equal to one of the bending moduli. Therefore, shapes are fitted with the value of $\kappa^{(1)}$ fixed to unity, and values of fitting parameters relative to one bending modulus are quoted (typically the bending modulus of the L_d phase, which has been estimated experimentally for a few pure lipidic systems).

As primarily bilayers made up of phospholipids and Chol are being considered, and the rate of Chol flip–flop is high, effects because of differences in number densities of lipids in the opposing monolayer leaflets can be neglected (*32,33*); spontaneous curvatures are also typically neglected, although in principle may be included. This procedure is accomplished using a computer program written in MATLAB. Column vectors containing S, R, ψ, and $d\psi/dS$ are determined from shape tracing algorithms (*see* above), and used as inputs to the program along with initial guesses at the shape parameters and the location of the phase boundary. The routine then searches using the fminsearch function for the minimum sum of $[H(S)]^2$ evaluated at all points along the equatorial section. This method has the advantage that it starts with the experimentally observed shape, which by definition incorporates the constraints on the total area, area fraction, and total volume for the shape.

5.2. Shape Fitting Through Full Integration of Differential Shape Equations

It was shown in the previous section that line tension estimates, lateral tension, normal pressure difference, and the relative bending stiffness in vesicles with fluid phase coexistence can be obtained from adjusting fit parameters in a conserved zero value Hamiltonian-like function of the experimentally obtained vesicle geometry, in terms of radius R, tangent angle ψ, and derivative of tangent angle with respect to arclength ψ'. An advantage of this approach is that a complete integration of differential shape equations to yield a fitted vesicle shape is not necessary, which minimizes computational efforts. A disadvantage of the Hamiltonian fit method is the high uncertainty of fit parameters that appear only in the jump conditions of the shape equations at the phase boundary (but not in the bulk differential shape equations). These parameters include the relative mean curvature bending stiffness, as well as the relative Gauss curvature stiffness difference.

The comparison of fully integrated theoretical vesicle shapes with experimentally obtained vesicle shapes can yield estimates for both mean curvature and Gauss curvature differences

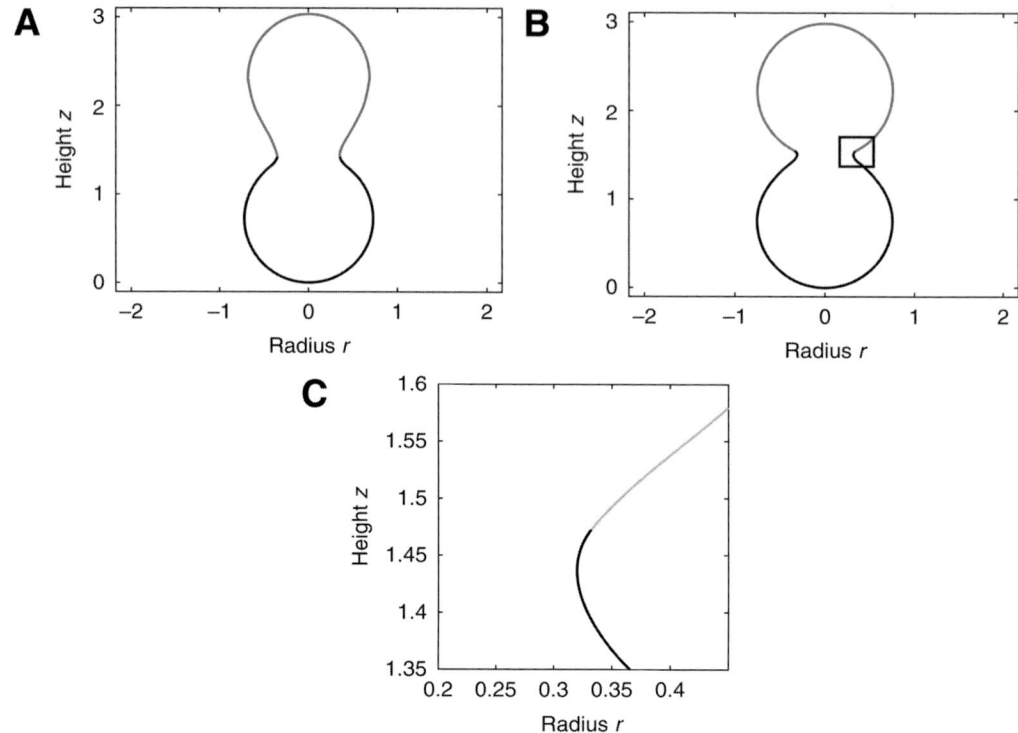

Fig. 11. Vesicles with fixed line tension, deformed by differing mechanical effects. All vesicles have the same volume of 76% of the volume of a sphere. Furthermore, all vesicles have the area of a sphere with unit radius. **(A)** The mean curvature stiffness of the gray phase is a factor of 100 higher compared with the black phase, but there is no difference in Gaussian bending stiffness. **(B)** The Gaussian bending stiffness of the black phase is higher [$\hat{\Delta}_g = \left(\kappa_g^{black} - \kappa_g^{gray}\right)/\kappa_g^{black} = 3$] than the gray phase, but the two phases have equal mean curvature stiffness. **(C)** Inset of the right phase boundary (vesicle neck) of the vesicle shown in **B**.

between fluid-ordered and -disordered phases *(17)*. This finding follows from the observation that differing mean and Gauss curvature moduli have characteristically differing effects on experimental vesicle shapes. **Figure 11A** shows a vesicle with equal area fraction and Gaussian bending stiffness, but a much higher mean curvature modulus in the gray membrane phase versus the black phase. The membrane phase with the higher bending stiffness forms a droplet or pear shape, whereas the phase with smaller bending stiffness shows the shape of a truncated sphere. In contrast to differing mean curvature moduli, a difference in Gaussian bending stiffness leads to a vesicle deformation that is localized to the phase boundary.

Figure 11 indicates that the phase with the smaller resistance toward Gauss curvature forms the high curvature neck, i.e., the phase boundary is shifted out of the neck. This phenomenon was theoretically described by Jülicher et al. *(34)*. Furthermore, increasing line tension leads to increase of curvature at the phase boundary of a phase separated vesicle, as indicated in **Fig. 12**, wherein a vesicle with fixed reduced volume of 0.76 is successively deformed by increasing line tension. Deformation leads to a limit shape that is described by truncated spheres with constant curvature, connected at the phase boundary.

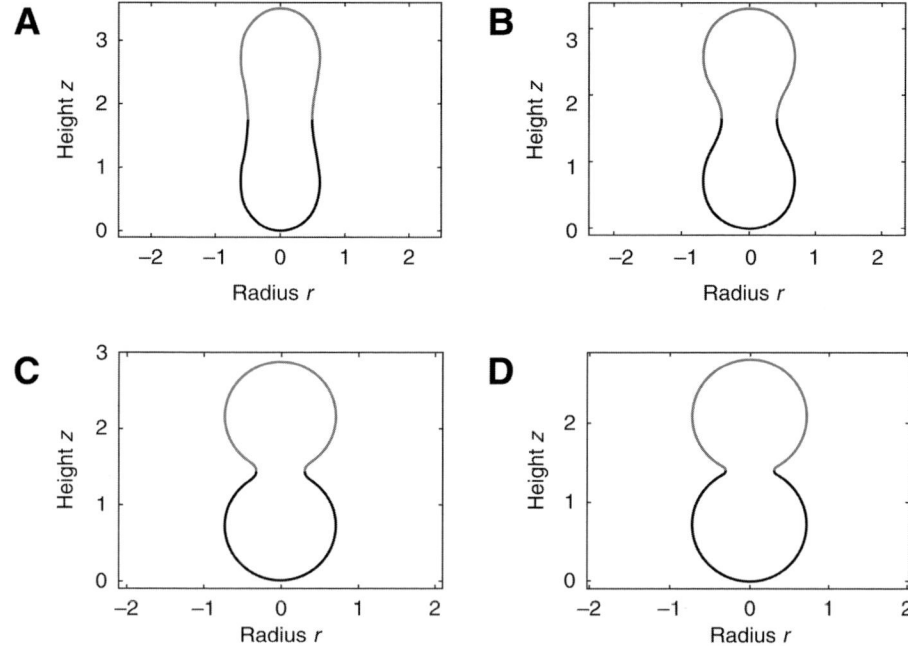

Fig. 12. Vesicles with the same bending stiffness (equal mean and Gauss curvature moduli), equal volume, and an area equal to a unit sphere, but increasing line tension. From **A–D** the line tension increases with values of 0, 2, 50, and 500, respectively. In all cases, the vesicle volume had a constant value of 76% from the volume of a sphere with the area of a unit sphere.

The following section describes, from a mechanical point of view, the terminology and shape equations that allow comparison between experimental and theoretical vesicle shapes. The shape equations obtained from balancing forces and moments in membrane area elements are equivalent to the shape equations described previously, which are derived from minimizing the global mechanical energy. The out-of-plane bulk force balance equation can be expressed as *(35–37)*

$$Q' + \frac{\cos\psi}{R}Q - T_m\psi' - \frac{\sin\psi}{R}T_p - p = 0 \qquad (13)$$

where Q is the transverse (i.e., along the surface normal) shear per unit length acting on an edge along the parallels, a prime indicates a derivative with respect to arclength, T_m and T_p are lateral stress components along the meridian and parallels, respectively, and p is a pressure difference across the membrane acting along the *inward* surface normal direction (i.e., p is an *outer excess* pressure). The in-plane balance of forces can be written as

$$T_m' + \frac{\cos\psi}{R}(T_m - T_p) + Q\psi' = 0 \qquad (14)$$

The jump conditions originally derived by Jülicher and Lipowsky *(32)* can be expressed as *(17)*

$$T_{mL_d} - T_{mL_o} - \sigma\frac{\cos\psi}{R} = 0, \qquad (15)$$

for the jump in lateral stress, where L_d indicates the value before and L_o the value after the phase jump (in the direction away from the north pole of the vesicle), respectively, in a vesicle with L_o phase oriented toward the north pole. The jump condition for transverse shear is

$$Q_{L_d} - Q_{L_o} - \sigma \frac{\sin \psi}{R} = 0 \tag{16}$$

Accordingly, the jump in lateral stress is equal to line tension multiplied by the geodesic curvature ($\cos \psi/R$) along the boundary, whereas the jump in transverse shear is the line tension multiplied by the principal curvature along the boundary. It can be shown that tensions T_m and T_p and transverse shear Q are related to curvature, mean curvature bending resistance, κ, and mean tension, d, in the membrane *(38)* by $T_m = -[d - 2\kappa h(\psi' + h)]$, $T_p = -[d + 2\kappa h(\psi' + h)]$, and $Q = -2\kappa h'$. The Gaussian curvature resistance modulus κ_G enters the condition for zero jump in moments across the boundary, $M_{mL_d} - M_{mL_o} = 0$, through the constitutive equation $M_m = -(2\kappa h - \kappa_G \sin \psi / R)$ *(32,38)*.

To solve the system of differential shape equations for a membrane with coexisting phases, all mechanical parameters are expressed relative to a given bending stiffness of one of the two coexisting phases *(see above) (32)*: $\varepsilon = \kappa^{L_o}/\kappa^{L_d}$ is the ratio between the mean curvature bending rigidities of the two regions, $\hat{\Delta}_g = \left(\kappa_g^{L_d} - \kappa_g^{L_o}\right)/\kappa^{L_d}$ provides a measure of the difference in Gaussian curvature rigidities between the two regions, and dimensionless transverse shear, line tension, mean lateral tension, and pressure, respectively are

$$\hat{Q} = \frac{QR_0^2}{\kappa^{L_d}}, \quad \hat{\sigma} = \frac{\sigma R_0}{\kappa^{L_d}}, \quad \hat{d} = \frac{dR_0^2}{\kappa^{L_d}}, \quad \text{and} \quad \hat{p} = \frac{pR_0^3}{\kappa^{L_d}}, \tag{17}$$

where R_0 is the radius of a spherical membrane (vesicle) with the same area as the particular deformed vesicle. All further quantities bearing the dimension of length are nondimensionalized by R_0 as well. The shape equations are simultaneously solved in the L_o and L_d regions of the membrane subject to appropriate boundary conditions.

The six quantities $\hat{Q}$, h, ψ, R, z, and $\hat{d}$, are chosen as dependent variables, where z is the vesicle height *(see* **Fig. 11**), whereas the arclength s is the independent variable. For each phase of a membrane with domains, a set of six coupled first order differential equations is obtained. Two of these are derived from **Eqs. 14** and **15**, and read for the L_o region (i.e., the region before the phase jump, viewed from the north pole of the vesicle)

$$\hat{Q}' = \frac{\cos \psi}{R}\hat{Q} + 2h\left[\hat{d} + \varepsilon h^2 + \varepsilon \frac{\sin \psi}{R}\left(2h + \frac{\sin \psi}{R}\right)\right] + \hat{p} \tag{18}$$

and

$$h' = -\hat{Q}/\varepsilon. \tag{19}$$

The four remaining equations are $\psi' = -2h - \sin \psi/R$,

$$R' = \cos \psi, \quad z' = \sin \psi, \quad \text{and} \quad \hat{d}' = 0.$$

The six equations for the L_d region are identical, except that the parameter ε in **Eqs. 18** and **19** does not appear (because of scaling with respect to κ^{L_d}). The 12 boundary conditions at north pole ($s = 0$), phase discontinuity ($s = s^*$), and south pole ($s = s_e$) are $Q_s(0) = Q_s(s_e) = 0$, $\psi(0) = 0$, $\psi(s_e) = \pi$, $r(0) = 0$, $z(0) = 0$, the continuity equations $\psi(s*^{L_o}) = \psi(s*^{L_d})$, $R(s*^{L_o}) = R(s*^{L_d})$, $z(s*^{L_o}) = z(s*^{L_d})$, and finally, the three jump conditions obtained from Eqs. 15 and 16, and $M_{mL_d} - M_{mL_o} = 0$:

$$\hat{Q}_{L_d} - \hat{Q}_{L_o} - \hat{\sigma}\frac{\sin\psi}{R} = 0, \tag{20}$$

$$\hat{d}^{L_d} - \hat{d}^{L_o} + h^{L_d 2} - \varepsilon h^{L_o 2} + (h^{L_d} - \varepsilon h^{L_o})\frac{\sin\psi}{R} + \hat{\sigma}\frac{\cos\psi}{R} = 0, \tag{21}$$

and

$$h^{L_d} - \varepsilon h^{L_o} - \hat{\Delta}_g \frac{\sin\psi}{R} = 0. \tag{22}$$

It can be shown that $R(s_e) = 0$ is automatically satisfied. The choice of a continuous tangent angle over the phase boundary is based on experimental evidence in vesicles with fluid phase coexistence (*see* ref. *18*). To account for the constraints on total membrane area and area fraction of the coexisting phases of a phase-separated lipid vesicle at constant temperature, the generalized variable t is introduced *(32)*, such that $s(t = 0) = 0$, $s(t^*) = s^*$, $s(t_e) = s_e$, and $\dot{s} \equiv ds/dt$. Derivatives with respect to s are thus expressed in terms of derivatives with respect to t, for example, $Q's = \dot{Q}$. The differential shape equations of the (L_o) and (L_d) regions are solved by mapping t^{L_o} in the range [0 t^*] and t^{L_d} in the range [t^* π] onto the common interval [0 π].

The arclength of the deformed vesicle is obtained by simultaneously integrating $\dot{s} = \sin(t)/R$, with the boundary condition $s(t = 0) = 0$. The singularities at the poles are approximated by expansions near the poles. The pressure difference, $\hat{p}$, is either prescribed or, in case of a fixed volume V_f, can be obtained as an eigenvalue of the boundary value problem, with introduction of the additional differential equation $\dot{V} = \pi r \sin\psi \sin t$, with boundary conditions $V(t = 0) = 0$ and $V(t = \pi) = V_f$. Two additional boundary values can be determined from the experimental shape. These are the radius R and tangent angle ψ at the phase boundary.

Defining these experimentally measured geometric parameters as boundary values in the numerical boundary value problem solver, allows determination of additional mechanical parameters as eigenvalues. These are the line tension at the phase boundary, and normalized difference between Gaussian curvature moduli $\hat{\Delta}_g = (\kappa_g^{L_d} - \kappa_g^{L_o})/\kappa^{L_d}$.

Vesicle geometries can thus be numerically generated, which have a vesicle volume, boundary radius, and tangent angle equal to the experimental shape. Varying the single fit parameter ε, the ratio between mean curvature bending modulus of L_o and L_d phase, respectively, allows to obtain a best fit to the experimental vesicle shape by LSQ fitting of numerical shapes to experimentally determined shapes.

6. Conclusion

In summary, the experimental acquisition and subsequent quantitative analysis of vesicle shape traces in terms of two shape-tracing algorithms have been described. Besides the vesicle trace, the intensity profile along the vesicle trace can be determined. Furthermore, numerical methods were developed to examine the mechanics of vesicle shapes. A Hamiltonian fit method was devised, which allows quantification of line tension, lateral tensions, pressure difference, and ratio of mean curvature bending moduli *(18)*. Full integration of the shape equations allows the determination of Gaussian bending stiffness differences between coexisting fluid phases. The experimental and numerical methods of vesicle shape analysis herein described are likely to provide additional insight into the mechanics and physical chemistry of membrane systems more complex than those described in this chapter. For example, an important and biologically relevant extension of the approach described herein is to consider membranes with intrinsic curvature preference (spontaneous curvature) that is caused, for example, by the binding of peripheral membrane proteins.

Acknowledgments

The authors thank Dr. Joshua Zimmerberg and Dr. Paul Blank for the use of the confocal microscope. The work described was supported in part by grant no. 1-K25-AI-65459-01 from the National Institute of Allergy and Infectious Diseases. Its contents are solely the responsibility of the authors and do not necessarily represent the official views of the National Institute of Health.

References

1. Anderson, R. G. W. and Jacobson, K. (2002) Cell biology—A role for lipid shells in targeting proteins to caveolae, rafts, and other lipid domains. *Science* **296**, 1821.
2. Sharma, P., Varma, R., Sarasij, R. C., et al. (2004) Nanoscale organization of multiple GPI-anchored proteins in living cell membranes. *Cell* **116**, 577.
3. Fujiwara, T., Ritchie, K., Murakoshi, H., Jacobson, K., and Kusumi, A. (2002) Phospholipids undergo hop diffusion in compartmentalized cell membrane. *J. Cell Biol.* **157**, 1071.
4. Nakada, C., Ritchie, K., Oba, Y., et al. (2003) Accumulation of anchored proteins forms membrane diffusion barriers during neuronal polarization. *Nat. Cell Biol.* **5**, 626.
5. Edidin, M. (2003) Lipids on the frontier: a century of cell-membrane bilayers. *Nat. Rev. Mol. Cell Biol.* **4**, 414.
6. Kwik, J., Boyle, S., Fooksman, D., Margolis, L., Sheetz, M. P., and Edidin, M. (2003) Membrane cholesterol, lateral mobility, and the phosphatidylinositol 4,5-bisphosphate-dependent organization of cell actin. *Proc. Natl. Acad. Sci. USA* **100**, 13,964.
7. Simons, K. and Ikonen, E. (1997) Functional rafts in cell membranes. *Nature* **387**, 569.
8. Manes, S., del Real, G., and Martinez, C. (2003) Pathogens: Raft Hijackers. *Nat. Rev. Immun.* **3**, 557.
9. Marks, D. L. and Pagano, R. E. (2002) Endocytosis and sorting of glycosphingolipids in sphingolipid storage disease. *Trends Cell Biol.* **12**, 605.
10. Pagano, R. E., Puri, V., Dominguez, M., and Marks, D. L. (2000) Membrane traffic in sphingolipid storage diseases. *Traffic* **1**, 807.
11. Takeda, M., Leser, G. P., Russell, C. J., and Lamb, R. A. (2003) Influenza virus hemagglutinin concentrates in lipid raft microdomains for efficient viral fusion. *Proc. Natl. Acad. Sci. USA* **100**, 14,610.
12. Kuzmin, P. I., Akimov, S. A., Chizmadzhev, Y. A., Zimmerberg, J., and Cohen, F. S. (2005) Line tension and interaction energies of membrane rafts calculated from lipid splay and tilt. *Biophys. J.* **88**, 1120.
13. Lipowsky, R. (1992) Budding of Membranes Induced by Intramembrane Domains. *J. Physique Ii* **2**, 1825.
14. Shi, Q. and Voth, G. A. (2005) Multi-scale modeling of phase separation in mixed lipid bilayers. *Biophys. J.* **89**, 2385.
15. Brown, D. A. and London, E. (2000) Structure and function of sphingolipid- and cholesterol-rich membrane rafts. *J. Biol. Chem.* **275**, 17,221.

16. Akimov, S. A., Kuzmin, P. I., Zimmerberg, J., Cohen, F. S., and Chizmadzhev, Y. A. (2004) An elastic theory for line tension at a boundary separating two lipid monolayer regions of different thickness. *J. Electroanal. Chem.* **564**, 13.
17. Baumgart, T., Das, S., Webb, W. W., and Jenkins, J. T. (2005) Membrane elasticity in giant vesicles with fluid phase coexistence, *Biophys. J.* **89**, 1067.
18. Baumgart, T., Hess, S. T., and Webb, W. W. (2003) Imaging coexisting fluid domains in biomembrane models coupling curvature and line tension. *Nature* **425**, 821.
19. Veatch, S. L. and Keller, S. L. (2005) Seeing spots: Complex phase behavior in simple membranes. *Biochim. Biophys. Acta*
20. Ipsen, J. H., Karlstrom, G., Mouritsen, O. G., Wennerstrom, H., and Zuckermann, M. J. (1987) Phase-Equilibria in the Phosphatidylcholine-Cholesterol System. *Biochim. Biophys. Acta* **905**, 162.
21. Filippov, A., Oradd, G., and Lindblom, G. (2003) The effect of cholesterol on the lateral diffusion of phospholipids in oriented bilayers. *Biophys. J.* **84**, 3079.
22. Veatch, S. L. and Keller, S. L. (2003) A closer look at the canonical raft mixture in model membrane studies. *Biophys. J.* **84**, 725.
23. Veatch, S. L. and Keller, S. L. (2002) Organization in lipid membranes containing cholesterol. *Phys. Rev. Lett.* **89**, art. no.
24. Feigenson, G. W. and Buboltz, J. T. (2001) Ternary phase diagram of dipalmitoyl-PC/dilauroyl-PC/cholesterol: Nanoscopic domain formation driven by cholesterol. *Biophys. J.* **80**, 2775.
25. Bacia, K., Schwille, P., and Kurzchalia, T. (2005) Sterol structure determines the separation of phases and the curvature of the liquid-ordered phase in model membranes. *Proc. Natl. Acad. Sci. USA* **102**, 3272.
26. Bagatolli, L. A. (2003) Thermotropic behavior of lipid mixtures studied at the level of single vesicles: giant unilamellar vesicles and two-photon excitation fluorescence microscopy. *Methods Enzymol.* **367**, 233.
27. Bagatolli, L. A. and Gratton, E. (2000) Two photon fluorescence microscopy of coexisting lipid domains in giant unilamellar vesicles of binary phospholipid mixtures. *Biophys. J.* **78**, 290.
28. Korlach, J., Baumgart, T., Webb, W. W., and Feigenson, G. W. (2005) Detection of motional heterogeneities in lipid bilayer membranes by dual probe fluorescence correlation spectroscopy. *Biochim. Biophys. Acta* **1668**, 158.
29. Korlach, J., Schwille, P., Webb, W. W., and Feigenson, G. W. (1999) Characterization of lipid bilayer phases by confocal microscopy and fluorescence correlation spectroscopy. *Proc. Natl. Acad. Sci. USA* **96**, 8461.
30. Mathivet, L., Cribier, S., and Devaux, P. F. (1996) Shape change and physical properties of giant phospholipid vesicles prepared in the presence of an AC electric field. *Biophys. J.* **70**, 1112.
31. Thompson, R. E., Larson, D. R., and Webb, W. W. (2002) Precise nanometer localization analysis for individual fluorescent probes. *Biophys. J.* **82**, 2775.
32. Julicher, F. and Lipowsky, R. (1996) Shape transformations of vesicles with intramembrane domains. *Phys. Rev. E* **53**, 2670.
33. Seifert, U., Berndl, K., and Lipowsky, R. (1991) Shape Transformations of Vesicles—Phase-Diagram for Spontaneous-Curvature and Bilayer-Coupling Models. *Phys. Rev. A* **44**, 1182.
34. Julicher, F. and Lipowsky, R. (1993) Domain-Induced Budding of Vesicles. *Phys. Rev. Lett.* **70**, 2964.
35. Evans, E. and Yeung, A. (1994) Hidden Dynamics in Rapid Changes of Bilayer Shape. *Chem. Phys. Lipids* **73**, 39.
36. Evans, E. A. and Skalak, R. (1980) *Mechanics and Thermodynamics of Biomembranes*. CRC Press, Boca Raton, FL.
37. Powers, T. R., Huber, G., and Goldstein, R. E. (2002) Fluid-membrane tethers: Minimal surfaces and elastic boundary layers. *Phys. Rev. E* **65**.
38. Jenkins, J. T. (1976) Static equilibrium configurations of a model red blood cell. *J. Math. Biol.* **4**, 149.

26

Laser Tweezer Deformation of Giant Unilamellar Vesicles

Cory Poole and Wolfgang Losert

Summary

Two methods are presented for deforming giant unilamellar vesicles with holographic optical tweezers. The first allows ultrahigh spatial- and temporal-resolution optical tracking of membrane deformations, by using embedded silica microspheres in a giant unilamellar vesicle as tracers. The vesicles are stretched by moving several beads with multiple optical tweezers and then are released from an elongated shape. Time constants of relaxation can be extracted by tracking the beads with 0.5-ms time resolution and 10 nm or better spatial resolution. The second method allows for direct deformation of the membrane into complex shapes using two solutions with different indices of refraction and holographic optical tweezer. Vesicle shapes are extracted directly with an active contour algorithm. Fourier analysis of the relaxation of the vesicle shape back to an equilibrium shape demonstrates a possible application of this technique.

Key Words: GUV; laser tweezers; particle tracking; active contour; Fourier analysis.

1. Introduction

Giant unilamellar vesicles (GUVs) range in size from 1 µm to sizes more than 100 µm. Because of their size and similarity of composition, vesicles are studied as a model system for cell membranes *(1)*. There are many experimental techniques for determining physical properties of GUVs. The elastic and bending moduli can be measured from the response to strong deformations driven by a large local decrease in pressure (micropipet aspiration) or by observation of thermal fluctuations *(2,3)*. Herein, a method to determine vesicle-shape deformations in response to intermediate forces of order several piconewton applied at multiple points using a holographic optical tweezer (HOT). The forces are one or two orders of magnitude larger than thermal forces, but generally smaller than forces resulting from micropipet aspiration. The time constants of the relaxation are measured back to equilibrium after the forces are stopped.

The methods presented focus on the use of multipoint laser tweezers to deform vesicles, incorporation of microspheres into GUVs, subpixel resolution tracking of microspheres, direct laser tweezing of vesicles, extraction of vesicle boundaries form images, and Fourier analysis of boundary shapes. The use of these tools is described in the context of measuring the relaxation to equilibrium vesicle shapes, but they should be of more general use for studies of mechanical properties of membranes. For example, shape deformations may prove useful as a way to assess vesicle properties, as additional components such as transmembrane proteins are added.

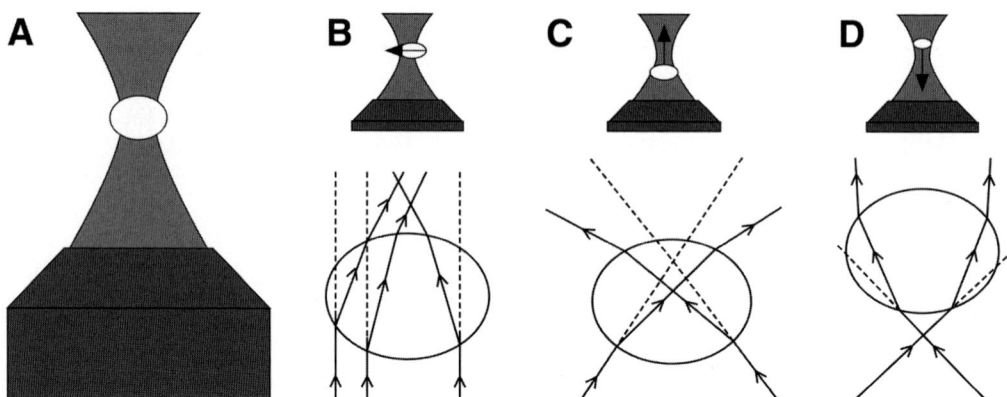

Fig. 1. Laser tweezers utilize the momentum of light to produce piconewton forces on dielectric materials. In (**A**) a dielectric sphere is seen in the focus of the laser beam that is coming through a microscope objective. This is the equilibrium position and displacements from it result in the forces depicted in (**B–D**). The dotted lines represent the path a photon would take if the sphere were not in its path. Because momentum is conserved, the bead must move in the opposite direction of the bending light. The bead in **B** has more photons hitting the left than the right because the beam is focused on the left side. This results in more photons bending to the right and so the bead must move to the left. In **C** and **D** it is seen how there is a restoring force in the vertical direction. This restoring force must be enough to overcome the scattering force from reflected photons. This is the reason why photons must come in at a steep angle, requiring a high numerical aperture objective.

1.1. Laser Tweezers

Laser Tweezers generate piconewton forces on micron-sized dielectric objects through the refraction of light *(4–7)*. Refraction at the interface between fluid and object changes the direction of light, and (because light carries momentum) also changes the light momentum. This results in a force pulling an object of higher index of refraction than the surrounding fluid toward regions of highest light intensity. The object is pulled toward the focal point when a laser beam is tightly focused through the microscope objective onto a focal point (*see* **Fig. 1**). This principle has been utilized to manipulate objects ranging from silica beads *(5)* to living cells *(8)* and will be used herein to deform GUV.

1.2. Holographic Laser Tweezers

Multipoint laser tweezers can be generated using a single laser. One method is time-sharing of the beam, wherein a beam is rapidly scanned between locations. This results in fluctuations of the laser tweezer intensity at each point on the time-scale of the scanning, generally in the order of milliseconds *(9–11)*. Instead, a technique based on HOT is used, which uses spatial light modulators (SLMs) to generate several laser tweezers from a single beam *(5,12,13)*.

A single beam is expanded to fill the surface area of a liquid crystal-based SLM. The SLM generates different phase lags for each pixel of the laser tweezer array, and acts as a programmable diffraction grating. This leads to simple interference patterns in the reflected light. The interference pattern can be tuned to create an array of traps wherein each trap can be moved independently several times per second (*see* **Figs. 2** and **3**). A commercial holographic laser tweezer system is used that is already integrated with a Nikon (www.nikonusa.com) inverted microscope, the Bioryx 200 from Arryx Inc.

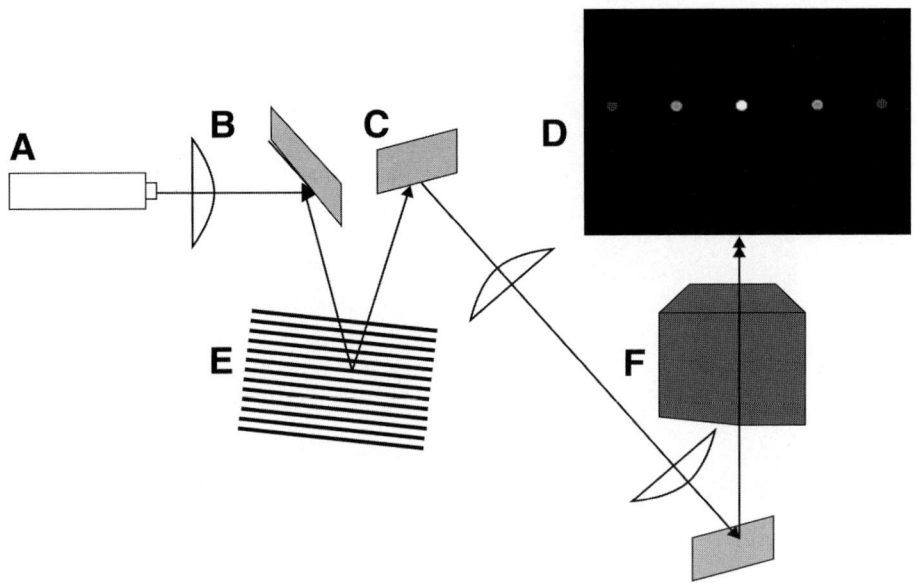

Fig. 2. Schematic of a HOT. Light from (**A**) a laser goes through (**B**) a beam expander before it is incident on (**C**) the SLM, which acts as a diffraction grating changing the phase of the light at each pixel. The light then passes through (**D**) a telescope set of lenses to ready the light to enter (**E**) the microscope objective, which focuses the light onto (**F**) a sample in which the resulting light field is seen. In this schematic what happens is observed when a simple parallel diffraction pattern is used. The result is a bright first-order spot and higher-order spots of diminishing intensity.

1.3. Particle Tracking

Particle Tracking consists of identifying objects in a time sequence of images, and tracking the motion of all objects through the sequence. Herein, particle tracking is utilized to determine time constants of a GUV's return to equilibrium after laser tweezer deformation, and also as a measurement of the force experienced by a vesicle as it is being deformed. Using information from all the pixels in an object, it is possible to find the location of the object to within much less than one pixel, if the shape of the object is known and fixed in time *(14)*. The particle tracking code shown herein can be used to track more than 1000 particles through thousands of images, and can find spheres with one-tenth of one pixel or better spatial resolution (<10 nm in this case).

1.4. Fourier Analysis

Fourier Analysis is a tool for decomposing functions into sinusoidal components. The Fourier transform is defined as

$$F(k) = \int_{-\infty}^{\infty} f(x)e^{-2\pi i k x}.$$

The discrete Fourier transform must be used for discrete data sets and is defined as

$$F_k = \sum_{x=0}^{N-1} f_x e^{-2\pi i k x / N}$$

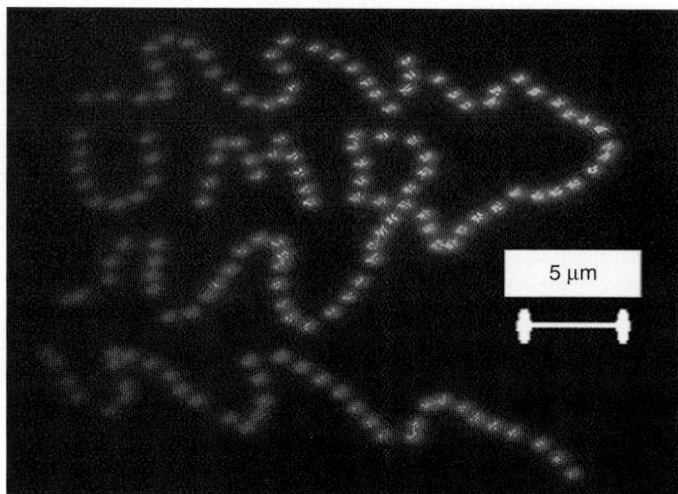

Fig. 3. HOTs allow generation of complex patterns of optical traps. Shown herein is the illumination of a fluorescent sample by a pattern of approx 160 laser traps. Each trap is independently movable in near real time.

where f_x is the xth point of the discrete data set to be transformed *(15,16)*. Herein, Fourier analysis is used to quantify the shape evolution of a GUV as it returns to equilibrium shape following a laser tweezer deformation.

2. Materials
2.1. Preparation of Electroformation Slides

1. ITO-coated glass microscope slides (ITO-coated Fusion Drawn Corning 1737F Aluminosilicate Glass 5–15 ohm $25 \times 75 \times 1.1$ mm³ from Delta Technologies Ltd., Stillwater, MN).
2. Digital multimeter for testing resistance of ITO slides.
3. 4-cm 14-gauge wire.
4. Silver conductive epoxy (Conductive Silver Epoxy Kit from Electron Microscopy Sciences, Hatfield, PA).
5. Hotplate for curing epoxy.

2.2. Electroformation of GUVs

1. Two Electroformation slides from **Subheading 2.1**.
2. Function Generator to provide 4–10 Hz electrical signal at 1.5 V.
3. Digital multimeter for measuring AC voltage and frequency of signal from **step 2**.
4. Dioleylphosphatidylcholine (DOPC) dissolved in 10:1 chloroform to methanol solution 1 mg lipids/mL (Avanti Polar Lipids, Alabaster, AL). Store in the freezer at −20°C (*see* **Note 1**).
5. −20°C freezer for storing DOPC from **step 4** to slow lipid degradation.
6. Refrigerator for storing GUVs produced in **Subheading 3.2**.
7. Pipeter and glass pipet tips (*see* **Note 1**).
8. Vacuum pump and vacuum desiccator for evaporating chloroform and methanol from electroformation slides.
9. Desiccant such as drierite.
10. 0.3 *M* sucrose and 0.3 *M* glucose solutions.
11. Chamber-making material (Seal Ease from Vacutainer Systems; Becton Dickinson, Franklin Lakes, NJ).
12. Three 1-mL syringe and three needles (22G 1 1/2).

2.3. Incorporation of Microspheres Into GUVs

1. Approx 2 µg of 1–4 µm silica microspheres (Bangs Labs, Fishers, IN) per 1 mL of prepared, vesicle-containing solution.
2. All materials listed in **Subheading 2.2**.

2.4. Sample Preparation of Vesicles With Internal Beads for Laser Tweezer

1. Microscope slides ($25 \times 75 \times 1.1$ mm^3).
2. Number 1 Microscope cover slips (0.13×0.17 mm^2) (*see* **Note 2**).
3. 0.3 M sucrose solution.

2.5. Deforming Vesicles Using Beads

1. A multipoint laser tweezer system (A Bioryx 200 from Arryx is used, which is a commercially available system.)
2. A camera for taking digital images of the experiment.

2.6. Particle Tracking

1. A computer running Windows, Linux, Sun Solaris, or Mac OS.
2. IDL virtual machine from RSI Inc. (Austin, TX), a free utility that allows compiled IDL code applets to run in many computing environments.
3. RyTrack.sav, a compiled IDL particle-tracking program.
4. Digital images of experiments such as those performed in **Subheading 3.5**.

2.7. Sample Preparation of Vesicles Containing Solution With Lower Index of Refraction

1. Microscope slides ($25 \times 75 \times 1.1$ mm^3)
2. Number 1 microscope cover slips (0.13×0.17 mm^2) (*see* **Note 2**).
3. 0.33–1.5 M glucose solution (*see* **Note 3**).

2.8. Deforming Vesicles Through Direct Interaction of Laser Tweezers With the Vesicle

1. A multipoint laser tweezer system (A Bioryx 200 from Arryx is used, which is a commercially available system) (*see* **Note 4**).
2. A camera for taking digital images of the experiment.

2.9. Image Extraction of Vesicle Boundary

1. A computer running Windows, Linux, Mac OS 9, or Mac OS X.
2. ImageJ, a free image processing and analysis program from National Institute of Health (NIH).
3. IDL Virtual Machine from RSI Inc. (Austin), a free utility that allows compiled IDL code applets to run in many computing environments.
4. A program such as Igor Pro (WaveMetrics), IDL, Matlab, or Microsoft Excel that can sort and perform mathematical operations, such as Fourier transforms on arrays of numbers.
5. Digital images of experiments, such as those performed in **Subheading 2.8**.

3. Methods

3.1. Preparation of Electroformation Slides

1. Use digital multimeter on two ITO-coated slides to determine conductive side (*see* **Note 5**). Place both slides conductive side up.
2. Place 0.1–0.25 cm^3 enough silver epoxy onto one end of both slides, so that there is a good contact area on the slide, and it is thick enough to hold a wire (*see* **Fig. 4**).
3. Place 0.75 cm of 14-gauge wire into the silver epoxy.

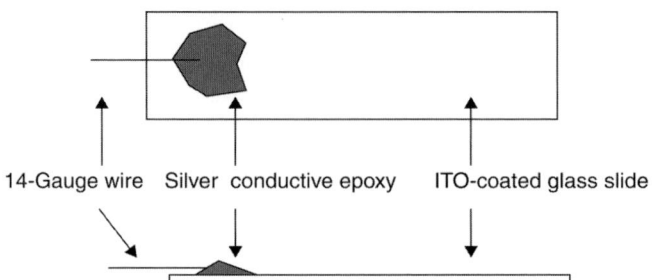

Fig. 4. Diagram of an ITO-coated glass slide with a wire attached *through* silver epoxy.

4. Place slides with wires onto hot plate and heat to 65–121°C for 5–10 min for maximum conductivity and adhesion. Remove slides from hot plate and allow slides to cool.
5. Test the slide resistance by attaching a multimeter to the wire and at the opposite end of the slide. The resistance should be only few to tens of ohms greater than resistance measured in **Subheading 3.1., step 1**.

3.2. Electroformation of GUVs

1. Make DOPC solution 9:1 chloroform to methanol. (1 mg lipids to 1 mL solution) Be sure to use all glass containers for storage and not allow plastics to contact the chloroform (*see* **Notes 1** and **6**).
2. Take two electroformation slides prepared in **Subheading 3.1.**, and using a pipeter and glass pipet tip, deposit 50 µL of DOPC solution from **Subheading 3.2., step 1.**, onto the ITO-coated side of each of the slides; be sure to deposit in a zigzag manner lightly touching the slide surface with the pipet tip and cover completely the section that is to become the electroformation chamber (*see* **Notes 6** and **7**).
3. Add desiccant to the vacuum desiccator. Place slides inside desiccator and turn on vacuum pump. Leave slides in chamber for a few hours or overnight allowing the chloroform and methanol to evaporate completely (*see* **Note 6**).
4. Turn off vacuum pump and remove slides from chamber.
5. Form 1-mm thick chamber between the two slides making sure that their conductive sides face each other using Seal Ease (*see* **Notes 5** and **6**).
6. Put two small holes into the side of the chamber by depressing a syringe needle into the Seal Ease. The first hole is for injecting the swelling solution and the second is to allow air to escape.
7. Using a syringe and a new syringe needle inject 0.3 *M* sugar solution (A mixture made up of 2/3 sucrose and 1/3 glucose) into chamber until it is full. Plug holes with a small amount of Seal Ease. This is the swelling solution (*see* **Note 8**).
8. Tune function generator to produce a 10-Hz sine wave at the lowest voltage the function generator can produce. Use a multimeter to measure the voltage and frequency of this electrical signal (*see* **Notes 5** and **6**).
9. Connect function generator to chamber and turn on generator.
10. Over the next hour gradually ramp the voltage up to 1.5 V.
11. After 2–8 h reduce frequency to 4 Hz while maintaining 1.5 V.
12. Leave at 4 Hz for 1–3 h. This facilitates the detachment of vesicles from the slides.
13. Extract vesicle-containing solution using a new syringe and tip. Look at a sample under the microscope to make sure vesicles have formed (*see* **Note 6**) (**Figs. 5–7**).

3.3. Incorporation of Microspheres into GUVs

1. Prepare a solution of microbeads in 0.3 *M* sugar solution. This solution should be in the range of 0.01–0.1% beads by volume (*see* **Note 9**). Perform electroformation as in **Subheading 3.2.** with this solution as the swelling solution (*see* **Subheading 3.2., step 7**) (**Figs. 8–10**).

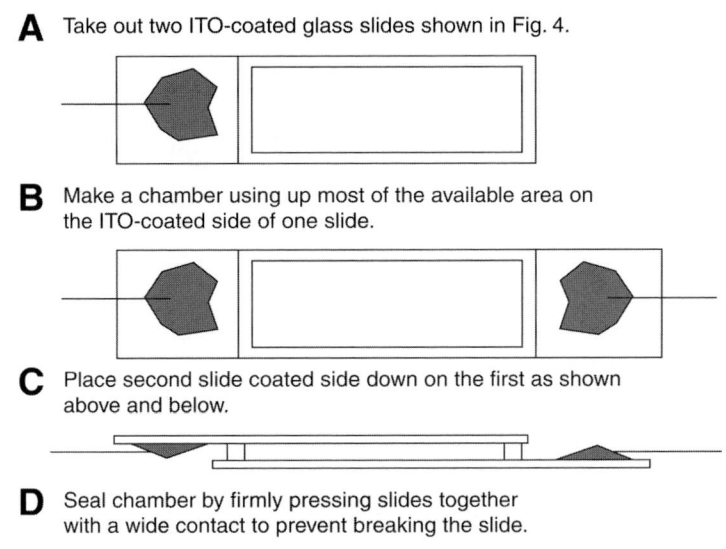

Fig. 5. Electroformation chamber construction.

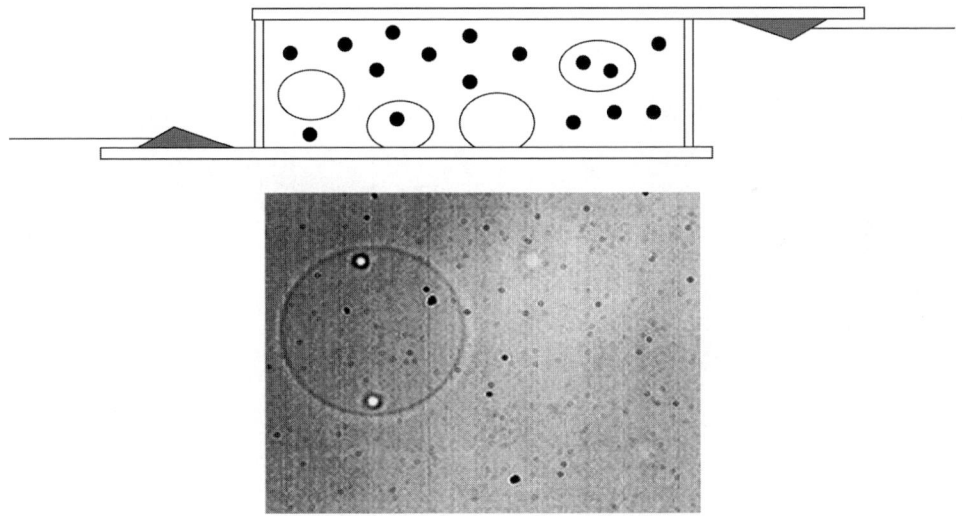

Fig. 6. **(A)** Schematic of electroformation chamber during electroformation with bead-containing solution. This illustrates how vesicles form around the beads, and that the number of beads incorporated in a particular vesicle is not controlled. **(B)** Pictures of vesicles with incorporated beads.

3.4. Sample Preparation of Vesicles With Internal Beads for Laser Tweezer

1. Prepare a slide using a no. 1 cover slip (*see* **Note 2**).
2. Using a pipeter extract a small amount of vesicle-containing solution created in **Subheading 3.3**. One should stir or vortex the solution before extraction because the vesicles will stick to the bottom.
3. Using the same pipet tip extract a small amount of 0.3 M sucrose solution (one-tenth of the amount of vesicle-containing solution from **Subheading 3.4., step 2**).
4. Stand pipeter upright for 10 min. Remove half of the contents of the pipet tip. This removes most of the beads in the external solution (*see* **Note 8**).

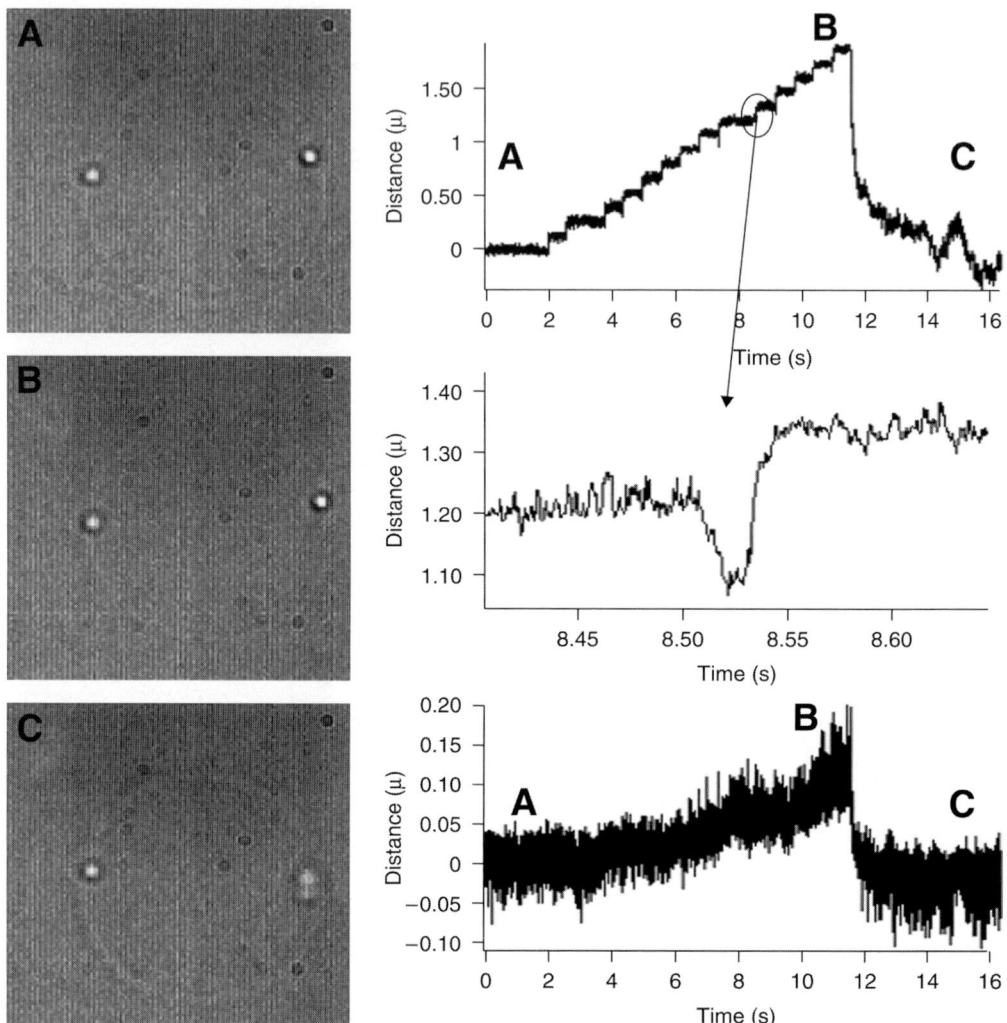

Fig. 7. The vesicle shown in the above image sequence is being stretched by two embedded microspheres. The sphere to the left is held stationary whereas the bead to the right is moved at uniform velocity. The upper right graph shows the x-position vs time of the moving microsphere. The graph in the middle shows the behavior of the bead during the trap switching time. At the lower left is a graph of the x-position vs time of the stationary bead. The bead is slowly pulled out of the trap by the force generated by membrane stress. As the traps are locally harmonic, this is a direct measure of the force. Note that subpixel particle tracking is necessary for the measurement.

5. Deposit remaining contents (GUVs in 0.3 M sucrose with very few external beads) of pipet tip onto sample slide.
6. If the laser tweezer objective is an oil immersion objective, place a drop of oil onto the objective. Most laser tweezer objectives are oil immersion owing to the need of a high numerical aperture (*see* **Note 2**).
7. Place sample on laser tweezers (**Fig. 11**).

3.5. Deforming Vesicles Using Beads

1. After letting the vesicles settle for 10 min, look for a vesicle near the glass surface that has beads on the inside. Vesicles that contain multiple beads will be rare. To find one might have to search

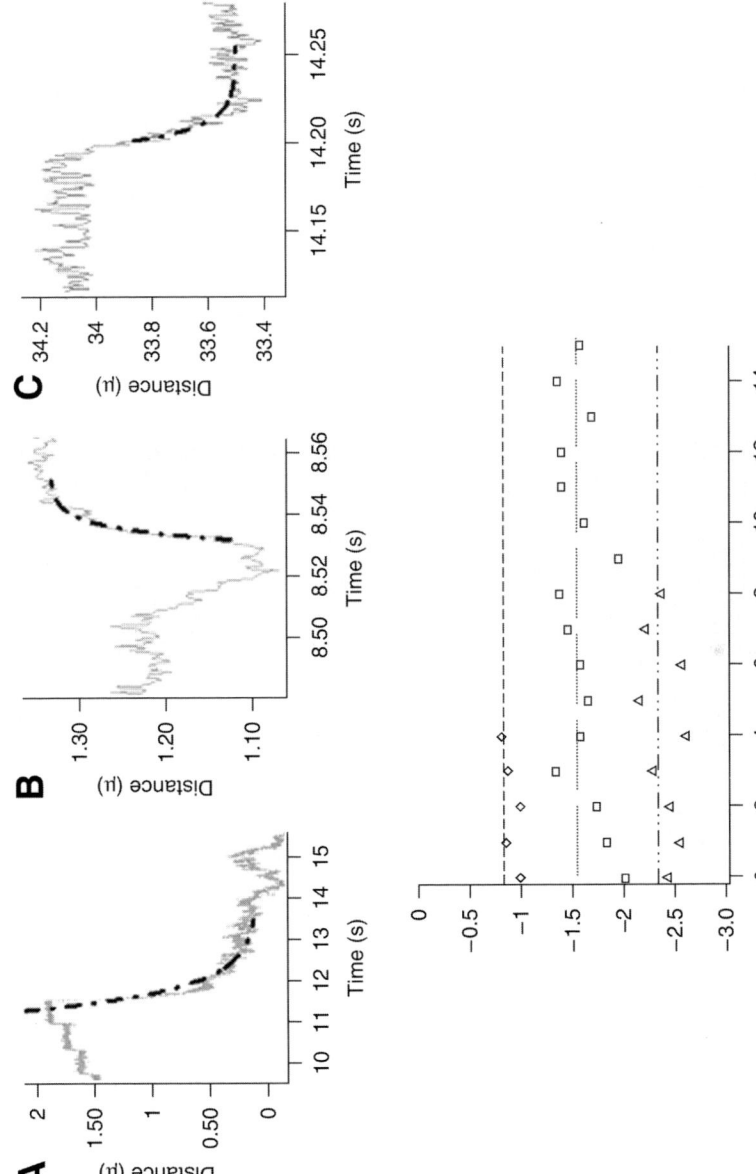

Fig. 8. Bead positions are fitted to exponentials in the cases of (**A**) falling out of a trap after pulled to maximal extension; (**B**) falling into a trap during stretching; and (**C**) falling into a trap when not in contact with a vesicle membrane.

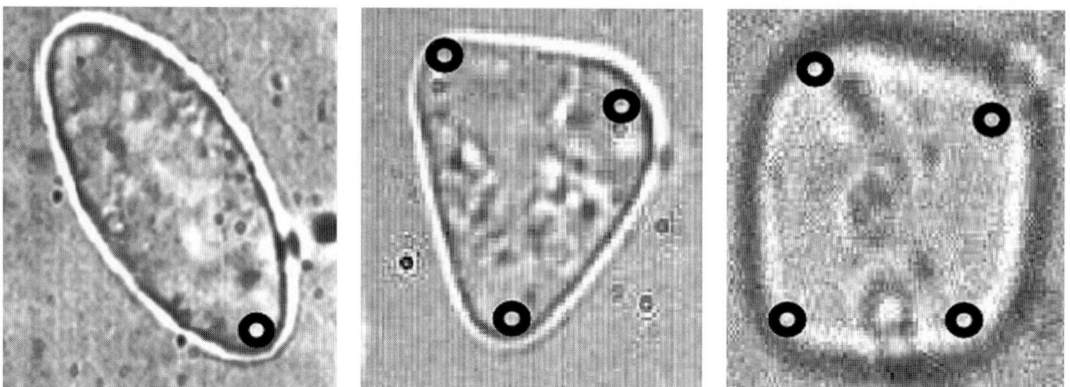

Fig. 9. Deformations of membranes by direct pulling of the fluid utilizing index of refraction mismatch between sucrose on the inside of the vesicle and glucose on the outside. Vesicles can be pulled into such shapes using only piconewton forces because the osmolarity of the outside solution was increased to induce osmotic shrinkage of the vesicle, resulting in an excess of membrane area. Black circles indicate locations of laser traps.

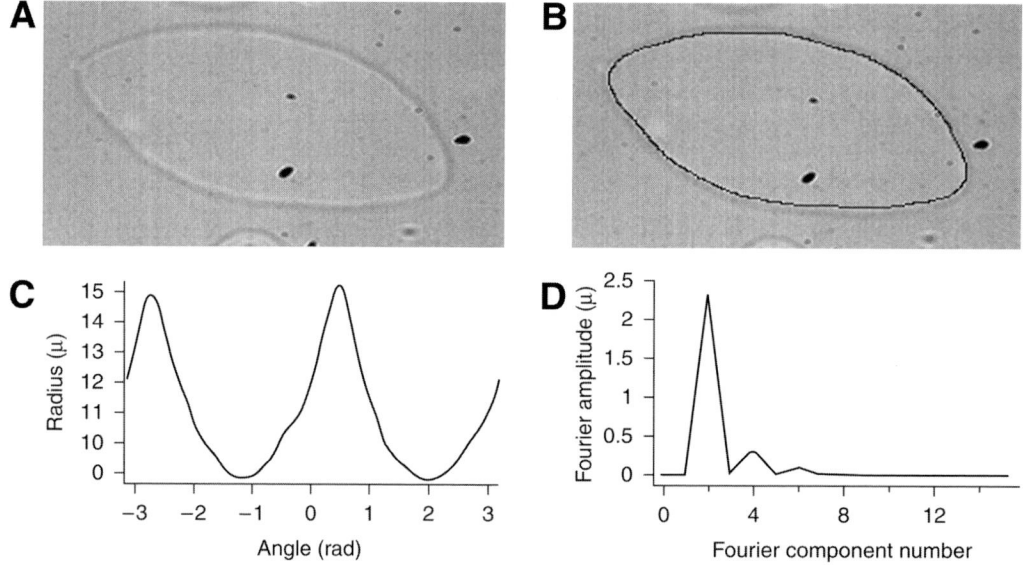

Fig. 10. Vesicle extraction sequence: (**A**) Raw image from the camera of a vesicle. (**B**) Adaptive contour algorithm (SplineSnake) in ImageJ has been used to extract the boundary of the vesicle. The boundary is shown in black overlaying the original image. (**C**) The boundary is then converted to polar coordinates. (**D**) The Fourier transform of **10C**.

 through tens to hundreds of vesicles. The vesicle should also be diffusing in the sample and not stuck to the glass surface to allow one to perform the experiment.
2. Adjust illumination so that there are no saturated pixels when imaging the beads. This will make the sample appear to be quite dark (*see* **Note 10**).
3. Set laser tweezer focal points (traps) at the bead locations and move traps until bead touch the membrane.
4. Use digital video camera to record time sequence of images as the beads are pushed further against the membrane using laser traps.

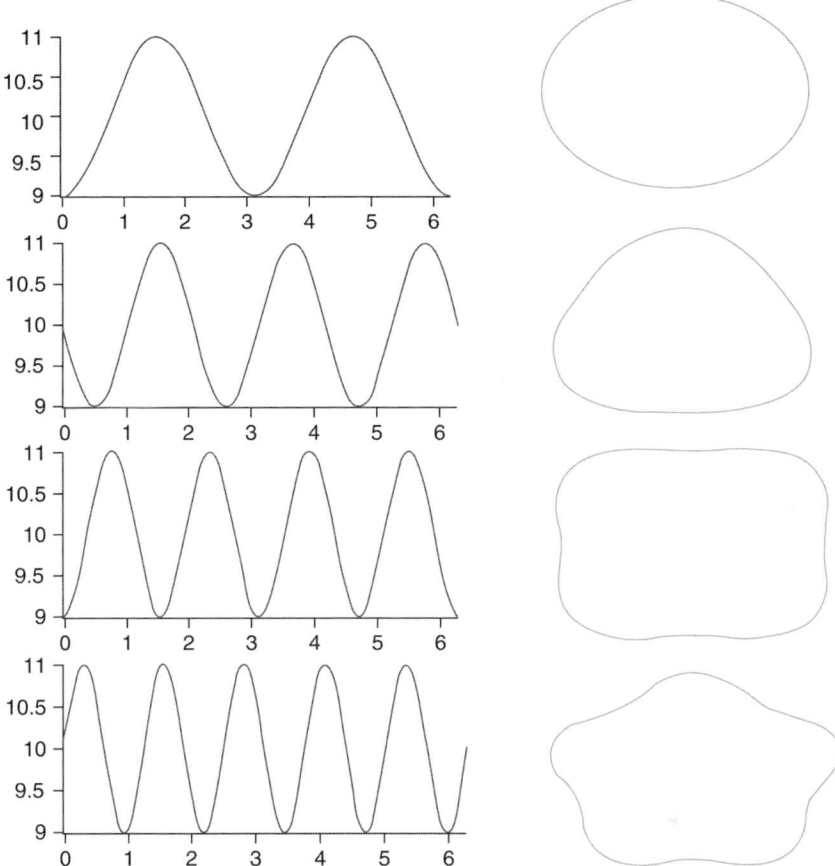

Fig. 11. The effect of Fourier components 2–5 in plots of radius vs angle are demonstrated. The plots of radius vs angle have a mean value of 10 and a single Fourier component with amplitude one. The pictures on the right show the resulting shapes plotted in Cartesian coordinates.

3.6. Particle Tracking

1. Open RyTrack.sav, a particle-tracking program written by Ryan Smith (available free of charge at http://titan.iwu.edu/~gspaldin/rytrack.html); based on tracking algorithms of David Grier, John Crocker, and Eric Weeks *(17,18)* (*see* **Note 11**). This program provides a graphical user interface for particle tracking and runs using IDL virtual machine (also available free of charge from RSI Inc. at http://www.rsinc.com).
2. Specify the image sequence to be used for particle tracking in the upper space labeled "Image Directory." One must include the name of the file and the file extension. Extensions that work with the program include: .jpg, .tif, .bmp, and others. An asterisk acts as a wildcard allowing the program to accept a set of images. An example set of files would be C:\mydata*.tif. This would track particles through all tif-formatted images ordered alphabetically. To load only tif-formatted images that start with an f, f*.tif could be used.
3. Using the vertical scroll bar, scroll down to the bottom of the user interface and click on the "Start" button. Now in the top right window one should see the images displaying one at a time. They may be warped as they are stretched to fit the window, but that will not affect particle tracking. In the bottom left window, one will see the filtered results of the image processing that one will perform in the next step.

4. Set image-processing parameters. The first two parameters are for band-pass filtering. "Bandpass1" filters out high-frequency noise. Usually, 1 pixel is a good starting point for this parameter. "Bandpass2" filters out the large-scale changes in brightness, larger than the object under study (e.g., variations in image brightness). This parameter should be set slightly larger than the radius of the largest object to be tracked. "Sobel Smooth" is a gradient-based edge-enhancement operator that should be used as a last resort as it also magnifies noise-making subpixel making its tracking very difficult. "Threshold" sets the minimum gray level of a pixel to be considered as a particle. A pixel below the threshold will be set to zero. A threshold, although many times necessary, should be applied carefully as it reduces information about particles to be tracked making subpixel tracking difficult. The parameter settings should be tuned until the processed image shown in the filtered results window shows all particles as white circles against a black background. If the particles to be tracked are dark particles against a white background then click "Invert Image."
5. Set particle identification parameters. The first parameter is "Particle Radius." This should be set slightly larger than the largest feature to be tracked. The second is "Particle Spacing." This sets the smallest distance (in pixels) between particles. This should be set larger than the particle radius. After setting an initial value for these two parameters scroll down to the bottom and check at least one of the boxes, "Overlay Original" or "Overlay Filtered." Now the results of particle identification on the images will be seen. At the bottom right it tells how many particles are being identified in each image and how many of these identifications are being retained. It will probably be observed that one is making false identification of particles. The next parameter is the "Mass Cut," a very useful parameter for eliminating these unwanted identifications. It sets a minimum for the integrated pixel value of a particle. The next two parameters can also be useful for eliminating excess particles. "Eccentricity" eliminates particles based on how circular they are. This is very useful when all the particles that one want to track are very circular or very elongated. A value of zero represents a circle whereas a value of 100 represents a line. After setting the value of this parameter check one of the boxes to eliminate particles either above or below this threshold. The last particle identification parameter is "Radius of Gyration," a measure of the width of the bright spot. This might prove to be a useful parameter to distinguish between particles of different size. Finally, the box at the bottom with the word "Field" should be checked if the image is one field of an interlaced image, such as that obtained from a VHS tape.
6. Set tracking parameters. These parameters are used to guide the program on how to assign the same number to the same particle in an image sequence. The first is "Maximum Displacement." This sets the farthest a particle can move between frames. If particles move further than the average distance between particles, neither the program nor a human could determine which particle is which from the previous frame. Therefore "Maximum Displacement" has to be set lower than the average distance between particles. The second is "Good Enough," the minimum length of trajectories. This filters out noise in the image, which may be misinterpreted as a particle in a single frame, but usually not in several subsequent frames. The last parameter is "Steps Memory" and represents the number of images for which a particle cannot be found in an image, yet still be tracked as the same particle if it reappears within this maximum number of images. The default is zero and this value should not be set very high.
7. Once all parameters are set satisfactorily click "ID and Track." Now the program will go through the directory and ID all the particles in each image. It will create a .gdf file that contains the particle locations for each image. All of particle data are combined into a file called pretrackoutput.gdf before the particle-tracking parameters are applied and the trajectories of the particles calculated. The final output of the program is trackoutput.gdf, trackplusvels.gdf, and their corresponding .dat files. The .gdf files are binary files with custom headers whereas the .dat files are tab-delimited text files. A params.txt file is also created, which contains all of the parameters used in RyTrack.sav on these images. If one already has a set of particle IDs click "Just Track" to calculate the trajectories from the existing .gdfs.

8. Open trackoutput.dat in a text editor, Microsoft Excel, or a graphing utility such as Igor Pro by WaveMetrics. Trackoutput.dat is an array containing information about each particle trajectory found in the image sequence. Each row represents x-, y-position, integrated brightness, square of the radius of gyration, eccentricity of a particle, frame number, and trajectory number (*see* **Note 12**).

3.7. Sample Preparation of Vesicles Containing Solution With Lower Index of Refraction

1. Take sample of GUVs in sucrose and dilute 1:1 with a 0.33–1.5 M glucose solution on a sample microscope slide (*see* **Note 8**).
2. Put sample onto laser tweezers using the method discussed in **Subheading 3.4**.
3. Let sample settle for 20 min.

3.8. Deforming Vesicles Through Direct Interaction of Laser Tweezers With the Vesicle

1. Using the microscope, find a vesicle that appears to be deflated and fluctuating. A spherical vesicle not exhibiting fluctuations is tense and not highly deformable *(19)* (*see* **Note 3**).
2. Place traps near the vesicle boundary.
3. Move traps slowly to pull out and stretch the membrane.
4. Take image sequences of the previous experiments.

3.9. Image Extraction of Vesicle Boundary

1. Open ImageJ, an image processing and analysis program written in Java developed by the NIH available free of charge at http://rsb.info.nih.gov/ij/.
2. Open the image to be analyzed. The image must be an 8-bit grayscale image. If it is not, convert it to 8-bit by clicking on "Image" and then "Type." Select "8-bit" from the list.
3. Start up the ImageJ plugin, SplineSnake, an adaptive contour algorithm written by Matthews Jacob available at http://ip.beckman.uiuc.edu/ *(20–22)*. Go to "Plugins" and then "SplineSnake."
4. The top icons in ImageJ now correspond to functions for SplineSnake. Select the upper left button. Pick a starting point on the image and click the right button on the mouse. Let go of the button and use the mouse to trace the vesicle boundary. When finished hit the left mouse button again. A series of control points with lines between them should appear in red. This will be the starting location for the iterative snaking algorithm. The second button from the left can be used to delete points and the third to delete all points.
5. Set snake parameters and constraints. The fourth and fifth buttons from the right are used to set and remove constraints for the curve. These constraints act as external forces on the locations where you set them. Constraints pulled far away from the curve represent a strong force. Use constraints as a last resort, as they strongly control what the contour extraction looks like and so introduce arbitrariness. The sixth icon, which is shaped like a heart, brings up a preferences selection box. The section on the left controls parameters for the initial curve. "Knot spacing" controls how far apart the knots for the spline are in pixels. Beneath these parameters is a sequence of check boxes. It is recommended to check "show mouse path", as this allows the initial curve to be seen as it is drawn. On the right are the parameters for controlling the snake. The top parameter controls how the edge-detection force is produced (*see* **Note 13**); "Spring weight" controls the strength of the internal forces produced by the snake.
6. Activate the snake by clicking on the icon that looks like a snake. A blue spline selection should appear. If this selection outlines the boundary of the vesicle well, click on the icon that looks like a play button and click on select. Otherwise, repeat **step 5** after changing snaking parameters or redrawing initial curve.
7. Click on the icon that looks like a microscope, the ImageJ icon.
8. Make the inside selection black by clicking on Edit and then Fill, and the outside white by clicking on Edit and then Clear Outside. Save the resulting image.

9. Use RyTrack.sav to create uniform points on the image boundary (*see* **Subheading 3.6.**). Set "Bandpass1" to 1 and "Bandpass2" to 2. Set "Radius" and "Separation" to 1. RyTrack will output the locations of all pixels on the boundary.
10. Convert the coordinates of the pixels from rectangular to polar (*see* **Note 14**).
11. Order the resulting coordinates and take their Fourier transform.

4. Notes

1. Lipids from Avanti Polar Lipids come in two forms. The first is pure powdered lipids. These must be dissolved in 9:1 chloroform to methanol solution (1 mg lipids per milliliter solution) to be used for electroformation. The second form, which the lab uses, comes dissolved in chloroform and is 25 mg lipids per mL of solution. To be used for electroformation, a sample must be diluted to 1 mg/mL in 9:1 chloroform to methanol solution. When making this solution, be sure to use glass containers and pipets and not let plastics come into contact with the chloroform. Plastic residue on the electroformation chamber appears to prevent proper vesicle formation as yield of GUVs drops.
2. Laser tweezers use high-numerical aperture lenses to tightly focus the laser for three-dimensional trapping. Such lenses have short working distances with standard no. 1 cover slips.
3. Water can move across the membrane but sugar and larger macromolecules cannot, on the time-scales of the experiments herein. Excess of macromolecules outside the vesicle will lead to osmotic pressure on the vesicle and deflation of the vesicle. Assuming that the volume of all the vesicles is negligible compared with the volume of the sample, the reduced volume (volume of the vesicle normalized to the volume of a sphere with equal area) of the deflated vesicles is given by the initial osmolarity of internal solution divided by the osmolarity of the external solution. The resulting vesicles are quite varied ranging from deflated fluctuating vesicles to pear shaped, to vesicles with internal and external budding of small vesicles *(23,24)*.
4. There are ways to carry out some of the experiments with a single laser tweezer if vesicles are attached (e.g., to the glass cover slip, *see* **ref. 25**).
5. Resistance of the electroformation slide should be low (<500 ohms across the entire length). The electroformation process requires an electric field on the order of 1 V/mm be produced across two conductive surfaces. The electric field produces undulations in lipid films eventually resulting in the creation of vesicles in the micron size range *(26,27)*.
6. In the authors' experience, problems with electroformation occur for the following reasons. Evaporation of chloroform and methanol is not complete. This usually results from a poor vacuum or insufficient time under vacuum. The electric field is too weak or too strong. (*See* **Note 5.**) This usually results from incorrect chamber thickness or poor connections to function generator. (*See* **Note 1.**) There are unwanted substances dissolved in lipid mixture (*see* **Note 1**).
7. Other types of electroformation chambers use two wires covered with lipids separated by a set distance in a solution such as sucrose. The advantage of the method used is more lipids can be deposited over the area of a slide, resulting in a higher yield of vesicles *(28)*.
8. Placing vesicles in solutions of lower density such as sucrose-swelled vesicles in glucose has the benefit of causing the vesicles to settle to the bottom *(1)*. This promotes easy viewing of the vesicles. It also allows for the manipulation of the solution. Because the vesicles are attached to the surface one can remove most of the solution without removing the vesicles. The solution can then be replaced by another solution. By placing vesicles in a higher-density solution, the vesicles will float to the top. This has proved useful for sorting objects according to their density and is used in **Subheading 3.4.** to sort silica beads from GUVs.
9. Bead solutions from Bangs Labs are typically 10% beads by volume. For incorporation of beads into GUVs the lab uses 0.1–0.01% beads by volume. A dilution of Bangs Labs beads as delivered into 0.3 M sucrose swelling solution at a ratio of 1:100 or 1:1000 accomplishes this. A higher-concentration solution results in less GUVs and a lower concentration causes bead-containing vesicles to be very rare.

10. Proper lighting is crucial for subpixel accuracy particle tracking. The bead should be illuminated so that no pixel is saturated. Saturation creates plateaus on the images indicating lost information about the bead shape.
11. There are other particle-tracking programs that run in other programming environments including an adaptation of the David Grier, John Crocker, and Eric Weeks software that runs in Matlab *(18)*.
12. Tracking randomly positioned particles with subpixel accuracy implies that the position should be random to the smallest digit. One test is to compute a histogram of the x- or y-positions of all the particles tracked in all images mode 1. A flat histogram indicates subpixel accuracy tracking *(18)*.
13. The energy term in SplineSnake can be set to be a combination of gradient and region energies. The gradient term pulls the snake toward regions of largest brightness gradients. In addition, the interior and exterior can also be distinguished in the average brightness or the amount of fluctuations in brightness between inside and outside. This is taken into account by the "region energy" term, which is some energy derived from the mean and variance in brightness differences between the inside and the outside. One advantage of using a regional energy term is that it results in a force on the snake even if the snake is far from the edge *(20,21)*. The inclusion of both of these terms (relative values seem unimportant) results in good vesicle-shape extraction from the images.
14. To convert to polar coordinates one must pick an origin of the coordinate system. The authors use the mean x $<x>$ and mean y $<y>$ positions of the vesicle boundary. For each point set $r = [(x - <x>)^2 + (y - <y>)^2]^{(1/2)}$ and $\theta = \arctan((y - <y>)/(x - <x>))$. Tangent is a multivalued function and this must be accounted for when determining θ, which is defined from 0 to 2π. One must add π to θ for points in the second and third quadrants and 2π for points in the fourth.

Acknowledgments

The authors would like to thank Peter Bradford, Joe Meszaros, and Kumar Senthil for help in working out the current methods, and Philippe Marmottant and Sascha Hilgenfeldt for the initial GUV-formation protocol. This work has been supported by NSF-MRSEC under grant no. DMR-00-80008 and by NIH under grant no. R21-EB-00328501.

References

1. Pier Luigi Luisi and Peter Walde (eds.) (2000) *Giant Vesicles*. Wiley and Sons, New York.
2. Evans, E. and Needham, D. (1987) Physical properties of surfactant bilayer-membranes: thermal transitions, elasticity, rigidity, cohesion, and colloidal interactions. *J. Phys. Chem.* **91,** 4219–4228.
3. Pécréaux, J., Döbereiner, H. -G., Prost, J., Joanny, J.-F., and Bassereau, P. (2004) Refined contour analysis of giant unilamellar vesicles. *Eur. Phys. J. E* **13,** 277–290.
4. Ashkin, A., Dziedzic, J. M., Bjorkholm, J. E., and Chu, S. (1986) Observation of a single beam gradient force optical trap for dielectric particles. *Opt. Lett.* **11,** 288–290.
5. Ashkin, A. (2000) History of optical trapping and manipulation of small-neutral particle, atoms, and molecules. *IEEE J. Sel. Top. Quantum Elec.* **6,** 841–856.
6. Ashkin, A. (1998) Forces of a single-beam gradient laser trap on a dielectric sphere in the ray optics regime. *Methods Cell Biol.* **55,** 1–27.
7. Grier, D. G. (2003) A Revolution in optical manipulation. *Nature* **424,** 810–816.
8. Ashkin, A., Dziedzic, J. M., and Yamane, T. (1987) Optical Trapping and Manipulation of Single Cells Using Infrared Laser Beams. *Nature* **330,** 769–771.
9. Sasaki, K., Koshio, M., Misawa, H., Kitamura, N., and Masuhara, H. (1991) Pattern formation and flow control of fine particles by laser-scanning micromanipulation. *Opt. Lett.* **16,** 1463–1465.
10. Sasaki, K., Fujiwara, H., and Masuhara, H. (1997) Optical manipulation of a lasing microparticle and its application to near-field microspectroscopy. *J. Vacuum Sci. Technol. B* **15,** 2786–2790.

11. Mio, C., Gong, T., Terray, A., and Marr, D. W. M. (2001) Morphological control of mesoscale colloidal models. *Fluid Phase Equilibria* **185,** 157–163.
12. Dufresne, E. R. and Grier, D. G. (1998) Optical tweezer arrays and optical substrates created with diffractive optics. *Rev. Sci. Instr.* **69,** 1974–1977.
13. Dufresne, E. R., Spalding, G. C., Dearing, M. T., Sheets, S. A., and Grier, D. G. (2001) Computer-generated holographic optical tweezer arrays. *Rev. Sci. Instr.* **72,** 1810–1816.
14. John, C. Crocker and David, G. Grier. (1996) Methods of Digital Video Microscopy for Colloidal Studies, *J. Colloid Interface Sci.* **179,** 298–310.
15. Eric, W. Weisstein. (1999) Discrete Fourier Transform. *MathWorld—A Wolfram Web Resource.* http://mathworld.wolfram.com/DiscreteFourierTransform.html
16. Folland, G. B. (1992) *Fourier Analysis and Its Applications.* Brooks/Cole Publishing Co., Pacific Grove, CA.
17. Ryan Smith and Gabe Spalding (2005) *User-Friendly, Freeware Image Segmentation and Particle Tracking.* http://titan.iwu.edu/~gspaldin/rytrack.html
18. John, C. Crocker and Eric, R. Weeks. (2001) *Particle tracking using IDL.* http://www.physics.emory.edu/~weeks/idl/index.html
19. Manouk Abkarian, Colette Lartigue, and Annie Viallat. (2002) Tank Treading and Unbinding of Deformable Vesicles in Shear Flow: Determination of the Lift Force *Phys. Rev. Let.* **88,** 8103–8107.
20. Mathews Jacob (2003) Parametric Shape Processing in Biomedical Imaging, Swiss Federal Institute of Technology Lausanne. EPFL *Thesis no. 2857.*
21. Jacob, M., Blu, T., and Unser, M. (2004) Efficient energies and algorithms for parametric snakes. *IEEE Trans. Image Processing,* **13,** 1231–1244.
22. Mathews Jacob (2003) SplineSnake, http://ip.beckman.uiuc.edu/Software/SplineSnake/
23. Jeremy Pencer, Gisèle F. White, and F. Ross Hallett (2001) Osmotically Induced Shape Changes of Large Unilamellar Vesicles Measured by Dynamic Light Scattering. *Biophys. J.* **81,** 2716–2728.
24. Udo Seifert, Karin Berndl, and Reinhard Lipowsky (1991) Shape Transformations of Vesicles: Phase Diagram for Spontaneous-Curvature and Bilayer-Coupling Models. *Phys. Rev. A* **44,** 1182–1202.
25. Mills, J. P., Qie, L., Dao, M., Lim, C. T., and Suresh, S. (2004) Nonlinear Elastic and Viscoelastic Deformation of the Human Red Blood Cell with Optical Tweezers. *Mech. Chem. Biosyst.* **1,** 169–180.
26. Angelova, M. I., Soléau, S., Méléard, P., Faucon, J. -F., and Bothorel, P. (1992) Preparation of giant vesicles by external ac electric fields: kinetics and applications. *Prog. Colloid Polym. Sci.* **89,** 127–131.
27. Philippe Marmottant and Sascha Hilgenfeldt. (2003) Personal Communication.
28. Bagatolli, L. A. and Gratton, E. (1999) Two-Photon Fluorescence Microscopy Observation of Shape Changes at the Phase Transition in Phospholipid Giant Unilamellar Vesicles. *Biophys. J.* **77,** 2090–2101.

27

Measurement of Lipid Forces by X-Ray Diffraction and Osmotic Stress

Horia I. Petrache, Daniel Harries, and V. Adrian Parsegian

Summary

Lipid suspensions in aqueous solutions most often form multilamellar vesicles of uniformly spaced bilayers. Interlamellar spacing is determined by the balance of attractive van der Waals (charge fluctuation) and repulsive forces. This balance of forces, as well as membrane elasticity, can be probed by applied osmotic stress. We describe how osmotic stress can be imposed on multilamellar lipid samples to study lipid interactions.

Key Words: Bending rigidity; cholesterol hydration; membrane fluctuations; osmotic pressure; spontaneous curvature.

1. Introduction
1.1. Interlamellar Forces

Because of mutually induced charge fluctuations, lipid membranes attract one another *(1)*. This charge fluctuation or van der Waals (vdW) force is responsible for the spontaneous formation of stable multilamellar structures: for example myelin sheets in vivo and multilamellar vesicles (MLVs) in vitro *(2–6)*. The vdW force is present regardless of the charged state of constituent lipids. Because attractive vdW interactions are counteracted by repulsive forces *(2)*, this attraction does not lead to the collapse of membrane stacks. Instead, an equilibrium spacing between lamellae is established to create equally spaced multilamellar structures of hundreds of layers. Measured by small-angle X-ray scattering, the interlamellar repeat spacing of neutral membranes made of phosphatidylcholines (PCs) and/or phosphatidylethanolamines (PEs) in aqueous solvents, is on the order of 50–100 Å, depending on acyl-chain composition, temperature, and other thermodynamic variables *(2)*. This repeat spacing represents the sum of membrane thickness and interlamellar water layer, with roughly equal contributions *(7)*.

Except for purely electrostatic forces, repulsive forces between membranes have an entropic origin. First, the work needed to remove hydrating waters opposes the membrane approach *(2)*. By direct measurement, this force (or free energy penalty) has been shown to be exponential with a decay length of about 2 Å. Second, an additional repulsive force, also of entropic origin, is due to confinement of membrane undulations *(8)*. As measured experimentally *(9)*, this force is also exponential with a decay length of 5–6 Å. The mutual restriction of thermally driven undulation (bending fluctuations) of approaching membranes incurs an entropic penalty. This penalty is proportional to a bending rigidity (elasticity), K_C, and mean-square fluctuations, σ. If the force (per area) needed to push membranes together is measured, the

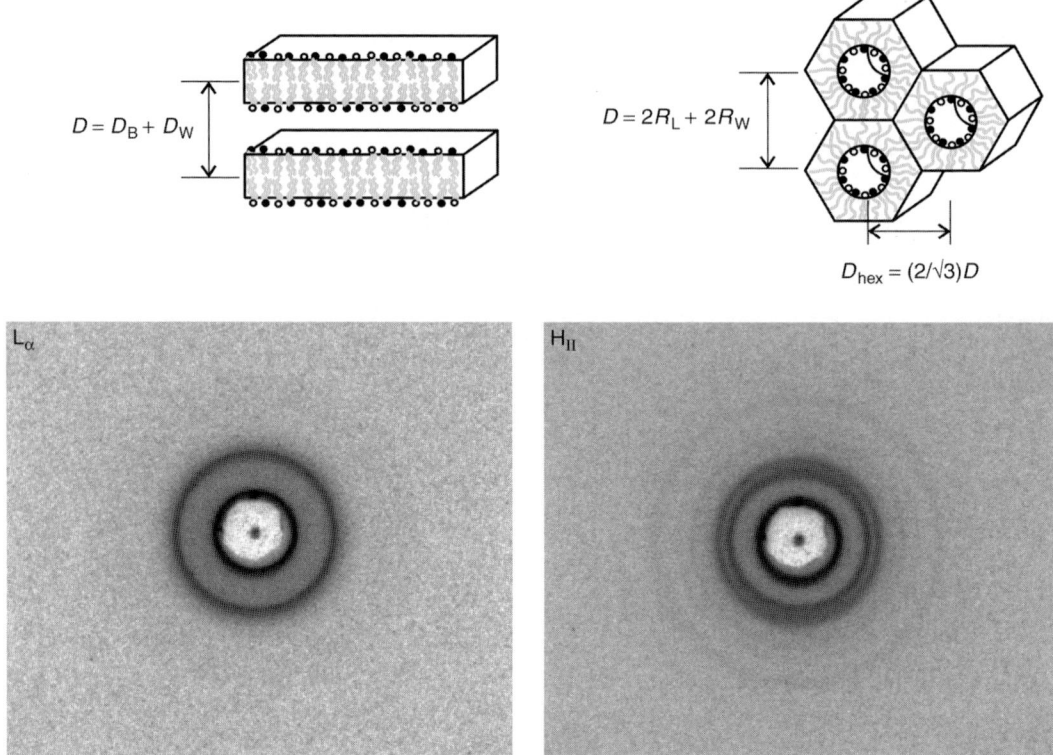

Fig. 1. X-ray scattering from multilamellar (L_α) DMPC and inverted hexagonal (H_{II}) DOPE, both fully hydrated at 35°C. Scattering rings of DMPC are equally spaced and reflect the regular spacing between stacked bilayers. The X-ray rings of DOPE index as 1, $1/\sqrt{3}$, $1/2$, $1/\sqrt{7}$, ... and reflect the honeycomb-like positioning of water cylinders in the H_{II} phase.

"spring constant" that restricts thermal motion is determined. How can it be done? By removing interlamellar water (i.e., by dehydrating the MLVs) using osmotic stress *(10)*. The chapter describes this approach.

1.2. X-Ray Images of Lipid Phases

Depending on acyl-chain and headgroup composition, lipids can form not only lamellar structures but also hexagonal phases (**Fig. 1**). Lipids with saturated acyl chains and PC headgroups tend to form bilayer phases at physiological temperatures. Conversely, lipids with unsaturated chains and PC (demethylated PC) tend to form inverse hexagonal phases. The 14-carbon disaturated dimyristoylphosphatidylcholine (DMPC) and the monounsaturated 18-carbon dioleoylphosphatidylethanolamine (DOPE) are common representatives of each class. DOPE has a negative spontaneous curvature and forms an inverted hexagonal phase when fully hydrated. DMPC, having a much smaller spontaneous curvature, forms a lamellar phase. The geometry of lipid aggregates, and corresponding X-ray pictures are shown in **Fig. 1**. Uniform scattering rings are obtained owing to random, "powder" orientations of lipid suspensions in the X-ray beam. The lattice (repeat) spacings are determined from the position of these rings. For lamellar structures, the rings are equally spaced and indexed simply as 1, 1/2, 1/3, ... (1/h,

Measurement of Lipid Forces

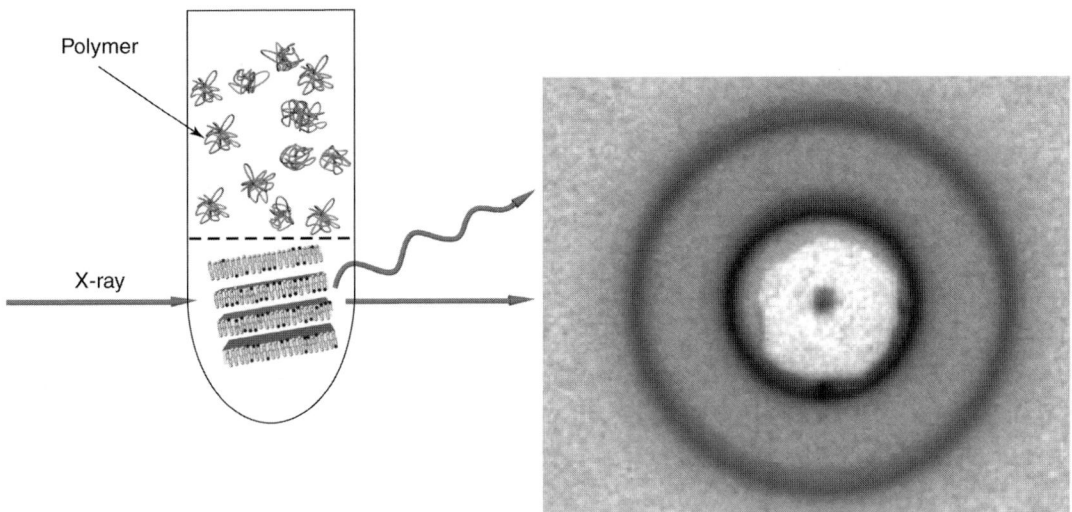

Fig. 2. Schematic of a multilamellar lipid domain under osmotic stress. Using large molecular stressors (polymers) that are excluded from the interlamellar space eliminates the need for a dialysis membrane (dashed line). A typical scattering pattern obtained from randomly oriented domains (MLVs) is shown.

$h = 1, 2, 3, \ldots$), whereas for the H_{II} phase the indexing is $1, 1/\sqrt{3}, 1/2, 1/\sqrt{7}, \ldots$ $(1/(h^2 + hk + k^2)^{1/2}, h, k = 0, \pm 1, \pm 2, \ldots)$.

For fully hydrated DMPC in water at 35°C, the interlamellar repeat spacing is $D = 63$ Å, which is decomposed into a bilayer thickness $D_B = 44$ Å, and a water spacing $D_W = 19$ Å (**Fig. 1**). For DOPE, with a hexagonal lattice spacing $D_{hex} = 64$ Å, $D = 74$ Å, with $2R_L = 36$ Å, and $2R_W = 38$ Å. The water content measured by D_W or R_W, can be reduced by osmotic stress. Although this chapter focuses on lamellar structures, the experimental techniques described apply to hexagonal structures as well.

2. Methods
2.1. Overview of the Osmotic Stress Method

The osmotic stress technique has become a popular tool for investigating the forces acting between lipid bilayers, as well as between other macromolecules *(11,12)*. Osmotic action represents the addition or removal of solvent molecules, most often water. Imposed by the presence of solutes (osmolytes), osmotic pressure acts on the water that hydrates the macromolecular or lipid aggregate. Glycerol, dextran, polyethylene glycol (PEG), and even common salts can be used as osmolytes *(13–18)*. Typically, the method uses either a semipermeable membrane (dialysis bag) that confines the large osmolytes yet allows water (and small osmolytes) to pass, or uses large solutes that are strongly excluded from the lipid aggregates, such as high molecular weight polymers (e.g., PEG) for which a dialysis membrane is not needed. By competing with lipid for available water, osmolytes reduce the equilibrium separation between lipid bilayers. Once equilibrated, the spacing between lipid bilayers at a specified osmotic pressure can be determined using X-ray scattering. This allows to measure the equation of state (force vs separation curves) for the bilayer stack *(9,19–21)*. (**Fig. 2**).

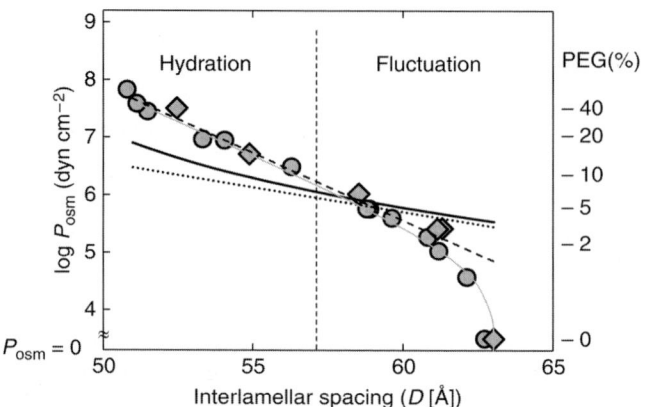

Fig. 3. Osmotic pressure vs interlamellar spacing for DMPC at 30°C (circles) and 35°C (diamonds), and theoretical decomposition into vdW (solid line), hydration (dashed), and fluctuation repulsion (dotted). Two regimes are distinguished: at low osmotic pressure wherein repulsion is dominated by fluctuations; and at high pressure wherein repulsion is mainly because of hydration. Reference PEG concentrations (MW = 20,000 g/mol) are shown on the right. The osmotic pressure vs PEG concentration is measured separately with a vapor pressure osmometer. Figure adapted from **ref. 41**.

2.2. Measurement of Interlamellar Forces

Results from an osmotic pressure experiment of (net neutral) DMPC multilayers are shown in **Fig. 3**. The interlamellar repeat spacing (D) from a series of DMPC samples equilibrated with PEG solutions of various concentrations are plotted for the corresponding pressures P_{osm}. Because relevant osmotic pressure values span many orders of magnitude data are typically plotted on a log scale. Note that the fully hydrated spacing corresponds to zero osmotic pressure and is arbitrarily placed on the log scale. Reference PEG concentrations (mylar window [MW] = 20,000 g/mol) are shown on the right. Osmotic pressure tables from Peter Rand and coworkers can be found at www.brocku.ca/researchers/peter_rand (also mirrored at lpsb.nichd.nih.gov/osmotic_stress.htm).

Because samples are in equilibrium, the applied osmotic pressure is exactly counterbalanced by the net interbilayer force. This force is, to a first approximation, a simple sum of the attractive and repulsive interactions. Interaction energies are given below (**9**).

$$F(D_W) = -\frac{H}{12\pi D_W^2} + P_h \lambda e^{-D_W/\lambda} + \left(\frac{k_B T}{2\pi}\right)\frac{1}{K_C \sigma^2} \quad (1)$$

The first term represents the vdW attraction with interaction strength quantified by the Hamaker parameter H, typically in the order of 1–2 $k_B T$. The second term represents the hydration repulsion with amplitude P_h and decay length λ. The last term represents the undulation (Helfrich) force depending on membrane bending rigidity (K_C) and mean-square fluctuation (σ); for measurement of σ, readers should *see* **refs. 9** and **22**. In **Eq. 1**, the derivative of free energy with respect to the interlamellar water spacing (D_W) equals the applied osmotic stress:

$$P_{osm} = -\frac{dF(D_W)}{dD_W}$$

Two distinct regimes are outlined in **Fig. 3**. At low applied pressures of a few atmospheres, membrane fluctuations serve as the dominant repulsive term, and the hydration force can be neglected. In contrast, at higher pressures, hydration acts as the main repulsive term counteracting vdW attraction. Therefore, it should be expected that low applied osmotic pressures mainly suppress repulsion owing to fluctuations *(9)*.

2.3. Work Done by Osmotic Stress

The applied osmotic pressure is in fact related to the amount of work required to change the separation between bilayers *(23)*. It is useful to realize that this work is the reversible work (or free energy) required to remove water from between the lipid layers and into a reservoir of pure water, conveniently expressed in terms of water's chemical potential (μ_w) that is related to water activity *(a_W)* through

$$\mu_W = \mu^0 + RT \ln a_W$$

where T is the absolute temperature and R is the gas constant; μ^0 is a reference "standard" chemical potential. The molar free energy of water in the lipid phase measured with respect to pure water is:

$$\Delta\mu_W = \mu_W - \mu_W^{bulk}$$

Because lipids form large aggregates that negligibly modify water's chemical potential ($\Delta\mu W$) is also simply a measure of the osmotic pressure (P_{osm}) of the water in lipid solution:

$$\Delta\mu_W = -V_W P_{osm}$$

where V_W is the volume of a water molecule (~30Å^3 at 30°C). Therefore, the incremental work needed to change membranes' spacing to release a small volume of water dV, translates to the chemical potential of water at that spacing,

$$\Delta\mu_W dn = -V_W P_{osm} d(V/V_W) = -P_{osm} dV$$

or expressed in terms of *D*-spacing per lipid with headgroup area A

$$\Delta\mu_W dn = -A P_{osm} dD.$$

At equilibrium, the net flow of water from the lipid phase to the bath is zero; the cost of removing water from one phase must equal the gain of adding it to the other, so that the force per unit area is:

$$P_{osm} = -dF/dD_W.$$

2.4. How to Impose and Measure Osmotic Stress

A simple and often practical way to change water activity in a stack is to limit the amount of available water (so-called gravimetric method) *(24)*. The stack will incorporate as much water as available (if this amount is less than the amount needed for maximal swelling). Once water is incorporated between lipid membranes it will generally be harder to remove than from a pure water phase. Measuring the vapor pressure in the gas phase over the equilibrated multilayers directly assesses this difference *(2)*. It is found that the vapor pressure of water is lower than over a pure water phase (increased osmotic pressure), indicating that added free energy is required to draw water from the lipid phase. The chemical potential of water is lower in the lipid phase compared with pure water.

The complementary strategy to limit the amount of water in the system is to equilibrate the aggregate against a bulk solution (bath) with a known osmotic pressure, allowing only water to move between the two phases *(2)*. For example, water can equilibrate through the vapor phase connecting the system and the bath (hydration chamber principle) *(17,25,26)*, or else through a partition allowing water but not solute to pass (dialysis). Here, any (nonvolatile) solute can be used as the osmotic stressor. Some solutes are so strongly excluded from the lipid phase that they can effectively serve as osmotic stressors even in the absence of a physical partition between the MLVs and the bath. This has been verified experimentally by comparing the interbilayer spacing equilibrated against high molecular weight PEG and other polymers, with and without partition (dialysis bag). However, care must be taken to ensure that solutes are indeed completely excluded. In particular, when smaller solutes are used or when several solutes are present in solution, the solutes can sometimes partition between the bath and aggregate phases as well, thus changing the water chemical potential whereas not always exerting their full osmotic effect *(27–30)*. For example, an added solute may happen to partition equally between the bath and lipid phase; in this case, the solute will exert no net osmotic stress on the bilayer stack, even though it significantly alters water activity.

A typical experiment starts with preparation of PEG solutions of known concentrations. Next, the osmotic pressure for each solution is measured, for example, by vapor pressure osmometry. This method gives an accurate determination of osmotic pressures in the range of just a few to almost 100 atm (~93% relative humidity [RH], *see* **Subheading 2.5.**). Finally, dry lipid is hydrated in excess PEG solutions and left to equilibrate. As the lipid hydrates by taking water from the PEG solution, PEG concentration in the bath increases. However, for small amount of lipid (e.g., 10 mg dry lipid/1 mL PEG solution), this perturbation can be neglected. In different situations, including the case of solute mixtures that can be partially incorporated into the MLVs, the osmotic pressure of the bath should be measured after MLVs have equilibrated. For some widely used solutes such as PEG and dextran, osmotic pressure values vs polymer concentration are available online at www.brocku.ca/researchers/peter_rand and lpsb.nichd.nih.gov/osmotic_stress.htm.

2.5. Osmotic Stress Units

The action of osmolytes can be reported either in pressure units (N/m^2 in SI and dyn/cm^2 in CGS) or corresponding RH with the equation

$$P_{osm} = -(k_B T / V_W) \log RH$$

where V_W represents the volume of a water molecule (~30Å^3 at room temperature). At room temperature, 1 atm then corresponds to RH = 99.9%, 50 atm to RH = 98%, and 100 atm to RH = 93%. In this range of pressures, one has $P_{osm} \cong (1.4 \times 10^9)(1-RH)$.

3. Sample Preparation

Most lipids are now available in high purity (>99%) from commercial sources such as Avanti Polar Lipids (Alabaster, AL) and Sigma-Aldrich (St. Louis, MO). Lipids can be ordered either in powder form or dissolved in chloroform. For work with unoriented MLVs, dry lipids between 3 and 15 mg of per sample (vial) have been used with the steps given below. Note that the equilibration time needed to obtain a homogeneous lipid sample can vary significantly with the lipid type and solvent used. Variations of the protocol are strongly recommended. With many suggestions from Stephanie Tristram-Nagle, Nola Fuller, Peter Rand, and Don Rau; the following procedures worked very well.

3.1. MLVs for X-Ray Scattering

1. Remove lipid bottle from freezer and let it equilibrate at room temperature before opening, to minimize hydration from air.
2. Use a spatula to transfer 3–15 mg of dry lipid into 1–2 mL plastic vials (*see* **Fig. 4**) and weigh. Because hydration is important, using vials provided with *O*-ring screw-caps is preferred. Beware that dry lipids can easily acquire electrostatic charge and fly off the spatula. Some others, such as DOPC, are sticky and require brushing of spatula on the inner walls of the vial to get the lipid off.
3. Add between 850 µL and 1 mL of premade solvent containing one or more of pure water, salt solutions, PEG solutions, and pH buffers.
4. Shake vial gently to dissolve all lipid powder.
5. Thermocycle the samples between 0°C and 5° to 10°C above the lipid melting temperature. Melting temperatures can be found in the online Lipidat database www.ldb.chemistry.ohio-state.edu. PC lipids with saturated chains melt at about 4, 24, 41, and 55°C for 12, 14, 16, and 18 carbons, respectively *(31)*. Phosphatidylserine (PS) and PE lipids melt about 10–20°C higher *(32,33)*. Lipids with unsaturated chains usually melt at lower temperatures. Samples can be shaken by hand or vortexed gently between temperature jumps from low-to-high and back, held at each temperature for about 5–10 min, and the cycle should be repeated at least three times. In principle, automatic thermocycling could also be used.
6. Store samples at 4°C for 48 h. Sedimentation of MLVs occur as shown in **Fig. 4**.
7. Before measurement, let vial equilibrate at the temperature of the measurement for 30 min to 2 h, with longer waiting times for more viscous samples. Gentle shaking or vortexing can also be performed to homogenize the samples. Centrifuge vials at low speed to repellet the MLVs. Note that, depending on lipid and solvent density, the MLVs can either sink or float *(34,35)*. In most cases the MLV aggregate is visible by eye even when not fully opaque.

3.2. Lipid Mixtures

1. Use chloroform or customized organic solvents (*see* Chapter 5) to codissolve measured quantities of lipid. Use standard glass test tubes.
2. Evaporate solvent under chemical hood using either an evaporator or manually by flowing nitrogen or argon gas. After evaporation, the lipid should form a thin film on the glass walls. Best results are obtained with lipid quantities less than 50 mg per test tube.
3. Cover test tubes with parafilm and punch needle holes for further evaporation. Place under high vacuum for at least 4 h or overnight if the sample is chemically stable.
4. There are many options at this point. Samples can be scraped off the glass, transferred into plastic vials, and hydrated according to the protocol in **Section 3.9**. Alternatively, samples can be hydrated in the original test tubes, thermocycled, and stored as such for short period. Seal with parafilm. Another alternative is to resuspend the lipid film in hexane or water and then lyophilize. This is done by quick freezing in dry ice followed by evaporation under high vacuum. After lyophilization, the fluffy powder can be hydrated as done previously.

3.2.1. Complications

The most delicate step is the weighing of dry lipid and the pipetting of small chloroform solution volumes. Errors in the estimate of lipid fractions will occur. It is best to deliver the

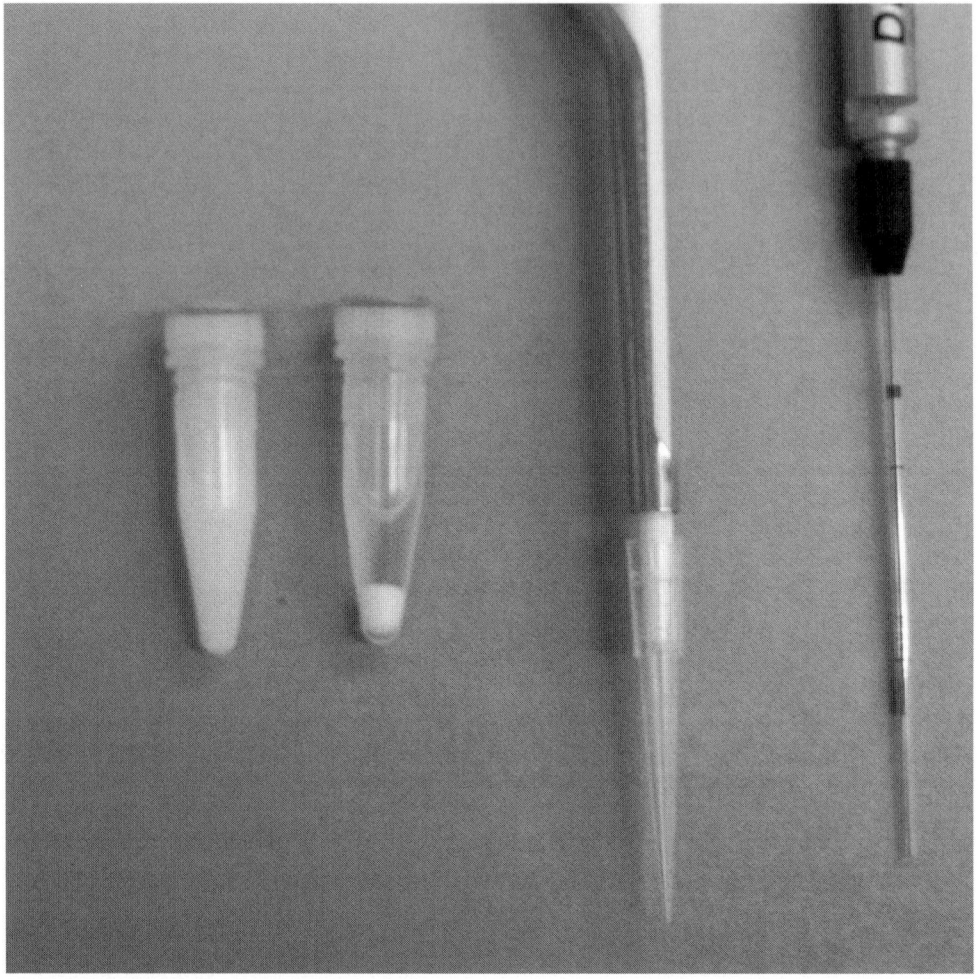

Fig. 4. Visual appearance of MLVs suspensions in aqueous solvents before and after sedimentation. The lipid pellet can be transferred into the sample holder using either a 10–100 µL micropipet, or a positive displacement 10–50 µL capillary pipet for viscous samples.

first chloroform drop on the bottom of test tubes, then the next containing the different lipid very close to it, but do so on the tube wall without touching the previous drop with the pipet tip.

3.3. Loading the X-Ray Sample Holder
3.3.1. The Sample Cell

Traditional sample cells (SCs) originally designed by Vittorio Luzzati and further modified by Peter Rand and coworkers are used. These cells contain a Teflon holder (TH) for the lipid pellet that is sandwiched between two mylar windows as shown in **Fig. 5**. The exact dimensions of the SC and holders are not critical as long as they match. Aluminum cells with a central hole of 2 mm in diameter were used. This central area will sit in the X-ray beam. The X-ray setup includes a matching cell crate sitting on a Peltier device that controls sample temperature. Once properly aligned, this setup allows remounting of the SCs without further adjustment.

Measurement of Lipid Forces

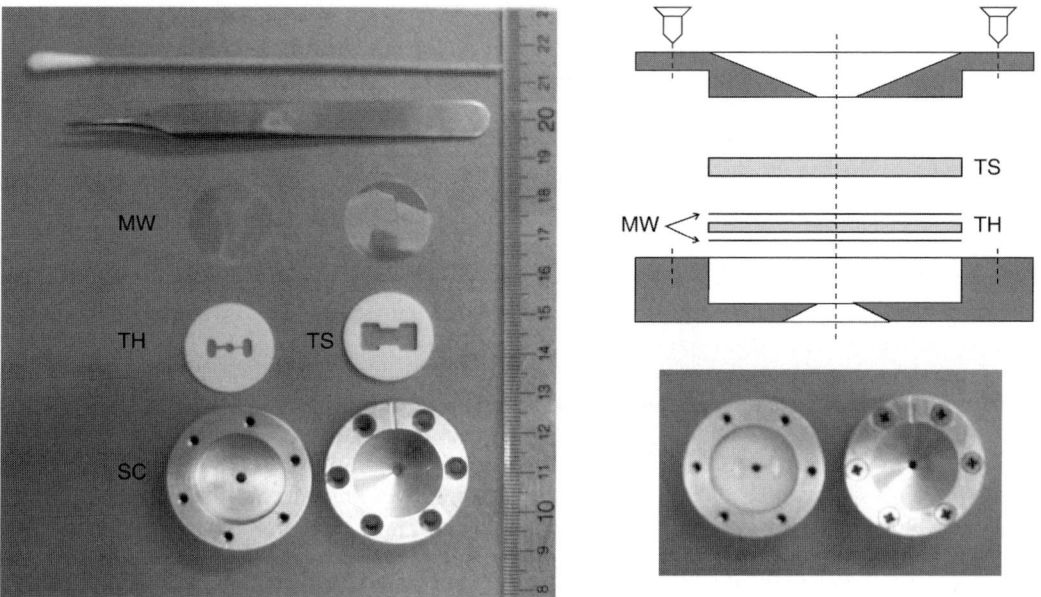

Fig. 5. The SC assembly. The lipid suspension is sequentially loaded into the teflon holder, sandwiched between mylar windows, placed inside the SC, and secured with screws. Tweezers and cotton swabs can be used to manipulate the MWs as described. The SC can be placed in the X-ray beam either way; however, the sample-to-film distance can differ between the two orientations and needs to be checked with a calibrator (e.g., AgBeh with interlamellar repeat spacing $D = 58.4$ Å (*see* **ref. 36**). Because the TS is put systematically on top of TH, the upper (shown) side of the cell is used as the X-ray exit side.

3.3.2. Teflon Holders

Holders are cut from 0.8–1 mm Teflon sheet. A central hole of 2-mm diameter is used to deposit the lipid pellet. Additional cutting can be made to hold extra PEG solution for multiple temperature measurements without reloading the cell.

3.3.3. Mylar Windows

Windows are cut from 100 μm mylar film by using a punch tool. The film is wrapped several times around a piece of cardboard so as to cut through many layers at once. Windows can also be cut by hand because they do not need to be perfectly circular. Before using, windows should be wiped with lens paper to remove dust and lint.

3.3.4. Loading the Sample Cell

1. Prepare a clean Sample cell, Teflon holder, Teflon spacer (TS), and two mylar windows.
2. Apply a thin film of vacuum grease on the TH to insure a good seal with the MWs. Avoid spreading grease on the holder center. If it does happen, grease can be removed using the wooden part of a cotton swab.
3. Using tweezers, stick one MW to one side of the TH (press circularly with the cotton swab), and set the holder inside the SC (**Fig. 5**).
4. Using a pipet, transfer between 10 and 2 μL of lipid pellet into the TH.

5. If the TH has lateral reservoirs, as in **Fig. 5**, fill them with solvent and make sure the solvent drop is continuous from side to side.
6. Using the tweezers, cover the sample with the second MW, avoiding trapping bubbles (slight overloading helps prevent bubbles). For a good seal use the cotton swab to press the mylar gently onto the TH.
7. Place the TS and the cell lid on top. The spacer is only needed if the height of the cell exceeds the width of the TH. Secure the lid with screws.
8. Re-equilibrate the loaded SC at the temperature of the measurement.

3.3.5. Complications

1. In principle, lipids might become contaminated with vacuum grease used to seal the MWs. No indication of such effects have been observed in the studies of PC, PE, and PS lipids at temperatures between 5° and 50°C while in aqueous buffer solutions. If such contamination is suspected, measurements can be done without vacuum grease by relying on a pressure seal between the mylar and Teflon. Dehydration of the sample should then be checked, for example, by repeating the X-ray scan.
2. In the absence of control X-ray data, lipid degradation resulting from oxidation or other chemical reactions should be checked, for example by thin-layer chromatography.

4. The X-Ray Measurement

4.1. A Basic X-Ray Setup

The basic X-ray setup permits quick sample replacement and accurate measurements of interlamellar spacings (**Fig. 6**).

The SC is held in the X-ray beam by sitting in a fixed crate, thermally regulated by a Peltier device, X-ray scattering is recorded on an image plate placed inside a 5 × 4 Fidelity Elite film holder (Fidelity, Sun valley, CA). The original entrance window was removed and replaced with a sheet of black paper usually used to protect photographic film. This paper is opaque at optical wavelengths but transparent to X-rays. To reduce background scattering, the X-ray beam travels between the sample and film through a helium-filled flight path (FP). Helium gas is flown continuously at very low pressure from a helium tank, through the flight path, and bubbled through an oil-filled vial out into the room. A semitransparent beamstop is fixed with double-stick tape on the exit window of the FP, just in front of the film. The shadow of the beamstop is visible in **Fig. 1**.

4.2. X-Ray Measurement and Analysis

Typical exposure times on the machine vary between 20 min and 3 h, depending on the concentration and ordering of MLVs. Fully hydrated samples tend to give weaker scattering than samples under osmotic stress. Phosphor imaging plates are used as detectors and a Fujifilm BAS-2500 plate reader (Fujifilm Life Science, Stamford, CA) capable of 50-μm pixel size. For analysis of X-ray spectra, the software FIT2D by Andy Hammersley (www.esrf.fr/computing/scientific/FIT2D) is very good and easy to use. To calculate the D-spacing, the sample-to-film distance is needed. This can be determined using silver behenate (AgBeh) powder as a calibrator. The D-spacing of AgBeh is 58.4 Å at room temperature *(36)*. For small angle scattering, the sample-to-film distance is

$$s = \Delta y (D / \lambda)$$

where Δy is the position of the first scattering ring in millimeters, D is the lattice spacing of AgBeh, and $\lambda = 1.54$ Å is the X-ray wavelength (from a Cu anode) (**Fig. 7**).

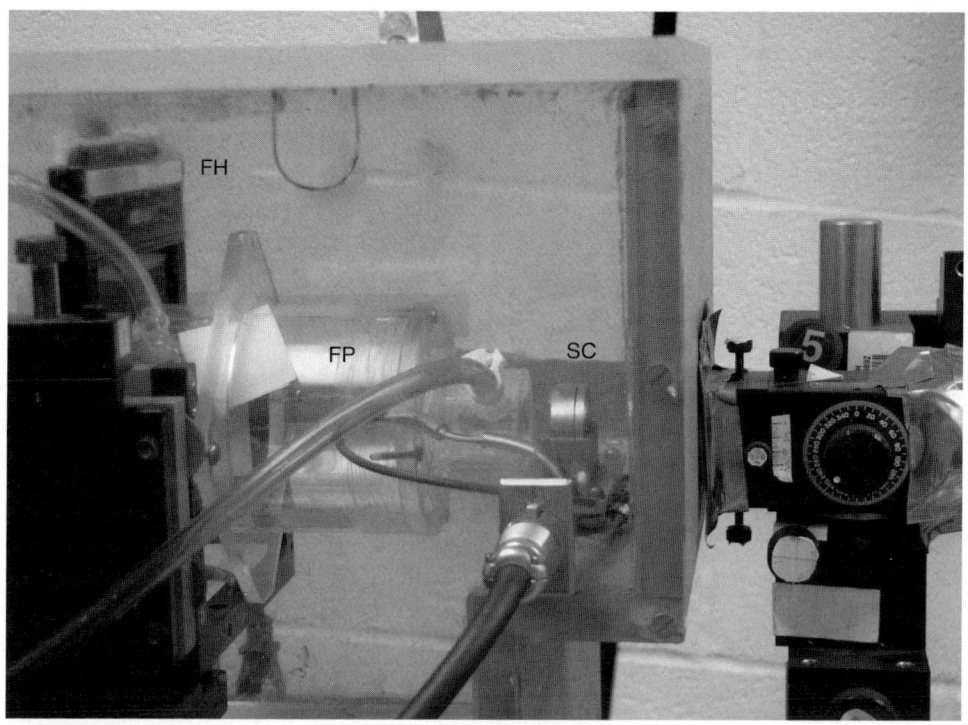

Fig. 6. Sample mounting on an X-ray port. The geometry is fixed. Sample cell, film holder, and flight path sit together inside a plexiglass box that can be oriented as needed to maximize X-ray intensity from the anode and through the focusing mirrors (*see* **ref. 42** for information on mirror alignment).

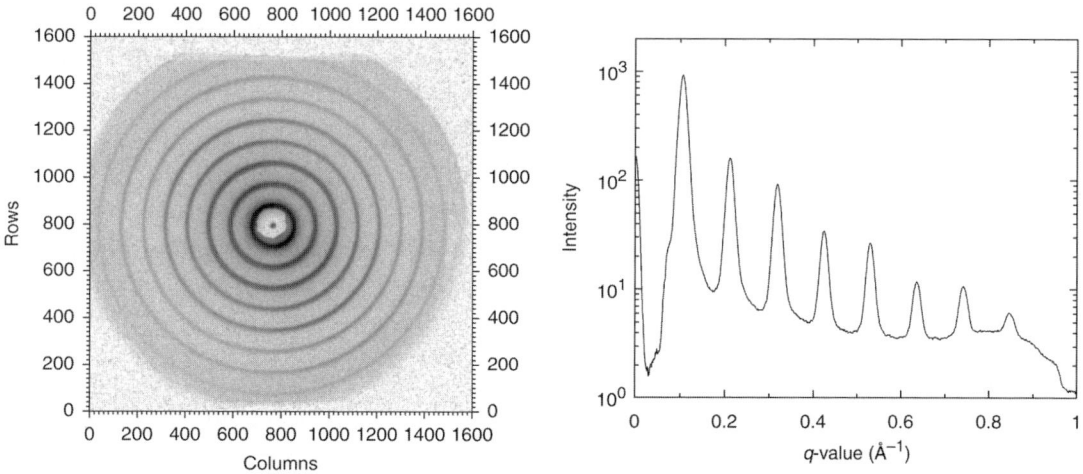

Fig. 7. X-ray scattering from AgBeh powder and radially integrated profile using the program FIT2D available from www.esrf.fr/computing/scientific/FIT2D. The known lattice spacing of 58.4 Å *(36)* allows calculation of the sample-to-film distance. AgBeh was kindly provided by Stephanie Tristram-Nagle.

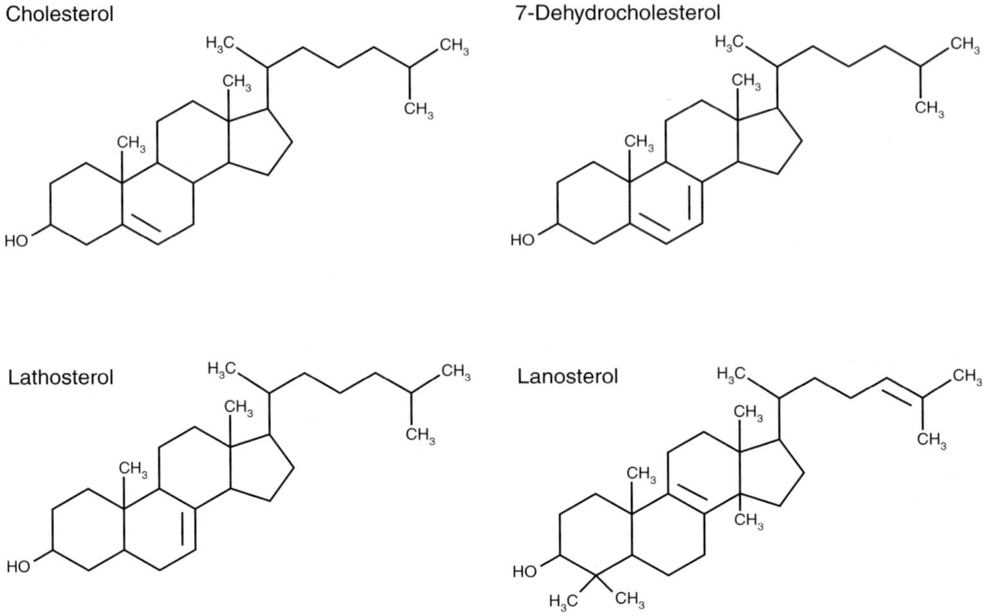

Fig. 8. Chemical structure of cholesterol and some of its metabolic precursors.

5. A Practical Example: Alteration of Membrane Rigidity by Sterols
5.1. A Cholesterol Problem

It is widely believed that an excess of cholesterol leads to atherosclerosis, cardiovascular diseases, and stroke. However, cholesterol deficiency is also dangerous, leading to serious congenital anomalies and mental retardation in newborns. These deficiencies occur when the chain of enzymatic reactions that lead to formation of cholesterol is broken *(37,38)*. Cholesterol precursors, such as the immediate precursor 7-dehydrocholesterol then accumulate in the cell instead of cholesterol. Although cholesterol and its precursors differ only by the number and position of double bonds and methyl groups (**Fig. 8**), it is found that bilayer rigidity conferred by cholesterol differ significantly from its precursors.

5.2. Applying Osmotic Stress to Sterol-Containing Bilayers

Using the repeat spacing as an indicator of modified interbilayer interaction, the modification of bilayer interactions by addition of sterols can be investigated. As shown in **Fig. 8**, there are marked differences between the effects of sterols on interlamellar repeat spacings at full hydration. For all bilayer compositions, the lamellar repeat spacing increases in the order: cholesterol < lathosterol < 7-dehydrocholesterol < lanosterol. However, differences vanish under only 0.26 atm of osmotic stress (5% PEG solutions, log P_{osm} = 5.4) when fluctuations are suppressed. From the modifications of interlamellar spacings, the modification of bending rigidities have been calculated using **Eq. 1**, as shown in **Fig. 9**.

Differences in sterol effects on membrane rigidity can be rationalized in terms of sterol location within the lipid bilayer. It is expected that sterols introduce an inhomogeneous

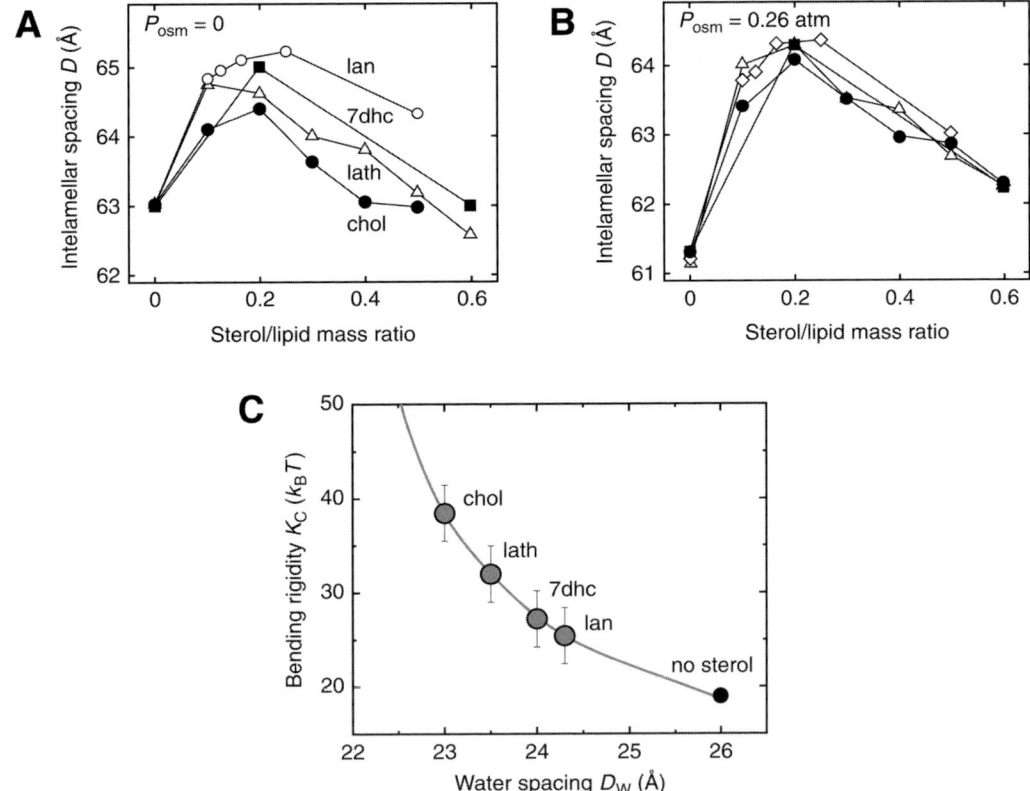

Fig. 9. Interlamellar repeat spacing vs sterol content for DMPC multilayers at 35°C. Differences between sterols measured at full hydration (**A**) vanish when fluctuations are suppressed by mild osmotic stress (**B**). (**C**) Membrane bending rigidity (K_C) vs equilibrium spacing obtained from **Eq. 1** (solid line). Symbols indicate K_C and D_W measured for 30 mol% sterols. Figure adapted from **ref. 41**.

modification of lateral forces within bilayers. For example, more polar sterols (owing to additional double bonds) might protrude further into the headgroup–water interface than cholesterol, and render the bilayer more flexible. The variation of D with sterol content in **Fig. 8** reflects the phase diagram of DMPC–sterol mixtures. At 35°C, a coexistence region exists between a liquid-disordered phase (with low sterol content) and a liquid-ordered phase (with high sterol content). The steep decline in the D-spacing values starting at a sterol/lipid mass ratio of 0.2 (30% mole sterol) corresponds to the transition to the liquid-ordered phase.

Although many questions regarding lipid–cholesterol interactions still remain, this example demonstrates the versatility of the osmotic stress technique. Used here primarily to analyze interlamellar interactions, osmotic stress also probes interactions within membranes. Because dehydration not only reduces intermembrane separation but also condenses lipids laterally (*23*), the work of lateral deformation can be measured (*39,40*) and related to forces acting in the membrane plane. Capable of revealing forces with immediate biological relevance, the osmotic stress technique should become the method of choice for investigating interactions between and within membranes.

Acknowledgments

We thank Stephanie Tristram-Nagle, Nola Fuller, Peter Rand, and Don Rau for teaching us the methods in this chapter and much more. This research was supported by the Intramural Research Program of the National Institutes of Health and National Institute of Child Health and Human Development.

References

1. Parsegian, V. A. (2005) *Van der Waals forces: A Handbook for Biologists, Chemists, Engineers, and Physicists*. Cambridge University Press, ISBN 0521547784 (New York, NY)
2. Rand, R. P. and Parsegian, V. A. (1989) Hydration forces between phospholipid bilayers. *Biochim. Biophys. Acta.* **988**, 351–376.
3. Luzzati, V. and Husson, F. (1962) The structure of the liquid-crystalline phases of lipid-water systems. *J. Cell Biol.* **12**, 207–219.
4. Luzzati, V., Reis-Husson, F., Rivas, E., and Gulik-Krzywicki, T. (1966) Structure and polymorphism in lipid-water systems, and their possible biological implications. *Annu. NY Acad. Sci.* **137**, 409–413.
5. Rand, R. P. and Luzzati, V. (1968) X-ray diffraction study in water of lipids extracted from human erythrocytes. *Biophys. J.* **8**, 125–137.
6. Luzzati, V., Benoit, E., Charpentier, G., and Vachette, P. (2004) X-ray scattering study of pike olfactory nerve. II-Elastic, thermodynamic and physiological properties of the axonal membrane. *J. Mol. Biol.* **343**, 199–212.
7. Nagle, J. F. and Tristram-Nagle, S. (2000) Structure of lipid bilayers. *Biochim. Biophys. Acta - Rev. Biomembr.* **1469**, 159–195.
8. Helfrich, W. (1973) Elastic Properties of Lipid Bilayers—Theory and Possible. Experiments. *Z. Naturforsch. C.* **28**, 693–703.
9. Petrache, H. I., Gouliaev, N., Tristram-Nagle, S., Zhang, R. T., Suter, R. M, and Nagle, J. F. (1998) Interbilayer interactions from high-resolution x-ray scattering. *Phys. Rev. E* **57**, 7014–7024.
10. Parsegian, V. A., Rand, R. P., Fuller, N. L., and Rau, D. C. (1986) Osmotic-Stress for the Direct Measurement of Intermolecular Forces. *Methods Enzymol.* **127**, 400–416.
11. Parsegian, V. A., Rand, R. P., and Rau, D. C. (2000) Osmotic stress, crowding, preferential hydration, and binding: A comparison of perspectives. *Proc. Nat. Acad. Sci. USA* **97**, 3987–3992.
12. Parsegian, V. A., Rand, R. P., and Rau, D. C. (1995) Macromolecules and water: probing with osmotic stress. *Methods Enzymol.* **259**, 43–94.
13. Harries, D., Rau, D. C., and Parsegian, V. A. (2005) Solutes probe hydration in specific association of cyclodextrin and adamantane. *J. Am. Chem. Soc.* **127**, 2184–2190.
14. Bonnet-Gonnet, C., Leikin, S., Chi, S., Rau, D. C., and Parsegian, V. A. (2001) Measurement offerees between hydroxypropylcellulose polymers: Temperature favored assembly and salt exclusion. *J. Phys. Chem. B* **105**, 1877–1886.
15. Chik, J., Mizrahi, S., Chi, S. L., Parsegian, V. A., and Rau, D. C. (2005) Hydration forces underlie the exclusion of salts and of neutral polar solutes from hydroxypropylcellulose. *J. Phys. Chem. B.* **109**, 9111–9118.
16. Deme, B. and Zemb, T. (2000) Measurement of sugar depletion from uncharged lamellar phases by SANS contrast variation. *J. Appl. Crystallogr.* **33**, 569–573.
17. McIntosh, T. J. and Simon, S. A. (1993) Contributions of hydration and steric (entropic) pressures to the interactions between phosphatidylcholine bilayers-Experiments with the subgel phase. *Biochemistry* **32**, 8374–8384.
18. Rau, D. C., Lee, B., and Parsegian, V. A. (1984) Measurement of the repulsive force between polyelectrolyte molecules in ionic solution: Hydration forces between parallel DNA double helices. *Proc. Nat. Acad. Sci. USA-Biol. Sci.* **81**, 2621–2625.

19. LeNeveu, D. M., Rand, R. P., and Parsegian, V. A. (1976) Measurement offerees between lecithin bilayers. *Nature* **259**, 601–603.
20. McIntosh, T. J. (2000) Short-range interactions between lipid bilayers measured by X-ray diffraction. *Curr. Opin. Struct. Biol.* **10**, 481–485.
21. Leikin, S., Parsegian, V. A., Rau, D. C., and Rand, R. P. (1993) Hydration forces. *Annu. Rev. Phys. Chem.* **44**, 369–395.
22. Lyatskaya, Y., Liu, Y. F., Tristram-Nagle, S., Katsaras, J., and Nagle, J. F. (2001) Method for obtaining structure and interactions from oriented lipid bilayers. *Phys. Rev. E* **63**, 011907.
23. Parsegian, V. A., Fuller, N., and Rand, R. P. (1979) Measured work of deformation and repulsion of lecithin bilayers. *Proc. Nat. Acad. Sci. USA* **76**, 2750–2754.
24. Gawrisch, K., Ruston, D., Zimmerberg, J., Parsegian, V. A., Rand, R. P., and Fuller, N. (1992) Membrane dipole potentials, hydration forces, and the ordering of water a membrane surfaces. *Biophys. J.* **61**, 1213–1223.
25. Katsaras, J. (1998) Adsorbed to a rigid substrate, dimyristoylphosphatidylcholine multibilayers attain full hydration in all mesophases. *Biophys. J.* **75**, 2157–2162.
26. Kucerka, N., Liu, Y. F., Chu, N. J., Petrache, H. I., Tristram-Nagle, S. T., and Nagle, J. F. (2005) Structure of fully hydrated fluid phase DMPC and DLPC lipid bilayers using X-ray scattering from oriented multilamellar arrays and from unilamellar vesicles. *Biophys. J.* **88**, 2626–2637.
27. LeNeveu, D. M., Rand, R. P., Parsegian, V. A., and Gingell, D. (1977) Measurement and modification offerees between lecithin bilayers. *Biophys. J.* **18**, 209–230.
28. Lis, L. J., McAlister, M., Fuller, N., Rand, R. P., and Parsegian, V. A. (1982) Interactions between neutral phospholipid bilayer membranes. *Biophys. J.* **37**, 657–665.
29. McDaniel, R. V., McIntosh, T. J., and Simon, S. A. (1983) Nonelectrolyte substitution for water in phosphatidylcholine bilayers. *Biochim. Biophys. Acta* **731**, 97–108.
30. Dubois, M., Zemb, T., Belloni, L., Delville, A., Levitz, P., and Setton, R. (1992) Osmotic pressure and salt exclusion in electrostatically swollen lamellar phases. *J. Chem. Phys.* **96**, 2278–2286.
31. Koynova, R. and Caffrey, M. (1998) Phases and phase transitions of the phosphatidylcholines. *Biochim. Biophys. Acta-Rev. Biomembr.* **1376**, 91–145.
32. Cevc, G. and Marsh, D. (1985) Hydration of noncharged lipid bilayer membranes. Theory and experiments with phosphatidylethanolamines. *Biophys. J.* **47**, 21–31.
33. Hauser, H., Paltauf, F., and Shipley, G. G. (1982) Structure and thermotropic behavior of phosphatidylserine bilayer membranes. *Biochemistry* **21**, 1061–1067.
34. Wiener, M. C., Tristram-Nagle, S., Wilkinson, D. A., Campbell, L. E., and Nagle, J. F. (1988) Specific volumes of lipids in fully hydrated bilayer dispersions. *Biochim. Biophys. Acta* **938**, 135–142.
35. Petrache, H. I., Kimchi, I., Harries, D., and Parsegian, V. A. (2005) Measured depletion of ions at the biomembrane interface. *J. Am. Chem. Soc.* **127**, 11,546–11,547.
36. Huang, T. C., Toraya, H., Blanton, T. N., and Wu, Y. (1993) X-ray powder diffraction analysis of silver behenate, a possible low-angle diffraction standard. *J. Appl. Crystallogr.* **26**, 180–184.
37. Lehninger, A. L. (1975) *Biochemistry*, Worth Publishers Inc., New York.
38. Porter, F. D. (2003) Human malformation syndromes due to inborn errors of cholesterol synthesis. *Curr. Opin. Pediatr.* **15**, 607–613.
39. Lis, L. J., McAlister, M., Fuller, N., Rand, R. P., and Parsegian, V. A. (1982) Measurement of the lateral compressibility of several phospholipid bilayers. *Biophys. J.* **37**, 667–672.
40. Petrache, H. I., Tristram-Nagle, S., Gawrisch, K., Harries, D., Parsegian, V. A., and Nagle, J. F. (2004) Structure and fluctuations of charged phosphatidylserine bilayers in the absence of salt. *Biophys. J.* **86**, 1574–1586.
41. Petrache, H. I., Harries, D., and Parsegian, V. A. (2004) Alteration of lipid membrane rigidity by cholesterol and its metabolic precursors. *Macromol. Symp.* **219**, 39–50.
42. Phillips, W. C. and Rayment, I. (1985) A systematic method for aligning double-focusing mirrors. *Methods Enzymol.* **114**, 316–329.

28

Micropipet Aspiration for Measuring Elastic Properties of Lipid Bilayers

Marjorie L. Longo and Hung V. Ly

Summary

Micropipet aspiration of giant unilamellar vesicles can be used to determine the mechanical properties of area compressibility modulus, bending modulus, and lysis tension of lipid bilayers. In this technique, giant (~25-μm diameter) single bilayered vesicles are aspirated into a pipet of inner diameter ≈8 μm. The changes in projection length of each vesicle inside of the pipet in response to changes in aspiration (suction) pressure are used to determine mechanical moduli. The suction pressure of vesicle rupture (lysis) is used to determine the membrane tension of lysis (lysis tension). Micropipet aspiration of giant unilamellar vesicles is highly specialized, requiring custom laboratory fabrication of most components (e.g., micropipets, chambers, and manometer). Herein, methods for fabrication of each of these components and instructions for measurements are described.

Key Words: Area compressibility modulus; bending modulus; giant vesicles; lysis tension; micromechanical properties; micropipet.

1. Introduction

The mechanics of lipid membranes play an important role in the stability, permeability, deformability, and viability of natural cells and drug-delivery liposomes. For example, activities of integral and anchored proteins depend on their native structures, which are strongly influenced by the surrounding lipid arrangement in the lipid bilayer (1). The lipid bilayer matrix is a dynamic system whose structure is regulated by short-range weak forces that result from interactions between lipid and water molecules and between lipid molecules themselves. These forces include hydrophobic forces between the lipid tails and water molecules, van der Waals interactions between the tails, and electrostatic interactions in the headgroups (2). The fine lipid microstructure gives rise to measurable macroscopic bilayer characteristics such as fluidity, permeability, and mechanical properties, these characteristics being interrelated. Mechanical properties include the membrane area expansion modulus (K_A), which represents the membrane resistance to isotropic area dilation, and the bending modulus (k_c), which represents the energetic penalty in deforming a flat surface into a curved structure.

Many techniques exist for measuring the mechanical stretch properties of bilayers including photon correlation spectroscopy (3), dynamic light scattering (4), and nuclear magnetic resonance and X-ray diffraction (5), but the most prominent for accurate and direct measurement of a single bilayer system is video microscopy micropipet aspiration (MPA). MPA has been used extensively to study mechanical and viscous properties of natural biological membranes, for example, red blood cells (6) and outer hair ear lateral wall (7); synthetic lipid vesicles (see **Fig. 1**), for example, egg lecithin (8), phosphatidylcholines with varying chain length and

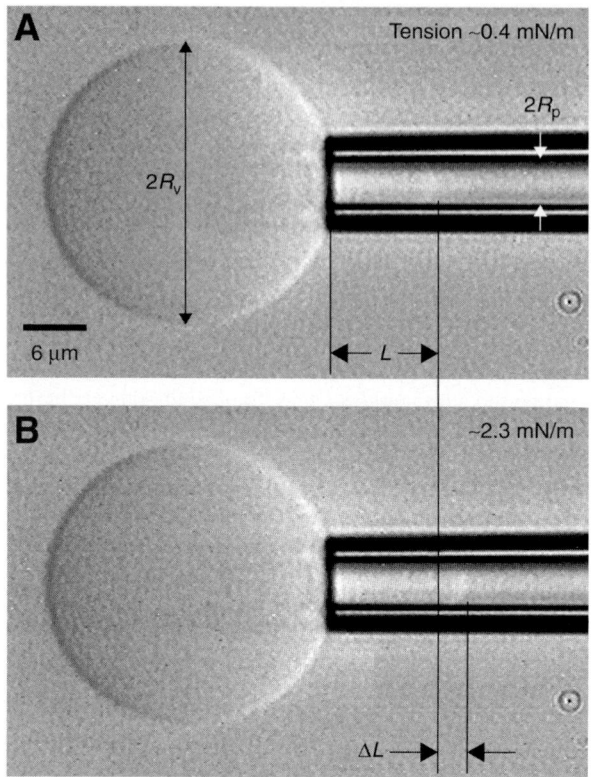

Fig. 1. Video micrograph of a GUV aspirated into a micropipet at (**A**) low and (**B**) medium tension. The change in projection length (ΔL) is proportional to the change in apparent surface area (ΔA).

degree of unsaturation *(9)*, and phosphatidylcholines/cholesterol *(10,11)*, and also diblock polymers *(12)*. MPA has been applied to measure adsorption of small molecules into the bilayer, such as the influenza hemagglutinin-fusion peptide *(13,14)*, alcohols *(15)*, and lysolipids *(16)*, study the effect of chain unsaturation on water permeability *(17)*, and measure interfacial tension of microscopic liquid–gas, liquid–liquid interfaces, and phospholipid monolayers *(18–20)*.

MPA requires custom laboratory fabrication of most components. Herein, methods for fabrication of each of these components and instruction for measurements are described. Sequentially, steps will be presented for fabricating and measuring pipets, fabricating needle probes for measuring pipet inner diameter (ID), coating glass cover slips, making giant vesicles by electroformation, setting up for a MPA experiment, measuring area compressiblity modulus, bending modulus and lysis tension, and analyzing the data.

2. Materials
2.1. Formation of Giant Vesicles and MPA
1. Glass reaction vials or V-vials (Fisher, Pittsburgh, PA) stored in chloroform when not in use. The vials should be equipped with PTFE screw caps or PTFE septa.
2. Stock solutions of lipids (Avanti Polar Lipids, Alabaster, AL) dissolved in chloroform at concentrations that range from 10 to 25 mg/mL housed in glass vials with PTFE-lined caps or PTFE septa. When storing lipids in organic solvent or as powder, use a freezer (not frost-free) and seal with PTFE tape around the closure.

3. 100 mM Sucrose solution; store in refrigerator for 3 d.
4. Glucose solution of concentration close to 100 mM; concentration adjusted for optimum vesicle projection length (*see* **Subheading 3.6., step 8**). Store in refrigerator for 3 d.
5. Bovine serum albumin (BSA) (fraction V, low heavy metals) from Calbiochem (San Diego, CA).
6. Gastight® glass syringes (10 μL, 250 μL, and 1 mL) each with attached needle, 10-mL luer lock with wide gauge needle, 1-mL luer lock with wide-gauge needle fitted with small amount of PTFE tubing (ID ~ 1mm) (Hamilton Syringe Company, Reno, NV).
7. 3-mL Disposable syringes, one filled with vacuum grease.
8. 1 Vol% Surfasil® (Pierce, Rockford, IL) solution in chloroform in an approx 250-mL glass container wide enough to vertically submerge a 45 × 50 mm^2 microscope cover slip. The container should be sealed with a screw cap and stored and used under a fume hood.
9. Chloroform, methanol, and ethanol each in the same type of screw cap container used in **step 8**, stored and used under a fume hood.
10. Diamond scribe (e.g., Glascribe® Pen, Bel-Art Products, Pequannock, NJ) and clean ruler (to act as straight edge) for cutting glass.
11. Pyrex 7740 glass capillary tubing (Frierich & Dimmock, Inc., Millville, NJ), external diameter (OD) ~0.9 mm, 0.2 mm wall, and 5 ft long.
12. Leaded glass powder (1 mL will last a lifetime).
13. Tungsten-carbide glass cutter for cutting pyrex tubing (e.g., from VWR Scientific Products Westchester, PA).
14. Large (e.g., 45 × 50 mm^2) no. 1 cover slips.
15. Cover slip holder for drying cover slips.
16. Flexible 34-gauge syringe needle for backfilling pipets (e.g., Microfil™, World Precision Instruments, Inc., Sarasota, FL).
17. 0.2-μm Disposable filter cassettes for use with syringes.
18. New condition, very sharp wire cutters.
19. Clear flexible laboratory tubing of various sizes and tubing connectors (e.g. from VWR Scientific Products, Westchester, PA).

2.2. Chambers

1. Electroformation chamber for making giant unilamellar vesicles (GUVs). **Figure 2A** shows a two-dimensional drawing of recommended design. Basically, it is a PTFE block with a large rectangular hole in the center and holes drilled on the sides. Two 7.5-cm sections of 1-mm diameter platinum wire (Aldrich, Milwaukee, WI) are threaded through the larger holes (distance between edges of wires ~2.5 mm); a narrow boar stainless steel syringe needle is pushed (just for tightness) into a smaller hole and a wider stainless steel syringe needle is pushed (just for tightness) into a larger hole. A Surfasil-coated cover glass (45 × 50 mm^2) fits over each large side of the PTFE block. Vacuum grease is used between the PTFE and each cover glass to seal each cover glass. Pushing down gently ensures a seal. This chamber is designed with a clear center so that, if desired, electroformation can be monitored by a microscope.
2. MPA chamber and humidity cover. **Figure 3** shows a three-dimensional drawing of recommended design. The MPA chamber is made of two small rectangular plexiglass (e.g. from United States Plastics Corp., Lima, OH) spacers, separated by 1.5 cm, sandwiched between two rectangular cut-cover-glasses (Surfasil protected) with vacuum grease (**Fig. 3B**). The result is a rigid chamber with an open transparent square cavity size of approx 0.7 cm^3. To minimize evaporation at the open ends of the sample chamber, the preparation is placed inside a humidified chamber on the microscope stage.

A humidified chamber can be made from two modified Petri dishes. A side slot for the micropipet and a center open square hole for the objective lens can be cut in a large Petri dish (e.g., 100 × 20 mm^2). A square hole can be cut into its top cover and sealed by a cover glass for uniform light entry through the condenser. A smaller Petri dish (e.g., 60 × 15 mm^2) with the same open square hole cut in the bottom can be glued-flush to the bottom of the larger

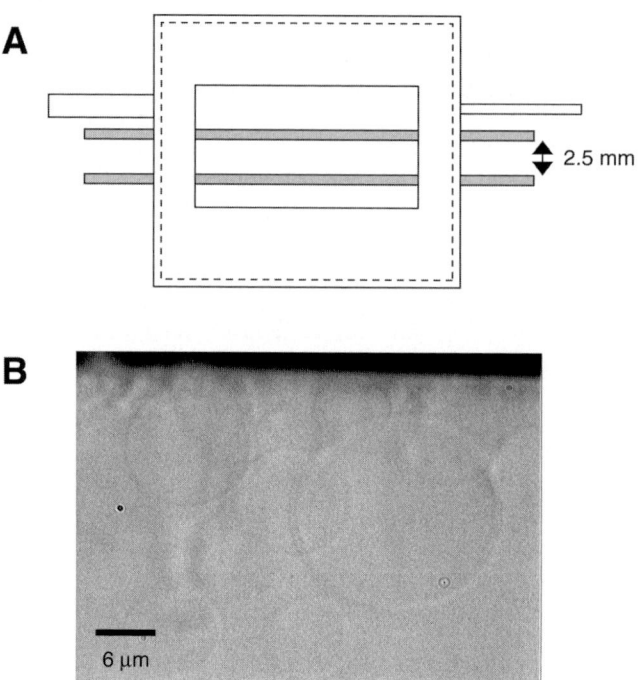

Fig. 2. **(A)** Electroformation chamber, enclosed by two glass cover slips (45 × 50 mm^2) sealed with vacuum grease. Side syringe ports are for exchanging fluid and two platinum electrodes (distance of separation 2.5 mm edge-to-edge) are attached to a function generator for desired voltage and frequency. **(B)** Giant vesicles budding on the platinum wires in the electroformation chamber.

Petri dish. Small slots to support the sample chamber and a side slot for the micropipet can be cut from the sides of the small Petri dish. Water is filled in the annular space to humidify the chamber.

2.3. Manometer

For a suggested design of a manometer that can be used to interrogate both the low- and high-pressure regimes, *see* **Fig. 4A**. This system includes a 70 cm ruled rapid advance unslide positioner with linear Acu-Rite® encoder (Velmex, Bloomfield, NY), mounted vertically to the side of the vibration isolation table. The top of the travel should be several centimeters above the microscope stage. A plate should be mounted onto the slider with screws. Two plexiglass water reservoirs (~50 mL each) with ports on top and bottom should be mounted on the plate. One of the water reservoirs should be attached to the plate through a small micrometer slider, allowing for the height of the water reservoir a total vertical travel of approx 2.5 cm relative to the plate. Tubing should go from the bottom port of each reservoir to a very low-pressure (~2 cm) transducer (e.g., model DP103, Validyne, Northridge, CA).

The low-pressure transducer can be mounted on the vibration table. A separate section of tubing should physically connect these two tubings and contain a valve. Tubing (e.g., Tygon tubing, ID = 1/16 in. and OD = 1/8 in.) should connect the micrometer-associated reservoir to the micropipet with a fitting. Using a small L-shaped polypropylene barbed fitting with a small (several mm) section of Tygon tubing (ID = 1/32 in. and OD = 3/32 in.) attached to the other end for insertion of the pipet is suggested. The L-fitting can be permanently mounted

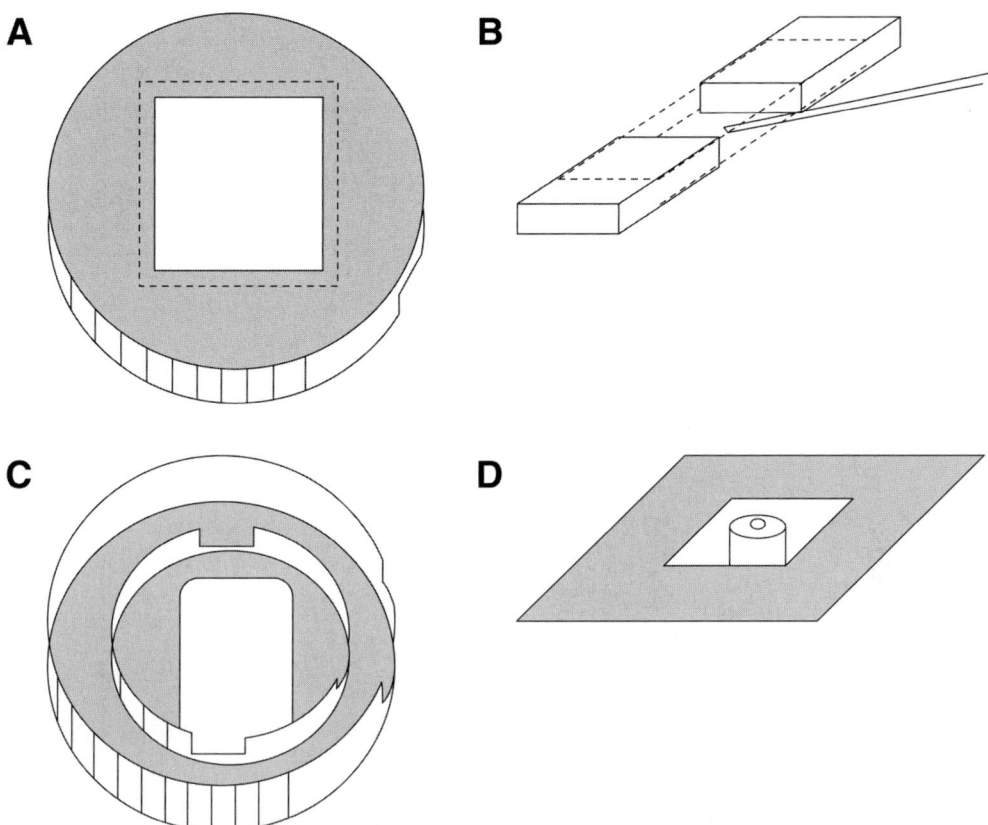

Fig. 3. **(A)** Top view of humidified chamber with glass-covered opening for condenser light and side opening for micropipet. **(B)** MPA sample chamber wherein GUV-containing solution resides between two cut glass cover slips. **(C)** Lid removed from humidifier chamber showing support for the sample chamber and openings for the micropipet and microscope lens. **(D)** Microscope stage and lens.

with an adjustable microclamp (model MM1-6, Technical Products International, St. Louis, MO) to control the angle of the pipet (*see* **Fig. 4B**) and for mounting on the micromanipulator. The manometer assembly should be filled with water.

Tapping on the tubing and transducer will generally eliminate all bubbles and should be performed fairly regularly. When not in use, the reservoirs should be stored at their uppermost distance of travel in order to maintain a positive pressure on the tubing and pressure transducer. The output of the pressure transducer and linear encoder generally need to go through output devices (e.g., Quick-Chek® by Metronics Incorporated; Bedford, NH, and digital indicator/demodulator by Validyne), which can then be directly attached to the overlay box for real-time overlay of pressure on the videotape. The manometer needs to be calibrated before use. The low-pressure transducer should accurately record a pressure change of 2 cm. Open the valve and close before this calibration in order to zero the pressure (valve will remain closed thereafter). The encoder on the slider should accurately record a pressure in the travel range that will be reached (e.g., 50 cm). Follow manufacturer instructions for adjusting if devices are not calibrated. Monitor the accuracy of the pressure measurement during MPA experiments as well.

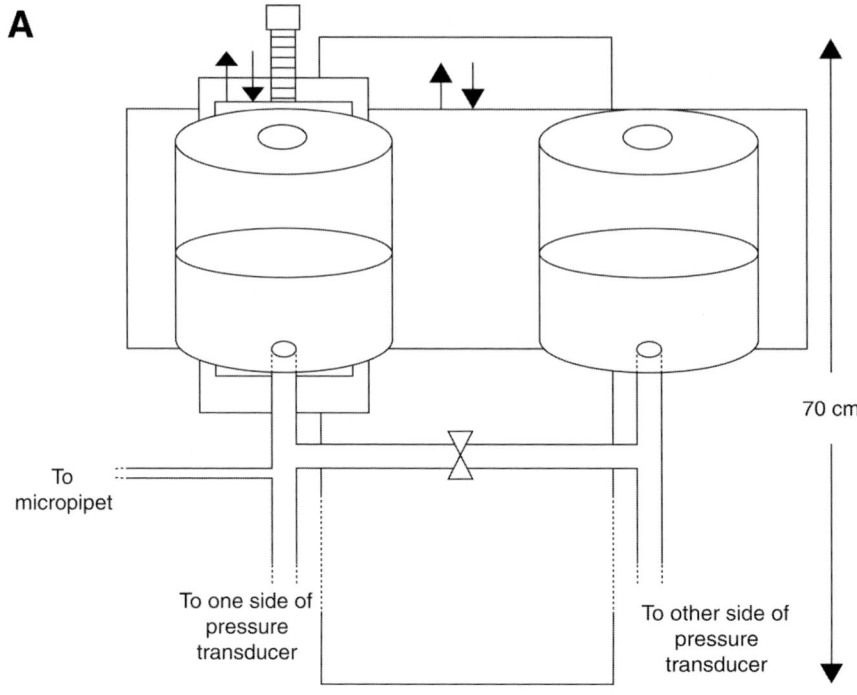

Fig. 4. **(A)** Manometer set-up for control of suction pressure as described in the text. **(B)** Connection between pipet and manometer tubing. Arrows (clockwise, starting with white arrow): manometer tubing, L-barbed fitting, pipet, knob for angle control of pipet.

2.4. Hardware and Software

1. Inverted optical microscope equipped with Hoffman-Modulation Optics (Modulation Optics, Greenvale, NY), a moveable stage, and a high-resolution black and white camera.
2. Attachment to the inverted microscope for mounting micromanipulators (e.g., Orbital Stage, Meridian Manufacturing Inc., Kent, WA).
3. One course micromanipulator (e.g., MMN-1, Narishige, Japan) and one fine micromanipulator (e.g., model MHW-3) mounted on a course micromanipulator.
4. Microscope and manometer should be mounted on an optical antivibration table.
5. Video-overlay hardware such as an overlay box (Polyvision, Western Australia).
6. Super VHS videotape recorder.

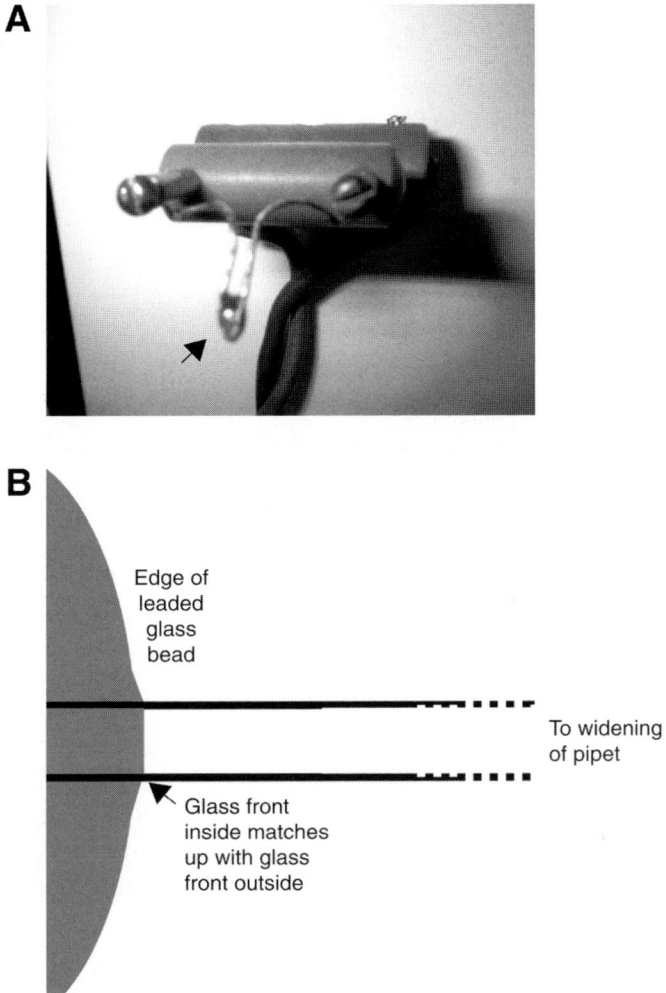

Fig. 5. **(A)** Proper shape of cylindrical platinum filament for the microforge. The leaded glass bead is also shown (arrow). **(B)** Sketch of shape of glass fronts inside and outside pipet for clean break.

7. Video capture card for capturing videotaped images (e.g. from ATI Multimedia center, Marlborough, MA).
8. Imaging software such as ATI Multimedia Center (Marlborough, MA) and Screen Ruler (Microfox Software, Columbus, OH) to measure dimensions on captured images.
9. Glass pipet puller (e.g., David Kopf Instruments, Tujunga, CA).
10. Pipet and needle storage containers. For example, a thin (1 cm) block of plexiglass with labeled holes drilled approx 1 mm diameter and 0.5 cm depth, and fitted with a plexiglass cover to avoid dust.
11. Microforge (e.g., Stoelting, Wood Dale, IL) for making a clean break of the micropipet. Mount the microforge electrode assembly **(Fig. 5A)** on the course manipulator on the microscope to get the best view of the micropipet break. Protect the objective lenses by placing a piece of cover glass over the opening on the microscope stage. Ensure proper grounding and electrical insulation from the micromanipulator. The microforge should have a cylindrical platinum electrode (diameter ~0.4 mm) mounted in a shape such as in **Fig. 5A**.
12. Function generator.

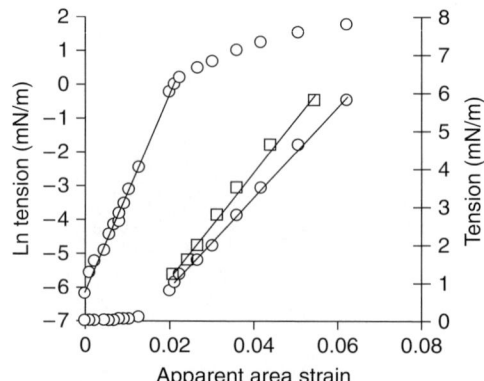

Fig. 6. Tension-strain measurements for a SOPC vesicle in a 30 vol% methanol/water solution. The points (circles in left curve) from plotting the natural log of the tension (τ) against area strain (α are linear in the low-tension regime (0.001–0.5 mN/m). The same points (circles in right curve) plotted with τ against α is nearly linear in the high-tension regime (>0.5 mN/m). Subtracting out contribution from smoothing out subvisible thermal shape undulations from α in the high tension regime gives the direct area strain (α_{dir}), and the replotted points (squares) shift the line to the left. $k_c = 4.4 \times 10^{-20}$ J, $K_{app} = 119$ dyn/cm, and $K_A = 144$ dyn/cm.

3. Methods

Aspiration of GUVs in a micropipet tip provides an excellent method for measuring elastic properties of lipid bilayers in fluid phases. In this method, a GUV is aspirated into a pipet of approx 8 µm ID (*see* **Fig. 1**). The increase in projected length (ΔL) inside the pipet in response to an increase in suction pressure (ΔP) is used to determine the observed or apparent area strain (α):

$$\alpha = (A - A_o)/A_o$$

where A_o is the membrane area of the vesicle measured at the lower suction pressure and A is the membrane area after the suction pressure is increased. The mechanics of thin materials show that increases in α comes from two modes of deformations of a vesicle under aspiration *(9,21)*:

$$\alpha = \left(\frac{k_B T}{8\pi k_c}\right) \ln\left(1 + \frac{c_o \tau A}{k_c}\right) + \frac{\tau}{K_A}$$

where A is the membrane area, τ is the induced isotropic membrane tension in direct relation to ΔP, k_c is the bending modulus, K_A is the direct area compressibility modulus, k_B is the Boltzmann's constant, T is absolute temperature, and c_o is a constant (~0.1) that depends on the type of modes (spherical harmonics or plane waves) used to describe surface undulations. Thermal shape undulations of flaccid vesicles are readily seen under microscope; the first term represents the smoothing out of these bending undulations. In the low-tension regime, the logarithmic term dominates, and shows that almost all increases in α come from smoothing out thermal undulations. The bending modulus is determined from plotting $\ln(\tau)$ vs α (**Fig. 6**) and is simply the product of the slope and $k_b T/8\pi$.

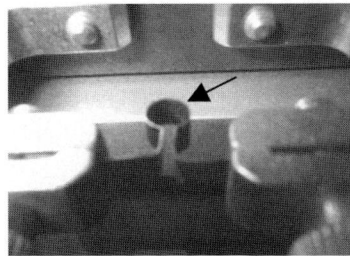

Fig. 7. Suggested shape of platinum ribbon electrode for pipet puller.

The second term represents direct stretching of the area per lipid molecule. The area compressibility modulus is determined by plotting τ vs α in the high-tension regime ($\tau > 0.5$ mN/m). In this regime, the linear term, τ/K_A dominates (see **Fig. 6**). By convention, the slope is known as the apparent area compressibility modulus, K_{app} *(9)*. This modulus includes a small contribution from the logarithmic term because even at the highest tension, subvisible thermal undulations persist. In **Subheading 3.7.** how the actual or "direct" area compressibility modulus is obtained will be discussed. Continued increases in suction pressure will eventually lead to vesicle rupture or "lysis." It is important to realize that the lysis tension (τ_{lyse}) is dependent on that rate of suction pressure increase (ramp-rate) *(22,23)*.

3.1. Fabricating and Measuring Pipets

1. Use appropriate gloves on hands to avoid placing fingerprints on the capillary glass and protect from organic solvents.
2. Use a glass cutter to cut 5 ft long sections of capillary glass into approx 30 sections of a length that is approximately double the desired pipet length, and set them aside in a container.
3. Wipe off a section of the capillary glass with a Kimwipe® (Kimberley Clark, Roswell, GA) soaked in ethanol.
4. Place the glass in a pipet puller equipped with a 3-mm width platinum ribbon-heating element shaped as shown in **Fig. 7**.
5. Adjust the pipet puller's clamp and heat setting to have a first pull that lengthens the capillary glass by 9–10 mm with an hourglass shape in the center.
6. Carefully slide the capillary glass to center the hourglass in the heating element. Adjust the pipet puller's heat settings to allow a second pull, such that the bottom-tapered pipet (if placed vertically) displays a long wispy end and is not attached to the top pipet. It is desirable to have the pipet narrow down rapidly to the desired ID rather than a gradual narrowing. Gradual narrowing will result in a vibration-prone pipet tip.
7. Use a glass cutter to trim the thick end of the pipet so the latter reaches approximately the length needed for MPA.
8. Store the pipet in a pipet storage container.
9. Repeat **steps 3–8** approx 20 times, whereas periodically performing a check on the pipet shape and ID by mounting them on a fine micromanipulator on a microscope. Use pen marks on the monitor as a guide to pipet ID.
10. Mount a pipet on the fine micromanipulator with the microscope in the microforge configuration.
11. The microforge should have a leaded glass bead of 1.5–2 mm diameter in the U-shaped section of the electrode (see **Fig. 5A**).
12. If there is no leaded glass bead, place some leaded glass powder on the base of the U-shape with a metal spatula, and heat the electrode to a temperature that melts the powder for a short period

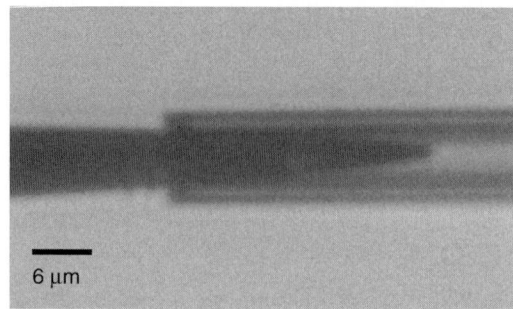

Fig. 8. Gold-coated glass probe inserted into pipet a distance of 28.6 µm. Diameter of micropipet is 6 µm.

(1 or 2 s), until bead appears red hot and spherical. Shut off the power, and the bead should appear clear. Periodically, replace the bead after a number of uses, or when it becomes difficult to microforge the tip of the pipet because the molten flow characteristics of the bead have changed significantly.

13. Set the bead's temperature to the melting point of the bead. (The power should be just high enough for the molten bead flow at a rate of ~10 µm/s into the pipet when it is touched). With the bead stationary, center the pipet and move it in the same plane as the bead. At this point, the wispy end can be removed by pushing the wispy end into the bead and quickly backing it away. Alternatively, very sharp wire cutters can be used to cut the pipet at an ID slightly smaller than the desired ID.
14. Move the pipet to touch the leaded glass bead. At this moment, the glass will instantly flow into the pipet. Keep inserting the pipet into the bead until the front end of the molten glass inside the pipet reaches the desired shaft ID.
15. Adjust the temperature of the bead such that the glass is soft, but it does not flow. Move the pipet (or bead) such that bead edge is even with the front of the glass inside the pipet (*see* **Fig. 5B**). This will give the best percentage of good breaks. Shut off the power to the electrode.
16. The thermal retraction of the bead may be enough for the pipet to break on its own. If not, move the pipet slowly up and down, and also move the bead slowly away (to put tension on the pipet) until the pipet breaks. About half of the time, the break occurs along the front end of the glass inside the pipet, and is clean and flat. If the break is poor, repeat **steps 14–18**.
17. Store the pipets vertically in a protected plexiglass pipet holder.
18. Measure the ID of each pipet. *See* **Subheading 3.2.**, for instructions on how to make needle probes necessary for this measurement.
19. Mount a metal-coated needle probe on a fine micromanipulator.
20. Mount a pipet on a course micromanipulator.
21. Bring both needle probe and pipet into the same plane using micromanipulators and viewing with the microscope. Ensure that their tips are relatively parallel and center with one another.
22. Insert the probe inside the bore of the pipet (**Fig. 8**) until movement is stopped. Do not overexert and break the pipet or probe. Capture this image by videotape or a framegrabber.
23. Repeat **steps 20–22** on all the pipets.
24. From each image, measure the distance from the pipet entrance to the tip of probe. Use the width–distance relationship calculated from scanning electron microscope (SEM) images of the probe to determine the ID of the pipet's entrance.
25. Coat each tip of the pipet with Surfasil to minimize vesicle adhesion.
26. Use a 10-mL gastight syringe with a large gauge leur-lock needle. Attach a short (~6 in.) length of Tygon tubing (ID = 1/32 in. and OD = 3/32 in.) to the needle. This tubing fits the pipet capillary tubing. Insert the pipet wide end inside of the tubing about 2 mm in depth to make a seal.

27. Perform the following steps (**steps 28–31**) in the fume hood.
28. Put the pipet tip into a solution of 1 vol% Surfasil in chloroform. By pushing down on the syringe plunger, blow air out of the syringe through the pipet, until air bubbles are seen coming out of the tip in the solution. Then, aspirate by pulling up on the plunger until the bubbling stops. Continue to aspirate until a liquid volume fills the inside tip, about 5 mm in length. Wait 15 s and then blow out air bubbles again. Remove the pipet tip from the Surfasil solution. Aspirate in air to dissipate any Surfasil solution that may remain within the tip. Replace screw cap immediately. Do not use excessive force on the plunger for any of these procedures.
29. Repeat **step 28** in chloroform. This step rinses off the unbound silanes.
30. Repeat **step 28** in methanol. This step neutralizes the sticky ends. Store each pipet in the plexiglass holder.
31. The pipet can be recleaned for continue use. Use the syringe set-up described in **step 26** and a 10-mL container filled with a 50 vol% ethanol/water mixture. Flow the mixture in and then out of the pipet tip by pulling up and pushing down the plunger while the pipet tip is dipped into the mixture. Repeat this motion several times. Finally, expel all the mixture from the pipet by pushing down the plunger until bubbles are expelled. Then repeat **steps 28–30**.

3.2. Fabricating Needle Probes for Measuring Pipet Inner Diameter

1. Repeat **steps 1–4** described in **Subheading 3.1.**
2. Decrease the heating setting from **step 5** described in **Subheading 3.1.**, and perform the first pull to stretch 15 mm.
3. Readjust the heat setting. Perform a second pull by gravity (or a similar pulling rate), which stretches and breaks the hourglass shape of the glass into long tapered ends with sharp points.
4. Use the glass cutter to trim the wide end of the needle probe to the length necessary for mounting on the fine micromanipulator.
5. Inspect the needle-shaped end by mounting in a manipulator on the microscope. The length:width ratio of the needle tip should be between 5:1 and 10:1 for a section starting at the tip and ending about 11 μm from the tip.
6. Check every few needles to make sure this ratio is maintained.
7. You will need to sputter coat the ends of these needles with a conductive metal (e.g., gold and/or platinum) for SEM. The coating should be approx 10 nm to avoid charging in the SEM.
8. Image each metal-coated needle probe in an SEM to obtain an image of the needle tip.
9. After SEM imaging, it may help to maintain the metal coating by coating the tips with formvar. Perform this procedure in a fume hood. Dip each tip in 1 vol% formvar solution in chloroform. Then store each needle probe in a plexiglass pipet holder.
10. The raw SEM images may be used to look up the width at the measured distance. Alternatively, image analysis may be performed to fit a polynomial to the SEM data in order to obtain an equation for width as a function of length for each needle probe (*see* **step 24** described in **Subheading 3.1.**).

3.3. Coating Glass Cover Slips

1. Use appropriate gloves on hands to avoid fingerprints on cover slips and protect from organic solvents.
2. Perform all steps in a fume hood.
3. From a box of 45 × 50 mm² (no. 1) cover slips, rinse each cover slip in ethanol. Hold each cover slip using plastic tweezers and gently swirl the cover slip in the ethanol held in a 250-mL wide-mouthed container. Replace the screw-top lid between cover slips.
4. Place each cover slip in a cover slip holder to dry in air.
5. Repeat **steps 3** and **4** for the box of cover slips.
6. Use plastic tweezers to grasp a dried cover slip and gently swirl in the 1 vol% Surfasil solution for about 15 s. Replace the screw-top lid immediately. Then, while still holding in tweezers allow the cover slip to air-dry. Wait for cloudiness on glass to dissipate.

7. Gently swirl the cover slip in chloroform for about 15 s. Replace the screw-top lid immediately. Air-dry while holding with tweezers.
8. Gently swirl the cover slip in methanol for about 15 s. Replace the screw-top lid immediately. Air-dry while holding with tweezers.
9. After repeating **steps 6–8** for each dried cover slip, put each cover slip in a cover slip holder. After completely air-drying, place all the coated cover slips back in the original manufacturer storage container as "Surfasil-coated."

3.4. Making Giant Vesicles by Electroformation

1. Clean and dry the electroformation chamber. The use of a small amount of chloroform and/or methanol under the hood in the rinsing procedure is helpful for removing any surfactants or remaining lipids and speeds drying. Wear appropriate gloves to protect from these organic solvents.
2. Remove your stock lipid solutions from the freezer and let them come up to room temperature.
3. Using concentrated stock solutions, make up a 2 mL solution of lipid (0.5 mg/mL) in 2:1 chloroform:methanol in a vial with PTFE or PTFE-lined septa or screw cap under a fume hood. Inclusion of 0.5% charged lipid will prevent vesicle–vesicle adhesion. An example of a mixture that makes very good giant vesicles is 99.5 mol% 1-stearoyl-2-oleoyl-phosphatidylcholine (SOPC) and 0.5% 1-stearoyl-2-oleoylphosphatidylserine. Use always clean gastight glass syringes with fixed needles for making up lipid solutions in chloroform. Never use a plastic pipeter. Stock solutions should only be uncapped while solution is being transferred from them. After use, they should be returned to the freezer.
4. If you are using a mixture, sonicate the loosely capped vial in a water bath sonicator for a minute or two to insure proper mixing of components. Perform this step and the next two steps in a fume hood.
5. Fill a 100-µL fixed-needle gastight glass syringe with 50 µL of 0.5 mg/mL lipid solution and spread (drop-by-drop) on the electrodes (platinum wires) using the syringe needle to make an even coating.
6. Briefly dry off the electrodes with a stream of dry nitrogen or argon.
7. Place the chamber under mild vacuum at room temperature for about 2 h.
8. After 2 h, degas about 10 mL of 100 mM sucrose solution at room temperature. For lipids with melting temperatures above room temperature *see* **step 9**.
9. For a lipid with a main phase-transition temperature (T_m) more than room temperature, heat the sucrose solution to a temperature about 10°C above the lipid T_m, and degas if there is excessive bubble formation.
10. Seal Surfasil-coated cover glasses to the electroformation chamber (*see* **Subheading 2.2.**).
11. Fill the chamber with degassed sucrose solution carefully using a 3-mL disposable syringe with attached 0.2-µm filter while chamber is in a vertical position. Keep the syringe attached during and after filling so that solution does not spill out. Fill until a drop of sucrose solution appears out of the opening of the syringe needle on the other side.
12. For a lipid with T_m above room temperature, ensure that the temperature of the water in the chamber is maintained at 10°C above T_m. For example, perform the next three steps with the body of the chamber in a Petri dish filled with warmed 100 mM sucrose solution.
13. With the function generator in the off position, attach the negative output to the end of one of the platinum electrodes and the positive output to the end of the other electrode.
14. Start the setting on the function generator at an amplitude of 3V (peak-to-peak), frequency of 10 Hz, and use the sine wave setting for 0.5 h. Then, lower the frequency sequentially to 3 Hz for 15 min, 1 Hz for 7 min, and 0.5 Hz for 7 min. It may be beneficial to lower the amplitude a bit as you lower the frequency to keep the vesicles from impacting with the electrodes. If viewing by microscope, GUVs should be visible (*see* **Fig. 2B**).
15. Shut off the function generator, then detach the outputs.
16. For the heated chamber, allow it to cool to room temperature if the GUVs are to be used below T_m or perform **step 17** above T_m if the GUVs are to be used above T_m.

17. Use the syringe to flow solution through the chamber into an Eppendorf tube. Remove the filter cartridge. Suck in and push out on the syringe to loosen GUVs from the electrodes. Use the syringe to ensure that all solution is collected. Two 1.5-mL Eppendorf tubes should be filled with GUV-containing solution. Store at room temperature for 3 d.

3.5. Setting Up for a MPA Experiment

1. Ensure that the manometers are calibrated (*see* **Subheading 2.3.**).
2. Put the humidified MPA chamber together (*see* **Subheading 2.2.**). Secure it on the microscope stage.
3. Pick out a Surfasil-coated and measured pipet from the plexiglass storage container.
4. Fill the pipet as follows in **steps 5** and **6**. BSA is included to coat the tip and thereby lower static charge.
5. Use the syringe set up described in **Subheading 3.1.**, **step 26**. Put the pipet tip in a 10-mL container filled with filtered glucose solution and 0.02 wt% BSA. By pushing down on the syringe plunger blow air out of the syringe through the pipet, until bubbles are seen coming out of the tip in solution. Then pull up on the plunger until the bubbling stops. Hold on this position until you see that the solution has been aspirated into the region wherein the pipet tip begins to widen out. Then, push down on the plunger such that solution is no longer forced into the tip. Remove the pipet tip from the glucose/BSA solution.
6. Backfill the pipet with the glucose/BSA solution using a syringe filled with the glucose/BSA solution and equipped with a very narrow flexible syringe needle (e.g., Microfil). Start by inserting the needle as far as it will go into the pipet and expelling a drop of solution out of the pipet tip. Then back the needle up, filling the entire pipet to slightly overfull. Glucose solution is naturally washed in and out of the tip during an experiment, dramatically lowering the actual unbound BSA concentration at the surface of the aspirated GUVs (*see* **Note 1**).
7. Attach the pipet to the manometer tubing as follows. Lower the level of the L-fitting in the microclamp until water drips out. Place the wide end of the pipet inside (~2 mm) the attached tubing; mount it on the fine micromanipulator.
8. Charging of the pipet tip can be problematic and will depend on the exact environment (*see* **Note 2**). If GUVs display lysis tensions consistently lower than the reported values, grounding the pipet during this step is suggested. Fine copper wire grounded on one end and shaped like a small hook on the other end, can be gently placed over the wide section of the micropipet.
9. Fill up the MPA chamber with filtered glucose solution. Surface tension will hold the water in the space between the cover slips. The air–water interface should be slightly concave to prevent dramatic changes in Laplace pressure inside of the chamber if evaporation should occur.
10. Add some GUVs as follows. Use a syringe with wide bore needle and a small section of PTFE tubing attached to the needle so that the tubing protrudes about 1 cm beyond the needle tip. Suck in some GUVs in 100 mM sucrose so that the 1-cm section of the PTFE tubing is filled. Deposit approx 6 µL with the 1-cm section through the air–water interface.
11. Use the micromanipulator to ensure that the pipet is positioned, so it will enter the chamber without hitting the walls. The tip of the pipet should be centered on the microscope and in focus and the MPA chamber off center. The pipet will need to be at a slight angle (tip end lowest) in order to pick up GUVs. This angle should not be so sharp that a change in focus occurs at the tip over the distance of several micrometer.
12. Move the MPA chamber over using the microscope stage so that the pipet is submersed in the chamber and focus on the tip.
13. If you do notice air in the tip of the pipet, remove it and repeat **steps 6** and **7**. This will usually suffice to remove the bubble.
14. Focus on the bottom cover glass of the chamber. Find a small vesicle (~5 µm in diameter). Aspirate the small vesicle into the pipet tip with very little suction and adjust the suction until the vesicle is not moving. The pipet should then be raised to the height used for the measurements below to ensure accuracy. The pressure at which the vesicle is stationary at this height is zero for analyzing data. Do this procedure more than once in a set of MPA experiments. This is the case

in particular for bending modulus measurements wherein it is necessary to know exact zero pressure. Evaporation in the chamber can change the exact zero.
15. Ensure that the Hoffmann optics is adjusted, so that the vesicle is dark on one side and light on the other side (i.e., light on right, dark on left). This may require unscrewing slightly the objective lens and readjusting the condenser polarizers. You are now ready to start your experiment.

3.6. Measuring Area Compressiblity Modulus, Bending Modulus, and Lysis Tension

1. This measurement is for GUVs that behave like an elastic fluid and does not apply for GUVs that behave like a solid. For example, fluid GUVs include SOPC GUVs and 1,2-dipalmitoylphosphatidylcholine GUVs with cholesterol concentrations ≥10 mol%.
2. Choose a vesicle between 25 and 40 μm in diameter. Look at the vesicle closely before picking it up to make sure that it does not have any vesicles inside or obvious vesicle tubes associated with it (*see* **Note 3**).
3. Pick up the vesicle with a suction pressure of about −2 cm.
4. Raise the height of the micropipet so that the vesicle is raised from the bottom cover glass. It is a good idea to translate the chamber a few hundred micrometers as well. If the vesicle tethers to the cover glass exist, this procedure should break them.
5. Increase the suction to about −15 cm. This step (prestress) will allow any small, attached vesicles to reincorporate or tethers to break (usually seen as a jump in the projection length). Begin to videotape.
6. Hold for about 3 s, then decrease the suction pressure in a stepwise manner to about −2 cm of suction (at each step, hold for about 3 s). This step will be used for the area compressibility modulus measurement.
7. At −2 cm, switch to the micrometer mechanism for changing pressure. Decrease the suction pressure in smaller steps holding at each step for about 3 s. This is for the bending modulus measurement.
8. At the suction pressure wherein the vesicle just gets sucked into the pipet, be careful not to push the vesicle projection all the way out of the pipet as this may cause small vesicles to form. There will be a small amount of suction pressure applied at which the projection goes from near the opening of the pipet to a significant projection (several μm). Practically, ΔL is measured with respect to that L for the bending modulus measurement. If this does not happen or the projection is too large, replace the cover glasses on the MPA chamber, and repeat steps starting at **step 9** described in **Subheading 3.5.**, with an adjustment to the glucose concentration (*see* **Note 4**).
9. Now reverse the above order (**steps 7–5** increasing suction pressure). Once a pressure of −15 cm is obtained, increase the suction pressure at a constant rate for measuring the lysis tension (determined from the pressure at which the vesicles rupture/lyse). Be mindful that the lysis tension is dependent on the rate at which the pressure is increased (ramp rate), so it needs to be in a narrow range.
10. After GUV lysis, maintain the suction pressure at the lysis pressure for several seconds to allow the lysed vesicle to be further aspirated into the pipet. This will ensure that GUV debris does not come into contact with and adhere to the micropipet tip. Then, lower the pressure to near 0 cm while looking for another GUV.
11. Practically, more careful measurements may be made when focusing on making only one measurement at a time (e.g., lysis tension). It is always necessary to perform a prestress for accurate measurements.

3.7. Analyzing the Data

1. Capture an image at each pressure held above.
2. Measure the distance (L) from the entrance of the pipet to the end of the vesicle projection (*see* **Fig. 1**).

3. Keep the first image on the screen with the measurement ruler. Referring occasionally to that image will ensure that the distance from the pipet entrance to the vesicle projection is performed in a repeatable manner.
4. The applied suction pressure (ΔP) is related to the induced isotropic membrane tension (τ), as follows *(24)*:

$$\tau = \frac{\Delta P \, R_p}{2\left(1 - \dfrac{R_p}{R_v}\right)} \qquad (2)$$

where R_p is the radius of the pipet and R_v is the radius of the vesicle. *See* **Fig. 1** for these dimensions. Geometrical arguments show that changes in R_v because of aspiration are similar to the error in measuring R_v for GUVs of size between 25 and 40 µm. One R_v measurement at low suction pressure (approx −0.1 and −2 cm) for the bending modulus and area compressibility modulus measurements should suffice.

The change in projection length from an increase in suction pressure (ΔL) is related to the observed or apparent area strain (α): $\alpha = (A - A_o)/A_o$ where A_o is the membrane area of the vesicle measured at an initial low tension state and A is the membrane area after pressurization to a higher tension state. Through simple geometric arguments, the relationship reduces to *(24)*:

$$\alpha = \frac{2\pi R_p \Delta L}{A_o}\left(1 - \frac{R_p}{R_v}\right) \qquad (3)$$

5. The bending modulus is determined from plotting $\ln(\tau)$ vs α in the tension range of 0.001 to generally 0.5 mN/m and is simply the product of the slope and $k_b T/8\pi$.
6. The area compressibility modulus is determined by plotting (τ vs α in the high-tension regime ($\tau > 0.5$ mN/m). By convention, the slope is known as the apparent area compressibility modulus, K_{app} *(9)*. The direct area compressibility modulus (K_A) is determined by subtracting the logarithmic contribution from α to get the direct area strain, α_{dir}. The appropriate contribution, $\Delta\alpha(i)$, for each i-th value of α that is subtracted out is *(9)*:

$$\Delta\alpha(i) = \left(\frac{k_B T}{8\pi k_{c,avg}}\right) \ln\left(\frac{\tau(i)}{\tau(1)}\right) \qquad (4)$$

where $k_{c,avg}$ is the average bending modulus and $\tau(1)$ is the initial low tension state of the high tension regime, usually experimentally set at approx 1 mN/m. K_A is the revised slope from plotting τ vs α_{dir} (*see* **Fig. 6**).

4. Notes

1. Although previous protocols call for small amounts of BSA in glucose solution in the MPA chamber, it has been eliminated because of inconsistent results in lysis tension when using BSA.
2. Static buildup can result in electroporation of the GUVs, and thus, low lysis tension or swelling of the GUVs is followed by bursting. Static buildup can be an erratic problem. When GUVs display these sort of behavior, assume first that a buildup of static charge is occurring and ground the pipet.
3. MPA of GUVs is a technique that requires practice and "feel." It usually takes 6 mo of playing, practice, and reworking equipment and methodology (everybody has their "pet" ways of doing certain things) before consistent data is obtained. Because lysis tension of these vesicles is

reasonably high and the mechanical moduli are well known, 99.5% SOPC:0.5% 1-stearoyl-2-oleoylphosphatidylserine make good practice GUVs.
4. If the projection length is too short, it tends to be difficult to measure changes because of interference fringes near the pipet entrance. If the projection length is too long, it will tend to pinch off under high suction pressure. Both "feel" and adjustment of the glucose concentration should minimize these problems.

Acknowledgments

We acknowledge other experimentalists who have substantial contributions in refining the techniques described herein; in particular, the laboratory of Richard E. Waugh, University of Rochester. MLL and HVL acknowledge partial funding from the Center for Polymer Interfaces and Macromolecular Assemblies (grant no. NSF DMR 0213618), the Nanoscale Interdisciplinary Research Teams Program of the National Science Foundation (under award no. CHE 0210807), and a generous gift of Joe and Essie Smith that endowed part of this work.

References

1. Gennis, R. B. (1989) *Biomembranes: Molecular Structure and Function.* Springer-Verlag, New York.
2. Israelachvili, J. N. (1992) *Intermolecular and Surface Forces.* Academic Press London, London.
3. Rutkowski, C. A., Williams, L. M., Haines, T. H., and Cummins, H. Z. (1991) The Elasticity of Synthetic Phospholipid Vesicles Obtained By Photon Correlation Spectroscopy. *Biochemistry* **30,** 5688–5696.
4. Hallett, F., Marsh, J., Nickel, B., and Wood, J. (1993) Mechanical properties of vesicles. II. A model for osmotic swelling and lysis. *Biophys. J.* **64,** 435–442.
5. Koenig, B., Strey, H., and Gawrisch, K. (1997) Membrane lateral compressibility determined by NMR and x-ray diffraction: effect of acyl chain polyunsaturation. *Biophys. J.* **73,** 1954–1966.
6. Waugh, R. E., Narla, M., Jackson, C. W., Mueller, T. J., Suzuki, T., and Dale, G. L. (1992) Rheologic Properties of Senescent Erythrocytes—Loss of Surface-Area and Volume with Red-Blood-Cell Age. *Blood* **79,** 1351–1358.
7. Sit, P., Spector, A., Lue, A., Popel, A., and Brownell, W. (1997) Micropipette aspiration on the outer hair cell lateral wall. *Biophys. J.* **72,** 2812–2819.
8. Kwok, R. and Evans, E. (1981) Thermoelasticity of large lecithin bilayer vesicles. *Biophys. J.* **35,** 637–652.
9. Rawicz, W., Olbrich, K. C., McIntosh, T., Needham, D., and Evans, E. (2000) Effect of chain length and unsaturation on elasticity of lipid bilayers. *Biophys. J.* **79,** 328–339.
10. Needham, D. and Nunn, R. (1990) Elastic deformation and failure of lipid bilayer membranes containing cholesterol. *Biophys. J.* **58,** 997–1009.
11. Tierney, K. J., Block, D. E., and Longo, M. L. (2005) Elasticity and phase behavior of DPPC membranes modulated by cholesterol, ergosterol, and ethanol. *Biophys. J.* **89,** 2481–2493.
12. Discher, B. M., Won, Y. Y., Ege, D. S., et al. (1999) Polymersomes: Tough vesicles made from diblock copolymers. *Science* **284,** 1143–1146.
13. Longo, M. L., Waring, A. J., and Hammer, D. A. (1997) Interaction of the influenza hemagglutinin fusion peptide with lipid bilayers: Area expansion and permeation. *Biophys. J.* **73,** 1430–1439.
14. Longo, M. L., Waring, A. J., Gordon, L. M., and Hammer, D. A. (1998) Area expansion and permeation of phospholipid membrane bilayers by influenza fusion peptides and melittin. *Langmuir* **14,** 2385–2395.
15. Ly, H. V. and Longo, M. L. (2004) The influence of short-chain alcohols on interfacial tension, mechanical properties, area/molecule, and permeability of fluid lipid bilayers. *Biophys. J.* **87,** 1013–1033.

16. Needham, D. and Zhelev, D. V. (1995) Lysolipid Exchange With Lipid Vesicle Membranes. *Annu. Biomed. Eng.* **23,** 287–298.
17. Olbrich, K., Rawicz, W., Needham, D., and Evans, E. (2000) Water permeability and mechanical strength of polyunsaturated lipid bilayers. *Biophys. J.* **79,** 321–327.
18. Lee, S., Kim, D. H., and Needham, D. (2001) Equilibrium and dynamic interfacial tension measurements at microscopic interfaces using a micropipet technique. 2. Dynamics of phospholipid monolayer formation and equilibrium tensions at water-air interface. *Langmuir* **17,** 5544–5550.
19. Lee, J. C. M., Bermudez, H., Discher, B. M., et al. (2001) Preparation, stability, and in vitro performance of vesicles made with diblock copolymers. *Biotechnol. Bioeng.* **73,** 135–145.
20. Lee, S., Kim, D. H., and Needham, D. (2001) Equilibrium and dynamic interfacial tension measurements at microscopic interfaces using a micropipet technique. 1. A new method for determination of interfacial tension. *Langmuir* **17,** 5537–5543.
21. Evans, E. and Rawicz, W. (1990) Entropy-driven tension and bending elasticity in condensed-fluid membranes. *Phys. Rev. Lett.* **64,** 2094–2097.
22. Evans, E. and Ludwig, F. (2000) Dynamic strengths of molecular anchoring and material cohesion in fluid biomembranes. *J. Phys. Condensed Matter* **12,** A315–A320.
23. Evans, E., Heinrich, V., Ludwig, F., and Rawicz, W. (2003) Dynamic tension spectroscopy and strength of biomembranes. *Biophys. J.* **85,** 2342–2350.
24. Evans, E. and Needham, D. (1987) Physical-Properties of Surfactant Bilayer-Membranes—Thermal Transitions, Elasticity, Rigidity, Cohesion, and Colloidal Interactions. *J. Phys. Chem.* **91,** 4219–4228.

29

Langmuir Films to Determine Lateral Surface Pressure on Lipid Segregation

Antonio Cruz and Jesús Pérez-Gil

Summary

Interfacial monolayers used as membrane models have become a practical technique to obtain detailed information about lateral processes taking place in the membrane. These monolayers are particularly useful to study the interactions and parameters governing lateral distribution of lipid and protein species and the association of different molecules with membrane surfaces. In the last few years, these classical models have been complemented by a whole collection of new techniques that are able to provide spatial information on the structure of the interfacial phospholipid-based films at both microscopic and nanoscopic scales. In the present chapter, some detailed protocols are described on how to prepare phospholipid Langmuir films, obtain structural information from their compression isotherms, and study their structure either *in situ* at the interface or on transfer onto solid supports by applying different microscopy techniques. The use of exogenous fluorescent probes and the extraction of qualitative and quantitative information from epifluorescence microscopy images are particularly addressed.

Key Words: Epifluorescence microscopy; interfacial monolayer; Langmuir films; Langmuir–Blodgett films; lipid domains; liquid-condensed; liquid-expanded; liquid-ordered; phase diagram; phase segregation; phase transition; rafts; surface tension.

1. Introduction

Lateral segregation of lipids in biological membranes is being recently proposed as a general mechanism governing different cellular processes, such as signal transduction and inter- and intracellular trafficking (for a comprehensive review, *see* **refs.** *1–3*). The development of techniques to evaluate lateral membrane structure has been highly demanded in this respect. Lipid monolayers were already at the basis of the historic observation by Benjamin Franklin that the waves in a pond could be reduced by spreading olive oil at the surface, but the pioneer work by Irving Langmuir during the first years of the 20th century was the one that set the fundamentals of modern surface balances, monolayer preparation, and the basis for the thermodynamic analysis of the behavior of such films. Langmuir films have been extensively used as membrane models to study lipid lateral organization and lipid–protein interactions, providing some advantages when compared with liposomes.

Interfacial monolayers allow a precise control of some factors affecting lipid structure, such as accurate composition, lateral pressure, and packing state. The precise spatial localization of monolayers at the air–water interface allows application of different techniques particularly suited to visualize Langmuir films at different scales, permitting the identification of lateral segregation on membrane-based lipid mixtures and the recognition of lipid–protein interaction sites. Transference of Langmuir films onto solid supports facilitates the application of other complementary techniques to the analysis of the structure of monolayers as membrane models.

2. Materials

The best available quality materials must be used to ensure good results when preparing Langmuir films. Water for the preparation of buffers must be, at least, double distilled. Deionized water (Milli-Q quality, Millipore, Billerica, MA) could be used, but a further distillation step performed in the presence of potassium permanganate to oxidize organic molecules is highly recommended. It is important to use freshly prepared water; nevertheless, water could be preserved at 4°C for a few days. Plastic recipients must be avoided to prepare or store materials.

1. Pure water is recommended as the aqueous subphase. When control of pH and pI is necessary, the experiment could be carried out using buffers made with nonsurface-active compounds. It is necessary to take into account that the presence of ions usually yields interactions at the films affecting the slope of the isotherms. Tris-HCl buffer at low concentrations (~5 mM) has been used to work at pH near neutrality. When preparing transferred films, the presence of salts may lead to crystal formation during the dried out period.
2. Solutions of the different film components must be prepared using high-performance-liquid chromatography grade solvents. Chloroform (Chl) and methanol (MetOH) mixtures are frequently preferred to prepare phospholipid-based solutions; Chl/MetOH 3:1 (v/v) is a broad solvent optimal for most phospholipids.
3. Lipids: analysis by thin-layer chromatography could be required to check for purity. Dipalmitoylphosphatidylcholine (DPPC), whose films show a well-defined Π-area isotherm, is a good standard phospholipid that could be used as training lipid or to check periodically that the system is clean enough and ready to work.
4. Fluorescent dyes: 7-nitrobenz-2-oxa-1,3-diazole-phosphatidylcholine (NBD-PC), rhodamine-PE, DiIC$_{18}$, Bodipy-PC (Molecular Probes, Junction City, OR), and so on. Dyes added as traces to the lipid solutions must be kept at proportions not higher than 0.5–1% M with respect to phospholipids (and preferably in the lowest range), to prevent spurious effects on the monolayer structure.
5. Paper or platinum flags may be used to measure surface tension. To get optimal contact angle, platinum flags must be scratched with sandpaper before taking them to the flame for cleaning. When using papers flags, they should be frequently replaced to avoid contamination between different experiments.
6. Glass cover slips: cut at the proper size (20 × 60 mm^2), are good substrates to transfer films for fluorescence applications. Molecularly flat substrates such as freshly exfoliated mica sheets or silicon wafers should be used for atomic force microscopy (AFM).

3. Methods

3.1. Langmuir Films

3.1.1. Layer Material and Solvent Selection

Any amphipathic material can be used to prepare an interfacial monolayer. This includes lipids, detergents, polypeptides, polymers, and so on. An amphipathic molecule contains a hydrophilic region, which orientates toward the water subfase, and a hydrophobic region exposed to the air side once at the water–air interface. Amphipathic compounds are able to form single-molecule thick layers at the interface. Molecules forming Langmuir films must be insoluble in water. Although water-soluble amphiphiles such as detergents could also form interfacial monolayers, these films sustaining a dynamic equilibrium with the bulk subphase.

The structure and properties of these solubilizable monolayers (termed Gibbs monolayers to differentiate them from true Langmuir films), are in tight dependence with the subphase composition and the bulk↔surface equilibrium conditions. On the other hand, Langmuir monolayers are characterized by the property of their molecules being strictly confined at the air–liquid interface. This feature is the first determinant to select the kind of solvents to use to manipulate monolayer components. To allow for interfacial lipid deposition without loss of material toward the subphase, the solvent should be also immiscible with water. Finally, the carrying solvent has to be eliminated from the surface by evaporation; thus, very volatile solvents are preferred. Chl/MetOH 3:1 (v/v) mixture, for instance, is a good solvent for dissolving phospholipid-based mixtures to prepare the kind of films frequently used as models of biological membranes.

Unless explicitly stated, most of the procedures described in this chapter make reference to the preparation and use of phospholipid interfacial phospholipid films. Preparation of layers of other amphiphiles different than phospholipids may require other solvent vehicles, such as ethyl-ether, benzene, or *n*-hexane. In some cases, when the influence of the hydrophilic part of the molecule makes it insoluble in nonpolar solvents, the proportion of alcohols like MetOH or ethanol could be increased. In these cases, care must be taken to avoid loss of material into the subphase. Proteins or peptides can be deposited into the interface through dissolution in polar solvents such as dimethyl-sulphoxide or acetone. However, the high miscibility of these solvents with water makes it difficult, if not impossible, to avoid a partial loss of the protein into the bulk phase.

Isopropanol-based solvent mixtures may be a good alternative *(4)*. It is important to take into account that the structure of proteins and peptides may be irreversibly affected by the exposure to some of these solvents, producing not easily explainable results.

To prepare the films, phospholipid organic solutions at concentrations between 0.1 and 2 mg/mL can be used. The concentration must be selected taking into account the amount of material to be initially applied, which is also related to the initial surface area of the trough at maximal opening of the balance. Diluted lipid solutions may lead to the application of large volumes of carrying solvent onto the surface, needing long times to complete solvent evaporation and increasing the risk of accumulation of solvent trace contamination. Concentrated lipid solutions, more than 2 mg/mL, could make it difficult to accurately manipulate the volume to be applied, increasing the experimental errors on isotherm calculations. Moreover, too concentrated solutions could yield an undesired increase in the solution density, potentially causing loss of material sunken toward the subphase during monolayer deposition.

3.1.2. Surface Balance

3.1.2.1. Trough Types

In general, troughs are typically made from polytetrafluoethylene (PTFE). This is a hydrophobic and highly inert material, allowing easy cleaning of the trough with organic solvents or inorganic acids. Some surface balances are differentiated by the mechanism set to compress and expand the film at the interface. The most common and simplest system is a PTFE bar, which is used as a single movable barrier that scans the subphase meniscus while maintaining the Langmuir film confined in the surface (*see* **Note 1**). This mechanism is good enough for most applications, but the subphase meniscus in this device acts as the monolayer container. Thus, special care is required to maintain the surface level constant against subphase evaporation **(Note 2)**. The use of balances equipped with bar meniscus barriers is usually enough to study monolayers in the pressure ranges thought to mimic lipid packing in free-standing membranes (in the order of ~30 mN/m). However, this type of barriers does not overcome some problems when the films are taken to high enough pressures (>40 mN/m), because surface tension may not be enough to sustain the meniscus, and leakage of the surface material through the barrier and through the edges of the trough may occur.

To avoid these problems, other compression mechanisms have been implemented. The solution of dipping the barrier inside the trough walls could be enough to avoid some inconveniences, but some leakage through the contact edges of barrier and trough may still take place if the contact zone is not tight enough. The best solution for applications requiring full access to the highest surface pressure regimes is the use of balances with a continuous perimeter barrier. In these balances, the whole film is confined inside an edge-free, continuous PTFE ribbon belt or an articulated single PTFE piece, which is deeply immersed into the subphase. Thus, the change of available surface during compression/expansion is achieved by modifying the shape of the area engulfed by the flexible barrier.

3.1.2.2. TROUGH CLEANING

The trough may be cleaned in four steps:

1. Dismount the balance; immerse the barriers and any mobile pieces into enough volume of Chl/MetOH 3:1 solvent solution.
2. Fill the trough with enough volume of Chl/MetOH 3:1 and wait for 5 min.
3. Remove the solution and fill the trough with MetOH to remove Chl traces.
4. Wash three times with ddH_2O.

Addition of 2% of 0.1 N HCl to the Chl/MetOH 3:1 cleaning solution could aid to eliminate protein remains. Wiping the trough with a clean paper impregnated with Chl/MetOH before proceeding to solvent washing could be required to remove material adhered to the trough.

3.1.3. Interfacial Deposition

3.1.3.1. FILMS PREPARED FROM ORGANIC SOLVENT SOLUTIONS

Surface pressure measurements and compression parameters in modern troughs are under control of a computer and proper software. Before starting acquisition of the isotherm, films can be formed by spreading the amphiphiles in solvent solutions according to the following protocol:

1. Fill the trough with clean subphase and wait until the chosen temperature of the subphase has been achieved.
2. Clean the flag. Platinum flags can be easily cleaned by flaming. Paper flags should be exhaustively washed with the subphase solution before wetting.
3. Calibrate the force transducer of the balance using known weights and adjust the program to monitor surface pressure (Π) (*see* **Note 3**). Determination of surface pressure requires calculation of the contact edge between flag and water. Flag weight is discounted at the initial step.
4. Hang the flag from the force transducer and adjust the balance to measure 0 mN/m surface pressure. Dip the flag 1 mm into the subphase and measure the apparent surface pressure (*see* **Note 4**). Because air is used as reference, the measurement in terms of surface pressure must read now approximately the negative value of the surface tension of the subphase (~72 mN/m in clean water), with some deviation from the theoretical value because of the effect of the floating force produced by the immersed part of the flag. For higher accuracy of this control measurement, the flag must touch the surface without any immersed volume; it is not so critical to obtain the pressure isotherm because floating force will be initially discounted. If the apparent baseline surface pressure value is far from the expected, check the subphase purity or reclean the system.

5. Set the balance to 0 mN/m and close the barrier to compress the interface. If the interface is really clean, no changes in surface pressure on compression should be observed. A limited increase in pressure is usually observed as a result of the presence of some trace of surface active molecules at the minimum area position; clean then the interface by sucking the surface with a pipet connected to a vacuum pump. Repeat this procedure as many times as necessary until no changes in surface pressure are observed on compression.
6. After those cleaning procedures, a lack of increase in surface pressure indicates that the air–liquid interface is free of any surface active molecule. If a systematic repetition of **step 5** does not end with good base lines, contamination of the subphase with some amphipathic material that is continuously adsorbing into the interface during the cleaning steps most likely occurred. In this case, check the trough cleaning, wash it again with double distilled water, and change the subphase with clean freshly prepared solution.
7. Once the surface is thoroughly clean, take the appropriate amount of the phospholipid organic solution with a microsyringe; deposit the solution drop-by-drop by placing the microsyringe tip end at a few millimetres from the water surface. Deposit lipid at the interface until the surface pressure increases from 0 to 0.1–0.2 mN/m, measuring accurately the volume that is finally applied. The volume measured may be used as a reference for successive experiments, and has to be strictly known to obtain accurate Π-area (Π-A) isotherms. An estimation of the amount of lipid required to start obtaining the isotherm can be derived from reference isotherms of the lipid. When a relatively large volume of lipid solution has to be spread, deposition of drops on different positions over the surface is recommended, waiting some time to allow for solvent evaporation between drop depositions.
8. Once all the solvent volume has been spread, wait 10 min to allow for solvent evaporation and lipid extension. However, longer times might be required, depending on the solvent and volume applied.

3.1.3.2. Films Prepared From Aqueous Suspensions

It may be useful to obtain interfacial phospholipid-based films that also contain proteins. Interaction of the proteins with lipid vesicles in a previous step permits forming films from these lipid–protein suspensions. Proteins could have been isolated or reconstituted in lipids for a better stabilization of their native structure. Preparation of films by spreading aqueous lipid/protein suspensions assumes that at least part of the lipid and protein molecules are transferred from the bilayered structures into the interface *(5)*. Such transfer is more efficient in some system than others. Some other considerations must be taken into account:

1. Vesicle suspensions are prepared in water; special care has to be taken during lipid deposition of these aqueous suspensions. Apply the material by forming a small drop on the syringe tip and let the drop touch the surface to allow for vesicle spreading. Because loss of some material into the subphase is unavoidable, more material is needed when using suspensions than when using organic solvent solutions *(6)*.
2. The subphase volume is much higher than the volume applied; thus, in practical terms, the material lost in the subphase during the deposition procedure does not further reach the interface. The amount of material that is finally transferred onto the surface then depends on the initial adsorption efficiency *(7)*. This means that only suspensions that adsorb quickly enough at the interface are useful to prepare lipid/proteins films. For example, vesicles made only with lipids bearing long hydrocarbon tails, such as DPPC, are very stable and do not practically adsorb into the interface. The morphology of the lipid vesicles used can also be important: small unilamellar vesicles prepared by strong sonication are intrinsically unstable and adsorb more efficiently into the interface than large unilamellar vesicles.
3. Partial loss of material because of incomplete transfer must be taken into account when comparing different isotherms, because the actual material on the surface cannot be easily calculated. An apparent area per lipid molecule could be estimated from the volume applied, but isotherms

calculated this way can only be compared when obtained from materials having comparable interfacial adsorption efficiencies. Interfacial adsorption kinetics may need to be analyzed in order to explain differences on area per lipid molecule data *(7)*.

3.1.4. Π-A Isotherms

The structure of an interfacial monolayer formed by a given lipid mixture depends basically on the temperature, the surface concentration of lipid molecules, and the lateral pressure of the film. Surface pressure vs area (Π-A) isotherms are usually recorded by changing the area occupied by the monolayer at the air-liquid interface, while the surface pressure is continuously monitored. Assuming that all applied lipid has been transferred and is confined to the air–liquid interface, the area per lipid molecule can be easily calculated from the known total area of the trough, the sample concentration, and the initial volume of spreading solution that was applied. This assumption is reasonable if the film has been formed carefully as described in **Subheading 3.1.3.1.**, but it can be difficult to confirm when the films are formed from aqueous vesicle suspensions or from lipid solutions containing high proportions of polar solvents.

All modern Langmuir troughs have motorized barriers controlled by software, allowing the selection of start and end positions, as well as the barrier speed. To obtain an isotherm, fill the trough with clean bulk solution and wait until the required temperature is attained. Clean the interface, open the barrier, and spread the lipid as described in **Subheading 3.1.3.1**. Select a proper compression rate and start the compression program (*see* **Note 5**). An incomplete equilibration of the film that may occur during relatively fast compression may affect the slope of the isotherms, depending on the barrier speed (**Note 6**). On the other hand, the apparent pattern of the compression-driven structural transitions occurring in the films is largely affected by the compression rate. Rapid compression usually produces higher number of condensed lipid domains smaller than those seen produced in slowly compressed isotherms *(8)*. Thus, the compression speed used to take the films to the desired pressures must be taken especially into consideration when comparing epifluorescence images.

A typical Π-A isotherm of a DPPC film is shown in **Fig. 1**. During compression, surface pressure increases as a function of area reduction and associated increase in lipid concentration at the interface. The different slopes in the segments of the curve indicate different compressibility degrees of the film at the different pressure regimes. Abrupt changes in the curve are commonly associated with compression-driven structural transitions occurring in the lipid films. Typically, four different two-dimensional (2D) phases have been described in isotherms from single lipid components: gas (G), liquid-expanded (LE), liquid-condensed (LC), and solid phase (S). At low pressures, the lipid behaves as a 2D G phase, characterized by very low density of disordered lipid molecules at the interface. As compression is increased, the available free surface is reduced until the molecules are forced to touch each other. At this point, surface pressure starts to rise and the lipid adopts a LE phase (*see* **Note 7**). Structurally, this phase shows up because of a reorganization of the molecules that form the monolayer, i.e., with the phospholipid head groups facing the aqueous bulk phase and the phospholipid hydrocarbon tails oriented almost perpendicularly to the interface plane (yet with some intrinsic flexibility because of *trans–gauche* isomerizations of the hydrocarbon chains). At higher pressures, the lipid may get into a LC phase, with a more compact ordering (*see* **Note 8**). The steeper slope of this segment of the isotherms reveals that this phase possesses low compressibility. In the LC phase, the lipid molecules are tightly packed, exhibiting highly limited molecular movements in the plane of the monolayer.

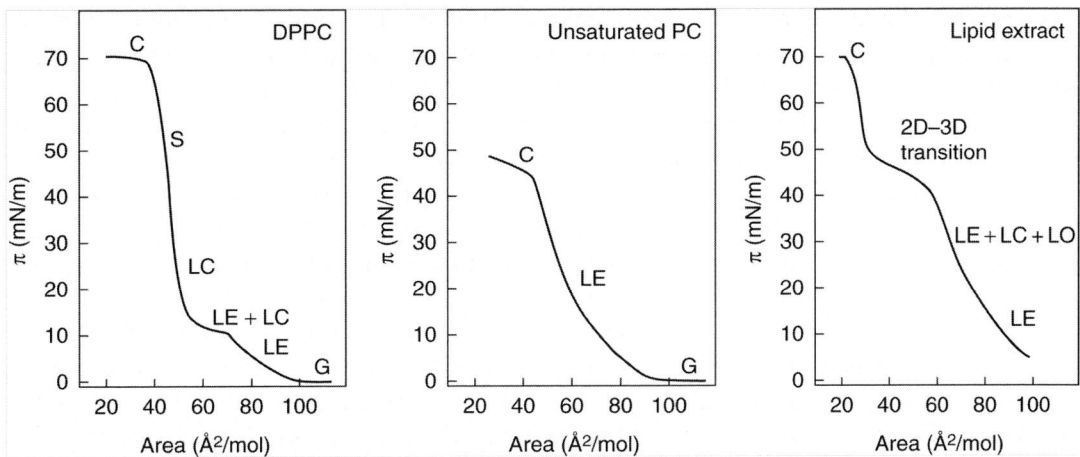

Fig. 1. Surface pressure–area (Π–A) compression isotherms, and the 2D phases associated, of some illustrative interfacial phospholipid films. Films made from a single saturated phospholipid species (DPPC) **(left panel)** transit at low pressures from a gas (G) to a LE phase. When compressed at temperatures lower than 40°C, a conspicuous plateau in the isotherm is the result of the LE/LC transition ending first in a pure LC phase, and then in a 2D solid-like (S) film, collapsing (C) at pressures higher than 70 mN/m. Films made from an unsaturated phospholipid species, such as egg yolk PC **(central panel)**, collapse directly from the LE phase. Isotherms obtained from films made up of a full lipid mixture such as, for instance, the organic-extracted fraction of pulmonary surfactant **(right panel)** are much more complex to interpret, with contribution of LE and LC phases but also of liquid-ordered-like (LO) regions. Beyond certain pressures, these complex isotherms may include squeeze-out plateaus originated by the partial exclusion of some of the components or the whole film away from the interface.

Coexistence of regions having LE and LC phases in the films is usually associated with a marked plateau in the isotherm, where compression produces very slight increases on surface pressure. Along this plateau, which behaves as a real phase transition edge of the phase diagram, the work of compression is used to promote the change from LE to LC phase. Morphologically (*see* **Fig. 3**) the coexistence of the two phases along this plateau is observed as the formation of lipid domains in LC phase surrounded by a background of LE phase. This constant pressure plateau is characteristic of a 2D phase transition and depends on temperature. Higher temperatures shift the transition pressure to higher values.

If surface pressure is increased further, a solid phase can be reached, observed as a kink in the isotherm to a segment with increased slope and practically no compressibility. There is some controversy about the significance of this phase because the translational movement of the molecules at the interface is similar to that observed in the LC phase. X-ray diffraction shows that the main difference between the two states can affect the tilt of the hydrocarbon chains of the phospholipid molecules *(9)*. Compression of the solid phase beyond a certain limit produces the collapse of the film with loss of material from the monolayer and formation of three-dimensional (3D) structures beneath or above the interface. At this point of the isotherm, surface pressure is no more affected by compression. This pressure is known as the collapse pressure of the film. A monolayer made of pure DPPC, for instance, is able to reduce surface tension to near 0 mN/m (reaching above 70 mN/m surface pressure) before collapsing. The area occupied per lipid molecule at LC or LE states can be obtained by extrapolating the curve segment corresponding to any of those phases to Π = 0 mN/m.

Isotherms from monolayers made of a mixture of lipids are more complex to interpret than those from single lipid films. Theoretically, the isotherm obtained from a mixture of two or more lipid components should be comparable with the algebraic sum of the isotherms of films made of the single components normalized according to their relative molar fraction. Any deviation of this theoretical curve is indicative of the existence of attractive or repulsive forces between the different molecular components. Typical curves could be obtained plotting the area occupied by the lipid mixture at a given pressure vs the molar fraction of one of the components. Ideally, a straight line must be found when plotting the area per lipid molecule of any of the components against its molar fraction. A negative deviation from this theoretical line could indicate a potential interaction between the different components *(10)*.

In mixed films, mutual interaction or segregation of the different components can yield structures or phases not existing in single lipid monolayers. Lateral immiscibility or segregation may not be easily detected in the isotherms of these mixed films in the form of plateaus, and sometimes it can only be observed when structure is analyzed. On the other hand, plateaus showing up in mixed systems could be owing to 2D to 3D structural transitions, including exclusion of part of the components of the layer during lateral compression. Methods that allow for microscopic observation of the monolayer usually help to interpret the data obtained from the isotherms.

3.2. Langmuir–Blodgett Films

Katharine Blodgett was the first to study the transfer of monolayers from the liquid surface onto a solid substrate crossing the liquid interface vertically. Blodgett's work, carried out under Langmuir supervision, included characterization of parameters that control monolayer transfer and some physical properties of the transferred films. Some later studies demonstrated that Langmuir–Blodgett (LB) films transferred at constant surface pressure may conserve most of the structural information of the Langmuir monolayers as they are at the interface.

3.2.1. Selection and Cleaning of Solid Supports

The LB procedure allows transferring the interfacial films either from the hydrophobic or the hydrophilic side of the monolayer by selecting the support material and the direction of the transfer. Hydrophilic substrates immersed into the subphase and raised through the monolayer plane adsorb the lipid monolayer, producing a film in which the lipid head groups interact with the substrate and the hydrocarbon chains are exposed to air (*see* **Fig. 2**). Films transferred with the opposite orientation may be obtained by using a hydrophobic material as substrate and reversing the transfer direction. The hydrophilic transfer is preferred because of inherent problems associated with transfer to hydrophobic supports. The adherence of lipid acyl chains to hydrophobic substrates is intrinsically weak, and as a consequence, the transfer of lipid films to those supports is efficient only from relatively dense monolayers compressed above certain pressures. Depending on the desired application, a variety of supports can be used, including polymers, metal, or silicon wafers. Inert substrates as glass, quartz, or mica are usually used as hydrophilic substrates for microscopy applications. Transparent quartz substrates may be required when samples are analyzed using ultraviolet light.

Mica is a flat substrate at the molecular scale, useful for AFM applications. Mica sheets can be easily cut to the desired size, and efficiently cleaned by just exfoliating the surface layers with adhesive tape (although formation of micrometer-sized *terraces* could affect the observation of the supported films at the optical microscopy scale). Microscopy cover slips are

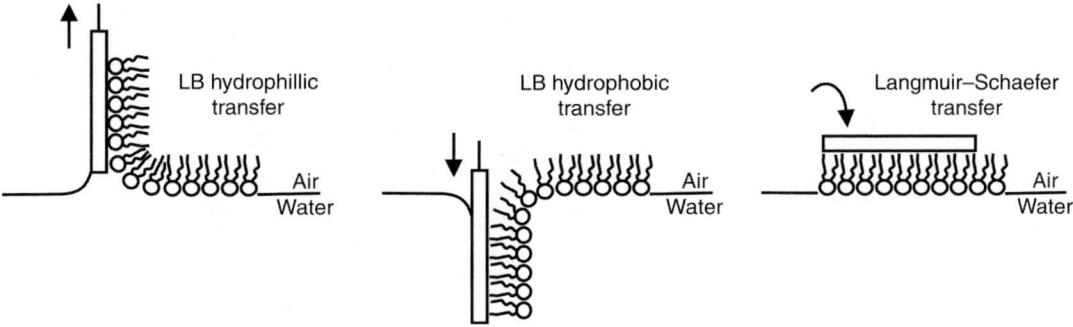

Fig. 2. Different modes of preparing supported phospholipid films on transfer of interfacial monolayers onto solid supports of different nature.

good substrates for films to be analyzed by epifluorescence microscopy. New cover slips can be washed with Chl/MetOH 3:1 (v/v) and dried in air to be clean enough. There is a whole variety of alternative methods to clean recycled substrates. Chromic acid or saturated solutions of KaOH in water or ethanol, combined with ultrasonic treatment and water wash, may be used for quartz or glass substrates. Rinse with a volatile solvent, such as isopropanol; later, N_2 flow facilitates a rapid and homogeneous dry off.

3.2.2. Transfer to Solid Supports

The transfer of lipid monolayers is a suitable technique for the observation of lipid lateral segregation in membrane-mimicking interfacial films. The major advantage of working with LB films is that the samples are immobilized and can be easily transported. This extends the possibilities for their study by a combination of techniques not specifically designed for studying monolayers *in situ*, including most spectroscopical techniques. One of the practical disadvantages of LB films is the obligation of preparing a different independent film to study the structure at each of the surface pressures of interest. The target pressures must be selected taking into account the information from isotherm curves.

Pressures producing kinks and plateaus in the isotherms are good candidates to be related to lateral segregation processes. To identify pressures producing lipid segregation, the transfer of the monolayer may be performed while the monolayer is compressed. A precise control of compression and transfer velocities allows obtaining LB films that contain all the structures occurring along the isotherm on a unique support, with pressure being a function of the position through all the transfer length.

A general protocol to transfer lipid monolayers to supports could be:

1. Clean carefully the surface balance.
2. Add pure water or buffer as subphase.
3. Clean the surface as indicated in **Subheading 3.1.3.1.** and the supports as suggested in **Subheading 3.2.1.**
4. With the barrier open at the largest area, mount the transfer plate on the lift system. Check that the support is completely out of the bulk phase in the upper position, and that the full transfer area is immersed into the subphase at the lower position. Then, move the lift to the lowest position.
5. Close the barrier and clean the surface by aspirating the interface with a pipet connected to a vacuum pump. If the substrate is perfectly clean, no changes must be observed on surface pressure during compression.

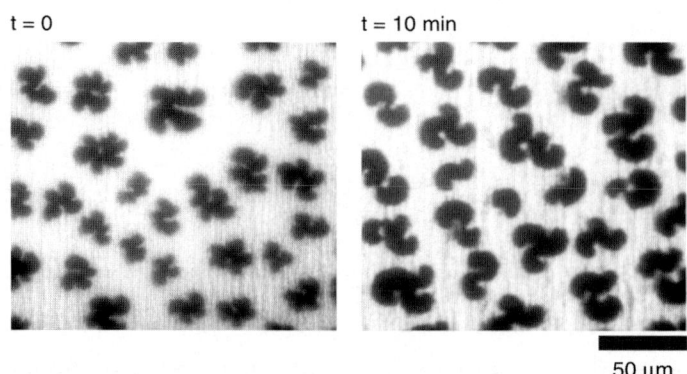

Fig. 3. Effect of equilibration on the shapes and morphology of condensed lipid domains. Epifluorescence microscopy images obtained just after finishing compression of a DPPC film up to 12 mN/m (left picture) and after a 10 min equilibration of the film while maintaining the same pressure (right picture). Notice the higher branching level and the blurry boundaries of the domains before equilibration.

6. Spread the solution as in **Subheading 3.1.3.1.** and wait until solvent is totally evaporated.
7. Select a proper compression speed and compress the monolayer to the desired target pressure. To maintain the surface pressure constant along the transfer process, most of the commercials surface balances include a feedback mechanism that automatically compensates with proper compression changes in the pressure owing to lipid extraction. The sensitivity and time-constant response of the feedback program must be carefully adjusted to avoid artefacts that can be observed in the microscopy images as scratches in the film, usually orientated in a perpendicular way to the transfer direction.
8. Once the target pressure is reached, wait 10 min to allow for re-equilibration of the film. When lipid domains have been generated on compression, their shape usually changes from dendritic-like shapes just at the end of compression to characteristic equilibrium shapes at longer times **(Fig. 3)**. Ten minutes is usually long enough to reach local equilibrium shapes in most systems. Liquid domains formed by lateral immiscibility (such as those formed by lipid mixtures containing cholesterol even at low pressures) are fluid enough to change substantially over time, often including domain coalescence *(11)*. Therefore, the equilibration time is a very important parameter when examining and comparing the morphology of films made of different mixtures or systems of this type.
9. While the feedback program is activated, raise the support until the lower edge is out of the bulk phase. A proper transfer speed is a crucial parameter to obtain well-transferred films without artefacts. For each experiment, the optimal speed depends on the nature of the material forming the monolayer and the support dimensions. Transfer speeds of 5 mm/min are commonly used to obtain LB films from lipid monolayers. The progress of film transfer can be followed by plotting the reduction in surface area that is required to maintain surface pressure constant on transfer against time.
10. Dismount the substrate from the lift and store the supported film at room temperature. To avoid lipid oxidation, storage of the LB's in an inert atmosphere is recommended. When light-sensible dyes are included into the films, protect the LB's in a dark environment.
11. To check for transfer efficiency, the deposition ratio may be calculated. This parameter is the rate between the final area reduction introduced at the interfacial monolayer to maintain constant pressure during transfer, and the area of the immersed support that has been lifted off the bulk phase. Films transferred with deposition ratios smaller than 0.9 must be discarded.

It is possible to stack more than one monolayer transferred into the same support, by repeating the process after the first deposition (*see* **Note 9**). To facilitate the second transfer, it is recommendable to let first the previous layer to dry out completely, in order to avoid the lipid of this first layer to be transferred again from the support back to the air–liquid interface. The structural information obtained from these multilayers is more difficult to interpret owing to overlapping in the structures of the different layers (*see* **Note 10**). To facilitate transfer of interfacial films onto hydrophobic substrates, the Langmuir–Schaefer method can be used (*see* **Fig. 3**). In this technique, the support is placed in contact with the monolayer that is compressed at the desired pressure, and the film is removed horizontally from the interface at the appropriate speed.

3.3. Observation of Film Lateral Structures: Fluorescence Microscopy

Different techniques are available to analyze the structure of phospholipid-based interfacial films. Some of the spectroscopic techniques, such as grazing incidence X-ray diffraction, neutron scattering, or infrared reflection-absorption spectroscopy *(9,12,13)*, report information that is averaged over the entire structure of the film. To obtain information on how lateral transitions take place or how the different lipid and protein components are laterally distributed, some optical techniques are available providing enough spatial resolution to produce images of the films. In this respect, fluorescence microscopy is one of the most used techniques *(14–17)*. This technique needs for the inclusion of traces of fluorescent dyes into the monolayers. Lateral inhomogeneities of the film structure are then revealed as far as they are associated with differential probe solubility. The inclusion of these fluorescent dyes may be considered a potential source of artefacts but the high sensitivity of the technique allows obtaining good results with very-low dye concentrations. Some other techniques such as Brewster angle microscopy (BAM), ellipsometry, or scanning force microscopy have no need for the use of dyes.

BAM takes advantage of the differences in the reflective properties of the interface and the different regions of the films to obtain images of the interfacial structures, without requiring the inclusion of exogenous probes *(18)*. However, BAM has a relatively low lateral spatial resolution. Ellipsometry is a more sophisticated technique producing images of films from the changes in polarization introduced on reflection of light by the interfacial surfaces. Apart from detecting segregated structures, ellipsometry may provide information about some physicochemical properties of the monolayers such as thickness, roughness, and so on *(19)*. Fluorescence microscopy has some advantages when compared with these other techniques, the most important being that it has very high sensitivity, plus the possibilities derived from the inclusion of multiple dyes, which allows obtaining simultaneous information from the distribution of different lipid and protein species.

3.3.1. Fluorescent Labels for the Observation of Lipid Segregation

Most dyes used for fluorescence microscopy of monolayer films are derivatives of membrane components, such as phospholipids, sphingolipids and cholesterol, or amphipathic molecules structurally similar to any of these membrane components *(20)*. In general, any membrane-partitioning dye could be used to label interfacial monolayers, but molecules that partition differentially into different phases are the most useful when observation of domain distribution is pursued. This differential partition of a given probe depends not only on its structure and intrinsic properties, but also on the potential interaction of the dye with the different components forming the monolayer and on the physical state of the different regions or phases existing in the film.

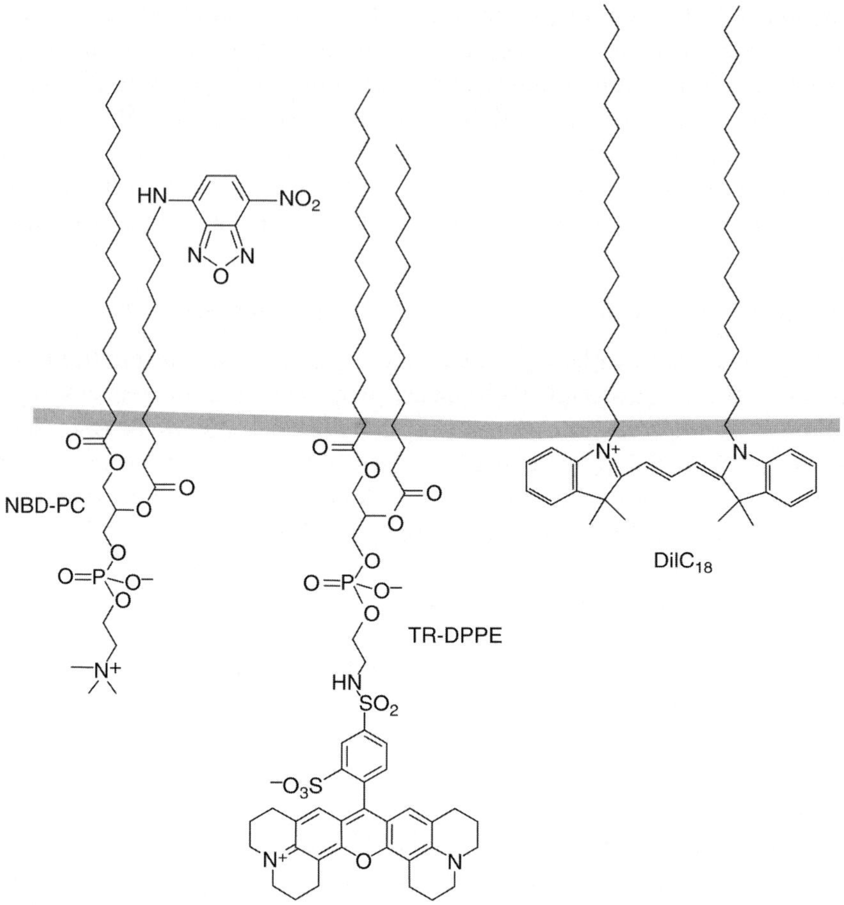

Fig. 4. Structure and typical orientation of different fluorescent phospholipid derivatives at the air–liquid interface.

Bodipy, NBD, Fluoresceine (FL), dansyl, or Texas Red© (TR) (available from Molecular Probes, Junction City, OR) (*see* **Note 11**) are dyes frequently used to synthesize covalently labeled fluorescent phospholipid molecules (**Fig. 4**). The distribution of these modified lipids in monolayers is highly dependent on where the dye is chemically attached at the phospholipid molecule. In principle, phospholipids bearing long saturated acyl chains and modified at the head group, can still be highly packed, remaining in LC or ordered phases at high pressures. In contrast, labels modified at the acyl chains of the molecule are frequently excluded from ordered phases. For instance, NBD-PC modified at one of the acyl chains of the molecule, is typically excluded from ordered phases of DPPC monolayers *(8,16)*. In contrast, a phospholipid modified at the headgroup such as NBD-dipalmitoylphosphatidylethanolamine (NBD-DPPE), can still remain in packed LC domains of DPPC monolayers.

However, the behavior of different probes cannot be easily predicted based only on their structural location. Probes such as TR-DPPE or FL-DPPE, which are also labeled at the head group, are effectively excluded from LC regions owing to the bulky structure of the fluorescent group *(21)*. In general, the distribution of the different dyes must be understood taking into account the properties of the different lipid phases analyzed and the "in-plane solubility" of the probe within each phase. As stated in **Subheading 2.1.**, the use of label concentrations more than 1% (*M*) is not recommended when avoiding effects on the monolayer structure is pursued (*see* **Note 12**).

Other available amphiphilic molecules such as dialkycarbocyanines, Laurdan, or Prodan (i.e., from Molecular Probes, Junction City, OR) may also be used. The octadecyl indocarbocyanine $DiIC_{18}$ is frequently used to label ordered phases. The emission spectra of prodan and laurdan depends on the orientation of the dye in the membranes, making it possible to obtain measurements of fluorescence anisotropy from defined regions of the layers *(22,23)*.

Some other probes could be used in more specific applications. Specific interactions of proteins with monolayers have been analyzed using fluorescent derivatives of certain proteins. Combination of ganglioside GM1 and fluorescein-conjugated cholera toxin B have been used to detect *raft-like* domains in monolayers and bilayers *(21,24,25)*.

Combination of two or more probes could be used to label different domains simultaneously *(26,27)*. In these experiments, the excitation and emission spectral properties of the different probes must be taken into consideration to avoid or to take advantage of fluorescent resonance energy transfer processes occurring in the films.

3.3.2. Capturing Images for Analysis

To obtain images from fluorescently labeled interfacial films, two possibilities can be considered: either a direct observation of the monolayer *in situ* at the air–liquid interface or observation of immobilized transferred monolayers.

3.3.2.1. Monolayers "In Situ"

The films can be directly observed at the interface *in situ,* directly focusing at the interface using a fluorescence microscope installed over the surface balance. The trough in this set-up must be well isolated from vibrations to avoid changes in the focus plane during image capture. The experiments have to be conducted in dark environments to protect fluorescent dyes from light bleaching. In some troughs, an inverted microscope could be used to watch the interface through a glass window installed at the base of the container, and located only at a few millimetres from the air–liquid interface to allow proper focusing. It is important to take into account that even in the absence of stirring, different phenomena including convection, evaporation, and bulk phase displacements initiated by the mechanical movements of the barrier, make the liquid surface to be in permanent motion. As a consequence of this, the interfacial film is usually experimenting fast lateral diffusion, except when the interface is occupied by highly packed material.

Most of the photographic cameras use exposure times that are too long to obtain good frozen images from the fluorescence of a monomolecular monolayer under continuous movement. Observation of films *in situ* therefore critically requires that a highly sensitive intensified CCD camera (i.e., from Andor Technology, South Windsor, CT, or Hamamatsu Photonics K.K., Hamamatsu, Japan) be available to record short videos from the interface that can be later used to extract independent frames. Simultaneous observation of different fluorescent probes in the same frames of those fast-moving films is a technical challenge. To minimize movements, the small region of the interface just under the microscope can be confined into a small compartment connected to the rest of the interface through a narrow channel; care must be taken to ensure that the connecting channel is allowing for equilibration of the most viscous films, such as phospholipid monolayers compressed to the highest pressures.

When image capturing is performed *in situ*, the changes in the structure of the film and the lateral distribution of fluorescence can be recorded from an entire Π-A or Π-time isotherm, using a unique film compressed to any desired pressure. Once compression has been stopped at the target pressure, some time of equilibration (2–5 min) is usually waited before starting video capturing.

3.3.2.2. TRANSFERRED MONOLAYERS

Once an interfacial film compressed to a given pressure is transferred to a solid support, the resulting LB film needs less technical requirements to be observed under fluorescence microscopy. These samples are completely immobilized allowing easy observation of overlapping regions using different filters, which facilitates the use of dye combinations in the films to localize specific regions or exploring for colocalization of different components. Transferred monolayers may also be subjected to different treatments after the transfer, such as blotting or staining. For instance, treatment with detergents of LB films made from dioleoylphosphatidylcholine (DOPC)/cholesterol/sphingomyelin mixtures has been used to detect formation of detergent-resistant "raft-like" domains *(21)*.

The main disadvantage of observing support-immobilized films is their rapid fluorescence photobleaching under the microscope as a consequence of sample immobilization. Therefore, it is important to limit the time of exposure during image collection. It may be enough to obtain a few images per frame of film, yet the observed region gets irreversibly burned and remains unusable for further observations.

It is also important to take into account that the LB method described in **Subheading 3.2.2.** produces films transferred at both sides of the substrate. Focusing on just one of the monolayers could be a problem. To ensure the observation of the right surface, it is recommended to carry out the transfer using two supports coupled side by side. Once the transfer has been finished, one of the two supports (preferably the one immobilizing the film that was behind the scan direction of the barrier) should be discarded.

3.3.3. Analysis of Lipid Segregation in Interfacial Films

3.3.3.1. QUALITATIVE EVALUATION OF LATERAL FILM REORGANIZATION

Information acquired from the isotherms of phospholipid films compressed to different pressures may be used to build phase diagrams. In the simplest case, Π A isotherms obtained at different temperatures may give enough information to construct Π vs temperature phase diagrams. Transition pressures obtained from horizontal plateaus or kinks observed in the slope of the isotherms are used to determine boundaries between the phases in the diagram.

Complementary information obtained from epifluorescence microscopy is useful to complete the structural information reported by the phase diagram. Observation of the apparent morphology of segregated lipid domains may allow, for instance, distinguishing between a LC/LE coexistence and one of liquid-ordered-like (LO)/LE type (*see* **Fig. 5**).

Phase diagrams, as isotherms, are very dependent on the structural properties of the phospholipid. The temperature and the surface pressure at which the different lateral transitions occur are parameters very much dependent on the length and/or the saturation state of the phospholipid acyl chains, in a similar way as they affect the fluid–gel phase transitions in membranes. Phospholipids with longer acyl chains have transitions at higher temperatures and lower pressures than those with shorter acyl chains. Increasing the length of the chain by a single methylene group is enough to increase by 5–10°C the temperature of its transition and decrease by about 10 mN/m the surface pressure *(9,28)*. Phase diagrams may also be

Lipid Segregation in Langmuir Films

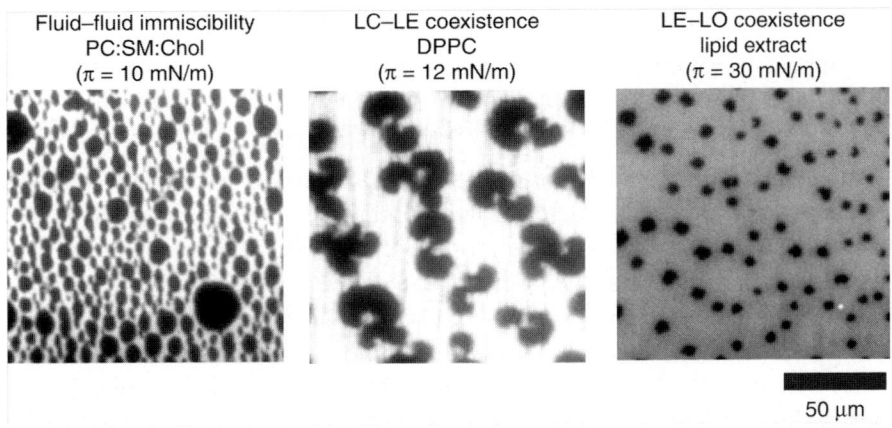

Fig. 5. Different examples of lipid phase and domain segregation as observed by epifluorescence microscopy of interfacial films. (**Left panel**) Typical round-shaped domains in PC/sphyngomyelin/Chol mixtures (24:45:30 w/w/w) exhibiting fluid/fluid immiscibility. (**Central panel**) chirality-shaped, LC, ordered domains of DPPC films compressed into the LE/LC plateau. (**Right panel**) DPPC and Chl-enriched liquid-ordered-like condensed domains segregated in pulmonary surfactant films compressed more than 30 mN/m.

affected by subphase properties, such as pH or presence of ions, which could affect interactions between charged lipid head groups in the monolayers.

To analyze lateral lipid segregation in monolayers, representation of composition-pressure phase diagrams is also useful. In this case, structure-sensitive techniques such as epifluorescence are the most useful because immiscibility is not easily detected exclusively from the information reported by isotherms (*see* **Note 13**). In binary systems, typical phase diagrams of miscibility pressures vs the mole fraction of one of the components are used (*see* **Note 14**). When systems containing three components are analyzed, ternary diagrams can be obtained from binary diagrams of two of the components at different constant molar fractions of the third one. As an example, liquid–liquid phase coexistence occurring in binary and ternary mixtures of dihydrocholesterol and phospholipids have been studied from their composition-pressure phase diagrams, as models to understand the lateral segregation thought to occur in real cell membranes *(29,30)*.

3.3.3.2. QUANTITATION OF COMPRESSION-DRIVEN LIPID-SEGREGATION

Additional quantitative information may be obtained from epifluorescence images to analyze the effect of different factors on the phase transition, along the compression isotherm. Quantitative analysis of lipid segregation can be made by computer-assisted estimation of the amount of fluorescent/nonfluorescent areas, with respect to the particular lateral distribution of any probe. Different image-analyzing programs are available to process digitally captured images and obtain quantitative data that include statistics of domain sizes or perimeters, the total fraction of condensed area, and the density of domains. NIH Image and ImageJ are free programs from the National Institute of Health for the Macintosh or Java platforms, respectively (*see* **Note 15**). A PC version of this program, Scion Image, is also available from Scion Corporation (MD). Data analysis requires processing of at least 10 frames per surface pressure to obtain statistical results.

Typical data reported from the quantitative analysis of images are:

1. Average percent of condensed area vs surface pressure.
2. Domain area vs surface pressure.
3. Number of domains per frame vs surface pressure.

Any component interacting with the phospholipid monolayer can affect one or several of these curves depending on the character of the interaction. Some components such as certain proteins or peptides may interact with the monolayer without altering the mean proportion of condensed phase observed along the transition. However, there is a reduction in size and increase in number of the condensed domains. This type of effect is usually interpreted as a consequence of the interaction of the added component with condensed domain boundaries, which produces a reduction in the line tension (*see* **Note 16**). An alternative explanation is that the additive may act as a center for nucleation of phospholipid condensates during compression. Molecules such as peptides or proteins interacting deep into the acyl chain region of phospholipid layers do usually affect both the domain distribution and the proportion of total condensed phase at any surface pressure along the compression isotherm. The interpretation of this effect is similar to that of proteins removing enthalpy from the gel-to-fluid thermotropic transition in lipid–protein suspensions studied by calorimetry. Interaction of proteins with the acyl chains of phospholipids subtracts lipid molecules from undergoing compression-driven condensation, in a proportion defined by the lipid/protein stoichiometry.

Care must be taken when quantitative data on lipid segregation are used to discuss membrane-relevant features because size, shape, and number of condensed domains are very dependent on the way the monolayer is prepared and compressed. As an example of this, the faster the monolayer is compressed, the smaller is the size of the lipid domains. Another consequence of this feature is that in monolayers prepared from adsorbed material, the size of the domains is also dependent on how rapid the material adsorbs into the air–liquid interface.

4. Notes

1. Alternatively, some balances permit a more isotropic compression by incorporating two PTFE barriers that reduce the surface available to the film in a more homogeneous way around the surface tension measurement point. This approach may be particularly useful when observing simultaneously the distribution of fluorescent probes at the surface, because it may largely reduce the lateral movements owing to displacement by the barrier of the upper layers of the subphase.
2. To maintain the subphase volume and avoid changes because of evaporation, a feedback system consisting of a buffer reservoir connected to the bulk phase, and an ellipsometric device detecting changes on the surface level may be used.
3. Surface pressure (Π) is defined as the reduction on surface tension of the clean liquid phase produced by the monolayer at the interface.
4. The Wilhelmy method uses a plate or flag partially immersed in the subphase and suspended from a balance to measure surface tension. The force measured depends on the weight of the plate, the buoyant force, and is a function of the surface tension (γ). The force produced by surface tension over the flag depends on the contact perimeter and on the cosine of the contact angle of the Wilhelmy plate with the interface. The contact perimeter can be easily determined; the contact angle is negligible when the flag is wet enough. To improve wetting of platinum Wilhelmy flags, these must be well scratched and completely immersed in water before raised to its final position. This step is not necessary for paper plates if they were previously immersed in water.
5. The compression rate of the film in Π-A isotherms should be always given as normalized area per molecule per time unit, for the isotherms to be directly comparable with those obtained in different troughs.

6. A very slow compression rate (in the order of 2 Å/mole/min) allows obtaining the isotherms under *quasi*-static equilibrium conditions. In thermodynamical terms, such isotherms are much better interpreted than isotherms recorded under dynamic conditions, but they may reflect poorly some of the phenomena occurring in living membranes. It is important to check how dependent is the structure of lipid films at defined pressures (i.e., at the membrane relevant pressure of ~30 mN/m) on the compression rate used to reach them.
7. The area per molecule at this "lift-off" pressure of the isotherm is a characteristic structural parameter, as it can be related to the maximal surface occupied by a single molecule (its molecular shape) in the absence of compression.
8. Ordered phases such as LC or the solid phase are typically reached by films made of phospholipids bearing saturated acyl chains at temperatures lower than their gel-to-fluid melting T_m. If temperature is higher than T_m, the lipid is intrinsically too disordered to be condensed and the films collapse when certain pressure is reached directly from the LE phase (*see* **Fig. 1**).
9. A second transferred lipid film could allow preparation, for instance, of supported bilayers, wherein the lateral distribution of lipids and proteins could be analyzed in a disposition that mimics free-standing membranes better than that of pure monolayer LB films. In these supported bilayers, the lipid and protein molecules taking part of the second layer diffuse in the plane of the film more freely than those directly in contact with the support.
10. When forming multilayer arrays, the orientation of the first layer determines the subsequent transferences. In "Y-type" arrays, which are the most common including those mimicking biological membranes, any layer has an opposite orientation to the surrounding ones. Less frequently, all the layers can be oriented in the same direction. These films are called X- or Z-type, depending on whether the outermost side is hydrophilic or hydrophobic, respectively.
11. Bodipy (4,4-difluoro-4-bora-3a,4a-diaza-*s*-indacene), NBD, fluorescein (spiro[isobenzofuran-1(3H), 9′-(9H)-xanthen]-3-one, 3′,6′-dihydroxy), dansyl (5-dimethylaminonaphthalene-1-sulfonyl), Texas Red (1H,5H,11H,15H-Xantheno[2,3,4-ij:5,6,7-i′j′]diquinolizin-18-ium,9-[2(or4)-(chlorosulfonyl)-4(or2)-sulfophenyl]-2,3,6,7,12,13,16,17-octahydro).
12. Recent experiments using high-resolution techniques such as AFM to compare the structure of probe-free and probe-containing phospholipid films at both micro- and nanoscopic scales, have allowed detection of effects produced by the dye on monolayer nanostructure even at low concentrations (0.5–1% *M*) *(14)*. This suggests that the proportion of exogenous probes should be maintained as low as possible, and their effect evaluated extensively by complementary techniques.
13. It is important to consider that detection of lipid segregation in compressed interfacial monolayers is only possible at the level of resolution provided by the particular microscopy technique used. Observation of fluorescent probes by epifluorescence microscopy *in situ* allows detection of lipid domains not much smaller than 1 µm. The actual size of the different types of membrane domains reported in the literature is a matter of discussion, but different authors argue that domains segregated in lipid mixtures mimicking the real complexity of membranes may have sizes ranging more in the nanometer than in the micrometer scale. Fluorescence observation of immobilized LB films improves resolution a bit, but AFM of films transferred onto mica or silicon is a much better technique to observe domains of only a few nanometers in diameter.
14. Miscibility pressure is the maximal surface pressure at which ordered domains are still segregated.
15. NIH Image and ImageJ are available for downloading at: http://rsb.info.nih.gov/. Scion Image may be downloaded from Scion Corporation home page at http://www.scioncorp.com/.
16. Line tension is the net component of forces experimented by the molecules at the boundaries of 2D segregated domains in interfacial films. Size and shape of segregated domains have been interpreted as a function of the dipolar interactions between the molecules taking part of the domains, and the line tension at their boundaries. Line tension is what makes fluid domains to adopt a circular shape in the films, as surface tension minimizes the exposed surface of soap micelles or oil drops in 3D emulsions.

Acknowledgments

Research in the laboratory of the authors is founded by grants from the Spanish Ministry of Science and Education (BIO2006-03130) and Community of Madrid (S-0505/MAT/0283). The research group at Universidad Complutense also participates in Marie Curie Networks EST-007931 and RTN-512229.

References

1. Harder, T. (2003) Formation of functional cell membrane domains: the interplay of lipid- and protein-mediated interactions. *Philos. Trans. R. Soc. London B* **358,** 863–868.
2. Helms, J. B. and Zurzolo, C. (2004) Lipids as targeting signals: lipid rafts and intracellular trafficking. *Traffic* **5,** 247–254.
3. Simons, K. and Toomre, D. (2000) Lipid rafts and signal transduction. *Nat. Rev. Mol. Cell Biol.* **1,** 31–39.
4. Taneva, S., McEachren, T., Stewart, J., and Keough, K. M. (1995) Pulmonary surfactant protein SP-A with phospholipids in spread monolayers at the air-water interface. *Biochemistry* **34,** 10,279–10,289.
5. Nag, K., Perez-Gil, J., Cruz, A., Rich, N. H., and Keough, K. M. (1996) Spontaneous formation of interfacial lipid-protein monolayers during adsorption from vesicles. *Biophys. J.* **71,** 1356–1363.
6. Launois-Surpas, M. A., Ivanova, T., Panaiotov, I., Proust, J. E., Puisieux, F., and Georgiev, G. (1992) Behavior of pure and mixed DPPC liposomes spread or adsorbed at the air-water interface. *Colloid Polym. Sci.* **270,** 901–911.
7. Cruz, A., Worthman, L. A., Serrano, A. G., Casals, C., Keough, K. M., and Perez-Gil, J. (2000) Microstructure and dynamic surface properties of surfactant protein SP-B/dipalmitoylphosphatidylcholine interfacial films spread from lipid-protein bilayers. *Eur. Biophys. J.* **29,** 204–213.
8. Nag, K., Boland, C., Rich, N., and Keough, K. M. (1991) Epifluorescence microscopic observation of monolayers of dipalmitoylphosphatidylcholine: dependence of domain size on compression rates. *Biochim. Biophys. Acta* **1068,** 157–160.
9. Kaganer, V. M., Möhwald, H., and Dutta, P. (1999) Structure and phase transitions in Langmuir monolayers. *Rev. Mod. Phys.* **71,** 779–819.
10. Marget-Dana, R. (1999) The monolayer technique: a potent tool for studying the interfacial properties of antimicrobial and membrane-lytic peptides and their interactions with lipid membranes. *Biochim. Biophys. Acta* **1462,** 109–140.
11. Samsonov, A. V., Mihalyov, I., and Cohen, F. S. (2001) Characterization of cholesterol-sphingomyelin domains and their dynamics in bilayer membranes. *Biophys. J.* **81,** 1486–1500.
12. Peng, J. B., Barnes, G. T., and Gentle, I. R. (2001) The structures of Langmuir-Blodgett films of fatty acids and their salts. *Adv. Colloid Interface Sci.* **91,** 163–219.
13. Wang, L., Cai, P., Galla, H. J., He, H., Flach, C. R., and Mendelsohn, R. (2005) Monolayer-multilayer transitions in a lung surfactant model: IR reflection-absorption spectroscopy and atomic force microscopy. *Eur. Biophys. J.* **34,** 243–254.
14. Cruz, A., Vazquez, L., Velez, M., and Perez-Gil, J. (2005) Influence of a fluorescent probe on the nanostructure of phospholipid membranes: dipalmitoylphosphatidylcholine interfacial monolayers. *Langmuir* **21,** 5349–5355.
15. Nag, K., Perez-Gil, J., Ruano, M. L., et al. (1998) Phase transitions in films of lung surfactant at the air-water interface. *Biophys. J.* **74,** 2983–2995.
16. Perez-Gil, J., Nag, K., Taneva, S., and Keough, K. M. (1992) Pulmonary surfactant protein SP-C causes packing rearrangements of dipalmitoylphosphatidylcholine in spread monolayers. *Biophys. J.* **63,** 197–204.
17. von Tscharner, V. and McConnell, H. M. (1981) An alternative view of phospholipid phase behavior at the air-water interface. Microscope and film balance studies. *Biophys. J.* **36,** 409–419.

18. Kaercher, T., Honig, D., and Mobius, D. (1993) Brewster angle microscopy. A new method of visualizing the spreading of Meibomian lipids. *Int. Ophthalmol.* **17,** 341–348.
19. Bakowsky, U., Rothe, U., Antonopoulos, E., Martini, T., Henkel, L., and Freisleben, H. J. (2000) Monomolecular organization of the main tetraether lipid from Thermoplasma acidophilum at the water-air interface. *Chem. Phys. Lipids* **105,** 31–42.
20. Maier, O., Oberle, V., and Hoekstra, D. (2002) Fluorescent lipid probes: some properties and applications (a review). *Chem. Phys. Lipids* **116,** 3–18.
21. Dietrich, C., Bagatolli, L. A., Volovyk, Z. N., et al. (2001) Lipid rafts reconstituted in model membranes. *Biophys. J.* **80,** 1417–1428.
22. Bagatolli, L. A. and Gratton, E. (2000) Two photon fluorescence microscopy of coexisting lipid domains in giant unilamellar vesicles of binary phospholipid mixtures. *Biophys. J.* **78,** 290–305.
23. Krasnowska, E. K., Bagatolli, L. A., Gratton, E., and Parasassi, T. (2001) Surface properties of cholesterol-containing membranes detected by Prodan fluorescence. *Biochim. Biophys. Acta* **1511,** 330–340.
24. Bacia, K., Scherfeld, D., Kahya, N., and Schwille, P. (2004) Fluorescence correlation spectroscopy relates rafts in model and native membranes. *Biophys. J.* **87,** 1034–1043.
25. de Almeida, R. F., Loura, L. M., Fedorov, A., and Prieto, M. (2005) Lipid rafts have different sizes depending on membrane composition: a time-resolved fluorescence resonance energy transfer study. *J. Mol. Biol.* **346,** 1109–1120.
26. Nag, K., Taneva, S. G., Perez-Gil, J., Cruz, A., and Keough, K. M. (1997) Combinations of fluorescently labeled pulmonary surfactant proteins SP-B and SP-C in phospholipid films. *Biophys. J.* **72,** 2638–2650.
27. Ruano, M. L., Nag, K., Worthman, L. A., Casals, C., Perez-Gil, J., and Keough, K. M. (1998) Differential partitioning of pulmonary surfactant protein SP-A into regions of monolayers of dipalmitoylphosphatidylcholine and dipalmitoylphosphatidylcholine/dipalmitoylphosphatidylglycerol. *Biophys. J.* **74,** 1101–1109.
28. Mohwald, H. (1990) Phospholipid and phospholipid-protein monolayers at the air/water interface. *Annu. Rev. Phys. Chem.* **41,** 441–476.
29. Keller, S. L., Anderson, T. G., and McConnell, H. M. (2000) Miscibility critical pressures in monolayers of ternary lipid mixtures. *Biophys. J.* **79,** 2033–2042.
30. McConnell, H. M. and Radhakrishnan, A. (2003) Condensed complexes of cholesterol and phospholipids. *Biochim. Biophys. Acta* **1610,** 159–173.

30

Detergent and Detergent-Free Methods to Define Lipid Rafts and Caveolae

Rennolds S. Ostrom and Xiaoqiu Liu

Summary

Lipid rafts and their related membrane vesicular structures, caveolae, are cholesterol- and sphingolipid-rich microdomains of the plasma membrane that have attracted considerable interest because of their ability to concentrate numerous signaling proteins. Efforts to define the proteins that reside in lipid rafts and caveolae as well as investigations into the functional role of these microdomains in signaling, endocytosis, and other cellular processes have led to the hypothesis that they compartmentalize or prearrange molecules involved in regulating these pathways. This chapter describes biochemical approaches for defining lipid rafts and caveolae. Included are detergent- and nondetergent-based fractionations on sucrose-density gradients that isolate buoyant lipid rafts and caveolae as well as caveolin antibody-based immunoisolation of detergent-insoluble membranes that selectively isolates caveolae and not lipid rafts. Also, a general method to disrupt lipid rafts and caveolae using β-cyclodextrin that is useful for probing the role of these microdomains in cellular processes is described. The advantages and disadvantages of the respective approaches are discussed. Taken together, these methods are useful for defining the role of lipid rafts and caveolae in cell signaling.

Key Words: β-cyclodextrin; caveolae; density gradient centrifugation; immunoisolation; lipid rafts; membrane microdomains.

1. Introduction

Lipid rafts are plasma membrane microdomains formed through the association of sphingolipid and cholesterol that have rapidly become recognized as important to many types of cellular signal transduction. Lipid rafts along with caveolae, which are thought to form from lipid rafts because of their similar lipid composition, appear to be signaling "hot spots" because of their ability to attract and retain numerous and diverse signaling molecules (1). Thus, caveolae and lipid rafts concentrate, and perhaps, promote the formation of signaling complexes that are essential for rapid and specific signal transduction (2,3). In this chapter, caveolae and lipid rafts will be introduced, their similarities and differences with respect to how they can be studied will be discussed, and several detailed methodologies for defining the proteins associated with lipid raft and caveolar structures will be presented.

Caveolae were originally identified in endothelial cells as 50–100-nm flask-like invaginations of the plasma membrane (4). Caveolae were later shown to be involved in the transcellular movement of molecules (potocytosis and endocytosis) (5). Endocytosis by caveolae represents a parallel but distinct pathway from clathrin-coated pits for the removal and destruction or recycling of plasma membrane receptors (5,6). Caveolae are similar to lipid rafts in that they are both enriched in sphingolipid and cholesterol, but caveolae also express a coat of caveolin proteins on the inner leaflet of the membrane bilayer (7). All mammalian cells appear to contain plasma membrane-lipid rafts because of the ubiquitous nature of sphingolipid and cholesterol, but only cells expressing one of the three isoforms of caveolin appear to contain caveolae (8). Of the three isoforms of caveolin, caveolin-1, -2, and -3 only caveolin-1 (the predominant isoform) and caveolin-3 (the striated muscle-specific isoform) are capable of inducing caveolar biogenesis (9,10). Caveolin-2, when expressed alone, cannot induce caveolar formation, but this isoform is found in hetero-oligomers with caveolin-1 and caveolin-3 (11–15). Whereas it is unclear whether different caveolin isoform compositions create functionally distinct caveolae (16), it is clear that lipid rafts and caveolae differ in a variety of ways (17–19). Approaches to differentiate between lipid rafts and caveolae as well as to manipulate caveolin expression in cells, will likely lead to much more information regarding the differences between these structures in the near future.

Despite recent advances in microscopic approaches (including atomic force microscopy to visualize lipid rafts [20]) lipid rafts and caveolae are most readily defined using biochemical approaches. Lipid rafts and caveolae can be extracted from other cellular material in cell homogenates based on their relative insolubility in particular detergent or nondetergent conditions and their high buoyancy when centrifuged on a density gradient. These approaches, a few of which are described in this chapter, rely on properties common to both caveolae and lipid rafts, thus cannot distinguish between these domains. Caveolar domains can be specifically isolated using immunological approaches to trap caveolin proteins from plasma membrane preparations (21,22). One method that is applicable for specifically isolating caveolae and not lipid rafts from numerous types of cells and tissues is described in this chapter.

The function of lipid rafts and caveolae in signaling or other cellular processes can also be inferred from studies in which the microdomains are disrupted. β-cyclodextrin, a chemical that does not enter cells but can bind cholesterol and remove it from the plasma membrane, disrupts lipid rafts and caveolae (23,24). Filipin, a polyene antibiotic and sterol-binding agent, also disrupts lipid rafts and caveolae (25,26). This chapter describes a method using β-cyclodextrin to disrupt lipid rafts and caveolae that can be applied in many experimental paradigms. For more detailed information regarding the manipulation of cellular cholesterol using cyclodextrins, refer to Christian et al. (27).

Microscopic as well as other types of nonbiochemical studies are often important for corroborating results from biochemical studies of lipid rafts and caveolae (see **Note 1**). Microscopic approaches for defining caveolae are limited to electron microscopic studies, because the resolution of light (including fluorescent microscopy) precludes the detection of 50–100 nm structures. However, laser confocal microscopy using deconvolution can be used to define the colocalization of two proteins, and when combined with detection of caveolins, can infer caveolar localization (28). Microscopy is also limited by the suitability of antibodies to detect native proteins of interest in fixed cells and tissues. One can utilize fluorescent or epitope tags to circumvent the antibody problem and take advantage of other powerful technologies such as fluorescence resonance energy transfer and bioluminescence energy transfer that define the interaction of components in living cells (29). However, the usefulness of tagged proteins is limited to the examination of exogenously expressed proteins, which can localize differently than their native counterparts (30). Thus, a combination of biochemical methods, several of which are described herein, and microscopic approaches

2. Materials

1. Phosphate-buffered saline (PBS) 137 mM NaCl, 10 mM Phosphate, 2.7 mM KCl, pH 7.4.
2. 500 mM Na$_2$CO$_3$ (should be ~pH 11.0, but do not adjust).
3. MES buffered saline (MBS): 25 mM 2-(N-Morpholino) ethanesulfonic acid 4-Morpholineethane sulfonic acid (MES), 150 mM NaCl, pH 6.0.
4. MBS/Na$_2$CO$_3$: MBS, 250 mM Na$_2$CO$_3$.
5. 90% Sucrose/MBS: dissolve 45 g sucrose with MBS until volume equals 50 mL. Heat in a microwave oven (in 10-s intervals) to dissolve/melt.
6. 35% Sucrose in MBS/Na$_2$CO$_3$: 5.83 mL 90% sucrose/MBS and 9.17 mL MBS/Na$_2$CO$_3$.
7. 5% Sucrose in MBS/Na$_2$CO$_3$: 0.83 mL 90% sucrose/MBS and 14.17 mL MBS/Na$_2$CO$_3$.
8. Triton X-100 Sigma-Aldrich, St. Louis, MO. Catalog # ×100 buffer: MBS, 1% Triton X-100, protease inhibitor mix (Sigma P-8340) (diluted 1:100).
9. 35% Sucrose in MBS/Triton X-100: 5.83 mL 90% sucrose/MBS and 9.17 mL Triton X-100 buffer.
10. 5% Sucrose in MBS/Na$_2$CO$_3$: 0.83 mL 90% sucrose/MBS and 14.17 mL Triton X-100 buffer.
11. Membrane buffer: 0.25 M sucrose, 1 mM ethylenediaminetetraacetic acid, 20 mM Tricine, pH 7.8.
12. 30% Percoll Sigma-Aldrich, St. Louis, MO. Catalog # 7737: 3 mL Percoll stock solution diluted in 9 mL PBS 137 mM NaCl, 10 mM Phosphate, 2.7 mM KCl, pH 7.4.
13. Modified lysis buffer: 50 mM Tris-HCl, pH 7.5, 150 mM NaCl, 1 mM ethylene glycolbis (2-aminoethylether)-N, N, N, N-tetraacetic acid (EGTA), 10 mM MgCl$_2$, 0.5% Triton X-100, and protease inhibitor mix (Sigma P-8340, diluted 1:100). Sigma-Aldrich, St. Louis, MO. Catalog # P-8340.
14. Protein A-agarose and protein G-agarose.
15. Immunoprecipitation (IP) wash buffer 1: 50 mM Tris-HCl, pH 7.5, 500 mM NaCl, and 0.2% Triton X-100.
16. IP wash buffer 2: 10 mM Tris-HCl, pH 7.5, 0.2% Triton X-100.
17. Methyl-β-cyclodextrin (MBCD) media: serum- and NaHCO$_3$-free Dubelco's Modified Eagle's Medium (DMEM), 20 mM HEPES, 2% (2-hydroxypropyl)-β-cyclodextrin, pH 7.4. Solution may require sonication to fully solubilize.
18. MBCD vehicle media: serum- and NaHCO$_3$-free DMEM, 20 mM HEPES, pH 7.4.
19. MBCD-cholesterol media: serum- and NaHCO$_3$-free DMEM, 20 mM HEPES, β-cyclodextrin (βCD)–cholesterol complexes (10 μg/mL cholesterol: β-cyclodextrin in 1:6 molar ratio, such as Sigma cat no. C4951), pH 7.4. Solution may require sonication to fully solubilize.

3. Methods

3.1. Isolation of Lipid Rafts and Caveolae by Sucrose-Density Centrifugation

The unique lipid composition (i.e., enrichment in sphingolipid and cholesterol) of lipid rafts and caveolae makes them resistant to solubilization in detergents and certain other conditions. Once other cellular material is dissolved, lipid rafts and caveolae can be separated from the rest of the cellular contents using sucrose-density centrifugation. The method for adherent cells in tissue culture is described, but it can be readily adapted for cells in suspension or for tissue samples. The authors have found that two 150-mm plates of cells are adequate for one preparation, but studies to optimize the amount of starting material are recommended.

3.1.1. Nondetergent Isolation of Lipid Rafts and Caveolae

1. Check that cells are at least 70% confluent. Aspirate medium and wash three times with ice-cold PBS. On the last wash, be sure to remove all PBS by tilting the plate at a steep angle for 30 s then aspirating all liquid. This step will ensure the lysis buffer is not overly diluted.
2. Apply 1 mL of 500 mM Na_2CO_3 to each 150-mm plate and make sure it covers the entire monolayer. Scrape cells from the plate with a cell scraper in a circular motion from top to bottom, making sure to retain as much cellular material as possible in a pool at the bottom of the tilted plate.
3. Transfer the cells and all liquid from two 150-mm plates (2 mL total) to a prechilled Dounce (glass–glass) homogenizer Wheaton Science Products, Millville, NJ: homogenize the cells with 20 strokes (one stroke is all the way down then all the way up) on ice.
4. Transfer the homogenate to a prechilled 50-mL conical tube and homogenize with a polytron three times for 10 s with intervals of 10–15 s. Rinse the polytron blade with 0.5 mL of 500 mM Na_2CO_3 into the sample to recover all possible material.
5. Homogenize the sample using an ultrasonic cell disruptor equipped with a stainless steel probe using high power three times for 20 s each with a full 60 s rest between each homogenization. Ultrasonic disruptors can vary from model to model in their power output. Thus, the power setting may need to be optimized (*see* **Note 2**).
6. Proceed to **Subheading 3.1.4**.

3.1.2. Detergent Isolation of Lipid Rafts and Caveolae

1. Check that cells are at least 70% confluent. Aspirate medium and wash three times with ice-cold PBS. On the last wash, be sure to remove all PBS by tilting the plate at a steep angle for 30 s and then aspirating all liquid. This step will ensure the lysis buffer is not overly diluted.
2. Apply 1 mL of 1% Triton-X 100 buffer to each 150-mm plate so that it covers the entire monolayer (*see* **Note 3**). Scrape cells from the plate with a cell scraper in a circular motion from top to bottom, making sure to retain as much cellular material as possible in a pool at the bottom of the tilted plate.
3. Transfer the cells from two plates (2 mL total) to a prechilled Dounce (glass–glass) homogenizer and incubate on ice for 20 min. Homogenize the cells with 20 strokes (one stroke is all the way down then all the way up) on ice.
4. Proceed to **Subheading 3.1.4**.

3.1.3. Variation: Isolation of Lipid Rafts and Caveolae From Plasma Membranes

1. Check that cells are at least 70% confluent. Aspirate medium and wash three times with ice-cold PBS. On the last wash, be sure to remove all PBS by tilting the plate at a steep angle for 30 s and then aspirating all liquid. This step will ensure the lysis buffer is not overly diluted.
2. Apply 1 mL of membrane buffer to each 150-mm plate so that it covers the entire monolayer. Scrape cells from the plate with a cell scraper in a circular motion from top to bottom, making sure to retain as much cellular material as possible in a pool at the bottom of the tilted plate.
3. Collect the cells from two plates (2 mL total) and homogenize cells with 20 strokes (one stroke is all the way down then all the way up) in a Dounce (glass–glass) or Teflon-glass homogenizer Wheaton Science Products, Millville, NJ. on ice then centrifuge at 300g for 5 min and collect the supernatant.
4. Layer the supernatant on top of 30% Percoll and centrifuge at 64,000g (19,000 rpm on a SW41 ultracentrifuge rotor [Beckman Coulter, Fullerton, CA]) for 30 min.
5. Collect the opaque band near the top of the Percoll layer as the plasma membrane fraction.
6. Adjust the plasma membrane fraction to a final concentration of 500 mM Na_2CO_3 by adding an equal volume of 1 M Na_2CO_3 and sonicate three times for 20 s with full 60 s rests between intervals (*see* **Note 2**).

Methods to Define Lipid Rafts and Caveolae 463

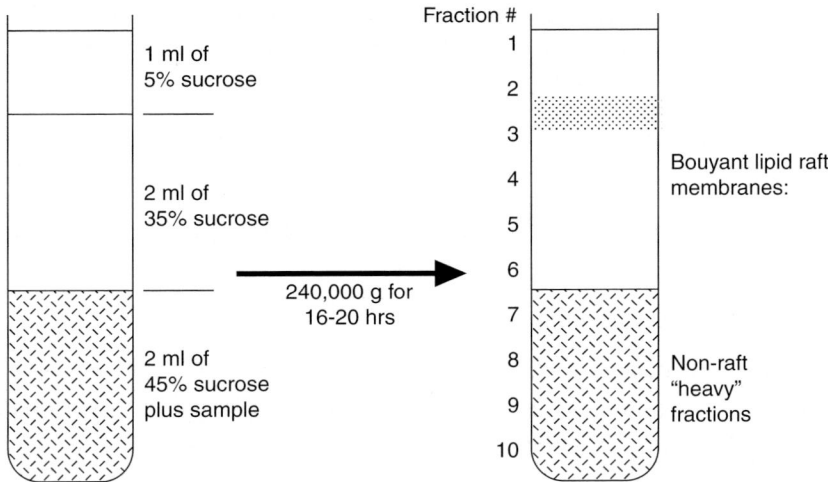

Fig. 1. Schematic diagram illustrating the isolation of buoyant lipid-raft fractions. The primary biochemical approach for defining lipid rafts involves the use of sucrose-density centrifugation. Following homogenization of cells using either detergent- or nondetergent- based approaches (*see* **Subheading 3.1.**), the sample is adjusted to 45% sucrose and bottom-loaded in an ultracentrifuge tube. A discontinuous sucrose gradient consisting of 35% sucrose and 5% sucrose is constructed on top of the sample and the entire gradient is centrifuged at approx 240,000*g* (maximum relative centrifugal force *rcf*) for 16–20 h. Buoyant lipid rafts (as well as caveolae, *see* **Subheading 1.**) "float" in the gradient, whereas the bulk of the "heavy" cellular material remains in the 45% sucrose layer. The gradient is typically collected in 0.5 mL fractions starting at the top and analyzed by immunoblot.

3.1.4. Sucrose-Density Centrifugation to Fractionate Cell Homogenates

Cells or tissues should be prepared and homogenized using one of the aforementioned approaches (*see* **Subheadings 3.1.1.**, **3.1.2.**, or **3.1.3.**, *see also* **Note 4**). Once the homogenate is prepared, the buoyant lipid-raft fraction can be isolated by floatation on a sucrose-density gradient. The method described here utilizes a discontinuous sucrose gradient (**Fig. 1**) but continuous gradients can also be used (*see* **Note 5**).

1. Mix 1 mL of homogenized sample (leaving any foam behind) with 1 mL of 90% sucrose/MBS in a 5-mL Beckman ultraclear ultracentrifuge tube. Save any remaining sample as whole cell lysate. Leftover lysates can also be frozen at −80°C and fractionated by sucrose density centrifugation at a later date.
2. Carefully layer 2 mL of either 35% sucrose in MBS/Na$_2$CO$_3$ (if sample was homogenized by nondetergent method, *see* **Subheadings 3.1.1.** or **3.1.3.**) or 35% sucrose in MBS/Triton X-100 buffer (if sample was homogenized by detergent method; *see* **Subheading 3.1.2.**) on top of the sample/90% sucrose/MBS layer. A visible interface should exist between the two density layers.
3. Carefully layer 1 mL of either 5% sucrose in MBS/Na$_2$CO$_3$ (if sample was homogenized by nondetergent method, *see* **Subheadings 3.1.1.** or **3.1.3.**) or 5% sucrose in MBS/Triton X-100 buffer (if sample was homogenized by detergent method; *see* **Subheading 3.1.2.**) on top of the 35% sucrose layer. A second interface should be visible between the 35% and the 5% sucrose layers, and the ultracentrifuge tube should be nearly full. Although the gradient is not highly sensitive, all movement of the gradient should be made carefully and deliberately in order to not disrupt the gradient interfaces.
4. Centrifuge for 16–20 h at 46,000 rpm at 4°C in a SW55Ti rotor (Beckman), equivalent to a maximum force (bottom of the tube) of approx 260,000*g* and an average force (middle of the tube) of approx 200,000*g* (*see* **Note 6**).

5. At the completion of the centrifugation, carefully remove the ultracentrifuge tube from the bucket. A faint light-scattering band, which consists of the buoyant lipid raft/caveolar material, is often visible at the 35% sucrose –5% sucrose interface.
6. Collect samples from the gradient from the top down in 0.5 mL volumes, putting each fraction in a labeled tube and yielding 10 fractions. One should be careful to keep the pipet at the top of the liquid in order to draw each fraction appropriately. If cellular material is visible at the upper gradient interface, care should be taken to collect this material in fractions two and three.
7. Fractions can then be analyzed by sodium dodecyl sulfate-polyacrylamide gel electrophoresis (SDS-PAGE) and immunoblotting (*see* **Note 7**).

3.2. Immunoisolation of Caveolae

This method takes advantage of the reduced solubility of lipid rafts and caveolae to detergent in order to isolate these domains from the rest of the cellular material (as in **Subheading 3.1.2.**) and then specifically "traps" caveolae (and not lipid rafts) by using an antibody to immunoprecipitate caveolin. This method will also pull down the caveolar lipids, cholesterol, and associated proteins. One needs to first define the caveolin isoform expression in a given cell or tissue type in order to select an appropriate caveolin antibody. For cells expressing multiple isoforms, an antibody to the most predominant or readily detectable caveolin can be used because coexpressed isoforms form hetero-oligomers *(31)*. Particular advantages of this method are that one can maintain enzyme and receptor-binding activity in the isolates, facilitating the assessment protein function. The authors as well as others have used this approach to assay adenylyl cyclase activity regulated by G protein-coupled receptors in caveolar domains isolated from cardiac myocytes *(22,28)*.

1. Check that cells are at least 70% confluent. Aspirate medium and wash three times with ice-cold PBS. On the last wash, be sure to remove all PBS by tilting plate at a steep angle for 30 s and then aspirating all liquid. This step will ensure the lysis buffer is not overly diluted.
2. Add 2 mL of modified lysis buffer to each 15-cm plate. Homogenize cells with 20 strokes (one stroke is all the way down then all the way up) in a Dounce (glass–glass) homogenizer.
3. Transfer to a 1.5-mL microtube and add 50 µL of either protein G- or protein A-agarose suspension (*see* **Note 8**) to preclear any native antibodies. Incubate at 4°C on a rocking platform for 1 h.
4. Centrifuge in a microcentrifuge at maximum speed (12,000–14,000 rpm) for 30 s to pellet the agarose and then transfer the supernatant to a new tube. The pellet can be discarded.
5. Add primary antibody (1–3 µL, depending on the antibody concentration and affinity) and continually mix (preferably by rocking) at 4°C for 1 h to allow antibody binding to epitope.
6. Add 50 µL protein A- or protein G-agarose to tube and continually mix (preferably by rocking) at 4°C for 1 h to allow binding to the antibody–epitope complexes.
7. Centrifuge in a microcentrifuge at maximum speed (12,000–14,000 rpm) for 30 s to pellet agarose. Supernatant should be saved as the IP supernatant. IP supernatant should be assessed for the amount of epitope not trapped in the immunoprecipitates and can behave as a control of nonprecipitated material in the assay of choice.
8. Wash the pellet by adding 1 mL of modified lysis buffer, mix and rock at 4°C for 5 min.
9. Centrifuge in a microcentrifuge at maximum speed (12,000–14,000 rpm) for 30 s to pellet agarose, remove supernatant and add 1 mL of wash buffer 1 to pellet, mix and rock at 4°C for 5 min.
10. Centrifuge in a microcentrifuge at maximum speed (12,000–14,000 rpm) for 30 s to pellet agarose, remove and discard supernatant.
11. Wash the pellet a second time by adding 1 mL of wash buffer 2, mix and rock at 4°C for 5 min.
12. Centrifuge in a microcentrifuge at maximum speed (12,000–14,000 rpm) for 30 s to pellet agarose, remove and discard supernatant.
13. The final pellet should then be suspended in a suitable assay buffer (if enzyme activity is to be measured) and/or in sample buffer for analysis by SDS-PAGE (for immunoblotting). Immunoblot

analysis should be performed on a portion of the immunoprecipitated pellet and the IP supernatant to confirm appropriate IP of caveolin and to assess which proteins have been coprecipitated.

3.3. Disruption of Lipid Rafts and Caveolae

Using β-cyclodextrin, a cholesterol-binding agent, one can remove cholesterol from the plasma membrane of living cells and disrupt both lipid rafts and caveolae. Cholesterol depletion is toxic to cells over time, thus one must be careful to optimize exposure to β-cyclodextrin as well as control for cell toxicity or stress. The method below describes using a 60-min treatment period (which have been found to be appropriate in several cell types). However, it is strongly suggested to perform a treatment time-course and determine cell viability (e.g., using trypan blue exclusion) in order to optimize the treatment for each cell type. The most critical control, which is described in **Subheading 3.3.4.**, involves the replenishment of cholesterol in cells following its extraction. This condition will control for nonspecific effects of β-cyclodextrin treatment and indicate the reversibility of any observed effects. Thus, all assays should be conducted in vehicle treated control cells, β-cyclodextrin treated cells, and cells that are treated with β-cyclodextrin followed by cholesterol replenishment. This method can be used before almost any type of signal transduction assay to ascertain the role of lipid rafts and caveolae in a given response.

1. Check that cells are at least 70% confluent. Aspirate medium and wash three times with warm (37°C) PBS. On the last wash, be sure to remove all PBS by tilting the plate at a steep angle for 30 s and then aspirating all liquid. This step will ensure the lysis buffer is not overly diluted.
2. To one plate or set of plates apply 15 mL of warm DMEH containing β-cyclodextrin (MBCD media) to each 150-mm plate, so that it covers the entire monolayer. Incubate cells at 37°C for 60 min. A CO_2-enriched environment (i.e., cell culture incubator) is not needed because of the buffering of this media with HEPES.
3. To a separate plate or set of plates apply 15 mL of warm DMEH containing MBCD vehicle (MBCD vehicle media) to each 150-mm plate following aspiration (**step 1**), so that it covers the entire monolayer. Incubate cells at 37°C for 60 min. This step will remove membrane cholesterol, disrupting lipid rafts, and caveolae.
4. To a third plate or set of plates apply 15 mL of warm MBCD Media to each 150-mm plate following aspiration (**step 1**), so that it covers the entire monolayer. Incubate cells at 37°C for 60 min. Aspirate MBCD media, wash cells three times with warm PBS (as in **step 1**), then add 15 mL of warm maintenance media containing β-cyclodextrin/cholesterol complexes (MBCD-cholesterol media). Incubate cells at 37°C for 60 min.
5. Aspirate media from cells, wash three times with warm PBS (as in **step 1**), and conduct assay of choice. Positive controls for this method include performing sucrose density centrifugation (*see* **Subheading 3.1.4.**) and/or assaying for cholesterol content.

4. Notes

1. Complementary approaches to defining caveolae and lipid rafts must be considered. Caveolins appear to also act as scaffolding proteins that can bind multiple signaling proteins; thus, caveolins may act to organize a signaling pathway or regulate signaling activity *(3)*. Therefore, IP of caveolin proteins and expression of peptides that interfere with the caveolin-binding motif are useful approaches for defining the role of caveolins in organizing signal-transduction cascades *(3)*. Overexpression or knockout of caveolins have also been used to examine the physiological role of these proteins *(32)*. However, altering caveolin expression should not be considered a pure probe of the compartmentation or organization of a signaling pathway in caveolae, because caveolins also act as direct regulators of several signal transduction pathways *(7)*.
2. In the nondetergent fractionation method, the ultrasonic disruption of cells is the most critical step, as this is the point at which membrane lipids are dissolved but lipid raft material remains

intact. Thus, the power of the sonication step is critical. Too much power can disrupt the raft structure, resulting in little material floating up in the sucrose gradient. Too little power can result in insufficient dissolution of nonraft lipids, resulting in raft material not being sufficiently freed from "heavy" material, which also results in less material floating up in the fractionation. Therefore, the power setting used will depend on the make and model of the ultrasonic disruptor used and may need to be optimized for individual cell types. In the authors' experience, the sonication time and rest period should not be significantly altered.

3. Each method for defining lipid rafts and caveolae has advantages and disadvantages. Experimentalists should consider these factors when choosing the experimental approaches they will use to answer their particular biological question. For example, the nondetergent fractionation of cells retains certain proteins in lipid raft fractions that are often lost in detergent-based methodologies *(33,34)*. However, detergent-based approaches allow for measuring protein function, such as enzyme activity, whereas the high-pH and -energy sonication of the nondetergent methods generally impair protein function. It is generally desirable to use a combination of different, complementary approaches in order to define signal transduction in lipid rafts and caveolae.

4. Nonionic detergents other than Triton X-100, including NP-40, octylglucoside, CHAPS, Lubrol, and Brij 98, can be used to solubilize cells and isolate lipid raft and caveolar domains *(35)*. In addition, some investigators have used concentrations of Triton X-100 lower than 1% in protocols similar to that described in **Subheading 3.1.2.**

5. The method described here utilizes a discontinuous gradient of sucrose. However, continuous gradients of sucrose or of Optiprep Sigma-Aldrich, St. Louis, MO. Catalog # D1556 (described in **refs. *33*** and ***36***) are capable of resolving proteins and structures with intermediate buoyancies.

6. Other rotors can be used for the sucrose density centrifugation, including a Beckman SW41Ti rotor with 12-mL buckets. In this case, 2 mL of cell homogenate is mixed with 2 mL of 90% sucrose and 4 mL of 35% sucrose and 4 mL of 5% sucrose is layered on top. The rotation speed is adjusted to maintain equivalent *g*-force (~39,000 rpm). Fractions are collected in 1 mL aliquots to yield 12 fractions. Other rotors can be used and the Beckman rotor resources web page (http://www.beckman.com/resourcecenter/labresources/centrifuges/rotorcalc.asp) is useful for calculating the appropriate rotational speed.

7. Each fractionation from a sucrose-density centrifugation should be carefully analyzed for markers of certain cellular organelles that can contaminate the buoyant fractions. Immunoblot analysis of fractions from the 5% sucrose/35% sucrose interface (fractions numbered 2 and 3 from the method described in **3.1.4.**) should contain the bulk of caveolin isoform immunoreactivity. At the same time, these fractions should largely exclude markers of clathrin-coated pits (such as adaptin-β) and Golgi apparatus (such as mannosidase II). One can also confirm the appropriateness of a fractionation by examining the total protein in each fraction. The buoyant fractions from most cells should contain approx 5% of the total cellular protein. When immunoblot analysis of fractions is planned, it is best to add SDS-PAGE sample buffer to each fraction and to denature at (70°C for 10 min immediately after collecting the gradient. This will ensure more reproducible results when storing frozen samples for extended periods. For detection of low-abundance proteins, samples can also be concentrated in a speed-vac (or similar type) concentrator Thermo Fisher Scientific, Waltham, MA before addition of sample buffer. However, the fractions from the bottom of the gradient will not concentrate as well because of the presence of higher concentrations of sucrose. Dialysis can also be used to remove sucrose and to concentrate the samples.

8. IP efficiency can be maximized by carefully selecting the most effective agarose bead conjugate based on the primary antibody being used. Protein A has high affinity for human, rabbit, guinea pig, and pig immunoglobulin G's (IgG). Protein G has high affinity for human, horse, cow, pig, and rabbit IgG's. When using mouse IgG's, both protein A and protein G have moderate affinity. However, protein A and protein G have further differences in affinities for the subclasses of IgG's. For more detailed information on the different affinities of protein A and protein G, refer to the manufacturer's product information sheet.

References

1. Shaul, P. W. and Anderson, R. G. (1998) Role of plasmalemmal caveolae in signal transduction. *Am. J. Physiol.* **275(5 Pt 1),** L843–L851.
2. Ostrom, R. S. and Insel, P. A. (2004) The evolving role of lipid rafts and caveolae in G protein-coupled receptor signaling: Implications for molecular pharmacology. *Br. J. Pharmacol.* **143(2),** 235–245.
3. Okamoto, T., Schlegel, A., Scherer, P. E., and Lisanti, M. P. (1998) Caveolins, a family of scaffolding proteins for organizing preassembled signaling complexes at the plasma membrane. *J. Biol. Chem.* **273(10),** 5419–5422.
4. Palade, G. (1953) Fine structure of blood capilaries. *J. Appl. Physiol.* **24,** 1424.
5. Anderson, R. G. (1998) The caveolae membrane system. *Annu. Rev. Biochem.* **67,** 199–225.
6. Rapacciuolo, A., Suvarna, S., Barki-Harrington, L., et al. (2003) Protein kinase A and G protein-coupled receptor kinase phosphorylation mediates beta-1 adrenergic receptor endocytosis through different pathways. *J. Biol. Chem.* **278(37),** 35,403–35,411.
7. Razani, B., Woodman, S. E., and Lisanti, M. P. (2002) Caveolae: from cell biology to animal physiology. *Pharmacol. Rev.* **54(3),** 431–467.
8. Hooper, N. M. (1999) Detergent-insoluble glycosphingolipid/cholesterol-rich membrane domains, lipid rafts and caveolae. *Mol. Membr. Biol.* **16(2),** 145–156 (review).
9. Song, K. S., Scherer, P. E., Tang, Z., et al. (1996) Expression of caveolin-3 in skeletal, cardiac, and smooth muscle cells. Caveolin-3 is a component of the sarcolemma and co-fractionates with dystrophin and dystrophin-associated glycoproteins. *J. Biol. Chem.* **271(25),** 15,160–15,165.
10. Tang, Z., Scherer, P. E., Okamoto, T., et al. (1996) Molecular cloning of caveolin-3, a novel member of the caveolin gene family expressed predominantly in muscle. *J. Biol. Chem.* **271(4),** 2255–2261.
11. Scherer, P. E., Okamoto, T., Chun, M., et al. (1996) Identification, sequence, and expression of caveolin-2 defines a caveolin gene family. *Proc. Natl. Acad. Sci. USA* **93(1),** 131–135.
12. Scherer, P. E., Lewis, R. Y., Volonté, D., et al. (1997) Cell-type and tissue-specific expression of caveolin-2. Caveolins 1 and 2 co-localize and form a stable hetero-oligomeric complex in vivo. *J. Biol. Chem.* **272(46),** 29,337–29,346.
13. Razani, B., Wang, X. B., Engelman, J. A., et al. (2002) Caveolin-2-deficient mice show evidence of severe pulmonary dysfunction without disruption of caveolae. *Mol. Cell Biol.* **22(7),** 2329–2344.
14. Rybin, V. O., Grabham, P. W., Elouardighi, H., and Steinberg, S. F. (2003) Caveolae-associated proteins in cardiomyocytes: caveolin-2 expression and interactions with caveolin-3. *Am. J. Physiol. Heart Circ. Physiol.* **285(1),** H325–H332.
15. Lahtinen, U., Honsho, M., Parton, R. G., Simons, K., and Verkade, P. (2003) Involvement of caveolin-2 in caveolar biogenesis in MDCK cells. *FEBS Lett.* **538(1–3),** 85–88.
16. Ostrom, R. S. (2005) Caveolins muscle their way into the regulation of cell differentiation, development, and function. Focus on "Muscle-specific interaction of caveolin isoforms: differential complex formation between caveolins in fibroblastic vs. muscle cells." *Am. J. Physiol. Cell Physiol.* **288(3),** C507–C509.
17. Sowa, G., Pypaert, M., and Sessa, W. C. (2001) Distinction between signaling mechanisms in lipid rafts vs. caveolae. *Proc. Natl. Acad. Sci. USA* **98(24),** 14,072–14,077.
18. Williams, T. M. and Lisanti, M. P. (2004) The caveolin proteins. *Genome Biol.* **5(3),** 214.
19. Oh, P. and Schnitzer, J. E. (2001) Segregation of Heterotrimeric G Proteins in Cell Surface Microdomains. G(q) binds caveolin to concentrate in caveolae, whereas g(i) and g(s) target lipid rafts by default. *Mol. Biol. Cell.* **12(3),** 685–698.
20. Henderson, R. M., Edwardson, J. M., Geisse, N. A., and Saslowsky, D. E. (2004) Lipid rafts: feeling is believing. *News Physiol. Sci.* **19,** 39–43.

21. Oh, P. and Schnitzer, J. E. (1999) Immunoisolation of caveolae with high affinity antibody binding to the oligomeric caveolin cage. Toward understanding the basis of purification. *J. Biol. Chem.* **274(33)**, 23,144–23,154.
22. Ostrom, R. S., Gregorian, C., Drenan, R. M., et al. (2001) Receptor number and caveolar co-localization determine receptor coupling efficiency to adenylyl cyclase. *J. Biol. Chem.* **276(45)**, 42,063–42,069.
23. Smart, E. J. and Anderson, R. G. (2002) Alterations in membrane cholesterol that affect structure and function of caveolae. *Methods Enzymol.* **353**, 131–139.
24. Ostrom, R. S., Bundey, R. A., and Insel, P. A. (2004) Nitric oxide inhibition of adenylyl cyclase type 6 activity is dependent upon lipid rafts and caveolin signaling complexes. *J. Biol. Chem.* **279(19)**, 19,846–19,853.
25. Orlandi, P. A. and Fishman, P. H. (1998) Filipin-dependent inhibition of cholera toxin: evidence for toxin internalization and activation through caveolae-like domains. *J. Cell Biol.* **141(4)**, 905–915.
26. Schnitzer, J. E., Oh, P., Pinney, E., and Allard, J. (1994) Filipin-sensitive caveolae-mediated transport in endothelium: reduced transcytosis, scavenger endocytosis, and capillary permeability of select macromolecules. *J. Cell Biol.* **127(5)**, 1217–1232.
27. Christian, A. E., Haynes, M. P., Phillips, M. C., and Rothblat, G. H. (1997) Use of cyclodextrins for manipulating cellular cholesterol content. *J. Lipid Res.* **38(11)**, 2264–2272.
28. Head, B. P., Patel, H. H., Roth, D. M., et al. (2005) G-protein-coupled receptor signaling components localize in both sarcolemmal and intracellular caveolin-3-associated microdomains in adult cardiac myocytes. *J. Biol. Chem.* **280(35)**, 31,036–31,044.
29. Zacharias, D. A., Violin, J. D., Newton, A. C., and Tsien, R. Y. (2002) Partitioning of lipid-modified monomeric GFPs into membrane microdomains of live cells. *Science* **296(5569)**, 913–916.
30. Ostrom, R. S., Liu, X., Head, B. P., et al. (2002) Localization of adenylyl cyclase isoforms and G protein-coupled receptors in vascular smooth muscle cells: expression in caveolin-rich and noncaveolin domains. *Mol. Pharmacol.* **62(5)**, 983–992.
31. Capozza, F., Cohen, A. W., Cheung, M. W., et al. (2005) Muscle-specific interaction of caveolin isoforms: differential complex formation between caveolins in fibroblastic vs. muscle cells. *Am. J. Physiol. Cell Physiol.* **288(3)**, C677–C691.
32. Razani, B. and Lisanti, M. P. (2001) Caveolin-deficient mice: insights into caveolar function in human disease. *J. Clin. Invest.* **108(11)**, 1553–1561.
33. Smart, E. J., Ying, Y. S., Mineo, C., and Anderson, R. G. (1995) A detergent-free method for purifying caveolae membrane from tissue culture cells. *Proc. Natl. Acad. Sci. USA* **92(22)**, 10,104–10,108.
34. Rybin, V. O., Xu, X., and Steinberg, S. F. (1999) Activated protein kinase C isoforms target to cardiomyocyte caveolae: stimulation of local protein phosphorylation. *Circ. Res.* **84(9)**, 980–988.
35. Pike, L. J. (2003) Lipid rafts: bringing order to chaos. *J. Lipid Res.* **44(4)**, 655–667.
36. Rybin, V. O., Xu, X., Lisanti, M. P., and Steinberg, S. F. (2000) Differential targeting of beta-adrenergic receptor subtypes and adenylyl cyclase to cardiomyocyte caveolae. A mechanism to functionally regulate the cAMP signaling pathway. *J. Biol. Chem.* **275(52)**, 41,447–41,457.

31

Near-Field Scanning Optical Microscopy to Identify Membrane Microdomains

Anatoli Ianoul and Linda J. Johnston

Summary

Near-field scanning optical microscopy (NSOM) allows optical imaging with a spatial resolution that is significantly better than the diffraction limited resolution achievable with conventional optical microscopy. NSOM has the potential to study the nanoscale organization of membrane surfaces and ultimately to resolve questions concerning lipid rafts in both model-supported membranes and cellular membranes. Supported phospholipid monolayers and bilayers of phase separated binary and ternary lipid mixtures that model the composition of lipid rafts in natural membranes have been studied by NSOM. The results illustrate the ability of NSOM measurements with 50–100 nm probe apertures to obtain detailed nanoscale information for small, closely spaced domains, and the utility of two-color experiments to probe localization of raft markers in supported membranes.

Key Words: Fluorescence; lipid rafts; microdomains; near-field scanning optical microscopy; phospholipid bilayers; phospholipid monolayers.

1. Introduction

Much of the detailed knowledge of the physical behavior of membranes comes from studies of model membranes such as phospholipid monolayers, planar lipid bilayers, and vesicles, all of which are amenable for study with a variety of spectroscopic and microscopic techniques. Phase separation in these models mimics some aspects of the behavior of cellular membranes wherein the formation of lipid domains plays an important role in the aggregation of protein complexes and initiation of signal transduction pathways *(1–3)*. The raft model recently proposed by Simons and Ikonen *(4)* suggests that the plasma membrane has small, dynamic microdomains (lipid rafts) that are rich in cholesterol, glycolipids, and saturated lipids such as sphingomyelin (SPM); specific proteins, particularly those involved in signaling cascades, are localized in these rafts. There is increasing evidence for the biological importance of rafts, although their postulated small size is beyond the resolution of conventional optical microscopy, making their direct detection difficult. This has led to a number of divergent views on the size, formation, and role of lipid rafts *(5–7)*. Advanced imaging techniques with the ability to detect nanosized domains have the potential to resolve some of the questions concerning lipid rafts in natural membranes.

Near-field scanning optical microscopy (NSOM) allows simultaneous topographic and fluorescence imaging using an optical fiber probe with a subwavelength aperture *(8–10)*. The optical resolution, determined by the probe aperture size, is approximately an order of magnitude better than the diffraction-limited resolution ($\lambda/2$) of conventional optical microscopy. The flexibility and specificity of using fluorescence as a contrast mechanism provides significant

From: *Methods in Molecular Biology, vol. 400: Methods in Membrane Lipids*
Edited by: A. M. Dopico © Humana Press Inc., Totowa, NJ

advantages over atomic force microscopy (AFM) for imaging complex samples. The high-spatial resolution and sensitivity allow for the independent observation of molecules at physiologically relevant packing densities, and NSOM is expected to be of particular value for studies of the spatial organization of membranes on the nanometer scale *(11,12)*. NSOM has been used to study the formation of domains in supported lipid monolayers and bilayers, as a model for the formation of raft domains in cellular membranes. Although the description herein refers exclusively to NSOM imaging of supported phospholipid membranes *(13–15)*, NSOM has also been used to study protein domains in supported bilayers and cellular membranes *(11,12,16)*.

2. Materials

2.1. Preparation of Phospholipid Monolayers and Bilayers by Langmuir–Blodgett Transfer

1. Stock solutions of dipalmitoylphosphatidylcholine (DPPC), dilauroylphosphatidylcholine (DLPC), dipalmitoylphosphatidylethanolamine (DPPE), dioleoylphosphatidylcholine (DOPC), and cholesterol (all from Avanti Polar Lipids, Alabaster, AL): approx 1 mg/mL in chloroform (*see* **Note 1**).
2. Stock solutions of *N*-palmitoyl-D-SPM from bovine brain and monosialoganglioside GM1 from bovine brain (from Sigma Chemical Co., Saint Louis, MO): approx 1 mg/mL in chloroform.
3. Solutions of Texas Red-DPPE (TR-DPPE) and GM1-Bodipy (Bodipy-FLC$_5$-GM1), both from Molecular Probes Invitrogen, Eugene, or, approx 1 mg/mL in chloroform.
4. Solution of 1:1:1 DOPC/SPM/cholesterol in chloroform (molar ratios; total lipid ~1 mg/mL) + 1% Texas Red-DPPE.
5. Solution of 1:1:1 DOPC/SPM/cholesterol in chloroform (molar ratios; total lipid ~1 mg/mL) + 1% Texas Red-DPPE + 1% GM1 + 0.005% GM1-Bodipy.
6. Solution of 7:3 DLPC/DPPC in chloroform.
7. Mica sheets, 25 × 25 mm^2 or 18-mm diameter (Ted Pella, Redding, CA), cleaved with tape immediately before use.

2.2. Preparation of Phospholipid Bilayers by Vesicle Fusion

1. Solution of 2:2:1 DOPC/egg SPM/cholesterol (molar ratios) in chloroform + 0.5% Texas Red-DPPE (total lipid ~1 mg/mL).
2. Solution of 15 m*M* CaCl$_2$.
3. Mica sheets, 25 × 25 mm^2 or 18-mm diameter (Ted Pella), cleaved with tape immediately before use.

2.3. Preparation of NSOM Test Samples

1. Solution of polymer spheres in polyvinyl alcohol (PVA): 1% wt aqueous solution of PVA (Molecular Weight 70,000–100,000, Sigma) containing 0.0002 wt% dye labeled polymer spheres (FluoSpheres®, nominally 40 nm in diameter, 350 dye molecules/sphere, Molecular Probes).
2. Glass cover slips, 18 mm (Fisher Scientific, Ottawa, ON, Canada), cleaned with methanol and water before use.

2.4. NSOM Probe Fabrication

1. High GeO$_2$-doped single mode optical fiber (Mitsubishi Cable Industries, Inc, Tokyo, Japan) with 3-µm core diameter, 125-µm cladding, and 1.45-µm cutoff wavelength.
2. Fiber-etching solutions: Solution 1 is a 14:4.7:6.7 (by volume) mixture of 6:1 buffered oxide etchant (Arch Chemicals, Norwalk, CT), HF (49% by weight), and distilled water. Solution 2 is 11:1 mixture of 10:1 buffered oxide etchant and distilled water.
3. Aluminum purity grade more than 99.999% (GoodFellow, Devon, PA).

3. Methods

Binary and ternary mixtures of lipids are used to prepare monolayers and bilayers for NSOM imaging. The lipid mixtures selected show clear phase separation to give coexisting gel and fluid phases (DLPC/DPPC) or coexisting liquid-ordered (raft) and fluid phases (DOPC/SPM/cholesterol). Phase separation is visualized by the addition of a dye-labeled lipid, Texas Red-DPPE, which has a strong preference for localization in fluid membrane domains (*see* **Note 2**). The distribution of an in vivo raft marker, ganglioside GM1, in supported membranes is also examined. In this case a combination of two labels and two-color NSOM measurements are used to conclusively identify the complex phase separation behavior observed.

It is recommended to use AFM to verify that sample preparation conditions are well optimized before NSOM imaging for new samples because the NSOM experiments are relatively time-consuming. As the sample holders are different for the two measurements, this is usually done using a duplicate sample prepared in an identical manner to that for NSOM experiments. Once the sample preparation conditions have been verified, the procedure is sufficiently reproducible that it is unnecessary to check samples by AFM for repeat experiments using the same lipid mixtures.

3.1. Preparation of Phospholipid Monolayers and Bilayers by Langmuir–Blodgett Transfer

Monolayers and bilayers are prepared on a Langmuir–Blodgett trough (Model 611, Nima Technology Coventry, UK) (surface area of 600 cm^2) equipped with a Wilhelmy balance (Nima) and a dipping apparatus and using Milli-Q (Millipore, Mississauga, ON, Canada) water as the subphase. Other Langmuir–Blodgett troughs will work equally well, with adjustment of volumes of water and lipid solutions to account for the difference in surface area (*see* **Note 3**).

1. Clean the trough before use by wiping with a 1:1 chloroform/2-propanol solution. Fill the trough with approx 250 mL water and then remove the water with an aspirator. Fill the trough a second time, taking care to ensure that the meniscus is slightly above the bottom of the barrier.
2. Clamp a freshly cleaved mica sheet in the dipping apparatus and then submerge it in the subphase so that the top edge of the mica is approx 3 mm below the water surface.
3. Open the barrier completely and zero the Wilhelmy balance and compress the barrier at a rate of 500 cm^2/min to within 10 cm of the mica sheet. An increase in surface pressure of more than 0.2 mN/m indicates that the concentration of impurities at the air–water surface is too high and the above cleaning procedure should be repeated. Otherwise, open the barrier and wait for the surface pressure to level off to a constant value (typically 15 min).
4. Set the parameters for deposition of the monolayer and then use a syringe to spread 25–30 μL of lipid solution on the water surface. Wait 10–15 min to ensure that all the solvent has evaporated. Compress the monolayer to the desired surface pressure (less than the collapse pressure for the lipid and equal to or more than the desired deposition pressure) at a rate of 100 cm^2/min and reopen the barrier. It is important to ensure that there is a baseline region wherein the surface pressure does not change with barrier position. The absence of a baseline region indicates that too much lipid has been added to the surface and the procedure should be restarted.
5. Expand and recompress the monolayer at least twice to anneal the sample before deposition. Then deposit the monolayer by vertical deposition to the mica using a dipping speed of 2 mm/min. The transfer ratio should be 100 ± 10%.
6. Dry the monolayer for 10 min before removing the mica from the dipper. Store the sample in a clean, dry container before imaging.
7. Bilayers are prepared by transferring a second monolayer to mica that has been precoated with a lipid monolayer (*see* **steps 1–6**, **Subheading 3.1.**). If hybrid bilayers with different compositions

for the top and bottom leaflet are required, the surface of the trough is cleaned and a new monolayer is prepared as in **steps 1**, **3**, and **4**, *see* **Subheading 3.1.** The monolayer-coated mica will already be in the dipper in the raised position (i.e., in air). The second monolayer is annealed twice before transferring to mica by dipping the slide down through the air–water interface. The resulting bilayer is kept under water in a small container before imaging.

3.2. Preparation of Phospholipid Bilayers by Vesicle Fusion

1. One to five milliliter of the DOPC/egg SPM/cholesterol lipid mixture containing 0.5% Texas Red-DPPE in chloroform is placed in a 20-mL vial and dried under a stream of nitrogen for 2 h to give a thin lipid film. The sample is further dried under high vacuum overnight to completely remove any traces of solvent.
2. Multilamellar vesicles are prepared by adding Milli-Q water to the lipid film to give a final lipid concentration of approx 1 mg/mL (*see* **Note 4**) and vortexing the sample for 2 min. The resulting multilamellar vesicles are sonicated using a bath-type sonicator at approx 45°C until a clear solution is obtained, typically for 30–60 min. This gives a solution of small unilamellar vesicles (50–100 nm in diameter). (*See* **Note 5**.)
3. Clamp a freshly cleaved sheet of mica in the NSOM fluid cell (a 20-mm Coverwell, Grace Bio-Labs, purchased from Sigma) and add 150 µL of vesicle solution and 300 µL $CaCl_2$ to the cell. The incubation time (room temperature) is varied from 15 to 40 min to test for complete formation of the lipid bilayer on mica. Bilayers are rinsed extensively with water to remove excess vesicles before imaging.

3.3. Preparation of NSOM Test Samples

1. Samples of 40-nm polymer spheres embedded in a 40-nm polymer film are prepared by depositing 5 µL of a PVA solution of polymer beads on an 18-mm circular glass slide and spin coating at approx 3415*g* for 30 s. Typically, several replicate samples are prepared, one for AFM analysis and the remainder for NSOM test patterns.
2. Samples are dried under nitrogen for 24 h at room temperature.
3. At least one sample slide from each batch is imaged by AFM to ensure that the density of spheres is appropriate and that the film thickness is correct. The polymer spheres should protrude slightly (1–5 nm) from the polymer film and there should be a significant number of individual (not aggregated) spheres that are well separated from each other. The film thickness can be evaluated by making a clean scratch across the slide to remove the polymer and then measuring the depth by AFM. (*See* **Note 6**.)
4. Test samples should be discarded after use for a number of images (10–20) as the spheres will photobleach and the cleanliness of the samples will deteriorate.

3.4. NSOM Probe Fabrication

The two main techniques for NSOM probe fabrication are laser pulling and chemical etching. The experiments described here used bent NSOM probes prepared from high GeO_2-doped optical fibers using a two-step chemical etching method followed by aluminum deposition and focused ion beam (FIB) milling to produce a flat circular aperture *(13)*. Commercially available bent optical fiber probes can in principle also be used. Fiber probes are examined using a microscope after each step to verify that the probe characteristics are within an acceptable range. Probes that do not meet specifications or have any visible damage are discarded. Scanning electron microscopy is used to assess the probe characteristics during optimization of the fabrication procedure.

3.4.1. Fiber Bending

Bent fibers are required to make cantilevered probes suitable for operation with the laser deflection distance regulation system of the AFM. Fibers are bent to approx 80° in a spark-discharge apparatus, which uses a stream of nitrogen to assist bending to the correct angle. A 1-m length of optical fiber is mounted in the bending apparatus, with a bend introduced at the end so that the length of fiber after the bend is approx 1.5 mm. The bend angle is selected to compensate for the angle in the holder of the microscope head and to ensure that the probe is perpendicular to the surface. The procedure produces bends with very little distortion owing to either twisting or changes in fiber diameter.

3.4.2. Fiber Etching

The fiber is first etched by immersing 2.5–3 mm of the probe in solution one for 60 min at constant temperature bath (28°C). This is followed by a second etch with solution two for 30 min at 28°C. This procedure generates a reproducible small and smooth conical structure with a height of approx 3.5 µm.

3.4.3. Aperture Formation

Apertures are created using FIB technology. The last few millimeters of each probe are completely coated with aluminum using an Edwards Auto 306 vacuum evaporator (BOC Edwards, Mississauga, ON, Canada). The aluminum coating is done in two steps to ensure that both the end of the probe and also the bend region are well covered with aluminum. Typically, an aluminum deposition rate of 7–9 mm/s is used to give a final thickness of 250 ± 50 nm. A Micrion 2500 FIB (Fibics Inc., Ottawa, Canada) operating in a multipass polish mill mode is used to produce an aperture by milling off the top of the tip. The size of the aperture is determined by the depth of milling and is typically in the range of 40–120 nm. (*See* **Note 7**.)

3.4.4. Probe Installation

Probes are glued to AFM stubs for insertion into the NSOM microscope. The end of the fiber is inserted in a fiber optic positioning system (Newport, Irvine, CA) for precise coupling of laser light into the probe. The end of the fiber is then cleaved with a fiber cleaver and the optics are adjusted for maximum coupling of light into the fiber.

3.5. NSOM Imaging

3.5.1. NSOM Description

NSOM experiments are carried out on a combined AFM/NSOM microscope, operated in transmission mode, with the optical fiber probe as excitation source and emission collected with an objective mounted directly under the sample. The microscope consists of a Digital Instruments Bioscope (Veeco, Santa Barbara, CA) mounted on an inverted fluorescence microscope (Zeiss Axiovert 100, Carl Zeiss, Chester, VA). A separate x–y piezo scanner with a 50-µm lateral scan range (Polytec PI, Auburn, MA) is used for sample scanning and is controlled by the Bioscope software. A continuous wave mixed gas ion laser (Innova 70 Spectrum, Coherent Inc, Santa Clara, CA) is used for excitation purposes (488 or 568.5 nm, 20 mW output, and linear polarization). A variable neutral density filter is used to attenuate the laser beam before coupling it with a ×10 objective into the optical fiber. The input power typically does not exceed 1 µW. The NSOM fluorescence signal is collected with either a ×100 oil immersion (1.3 numerical aperture [NA]) or a ×40 (0.65 NA) objective, with appropriate notch (Kaiser Optical Systems, Ann Arbor, MI) and band pass (Omega Optical, Brattleboro, VT) filters to remove residual excitation and red alignment laser light. An avalanche photodiode detector (Perkin Elmer Optoelectronics, SPCM-AQR-15, Vaudreuil, Canada) is used for signal detection. The detector is mounted on a three-dimensional positioning stage to allow adjustment of its position for maximum signal collection efficiency.

3.5.2. Sample Handling

Samples are stored in closed containers and protected from light to prevent any photobleaching before imaging (*see* **Note 8**). For NSOM measurements on dry samples, the sample slide is mounted on the microscope stage. For NSOM measurements in liquid, the bilayer sample in the NSOM fluid cell is covered with a HybriWell (Molecular Probes, www.invitrogen.com) hybridization sealing system. The sample is kept in liquid during the transfer and imaging.

3.5.3. Scanning Conditions

Imaging is performed at room temperature, $23 \pm 2°C$. Images are recorded in tapping mode at a scan rate of 0.25–0.5 Hz and a resolution of 512×512 pixels for images lager than 10×10 µm^2, and 256×256 for images smaller than 5×5 µm^2. This gives a pixel size between 10 and 80 nm, which is smaller than or comparable with the diameter of the aperture of the NSOM probe.

3.5.4. Two-Color Imaging

Two-color imaging is performed as a sequence of two independent NSOM scans using two different excitation wavelengths and the appropriate notch and band pass filters. Between the two scans the NSOM probe is disengaged from the sample surface to prevent damage.

3.5.5. Imaging Test Samples

1. Each new NSOM probe is characterized by imaging a test sample of dye-labeled polymer spheres. This provides an estimate of the probe aperture size and shape and thus the lateral resolution achievable with this probe. An example image for a test sample of polymer spheres recorded with an approx 60-nm aperture probe (estimated from images obtained during focus ion-beam milling) is shown in **Fig. 1A**. Note that there is some variation in both size and intensity for individual spheres. The measured size of the individual spheres (*see* the section analysis in **Fig. 1A**) reflects a convolution of the sphere size, the probe aperture diameter, and the pixel size for a particular measurement.
2. The *z*-resolution of the instrument is determined by imaging test samples of dye-labeled polymer spheres using the LiftMode™ feature of the Bioscope. The LiftMode measurement allows the simultaneous acquisition of two optical images: one with the tip engaged at the surface and the second with the probe tip at a desired height above the surface. The two 5×5 µm^2 LiftMode images shown in **Fig. 1A,B** were recorded for the same area of the sample with zero and 100 nm tip-sample separations, respectively. The image for 100 nm separation shows a large decrease of the fluorescence signal intensity and a significant broadening of feature size. As shown in the two cross sections **(Fig 1A,B)**, the size of an individual sphere increases from 97 to 210 nm when the tip is 100 nm above the sample surface. The plot of signal intensity vs tip-sample separation **(Fig. 1C)** indicates that the signal has decreased to less than 20% of its initial value for a probe-sample separation of 100 nm *(11)*.
3. Test samples are routinely imaged before and after each series of experiments to ensure that the probe aperture does not change during the experiment. It is also useful to image the test sample if one suspects a problem with either the sample or with a contaminated or damaged tip. (*See* **Notes 9 and 10**.)

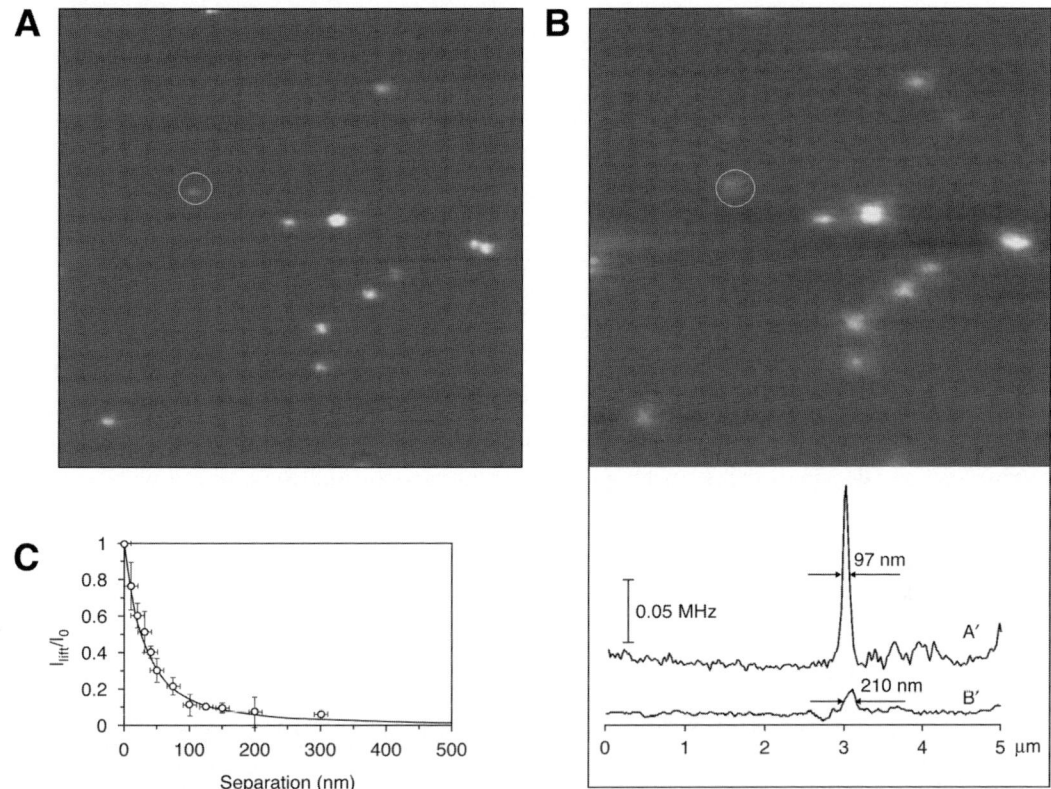

Fig. 1. NSOM images of a polymer sphere test pattern at different NSOM probe-sample separations. Images were obtained using an approx 60-nm diameter aperture probe with a ×100 oil-immersion objective (NA = 1.3). Image (**A**) was obtained with the tip engaged at the surface (separation is zero), and image (**B**) was measured in interleave mode with the tip lifted by 100 nm. The cross sections show the difference in feature size for the same sphere (circled in each image) at the two tip-sample distances. Graph (**C**) shows the dependence of the LiftMode relative intensity of the maximum fluorescence signal across a sphere (averaged over many spheres) on the tip to sample distance. The solid line shows the decay predicted by a $(d/[d + h])^2$ function, where d is NSOM probe aperture diameter and h is tip to sample separation. Reproduced with permission from **ref. 11**.

3.5.6. Imaging Lipid Monolayers

1. A typical NSOM fluorescence image of a phase-separated lipid monolayer is presented in **Fig. 2**, for a DOPC/SPM/cholesterol sample containing 1% TR-DPPE and imaged with a probe with a 120-nm diameter aperture. Dark, elliptical domains of an ordered SPM/cholesterol-rich phase are surrounded by a fluid fluorescent phase. Small, closely separated nanodomains of the ordered phase are also detected in some areas (cross section for a 117-nm domain), clearly illustrating the high-spatial resolution achievable with NSOM. The domain boundaries are sharp and well resolved, giving a transition between phases of 136 nm (for 10–90% change in intensity) that is close to the diameter of the probe aperture.
2. A second example of monolayer imaging is shown in **Fig. 3** for a similar DOPC/SPM/cholesterol monolayer containing ganglioside GM1, a raft marker, and imaged with a probe with an aperture diameter of approx 85 nm. In this case TR-DPPE is used to stain the fluid membrane phase (**Fig. 3A**, excitation at 568 nm) and GM1-Bodipy is used to visualize the ganglioside (**Fig. 3B**,

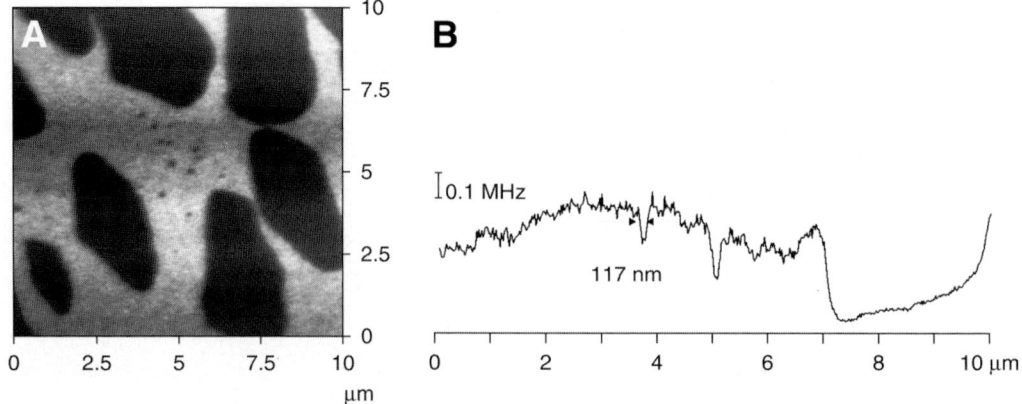

Fig. 2. NSOM image (**A**) for a DOPC/SPM/cholesterol (1/1/1) monolayer containing 1% TR-DPPE. The cross section (**B**) shows the diameter for the circled nanodomain. Reproduced with permission from **ref. 13**.

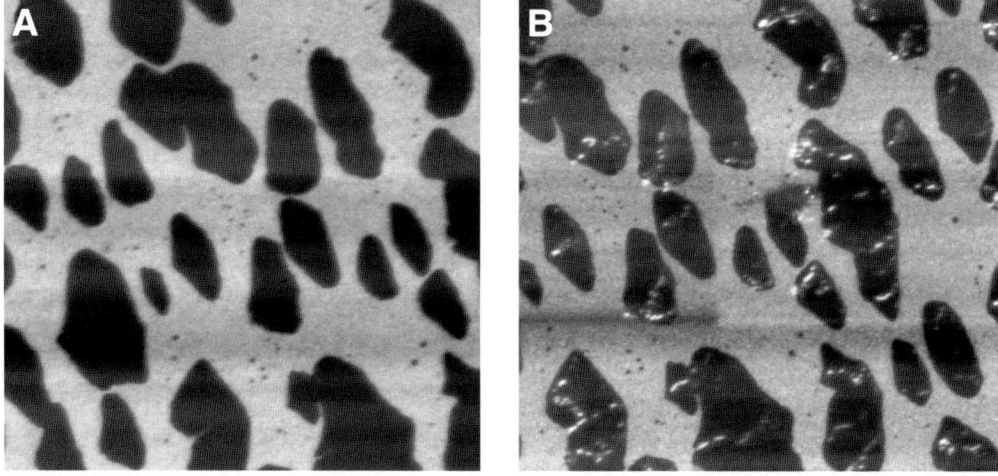

Fig. 3. NSOM images for a DOPC/SPM/cholesterol monolayer (1/1/1) containing 1% Texas Red-DPPE, 1% GM1, and 0.005% GM1-Bodipy, with excitation at 568 nm (**A**) and 488 nm (**B**). Reproduced with permission from **ref. 14**.

excitation at 488 nm where both TR and Bodipy absorb), which is located in small islands in the condensed domains. These results illustrate the possibility of two-color imaging using NSOM and the advantages of the high spatial resolution for observing small nanodomains that localize raft components.

3.5.7. Imaging Lipid Bilayers in Water

1. For imaging a lipid bilayer in liquid a two-step approach of the NSOM probe to the surface is used: approach to the water surface and approach to the sample surface. After initial engagement to the point when the tip contacts the water surface, engagement is stopped and the resonance frequency of the tip is adjusted and the drive voltage is correspondingly increased to maintain the root mean square amplitude. The final engagement of the tip is performed in several steps, with continuous adjustment of the resonance frequency and drive voltage until the sample surface is reached. During imaging only a fraction of the tip (~0.5 mm) is immersed in liquid, whereas

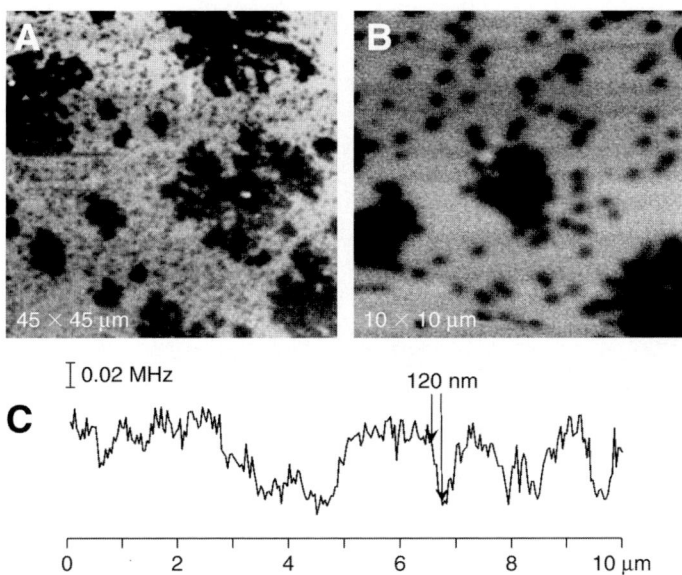

Fig. 4. Large- and small-scale NSOM images (**A,B**) of a DPPE/DLPC-DPPC-DHPE-TR bilayer obtained in aqueous solution. The lower DPPE leaflet was deposited on mica at a surface pressure of 45 mN/m. The upper leaflet (DLPC/DPPC (7/3) with 2 mol of DHPE-TR) was deposited at 30 mN/m. The cross section shown in (**C**) was used to estimate the NSOM resolution. Reproduced from **ref. 15** with permission.

the rest is kept in air. Images are recorded in tapping mode at a scan rate of 0.5 Hz and a resolution of 256 × 256.

2. **Figure 4** shows NSOM images recorded in water for a hybrid phospholipid bilayer with the bottom DPPE leaflet transferred at 45 mN/m and the top leaflet (7:3 DLPC/DPPC + 2% TR-DPPE) deposited at a surface pressure of 30 mN/m. The large-scale image shows a number of large, dark micron-sized areas that correspond to solid DPPC domains from which the dye is excluded. These are surrounded by a bright fluorescent DLPC + TR-DPPE phase. The fluid phase also contains a number of small-dark islands (as illustrated more clearly in the smaller-scale image in **Fig. 4B**) that correspond to holes in the bilayer. The images show excellent contrast between the two phases and are reproducible. The resolution is estimated by using the transition between the bright and the dark phases in the image, and is on the order of 120 nm, consistent with the diameter of the probe aperture (**Fig. 4C**). (*See* **Note 11**.)

3. **Figure 5** shows an NSOM image of a phospholipid bilayer prepared by vesicle fusion from 2:2:1 DOPC/egg SPM/cholesterol with 0.5% TR-DPPE and imaged in water with a 70-nm probe aperture. The dye is excluded from the liquid-ordered domains. The lack of areas of intermediate fluorescence intensity indicates that domains are coupled between upper and lower membrane leaflets. The ability to resolve small, closely spaced domains for bilayers in aqueous solution and the sharp transition between the domains and fluid phase (**Fig. 5C**), highlight the potential of NSOM for imaging under physiologically relevant conditions.

3.6. Image Treatment

Images are processed using standard Digital Instruments/Veeco software as well as Image J software (NIH, Bethesda, MD). Sphere size analysis is performed using original nonprocessed NSOM images with custom-made software. The software allows the determination of the number of clusters, their location in the image, as well as their height (intensity) and halfwidth.

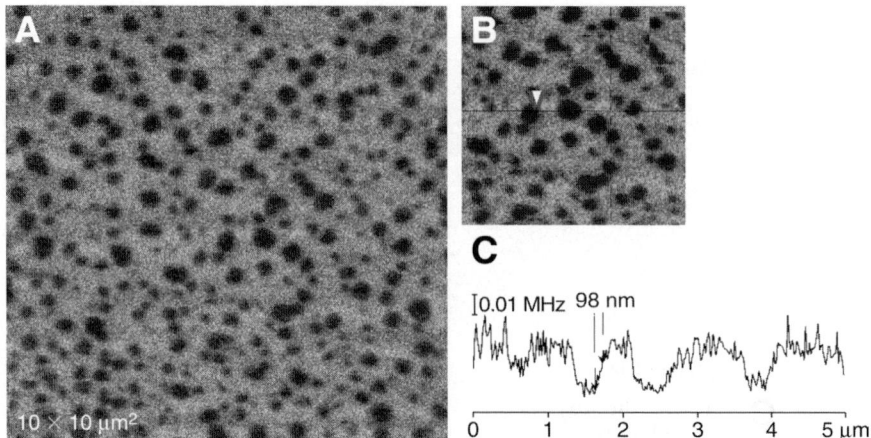

Fig. 5. NSOM images (**A,B**) and cross section (**C**) for a phospholipid bilayer prepared by vesicle fusion and imaged in water. The lipid mixture was 2/2/1 DOPC/egg SPM/cholesterol with 0.5% TR-DPPE.

4. Notes

1. Lipid solutions for preparing monolayers and bilayers and vesicle solutions should be prepared immediately before use wherever possible. This is particularly important for unsaturated lipids. In cases wherein storage of lipid solution is unavoidable, samples should be placed under a nitrogen atmosphere and stored in the freezer (organic solutions) or refrigerator at 4°C (vesicles).
2. It was found that Texas Red gives excellent contrast for NSOM measurements, can be used at relatively low concentrations compared with other dyes, and is somewhat more resistant to rapid photobleaching than other dyes. However, Bodipy, fluorescein, and (7-nitrobenz-2-oxa-1,3-diazole) (NBD) labeled lipids have also been used in some studies. The majority of commercially available dyes localize preferentially in fluid membrane domains.
3. All experiments described here are carried out in air. However, in some cases it is necessary to prepare lipid monolayers in a nitrogen atmosphere to minimize oxidation of lipids at the air–water interface. The procedure used is similar to that described herein, except that the Langmuir–Blodgett trough is enclosed in a glove box that can be kept in a nitrogen atmosphere. For the monolayers described herein (ternary lipid mixtures containing dye or ganglioside and deposited at 10 mN/m pressure) no difference between an air or nitrogen atmosphere is observed. However, significant changes in monolayer morphology are observed for other mixtures, particularly at high-surface pressures (e.g., 30 mN/m).
4. Vesicles are also prepared in buffered aqueous solution for many applications. Alternately, vesicles can be prepared in water as described herein, with the buffer being added during or after formation of the supported bilayer.
5. Vesicle solutions are sonicated at a temperature close to the phase transition temperature for the lipid with the highest T_m in the mixture. The time required will depend on the sonicator. If a probe type sonicator is used, it is essential to filter the vesicle solution after sonication to remove any particulate matter from the probe. The size of the vesicles produced by sonication can be determined by light scattering. Extrusion of the multilamellar vesicle solution can also be used to give a well-controlled distribution of small unilamellar vesicles. However, the size is not critical for preparation of supported bilayers.
6. It may be necessary to adjust the concentrations of the polymer or spheres, the rate of spin coating or the amount of solution used in order to optimize the density of spheres and the thickness of the film for the test samples. Either larger or smaller diameter polymer spheres can be used. Larger spheres with more dye are easier to image in case of low light levels, whereas smaller

polymer spheres provide a better test of the ability to image weakly fluorescent samples and will give a more accurate assessment of the size of the probe aperture.
7. A smooth uniform aluminum coating is important to prevent light leakage near the end of the probe. In general faster rates for aluminum deposition will favor smoother aluminum surfaces with fewer pinholes or large grains.
8. Samples containing dye-labeled lipids should be protected from light during storage.
9. When imaging soft samples, such as lipid bilayers in liquid, NSOM tip contamination with the lipid is a frequent problem. The tip contamination results in a decrease in the light intensity from the probe and also prevents the tip from approaching the sample surface. As a result, decreased resolution and low fluorescence signal intensity are observed. Therefore, it is important to verify the NSOM tip quality before and after imaging lipid bilayers, using the polymer sphere NSOM test pattern.
10. Sudden changes in the position of the NSOM probe with respect to the sample surface result in a change of the optical signal background intensity—so-called topography-induced artifacts. The topography-induced artifacts are particularly strong when studying membrane protein complexes and cells. Therefore, in order to avoid misinterpretation of NSOM signal, it is necessary to perform additional control experiments using a different excitation wavelength to prove that NSOM signal is "real."
11. Because of the high spring constant of the tip (~100 N/m) and high excitation power densities used in NSOM experiments, it is important to optimize scan parameters to minimize photo-bleaching and physical damage to the sample, particularly when imaging lipid bilayers. Parameters such as force set point, scanning speed, laser excitation intensity, and gains are adjusted in order to achieve the best image quality and reproducibility.

Acknowledgments

We thank Drs. Rod Taylor and Zhengfang Lu for their ongoing contributions to our NSOM development work and the coauthors listed in **refs. *11–16*** for their contributions to some of the experiments reviewed herein.

References

1. Binder, W. H., Barragan, V., and Menger, F. M. (2003) Domains and rafts in lipid membranes. *Angew. Chem. Int. Ed.* **42,** 5802–5827.
2. Simons, K. and Vaz, W. L. C. (2004) Model systems, lipid rafts and cell membranes. *Annu. Rev. Biophys. Biomol. Struct.* **33,** 269–295.
3. London, E. (2002) Insights into lipid raft structure and formation from experiments in model membranes. *Curr. Opin. Struct. Biol.* **12,** 480–486.
4. Simons, K. and Ikonen, E. (1997) Functional rafts in cell membranes. *Nature* **387,** 589–572.
5. Anderson, R. G. W. and Jacobson, K. (2002) A role for lipid shells in targeting proteins to caveolae, rafts and other lipid domains. *Science* **296,** 1821–1825.
6. Pike, L. J. (2004) Lipid rafts: heterogeneity on the high seas. *Biochem. J.* **378,** 281–292.
7. Munro, S. (2003) Lipid rafts: elusive or illusive? *Cell* **115,** 377–388.
8. Dunn, R. C. (1999) Near-field scanning optical microscopy. *Chem. Rev.* **99,** 2891–2927.
9. Edidin, M. (2001) Near-field scanning optical microscopy, a siren call to biology. *Traffic* **2,** 797–803.
10. Lewis, A., Taha, H., Strinkovski, A., et al. (2003) Near-field optics: from subwavelength illumination to nanometric shadowing. *Nat. Biotechnol.* **21,** 1378–1386.
11. Ianoul, A., Street, M., Grant, D., Pezacki, J., Taylor, R. S., and Johnston, L. J. (2004) Near-field scanning fluorescence microscopy study of ion channel clusters in cardiac myocyte membranes. *Biophys. J.* **87,** 3525–3535.

12. Ianoul, A., Grant, D., Rouleau, Y., Bani, M., Johnston, L. J., and Pezacki, J. P. (2005) Imaging nanometer domains of β-adrenergic receptor complexes on the surface of cardiac myocytes. *Nat. Chem. Biol.* **1,** 196–202.
13. Burgos, P., Lu, Z., Ianoul, A., et al. (2003) Near-field scanning optical microscopy probes: a comparison of pulled and double-etched bent NSOM probes for fluorescence imaging of biological samples. *J. Micros.* **211,** 37–47.
14. Burgos, P., Yuan, C., Viriot, M.-L., and Johnston, L. J. (2003) Two-color near-field fluorescence microscopy studies of microdomains ('rafts') in model membranes. *Langmuir* **19,** 8002–8009.
15. Ianoul, A., Burgos, P., Lu, Z., Taylor, R. S., and Johnston, L. J. (2003) Phase separation in complex supported phospholipid bilayers visualized by near-field scanning optical microscopy in liquid. *Langmuir* **19,** 9246–9254.
16. Murray, J., Cuccia, L., Ianoul, A., Cheetham, J. J., and Johnston, L. J. (2004) Imaging the selective binding of synapsin to anionic membrane domains. *Chem. Biol. Chem.* **5,** 1489–1494.

32

Fluorescence Microscopy to Study Domains in Supported Lipid Bilayers

Jonathan M. Crane and Lukas K. Tamm

Summary

Fluorescence microscopy of model membranes is a powerful tool for identifying the nature and extent of coexisting phases in biologically relevant lipid mixtures. Planar supported bilayers offer the advantage over spherical model membranes in that both overall composition and lipid asymmetry can be controlled. In addition, the membrane can be easily accessed by perfusion of soluble components. Here, the necessary techniques for reconstituting bilayers of complex composition and phase behavior in planar systems are outlined. Effective methods for the formation of both symmetric and asymmetric bilayers are described. Considerations that must be taken into account in choosing suitable lipid compositions, effective fluorescent lipid dyes, and adequate microscope setups are discussed.

Key Words: Cholesterol; domain; fluorescence microscopy; Langmuir–Blodgett; lipid raft; liquid-disordered; liquid-ordered; phospholipid asymmetry; planar supported bilayers; self-assembly; vesicle fusion.

1. Introduction

Lipid "rafts" were originally defined as detergent-resistant cell membrane patches that are made up of a cholesterol-rich liquid-ordered (L_o) phase and surrounded by a more fluid, liquid-disordered (L_d) phase. Their name reflects the belief that they serve as platforms for the transport of certain lipids and proteins from the Golgi to specific parts of the plasma membrane (PM), or for the sequestration of signaling components within the PM *(1)*. A major caveat of defining lipid rafts by detergent resistance is the necessary assumption that the phase behavior of the lipid bilayer is not altered by the addition of the detergent. This assumption is probably incorrect (*see* Chapter 13). Fluorescence techniques are a favorite alternative to detergent extraction when looking for lipid cluster or raft formation in cell membranes. For example, single particle tracking (SPT) of fluorescent lipids and proteins has revealed "transient confinement zones" in the PM of living cells that are related to cholesterol concentration *(2–4)*.

Spherical model membranes, in the form of giant unilamellar vesicles, are also useful systems in which fluorescence microscopy and fluorescence correlation spectroscopy have revealed much about the existence and nature of ordered lipid domains *(5–9)*. However, the known phospholipid asymmetry in biological membranes *(10)* cannot be mimicked in giant unilamellar vesicles. This asymmetry remains an issue that impedes the understanding of the structure and function of lipid rafts. The majority of the inner leaflet of the mammalian PM is made up of phospholipids that are found in the detergent-soluble fraction of cell extracts *(11–13)*. Still, many proteins that associate exclusively with the inner leaflet of the PM have been found in detergent resistant fractions *(14–16)*. The nature of this apparent transbilayer interaction remains a central open question in the field of molecular membrane physiology.

Planar supported bilayers are attractive systems for studying the formation, nature, and influences of lipid microdomains because the lipid composition and the lipid asymmetry can be controlled *(17)*. Symmetric planar supported bilayers, which sometimes exhibit large lipid domains that are visible by fluorescence microscopy *(18)*, are made by the combined Langmuir–Blodgett (LB) and Langmuir–Schäfer (LS) method *(19)*. A combined LB/vesicle-fusion technique *(20)* is used to construct asymmetric planar supported bilayers that may contain rafts in only one leaflet. The planar geometry of these bilayers facilitates the simple interpretation of structural data through fluorescence microscopy. The shapes of lipid domains can provide information on their phase structure. For example, solid-ordered domains consisting of sphingomyelin (SM) in the absence of cholesterol have a snowflake-like shape, reflecting the gel-like, semicrystalline nature of this phase. However, on the addition of cholesterol, the domains become rounded, indicating liquid–liquid phase coexistence **(Fig. 1)**. The point of percolation, where the L_o phase becomes connected and the L_d phase disconnected may also be of interest biologically *(21)*. In addition to simple microscopy, lateral mobility and molecular interactions between lipids and/or proteins can be measured by fluorescence recovery after photobleaching, fluorescence correlation spectroscopy, or SPT.

2. Materials

2.1. Preparation of Lipid Samples and Solid Supports

1. Synthetic phospholipids: 1-palmitoyl-2-oleoyl phosphatidylcholine, 1,2-dipalmitoyl phosphatidylcholine, and N-stearoyl sphingomyelin (SM) (Avanti Polar Lipids, Alabaster, AL).
2. Natural phospholipid mixtures: porcine-brain extracts of phosphatidylcholine (PC), phosphatidylethanolamine (PE), phosphatidylserine, and SM (Avanti Polar Lipids).
3. Fluorescent phospholipids: 1,2-dioleoyl PE-N-(lissamine rhodamine B) (Rh-DOPE), 1,2-dipalmitoyl PE-N-(lissamine rhodamine B) (Rh-DPPE), 1,2-dioleoyl PE-N-(7-nitro-2-1,3-benzoxadiazol-4-yl) (NBD-DOPE), 1,2-palmitoyl PE-N-(7-nitro-2-1,3-benzoxadiazol-4-yl) (NBD-DPPE) (Avanti Polar Lipids).
4. Cholesterol (Sigma Chemical Co., St. Louis, MO).
5. Chloroform and methanol (Fisher Scientific, Fair Lawn, NJ).
6. Glass vials with Teflon-coated caps (Fisher).
7. Quartz slides (Quartz Scientific, Inc., Fairport Harbor, OH).
8. Contrad 100 detergent (Cole-Palmer, Vernon Hills, IL), 95% sulfuric acid, 30% hydrogen peroxide (Fisher).

2.2. LB/LS Technique

1. Glass cover slips (Fisher).
2. Water-resistant double-sided tape (3 *M*, St. Paul, MN).
3. Hamilton syringes with Teflon-tipped plungers (Fisher).
4. Soaking buffer: 10 m*M* HEPES, 150 m*M* NaCl, and pH 7.4.

2.3. Combined LB/Vesicle Fusion Technique

1. Glass test tubes (Fisher).
2. Reconstitution buffer: 25 m*M* HEPES, 100 m*M* KCl, 1 m*M* $CaCl_2$, and pH 7.4 (*see* **Note 1**).
3. 1.5-mL polypropylene conical tubes with cap (Fisher).

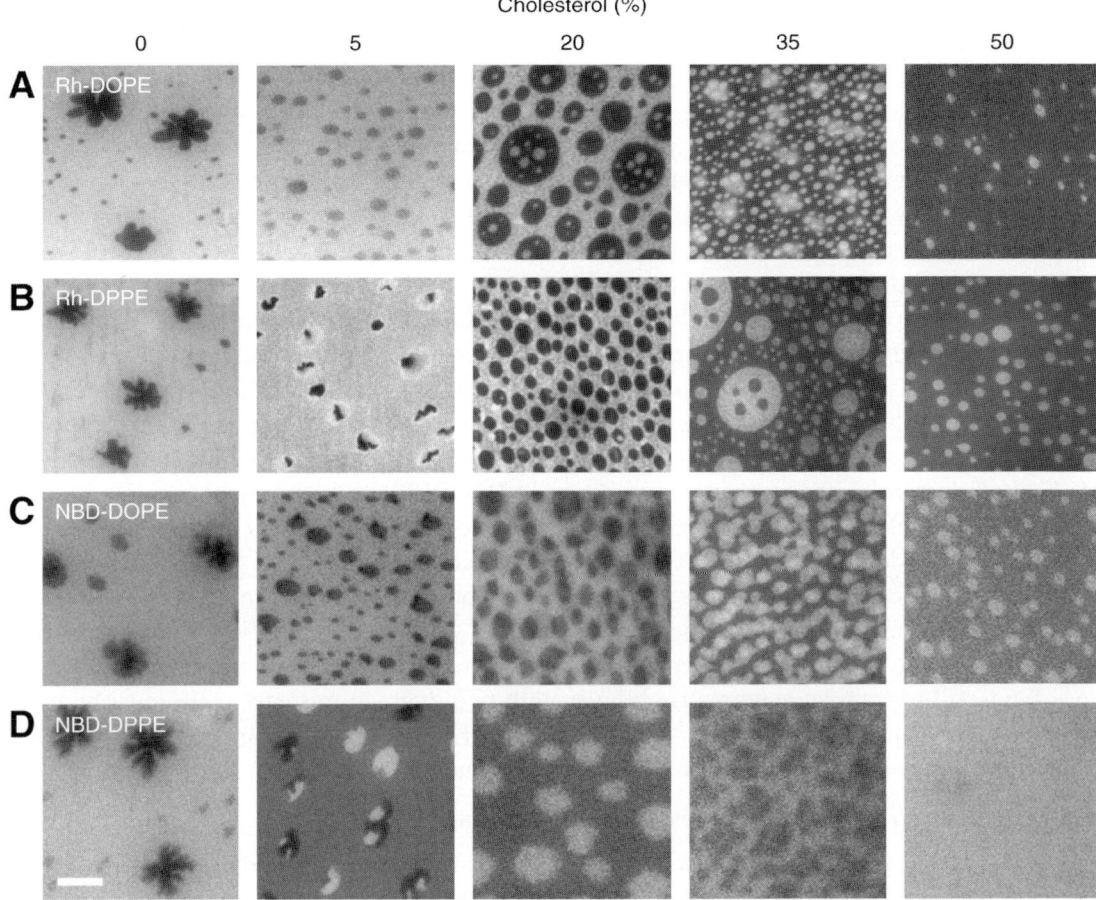

Fig. 1. Fluorescence micrographs of planar supported bilayers made up of PC/SM (1:1) with increasing cholesterol concentrations, and stained with four different fluorescently labeled lipids. Bilayers were formed by the LB/LS method on quartz slides. Each bilayer was stained with one of the following dyes in the LS monolayer only: **(A)** 0.5% Rh-DOPE, **(B)** 0.5% Rh-DPPE, **(C)** 1% NBD-DOPE, **(D)** 1% NBD-DPPE. Experiments were performed at room temperature. The white bar represents 10 µm.

4. LiposoFast-basic membrane extruder (Avestin, Ottawa, Canada).
5. Polycarbonate extruding filters, 100-nm pore size (Avestin).
6. 1,2-dimyristoyl PE-N-[poly(ethylene glycol)-triethoxysilane] (DPS) (Nektar Therapeutics, Huntsville, AL).
7. Disposable syringes and needles (Becton, Dickinson and Co., Franklin Lakes, NJ).

3. Methods

Lipid composition is the key factor when designing bilayers with complex phase behavior. For the reconstitution of L_d–L_o phase coexistence in planar supported bilayers, the required elements are (1) an unsaturated PC species such as 1-palmitoyl-2-oleoyl PC, (2) a saturated SM species such as N-stearoyl SM, and (3) cholesterol. Alternatively, a doubly saturated long-chain PC such as 1,2-dipalmitoyl PC may be used in place of SM. Natural mixtures of phospholipids such as porcine-brain extracts of PC and SM may also be used and are commercially available. Cholesterol content can be varied from 0 up to 66% of the total lipid *(22)* (*see* **Note 2**) depending on the extent of L_o phase desired (**Fig. 1**). The unsaturated/saturated, or PC/SM, ratio can also be varied, but keep in mind that the PC/SM ratio is about 50/50 in the extracellular leaflet of the PM *(13)*. Concerns have been raised about the possible oxidation of unsaturated lipid species in these preparations. Recent results from our laboratory show that qualitatively very similar results are obtained with non-oxidizable completely saturated lipid species.

For membrane visualization, commercially available fluorescent headgroup-labeled PEs are the best choice (*see* **Note 3**). The chemical structure of the chromophore and acyl chain composition of the dye molecule must also be considered, as different dyes have different optical properties and stain different lipid phases (**Fig. 1**). It is important to repeat experiments using multiple dyes when examining L_o–L_d phase coexistence in planar supported bilayers. For example, if one wanted to determine the cholesterol concentration at which a PC/SM/cholesterol bilayer becomes uniform, NBD-DPPE, which partitions equally into both phases at 50% cholesterol, would be a poor choice of dye (**Fig. 1D**). The use of multiple dyes shows that in fact the required concentration is higher than 50% (**Fig. 1A–C**).

When constructing bilayers of asymmetric lipid composition, the dye must also be asymmetrically distributed. The experimenter must decide which monolayer is to contain the dye, depending on the type of fluorescence experiment that is to be conducted (**Fig. 2**). Often it may be useful to stain an asymmetric bilayer with two dyes. In this case, care must be taken to avoid dyes that have significant spectral overlap. For example, Rh and NBD would be a poor choice because NBD emission will be locally quenched by the presence of Rh.

3.1. Preparation of Lipid Samples and Solid Supports

1. 40×25 mm^2 glass or quartz microscope slides are boiled in 1% Contrad detergent for 10 min, then placed in a hot bath sonicator while still in detergent for 30 min, followed by extensive rinsing in deionized water, methanol, and then water again.
2. Any remaining organic residue is removed by immersion in 3 vol of 95% sulfuric acid to 1 vol of 30% peroxide for 10–20 min, followed by extensive rinsing in water. At this point, slides can be covered and stored in 18 MΩ water for several days.
3. Immediately before use, slides are dried and further cleaned for 10 min in an argon plasma sterilizer (Harrick Scientific, Ossining, NY).
4. Phospholipids and cholesterol are purchased in powder form and can be stored at –20°C for several months. Stock solutions of the individual lipids are made in chloroform (*see* **Note 4**) at a concentration of 10 m*M* and can be stored at –20°C for several weeks. Before use, stock solutions are warmed to room temperature and mixed at the desired ratios.

3.2. LB/LS Technique

1. These instructions assume the use of a LB trough with a Wilhelmy plate to measure surface pressure, a computer controlled barrier and dipping mechanism, and a surface area of about 500 cm^2. The trough is filled with 18 MΩ water and the surface is cleaned by suction upon closing of the barrier. We use a Nima 611 trough (Nima Technology, Coventry, England)
2. Two strips (25×5 mm^2) of water-resistant double-sided tape are placed approx 20 mm apart on a new rectangular glass cover slip (35×50 mm^2) and rinsed with water. The cover slip is then submerged and placed in the front corner of the trough, near the dipping mechanism. The barrier is expanded to an area of 350 cm^2 and the surface of the trough is again cleaned by suction. More water is added if needed, and the offset on the surface pressure transducer is adjusted to read zero.

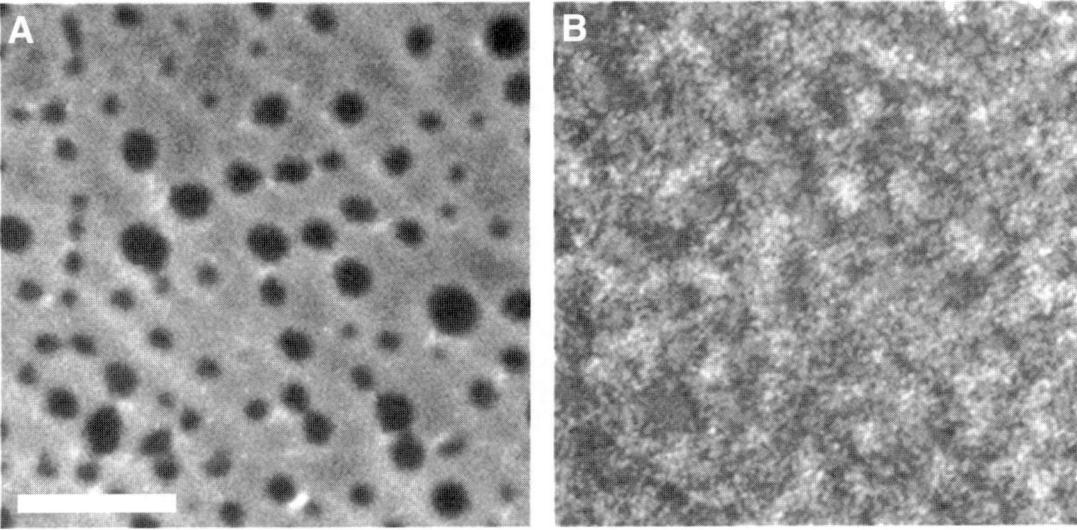

Fig. 2. Fluorescence micrographs of tethered polymer-supported planar supported bilayers of asymmetric lipid composition. Bilayers were made by the combined LB/vesicle fusion technique. In each case, the LB monolayer was made up of PC/SM/cholesterol (2:2:1) and 3% DPS, a mixture that forms domains of coexisting liquid-ordered and liquid-disordered phases. The vesicles were made up of PE/PC/phosphatidylserine/cholesterol (33:22:15:30), a mixture designed to mimic the inner leaflet of the PM. The bilayers were stained with 0.5% Rh-DPPE in either the LB monolayer only (**A**) or the vesicles only (**B**). Experiments were performed at room temperature. The white bar represents 20 µm. Domains are clearly visible in the SM-containing first layer. Domains are also induced in the PE/PS-containing second layer by the underlying domains in the first layer. For more information on induced domains and single molecule tracking in such domains, *see* **ref. 23**.

3. A clean Hamilton syringe is filled with 10 µL of the desired lipid solution in chloroform. The solution is then delicately placed on the surface until the pressure reading is steadily above zero. The solvent is allowed to evaporate for about 10 min.
4. A clean glass or quartz slide is removed from the plasma sterilizer and placed in the dipping mechanism above the surface of the monolayer in a vertical orientation. The barrier is expanded until the surface pressure reaches zero, then compressed at a rate of 10 cm^2/min until the surface pressure reaches 35 mN/m, then expanded back to a final pressure of 32 mN/m. The monolayer is allowed to equilibrate for 5–10 min. The slide is rapidly dipped through the surface at a rate of 200 mm/min and then slowly withdrawn back up through the surface at 5 mm/min, while the surface pressure is maintained at 32 mN/m (*see* **Note 5**). This results in the deposition of a single monolayer of lipids, known as the LB monolayer, on the surface of the slide. Because the slide is hydrophilic, the headgroups of the lipids will be oriented proximal to the surface with the acyl chains extending away.
5. Once the dipping of the slide is complete, the remaining monolayer on the trough is removed by suction, and a second monolayer is deposited and compressed to a surface pressure of 32 mN/m in exactly the same fashion as the first. This second monolayer is called the LS monolayer.
6. The coated slide is removed from the dipping mechanism, and a suctioning tip is attached to the center of one face. The slide is oriented above the trough with the face opposite the suctioning tip facing down. The LB monolayer is gently lowered to contact the LS monolayer at the air/water interface for a few seconds. The slide is then pushed through the interface and placed on the taped cover slip below. This completes the bilayer.
7. The remaining monolayer is removed from the trough by suction. The assembled bilayer can be removed from the trough at this time. The tape provides a spacer between the slide and the cover slip, and the water will remain between the surfaces by capillary forces. The water is then exchanged with

buffer by soaking the assembly for about 20 min (*see* **Note 6**). Buffers may also be exchanged by flowing the desired buffer through the sandwich, while always maintaining the hydration of the bilayer.

8. The assembly is transported to an inverted epifluorescence microscope for examination of the resulting planar supported bilayer (*see* **Note 7**). Micrographs of several bilayer samples made by the LB/LS technique are shown in **Fig. 1**.

3.3. Combined LB/Vesicle Fusion Technique

1. 50 µL of a 10 m*M* solution of the desired lipid mixture is placed in a glass test tube and the solvent is evaporated under a stream of nitrogen gas, followed by vacuum for at least 1 h.
2. The dried lipid film is resuspended in 0.5 mL of reconstitution buffer and vortexed rapidly until it appears homogeneously mixed and no lipid is visibly stuck to the glass.
3. The resulting liposome suspension is transferred into a 1.5-mL polypropylene tube, capped, flash frozen by immersion in liquid nitrogen for 30 s, then thawed by incubation in a water bath at 40°C. The freeze-thaw process is repeated four more times.
4. The liposomes are passed 11 times through two polycarbonate filters with a pore diameter of 100 nm in a LiposoFast-basic membrane extruder. This results in a uniform suspension of large unilamellar vesicles (LUVs), which may be stored at 4°C for up to 5 d.
5. A LB monolayer containing 3% DPS is prepared on a quartz slide as described in **Subheading 3.2.** The slide is then dried in a desiccator overnight and incubated at 70°C for 40 min to covalently tether the silane moiety of DPS to the SiO_2 surface (*see* **Note 8**). After removing the slide from the oven, it is allowed to cool to room temperature in a desiccating chamber.
6. The coated slide is placed in a custom-built chamber and the system is sealed. The chamber has a volume of 1 mL and has inlet and outlet ports at either end to allow for fluid exchange.
7. 100 µL of the 1 m*M* suspension of LUVs is diluted into 1 mL of reconstitution buffer, mixed thoroughly, and slowly injected into the chamber until full (*see* **Note 9**). The vesicles are allowed to incubate with the monolayer for at least 30 min at room temperature (*see* **Note 1**).
8. During the incubation, LUVs fuse with the LB monolayer, forming a single asymmetric bilayer *(20)*. Excess vesicles are washed out by pushing 10 mL of buffer through the chamber (*see* **Note 10**).
9. The chamber is then mounted on an inverted fluorescence microscope for examination. Examples of asymmetric planar supported bilayers made by this method are shown in **Fig. 2**.

3.4. Fluorescence Microscopy

1. Liquid-ordered phase domains in planar supported bilayers are large enough to be viewed with either a ×40 or ×63 water-immersion objective on an inverted epi-fluorescence microscope with a ×10 eyepiece and a mercury lamp for a light source. Image acquisition is accomplished with a cooled CCD camera. [We use a Cooke Sensicam QE (Cooke, Auburn Hills,, MI), but many other suitable products may perform equally well].
2. For viewing NBD-labeled lipid dyes, a filter set made up of a 470 nm bandpass excitation filter, a 510 nm long-pass dichroic mirror, and a 532 nm bandpass emission filter is used. For viewing Rh-labeled lipids, a 546 nm bandpass excitation filter, a 580 nm long-pass dichroic mirror, and a 610 nm bandpass emission filter is used.

4. Notes

1. This buffer is used for reconstitution of some integral proteins. Other buffers may also work. Incubation times may increase for low-ionic strength buffers *(20)*.
2. Bilayers with higher cholesterol content than 50% have not been tested, but according to Feigenson and coworkers *(22)*, a PC bilayer can contain up to 66% cholesterol, whereas PE bilayers can only tolerate up to 51%.
3. Carbocyanine dyes such as dialkyl-tetramethylindocarbocyanine perchlorate (DiI) are also popular fluorescent lipid markers. However, it has been generally found that fluorescent PEs give more reproducible results, particularly in the presence of cholesterol.
4. Some lipids require a small amount of methanol to dissolve completely.

5. Once the surface pressure reaches 32 mN/m, the computer takes over control of the barrier and dipping mechanism. Preprogrammed protocols are used to ensure the proper rates of dipping and withdrawal. Feedback from the pressure transducer to the computer ensures that a constant surface pressure is maintained throughout the dipping process.
6. Buffered saline is used to mimic either the extracellular fluid or cytoplasm. The authors have found that changing the buffer or leaving the bilayer in pure water has no effect on the appearance of L_o phase domains. However, if fluorescence recovery after photobleaching or SPT experiments are to be conducted, cations in the buffer may affect the diffusion rates of negatively charged lipids.
7. Bilayers should be viewed as soon as possible after completion to reduce the possible effects of lipid flip–flop or degradation. The authors have found that the appearance of L_o phase domains is stable for several hours.
8. DPS provides a cushion between the LB monolayer and the solid support *(24,25)*. DPS-supported LB monolayers are much more stable when exposed to aqueous solutions and vesicle suspensions than directly supported monolayers. Whereas tethering the polymer by desiccating and baking the monolayer covalently links the polymer cushion to the support, the increased stability of the initial monolayer is likely because of improved hydration. If one is worried about lipid oxidation, the 70°C baking step may be omitted. In many experiments, the authors found that DPS-supported bilayers without baking behave just as well as tethered DPS-supported bilayers with baking.
9. A slow, steady injection is crucial at this point to avoid washing material out of the LB monolayer. Some monolayers are more tolerant than others. Depending on the nature of the LB monolayer present, this step can be difficult to reproducibly perform. The first time this is attempted, it is a good idea to add a fluorescent dye to the LB monolayer so that its stability can be monitored visibly immediately after the injection.
10. Avoid pushing air bubbles into the chamber at this point. An air bubble will instantly ruin any part of a bilayer that it contacts.

References

1. Simons, K. and Ikonen, E. (1997) Functional rafts in cell membranes. *Nature* **387**, 569–572.
2. Simson, R., Sheets, E. D., and Jacobson, K. (1995) Detection of temporary lateral confinement of membrane proteins using single-particle tracking analysis. *Biophys. J.* **69**, 989–993.
3. Sheets, E. D., Lee, G. M., Simson, R., and Jacobson, K. (1997) Transient confinement of a glycosylphosphatidylinositol-anchored protein in the plasma membrane. *Biochemistry* **36**, 12,449–12,458.
4. Dietrich, C., Yang, B., Fujiwara, T., Kusumi, A., and Jacobson, K. (2002) Relationship of lipid rafts to transient confinement zones detected by single particle tracking. *Biophys. J.* **82**, 274–284.
5. Dietrich, C., Bagatolli, L. A., Volovyk, Z. N., et al. (2001) Lipid rafts reconstituted in model membranes. *Biophys. J.* **80**, 1417–1428.
6. Feigenson, G. W. and Buboltz, J. T. (2001) Ternary phase diagram of dipalmitoyl-PC/dilauroyl-PC/cholesterol: nanoscopic domain formation driven by cholesterol. *Biophys. J.* **80**, 2775–2788.
7. Veatch, S. L. and Keller, S. L. (2002) Organization in lipid membranes containing cholesterol. *Phys. Rev. Lett.* **89**, 268101.
8. Baumgart, T., Hess, S. T., and Webb, W. W. (2003) Imaging coexisting fluid domains in biomembrane models coupling curvature and line tension. *Nature* **425**, 821–824.
9. Kahya, N., Scherfeld, D., Bacia, K., Poolman, B., and Schwille, P. (2003) Probing lipid mobility of raft-exhibiting model membranes by fluorescence correlation spectroscopy. *J. Biol. Chem.* **278**, 28,109–28,115.
10. Bretscher, M. (1972) Asymmetrical lipid bilayer structure of biological membranes. *Nat. New Biol.* **61**, 11–12.
11. Devaux, P. F. (1991) Static and dynamic lipid asymmetry in cell membranes. *Biochemistry* **30**, 1163–1173.
12. Brown, D. A. and Rose, J. K. (1992) Sorting of GPI-anchored proteins to glycolipid-enriched membrane subdomains during transport to the apical cell surface. *Cell* **68**, 533–544.

13. Quinn, P. J. (2002) Plasma membrane phospholipid asymmetry. *Subcel. Biochem.* **36,** 39–60.
14. Rodgers, W., Crise, B., and Rose, J. K. (1994) Signals determining protein tyrosine kinase and glycosyl-phosphatidylinositol-anchored protein targeting to a glycolipid-enriched membrane fraction. *Mol. Cell Biol.* **14,** 5384–5391.
15. Baird, B., Sheets, E. D., and Holowka, D. (1999) How does the plasma membrane participate in cellular signaling by receptors for immunoglobulin E? *Biophys. Chem.* **82,** 109–119.
16. Simons, K. and Toomre, D. (2000) Lipid rafts and signal transduction. *Nat. Rev. Mol. Cell Biol.* **1,** 31–39.
17. Crane, J. M., Kiessling, V., and Tamm, L. K. (2005) Measuring lipid asymmetry in planar supported bilayers by fluorescence interference contrast microscopy. *Langmuir* **21,** 1377–1388.
18. Crane, J. M. and Tamm, L. K. (2004) Role of cholesterol in the formation and nature of lipid rafts in planar and spherical model membranes. *Biophys. J.* **86,** 2965–2979.
19. Tamm, L. K. and McConnell, H. M. (1985) Supported phospholipid bilayers. *Biophys. J.* **47,** 105–113.
20. Kalb, E., Frey, S., and Tamm, L. K. (1992) Formation of supported planar bilayers by fusion of vesicles to supported phospholipid monolayers. *Biochim. Biophys. Acta* **1103,** 307–316.
21. Thompson, T. E., Sankaram, M. B., Biltonen, R. L., Marsh, D., and Vaz, W. L. (1995) Effects of domain structure on in-plane reactions and interactions. *Mol. Membr. Biol.* **12,** 157–162.
22. Huang, J., Buboltz, J. T., and Feigenson, G. W. (1999) Maximum solubility of cholesterol in phosphatidylcholine and phosphatidylethanolamine bilayers. *Biochim. Biophys. Acta* **1417,** 89–100.
23. Kiessling, V., Crane,, J. M., and Tamm, L. K. (2006) Transbilayer effects of raft-like lipid domains in asymmetric planer bilayers measured by single molecule tracking. *Biophys. J.* **91,** 3313–3326.
24. Wagner, M. L. and Tamm, L. K. (2000) Tethered polymer-supported planar lipid bilayers for reconstitution of integral membrane proteins: silane-polyethyleneglycol-lipid as a cushion and covalent linker. *Biophys. J.* **79,** 1400–1414.
25. Kiessling, V. and Tamm, L. K. (2003) Measuring distances in supported bilayers by fluorescence interference-contrast microscopy: polymer supports and SNARE proteins. *Biophys. J.* **84,** 408–418.

33

Fluorescence Resonance Energy Transfer to Characterize Cholesterol-Induced Domains

Luís M. S. Loura and Manuel Prieto

Summary

Cholesterol is a major component of mammalian cell membranes. It has remarkable effects on the properties of phospholipids bilayers, and is implicated in the lipid raft model. Depending on the membrane composition, cholesterol-containing bilayers can exist either as single phase or as mixture of coexisting phases. These are organized in domains of variant size determined by composition, but can be smaller than the optical microscopy resolution limit. This chapter describes a methodology based on fluorescence resonance energy transfer, which is sensitive to phase separation and can provide estimates of domain size in these usually hard to characterize systems.

Key Words: Fluorescence spectroscopy; FRET; lipid mixture; phase diagram; raft; liquid ordered phase.

1. Introduction
1.1. Cholesterol in Phospholipid Bilayers

Cholesterol (Chol) is a major component of mammalian cells, and its effects on the physical properties of lipid bilayers have been studied actively in the last three decades. The notable effects of Chol include condensation of area/lipid molecule (e.g., see **ref. 1**), reduction in passive permeability of the bilayer (e.g., see **ref. 2**), increase in orientational order of the phospholipid acyl chains (e.g., see **ref. 3**), and increase in bending elasticity (e.g., see **ref. 4**), relative to the values in pure phospholipids fluid membranes. A most remarkable property of Chol is its ability to induce phase separation both above and below the main transition temperature (T_m) of saturated phosphocholines, such as 1,2-dimyristoyl-3-sn-glycerophosphocholine (DMPC) and 1,2-dipalmitoyl-3-sn-glycerophosphocholine (DPPC), and also above the T_m of mixed-chain 1-palmitoyl-2-oleyol-3-sn-glycerophosphocholine (POPC), the latter being observed with Chol mole fractions of approx 0.1–0.3. The two phases coexisting above T_m, the Chol-rich or liquid-ordered (β or L_o, after the nomenclature introduced in **ref. 5**) and Chol-poor or liquid-disordered (α or L_d) phases, have been thoroughly characterized in terms of physical properties.

Although L_d resembles the pure lipid fluid, L_o has intermediate properties between those of pure phospholipid fluid and gel. Similarly, below T_m, there is coexistence of L_o and a phase essentially identical to pure phospholipid gel, termed solid-ordered (so) phase. For very high Chol content (mole fractions higher than ~0.4–0.5), the phase transition is eliminated, and L_o is the sole phase stable both below and above the pure phospholipid T_m.

Most of the results described above were carried out in binary phosphatidylcholine (PC)–Chol mixtures. Determination of the phase diagrams of Chol-containing systems is not

From: *Methods in Molecular Biology, vol. 400: Methods in Membrane Lipids*
Edited by: A. M. Dopico © Humana Press Inc., Totowa, NJ

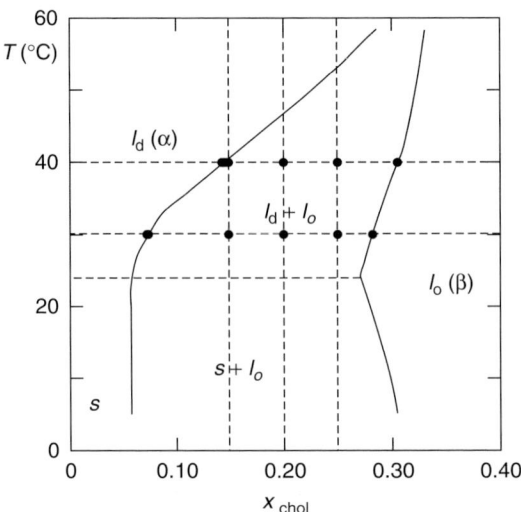

Fig. 1. Phase diagram of DMPC/Chol (see **ref. 17**). Points indicate the mixtures and temperatures addressed in **Subheading 3.1**. Reprinted with permission. © 2001 Biophysical Society.

an easy task, because the difference in properties between the coexisting phases is much less pronounced than in typical gel–fluid phospholipid mixtures. Nevertheless, several temperature/composition phase diagrams were obtained in the 1990s, namely for DPPC/Chol (**6**), DMPC/Chol (**7**) (see **Fig. 1**), and POPC–Chol (**8**) mixtures, using techniques such as ^{2}H nuclear magnetic resonance, electron spin resonance, and fluorescence.

More recently, Chol has been implicated in the lipid raft model (**9**). The proposal of this model was based mainly on the fact that detergent-resistant membranes isolated from cellular membranes were enriched in (glyco)sphingolipids and Chol, and depleted in (unsaturated) PC, and also that the simultaneous presence of Chol and sphingolipids was necessary to confer detergent insolubility to some proteins. On the other hand, it was known that liposomes with resistance to Triton X-100 (TX-100) solubilization are in ordered phases (so and L_o), whereas in the L_d phase they are solubilized. Given that the so phase is found in cell membranes only in exceptional cases and that detergent-resistant membranes isolated from cell membranes are rich in Chol, it is currently accepted that rafts are the cellular equivalent of in vitro L_o domains.

The simplest model for lipid rafts is a ternary mixture of the type low-T_m lipid/high-T_m lipid/Chol. Unsaturated PCs are often used as the low-melting lipid component, whereas sphingomyelin (SM) is frequently used as the high-melting component. Phase diagrams of mixtures of this type are naturally hard to obtain, because, in addition to the aforementioned difficulties for PC/Chol binary systems, there is the obvious increase in complexity inherent to the addition of a third component and the appearance of a variety of new interactions among components.

The size of rafts and other microdomains in membranes of cells has been usually detected by indirect methods, such as detergent extraction, yet this can increase the size of the preexisting domains by at least one order of magnitude, as verified *in situ* in TX-100 treated fibroblast-like cells (**10**). They can also be detected directly, for example, after cross-linking of membrane components. This leads to the formation of larger, otherwise undetectable, aggregates *(11,12)*. Some studies rely on ganglioside G_{M1} as one of the most popular raft markers *(13)*, which are usually detected by addition of labeled cholera toxin subunit B (CTB) *(14,15)*. CTB forms stable pentamers, and in this oligomeric state, are able to bind 5 G_{M1} molecules *(16)*. It is possible that this binding has an effect similar to cross-linking, thus changing the size of the G_{M1}-enriched domains. In this chapter, the fluorescence resonance energy transfer (FRET) methodology developed by the authors to characterize Chol-containing lipid mixtures is described.

1.2. FRET in Phase-Separated Lipid Membranes

In this section, a simplified version of the formalism of time-resolved FRET in biphasic membranes is presented (the full formalism description is given in **ref. 17**). Consider a system made up of two separated phases, 1 and 2. The donor decay in the absence of acceptor is simply given by:

$$i_D(t) = A_1 \rho_{D1}(t) + A_2 \rho_{D2}(t) \tag{1}$$

In this equation, ρ_{Di} is the donor decay in phase i, and for an exponentially decaying fluorophore, it is given by $\exp(-t/\tau_{Di})$, where τ_{Di} is the donor lifetime in phase i. On the other hand, A_i is proportional to the number of donor molecules in this phase. In presence of acceptor, the donor decay changes (becomes faster), and is now given by:

$$i_{DA}(t) = A_1 \rho_{DA1}(t) + A_2 \rho_{DA2}(t) \tag{2}$$

The donor decay inside phase i, ρ_{DAi}, is the product of the donor intrinsic decay ρ_{Di} and the FRET term. From the latter, eventually topographic information is recovered (distance between donor and acceptor planes in the bilayer), as well as the c_i parameter, which is proportional to the acceptor concentration in phase i, as follows:

$$c_i = 1.354 n_i \pi R_{0i}^2 \tau_{Di}^{-1/3} \tag{3}$$

where n_i is the number of acceptor per unit area in phase i, whereas R_{0i} is the Förster radius (characteristic distance for FRET between a given donor/acceptor pair).

The parameters A_i and c_i contain information regarding the partition of donor and acceptor between phases 1 and 2. If one considers a partition equilibrium of both probes between the coexisting phases,

$$(\text{probe})_1 \leftrightarrows (\text{probe})_2 \tag{4}$$

then the amounts of the probes in each phase are related through their partition coefficients (e.g., *see* **ref. 18**):

$$K_p = (P_2 / X_2) / (P_1 / X_1) \tag{5}$$

where P_1 is the probe mole fraction in lipid phase 1, and X_1 is the lipid phase mole fraction (therefore, $P_2 = 1 - P_1$ and $X_2 = 1 - X_1$). It is easy to show that the partition coefficients of donor (K_{pD}) and acceptor (K_{pA}) probes can be calculated straightforwardly from the FRET decay parameters,

$$K_{pD} = (A_2 / X_2) / (A_1 / X_1) \tag{6}$$

$$K_{pA} = (c_2 a_2) / (c_1 a_1) \qquad (7)$$

where a_i is the area per lipid molecule in phase i. It can also be shown that, from the recovered values of c_1 and c_2 for a sample inside the phase coexistence range (together with the overall acceptor surface concentration), X_1 and X_2 can be calculated even without previous knowledge of the phase boundary compositions x_1 ($X_2 = 0$) and x_2 ($X_2 = 1$) (**ref. 17**). In this situation, if the mixture phase diagram is unknown yet, at a given temperature T, X_1, and X_2 have been thus calculated for two points (x_A, T) and $B(x_B, T)$ inside the phase coexistence range, then x_1 and x_2 are given by:

$$x_1 = (x_A X_{2B} - x_B X_{2A}) / (X_{1A} - X_{1B}) \qquad (8)$$

$$x_2 = (x_B X_{1A} - x_A X_{1B}) / (X_{1A} - X_{1B}) \qquad (9)$$

which allows us to calculate the compositions of phases 1 and 2 at that temperature from time-resolved FRET data. If this procedure is repeated for several temperatures, the phase diagram is obtained.

These simple relationships are strictly valid only for very large domains (>5–10 R_0). Otherwise, boundary effects become important, i.e., if the acceptor prefers to incorporate in the minority phase 1 ($K_{pA} <1$) but the domains of this phase are very small, donors inside these domains are still sensitive to the region outside them and "see" a local concentration of acceptor that is smaller than the domain value c_1. The opposite happens for donors outside the domains. As a consequence, when analyzing decays with **Eq. 2**, c_1 is underestimated, c_2 is overestimated, and K_{pA} calculated from **Eq. 7** is overestimated. Conversely, if $K_{pA} > 1$, but the domains of phase 1 are very small, K_{pA} calculated from time-resolved FRET parameters is underestimated.

Of course, K_p values can be obtained by a plethora of established methods, including other photophysical techniques (*18*). The uniqueness of FRET in this respect resides in the dependence of the "apparent K_p" (the value recovered after analysis) on the size of the phases. Other fluorescent properties often used for calculation of K_p, such as fluorescence intensity, lifetime, and anisotropy are dependent solely on the immediate environment of the probe (at least for common dyes, with lifetimes shorter than 10 ns) and are insensitive to domain size.

2. Materials

1. Nonfluorescent lipids: Chol, DMPC, POPC, and N-palmitoyl-D-SM (PSM). Prepare stock solutions in chloroform and store at –20°C.
2. Raft markers: ganglioside G_{M1} from bovine brain, CTB from *Vibrio cholerae*.
3. Fluorescent probes N-(7-nitrobenz-2-oxa-1,3-diazol-4-yl)-dimyristoylphosphatidylethanolamine (NBD-DMPE), N-(lissamine™[Molecular Probes, Eugene, OR]-rhodamine B)-dipalmitoylphosphatidylethanolamine (Rh-DMPE), N-(7-nitrobenz-2-oxa-1,3diazol-4-yl)-dipalmitoylphosphatidylethanolamine (NBD-DPPE), N-(lissamine-rhodamine B)-dioleoylphosphatidylethanolamine (Rh-DOPE), 1,6-diphenylhexatriene (DPH), and *trans*-parinaric acid (*t*-PnA). Prepare stock solutions using chloroform as solvent for the NBD and Rh probes, and ethanol for the other probes. Store at –20°C.
4. Spin-labeled probe 5-doxyl stearic acid (5-NS). Prepare stock solution in methanol. Store at –20°C.
5. Buffer I: 50 mM Tris-HCl, 100 mM NaCl, and 0.2 mM ethylenediaminetetraacetic acid, pH 7.4.
6. Buffer II: 10 mM sodium phosphate, 150 mM NaCl, and 0.1 mM ethylenediaminetetraacetic-acid, pH 7.4.
7. Buffer III: 0.05 M Tris-HCl, 200 mM NaCl, and 3 mM NaN$_3$, 1 mM, pH 7.5.

3. Methods
3.1. Domains in Binary PC–Chol Mixtures
3.1.1. Vesicle Preparation

1. Mix adequate volumes of DMPC and Chol stock solutions to prepare 1 µmol of DMPC + Chol, with Chol mole fraction $x_{chol} = 0.075$. Add adequate volume of NBD-DMPE stock solution to obtain 0.1 mol% of this probe relative to total nonfluorescent lipid (sample 1D).
2. Repeat previous step, also adding adequate volume of Rh-DMPE stock solution to obtain 0.5 mol% of this probe relative to total nonfluorescent lipid (sample 1DA).
3. Repeat **steps 1** and **2**, but varying the volumes of DMPC and Chol stock solutions, rendering samples with $x_{chol} = 0.14$ (samples 2D and 2DA), $x_{chol} = 0.15$ (samples 3D and 3DA), $x_{chol} = 0.20$ (samples 4D and 4DA), $x_{chol} = 0.25$ (samples 5D and 5DA), and $x_{chol} = 0.28$ (samples 6D and 6DA).
4. After thorough mixing of each sample, dry all samples under a gentle stream of nitrogen until complete evaporation.
5. Dry all samples further by leaving them in vacuum for 6 h.
6. Suspend each sample in 10 mL of buffer I above the phospholipid main T_m (e.g., at 30°C for DMPC).
7. Prepare large unilamellar vesicles by the extrusion technique. To this effect, extrude the lipid dispersions 10 times through 0.1-µm pore diameter polycarbonate filters (*see* **Note 1**).

3.1.2. Donor (NBD-DMPE) Partition Coefficient

1. In a spectrofluorimeter with polarizers, measure steady-state anisotropy <r> (*see* details in **ref. 18**) of samples 1D, 3D, 4D, 5D, and 6D at 30°C and 1D, 2D, 3D, 4D, and 5D at 40°C. Use $\lambda = 470$ nm as excitation wavelength and $\lambda = 540$ nm as emission wavelength.
2. Calculate the fraction of L_o phase, X_β or X_{lo}, for each composition from the lever rule,

$$x_\beta = \frac{x_{chol} - x_\alpha}{x_\beta - x_\alpha} \tag{10}$$

Take $x_\alpha(30°C) = 0.075$, $x_\alpha(40°C) = 0.14$, $x_\beta(30°C) = 0.28$, and $x_\beta(40°C) = 0.31$ (*7*).
3. Analyze the (<r>, X_β) data for each temperature with **Eq. 11**, using nonlinear least-squares software (see **Note 2**). For NBD-DMPE in the described system, take as quantum yields $\Phi_\alpha(30°C) = 0.26$, $\Phi_\beta(30°C) = 0.29$, $\Phi_\alpha(40°C) = 0.21$, and $\Phi_\beta(40°C) = 0.25$. <r>$_\alpha$ and <r>$_\beta$ are best fixed at the values measured for the samples with lowest and highest (respectively) x_{chol} for each temperature. Obtain K_{pD} as the sole fitting parameter.

$$\langle r \rangle = \frac{K_{pD}^{\beta/\alpha} \cdot X_\beta \cdot \Phi_\beta \langle r \rangle_\beta + (1 - X_\beta) \cdot \Phi_\alpha \cdot \langle r \rangle_\alpha}{K_{pD}^{\beta/\alpha} \cdot X_\beta \cdot \Phi_\beta + (1 - X_\beta) \cdot \Phi_\alpha} \tag{11}$$

3.1.3. Acceptor (Rh-DMPE) Partition Coefficient

1. Using a spectrofluorimeter, measure steady-state intensity I_F of samples 1DA, 3–6DA at 30°C and 1–5DA at 40°C. Use as $\lambda = 570$ nm as excitation wavelength and $\lambda = 593$ nm as emission wavelength. Choose narrow bandwidths to minimize detection of scattered excitation light.
2. Calculate X_β as described in **step 2**, *see* **Subheading 3.1.2.**
3. Analyze the (I_F, X_β) data for each temperature with **Eq. 12**, using nonlinear least-squares software (*see* **Note 2**). $I_{F\alpha}$ and $I_{F\beta}$ are best fixed at the values measured for the samples with lowest and highest (respectively) x_{chol} for each temperature. Obtain K_{pA} as the sole fitting parameter.

$$I_F = \frac{K_{pD}^{\beta/\alpha} \cdot X_\beta \cdot I_{F\beta} + (1 - X_\beta) \cdot I_{F\alpha}}{K_{pD}^{\beta/\alpha} \cdot X_\beta + (1 - X_\beta)} \tag{12}$$

3.1.4. Donor Decay Measurement and Analysis

1. Using instrumentation for measurement of fluorescence decays, obtain decays for NBD-DMPE fluorescence for samples 1D, 3–6D, 1DA, 3–6DA at 30°C, and 2–6D, 2–6DA at 40°C (*see* **Note 3**).
2. Analyze each decay with an empirical sum-of-exponentials equation (*see* **Note 4**). Typically, three exponentials will suffice for both donor only (jD) and donor plus acceptor (jDA) samples:

$$i_k(t) = \sum_{n=1}^{3} \alpha_{kn} \exp(-t/\tau_{kn}) \quad (13)$$

This will be used to calculate the experimental FRET efficiency, E:

$$E = 1 - \frac{\int_0^\infty i_{jDA}(t)dt}{\int_0^\infty i_{jD}(t)dt} = 1 - \frac{\alpha_{jDA1}\tau_{jDA1} + \alpha_{jDA2}\tau_{jDA2} + \alpha_{jDA3}\tau_{jDA3}}{\alpha_{jD1}\tau_{jD1} + \alpha_{jD2}\tau_{jD2} + \alpha_{jD3}\tau_{jD3}} \quad (14)$$

3. Analyze each pair of decays globally with **Eqs. 1–2** (*see* **Note 4**). The parameters are described in detail in **ref. 17**. **Figure 2** shows results for the variation of the recovered values for c in the L_d and L_o phases (c_α and c_β, respectively) with the fraction of L_o phase together with that expected from the K_{pA} values obtained from fluorescence intensity measurements, at both temperatures, obtained in the DMPC/Chol system.
4. From the recovered parameters, calculate the FRET K_{pD} and K_{pA} using **Eqs. 6** and **7** (taking phase 1 = α = L_d, phase 2 = β = L_o, so that $K_p > 1$ denotes preference for the L_o phase). These values are shown and compared with the values recovered from distance-independent methods (*see* **Subheadings 3.1.2.** and **3.1.3.**) in **Table 1**. There is fair agreement between the FRET global analysis and fluorescence anisotropy values for the donor NBD-DMPE whereas there is a systematic decrease of the FRET K_{pA} with increasing X_β. For lower x_{chol}, K_{pA} is closer to unity than the fluorescence intensity value for the K_p of Rh-DMPE. This was already apparent in **Fig. 2**, as seen from the closeness of the c_α or c_β values (inside the curves plotted using the fluorescence intensity K_{pA}) at low X_β. On the other hand, for $x_{chol} = 0.25$, the FRET K_{pA} are lower and further from unity than the distance-independent values (there is no complete agreement probably because of some degree of acceptor aggregation in the L_o phase). Following the previous section, this points to small (<3–5 R_0 or 18–30 nm) L_o domains in the low-Chol end, and large (>5R_0 or ~30 nm) L_d domains in the high-Chol end of the tie-lines, at both temperatures.
5. Estimate the phase boundaries using **Eqs. 8** and **9**. For the DMPC/Chol system, when the c_α and c_β values for $x_{chol} = 0.20$ and 0.25 are used to estimate the phase boundaries using **Eqs. 8** and **9**, $x_\alpha = 0.178$ and $x_\beta = 0.277$ are obtained at 30°C (compared with 0.075 and 0.28, respectively, as estimated from the phase diagram of **Fig. 1** (*see* **ref. 7**), $x_\alpha = 0.187$ and $x_\beta = 0.270$ are obtained at 40°C (0.14 and 0.31, respectively, estimated from **Fig. 1**). The fair agreement between the FRET calculated values for x_β and those obtained from domain size-independent methodologies confirms the fact that, in the high-Chol end of the phase coexistence range, the domains are large. On the other hand, the higher values for the FRET x_β, compared with those of the published diagram, are another manifestation that FRET only "begins to see" phase separation for relatively high x_{chol}. However, for lower Chol content, phase separation takes the shape of tiny nanodomains that are undetectable by FRET.

3.2. Ternary PC/SM/Chol Phase Diagram

Contrary to the binary mixture, the phase diagram for the PC/SM/Chol ternary system was previously unknown. As it should be clear from **Subheading 1.**, the use of FRET for domain size estimation requires the knowledge of the phase boundaries, and FRET measurements

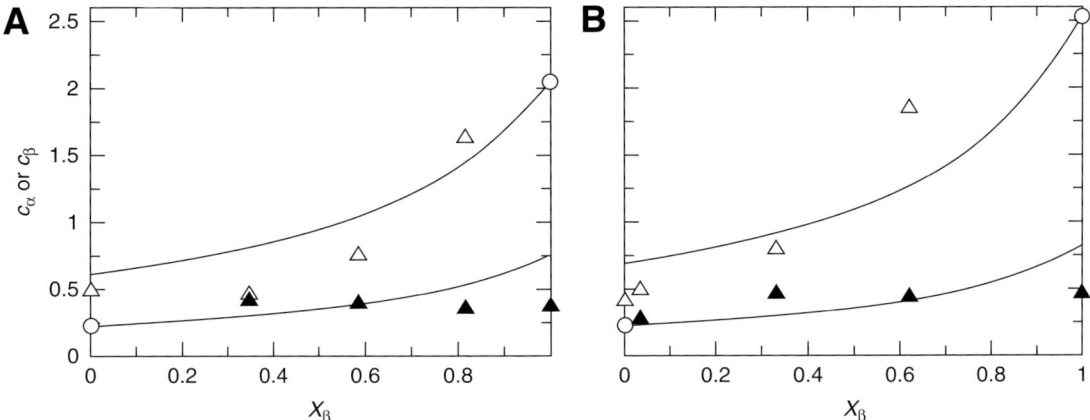

Fig. 2. Theoretical (——— and – – –) and experimental fitting values (△ and ▲) for c_α and c_β, the c parameters associated to L_o and L_d phases, respectively, for NBD-DMPE/Rh-DMPE in DMPC/chol LUV. The open circles represent points where either c_α or c_β is not defined. $X_\beta = X_{l_o}$. (**A**) $T = 30°C$. (**B**) $T = 40°C$. Reprinted with permission from **ref. 17**. © 2001 Biophysical Society.

Table 1
Comparison of K_p Values Obtained from FRET Global Decay Analysis (Second, Third, and Fourth Columns) With Those Obtained from Variations of Fluorescence Anisotropy, r (K_{pD}) or Fluorescence Intensity, I_F (K_{pA})

	$x_{chol} = 0.15$	$x_{chol} = 0.20$	$x_{chol} = 0.25$	Values from r or I_F
30°C				
K_{pD}	1.1	1.5	1.2	1.1
K_{pA}	0.73	0.42	0.18	0.30
40°C				
K_{pD}	1.5	3.2	2.6	2.6
K_{pA}	0.47	0.49	0.20	0.27

Reprinted with permission from **ref. 17**. © 2001 Biophysical Society.

should be carried out along a tie-line, so that the composition of each phase does not vary. Whereas for binary phase diagrams the tie-lines are always horizontal lines, which begin and end at the phase boundaries, for their ternary counterparts the determination of tie-lines is not a trivial issue *(19)*.

3.2.1. PSM/POPC Binary Phase Diagram

1. Prepare PSM/POPC liposomes, as described in **Subheading 3.1.1.** for DMPC/Chol, but now using buffer II instead of I, and adding DPH (1:200 probe:total lipid ratio) instead of NBD-DMPE and Rh-DMPE. Choose at least seven different compositions.
2. In a spectrofluorimeter with polarizers and temperature control, measure steady-state anisotropy <r> of each sample as a function of temperature, in the $T \in [3°C, 60°C]$ range. Use λ = 358 and 430 nm as excitation and emission wavelengths, respectively.
3. Determine the so/L_d phase boundaries at each temperature from variations in the slope of the measured <r>(T) curves (*see* **Note 5**).

3.2.2. PSM/Chol Binary Phase Diagram

1. Prepare PSM/Chol liposomes, as described in **Subheading 3.1.1.**, for DMPC/Chol, but now using buffer II instead of I, and adding DPH (1:200 probe:total lipid ratio) instead of NBD-DMPE and Rh-DMPE. Choose at least seven different compositions.
2. For temperatures above the T_m of PSM (in practice, for $T > 45°C$), determine the L_d/L_o phase boundaries using $<r>$ of DPH, as described in **Subheading 3.2.1.** for the so/L_d phase boundaries of PSM/POPC.
3. For temperatures below the T_m of PSM (in practice, for $T < 37°C$), add 5-NS (quencher; 1:25 probe:lipid ratio) to aliquots of the DPH-labeled vesicles. Using instrumentation for measurement of fluorescence decays, measure the fluorescence lifetime of DPH in presence ($\bar{\tau}$) and absence ($\bar{\tau}_0$) of quencher. Represent $\left(\bar{\tau}\right)/\left(\bar{\tau}_0\right)$ vs x_{chol}. A steep increase and a steep decrese reveal the so/L_o phase boundaries (onset and end of formation of L_o, respectively) (*see* **Note 3**).

3.2.3. POPC/Chol Binary Phase Diagram

1. Prepare POPC/Chol liposomes, as described in **Subheading 3.1.1.** for DMPC/Chol, but now using buffer II instead of I, and adding DPH (1:200 probe:total lipid ratio) instead of NBD-DMPE and Rh-DMPE. Choose at least seven different compositions (*see* **Note 6**).
2. In a spectrofluorimeter with polarizers and temperature control, measure steady-state anisotropy $<r>$ of each sample as a function of temperature, in the $T \in [15°C, 40°C]$ range. Use $\lambda = 358$ and 430 nm as excitation and emission wavelengths, respectively.
3. Determine the L_o/L_d phase boundaries at each temperature from variations in the slope of the measured $<r>(T)$ curves (*see* **Note 6**).

3.2.4. PSM/POPC/Chol Ternary Phase Diagram

For the construction of the ternary phase diagram, besides the knowledge of the phase boundaries for the three binary subsystems, one has to obtain phase boundaries for ternary mixtures. For moderate amounts of PSM ($x_{PSM} < 0.3$–0.4), one expects that there will be coexistence of L_o and L_d phases, and the phase boundaries in this region are determined using $<r>$ of DPH, as described in **Subheading 3.2.3.** For moderate amounts of POPC ($x_{POPC} < 0.2$–0.3), one expects that there will be coexistence of so and L_o phases, and the phase boundaries in this region are determined using quenching of DPH by 5-NS, as described in **Subheading 3.2.2.** The end result for room temperature is shown in **Fig. 3**.

As aforementioned, correct ascertainment of a tie-line is of paramount importance for the purpose of characterization of the system using FRET. Based on thermodynamic reasoning, estimates for the phase compositions along the tie-line containing the often-used 1:1:1 composition are obtained at both 23°C and 37°C. For the former, which is the temperature relevant for the FRET study described in **Heading 3.3.**, these are $(x_{POPC}, x_{PSM}, \text{and } x_{chol}) = (0.74–0.69, 0.19–0.27, 0.06–0.04)$ for the L_d phase, and $(x_{POPC}, x_{PSM}, \text{and } x_{chol}) = (0.25–0.26, 0.36–0.35, 0.39–0.40)$ for the L_o phase. On the same grounds, it can be argued that the slope of this tie-line must lie between 1.0 and 1.3; it is in fact expected to be closer to the larger value *(19,20)*. Thus, the compositions chosen for the FRET study lie along the gray line in **Fig. 3**, which has slope equal to 1.2, and contains the 1:1:1 point.

3.3. Domains in the Ternary PC/SM/Chol Phase System

1. Prepare POPC/PSM/Chol vesicles, as described in **Subheading 3.1.1.**, but now using buffers II or III instead of buffer I. Choose compositions (at least five) along the tie-line discussed previously

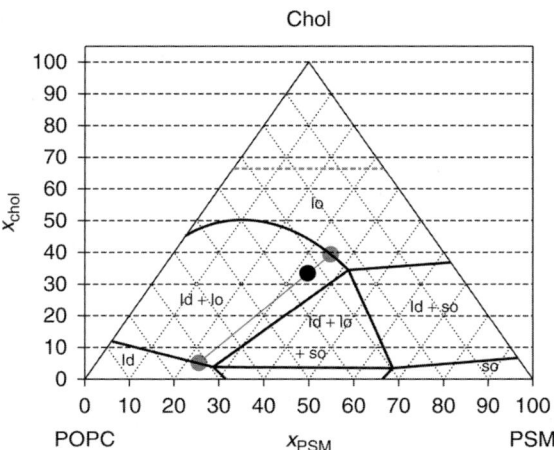

Fig. 3. PSM/POPC/Chol phase diagram at 23°C (*19*). The tie-line (gray line) contains the PSM/POPC/Chol (1:1:1) mixture (black point), which is used in the present FRET study. The extremes of the tie-line (gray points) give the compisition of the coexisting L_d (low Chol) and L_o (high Chol) phases. The dashed line for a Chol mol fraction of $x_{chol} = 0.66$ represents the solubility limit of Chol in lipid bilayers. Reprinted with permission from **ref. 20**. © 2005, from Elsevier.

and depicted in **Fig. 3**. For each composition, prepare two samples, one labeled with donor (NBD-DPPE) only and the other labeled with both donor and acceptor (Rh-DOPE). Donor probe:total lipid mole ratio should be not larger than 1:1000, whereas appropriate acceptor:total lipid mole ratios can lie in the 1:500–1:200 range (but must be the same for all compositions).

2. Determine the donor partition coefficient, as described in **Subheading 3.1.2**. Verify the preference of this probe for the L_o phase ($K_{pD} = 4.3 \pm 1.2$) (*see* **Note 7**).
3. Determine the acceptor partition coefficient, as described in **Subheading 3.1.3**. (*see* **Note 8**). Verify the preference of this probe for the L_d phase ($K_{pA} = 0.37 \pm 0.06$).
4. Measure the fluorescence decay of all samples and calculate the FRET efficiency E, as described in **steps 1–2**, *see* **Subheading 3.1.4**. (*see* **Note 9**).
5. To study the effect of raft markers ganglioside G_{M1} and CTB, repeat **steps 1–4**, adding up to 4 mol% G_{M1}, with or without CTB, to each composition.

Given the choice of a L_o-preferring donor and a L_d-preferring acceptor, FRET is expected to be less efficient in the phase coexistence region than in pure L_d and L_o phases. As shown in **Fig. 4**, this is indeed the case; E decreases from the value measured for $X_\beta = 0$ as a consequence of phase separation, before rising again near the high-Chol end of the tie-line depicted in **Fig. 3**. **Figure 4** shows several data sets: acceptor:lipid mole ratio of 1:200 in absence of other lipid species **(Fig. 4A)**, and in the presence of different amounts of the raft marker ganglioside G_{M1} (up to 4 mol%, with or without CTB) **(Fig. 4B)**, and acceptor:lipid mole ratio of 1:500 in the absence of raft markers **(Fig. 4A)**.

The curves in the absence of raft markers are qualitatively very similar, showing that a significant drop in FRET efficiency only occurs at $X_\beta \approx 0.35$. This is in contrast with the theoretical curve assuming infinite phase separation (large domains), for which a steep drop was expected already at very low X_β. Thus, domains in the low X_β range, although very small, are definitely forming (in the absence of phase separation, E would actually *increase* owing to the Chol condensing effect, as apparent on the thin lines in **Fig. 4A**). On the other hand, the theoretical curve for infinite phase separation describes well the experimental data for high

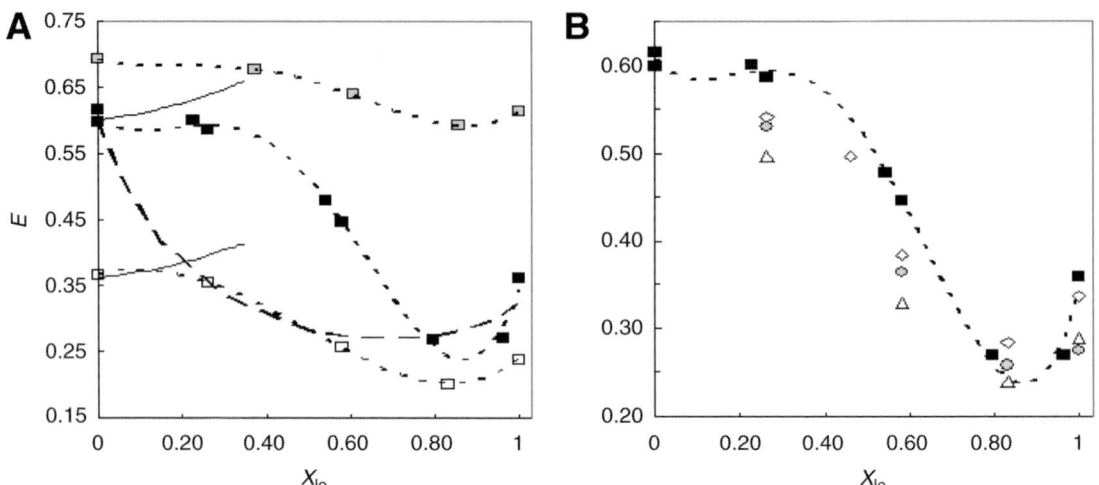

Fig. 4. Variation of E of the donor/acceptor pair NBD-DPPE/Rh-DOPE in PSM/POPC/Chol LUV as a function of X_{lo}, along the gray tie-line of **Fig. 3**. **(A)** Effect of acceptor concentration and comparison with the binary system DMPC/Chol. The dotted lines are merely visual guides. The thin continuous lines are the values of E calculated for a uniform distribution of donor and acceptor molecules in hypothetical pure L_d phase for acceptor/lipid ratios of 1:200 and 1:500. The different data sets correspond to 1:200 acceptor lipid ratio (■); 1:500 acceptor/lipid ratio (□); for comparison, the values for the binary system DMPC/Chol along the L_d/L_o tie-line at 30°C are also shown (■; *see also* **ref. 17**). The theoretical line for infinite phase separation (large domains) for the 1:200 acceptor/total lipid mole ratio is also shown (thin broken line). **(B)** Effect of raft markers. The different data sets correspond to 1:200 acceptor lipid ratio with 0 mol% G_{M1} and 2 mol% G_{M1}/no CTB indistinguishable (■), 2 mol% G_{M1} and excess CTB (◇), 4 mol% G_{M1} and no CTB (●), 4 mol% G_{M1} and excess CTB (△). Reprinted with permission from **ref. 20**. © 2005 Elsevier.

L_o phase fraction. There is also qualitative agreement with the data from the DMPC/Chol binary system at 30°C *(17)*.

However, the magnitude of the E drop is much higher in the ternary system, indicating that phase separation is more pronounced in the latter. The domains in the binary system (even in the large domain side of the tie-line) probably do not reach the size of the domains in the ternary system. The E vs X_{lo} curve predicted for very large domains (infinite phase separation) and acceptor/total lipid at a 1:200 mole ratio thin dashed line in **Fig. 4A** crosses the trendline of the experimental data at X_{lo} approx 0.75. This corresponds to a situation wherein rafts occupy approx 50% of the membrane surface area, in agreement with recent sophisticated studies in cell membranes (*see* **ref. 20** for a detailed discussion).

The presence of 2 mol% G_{M1} does not produce any significant changes in E. However, a drop is apparent in the low Chol region of the tie-line ($X_\beta < 0.6$) for 4 mol% G_{M1}. This decrease is enhanced if excess CTB is added. Addition of excess CTB in conjunction with 2 mol% G_{M1} produces a drop in E similar to that obtained with 4 mol% G_{M1} but no CTB. Overall, the curves' shapes are similar to those measured without raft markers, and the values obtained between the various data sets for $X_\beta = 0$ are almost identical, as is the case for $X_\beta = 1$. This indicates that G_{M1} and CTB do not affect the general phase behavior of the lipid system. This is also confirmed by verifying that these components do not change the phase boundaries (for the amounts used) determined by DPH anisotropy or *t*-PnA lifetime measurements. However, they

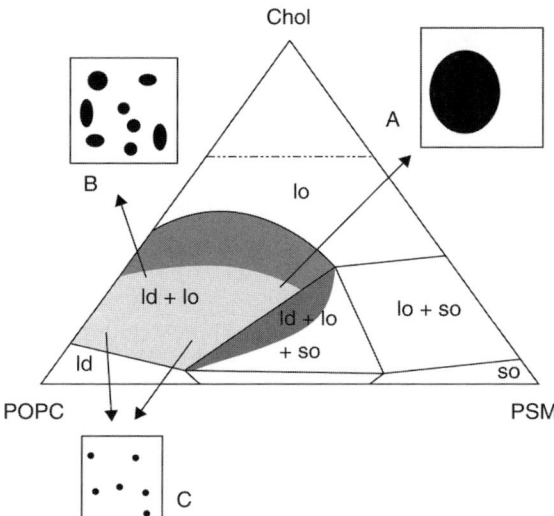

Fig. 5. PSM/POPC/Chol phase diagram at 23°C, which also shows the boundaries and size of lipid rafts. In the dark area, L_o predominates over L_d, and the reverse occurs in the light-shaded area. Rafts are present in the blue-shaded area (L_d/L_o coexistence). They can also exist in the green-shaded area, wherein there is coexistence of three-phases, yet the so phase is present only in very low amounts. **(Insets) (A)** region of large rafts, detected by microscopy and FRET (>75100 nm); **(B)** region of intermediate size rafts, detected by FRET but not microscopy (between ~20 nm and ~75–100 nm); **(C)** region of small rafts, undetected by FRET or microscopy (<20 nm). Reprinted with permission from **ref. 20**. © 2005 Elsevier.

can increase the size of the rafts in the small-to-intermediate domain region (and possibly in the large domain region. However, in this range FRET would not detect any further domain size increase). Because 2 mol% G_{M1} has no influence on E, and the presence of CTB or higher amounts of ganglioside only affects the domains' size when these are relatively small, the domains detected with those raft markers in giant unilamellar vesicles by microscopy are probably not induced by the presence of the markers: the domains observed by such techniques are larger and the amounts of G_{M1} used are usually smaller.

In **Fig. 5**, along with the phase diagram, sketches of the domains structures are represented with the best estimates for raft sizes in different regions of the diagram. An upper limit of 20 nm can be given for the L_o phase domains (rafts) when these represent less than 35 mol%, obtained from numerical simulations *(17)*.

Using the described FRET methodology, it is shown that the size of lipid rafts depends on the membrane composition in, at least, three ways: (1) the complexity of the system (e.g., binary vs ternary, absence vs presence of ganglioside), (2) very importantly, the fraction of the system in L_o phase, and (3) the presence of multivalent lipid-binding proteins. The first two involve only lipid–lipid interactions, and suffice to generate rafts with very different sizes.

4. Notes

1. Home-made extruder devices are used. There are also commercially available models (e.g., Avanti Polar Lipids Inc., Alabaster, AL; Avestin Inc., Ottawa, Canada; Lipex Biomembranes Inc., Vancouver, Canada).
2. This can be achieved using commercial software, for example, Origin (OriginLab Corp., Northampton, MA).

3. In these studies, several different systems are used. NBD-PE dyes can be conveniently excited at 340 nm by a frequency-doubled dye (4-dicyanomethylene-2-methyl-6-[p-dimethylaminostyryl]-4H-pyran or DCM) laser or at 428 nm by a Ti:Sapphire laser. In the lab, these secondary lasers are pumped by an Ar$^+$ ion laser and diode-pumped solid-state Neodymium dopedyttrium orthovanadate (ND:YVO$_4$) laser, respectively. DPH excitation is carried out with the dye laser setup at 340 nm. t-PnA excitation is carried out with a frequency-doubled rhodamine 6G laser dye (pumped by the Ar$^+$ ion laser), at 303 nm. Rh-DOPE is excited at 575 nm by the rhodamine 6G laser dye (Coherent 701-2 dye laser, Coherent, Santa Clara, CA) without frequency doubling. Emission wavelengths are 536 nm for the NBD dyes, 430 nm for DPH, 405 nm for t-PnA, and 610 nm for Rh-DOPE.
4. In these studies, all decay analyses are carried out using home-developed software using the Marquardt algorithm, as described, for example, in **ref. 21**.
5. For the gel phase, <r> is usually more than 0.3. For the fluid phase, for T above room temperature, <r> is typically less than 0.1. At low temperatures, one can observe higher <r> in the fluid phase (e.g., 0.2 for POPC at 3°C); therefore, an intermediate <r> value does not necessarily indicate gel/fluid coexistence. However, from breaks in the slope of the <r> (T) curves, and from a sensible analysis that takes into account the behavior at multiple temperatures, reasonable estimates of the phase boundaries are obtained.
6. An alternate procedure is the use of the average lifetime of t-PnA instead of DPH anisotropy. To this effect, t-PnA (1:500 probe:total lipid ratio) is added to blank vesicles by injection of an adequate small amount of stock solution.
7. K_{pD} can be determined either using <r> as described in **Subheading 3.1.2.**, or using the variation of NBD-DPPE average lifetime, $\bar{\tau}$ (*see* **ref. 17**). The equation analysis for the latter is identical to **Eq. 12**, with $\bar{\tau}$ in the place of I_F.
8. K_{pA} can be determined using either I_F as described in **Subheading 3.1.3.**, or the variation of Rh-DOPE average lifetime, $\bar{\tau}$ (*see also* **Note 7**).
9. Alternatively, global analysis of the FRET decays can be carried out, as described in **Subheading 3.1.4.** for the DMPC/Chol system. However, possibly because of the increased complexity of the ternary mixture, satisfactory global fitting statistics are not achieved for PSM/POPC/Chol.

Acknowledgments

The authors thank Dr. Rodrigo de Almeida for his contribution to the work on the ternary system. This work was supported by Programa Operacional Ciência e Inovação (POCI) projects, Fundação Para a Ciência e Tecnologia (FCT) (Portugal).

References

1. Smaby, J. M., Momsen, M. M., Brockman, H. L., and Brown, R. E. (1997) Phosphatidylcholine acyl unsaturation modulates the decrease in interfacial elasticity induced by cholesterol. *Biophys. J.* **73,** 1492–1505.
2. Xiang, T. -X. and Anderson, B. D. (1997) Permeability of acetic acid across gel and liquid–crystalline lipid bilayers conforms to free-surface-area theory. *Biophys. J.* **72,** 223–237.
3. Lafleur, M., Cullis, P. R., and Bloom, M. (1990) Modulation of the orientational order profile of the lipid acyl chain in the L alpha phase. *Eur. Biophys. J.* **19,** 55–62.
4. Méléard, P., Gerbeaud, C., Pott, T., et al. (1997) Bending elasticities of model membranes: influences of temperature and sterol content. *Biophys. J.* **72,** 2616–2629.
5. Ipsen, J. H., Karlström, G., Mouritsen, O. G., Wennerström, H., and Zuckermann, M. J. (1987) Phase equilibria in the phosphatidylcholine-cholesterol system. *Biochim. Biophys. Acta* **905,** 162–172.
6. Vist, M. R. and Davis, J. H. (1990) Phase equilibria of cholesterol/DPPC mixtures: ^{2}H-NMR and differential scanning calorimetry. *Biochemistry* **29,** 451–464.

7. Almeida, P. F. F., Vaz, W. L. C., and Thompson, T. E. (1992) Lateral diffusion in the liquid phases of dimirystoylphosphatidylcholine/cholesterol bilayers: a free volume analysis. *Biochemistry* **31,** 6739–6747.
8. Mateo, C. R., Acuña, A. U., and Brochon, J. -C. (1995) Liquid-crystalline phases of cholesterol/lipid bilayers as revealed by the fluorescence of trans-parinaric acid. *Biophys. J.* **68,** 978–987.
9. Simons, K. and Ikonen, I. (1997) Functional rafts in cell membranes. *Nature* **387,** 569–572.
10. Giocondi, M. -C., Vié, V., Lesniewska, E., Goudonnet, J. -P., and Le Grimellec, C. (2000) In situ imaging of detergent-resistant membranes by atomic force microscopy. *J. Struct. Biol.* **131,** 38–43.
11. Thomas, J. L., Howloka, D., Baird, B., and Webb, W. W. (1994) Large-scale co-aggregation of fluorescent lipid probes with cell surface proteins. *J. Cell Biol.* **126,** 795–802.
12. Zurzolo, C., van Meer, G., and Mayor, S. (2003) The order of rafts. Conference on microdomains, lipid rafts and caveolae. *EMBO Rep.* **4,** 1117–1121.
13. Pralle, A., Keller, P., Florin, E. -L., Simons, K., and Hörber, J. K. H. (2000) Sphingolipid-cholesterol rafts diffuse as small entities in the plasma membrane of mammalian cells. *J. Cell Biol.* **148,** 997–1007.
14. Dietrich, C., Bagatolli, L. A., Volovyk, Z. N., et al. (2001) Lipid rafts reconstituted in model membranes. *Biophys. J.* **80,** 1417–1428.
15. Dietrich, C., Volovyk, Z. N., Levi, M., Thompson, N. L., and Jacobson, K. (2001) Partitioning of Thy-1, GM1 and crossed-linked phospholipid analogs into lipid rafts reconstituted in supported model membrane monolayers. *Proc. Natl. Acad. Sci. USA* **98,** 10,642–10,647.
16. Ribi, H. O., Ludwig, D. S., Mercer, K. L., Schoolnik, G. K., and Kornberg, R. D. (1988) Three-dimensional structure of cholera toxin penetrating a lipid membrane. *Science* **239,** 1272–1276.
17. Loura, L. M. S., Fedorov, A., and Prieto, M. (2001) Fluid-fluid membrane microheterogeneity: a fluorescence resonance energy transfer study. *Biophys. J.* **80,** 776–788.
18. Davenport, L. (1997) Fluorescence probes for studying membrane heterogeneity. *Methods Enzymol.* **278,** 487–512.
19. de Almeida, R. F. M., Fedorov, A., and Prieto, M. (2003) Sphingomyelin/phosphatidylcholine/cholesterol phase diagram: boundaries and composition of lipid rafts. *Biophys. J.* **85,** 2406–2416.
20. de Almeida, R. F. M., Loura, L. M. S., Fedorov, A., and Prieto, M. (2005) Lipid rafts have different sizes depending on membrane composition: a time-resolved fluorescence resonance energy transfer study. *J. Mol. Biol.* **346,** 1109–1120.
21. O'Connor, D. V. and Philips, D. (1984) *Time-correlated single photon counting.* Academic Press, New York, NY.

34

Lipid Domains in Supported Lipid Bilayer for Atomic Force Microscopy

Wan-Chen Lin, Craig D. Blanchette, Timothy V. Ratto, and Marjorie L. Longo

Summary

Phase-separated supported lipid bilayers have been widely used to study the phase behavior of multicomponent lipid mixtures. One of the primary advantages of using supported lipid bilayers is that the two-dimensional platform of this model membrane system readily allows lipid-phase separation to be characterized by high-resolution imaging techniques such as atomic force microscopy (AFM). In addition, when supported lipid bilayers have been functionalized with a specific ligand, protein–membrane interactions can also be imaged and characterized through AFM. It has been recently demonstrated that when the technique of vesicle fusion is used to prepare supported lipid bilayers, the thermal history of the vesicles before deposition and the supported lipid bilayers after formation will have significant effects on the final phase-separated domain structures. In this chapter, three methods of vesicle preparations as well as three deposition conditions will be presented. Also, the techniques and strategies of using AFM to image multicomponent phase-separated supported lipid bilayers and protein binding will be discussed.

Key Words: AFM; gel-fluid coexisting; lipid domains; phase separation; rafts; supported lipid bilayer; transbilayer symmetry; vesicle fusion.

1. Introduction

The formation of ordered phase domains in the plasma membrane has elicited extensive attention for more than 30 yr. It has been shown that many important functions of cellular membranes are closely associated with their compositional and structural heterogeneity *(1,2)*. The idea of phase separation and membrane heterogeneity in cellular membranes originated from known biophysical properties of multicomponent lipid bilayers. Lipids can exist in multiple phases but the two phases that are pertinent to domain formation are the fluid and gel phases. Each lipid has a characteristic melting temperature (T_m), and the phase of the lipid depends on the temperature of the bilayer and the T_m. When the temperature of the bilayer is lowered past the T_m the lipids will undergo a phase transition from the fluid to the gel state *(3)*. When a bilayer contains multiple lipid components and exists at a temperature wherein one component is in the gel phase and the other in a fluid phase, the gel-phase lipids will protrude above the fluid component, resulting in a hydrophobic mismatch *(4)*. The unfavorable energy associated with the hydrophobic mismatch will cause gel phase lipids to aggregate into domains, thereby reducing the hydrocarbon surface area exposed to the solvent media *(5,6)*.

Model membrane systems such as giant unilamellar vesicles and supported lipid bilayers have been extensively used in understanding the fundamental properties of heterogeneity and lipid phase separation in biological membranes *(7–10)*. These model systems have successfully demonstrated the coexistence of ordered and disordered phases for a variety of different lipid compositions. In recent years, many studies have been conducted using supported lipid bilayers to study membrane biophysical properties such as lipid domain formation *(9,11,12)*, lipid lateral diffusion *(13)*, and ligand–protein interaction *(14,15)*. One of the primary advantages of using this model membrane system is that lipid-phase separation can be characterized through high resolution scanning probe techniques such as atomic force microscopy (AFM). Therefore, structural details that are beyond the diffraction limit of conventional light microscopy can be obtained.

Vesicle fusion is a common method of preparing supported lipid bilayers *(16,17)*. This method involves depositing small unilamellar vesicles (SUVs) on to a hydrophilic surface resulting in the spontaneous formation of a uniform two-dimensional supported lipid bilayer. The fundamental processes of forming supported lipid bilayers through vesicle fusion are usually described in four steps:

1. Adsorption of SUVs on the solid surface.
2. Fusion between intact SUVs to form larger vesicles.
3. Larger vesicles deforming and rupturing to form bilayer patches on the support.
4. Coalescences of bilayer patches to form a continuous bilayer (for review, *see* **ref. *18***).

It is generally accepted that the whole vesicle-fusion process takes place within 15 min *(19,20)*. It has been found that the thermal history of the vesicles before deposition as well as different deposition conditions, result in different types of phase-separated supported lipid bilayer: ordered domains in these bilayer can be transbilayer symmetric or asymmetric *(21)*, whereas the size of the domains can range from tens of nanometers to tens of microns. In this chapter, the different methods that can be used to control phase-separated domain transbilayer symmetry and size will be discussed in detail. In addition, the methods and strategies of using AFM to obtain surface topology images of these supported lipid bilayers with subnanometer resolution in height will be discussed. Physical properties of domains, such as size, area, perimeter, as well as domain symmetry can be easily obtained by analyzing topographical AFM images. Besides using AFM to image phase-separated domains in supported lipid bilayers, also using this method to image and characterize specific protein binding to supported lipid bilayers containing ligand-conjugated lipids will be discussed.

2. Materials
2.1. Supported Lipid Bilayer

1. Stock solutions of lipids (Avanti Polar Lipids, Birmingham, AL and Matreya, Pleasant Gap, PA) are dissolved in chloroform at concentrations that range from 10 to 25 mg/mL, and are housed in glass vials with Teflon® caps or Teflon septa at –20°C.
2. Glass reaction vials or V-vials (Fisher, Pittsburgh, PA) are stored in chloroform when not in use. The vials should be equipped with Teflon screw caps or Teflon septa.
3. All water used in the experiments is purified in a Barnstead Nanopure System (Dubuque, IA), with resistivity ≥17.9 MΩ and pH 5.5.
4. Extruder (includes a polycarbonate membrane holder and two gas-tight, glass syringes) (Avestin Inc., Ottawa, Canada) is stored in ethanol when not in use.
5. Mica substrates (Ted Pella Inc., Redding, CA) are freshly cleaved using scotch tape until a completely flat surface is obtained.

2.2. AFM and Bilayer-Protein Binding

1. AFM probes: silicon nitride microlever probes gold coated on the reflex side with spring constants ranging from 0.01 to 0.1 N/m (Veeco, Santa Barbara, CA).
2. Lipids functionalized with a ligand for specific protein-receptor binding, for example, 1,2-dioleoyl-sn-glycero-3-phosphoethanolamine-N-(cap biotinyl) (biotin-DOPE) (Avanti Polar Lipids).
3. Protein for specific binding to ligand conjugated lipid, for example, Neutravidin (biotin-binding protein) (Molecular Probes, Carlsbad, CA). Follow the manufactory data sheet and protocols for protein storage, i.e., buffer conditions and storage temperature.

3. Methods

Vesicle fusion involves depositing SUVs (which range from 25 to 200 nm in diameter) on to a hydrophilic surface. This method results in the spontaneous formation of a uniform two-dimensional supported lipid bilayer on the surface. Mica is widely used as a substrate because it is atomically flat for AFM imaging of phase-separated, supported lipid bilayers, which often require subnanometer topographical resolution.

SUVs are typically prepared by tip-sonification or extrusion. In the tip-sonification method, large multilamellar vesicles (MLVs) are broken down to SUVs (25–100 nm in diameter) by high-energy ultrasonic waves. In the extrusion method, large MLVs are pushed through a polycarbonate membrane of defined pore size (50–200 nm), so that they rupture into small vesicles dictated by the polycarbonate pore size. The characteristics of the bilayer (e.g., lateral mixing, leaflet asymmetry) in the vesicles used for fusion are affected by the thermal history of the SUV suspension, which can be varied through the method of preparation. Therefore, different methods of SUV preparation can have large effects on the structural properties of phase-separated supported lipid bilayer. For example, in 1,2-dilauroyl-sn-glycero-3-phosphocholine (DLPC—12:0, $T_m = (5°C)$/1,2-distearoyl-sn-glycero-3-phosphocholine (DSPC—18:0, $T_m = 55°C$) mixtures, it is possible to control transbilayer symmetry by controlling the thermal history of the vesicle suspension and vesicle deposition methods (**Fig. 1**). It is worth noting that the results presented in **Fig. 1** are specific to the phase-separated DLPC/DSPC binary mixture. Because the mechanism of domain formation strongly depends on the physical properties of the lipid mixture, such as the hydrophobic mismatch, different results should be expected with different lipid compositions when using the same preparation methods. An example is shown in **Fig. 2**.

Three types of deposition methods have been explored, vesicle fusion, quenched vesicle fusion, and slow-cooled vesicle fusion, when forming supported lipid bilayers. Thermal history is the main parameter that is varied in each method. In vesicle fusion (i.e., room temperature vesicle—room temperature substrate deposition), lipids experience no temperature change during supported lipid bilayer formation. On the other hand, when using quenched vesicle fusion (i.e., heated vesicle—room temperature substrate deposition) and slow-cooled vesicle fusion (i.e., heated vesicle—heated substrate deposition followed by slow cooling to room temperature) methods, lipids experience a fast and slow temperature drop, respectively, during the phase transition from the fluid state to the solid–fluid coexisting state. In general, slower cooling rates allow more time for domain growth. Thus, domain size can be controlled by the rate of cooling during the phase-separation process. These effects are demonstrated in **Fig. 3**. In summary, it is important to maintain consistency, especially with the thermal history, during both vesicle preparation and deposition to ensure that results are reproducible.

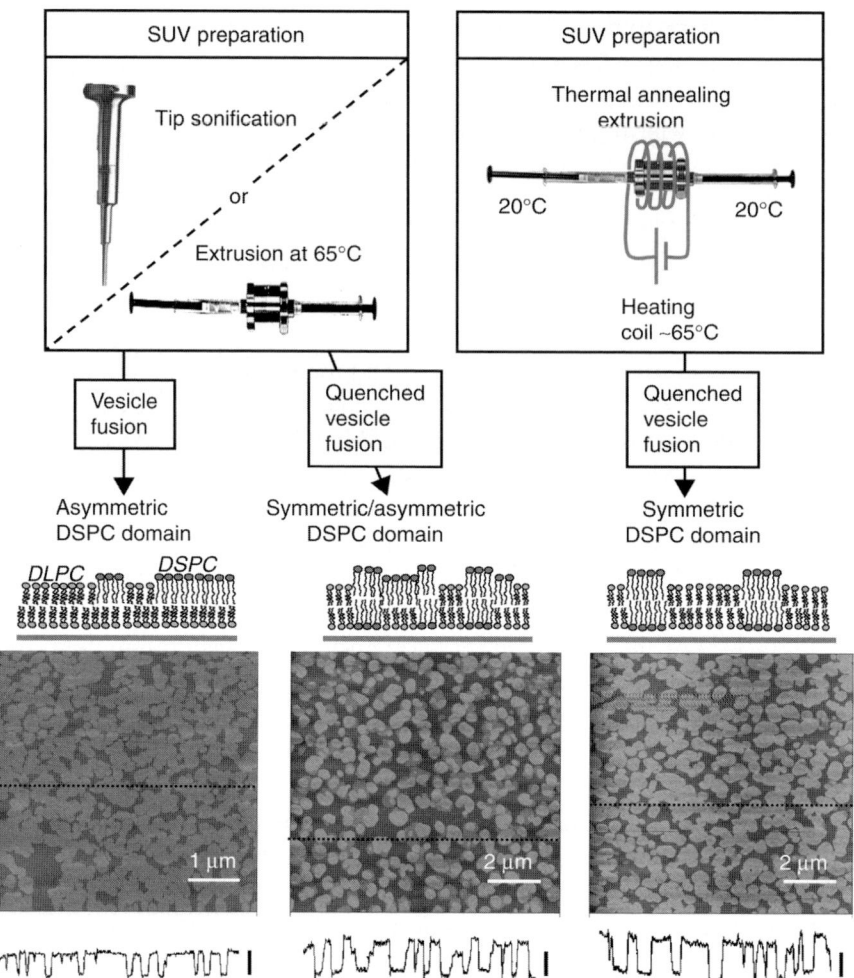

Fig. 1. Controlling transbilayer symmetry of DSPC domains in phase-separated DSPC/DLPC supported lipid bilayers. Three vesicle preparation methods (*see* **Subheading 3.1.2.**) combined with two deposition conditions (*see* **Subheading 3.1.3.**) were used to form supported lipid bilayers displaying three types of domain symmetry: asymmetric (domains in one leaflet), symmetric (domains superimposed in both leaflets), and symmetric/asymmetric (unequal DSPC concentration in each leaflet). AFM height images of the supported lipid bilayers are shown here. The dotted lines denote the location of the section analyses (bar = 1 nm). Asymmetric and symmetric domains extended approx 1 and 1.8 nm, respectively, above the fluid phase, whereas symmetric/asymmetric domains displayed both heights.

To image a supported lipid bilayer using AFM, it is necessary that the imaging system deliver resolution on the nanometer (lateral) and angstrom (vertical) scales in the fluid environment. In order to protect the delicate electronic parts such as the scanner, which is made from piezoelectric material, most commercialized AFMs are equipped with a fluid-imaging cell or a fluid-cantilever holder to operate the system in fluid environment. In general, it is possible to image lipid-phase separation in contact mode when a scanning probe containing a low spring constant (0.01–0.1 N/m) is used. Besides imaging lipid domains in supported lipid bilayers, AFM can also image and characterize protein-bilayer binding when the bilayer is functionalized with

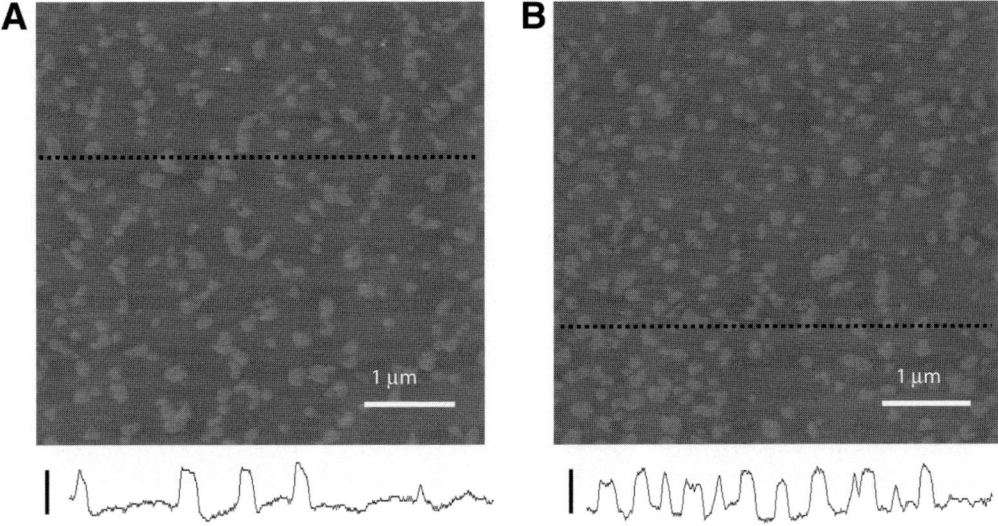

Fig. 2. Effect of changing lipid composition on domain symmetry. (**A**) AFM image and section analysis of a DPPC (16:0)/DLPC (12:0) phase-separated supported lipid bilayer. DPPC domains are approx 1.1 nm above the fluid phase, which correspond to an eight-carbon mismatch indicating the domains are transbilayer symmetric. (**B**) AFM image and section analysis of a DSPC (18:0)/DLPC (12:0) phase-separated supported lipid bilayer. DSPC domains are approx 1 nm above the fluid phase, which indicates the domains are transbilayer-asymmetric (*see* **Fig. 1**). Both bilayers were prepared through the same method: SUV suspensions were made through tip sonification (*see* **Subheading 3.1.2.1.**) followed by vesicle fusion (*see* **Subheading 3.1.3.1.**). The scale bar in section analysis is 1 nm.

protein receptors or ligands. This is demonstrated in **Fig. 4**, wherein specific binding between fluid phase biotin-DOPE and Neutravidin is observed.

3.1. Preparation of Supported Lipid Bilayer

3.1.1. Preparation of MLVs

1. Calculate the amount of lipid needed to make the vesicle suspension. In general, the concentration ranges from 0.05 to 0.5 mg/mL.
2. Stock lipid solutions must then be removed from the freezer and brought to room temperature before use.
3. Prepare a hot water bath. In general, the temperature of the bath depends on the lipids being used to make the supported lipid bilayer. The temperature must be higher than the phase-transition temperature of every lipid constituent.
4. Clean a glass reaction vial or V-vial with chloroform several times.
5. Clean the glass syringes with chloroform. The outside of the syringe should be rinsed with chloroform. The inside of the syringe can be cleaned by pumping chloroform in and then out 10–20 times.
6. Vortex the stock lipid solution for 20–30 s before appropriate volumes of lipid solution are added to the reaction vial. Then, the calculated volume of lipids should be added to the reaction vial.
7. Once the appropriate amount of lipid solution is added to the reaction vial, cap the stock solution of lipids to ensure consistency of stock lipid concentration. It is important that the stock solution does not stay unsealed for extended periods of time because chloroform rapidly evaporates, thus changing the concentration of the stock solution.
8. Seal the cap of the stock solution with parafilm and place lipid solution back into the freezer.
9. Then, clean the syringes as described in **subheading 3.1.1.5.** and put away; at this point check the temperature of the hot water bath. Do not proceed until the hot water bath is at the required temperature.

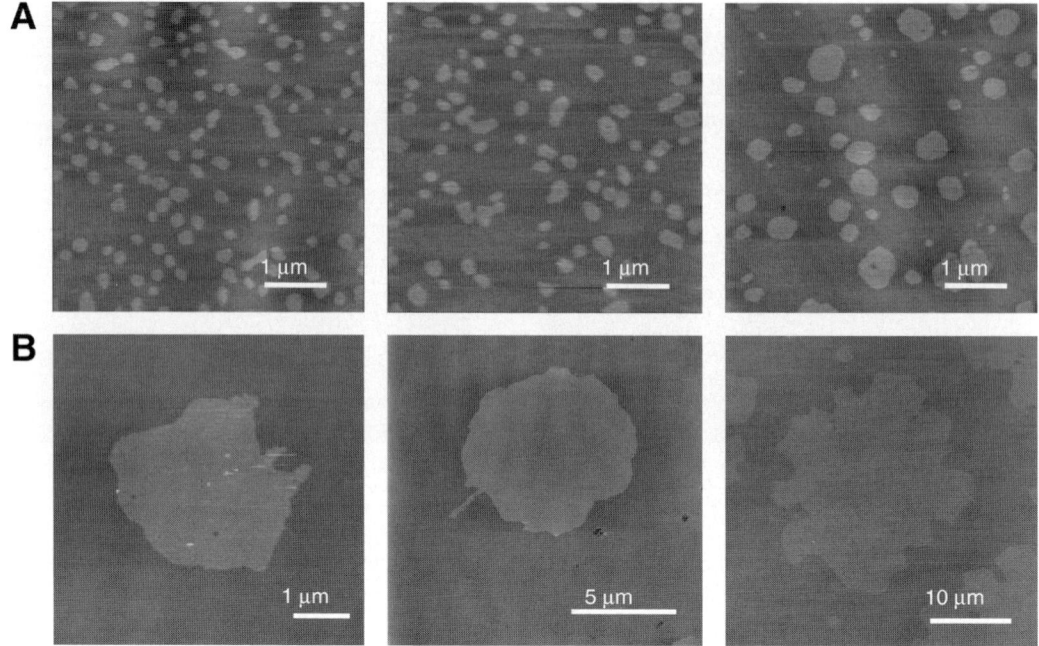

Fig. 3. Controlling GalCer (galactosylceramide) gel phase domain size at the nanometer and micrometer scale. (**A**) Quenched vesicle fusion: quenching rates were decreased from left to right, resulting in a gradual increase in GalCer domain size from 100 nm at the fastest quenching rate to approx 500 nm at the slowest quenching rate. (**B**) Slow cooled vesicle fusion: cooling rates were decreased from left to right (2–5 h cooling times, from 90 to 25°C), resulting in an increase in GalCer domain size from 4 to 25 μm.

10. Evaporate the chloroform from the reaction vial under a slow stream of N_2 while vortexing at the highest setting. This will create a dry lipid film on the surface of the reaction vial.
11. Add water to the reaction vial to bring the lipid–water suspension to the desired final concentration (range from 0.05 to 0.5mg/mL).
12. Then, place the vial in the hot water bath for approx 15 min (or until there are no more lipids dried to the bottom of the reaction vial) with periodically shaking or vortexing for periods of 15 s.
13. Transfer the milky lipid suspension (containing MLVs) to a plastic tube at room temperature before further treatment.

3.1.2. SUV Preparation Methods

3.1.2.1. METHOD A: TIP SONIFICATION

1. Allow the MLV suspension to cool to room temperature and clean the tip sonifier.
2. Sonicate the MLV suspension until the suspension reaches clarity. The sonification time may vary depending on the lipid content (*see* **Note 1**).
3. Incubate the resulting SUV suspension in a water bath at the desired temperature before further use (*see* **Note 2**).

3.1.2.2. METHOD B: EXTRUSION

1. Clean the extruder with ethanol and then rinse extensively with water.
2. Place a polycarbonate membrane of defined pore size (it can range from 50 to 200 nm in diameter) in the extruder.

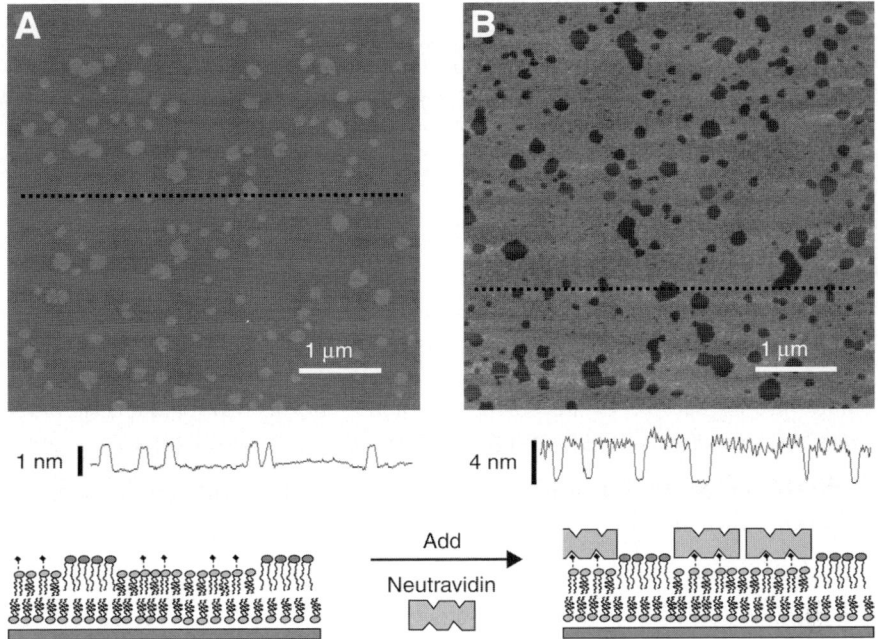

Fig. 4. Imaging membrane (biotin-DOPE)—protein (Neutravidin) interactions with AFM. **(A)** AFM image and section analysis of a DSPC/DLPC supported lipid bilayer doped with 1% biotin-DOPE (fluid-phase biotin-conjugated lipid). Bilayers were formed through methods resulting in asymmetric DSPC domains with a 1 nm domain height (*see* **Fig. 1**). **(B)** AFM image and section analysis of the bilayer after addition of Neutravidin. DSPC domains now appear as darker regions that are about 4 nm lower than the fluid region. The observed contrast reversal indicates that protein binding was directed to the fluid phase. The measured protein thickness (~5 nm) is consistent with the known protein dimensions.

3. Wet the polycarbonate membrane by pushing water through the extruder with a gas-tight, glass syringe.
4. Assemble the extruder with one empty syringe and other syringe containing the MLV suspension.
5. Place the extruder into a hot water bath or on a heating block for 10 min. The temperature of the heating device should be higher than the phase-transition temperatures of all lipid constituents.
6. Using the syringes slowly push the MLV suspension through the center part of the extruder 20 times or until the suspension reaches clarity (for thermal annealing method, *see* **Note 3**).
7. Incubate the resulting SUV suspension in a water bath at the desired temperature before further use.

3.1.3. Supported Lipid Bilayer Deposition

3.1.3.1. Vesicle Fusion

1. Add a droplet of the room temperature SUV suspension on to a freshly cleaved room temperature mica disk, chose a volume to ensure the mica surface remains hydrated during supported lipid bilayer formation.
2. Incubate the vesicle droplet on the mica disk for 30 min and then pump-rinse 40 times using a handheld pipeter with aliquots of water to remove excess vesicles; again, this procedure must be conducted to ensure the supported lipid bilayer remains hydrated (*see* **Note 4**).

3.1.3.2. QUENCHED VESICLE FUSION

1. Add a droplet of heated SUV suspension onto a freshly cleaved room temperature mica disk. Thus, the formation of the supported lipid bilayer occurs during a thermal quench from high-to-room temperature (*see* **Note 5**).
2. The vesicle droplet is incubated on the mica disk for 30 min and then rinsed 40 times using a handheld pipeter with aliquots of water to remove excess vesicles while keeping the supported lipid bilayer hydrated (*see* **Note 4**).

3.1.3.3. SLOW-COOLED VESICLE FUSION

1. Preheat a clean mica disk in a temperature-controlled oven (*see* **Note 6**).
2. Add a droplet of the heated SUV suspension on to the heated mica disk.
3. Slowly cool the mica disk/supported lipid bilayer to room temperature in the oven (*see* **Note 7**).
4. Then pump-rinse the sample 40 times using a handheld pipeter with aliquots of water to remove excess vesicles, whereas ensuring the supported lipid bilayer remains hydrated (*see* **Notes 4** and **8**).

3.2. Image Phase-Separated Supported Lipid Bilayers and Bilayer-Protein Binding in Fluid (AFM Study)

3.2.1. Image Phase-Separated Supported Lipid Bilayers

1. Prepare a phase-separated supported lipid bilayer as described previously (*see* **Subheading 3.1.**). Place the bilayer onto the AFM sample holder.
2. To clean the AFM probe, expose the probe to UV light for 60 s or use a plasma cleaner.
3. Mount a probe into the cantilever holder.
4. Place a small drop of water at the end of the probe then mount the cantilever holder onto the AFM (*see* **Note 9**).
5. Merge the probe into the fluid cell or hydrated sample container containing the supported lipid bilayer, then follow the manufactory manual to align the laser and adjust the photodetector (*see* **Note 10**).
6. Follow the manufactory manual to set initial scan parameters, then engage the probe.
7. Once the probe is in contact with the sample, decrease the set point so that the scanning force is 0.2–1 nN (*see* **Notes 11** and **12**).
8. Set desired scan size and acquire an image at the lowest scanning force.

3.2.2. Image Bilayer-Protein Binding

1. Prepare a phase-separated supported lipid bilayer as described previously (*see* **Subheading 3.1.**). Instead of using purified water, use buffer/water mixture to make a MLV suspension (*see* **Note 13**).
2. After the supported lipid bilayer forms, exchange the buffer/water mixture to 100% buffer gradually.
3. Image the supported lipid bilayer as described previously (*see* **Subheading 3.2.1.**).
4. Deposit protein stock solution directly on top of the bilayer surface.
5. After a 30-min equilibration period, rinse the bilayer gently to remove excess unbound protein.
6. Image the supported lipid bilayer with bound protein as described previously (*see* **Subheading 3.2.1.**). An example result is shown in **Fig. 4**.

4. Notes

1. To avoid overheating the suspension, sonicate the vesicle suspension at the lowest power, in two 30-s time intervals with 20-s pause in between. In addition, the condition of the sonifier tip can largely affect the performance of the sonifier. Polishing the tip with cloth regularly may help maintain the condition of the tip.
2. It is suspected that the tip sonifier may release metal particles to the SUV suspension during the sonification. The metal particles can be removed by centrifuging the suspension before any further use.

3. Thermal annealing extrusion: instead of the whole extruder, only the center part of the extruder (i.e., the polycarbonate membrane and the membrane holder) is heated (by a heating blanket), whereas the syringes are kept at room temperature. During the extrusion process the vesicle suspension is thermally annealed, i.e., as it passes through the center of the extruder, the suspension is heated above the phase-transition temperatures of the lipids that are used, and then cooled in the room temperature syringe.
4. Dropping the mica disk into a water reservoir and then oscillating by hand while submerged in the water can further rinse off extra vesicles that are absorbed to the supported lipid bilayer surface. Incubating the sample in clean bulk water for several hours can also help remove vesicles absorbed onto the supported lipid bilayer surface.
5. The temperature of the heated SUV suspension depends on the lipids that are used. Usually, the temperature is higher than phase-transition temperatures of all lipids. The temperature difference between the SUV suspension and the mica substrate will affect the resulting domain size. For example, depositing a 150 µL 65°C droplet of a DSPC ($T_m = 55°C$)/DLPC ($T_m = 5°C$) suspension onto 20°C mica surface results in about 100 nm DSPC domains **(Fig. 1)**.
6. The temperature of the oven depends on the lipids that are used. Usually, the temperature should be higher than phase-transition temperatures of every lipid. It is known that the phase-transition temperatures of lipids are lower when the bilayer is on a solid substrate, although the mechanism is unclear. Therefore, it is suggested to determine the exact oven temperature by trial and error. It is worth noting that if the oven temperature is too high (e.g., 90°C), the deposition of the supported lipid bilayer may be problematic.
7. The droplet may evaporate rapidly in the oven during the slow cooling process. Keep the sample in a humid container to prevent dehydration. The cooling rate affects the final domain size; in general, faster cooling results in smaller domains **(Fig. 3)**.
8. During slow cooling, the bilayer will be incubated with a high concentration of vesicles in solution. This may result in lots of vesicles absorbed to the bilayer surface, which cannot be washed off. If this is the case, first form the bilayer through vesicle fusion followed by a thorough rinsing. Then, heat the whole sample in the oven to the desired temperature and slowly cool to room temperature as described in **Subheading 3.1.3.3.** Using this alternative method ensures that no vesicles will be in solution during cooling to be absorbed to the bilayer surface. It is worth noting that this may result in defects because of bilayer tearing as the bilayer phase separates.
9. This small drop of water reduces the chance of air bubbles adsorbing to the AFM cantilever when the probe is merged into fluid. This technique also helps the probe break through the air–water interface.
10. The signal may drift at the beginning of the experiment owing to thermal unbalance. Usually, the system will reach thermal equilibrium within 30 min.
11. To image at the lowest possible scanning force, first decrease the setpoint until the probe loses contact with sample and then increase the set point slowly until the probe is barely in contact with the sample surface.
12. Larger scanning forces may be used to remove or dig out a region of the bilayer; this is generally used to determine the existence of a supported lipid bilayer and bilayer thicknesses. The force that is needed to remove a bilayer is generally more than 20 nN. An AFM probe with a higher spring constant may be needed.
13. The ratio between buffer and water depends on the type of buffer. In general, the total salt concentration has to be below 50 mM to successfully form a supported lipid bilayer on a mica substrate. For example, a PBS:water ratio of 1:2 has been used to make vesicle suspensions.

Acknowledgments

Marjorie L. Longo (MLL) acknowledges funding by the Center on Polymer Interfaces and Macromolecular Assemblies (NSF DMR-0213618), the Nanoscale Interdisciplinary Research Teams Program of the National Science Foundation (NSF CHE-0210807 and BES-0506602), and a generous endowment from Joe and Essie Smith. CDB acknowledges funding from the National Institutes of Health Biotechnology Training Grant of the University of California, Davis, and the Student-Employee Graduate Research Fellowship (SEGRF) Program of Lawrence Livermore National Laboratory. This work was performed under the auspices of the US Dept. of Energy by the University of California/Lawrence Livermore National Laboratory under Contract No. W-7405-Eng-48.

References

1. Brown, D. A. and London, E. (2000) Structure and function of sphingolipid- and cholesterol-rich membrane rafts [Review]. *J. Biol. Chem.* **275,** 17,221–17,224.
2. Devaux, P. F. and Morris, R. (2004) Transmembrane asymmetry and lateral domains in biological membranes. *Traffic* **5,** 241–246.
3. Jacobs, R. E., Hudson, B. S., and Andersen, H. C. (1977) Theory of Phase-Transitions and Phase-Diagrams for One-Component and 2-Component Phospholipid Bilayers. *Biochemistry* **16,** 4349–4359.
4. Mouritsen, O. G. and Bloom, M. (1984) Mattress Model of Lipid-Protein Interactions in Membranes. *Biophys. J.* **46,** 141–153.
5. Mabrey, S. and Sturtevant, J. M. (1976) Investigation of Phase-Transitions of Lipids and Lipid Mixtures by High Sensitivity Differential Scanning Calorimetry. *Proc. Natl. Acad. Sci. USA* **73,** 3862–3866.
6. Wu, S. H. W. and Mcconnell, H. M. (1975) Phase Separations in Phospholipid Membranes. *Biochemistry* **14,** 847–854.
7. Bagatolli, L. A. and Gratton, E. (2000) A correlation between lipid domain shape and binary phospholipid mixture composition in free standing bilayers: A two-photon fluorescence microscopy study. *Biophys. J.* **79,** 434–447.
8. de Almeida, R. F. M., Loura, L. M. S., Fedorov, A., and Prieto, M. (2002) Nonequilibrium phenomena in the phase separation of a two-component lipid bilayer. *Biophys. J.* **82,** 823–834.
9. Tokumasu, F., Jin, A. J., Feigenson, G. W., and Dvorak, J. A. (2003) Nanoscopic lipid domain dynamics revealed by atomic force microscopy. *Biophys. J.* **84,** 2609–2618.
10. Veatch, S. L. and Keller, S. L. (2003) Separation of liquid phases in giant vesicles of ternary mixtures of phospholipids and cholesterol. *Biophys. J.* **85,** 3074–3083.
11. Giocondi, M. C., Vie, V., Lesniewska, E., Milhiet, P. E., Zinke-Allmang, M., and Le Grimellec, C. (2001) Phase topology and growth of single domains in lipid bilayers. *Langmuir* **17,** 1653–1659.
12. Reviakine, I., Simon, A., and Brisson, A. (2000) Effect of Ca^{2+} on the morphology of mixed DPPC-DOPS supported phospholipid bilayers. *Langmuir* **16,** 1473–1477.
13. Ratto, T. V. and Longo, M. L. (2002) Obstructed diffusion in phase-separated supported lipid bilayers: A combined atomic force microscopy and fluorescence recovery after photobleaching approach. *Biophys. J.* **83,** 3380–3392.
14. Kaasgaard, T., Mouritsen, O. G., and Jorgensen, K. (2002) Lipid domain formation and ligand-receptor distribution in lipid bilayer membranes investigated by atomic force microscopy. *FEBS Lett.* **515,** 29–34.
15. Hussain, M. A., Agnihotri, A., and Siedlecki, C. A. (2005) AFM imaging of ligand binding to platelet integrin alpha(IIb)beta(3) receptors reconstituted into planar lipid bilayers. *Langmuir* **21,** 6979–6986.
16. Watts, T. H., Brian, A. A., Kappler, J. W., Marrack, P., and Mcconnell, H. M. (1984) Antigen Presentation by Supported Planar Membranes Containing Affinity-Purified I-Ad. *Proc. Natl. Acad. Sci. USA* **81,** 7564–7568.
17. Mcconnell, H. M., Watts, T. H., Weis, R. M., and Brian, A. A. (1986) Supported Planar Membranes in Studies of Cell-Cell Recognition in the Immune-System. *Biochim. Biophys. Acta* **864,** 95–106.

18. Reviakine, I. and Brisson, A. (2000) Formation of supported phospholipid bilayers from unilamellar vesicles investigated by atomic force microscopy. *Langmuir* **16,** 1806–1815.
19. Keller, C. A. and Kasemo, B. (1998) Surface Specific Kinetics of Lipid Vesicle Adsorption Measured with a Quartz Crystal Microbalance. *Biophys. J.* **75,** 1397–1402.
20. Richter, R. P. and Brisson, A. R. (2005) Following the formation of supported lipid bilayers on mica: A study combining AFM, QCM-D, and ellipsometry. *Biophys. J.* **88,** 3422–3433.
21. Lin, W. -C., Blanchette, C. D., Ratto, T. V., and Longo, M. L. (2006) Lipid asymmetry in DLPC/DAPC supported lipid bilayers, a combined AFM and fluorescence microscopy study. *Biophys. J.* **90,** 228–237.

35

Nuclear Magnetic Resonance Structural Studies of Membrane Proteins in Micelles and Bilayers

Xiao-Min Gong, Carla M. Franzin, Khang Thai, Jinghua Yu, and Francesca M. Marassi

Summary

Nuclear magnetic resonance (NMR) spectroscopy enables determination of membrane protein structures in lipid environments, such as micelles and bilayers. This chapter outlines the steps for membrane-protein structure determination using solution NMR with micelle samples, and solid-state NMR with oriented lipid-bilayer samples. The methods for protein expression and purification, sample preparation, and NMR experiments are described and illustrated with examples from γ and CHIF, two membrane proteins that function as regulatory subunits of the Na^+- and K^+-ATPase.

Key Words: Bilayer membrane; expression; FXYD; lipid; micelle; NMR; protein; structure.

1. Introduction

Integral membrane proteins constitute approx 30% of all expressed genes, and are the major regulators of the most basic cellular functions as well as the major targets for drug discovery initiatives. Despite membrane protein prevalence and importance, the Protein Data Bank (www.rcsb.org/pdb) contains only hundreds of membrane-protein structures, compared with the tens of thousands deposited for globular proteins to-date. This disparity is because of the lipophilic character of membrane proteins, which makes them difficult to overexpress and purify, and complicates their crystallization for X-ray analysis. The examples of membrane proteins whose structures have been determined with atomic resolution are exceptional, and highlight the importance of developing new methods for experimental structure determination.

Because the physical interactions of the lipid bilayer with membrane proteins are more important in determining protein stability and fold than specific lipid-binding interactions, it is desirable to determine protein structures within the lipid-bilayer environment. Nuclear magnetic resonance (NMR) spectroscopy has the potential to accomplish this goal because it can be applied to molecules in all physical states, including the liquid-crystalline bilayer and the micelle environments provided by the lipids that associate with membrane proteins. Solution NMR methods can be used on samples of proteins in lipid micelles, while solid-state NMR methods can be applied to samples of membrane proteins in lipid bilayers, enabling structures to be determined in a native-like environment. The two approaches are complementary, and can be used in combination as a unified method to membrane-protein structure determination.

High-quality solution NMR spectra can be obtained for some fairly large-membrane proteins in micelles, however, for helical proteins it is very difficult to measure and assign a sufficient number of long-range nuclear overhauser effect (NOE) restraints to determine protein folds. This limitation can be overcome by preparing weakly aligned micelle samples for the measurement of residual dipolar couplings (RDCs) and residual chemical shift anisotropies. High-resolution solid-state NMR spectra can be obtained for membrane proteins that are expressed, isotopically labeled, and reconstituted in uniaxially oriented planar lipid bilayers. The spectra have characteristic resonance patterns that directly reflect protein structure and topology, and this direct relationship between spectrum and structure provides the basis for methods that enable the simultaneous sequential assignment of resonances and the measurement of orientation restraints for protein structure determination.

Recent developments in sample preparation, recombinant bacterial expression systems for the preparation of isotopically labeled membrane proteins, pulse sequences for high-resolution spectroscopy, and structural indices that guide the structure assembly process, have greatly extended the capabilities of these NMR techniques. Thus, structures of a variety of membrane proteins have been determined by NMR in both micelles and bilayers *(1–9)*. In this chapter, the methods are illustrated with examples from γ (FXYD2) and CHIF (FXYD4, channel-inducing factor, corticosteroid hormone-induced factor), two homologous membrane proteins that function as regulatory subunits of the Na^+-, K^+-ATPase, the primary enzyme responsible for maintaining the distribution of Na^+ and K^+ concentrations across animal cell membranes *(10–12)*.

FXYD2 and CHIF belong to the FXYD family of Na^+-, K^+-ATPase regulatory membrane proteins, and are each expressed in distinct, specialized segments of the kidney, with unique expression patterns that help explain the physiological differences in Na^+-, K^+-ATPase activity among the nephron segments *(13–17)*. The FXYD protein sequences are highly conserved through evolution, and characterized by a 35-amino acid FXYD homology domain, which includes the short signature motif of the family (Pro, Phe, X, Tyr, and Asp) and a single transmembrane domain. Conserved basic residues flank the transmembrane domain, the extracellular N-termini are acidic, and the cytoplasmic C-termini are basic. Despite their relatively small sizes ranging from about 60 to 160 amino acids, all FXYD proteins are encoded by genes with six to nine small exons, and NMR has shown that the protein structures reflect the structures of their corresponding genes, suggesting that they were assembled from modules through exon shuffling *(18)*.

2. Materials

The specialized materials used for the experiments described in this chapter, and their sources, are listed in **Table 1**. They include lipids for protein reconstitution, and *Escherichia coli* cells, isotopically labeled salts, sugars, and amino acids, used to produce ^{15}N-, ^{13}C-, and ^{2}H-proteins by bacterial expression. The pBCL plasmids for protein expression **(Fig. 1)** were developed in our laboratory and are available on request. The sources of the free programs (NMRPipe, TALOS, XPLOR-NIH, Sparky, and REDCAT) used to process and analyze the NMR data are listed in **Table 1**.

3. Methods

3.1. Protein Expression and Purification

The FXYD proteins γ and CHIF, are expressed using the pBCL plasmid vector, which we have developed for the large-scale expression of membrane proteins *(19)*. This plasmid directs the expression of a target polypeptide fused to the C-terminus of a mutant form of the antiapoptotic protein Bcl-XL, where the hydrophobic C-terminus has been deleted, to be

Table 1
Specialized Materials Used for the Experiments Described in This Chapter, and Their Sources

Material	Source
Reagents	
DHPC	Avanti Polar Lipids (www.avantilipids.com)
DOPC (di-oleoyl-phosphatidyl-choline)	Avanti Polar Lipids (www.avantilipids.com)
DOPG (di-oleoyl-phosphatidyl-glycerol)	Avanti Polar Lipids (www.avantilipids.com)
LPPG	Avanti Polar Lipids (www.avantilipids.com)
OG	Fluka (www.sigmaaldrich.com)
^{2}H-SDS (^{2}H-sodiumdodecylsulfate)	Cambridge Isotopes laboratories (www.isotope.com)
^{2}H-DPC (^{2}H-dodecyl-phosphocholine)	Cambridge Isotopes laboratories (www.isotope.com)
^{2}H$_2$O	Cambridge Isotopes laboratories (www.isotope.com)
(^{15}NH$_4$)$_2$SO$_4$	Cambridge Isotopes laboratories (www.isotope.com)
^{13}C-glucose	Cambridge Isotopes laboratories (www.isotope.com)
E. coli C41(DE3) cells	Avidis (www.overexpress.com)
FF-S ion-exchange chromatography column	Amersham (www.amershambiosciences.com)
Delta-Pak C4 reverse-phase chromatography column	Waters (www.waters.com)
Programs	
NMRPipe	(spin.niddk.nih.gov/bax/software)
TALOS	(spin.niddk.nih.gov/bax/software)
Sparky	www.cgl.ucsf.edu/home/sparky/
XPLOR-NIH	(nmr.cit.nih.gov)
REDCAT	(tesla.ccrc.uga.edu/software)

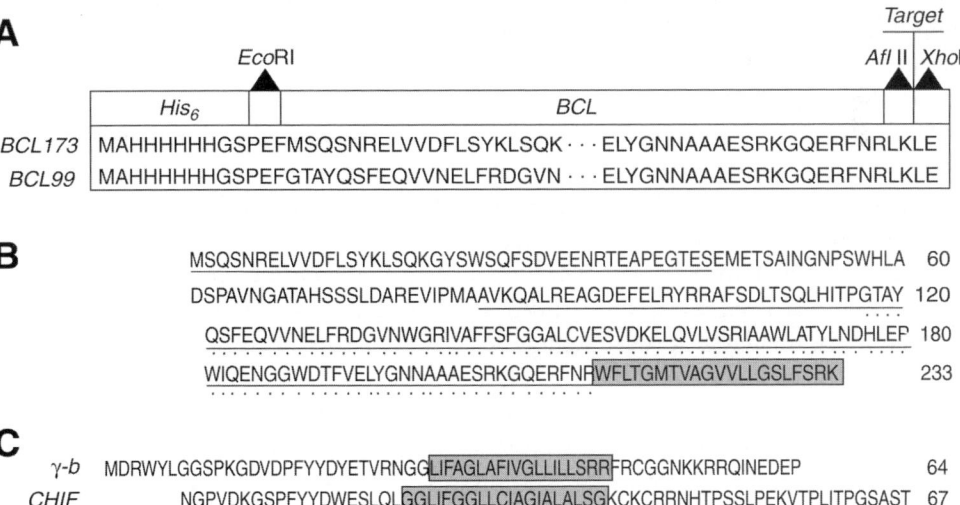

Fig. 1. **(A)** Construction of the pBCL173 and pBCL99 fusion protein expression plasmids. The target sequence with N-terminal Met is inserted between the *Afl* II and *Xho* I cloning sites. BCL173 has a cleavable Met after the His tag, whereas BCL99 does not. **(B)** Amino acid sequence of the Bcl-XL fusion protein. Both the BCL173 and BCL99 fusion proteins lack the Bcl-XL hydrophobic C-terminus (highlighted in the gray box). BCL173 (solid underline) also lacks the flexible loop of Bcl-XL, while BCL99 (dotted underline) lacks the first 116 residues. **(C)** Amino acid sequences of the FXYD proteins γ-b and CHIF. The transmembrane domains are in the gray box.

replaced with a hydrophobic polypeptide gene of interest by insertion at an engineered cloning site (*Afl* II/*Xho* I), and Met residues have been mutated to Leu to facilitate CNBr cleavage after a single Met inserted at the beginning of the target sequence (**Fig. 1A**). In cases where the target protein contains Met residues that cannot be mutated, separation from BCL can be obtained by introducing amino acid sequences specific for cleavage by other chemical means, such as hydroxylamine (Asn-Gly), or for cleavage by one of the commonly used proteases: thrombin, factor Xa, enterokinase, and tobacco-etch virus protease. Chemical cleavage is an attractive option because it eliminates the difficulties—poor specificity and enzyme inactivation— often encountered with protease treatment of hydrophobic proteins in detergents. The plasmid utilizes a T7 expression system *(20)*, and the fusion tag has an N-terminal $(His)_6$ sequence for protein purification by Ni-affinity chromatography.

For the FXYD proteins, the expression levels obtained using pBCL are greater than those obtained using the TrpΔLE (pTLE) *(21–23)*, or the ketosteroid isomerase (pKSI) *(24)* fusion protein expression systems. Thus, we could obtain milligram quantities of pure, isotopically labeled protein easily and quickly. After cleavage from the fusion partner, highly pure FXYD proteins were obtained using a combination of Ni-affinity, size exclusion, and reverse-phase chromatography, with yields in the range of 10 mg of purified protein per liter of culture in M9 minimal medium.

For protein expression, 5–10 µL of transformed C41(DE3) cells from a frozen glycerol stock were used to inoculate 10 mL of LB media, and grown for 5 h at 37°C with vigorous shaking. Then, 1 mL of this starter culture was added to 100 mL of minimal M9 media and grown overnight. All media contained 100 µg/mL of ampicillin. In the morning, 1 L of fresh M9 media was inoculated with the overnight culture, and the cells were grown to a cell density of $OD_{600} = 0.7$. Protein expression was induced by the addition of 1 mM IPTG for 4–5 h at 37°C. The cells were subsequently harvested by centrifugation and stored at −20°C overnight. For uniformly ^{15}N-labeled proteins, $(^{15}NH_4)_2SO_4$ was supplied to the M9 salts as the sole nitrogen source.

Frozen cells from 1 L of culture were lysed by French press in 30 mL of buffer A (50 mM Tris-HCl, pH 8.0, 15% glycerol). The soluble fraction was removed by centrifugation (48,000g, 4°C, 30 min), and the pellet was washed twice by resuspension in 30 mL of buffer A, followed by centrifugation (48,000g, 4°C, 30 min) to remove the soluble fraction. The resulting pellet was dissolved in 30 mL of 6 M GdnHCl, and again centrifuged (48,000g, 4°C, 2 h) to remove any insoluble materials. The 6 M GdnHCl protein solution was adjusted to 0.1 N HCl (pH 0.2), a 100-fold molar excess of solid CNBr was added, and the mixture was allowed to react overnight, in the dark, at room temperature.

CNBr is extremely toxic by inhalation and must be weighed and handled in the fume hood. In the morning, the reaction mixture was dialyzed against water until the pH reached about 5.0 (6 h with several changes of 4 L of water, in a dialysis membrane with a molecular weight cutoff of 1 kD), lyophilized to powder, and dissolved in buffer B (20 mM Tris-HCl, pH 7.0, 8 M urea). Proteins were purified by ion-exchange chromatography with a NaCl gradient (FF-S column), or by size-exclusion chromatography in buffer C (20 mM Tris-HCl, pH 7.0, 4 mM sodium dodecyl-sulfate [SDS]), followed by preparative reverse-phase high-pressure liquid chromatography (C4 column) with a gradient of acetonitrile in water and 0.1% trifluoroacetic acid. Purified proteins were stored as lyophilized powder at −20°C. Alternatively, the lyophilized cleavage mixture could be dissolved directly in buffer C, and the protein purified with reverse-phase high-pressure liquid chromatography.

3.2. Structural Studies in Micelles

3.2.1. Solution NMR Experiments

The NMR experiments were performed on a Bruker AVANCE 600 MHz spectrometer (Billerica, MA) using a triple-resonance ^{1}H/^{13}C/^{15}N-probe equipped with three-axis pulsed field gradients (www.bruker-biospin.com). All NMR experiments were performed at 40°C using a 1-s recycle delay. The chemical shifts were referenced to the ^{1}H$_2$O resonance, set to its expected position of 4.5999 ppm at 40°C *(25)*. The NMR data were processed using NMRPipe *(26)*, and the spectra were assigned and analyzed using Sparky *(27)*. These programs are free and available for a variety of platforms.

The standard fast heteronuclear single-quantum correlation (fHSQC) experiment was used for isotropic samples with 1024 points in $t2$ and 256 in $t1$ *(28)*. Backbone resonance assignments were made using a standard HNCA experiment with constant time evolution for ^{15}N, and solvent suppression was accomplished with a water flip-back pulse after the original ^{1}H–^{15}N magnetization transfer *(29–31)*. The spectra from selectively ^{15}N-labeled protein samples were necessary to resolve assignment ambiguities because of the extensive overlap among the Cα-resonances that is typical of helical membrane proteins in micellees. ^{1}H–^{15}N heteronuclear NOE measurements were made using difference experiments with and without 3 s of saturation of the ^{1}H resonances between scans *(32)*.

The ^{1}H–^{15}N RDCs were measured using a sensitivity-enhanced ^{1}H–^{15}N IPAP (in-phase–antiphase) experiment modified for suppression of the NH$_2$ signals from the acrylamide in the gel *(33–35)*. The contribution to the RDC splitting from the isotropic scalar coupling was determined by performing the same experiment on an isotropic micelle sample, and subtracting the value of the isotropic *J*-coupling obtained from that measured for the weakly aligned gel sample.

3.2.2. Choosing the Right Detergent

The first step in our structural studies of the FXYD family proteins was to find a detergent that would be suitable for high-resolution NMR spectroscopy. We examined the ^{1}H/^{15}N HSQC spectra of the proteins in the best-characterized micelle-forming detergents: dodecylphosphocholine (DPC), diheptanoyl-phosphocholine (DHPC), lyso-palmitoyl-phosphoglycerol (LPPG), octylglucopyranoside (OG), and SDS, at various conditions (protein, detergent, salt concentration, pH, and temperature) **(Fig. 2)**. The highest quality spectra were obtained in SDS micelles, at 40°C **(Fig. 2E,F)**.

Although SDS is widely assumed to be an universal protein-denaturing detergent because of its common use in protein electrophoresis, many hydrophobic membrane proteins actually retain their structures in SDS micelles *(36)*. Furthermore, since all functional studies of Na$^+$, K$^+$-ATPase have been done on enzymes purified in the presence of SDS, and the noncovalent associations of the α, β and FXYD subunits are maintained through the SDS purification process *(37–39)*, we reasoned that this detergent would also be a good choice for FXYD structural studies.

3.2.3. Protein Structure Determination

The HSQC spectra show that γ and CHIF adopt unique folded structures in SDS. The 1.5 ppm dispersion of the amide ^{1}H chemical shifts is typical of native helical membrane proteins in micelles, and the HSQC spectra obtained in D$_2$O revealed a core region in each protein, with

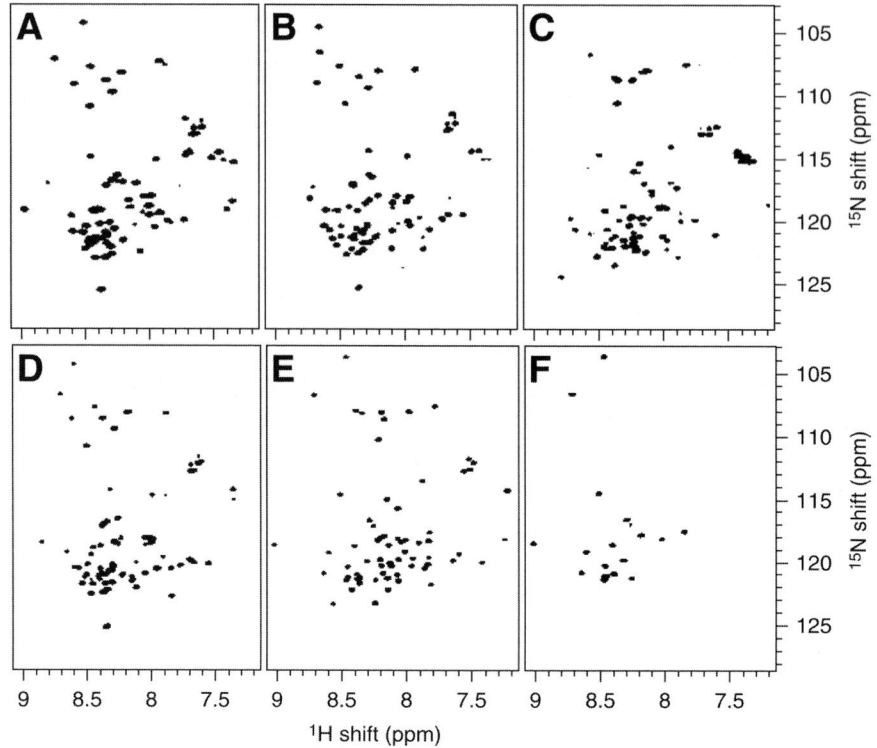

Fig. 2. 2D ^{1}H/^{15}N HSQC spectra of uniformly ^{15}N-labeled γ-b in detergent micelles. The samples were prepared by dissolving each protein in buffer (20 mM sodium citrate, pH 5.0, 10 mM DTT, and 1 mM sodium azide, in 90% H$_2$O, 10% D$_2$O) plus detergent. **(A)** 500 mM DPC; **(B)** 180 mM DHPC; **(C)** 100 mM LPPG; **(D)** 200 mM OG; and **(E)** 500 mM SDS. In **(F)** the spectrum of γ-b in 500 mM SDS was acquired 3 h after the addition of 100% D$_2$O.

amide protons that exchange very slowly with the surrounding aqueous solvent (**Figs. 2F** and **3B**). These regions match the hydrophobic transmembrane segments of the proteins identified by the hydropathy plots (**Fig. 3A**), and reflect the strong intramolecular hydrogen bonds present in transmembrane helices in the low dielectric environment of the membrane, or, in this case, the micelle interior. Hydrogen exchange experiments were performed by dissolving the lyophilized protein in SDS buffer with 100% D$_2$O, and then acquiring HSQC spectra at 0.5-h time intervals.

To determine the protein secondary structures we relied primarily on ^{1}H–^{15}N RDCs, measured from weakly aligned samples and analyzed in terms of dipolar waves *(40)* (**Fig. 3C**). These were supplemented with chemical shifts, analyzed in terms of chemical shift indices *(41)* (**Fig. 3D**), and with the program TALOS *(42)*. For the measurement of RDCs, the protein-SDS micelles were weakly aligned in 7% polyacrylamide gels, using either vertical compression *(34,43)* or expansion *(44)*. These samples were prepared by soaking the protein micelle solution into a column of acrylamide that had been cast inside an NMR tube, and then rinsed and dried. After soaking, the acrylamide column was inserted inside an NMR tube with smaller diameter for stretched gels, as described by Bax and coworkers *(44)*. The apparatus for

NMR of Proteins in Micelles and Bilayers

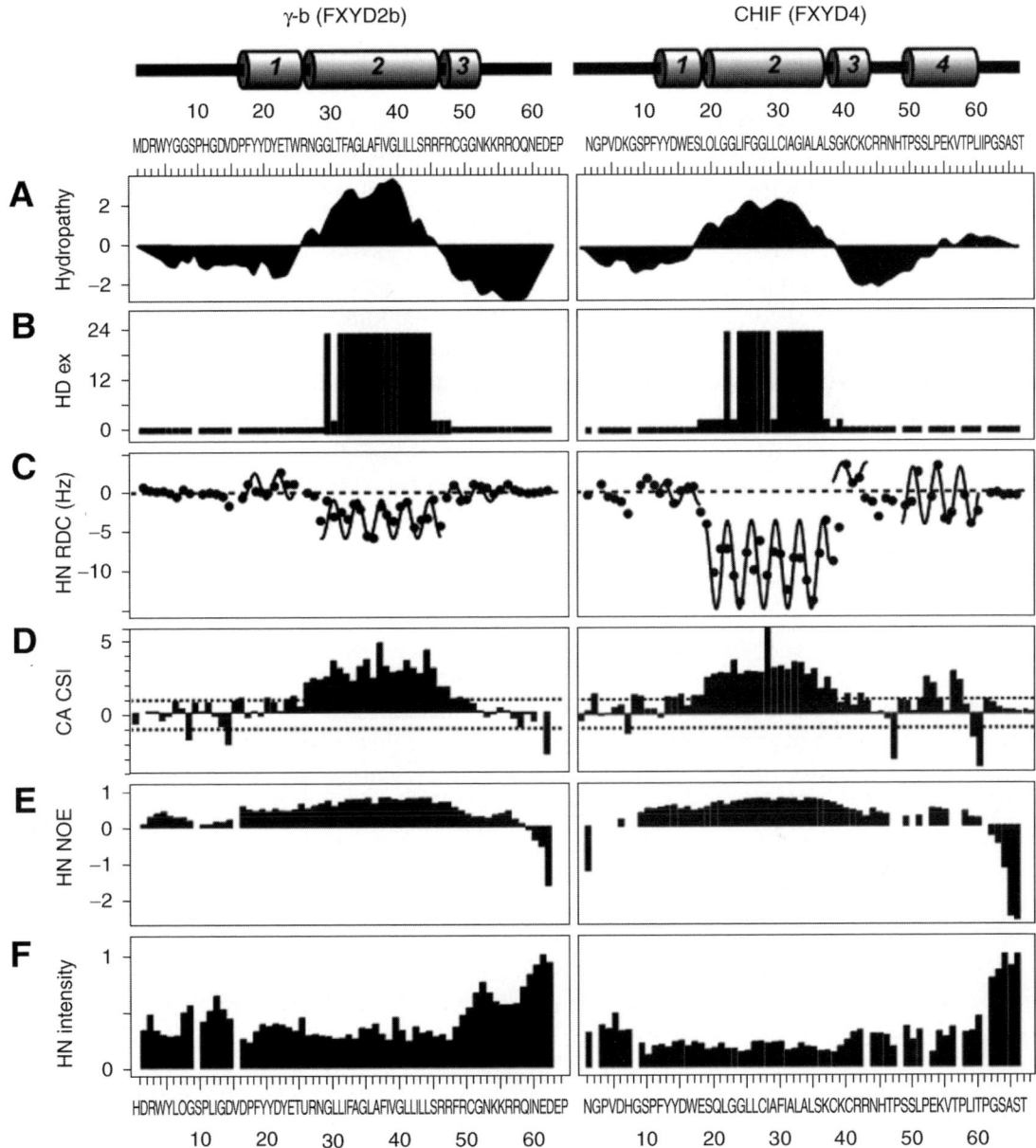

Fig. 3. Summary of NMR parameters for γ-b and CHIF, plotted as a function of residue number. The protein secondary structures are shown at the top of the figure (helices 1–4 are numbered). The residue numbers for the three proteins begin after the signal sequences. **(A)** Kyte-Doolittle hydropathy plot *(60)*; **(B)** amide hydrogen/deuterium exchange profiles, with exchange rates classified as rapid (<1 h), short (<3 h), medium (<12 h), or long (>24 h); **(C)** Values of ^{1}H–^{15}N RDCs with the characteristic periodicity in the α-helical regions of the proteins fitted to sinusoids; **(D)** ^{13}Cα-chemical shift index; **(E)** ^{1}H/^{15}N heteronuclear NOEs; and **(F)** Normalized ^{1}H/^{15}N HSQC peak intensities. Positions that are left blank correspond to prolines or overlapped resonances.

preparing stretched gels is available from New Era Enterprises (www.newera-spectro.com). Alternatively, for compressed gels, the soaked acrylamide was inserted in a NMR tube with larger diameter and compressed with a glass plunger (www.shigeminmr.com).

The RDCs were analyzed using MATLAB scripts as described in **refs. 40, 45,** and **46**. Helical regions were identified by applying a sliding window algorithm to fit the experimental RDCs. The RDCs within a five-residue window were fit to a sinusoid of periodicity 3.6, and the RMSD between the sinusoid and the data were plotted as a function of residue number. Continuous stretches of amino acids with low RMSD (less than the experimental error of 1.5 Hz) were identified as helices and fitted to a single sinusoid. Higher RMSDs were generally interpreted as deviations from ideality, including kinks, curvature, and loops. This analysis relating the orientation of the helix to the amplitude, average value, and phase of the sinusoid is an initial step toward structure determination, and can determine the relative orientations of helices to within four degenerate solutions *(45,47)*.

The protein backbone dynamics are characterized with measurements of the ^1H–^{15}N heteronuclear NOE, and resonance intensities. Within the core-helical regions of the three proteins, all residues have similar positive values of ^1H–^{15}N NOE, reflecting similar rotational correlation times, and indicating that the three helices are rigidly connected (**Fig. 3E**). Lower negative values of the ^1H–^{15}N NOE, reflecting additional backbone motions, are present in residues near the N- and C-termini, and at the helix boundaries that are also marked by exon junctions. Furthermore, all of the resonances from amino acids within the central helical regions of the three proteins have similar peak intensities that plateau at minimum values (**Fig. 3F**), indicating that the helices are rigidly connected. In contrast, residues at the core module boundaries and at the terminal regions of the proteins have greater intensities, reflecting narrower linewidths that result from the increased dynamics in these regions.

An important finding of these structural studies is that the helical secondary structures of the FXYD family proteins reflect the structures of their corresponding genes *(18)*. The coincidence of intron–exon junctions with helical structures and flexible connecting segments, support the hypothesis that the FXYD proteins may have been assembled from discrete structural modules through exon shuffling. Despite their relatively small sizes (60–160 residues), the FXYD family proteins are all encoded by genes with six to nine small exons, and this has been previously suggested to reflect modular gene assembly *(10)*. The presence of conserved modules in different FXYD family members suggests that they are derived from a common ancestor gene, from which FXYD5 appears to have been the first to diverge *(12)*. The multiple exon organization of the FXYD genes could serve to confer high structural and functional diversity among the family members.

3.4. Structural Studies in Lipid Bilayers

3.4.1. Solid-State NMR Experiments

The solid-state NMR spectra of membrane proteins in oriented lipid bilayers have frequencies that reflect the orientation of their respective sites relative to the direction of the magnetic field. Since the lipid-bilayer plane is perpendicular to the magnetic field direction, each resonance frequency reflects the orientation of its corresponding protein site in the membrane *(48)*. The spectra were obtained at 23°C on a Bruker AVANCE 500 (Billerica) spectrometer with a wide-bore 500/89 Magnex magnet (Yarnton, UK). The double-resonance ($^1H/^{15}N$ or $^1H/^{31}P$) probes had square radiofrequency (rf) coils wrapped directly around the samples. The NMR data were processed using the programs NMRPipe *(26)*. The ^{15}N spectra were obtained with single contact 1-ms cross polarization with mismatch-optimized IS polarization transfer (CPMOIST) *(49,50)*, and the ^{31}P spectra with a single pulse. Both types of spectra were acquired with continuous ^{1}H irradiation (rf field strength 63 kHz) to decouple the ^{1}H–^{15}N and ^{1}H–^{31}P dipolar interactions. The ^{15}N and ^{31}P chemical shifts were referenced to 0 ppm for liquid ammonia and phosphoric acid.

The polarization inversion with exchange at the magic angle (PISEMA) experiment *(51)* gives high-resolution, two-dimensional (2D), ^{1}H–^{15}N dipolar coupling and ^{15}N chemical shift correlation spectra of oriented membrane proteins where the individual resonances contain orientation restraints for structure determination *(4,52)*. The spectra were obtained with a cross polarization contact time of 1 ms, a ^{1}H 90° pulse width of 5 µs, and continuous ^{1}H decoupling of 63 kHz rf field strength. The 2D data were acquired with 512 accumulated transients and 256 complex data points, for each of 64 real $t1$ values incremented by 32.7 µs. The recycle delay was 6 s.

3.4.2. Sample Preparation

The samples were prepared by first dissolving 2 mg of ^{15}N-labeled FXYD protein in 0.5 mL of trifluoroethanol with 50 µL of β-mercaptoethanol, and then adding 100 mg of lipid, di-oleoyl-phosphatidyl-choline/di-oleoyl-phosphatidyl-glycerol (8/2 molar ratio), in 1 mL of chloroform. After spreading this solution on the surface of 35 glass slides (dimensions 11 × 20 × 0.06 mm³) (Paul Marienfeld GmbH, Germany), the solvents were removed under vacuum overnight, and the slides were stacked. Oriented-lipid bilayers were formed by equilibrating the stacked slides for 24 h, at 40°C, in a chamber containing a saturated solution of ammonium phosphate, which provides an atmosphere of 93% relative humidity. The samples were wrapped in parafilm and then sealed in thin polyethylene film before insertion in the NMR probe. Hydrogen exchanged samples were prepared by exposing the stacked-oriented bilayer samples to an atmosphere saturated with ^{2}H$_2$O. This was achieved by placing the sample in a closed chamber containing ^{2}H$_2$O and incubating at 40°C for 24 h. Protein purity is crucial for obtaining highly oriented lipid-bilayer samples that give high-resolution NMR spectra.

3.4.3. 1D Solid-State ^{15}N NMR Spectra

The spectrum of ^{15}N-labeled CHIF in oriented lipid bilayers (**Fig. 4B**) reflects the presence of transmembrane and C-terminal helical segments similar to those found in micelles *(53)*. It displays significant resolution throughout the frequency range of the ^{15}N amide chemical shift. The resonance intensity near 200 ppm is from backbone-amide sites in the transmembrane helix, with the amide-NH bonds being located nearly perpendicular to the plane of the membrane. On the other hand, the intensity near 80 ppm is from sites in the N- and C-termini of the protein, with the NH bonds being located nearly parallel to the membrane surface. The peak near 35 ppm is from the amino groups of the lysine sidechains and the N-terminus. The narrow chemical shift dispersion in the frequency range near 200 ppm associated with transmembrane helices indicates that the protein crosses the membrane with only a very small tilt angle, which is estimated to be around 15° from the 2D ^{1}H/^{15}N PISEMA spectrum (**Fig. 5**).

The spectrum of the oriented sample is strikingly different from that of the unoriented sample, which provides no resolution among resonances (**Fig. 4A**). Most of the backbone sites are structured and immobile on the time-scale of the ^{15}N chemical shift interaction (10 kHz), contributing to the characteristic amide powder pattern between about 220 and 60 ppm. Some of the backbone sites, in the loop and terminal regions, are mobile, giving rise to the

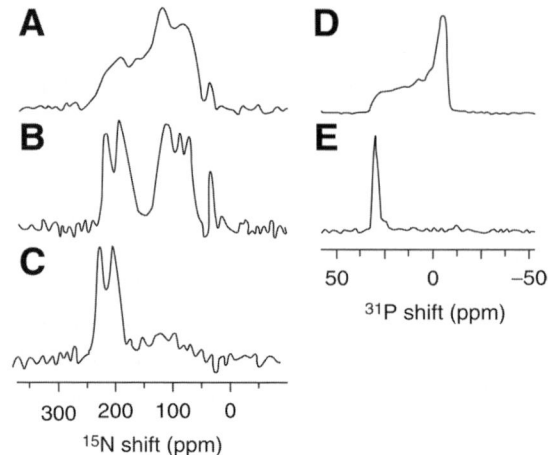

Fig. 4. Solid-state NMR ^{15}N and ^{31}P chemical shift spectra of uniformly ^{15}N-labeled CHIF in unoriented (**A,D**) and oriented (**B,C,E**) lipid bilayers. Resonances near 200 ppm are from amino acid residues in the CHIF transmembrane helix; (**C**) amide hydrogens in the transmembrane helix of CHIF are resistant to hydrogen exchange, and their resonances remain after exposure to D$_2$O while the resonances from exchangeable hydrogens disappear. The ^{15}N and ^{31}P chemical shifts are referenced to 0 ppm for liquid ammonia and phosphoric acid.

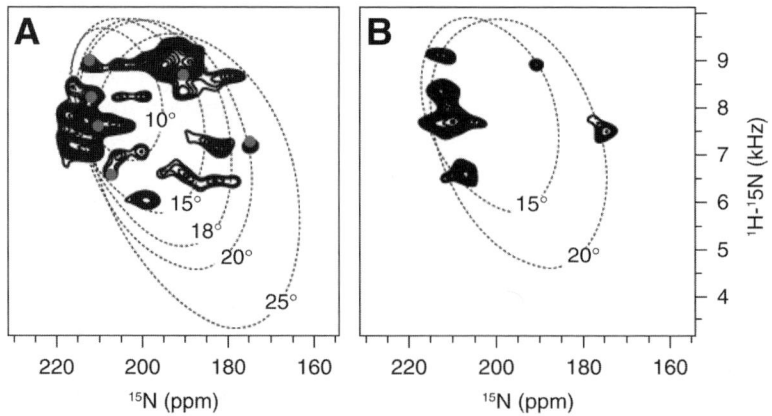

Fig. 5. ^{1}H/^{15}N solid-state NMR PISEMA spectra of uniformly ^{15}N-labeled (**A**) and Leu ^{15}N-labeled (**B**) CHIF in oriented lipid bilayers. In (**A**) the spectrum is superimposed on the Pisa wheels calculated for ideal α-helices with different tilts in the lipid bilayer, and on the spectrum of ^{15}N-Leu labeled CHIF, which is represented as red dots. ^{15}N chemical shifts are referenced to 0 ppm for liquid ammonia.

resonance band centered near 120 ppm (the isotropic frequency). Some resonances near 120 ppm, however, may reflect specific orientations of their corresponding sites.

Amide hydrogen exchange rates are useful for identifying residues that are involved in hydrogen-bonding and exposed to water. Although lipid bilayers are permeable to water and other small polar molecules, the amide hydrogens in transmembrane helices can have very slow exchange rates because of strong hydrogen bonds in the low dielectric of the lipid-bilayer environment, and their ^{15}N chemical shift NMR signals persist for days after exposure to D$_2$O. Faster exchange rates are observed for transmembrane helices that are not tightly hydrogen-bonded and are exposed to bulk water because they participate in channel pore formation *(54)*. Faster exchange rates are also observed for other water-exposed helical regions of proteins with weaker hydrogen-bonded networks. Many of the amide hydrogens exchange when the CHIF sample is exposed to D_2O. Since the signals are generated by cross polarization from ^1H, the amide resonances, thus, disappear from the spectrum (**Fig. 4C**). However, the amide hydrogens in the transmembrane helix did not exchange, and their signals persisted in the spectrum, indicating that the CHIF transmembrane helix forms a tight hydrogen bonding network that is resistant to hydrogen exchange.

3.4.4. ^{31}P NMR Spectra of the Membrane Lipids

The phospholipid phase and the degree of phospholipid-bilayer alignment can be assessed with ^{31}P NMR spectroscopy of the lipid phosphate headgroup. The ^{31}P NMR spectrum obtained for unoriented bilayer vesicles containing CHIF (**Fig. 4D**) is characteristic of lipids in a bilayer arrangement, while the spectrum for oriented lipids with CHIF has a single resonance near 30 ppm that is characteristic of oriented lipid-bilayer membranes (**Fig. 4E**). The presence of a single peak demonstrates that the samples are highly oriented, as required for NMR structure determination. Thus, taken together, the ^{15}N and ^{31}P spectra provide evidence that the FXYD proteins insert in membranes without disruption of the membrane structure.

3.4.5. 2D Solid-State $^1H/^{15}N$ Correlation Spectra: PISEMA

The PISEMA spectra of membrane proteins in oriented lipid bilayers provide sensitive indices of protein secondary structure and topology because they exhibit characteristic wheel-like patterns of resonances, called Pisa wheels, that reflect helical wheel projections of residues in both α-helices and β-sheets *(55–57)*. When a Pisa wheel is observed, no assignments are needed to determine the tilt of a helix, and a single resonance assignment can be sufficient to determine the helix rotation in the membrane. This information is extremely useful for determining the supramolecular architectures of membrane proteins and their assemblies. The shape and position of the Pisa wheel in the spectrum depends on the protein secondary structure and its orientation relative to the lipid-bilayer surface, as well as the amide N–H bond length and the magnitudes and orientations of the principal elements of the amide ^{15}N chemical shift tensor. This direct relationship between spectrum and structure makes it possible to calculate solid-state NMR spectra for specific structural models of proteins, and provides the basis for a method for backbone structure determination from a limited set of uniformly and selectively ^{15}N-labeled samples *(4,58)*.

The 2D $^1H/^{15}N$ PISEMA spectra of uniformly and selectively Leu ^{15}N-labeled CHIF in lipid bilayers are shown in **Fig. 5**. The Pisa wheel that is observed in the region from 6 to 10 kHz and 180 to 220 ppm in the spectrum, provides definitive evidence that the protein associates with the lipid bilayer as a transmembrane helix. To estimate the tilt of the CHIF transmembrane helix we compared the experimental spectrum with those calculated for an ideal α-helix, with 3.6 residues per turn and identical backbone dihedral angles for all residues (ϕ, φ = 57°, –47°), tilted at 10°, 15°, and 20° relative to the lipid-bilayer normal. This comparative analysis demonstrates that the CHIF helix is tilted by about 15° in the membrane (or 75° from the membrane surface). According to the solution NMR data in micelles, the peaks in the spectrum of ^{15}N-Leu-labeled CHIF should account for Leu 17 and 19 in helix 1 preceding the transmembrane helix (helix 2), and Leu 22, 27, 28, 35, and 37, in the transmembrane helix. The data suggest that the peaks in the PISEMA spectrum will have to be

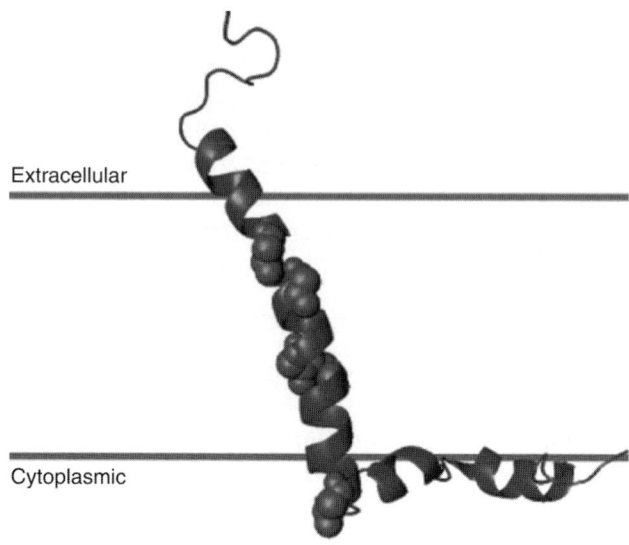

Fig. 6. Structure of CHIF determined by combining the NMR restraints obtained in micelles with those obtained in lipid bilayers. Leucine residues are shown as spheres.

fitted with Pisa wheels of different tilts, in agreement with the results obtained in micelles showing that the CHIF helices 1 and 2 have different orientations.

4. Conclusions

The structure of CHIF (**Fig. 6**) was determined by combining the restraints from measurements of RDCs, chemical shift, and H/D exchange, obtained by solution NMR in micelles, with the 15° tilt of the transmembrane helix obtained from solid-state NMR experiments in oriented lipid bilayers. Protein structures were calculated from the experimental data using a basic simulated annealing protocol in the program X-PLOR-NIH *(9,59)*. Because helices 1, 2, 3, and 4 are rather rigidly connected, their relative orientations could be obtained from the combined measurements of RDCs and chemical shifts, and their analysis using the programs REDCAT *(47)* and TALOS *(42)*. Additional measurement of RDCs from a sample with a different alignment would yield the helix orientations unambiguously.

The combination of solution NMR with lipid micelle samples, and solid-state NMR with lipid-bilayer samples, is a powerful approach for determining the structure of a membrane protein in an environment that closely resembles the biological membrane. New and ongoing developments in protein-expression systems, methods for reconstitution, NMR experiments and equipment, and computational methods, are bringing ever more complex membrane proteins into the range of molecules whose structures can be determined by NMR.

Acknowledgments

This research was supported by grants from the National Institutes of Health (CA082864, GM065374). The NMR studies utilized the Burnham Institute NMR Facility and the University of California San Diego (UCSD) Resource for Molecular Imaging of Proteins, each supported by grants from the National Institutes of Health (P30CA030199, P41EB002031).

References

1. Ketchem, R. R., Hu, W., and Cross, T. A. (1993) High-resolution conformation of gramicidin A in a lipid bilayer by solid-state NMR. *Science* **261,** 1457–1460.
2. Opella, S. J., Marassi, F. M., Gesell, J. J., et al. (1999) Structures of the M2 channel-lining segments from nicotinic acetylcholine and NMDA receptors by NMR spectroscopy. *Nat. Struct. Biol.* **6,** 374–379.
3. Wang, J., Kim, S., Kovacs, F., and Cross, T. A. (2001) Structure of the transmembrane region of the M2 protein H(+) channel. *Protein Sci.* **10,** 2241–2250.
4. Marassi, F. M. and Opella, S. J. (2003) Simultaneous assignment and structure determination of a membrane protein from NMR orientational restraints. *Protein Sci.* **12,** 403–411.
5. Park, S. H., Mrse, A. A., Nevzorov, A. A., et al. (2003) Three-dimensional structure of the channel-forming trans-membrane domain of virus protein "u" (Vpu) from HIV-1. *J. Mol. Biol.* **333,** 409–424.
6. Oxenoid, K. and Chou, J. J. (2005) The structure of phospholamban pentamer reveals a channel-like architecture in membranes. *Proc. Natl. Acad. Sci. USA* **102,** 10,870–10,875.
7. Zamoon, J., Mascioni, A., Thomas, D. D., and Veglia, G. (2003) NMR solution structure and topological orientation of monomeric phospholamban in dodecylphosphocholine micelles. *Biophys.J.* **85,** 2589–2598.
8. Sorgen, P. L., Cahill, S. M., Krueger-Koplin, R. D., Krueger-Koplin, S. T., Schenck, C. C., and Girvin, M. E. (2002) Structure of the Rhodobacter sphaeroides light-harvesting 1 beta subunit in detergent micelles. *Biochemistry* **41,** 31–41.
9. Howell, S. C., Mesleh, M. F., and Opella, S. J. (2005) NMR structure determination of a membrane protein with two transmembrane helices in micelles: MerF of the bacterial mercury detoxification system. *Biochemistry* **44,** 5196–5206.
10. Sweadner, K. J. and Rael, E. (2000) The FXYD gene family of small ion transport regulators or channels: cDNA sequence, protein signature sequence, and expression. *Genomics* **68,** 41–56.
11. Crambert, G. and Geering, K. (2003) FXYD proteins: new tissue-specific regulators of the ubiquitous Na,K-ATPase. *Sci. STKE* **2003,** RE1.
12. Garty, H. and Karlish, S. J. (2006) Role of FXYD Proteins in Ion Transport. *Annu. Rev. Physiol.* **68,** 431–459.
13. Mercer, R. W., Biemesderfer, D., Bliss, D. P., Jr., Collins, J. H., and Forbush, B., 3rd. (1993) Molecular cloning and immunological characterization of the gamma polypeptide, a small protein associated with the Na,K-ATPase. *J. Cell Biol.* **121,** 579–586.
14. Attali, B., Latter, H., Rachamim, N., and Garty, H. (1995) A corticosteroid-induced gene expressing an "IsK-like" K+ channel activity in *Xenopus* oocytes. *Proc. Natl. Acad. Sci. USA* **92,** 6092–6096.
15. Arystarkhova, E., Wetzel, R. K., Asinovski, N. K., and Sweadner, K. J. (1999) The gamma subunit modulates Na(+) and K(+) affinity of the renal Na, K-ATPase. *J. Biol. Chem.* **274,** 33,183–33,185.
16. Shi, H., Levy-Holzman, R., Cluzeaud, F., Farman, N., and Garty, H. (2001) Membrane topology and immunolocalization of CHIF in kidney and intestine. *Am. J. Physiol. Renal Physiol.* **280,** F505–F512.
17. Wetzel, R. K. and Sweadner, K. J. (2001) Immunocytochemical localization of Na-K-ATPase alpha- and gamma-subunits in rat kidney. *Am. J. Physiol. Renal Physiol.* **281,** F531–F545.
18. Franzin, C. M., Yu, J., Thai, K., Choi, J., and Marassi, F. M. (2005) Correlation of Gene and Protein Structures in the FXYD Family Proteins. *J. Mol. Biol.* **354,** 743–750.
19. Thai, K., Choi, J., Franzin, C. M., and Marassi, F. M. (2005) Bcl-XL as a fusion protein for the high-level expression of membrane-associated proteins. *Protein Sci.* **14,** 948–955.
20. Studier, F. W., Rosenberg, A. H., Dunn, J. J., and Dubendorff, J. W. (1990) Use of T7 RNA polymerase to direct expression of cloned genes. *Methods Enzymol.* **185,** 60–89.

21. Miozzari, G. F. and Yanofsky, C. (1978) Translation of the leader region of the *Escherichia coli* tryptophan operon. *J. Bacteriol.* **133,** 1457–1466.
22. Kleid, D. G., Yansura, D., Small, B., et al. (1981) Cloned viral protein vaccine for foot-and-mouth disease: responses in cattle and swine. *Science* **214,** 1125–1129.
23. Staley, J. P. and Kim, P. S. (1994) Formation of a native-like subdomain in a partially folded intermediate of bovine pancreatic trypsin inhibitor. *Protein Sci.* **3,** 1822–1832.
24. Kuliopulos, A., Nelson, N. P., Yamada, M., et al. (1994) Localization of the affinity peptide-substrate inactivator site on recombinant vitamin K-dependent carboxylase. *J. Biol. Chem.* **269,** 21,364–21,370.
25. Cavanagh, J. (1996) *Protein NMR spectroscopy: principles and practice*. Academic Press, San Diego.
26. Delaglio, F., Grzesiek, S., Vuister, G. W., Zhu, G., Pfeifer, J., and Bax, A. (1995) NMRPipe: a multidimensional spectral processing system based on UNIX pipes. *J. Biomol. NMR* **6,** 277–293.
27. Goddard, T. D. and Kneller, D. G. Sparky 3, University of California, San Francisco. (2004).
28. Mori, S., Abeygunawardana, C., Johnson, M. O., and Vanzijl, P. C. M. (1995) Improved Sensitivity of HSQC Spectra of Exchanging Protons at Short Interscan Delays Using a New Fast HSQC (FHSQC) Detection Scheme That Avoids Water Saturation. *J. Magn. Reson. B* **108,** 94–98.
29. Ikura, M., Kay, L. E., and Bax, A. (1990) A novel approach for sequential assignment of 1H, 13C, and 15N spectra of proteins: heteronuclear triple-resonance three-dimensional NMR spectroscopy. Application to calmodulin. *Biochemistry* **29,** 4659–4667.
30. Sattler, M., Schleucher, J., and Griesinger, C. (1999) Heteronuclear multidimensional NMR experiments for the structure determination of proteins in solution employing pulsed field gradients. *Prog. Nucl. Magn. Reson. Spectrosc.* **34,** 93–158.
31. Grzesiek, S. and Bax, A. (1992) Correlating backbone amide and side chain resonances in larger proteins by multiple relayed triple resonance NMR. *J. Am. Chem. Soc.* **114,** 6291–6293.
32. Farrow, N. A., Zhang, O., Forman-Kay, J. D., and Kay, L. E. (1994) A heteronuclear correlation experiment for simultaneous determination of 15N longitudinal decay and chemical exchange rates of systems in slow equilibrium. *J. Biomol. NMR* **4,** 727–734.
33. Ottiger, M., Delaglio, F., and Bax, A. (1998) Measurement of J and dipolar couplings from simplified two-dimensional NMR spectra. *J. Magn. Reson.* **131,** 373–378.
34. Ishii, Y., Markus, M. A., and Tycko, R. (2001) Controlling residual dipolar couplings in high-resolution NMR of proteins by strain induced alignment in a gel. *J. Biomol. NMR* **21,** 141–151.
35. Ding, K. and Gronenborn, A. M. (2003) Sensitivity-enhanced 2D IPAP, TROSY-anti-TROSY, and E.COSY experiments: alternatives for measuring dipolar 15N-1HN couplings. *J. Magn. Reson.* **163,** 208–214.
36. Tanford, C. and Reynolds, J. A. (1976) Characterization of membrane proteins in detergent solutions. *Biochim. Biophys. Acta* **457,** 133–170.
37. Jorgensen, P. L. (1988) Purification of Na+, K+-ATPase: enzyme sources, preparative problems, and preparation from mammalian kidney. *Methods Enzymol.* **156,** 29–43.
38. Maunsbach, A. B., Skriver, E., and Jorgensen, P. L. (1988) Analysis of Na+,K+-ATPase by electron microscopy. *Methods Enzymol.* **156,** 430–441.
39. Ivanov, A. V., Gable, M. E., and Askari, A. (2004) Interaction of SDS with Na+/K+-ATPase: SDS-solubilized enzyme retains partial structure and function. *J. Biol. Chem.* **279,** 29,832–29,840.
40. Mesleh, M. F., Lee, S., Veglia, G., Thiriot, D. S., Marassi, F. M., and Opella, S. J. (2003) Dipolar waves map the structure and topology of helices in membrane proteins. *J. Am. Chem. Soc.* **125,** 8928–8935.
41. Wishart, D. S. and Sykes, B. D. (1994) Chemical shifts as a tool for structure determination. *Methods Enzymol.* **239,** 363–392.
42. Cornilescu, G., Delaglio, F., and Bax, A. (1999) Protein backbone angle restraints from searching a database for chemical shift and sequence homology. *J. Biomol. NMR* **13,** 289–302.

43. Sass, H. J., Musco, G., Stahl, S. J., Wingfield, P. T., and Grzesiek, S. (2000) Solution NMR of proteins within polyacrylamide gels: diffusional properties and residual alignment by mechanical stress or embedding of oriented purple membranes. *J. Biomol. NMR* **18,** 303–309.
44. Chou, J. J., Gaemers, S., Howder, B., Louis, J. M., and Bax, A. (2001) A simple apparatus for generating stretched polyacrylamide gels, yielding uniform alignment of proteins and detergent micelles. *J. Biomol. NMR* **21,** 377–382.
45. Mesleh, M. F., Veglia, G., DeSilva, T. M., Marassi, F. M., and Opella, S. J. (2002) Dipolar waves as NMR maps of protein structure. *J. Am. Chem. Soc.* **124,** 4206–4207.
46. Mesleh, M. F. and Opella, S. J. (2003) Dipolar Waves as NMR maps of helices in proteins. *J. Magn. Reson.* **163,** 288–299.
47. Valafar, H. and Prestegard, J. H. (2004) REDCAT: a residual dipolar coupling analysis tool. *J. Magn. Reson.* **167,** 228–241.
48. Marassi, F. M. (2002) NMR of peptides and proteins in membranes. *Concepts Magn. Reson.* **14,** 212–224.
49. Pines, A., Gibby, M. G., and Waugh, J. S. (1973) Proton-enhanced NMR of dilute spins in solids. *J. Chem. Phys.* **59,** 569–590.
50. Levitt, M. H., Suter, D., and Ernst, R. R. (1986) Spin Dynamics and Thermodynamics in Solid-State NMR Cross-Polarization. *J. Chem. Phys.* **84,** 4243–4255.
51. Wu, C. H., Ramamoorthy, A., and Opella, S. J. (1994) High-resolution heteronuclear dipolar solid-state NMR spectroscopy. *J. Magn. Reson. A* **109,** 270–272.
52. Marassi, F. M., Ramamoorthy, A., and Opella, S. J. (1997) Complete resolution of the solid-state NMR spectrum of a uniformly 15N-labeled membrane protein in phospholipid bilayers. *Proc. Natl. Acad. Sci. USA* **94,** 8551–8556.
53. Franzin, C. M., Yu, J., and Marassi, F. M. (2005) Solid-state NMR of the FXYD family membrane proteins in lipid bilayers, in *NMR Spectroscopy of Biological Solids*, (Ramamoorthy A., ed.), Tylor and Francis, CRC Press, Vol. **34,** pp. 191–214.
54. Tian, C., Gao, P. F., Pinto, L. H., Lamb, R. A., and Cross, T. A. (2003) Initial structural and dynamic characterization of the M2 protein transmembrane and amphipathic helices in lipid bilayers. *Protein Sci.* **12,** 2597–2605.
55. Marassi, F. M. and Opella, S. J. (2000) A solid-state NMR index of helical membrane protein structure and topology. *J. Magn. Reson.* **144,** 150–155.
56. Wang, J., Denny, J., Tian, C., et al. (2000) Imaging membrane protein helical wheels. *J. Magn. Reson.* **144,** 162–167.
57. Marassi, F. M. (2001) A simple approach to membrane protein secondary structure and topology based on NMR spectroscopy. *Biophys. J.* **80,** 994–1003.
58. Marassi, F. M. and Opella, S. J. (2002) Using pisa pies to resolve ambiguities in angular constraints from PISEMA spectra of aligned proteins. *J. Biomol. NMR* **23,** 239–242.
59. Schwieters, C. D., Kuszewski, J. J., Tjandra, N., and Marius Clore, G. (2003) The Xplor-NIH NMR molecular structure determination package. *J. Magn. Reson.* **160,** 65–73.
60. Kyte, J. and Doolittle, R. F. (1982) A simple method for displaying the hydropathic character of a protein. *J. Mol. Biol.* **157,** 105–132.

36

Laurdan Studies of Membrane Lipid-Nicotinic Acetylcholine Receptor Protein Interactions

Silvia S. Antollini and Francisco J. Barrantes

Summary

The extrinsic fluorescent probe Laurdan (6-dodecanoyl-2-dimethylamino naphthalene) exhibits extreme sensitivity to the polarity and to the molecular dynamics of the dipoles in its environment. Dipolar relaxation processes are reflected as relatively large spectral shifts. Steady-state measurements of the so-called general polarization (GP) of Laurdan exploit the advantageous spectral properties of Laurdan. Since the main solvent dipoles surrounding Laurdan in biological membranes are water molecules, when no relaxation occurs GP values are high, indicating low water content in the hydrophilic/hydrophobic interface region. Laurdan fluorescence can also be used to obtain topographical information. A hitherto unexploited property of Laurdan, namely its ability to act as a Förster-type resonance energy transfer (FRET) acceptor of tryptophan emission, was used to learn about the physical state of lipids within Förster distance from donor tryptophan residues in integral membrane proteins. The application of this technique to the paradigm integral membrane protein, the nicotinic acetylcholine receptor, is described in this chapter.

Key Words: Cholinergic receptor; fluorescence spectroscopy; Förster resonance energy transfer; generalized polarization; Laurdan; intrinsic fluorescence.

1. Introduction

Despite constituting roughly one third of all gene products, only very few high-resolution structures of membrane proteins have been obtained to date. This is partly due to the inherent difficulty of obtaining three-dimensional crystals of sufficient quality. Drawing structural–functional correlations for many important membrane proteins has thus relied on studies in solution employing various spectroscopic techniques (e.g., electron spin resonance, nuclear magnetic resonance, fluorescence), site directed mutagenesis, cryoelectron microscopy and electron diffraction studies of two-dimensional crystals of membrane fragments and purified, reconstituted membrane proteins; and in the case of ion channels, high resolution single-channel patch-clamp studies.

Ligand-gated ion channels constitute an important superfamily of integral membrane proteins that mediate signal transduction. The nicotinic acetylcholine receptor (AChR) is one of the best characterized members of this superfamily *(1)*. Together with the serotonin receptor, the AChRs comprise two families of cation-selective channels, whereas glycine and gamma-amino butyric acid type A receptors are anion-selective channels. Signal transduction is relatively fast and results from similar mechanisms: binding of the neurotransmitter followed by conformational transitions in the receptor proteins that lead to changes in the ionic permeability of the postsynaptic membrane *(2)*.

From: *Methods in Molecular Biology, vol. 400: Methods in Membrane Lipids*
Edited by: A. M. Dopico © Humana Press Inc., Totowa, NJ

The AChR is composed of five homologous subunits organized pseudo-symmetrically around a central pore. Each subunit consists of a relatively large extracellular domain, four hydrophobic transmembrane (TM) segments (M1-M4) connected by loops of varying length and ending with a short extracellular carboxyl terminal domain *(1)*. The membrane itself, although relatively distant from both the ligand binding domain in the extracellular moiety of the AChR and the ion permeation pathway in the TM region, plays an important role in AChR function *(3–4)*. Cryoelectron microscopy studies have recently provided structural information on the TM organization of the AChR *(5)*. Three concentric rings can be distinguished in the AChR TM region *(3)*: a) an inner ring, formed by the M2 transmembrane segments of each subunit, and lining the walls of the ion channel proper; b) a middle ring, defined by Ml and M3; and c) an outer ring, which is in closest contact with the boundary- or belt-lipid region surrounding the AChR protein.

The composition of this singular lipid domain has recently been determined *(6)*. Cholesterol and acidic phospholipids are an absolute requisite for proper AChR function *(7–10)*. Hydrophobic compounds, including certain local anesthetics, steroids and fatty acids (FA), belong to a large family of non-competitive inhibitors of the AChR, affecting receptor function *(4)*. They act most likely at the lipid-protein interface, as postulated in early work *(11)*. In order to characterize the mechanism and site of action (i.e., annular vs. non-annular) of this type of molecules we introduced *(12,13)* the use of the extrinsic probe Laurdan (6-dodecanoyl-2-dimethylamino naphthalene) *(14,15)* and certain modalities of fluorescence spectroscopy, in particular Förster resonance energy transfer (FRET) *(16)*.

The spectral properties of Laurdan have their physical origin in the probe's extreme sensitivity to the polarity and molecular dynamics of the dipoles in its environment, owing to the effect of dipolar relaxation processes reflected as relatively large spectral shifts *(14,15)*. Laurdan molecules surrounded by gel-phase lipids, which provide an environment with low polarity and high order or "rigidity," undergo a lower relaxation process than those localized in an environment of high polarity and "fluidity," such as a liquid–crystalline phase. Since the main solvent dipoles surrounding Laurdan in the membrane are water molecules, when no relaxation occurs the generalized polarization (GP) values are high, indicating low water content in the hydrophilic/hydrophobic interface region of the membrane. A hitherto unexploited property of Laurdan, namely its ability to act as a FRET acceptor of tryptophan emission, was introduced to measure some physical properties of the AChR protein–vicinal lipid. FRET between an AChR-rich membrane protein and Laurdan was observed upon excitation at 290 nm (**Fig. 1**). The Trp-Laurdan pair has been characterized as a good donor-acceptor FRET pair, with a Förster distance of 31 ± 1 Å, and with a minimum donor-acceptor distance of 14 ± 1 Å, which corresponds roughly to the diameter of the first-shell protein-associated lipid *(12)*. Using this donor-acceptor pair it was possible to describe the occurrence of one type of site for cholesterol and another type for phospholipids, both accessible to FA, at the lipid-protein interface in native AChR-rich membranes, irrespective of their structure (i.e., number of carbon atoms and/or degree of saturation) *(17,18)*. This chapter describes the use of these techniques to exploit the advantageous spectroscopic properties of Laurdan, including its ability to act as a suitable acceptor for the AChR intrinsic protein fluorescence donor, to measure changes in FRET efficiency (E) caused by the addition of exogenous molecules that displace Laurdan from protein-vicinal sites. This strategy can be extrapolated to other membrane proteins for which solid-state structural information is still lacking.

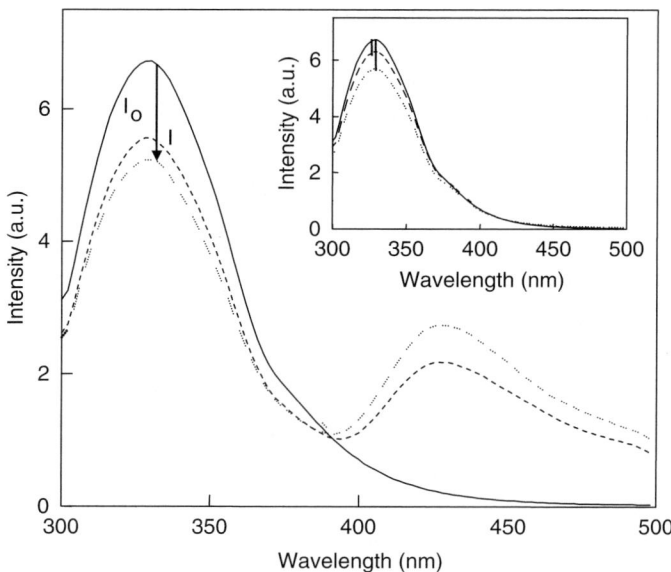

Fig. 1. Emission spectra of *T. californica* AChR-rich membranes under FRET conditions before and after the addition of Laurdan and FA (arachidic acid): (—) emission spectrum before Laurdan addition (I_o), (----) emission spectrum after Laurdan addition (I), and (······) emission spectrum after addition of 5 µM arachidic acid. The intrinsic fluorescence displays a maximum at 330 nm, and Laurdan emission shows two maxima between 400 and 500 nm. *Inset:* the same spectra obtained with the cuvette without Laurdan.

2. Materials

2.1. Chemicals

Most of the chemicals can be obtained from Sigma Chemical Co., St. Louis, MO. Laurdan can be purchased from Molecular Probes (Eugene, OR).

2.2. Materials for the Preparation of AChR-rich Membranes From Electric Tissue

1. *Torpedo californica* fish are obtained from the Pacific coast of California (U.S.A.). Fish are killed by pithing, and the electric organs are dissected and either used fresh or stored at –80°C until use.
2. Buffer A: 10 mM sodium phosphate buffer, pH 7.4 with 0.4 mM NaCl; 5 mM ethylenediaminetetraacetic acid; 5 mM ethylene glycol tetraacetic acid; 3 mM phenylmethanesulfonylfluoride and 0.02% sodium azide.
3. Buffer B: Buffer A + 25–30% sucrose (*see* **Note 1**).
4. Buffer C: 10 mM buffer sodium phosphate pH 7.4; 0.4 mM NaCl; 5 mM ethylenediaminetetraacetic acid.
5. Buffer D, for fluorescence studies: 20 mM HEPES buffer, pH 7.4, containing 150 mM NaCl and 0.25 mM $MgCl_2$.

2.3. Materials for Fluorescence Measurements

1. Quartz cuvette with four polished walls can be obtained from Starna, Germany. The cuvettes must be kept very clean, for which the customary protocol is to i) wash with distilled water + 0.1% mild detergent; ii) rinse with abundant distilled water; iii) rinse with methanol; and iv) dry the external walls with soft lens-tissue paper to avoid the formation of spots on the walls of the cuvette (*see* **Note 2**).
2. Laurdan crystals are dissolved in ethanol to obtain a 1.0–1.2 mM stock solution. Do not expose to light. It is necessary to vigorously vortex the sample until the crystals disappear completely (*see* **Note 3**). This solution can be stored in a vial at –20°C protected from light for up to 4 mo.

3. Free fatty acids (FA) are dissolved in ethanol (in all cases the amount of ethanol added to the samples must be kept below 0.5%) and protected from peroxidation by flushing the solutions with N_2. Stock sodium salts of FA are dissolved in the Buffer for fluorescence measurements with a bath sonicator. In the case of the more hydrophobic FA, it is recommended first to prepare a sodium salt solution in 4 mM NaOH, subsequently dilute an aliquot of this solution into the Buffer for fluorescence measurements (*see* **Note 4**).

3. Methods
3.1. Preparation of AChR-rich Membranes

1. Approximately 200 g of frozen (thawed) or fresh electric organ are dissected into small pieces (ca. 1 × 1 cm) on a sheet of metal foil placed on ice.
2. The tissue is homogenized at 4°C using a high speed metal blender (Virtis 60 glass homogenizer, The Virtis Company Inc.) until a homogeneous suspension is obtained. Usually three or four strokes of 30 s at 30,000 rpm (*see* **Note 5**) suffice. Tissue debris and large clumps can be eliminated by filtration through a nylon mesh.
3. The homogenate is subjected to a first centrifugation at 2500g for 10 min at 4°C. Connective tissue, clumps of cells and other large particles are pelleted and discarded. The supernatant is filtered through a fine nylon mesh and centrifuged for 1 h at 30,000g and 4°C. The pellet is resuspended in buffer B by gently passing through a needle followed by three strokes in a glass homogenizer.
4. 8 mL of the suspension are layered on a discontinuous sucrose gradient consisting of 8 mL of 50%, 9 mL of 39% and 8 mL of 35% sucrose in buffer C (*see* **Note 1**) and centrifuged at 80,000g for 3 h at 4°C in a 28 Beckman swinging bucket rotor (Beckman Coulter, Inc., Fullerton, CA) with slow acceleration and no de-acceleration in the range of 0–3300g. After centrifugation three light-scattering bands are formed at the interfaces of the sucrose solutions: top, middle and bottom fractions.
5. Each band is collected by aspiration and placed in a separate tube maintained at 4°C, diluted twofold in buffer C (without sucrose), divided among centrifuge tubes, and centrifuged at 138,000g for 45 min at 4°C in a Beckman 70.1 Ti rotor (Beckman Coulter, Inc., Fullerton, CA). The pellet is resuspended in a small volume of buffer for fluorescence by repeatedly passing gently through a 0.3-mm gauge hypodermic needle.
6. Small aliquots are separated to determine protein concentration and specific activity by the method of Lowry *(19)* and ^{125}I-labelled α–BTx binding assay, respectively. The rest is kept at −80°C until further use (*see* **Note 6**). Following this procedure, the MIDDLE fraction corresponds to the AChR-rich membrane fraction with highest specific activity, in the order of 2.0–2.8 nmol α-bungarotoxin sites per milligram protein, implying that more than 50% of the total protein is AChR protein.

3.2. Fluorescence Measurements

1. All fluorimetric measurements are performed in a steady-state fluorimeter (SLM Instruments, Urbana, IL) using a vertically polarized light beam from a mercury/xenon arc obtained with a Glan-Thompson polarizer (4-nm excitation and emission slits) and quartz cuvettes (*see* **Note 7**). The temperature is set at 25°C with a thermostated circulating water bath (Haake, Darmstadt, Germany).
2. Membrane samples in buffer D are prepared in a glass tube from concentrated AChR-rich membrane preparations to give final concentrations of approx 50 μg protein/mL (0.2 μM). The optical density of the membrane suspension is kept below 0.1 to minimize light scattering. The sample in the quartz cuvette is allowed to equilibrate at 25°C. For each experimental condition two cuvettes are prepared, one serving as the control and the other for Laurdan labeling.

3.3. Förster Resonance Energy Transfer (FRET) Measurements (see Note 8) and Laurdan Generalized Polarization (see Note 9)

1. The fluorescence of the donor (intrinsic fluorescence of the protein, Trp residues) in the absence of the acceptor is acquired before the addition of Laurdan to the samples. A first emission spectrum for

each cuvette is obtained by exciting the sample at 290 nm and collecting the emission spectrum from 300 to 520 nm (*see* **Notes 10–12**). The resulting spectrum is typical of Trp fluorescence emission with a clear maximum at approx 330 nm (*see* **Fig. 1A**) (*see also* **Note 13**). From this spectrum one obtains I_0 values at the maximum emission intensity. It is convenient to collect an additional emission spectrum by direct excitation of the probe Laurdan at 360 nm and collection of the emission from 390 to 520 nm. This provides a measurement of the background signal.

2. Laurdan is added next to the membrane samples from an ethanol solution, to give a final probe concentration of 0.6 µM. The amount of organic solvent should be kept ≤ 0.1%. The solvent (ethanol) is added to the control cuvette. Both samples are incubated in the dark for 60 min at 25°C.
3. The fluorescence of the donor in the presence of the acceptor is obtained. An emission spectrum for each cuvette is obtained by exciting the samples at 290 nm (FRET conditions) and at 360 nm (direct Laurdan excitation), respectively, as described under **step 1** of this subheading. In the case of samples excited under FRET conditions, the spectra of the samples with Laurdan show a decrease in the intensity at 330 nm and the appearance of a second maximum at a longer wavelength, which corresponds to FRET-excited Laurdan molecules (*see* **Fig. 1A**) (*see* **Note 14**). From these, spectra I values are obtained (*see* **step 1** of **Subheading 3.5.1.**). In the case of direct excitation conditions, the Laurdan spectrum exhibits two characteristic maxima, one near 430 nm, corresponding to the Laurdan molecules in the lipid gel phase (i.e., rigid domains) and a second maximum near 490 nm, corresponding to Laurdan molecules in the liquid-crystalline phase (i.e., in fluid domains) (*see* **Note 15**).

3.4. Titration Experiments

The location of the specific molecule under study (e.g., an endogenous lipid and/or an exogenously added hydrophobic ligand) in the AChR-vicinal region is studied by following the diminution of the FRET efficiency (E) between the AChR and Laurdan caused by the addition of the molecule being tested. The diminution of E reflects the displacement of the FRET acceptor molecules by the added molecule (see **Fig. 2**). Here we present an example using AChR-rich membranes from *T. californica* titrated with different FA.

1. After **step 3** of **Subheading 3.3.**, an aliquot of the FA stock solution is added to each cuvette (*see* **Notes 16–18**). Each experimental condition (i.e., for Each FA) comprises a pair of cuvettes (one with and one without Laurdan), and both must be treated in exactly the same way. Furthermore, one pair of cuvette is filled only with the solvent of the FA solution, which is taken into account for correction of dilution effects and for unspecific background subtraction. For example, if the effect of two different FA is studied, there will be at least six cuvettes: one pair for background correction, one for FA_1, and one for FA_2; within each pair one sample has Laurdan and the other does not. Incubate for 30 min in order to allow the partition of the exogenous molecule to reach equilibrium.
2. Repeat **step 3** of **Subheading 3.3.**, this time in the presence of FA.
3. Add another aliquot of the FA, wait for 30 min, and collect a new spectrum. Repeat this process until the desired final FA concentration is reached (*see* **Fig. 3** after data analysis).
4. It is also possible to perform competition experiments between two exogenous compounds in order to verify whether the two molecules compete for the same sites at the lipid-protein interface. To perform this kind of experiment, first add one molecule until saturation is reached, and then continue with the addition of the second molecule (*see* **Note 18**). Wait unit equilibrium is reached after each addition (time to equilibration varies among different ligands) and then collect the corresponding spectrum (*see* **Fig. 4** after analysis).

3.5. Data Analysis

3.5.1. Förster Resonance Energy Transfer (FRET) Measurements

1. The energy transfer efficiency (E) in relation to all other deactivation processes of the excited donor depends on the sixth power of the distance between donor and acceptor. According to Förster's theory *(16)* E can be calculated as follows:

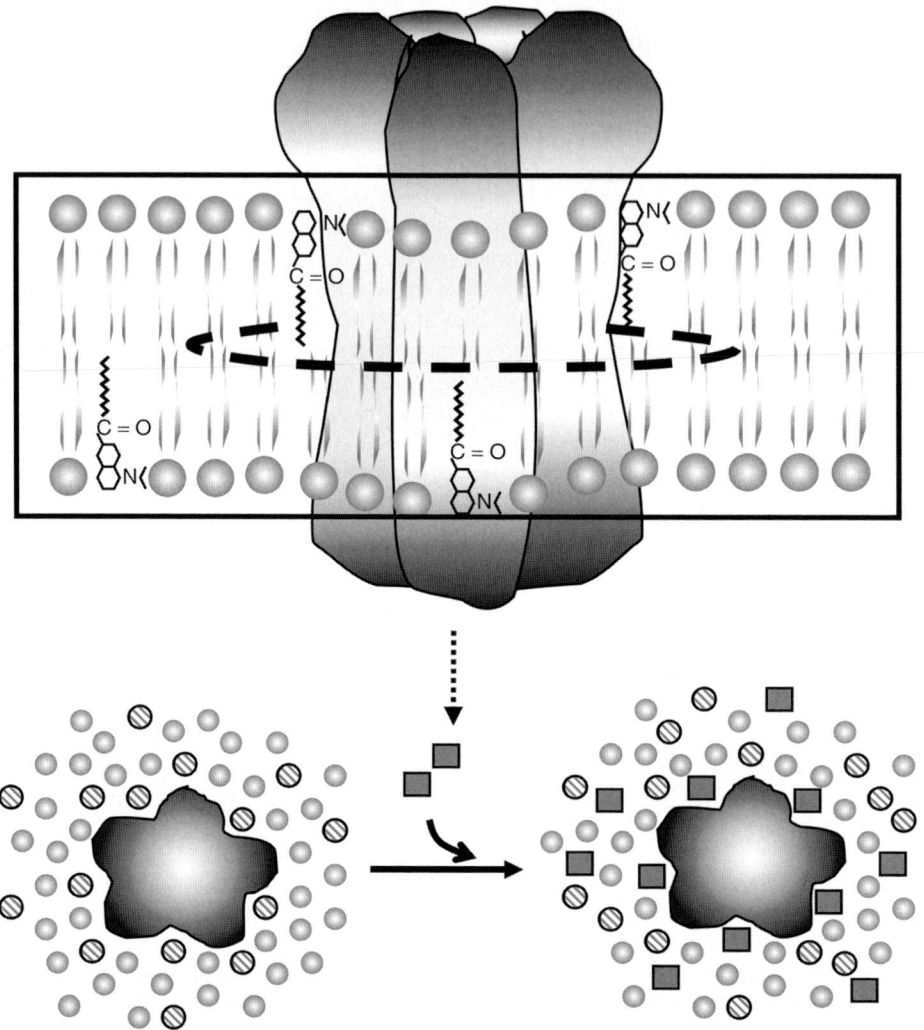

Fig. 2. Topographical relationship between the membrane-bound AChR, the AChR lipid-belt region in immediate contact with the receptor protein, and the fluorescent probe Laurdan in cross-sectional representation. The upper diagram illustrates the distribution of Laurdan molecules in the belt (outlined by a broken line) and bulk lipid regions, respectively. The bottom diagram depicts the displacement of Laurdan molecules (hatched circles) from the AChR belt region by an exogenously added hydrophobic compound (grey squares). Reproduced from Antollini and Barrantes, 2002 *(18)*.

$$E = 1 - (I/I_0) \tag{1}$$

where I and I_0 are the emission intensities in the presence and absence of the acceptor, respectively. Here, I corresponds to the maximal intrinsic protein emission intensity, at approx 330 nm.

2. For each condition, the energy transfer efficiency calculated in the presence of Laurdan (obtained from the Laurdan-doped cuvette of the pair) is corrected for E, calculated in the absence of Laurdan (i.e., obtained from the control cuvette of the pair) as follows:

$$E_{corr} = E_{(+\text{Laurdan})} - E_{(-\text{Laurdan})} \tag{2}$$

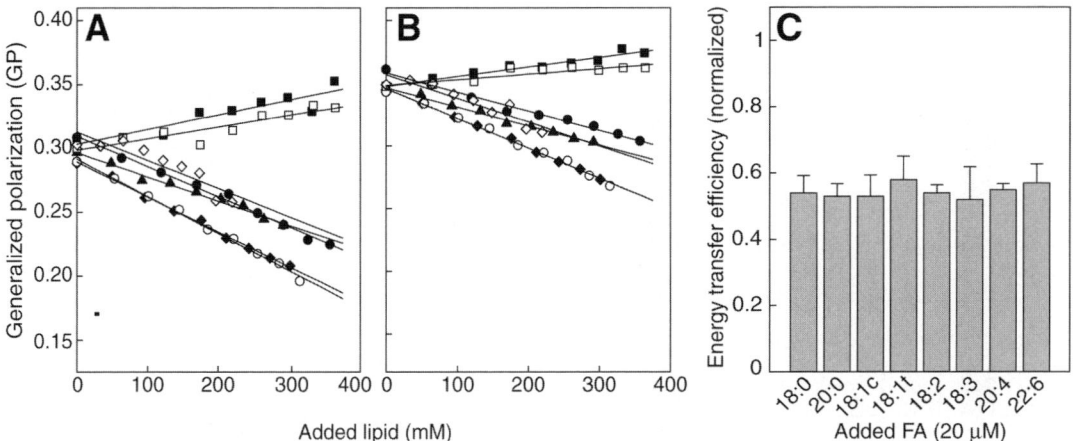

Fig. 3. Laurdan exGP using direct excitation of the probe (**A**) or FRET conditions (**B**). In the latter case the intrinsic protein emission from *T. californica* AChR-rich membranes is used as the donor in the presence of increasing concentrations of FA: 18:0 (■), 20:0 (□), 18:1c (●), 18:2 (▲), 18:3 (◇), 20:4 (◆), and 22:6 (○). The abscissa indicates effective FA concentrations in the membrane, calculated using the partition coefficient, K_p of each FA, obtained with ADIFAB as described under Methods. Each point is the average of at least four independent experiments (reproduced from Antollini and Barrantes, 2002 *(18)*. (**C**) Decrease in normalized FRET efficiency (*E*) between intrinsic fluorescence in AChR-rich membranes and Laurdan in the presence of 20 μM FA. Each point corresponds to the average ± S.D. of four independent measurements (reproduced from Antollini and Barrantes, 2002 *[18]*). Although the physical properties of the membrane change depending on the structure of the FA (unsaturated FA increase the water content in the membrane and disorder the bilayer, whereas saturated FA induce only a small decrease in the amount of water in the membrane), all FA, independently of their physical properties, bind to similar sites at the lipid-protein interface in native AChR-rich membranes.

3. Once the E_{corr} values are obtained for the different experimental conditions, they are normalized with respect to the control condition (obtained from the pair of cuvettes without the exogenous molecule) (*see* **Fig. 3C** for a displacement experiment and **Fig. 4C–F** for a competition experiment).

3.5.2. Generalized polarization (GP) of Laurdan

The GP for the Laurdan emission spectrum, which is called excitation GP (exGP), is calculated as follows,

$$exGP = (I_{430} - I_{490}) / (I_{430} + I_{490}) \qquad (3)$$

where I_{430} and I_{490} are the emission intensities at the characteristic wavelength of the gel phase (430 nm) and the liquid-crystalline phase (490 nm), respectively. The exGP values are obtained from the emission spectra acquired under FRET conditions (thus reporting on the lipid-protein interface region, within Förster distance) and by direct excitation (yielding information on bulk lipid properties) (*see* **Fig. 3A,B**, where exGP is obtained under the addition of different FA to *T. marmorata* AChR-rich membranes under both excitation conditions; and **Fig. 4**, where GP values were used to verify the incorporation of the different exogenous compounds added to the membrane).

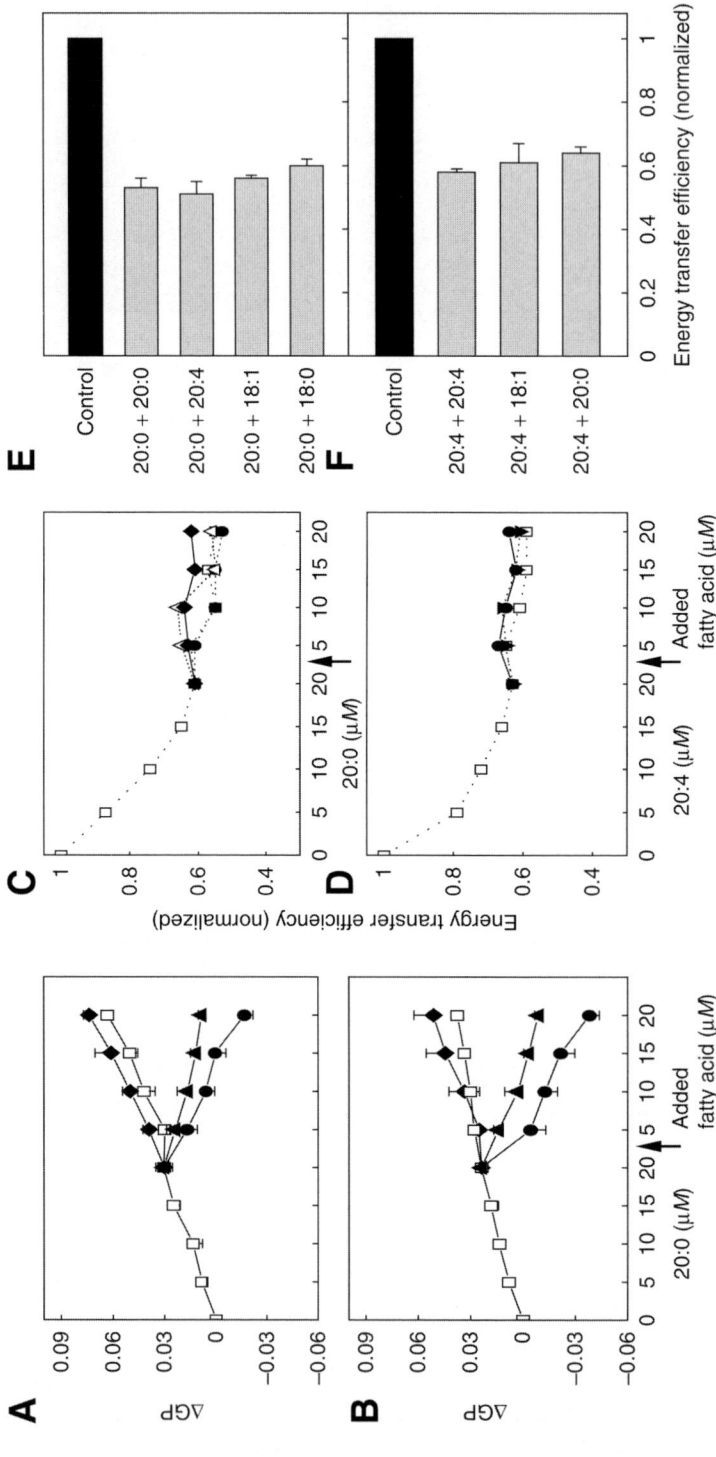

Fig. 4. Net variation in Laurdan GP using direct excitation of Laurdan (**A**) and FRET conditions (**B**) in the presence of increasing concentrations of arachidic acid (□) plus 18:0 (◆), 18:1c (▲), or 20:4 (●). (**C,D**) normalized FRET efficiency (*E*) for the AChR/Laurdan pair. Two different conditions are shown: increasing concentrations of 20:0 up to 20 μ*M* (**C**) and increasing concentrations of 20:4 up to 20 μ*M* (**D**). A subsequent addition of a second FA up to a concentration of 20 μ*M* is further shown in (**C**): 20:0 (□), 20:4 (●), 18:1c (△), and 18:0 (◆) and in (**D**): 20:4 (□), 18:1c (▼), and 20:0 (●). (**E,F**) depict final values of *E* attained upon addition of a second FA. The differences between experimental conditions were not statistically significant. These results correspond to the average ± S.D. of at least four independent experiments (reproduced from Antollini and Barrantes, 2002 *[18]*). The fact that Laurdan GP was affected under FRET conditions is a clear indication that the second FA effectively partitions in the membrane and localizes in the AChR-vicinal, lipid belt region. The decrease of *E* obtained with the first FA (saturated or unsaturated) remained constant in the presence of a second FA, suggesting that all FA share the same site at the protein-lipid interface.

4. Notes

1. When making the sucrose solutions take care to avoid sucrose hydrolysis by keeping the vessel close to 4°C.
2. Avoid touching the walls of the cuvette with bare hands and be meticulous about using clean cuvettes, otherwise they can be a major source of experimental variability.
3. Whenever a Laurdan stock solution is withdrawn from the freezer, wait until it reaches room temperature to avoid the appearance of Laurdan crystals. Always verify that the solution is homogeneous before proceeding.
4. In order to compare the effects caused by the presence of different exogenous FA it is necessary to take into consideration that different FA, with distinct structural characteristics, have different partition coefficients in membranes. It is therefore important to know –if available- or to experimentally determine their partition coefficient. There are several methods for carrying out this task. We use a spectroscopic method with an Acrylodan-Derivatized Intestinal Fatty Acid-Binding Protein (ADIFAB), which does not require physical separation of free and membrane-bound FA *(20,21)*.
5. During the homogenization procedure the inner glass vessel of the Virtis homogenizer should be properly refrigerated; an outer plastic cylinder is provided for this purpose and the space between the two vessels should be filled with ice. It is important that the high speed homogenization process does not warm up the sample; it suffices to wait for a couple of minutes in between homogenization cycles to avoid warming up.
6. Additional characterization of the membrane preparations can be mandatory for certain experiments. For instance, right-side-out membranes may be needed. The orientation of the AChR protein in the vesicles can by determined by measuring the total number of toxin-binding sites in the presence and absence of 1% Triton X-100 (Sigma Chemical Co., St. Louis, MO), respectively *(22)*. The integrity of the AChR subunits and the purity of the protein can be verified using Sodium Dodecyl Sulfate Polyacrylamide Gel **Electrophoresis** (SDS-PAGE) (the apparent molecular weights in SDS-PAGE are 40,000, 48,000, 60,000, and 65,000 for the α, β, γ, and δ subunit of electric fish AChR, respectively). The gels also provide information on the degree of protein degradation that may have occurred during membrane purification.
7. The size of the quartz cuvettes depends on the amount of sample available. One centimeter pathlength quartz cuvettes (2 mL sample or more) are excellent for titration experiments because these volumes allow several additions of solutions without a significant change of the total volume (i.e., 10 addition of 1 µL each to a 2 mL initial sample volume corresponds to a total added solvent of 0.5%). This is especially important when the solvent is not aqueous but organic, such as in the case of methanol, ethanol or DMSO. When the amount of sample is small, 5-mm quartz cuvettes are optional (up to 0.7 mL total volume) but one should bear in mind that the dilution effect is not trivial after a few µL additions. The same applies to 3-mm quartz cuvettes (130–150 µL of samples), which are quite useful in the case of precious samples but difficult to use in titration experiments.
8. FRET occurs when the emission spectrum of a fluorophore (donor) overlaps with the absorption spectrum of another molecule (acceptor) *(16)*. In this process there is no intermediate photon but the donor and the acceptor are coupled by dipole-dipole interactions (resonance). FRET is considered a "spectroscopic ruler" as the distance at which it occurs with 50% efficiency (R_0, Förster distance) is typically in the range of 20–60 Å. The distance range values are extremely convenient for studies of interrelations between molecules inside a membrane, and also for measurements of macromolecular associations/dissociations: E depends only on the donor-acceptor distance, and hence changes in such parameter are reflected dramatically in the efficiency of the process *(23)*.
9. Laurdan has no preference for a particular physical state of the membrane, but has an exquisite spectral sensitivity to the phase state of the membrane. This spectroscopic property of Laurdan has been integrated into a parameter called generalized polarization (GP) *(12,13)*. The spectral properties of Laurdan have their physical origin in the probe's capacity to sense the polarity and the molecular dynamics of the dipoles in its environment due to the effect of dipolar relaxation

processes *(12,13)*. The main dipoles sensed by Laurdan in the membrane are water molecules. Thus, GP values depend on the extent of water penetration allowed by the local membrane packing and constitute a faithful reflection of solvent dynamics in the immediate environment of membrane proteins.

10. The membrane sample is a suspension that needs to be maintained as such by gentle agitation immediately before acquisition of the spectrum. The acquisition time is obviously dependent on the instrumental sensitivity and gain, quantum yield of the fluorophore, and so on. The quantum yield of Laurdan is much higher in membranes than in aqueous environments, and the sensitivity of any state of the art fluorimeter allows one to acquire a spectrum within a reasonably short time (seconds), using 2 nm steps between each emission wavelength and no averaging.
11. The emission spectrum is taken with a bandwidth of $\lambda_{em} \pm 2$ nm. Acquired emission spectra are not perfect, because of the non-uniform spectral output of the light sources and the wavelength-dependent efficiency of the monochromators and detectors (photomultiplier tubes). Since accurately corrected spectra are difficult to obtain, it is not necessary to use corrected data when comparisons between spectra are made, technical spectra on the wavelength scale suffice in most cases.
12. The emission spectra are collected from 300 to 520 nm to obtain the emission spectrum of donor and acceptor. The Trp emission is centered at 330 nm, and at 400 nm the intrinsic fluorescence signal is no longer apparent. The Laurdan fluorescence emission is observed between 400 nm to 500 nm.
13. The fluorescence emission spectrum for Trp in water is centered at about 360 nm, whereas that of a Trp residue of a peptide located in the hydrophobic region of a lipid bilayer is centered at 323–330 nm depending on bilayer thickness, the position of the Trp residue along the peptide sequence, and peptide length *(24)*.
14. The spectra of the control cuvettes show a small diminution of the emission intensity at 330 nm without the appearance of the maximum corresponding to the Laurdan molecules. This diminution corresponds to an unspecific fluorescence quenching by oxygen that must be taken into account in this kind of experiment.
15. By exciting Laurdan molecules at 290 and 360 nm, it is possible to differentiate between the bulk bilayer properties (360 nm) and the AChR-vicinal lipid properties (290 nm excitation photoselects Laurdan molecules that are very close to the protein).
16. When a highly hydrophobic compound is added, it is important to know its critical micellar concentration (CMC), and work below this value. Added concentrations higher than the CMC give rise to errors and non-reproducible results as the incorporation of an exogenous molecule into the membrane in a micellar form is not as effective and quick as when the exogenous molecule inside the cuvette is in the monomeric form (i.e., add concentration below the CMC).
17. Titrations are usually carried out by addition of very small volumes (in general less than 1 µL). It is therefore important to exercise caution in wiping the external part of the tip to avoid "oversampling". At the same time, care should be taken not to extract sample from inside the pipette tip. It is also recommended to make a "routine" series of steps for each round of titrations (sometimes involving several cuvettes and long equilibration periods) to maximize reproducibility. Tip: always pipete the solution to be added from the top of the sample, add it to the cuvette in the same manner, mix all samples with the same technique, and so on.
18. It is also important to take into account the absorption spectrum of the exogenous molecule. If the added compound absorbs at the same wavelengths as the excitation and/or emission of the fluorophore, as is the case with some steroids, it is necessary to correct for the intensity of the donor. This correction, the so-called inner filter effect, is given by the following equation:

$$I_{corr} = I \times \text{antilog}\,[(OD_{290} + OD_{330})/2] \qquad (4)$$

where I is the measured fluorescence intensity of the Trp (330 nm) excited at 290nm, and OD_{290} and OD_{330} are the optical densities of the sample at the excitation and emission wavelength, respectively *(23)*.

Acknowledgments

The experimental work reported in this chapter was supported in part by grants from the Universidad Nacional del Sur, the Consejo Nacional de Investigaciones Científicas y Técnicas, the Comisión de Investigaciones Científicas de la Provincia de Buenos Aires, the Agencia Nacional de Promoción Científica (Fondo Nacional de Ciencia y Técnica), Argentina, the Ministerio de Salud, Argentina, Fogarty, International Center Research Collaboration Award, National Institutes of Health Grant 1-R03-TW01225-01, and Antorchas/British Council to F. J. Barrantes.

References

1. Karlin, A. (2002) Emerging structure of the nicotinic acetylcholine receptors. *Nat. Rev. Neurosci.* **3,** 102–114.
2. Barrantes, F. J. (1998) The Nicotinic Acetylcholine Receptor: current views and future trends. Springer Verlag/Heidelberg and Landes Publishing, Berlin/Georgetown, TX 226 pp, ISBN 3-540-64258-7.
3. Barrantes, F. J. (2003) Transmembrane modulation of nicotinic acetylcholine receptor function. *Curr. Opinion Drug Disc. & Develop.* **6,** 620–632.
4. Barrantes, F. J. (2004) Structural basis for lipid modulation of nicotinic acetylcholine receptor function. *Brain Res. Brain Res. Rev.* **47,** 71–95.
5. Miyazawa, A., Fujiyoshi, Y., and Unwin, N. (2003) Structure and gating mechanism of the acetylcholine receptor pore. *Nature* **423,** 949–955.
6. Mantipragada, S. B., Horvath, L. I., Arias, H. R., Schwarzmann, G., Sandhoff, K., Barrantes, F. J., and Marsh, D. (2003) Lipid-protein interactions and effect of local anesthetics in acetylcholine receptor-rich membranes from *Torpedo marmorata* electric organ. *Biochemistry* **42,** 9167–9175.
7. Criado, M., Eibl, H., and Barrantes, F. J. (1982) Effects of lipids on acetylcholine receptor. Essential need of cholesterol for maintenance of agonist-induced state transitions in lipid vesicles. *Biochemistry* **21,** 3622–3629.
8. Criado, M., Eibl, H., and Barrantes, F.J. (1984) Functional properties of the acetylcholine receptor incorporated in model lipid membranes. Differential effects of chain length and head group of phospholipids on receptor affinity states and receptor-mediated ion translocation. *J. Biol. Chem.* **259,** 9188–9198.
9. daCosta, C.J., Ogrel, A. A., McCardy, E. A., Blanton, M. P., and Baenziger, J.E. (2002) Lipid-protein interactions at the nicotinic acetylcholine receptor. A functional coupling between nicotinic receptors and phosphatidic acid-containing lipid bilayers. *J. Biol. Chem.* **277,** 201–208.
10. daCosta, C. J., Wagg, I. D., McKay, M. E., and Baenziger, J. E. (2004) Phosphatidic acid and phosphatidylserine have distinct structural and functional interactions with the nicotinic acetylcholine receptor. *J. Biol. Chem.* **279,** 14, 967–14, 974.
11. Marsh, D., and Barrantes, F. J. (1978) Immobilized lipid in acetylcholine receptor-rich membranes from *Torpedo marmorata*. *Proc. Natl. Acad. Sci. U.S.A.* **75,** 4329–4333.
12. Parasassi, T., De Stasio, G., d'Ubaldo, A., and Gratton, E. (1990) Phase fluctuation in phospholipid membranes revealed by Laurdan fluorescence. *Biophys. J.* **57,** 1179–1186.
13. Parasassi, T., De Stasio, G., Ravagnan G., Rusch, R. M., and Gratton, E. (1991) Quantitation of lipid phases in phospholipid vesicles by the generalized polarization of Laurdan fluorescence. *Biophys. J.* **60,** 179–189.
14. Antollini, S. S., Soto M. A., Bonini de Romanelli, I., Gutierrez-Merino, C., Sotomayor, P., and Barrantes, F. J. (1996) Physical state of bulk and protein-associated lipid in nicotinic acetylcholine receptor-rich membrane studied by laurdan generalized polarization and fluorescence energy transfer. *Biophys. J.* **70,** 1275–1284.

15. Zanello, L. P., Aztiria E., Antollini S., and Barrantes, F. J. (1996) Nicotinic acetylcholine receptor channels are influenced by the physical state of their membrane environment. *Biophys. J.* **70,** 2155–2164.
16. Föster, Th. (1948) Intermolecular energy migration and fluorescence. *Ann. Phys. (Leipzig)* **2,** 55–75.
17. Antollini, S. S., and Barrantes, F. J. (1998) Disclosure of discrete sites for phospholipid and sterols at the protein-lipid interface in native acetylcholine receptor-rich membrane. *Biochemistry* **37,** 16, 653–16, 662.
18. Antollini, S. S., and Barrantes F. J. (2002) Unique effects of different fatty acid species on the physical properties of the *Torpedo* acetylcholine receptor membrane. *J Biol. Chem.* **277,** 1249–1254.
19. Lowry, O. H., Rosebrough, N. J., Farr A. L., and Randall, R. J. (1951) Protein measurement with the Folin phenol reagent. *J. Biol. Chem.* **193,** 265–275.
20. Anel, A., Richieri, G. V., and Kleinfeld, A. M. (1993) Membrane partition of fatty acids and inhibition of T cell function. *Biochemistry* **32,** 530–536.
21. Richieri, G. V., Ogata, R. T, and Kleinfeld, A. M. (1992) A fluorescently labeled intestinal fatty acid binding protein. Interactions with fatty acids and its use in monitoring free fatty acids. *J. Biol. Chem.* **267,** 23, 495–23, 501.
22. Hartig, P. R., and Raftery M. A. (1979) Preparation of right-side-out, acetylcholine receptor enriched intact vesicles from *Torpedo californica* electroplaque membranes. *Biochemistry* **18,** 1148–1150.
23. Lakowicz, J. R. (1999) Principles of Fluorescence Spectroscopy, 2nd Edition, Kluwer Academic/Plenum Publishers, New York, 679 pp, ISBN: 0-306-46093-9.
24. Webb, R. J., East, J. M., Shanna, R. P., and Lee, A. G. (1998) Hydrophobic mismatch and the incorporation of peptides into lipid bilayers: a possible mechanism for retention in the Golgi. *Biochemistry* **37,** 673–679.

37

Single-Molecule Methods for Monitoring Changes in Bilayer Elastic Properties

Olaf S. Andersen, Michael J. Bruno, Haiyan Sun, and Roger E. Koeppe II

Summary

Membrane-spanning proteins perturb the organization and dynamics of the adjacent bilayer lipids. For example, when the hydrophobic length (l) of a bilayer-spanning protein differs from the average thickness (d_0) of the host bilayer, the bilayer thickness will vary locally in the vicinity of the protein in order to "match" the length of the protein's hydrophobic exterior to the thickness of the bilayer hydrophobic core. Such bilayer deformations incur an energetic cost, the bilayer deformation energy (ΔG_{def}^0), which will vary as a function of the protein shape, the protein-bilayer hydrophobic mismatch ($d_0 - l$), the lipid bilayer elastic properties, and the lipid intrinsic curvature (c_0). Thus, if the membrane protein conformational changes underlying protein function involve the protein/bilayer interface, the ensuing changes in ΔG_{def}^0 ($\Delta\Delta G_{def}^0$) will contribute to the overall free-energy change of the conformational changes (ΔG_{tot}^0)—meaning that the host lipid bilayer will modulate protein function. For a given protein, $\Delta\Delta G_{def}^0$ varies as a function of the bilayer geometric properties (thickness and intrinsic curvature) and the elastic (bending and compression) moduli, which vary as a function of changes in lipid composition or with the adsorption of amphiphiles at the bilayer/solution interface.

To understand how changes in bilayer properties modulate the function of bilayer-spanning proteins, single-molecule methods have been developed to probe changes in bilayer elastic properties using gramicidins as molecular force transducers. Different approaches to measuring the deformation energy are described: (1) measurements of changes in channel lifetimes and appearance rates as the lipid bilayer thickness or channel length are varied, (2) measurements of the equilibrium distribution among channels of different lengths, formed by homo- and heterodimers between gramicidin subunits of different lengths, and (3) measurements of the ratio of the appearance rates of heterodimer channels relative to parent homodimer channels formed by gramicidin subunits of different lengths.

Key Words: Bilayer deformation energy; bilayer elastic moduli; bilayer stiffness; gramicidin channels; hydrophobic coupling; hydrophobic mismatch; lipid intrinsic curvature.

1. Introduction

Biological membranes are mosaic structures, made up of lipid bilayers that are spanned by integral membrane proteins *(1)*, with an overall organization as illustrated in **Fig. 1**. The bilayer serves as a barrier against nonregulated transmembrane solute movement; membrane proteins catalyze transfer of material and information across the membrane.

1.1. Bilayer Barrier Function

The lipid-barrier function is well understood, and reflects the low solubility of polar solutes—whether ions or nonelectrolytes—into the hydrophobic bilayer core *(2–4)*. The chemical identity of the bilayer lipids is, to a first approximation, irrelevant, although the water permeability of

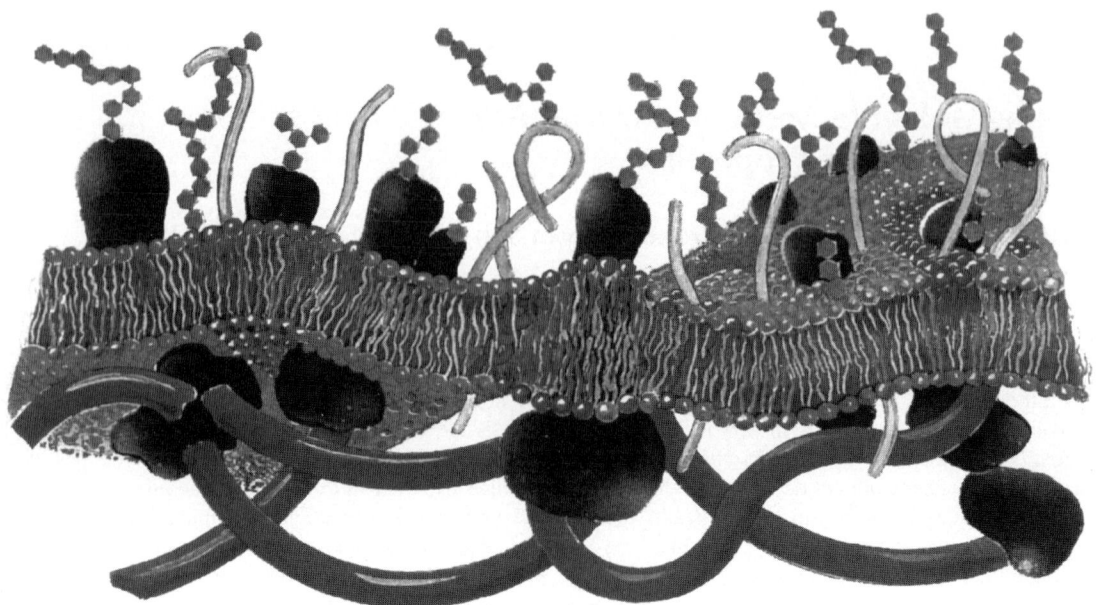

Fig. 1. Cartoon representation of a biological membrane. The membrane is oriented with the extracellular compartment located above and the intracellular compartment below. The lipid bilayer with its imbedded proteins separates the two compartments. The extracellular protein domains have covalently linked carbohydrate chains; the intracellular domains interact with the submembraneous cytoskeleton. Reflecting its elastic properties, the lipid bilayer component displays undulations and thickness fluctuations, as well as lateral heterogeneity and domain formation. For clarity, the membrane is depicted with a relatively low protein/bilayer ratio. Modified after **ref. *120***.

(liquid-crystalline) lipid bilayers can vary by more than two orders of magnitude as the lipid composition is varied *(5)*. The bilayer component is not just a "simple" barrier; it also is a regulator of membrane protein function *(6–12)*. However, despite recent progress, there is no consensus about the mechanism(s) underlying this bilayer control of protein function—a situation that arises, in part, because of the diversity of the lipids in an "average" cell membrane *(13–15)*.

1.2. Regulation of Protein Function

The bilayer regulation of a protein can be *direct*, when the function of the protein of interest is directly altered by changes in bilayer lipid composition, or *indirect*, when the changes in protein function are secondary to (direct) bilayer-dependent changes in some other membrane protein(s), for example, proteins that activate intracellular signal transduction cascades. In the case of direct bilayer regulation of membrane protein function, the regulation can be *specific*, because of specific (stoichiometric) lipid–protein interactions, as is the case for phosphatidylinositol 4,5-bisphosphate binding to pleckstrin homology domains *(16,17)*, or *nonspecific*, because of changes in global bilayer properties. The latter encompasses changes in interfacial (double layer or dipole) potential, bilayer thickness, monolayer intrinsic curvature, or bilayer elastic (compression and bending) moduli *(18)*.

1.3. Membrane Protein Function and Conformational Changes

There is ample evidence that membrane protein function depends on conformational changes involving the protein's transmembrane domains. Early low-resolution structures of gap junction channels *(19)* and the nicotinic acetylcholine receptor *(20)* showed that changes in channel state were associated with alterations in subunit tilt within the bilayer. Recent high-resolution structures of the transmembrane domains of three bacterial potassium channels, including H$^+$-, Ca^{2+}-, and voltage-activated channels *(21–24)*, of the sarcoplasmic reticulum Ca^{2+}-ATPase *(25,26)*, and of the large-conductance mechano-sensitive channel MscL *(27–29)* all provide evidence for reorganization within the protein's transmembrane domains, including changing the tilt of bilayer-spanning α-helices and the domain lengths.

Less direct evidence for reorganization within the transmembrane domain has been obtained in other membrane proteins, most notably rhodopsin. Light activation of mammalian rhodopsin causes a structural rearrangement *(30)* in which the bilayer-spanning α-helices move relative to each other *(31)*. The (pseudo)equilibrium between metarhodopsin I and II depends on the bilayer lipid composition *(32)*, and light-activation changes bilayer lipid organization *(33)*.

2. Lipid Bilayer–Membrane Protein Interactions

How does the lipid bilayer respond to changes in protein structure? And how do changes in bilayer properties alter protein function? There is a voluminous literature on bilayer–protein interactions. A central theme has been bilayer fluidity *(34,35)*. A liquid-crystalline lipid bilayer is necessary in order to allow for protein conformational changes. Nevertheless, bilayer fluidity *per se* cannot be the major factor regulating membrane protein function because changes in bilayer fluidity *cannot* shift the conformational preference of membrane proteins *(36)* (*see* **Note 1**). What then? Bienvenüe and Marie (*see* p. 325 in **ref. 9**) summarized the situation as follows: ".., modulation of many protein properties has been demonstrated in several different models. However, no clear rule exists that rationalizes the effect of lipid composition on ligand binding or enzyme activity. Considering each protein as a different model interacting with its own lipid environment, a theoretical breakthrough is necessary to explain how lipids regulate membrane protein activities."

Given the diversity of lipids in cell membranes, it is unlikely that there is just a single organizing principle that can account for the variety of lipid-dependent changes in membrane protein function. The available evidence shows that membrane lipids can serve as both specific and nonspecific regulators of membrane protein function. Some proteins interact directly with polyphosphatides *(17)* in "classic" ligand–protein interactions. Phospholipid molecules are also seen in high-resolution membrane protein structures *(37–40)*; in these cases, the lipid structure usually is disorganized, which could denote conformational heterogeneity or chemical heterogeneity among the acyl chains of the bound lipids. However, in many cases membrane protein function is regulated by "bulk" bilayer properties, such as bilayer thickness—red blood cell glucose transporter *(41)*, rhodopsin *(32)*, sarcoplasmic reticulum Ca^{2+}-ATPase *(42)*, Na$^+$, K$^+$-ATPase *(43)* and the large-conductance, mechanosensitive channel MscL *(44)*—and intrinsic lipid curvature—sarcoplasmic reticulum Ca^{2+}-ATPase *(45,46)*, rhodopsin *(32)*, insulin receptor *(47)*, Ca^{2+}-activated potassium channels *(48)*, and MscL *(44)*. Moreover, the membrane lipid composition is actively regulated to maintain some physical bilayer property (or properties) invariant *(49,50)*.

Many of the aforementioned proteins are known to undergo conformational changes that involve their bilayer-spanning domains in the course of their normal function cycle (*see* **Subheading 1.3.**). It therefore is not surprising that their function may be regulated by changes in the composition (physical properties) of the host lipid bilayer. Specifically, the bilayer control of membrane protein function can be understood by noting that the energetics of membrane protein conformation changes is related to the energetics of bilayer–protein interactions: to the bilayer deformation energy associated with membrane protein conformational changes that involve the protein–lipid interface, which are coupled through hydrophobic interactions (*see* **Fig. 2**).

2.1. Protein-Induced Bilayer Deformations and Deformation Energies

Protein-induced bilayer perturbations, or deformations, will in general incur an energetic cost (ΔG_{def}^0), which contributes to the free energy difference between different protein conformations I and II $(\Delta G_{tot}^{I \to II})$:

$$\Delta G_{tot}^{I \to II} = \Delta G_{prot}^{I \to II} + \Delta \Delta G_{def}^{I \to II} \tag{1}$$

where $\Delta G_{prot}^{I \to II}$ denotes the free energy difference of the protein conformational change *per se* (including contributions from interactions with the environment, such as changes in the protein/solution interface, not considered in the protein–bilayer interactions) and $\Delta \Delta G_{def}^{I \to II}$, the difference in bilayer deformation energy between protein conformations I and II ($\Delta \Delta G_{def}^{I \to II} = \Delta G_{def}^{II} - \Delta G_{def}^{I}$). The equilibrium distribution between the different protein conformations, therefore, is given by:

$$\frac{II}{I} = K_{II}^{I} = \exp\left[\frac{-\left(\Delta G_{prot}^{I \to II} + \Delta \Delta G_{def}^{I \to II}\right)}{k_B T}\right] \tag{2}$$

where K_{II}^{I} denotes the equilibrium distribution coefficient between protein states I and II (the superscript and subscript in K_{II}^{I} denote the initial and final state, respectively), T the temperature and k_B Boltzmann's constant. If ΔG_{def}^0 is significant, meaning that $|\Delta G_{def}^0| > k_B T$, then the equilibrium distribution between different membrane protein conformations (and the kinetics of the conformational changes) could be modulated by the bilayer in which the proteins are embedded *(7,32,51,52)*.

2.2. Energetics of Elastic Bilayer Deformations

In the case of a cylindrical membrane protein of hydrophobic length l, embedded in a bilayer of average thickness d_0, intrinsic monolayer curvature c_0, and bilayer compression and bending moduli K_a and K_c, the bilayer deformation can be decomposed *(53)* into local bilayer compression with an associated energy density $K_a \cdot (2u/d_0)^2$ (*see* **ref. 54**) and monolayer bending with an associated energy density $K_c \cdot (\nabla^2 u - c_0)^2$ (*see* **ref. 55**) (*see also* **Note 2**), where $u = (d_0 - d)/2$ with d being the local bilayer thickness (*see* **Fig. 2**). Combining these

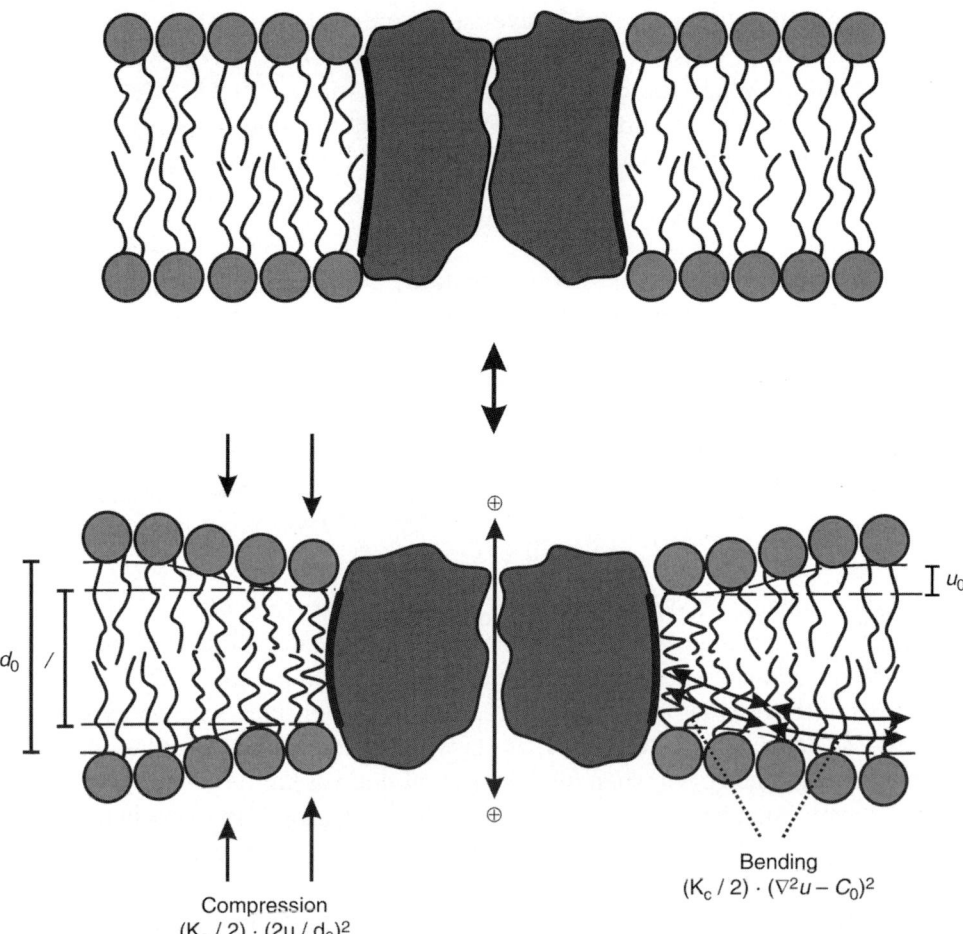

Fig. 2. Hydrophobic coupling between membrane protein conformational changes and lipid bilayer deformations/perturbations. The hydrophobic coupling between a membrane protein and the surrounding bilayer causes protein conformational changes that involve the hydrophobic protein/bilayer boundary (indicated by the heavy black lines) to be associated with local bilayer deformation. When the protein-induced bilayer deformation can be approximated as a change in bilayer thickness, the deformation can be described in terms of the compression and bending of the two bilayer leaflets, with energy densities $(K_a/2)\cdot(2u/d_0)^2$ and $(K_c/2)\cdot(\nabla^2 u - c_0)^2$, respectively, where K_a and K_c denote the bilayer compression and bending moduli, $2u$ the local bilayer deformation, that is the difference between the actual bilayer thickness (d) and the (average) thickness of the unperturbed bilayer (d_0), and c_0 the monolayer equilibrium curvature. Note, the bilayer itself is flat; it is the two bilayer leaflets that bend toward each other.

contributions, one can estimate ΔG^0_{def} using the theory of elastic bilayer deformations *(53,56–60)*:

$$\Delta G^0_{\text{def}} = \int_{r_0}^{\infty}\left[K_a\cdot\left(\frac{2u}{d_0}\right)^2 + K_c\cdot(\nabla^2 u - c_0)^2 + \alpha\cdot\left(\frac{du}{dr}\right)^2\right]\cdot\pi\cdot r\,dr - \int_{r_0}^{\infty} K_c\cdot c_0^2\cdot\pi\cdot r\cdot dr \quad (3)$$

where r_0 is the protein radius and α the bilayer/solution interfacial tension. (The second integral in **Eq. 3** arises from the curvature stress in an unperturbed bilayer with finite intrinsic monolayer curvature). When the contributions to ΔG_{def}^0 are evaluated, the surface tension component turns out to be negligible *(53,56,59)*; it will not be considered further here.

When $c_0 = 0$, **Eq. 3** reduces to a particularly user-friendly expression:

$$\Delta G_{def}^0 = H_B \cdot (2u_0)^2 = H_B \cdot (d_0 - l)^2 \qquad (4)$$

where H_B is a phenomenological spring constant, and $2u_0$ the hydrophobic mismatch $(2u_0 = d_0 - l)$. H_B describes the energetic consequences of the hydrophobic channel-bilayer coupling; it is determined by the bilayer thickness and elastic moduli (K_a and K_c) and r_0, as well as by the choice of boundary conditions used to describe the lipid packing adjacent to the protein *(59)*. When $c_0 \neq 0$, the unperturbed bilayer (where $2u_0 = 0$) possesses a curvature frustration energy density, which also contributes to ΔG_{def}^0 *(60)*, and **Eq. 3** reduces to (*see* **ref. 60**) (*see also* **Eqs. 17** or **28**):

$$\Delta G_{def}^0 = H_B \cdot (d_0 - l)^2 + H_X \cdot (d_0 - l) \cdot c_0 + H_C \cdot c_0^2 \qquad (5)$$

where H_X and H_C, again, are determined by d_0, K_a, K_c, and r_0 (**ref. 60**) (*see also* **Note 3**).

The second-order approximation used in elastic (liquid crystal) models of bilayer behavior *(53,61)* may be questioned when the radii of curvature are comparable with the bilayer thickness, and it is uncertain whether "macroscopic" material constants can be used to describe such systems, (*see* **refs. 62,63**). However, as shown by Partenskii and Jordan *(63)*, the structure of **Eq. 4** is maintained even when the material moduli vary as a function of distance from the protein/bilayer boundary. Moreover, when **Eq. 4** is used to quantitatively evaluate channel–bilayer interactions using gramicidin channels as molecular force transducers *(64)*, the experimentally determined H_B values are in agreement with those predicted based on previously determined values of K_a and K_c. Thus, **Eqs. 4** and **5** should provide a fairly robust framework for understanding the energetic consequences of protein-induced bilayer deformations.

The energetics of bilayer–protein interactions also have been described using other terms, such as: changes in bilayer compression *(54)* or curvature frustration *(55)* energy, which were combined using the model of elastic bilayer deformations *(53)*; changes in the lateral pressure profile across the bilayer *(65)*, or in lipid packing stress *(66)*; and changes in bilayer free volume *(67,68)*. Although couched in different terms, all of these different descriptions represent different approaches to parameterize the profile of intermolecular interactions across the bilayer, i.e., between the bilayer-forming lipids and between the lipids and the embedded proteins. We also note that, although changes in intrinsic curvature have long been emplicated in the bilayer control of membrane protein function *(32,55,69,70)*, the curvature stress is only one contribution to ΔG_{def}^0 (*see* **Eqs. 3** and **5**). Indeed, pharmacologically induced changes in curvature *(71,72)* usually alter (lower) the bilayer elastic moduli *(73–77)*, and thus, the coefficients in **Eqs. 4** and **5**. These changes in bilayer elastic moduli are likely to be of central importance for the bilayer regulation of membrane protein function.

Within the limits of the second-order description of elastic bilayer deformations, as embodied in **Eq. 3**, **Eqs. 4** and **5** provide exact expressions for the bilayer deformation energy associated with a protein-induced hydrophobic mismatch (*see* **Note 4**). Although the coefficients in **Eqs. 4** and **5** can be evaluated only for idealized geometries, the elastic bilayer model provides a basis for evaluating different scenarios. Thus, the advantages of the elastic bilayer model are the following: first, conceptual simplicity, at least when formulated as in **Eqs. 4** and **5**; second, although H_B (and H_X and H_C) initially were defined based on **Eq. 3**, they can equally well be regarded as empirical descriptors of protein–bilayer interactions; third, assuming d_0 and r_0 are known, one can compare an experimentally determined H_B with estimates based on measured values of K_a and K_c *(78–81)* obtained using the scaling relations developed by Nielsen et al. *(59,60)* (or predict H_B *de novo* from the material moduli using these scaling relations); and fourth (and perhaps most useful), one can estimate the value of H_B for a protein (and bilayer) of interest based on experimental values obtained with a simple channel using the scaling relations derived in **refs. 59** and **60**.

2.3. Application to Ion Channels

To implement the elastic bilayer model to understand the bilayer control of membrane protein function, and develop probes to assess how changes in bilayer properties alter protein function, the focus was on the interconversion between two different states of an ion channel: closed (C) and open (O),

$$C \leftrightarrow O$$

$$\frac{[O]}{[C]} = K_O^C = \exp\left(\frac{-\Delta G_{tot}^{C \to O}}{k_B T}\right); \quad \Delta G_{tot}^{C \to O} = \Delta G_{prot}^{C \to O} + z_a \cdot eV_m + \Delta\Delta G_{def}^{C \to O} + \ldots \quad (6)$$

where $\Delta G_{tot}^{C \to O}$ is the standard free energy change for the C↔O equilibrium. As noted previously (*see* **Subheading 2.1.**), $\Delta G_{tot}^{C \to O}$ is the sum of contributions from the protein *per se* ($\Delta G_{prot}^{C \to O}$) and terms that arise from the protein interactions with the environment. These latter terms include the electrostatic energy ($z_a \cdot eV_m$, where z_a is the apparent gating valence, e is the elementary charge, and V_m is the membrane potential difference) and $\Delta\Delta G_{def}^{C \to O}$, as well as other terms such as the one arising from a change in the number of water molecules associated with the channel *(82,83)*. $\Delta G_{prot}^{C \to O}$ and $z_a \cdot eV_m$ are generally recognized contributors to $\Delta G_{tot}^{C \to O}$, whereas $\Delta\Delta G_{def}^{C \to O}$ is not. $\Delta\Delta G_{def}^{C \to O}$ is not negligible, however, as demonstrated in experiments with alamethicin and gramicidin channels *(69,84–88)*. This is important, because when $\Delta\Delta G_{def}^{C \to O}$ is significant (meaning more than $k_B T$), the bilayer becomes an allosteric modulator of membrane protein function.

3. Gramicidin Channels as Probes of Bilayer Properties
3.1. Basic Principles

To understand better how the lipid bilayer can regulate protein function it is important to study proteins that undergo well-defined changes in structure *(51)*. These proteins preferably should be channels or receptors, so that the measured changes in function provide information about the pseudoequilibrium distribution between different protein conformations in a fairly straightforward manner *(32,89)*—as opposed to carriers or pumps, wherein the changes in function reflect the underlying changes in transport kinetics, which are more complex to interpret. Presently, it is not possible to characterize fully the bilayer–channel interactions for a complex integral membrane protein (*see* **ref. 29**), although changes in MscL function caused by changes in bilayer thickness and maneuvers that alter lipid intrinsic curvature, for example, can be understood by considering the changes in $\Delta\Delta G_{\text{def}}^{C\to O}$.

Among simple channels, the channels formed by the linear gramicidins are particularly useful as probes to study bilayer–channel interactions *(52,64,89–93)*. In this context, the gramicidin channels serve two different purposes: (1) as prototypical ion channels, which allow for detailed mechanistic studies on how changes in bilayer thickness, lipid intrinsic curvature, and bilayer elastic moduli can later function for bilayer-spanning channels, and (2) as probes of changes in these bilayer properties. With time, the latter purpose is likely to become the most important.

3.2. Primer on Gramicidin Channels

Gramicidin channels are miniproteins, formed by the transmembrane dimerization *(94)* of two single-stranded, right-handed $\beta^{6.3}$-helical subunits *(95,96)*. Several different gramicidins are produced by the soil bacterium *Bacillus brevis*, the most common form being valine gramicidin A, [Val^1]gA, which has the sequence *(97)*:

formyl-L-Val-Gly-L-Ala-D-Leu-L-Ala-
D-Val-L-Val-D-Val-L-Trp-D-Leu-L-Trp-
D-Leu-L-Trp-D-Leu-L-Trp-ethanolamine.

Gramicidin channels are symmetric antiparallel dimers that form by the reversible, transmembrane dimerization *(94)* of two $\beta^{6.3}$-helical subunits, whose structure is known at atomic resolution *(98–100)*. The channels can be approximated as cylinders, and the three horizontal rows in the sequence above denote, approximately, three helical turns in the structure. The subunit interface is formed by the formyl-L-Val-Gly-L-Ala-D-Leu-L-Ala segment. For a recent overview of gramicidin channel structure and function, *see* Andersen et al. *(101)*.

The hydrophobic length of [Val^1]gA channels is approx 2.3 nm *(53,102)*, which is less than the hydrophobic thickness of lipid bilayers *(103,104)*, such that the hydrophobic coupling between the channel and the surrounding bilayer causes channel formation to perturb the bilayer locally around the channel *(53,102)*. Essentially, all $\beta^{6.3}$-helical gramicidin dimers are bilayer-spanning channels *(105)*, and the nonconducting gramicidin subunits are inserted into each leaflet of the bilayer *(94)*. Consequently, the gramicidin monomer↔dimer equilibrium can be regarded as a special type of "conformational" transition in a membrane "protein," one that is associated with a well-defined bilayer deformation, which involves a compression and bending of the two monolayers (**Fig. 3**), similar to the situations for integral membrane proteins (**Fig. 2**).

Gramicidin channels do not open or close, they appear or disappear (**Fig. 3**). The channels form by the transmembrane dimerization of two $\beta^{6.3}$-helical subunits, one from each leaflet *(106)*. Given the mismatch between the channel hydrophobic length (~2.2 nm) and the thickness of the bilayer hydrophobic core (3–4 nm), channel formation usually involves a local bilayer thinning. Not surprisingly, the association rate constant is about two orders of magnitude less than that predicted for diffusional encounters in the plane of the bilayer *(107)*, which is consistent with the existence of an energetic penalty associated with the bilayer deformation.

3.3. Gramicidin Channels are Bilayer-Embedded Force Transducers

The distribution between gramicidin monomers (M) and dimers (D) is given by:

$$\frac{[D]}{[M]^2} = K_D^M = \exp\left(\frac{-\Delta G_{\text{tot}}^{M\to D}}{k_B T}\right) = \exp\left(\frac{-(\Delta G_{\text{prot}}^{M\to D} + \Delta\Delta G_{\text{def}}^{M\to D})}{k_B T}\right) \quad (7)$$

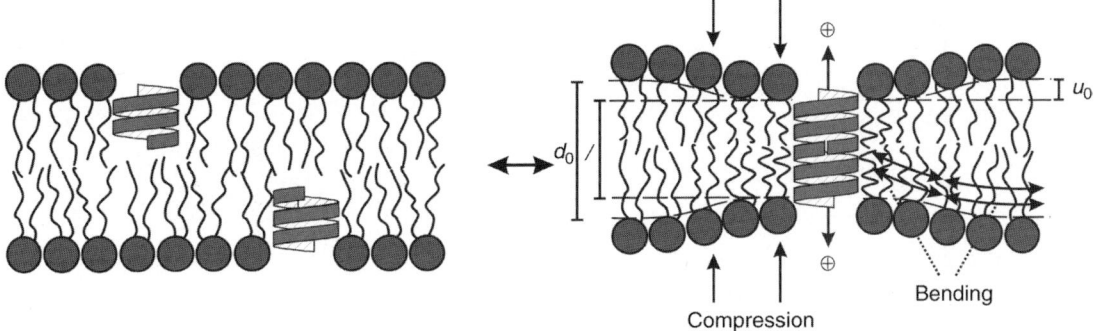

Fig. 3. Schematic representation of the bilayer deformation associated with the formation of gramicidin channels. Nonconducting $\beta^{6.3}$-helical gramicidin subunits are embedded in each bilayer leaflet *(106)*. The subunits diffuse in the plane of each leaflet and thereby encounter subunits in the opposite leaflets (still being on opposite sides of the bilayer). A small fraction of these encounters lead to the formation of bilayer-spanning channels *(107,121,122)*, which causes a bilayer deformation that involves the compression and bending of the bilayer leaflets adjacent to the channel.

where (*see* **Eq. 5**):

$$\Delta\Delta G_{\text{def}}^{M\to D} = \Delta G_{\text{def}}^{D} - \Delta G_{\text{def}}^{M} = H_B \cdot (d_0 - l)^2 + H_X \cdot (d_0 - l) \cdot c_0 \tag{8}$$

because $\Delta G_{\text{def}}^{M} = H_C \cdot c_0^2$.

$\Delta G_{\text{def}}^{D}$ varies as a function of $d_0 - l$, and the bilayer will respond to the deformation by imposing a disjoining force (F_{dis}) on the bilayer-spanning channels:

$$F_{\text{dis}} = -\left(-\frac{\partial(\Delta\Delta G_{\text{def}}^{D\to M})}{\partial(d_0 - l)}\right) = 2 \cdot H_B \cdot (d_0 - l) + H_X \cdot c_0 \tag{9}$$

Changes in F_{dis} will be observable as changes in channel lifetime (**Fig. 4**).

The channel lifetimes decrease as the hydrophobic mismatch increases, whether the mismatch is because of changes in bilayer thickness (**Fig. 4, left panel**) or channel length (**Fig. 4, right panel**), which means that gA channels can be used as *in situ* molecular force transducers—embedded in lipid bilayers!

3.4. Pharmacological Modification of Bilayer Properties; Changes in Bilayer Stiffness

When the bilayer lipid composition is varied, or small amphiphiles adsorb at the bilayer/solution interface, the bilayer geometric (thickness and intrinsic curvature) and elastic (the elastic moduli) properties are altered *(71–77)*. Consequently, the deformation energy associated with a given protein conformational change, and the disjoining force the bilayer imposes on bilayer-spanning gramicidin channels (and other proteins), are altered as well. Such changes in bilayer properties alter the gramicidin channels appearance rate and lifetime, usually by changing both the appearance rate and lifetime in the same direction—meaning that the time-averaged channel concentration varies.

Figure 5 shows results obtained when 3 μ*M* eicosapentenoic acid (EPA) is added to both sides of a bilayer that has been doped with two different gramicidin analogs, the sequence-substituted [Ala^1]gA with 15 residues in the sequence, and the sequence-shortened des-Val^1-Gly^2-gA$^-$ with

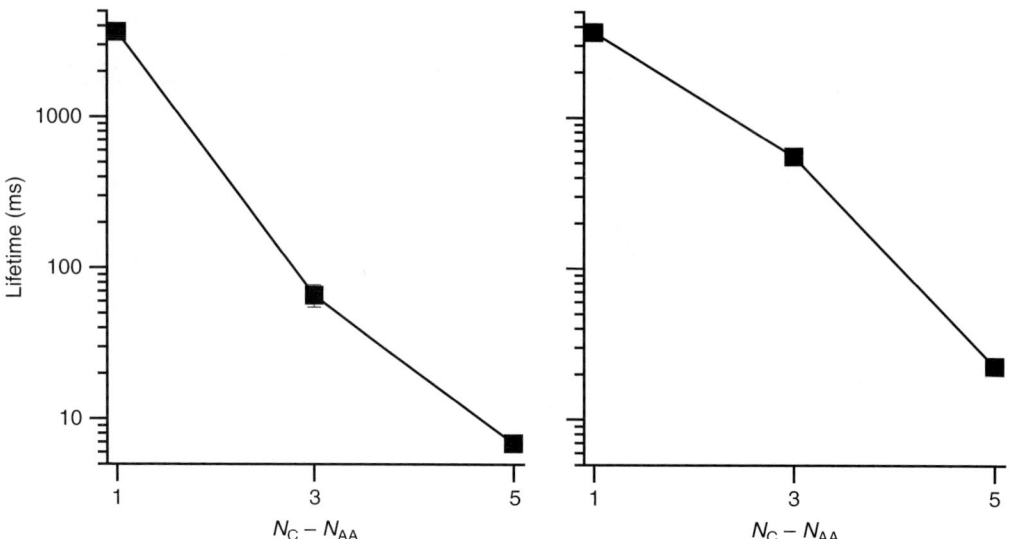

Fig. 4. Gramicidin channel lifetimes vary as a function of the channel-bilayer hydrophobic mismatch. The mismatch is characterized by $N_C - N_{AA}$, where N_C denotes the number of carbon atoms in the acyl chains of the bilayer-forming symmetric, mono-unsaturated phosphatidylcholines, and N_{AA} denotes the number of amino acids in the gramicidin sequence. In each panel the parameter that is changed is in **bold**. The left panel shows the lifetimes of endo-Gly^{0b}-D-Ala^{0a}-gA channels in bilayers of different thickness. The right panel shows the lifetimes of des-Val^1-Gly^2-gA, [Val^1]gA, and endo-Gly^{0b}-D-Ala^{0a}-gA channels in dioleoylphosphatidylcholine/n-decane bilayers. Records obtained in 1 M CsCl 10 mM HEPES buffered to pH 7.0, at 25°C; transbilayer voltage = 200 mV. (Experimental results from Hwang et al. *[91]*).

13 residues in the sequence. These two gramicidins differ in length, and thus in the hydrophobic mismatch they impose on the host bilayer when they form. They also differ in the amplitude of the current transitions that occur when a channel forms or disappears, as denoted by the two horizontal lines (surprisingly, the shorter des-Val^1-Gly^2-gA$^-$ channels have the smaller current transitions; it is not clear why).

Figure 6 shows current transition amplitude histograms and lifetime distributions obtained in the absence and presence of 3 µM EPA.

Two different channel types can be identified in the current transition amplitude histograms (left two panels): the peaks at about 2.1 pA (in the absence of EPA) and approx 2.5 pA (in the presence of EPA, which imparts a negative surface charge to the bilayer/solution interface) represent the des-Val^1-Gly^2-gA$^-$ channels; the peaks at about 3.5 pA (in the absence of EPA) and about 4.0 pA (in the presence of EPA) represent the [Ala^1]gA channels. The lifetime distributions (right two panels) show how the average lifetime of the shorter des-Val^1-Gly^2-gA$^-$ channels is 10-fold less than those of the longer [Ala^1]gA channels, and that the average lifetime of both channels are increased in the presence of EPA.

Although EPA is likely to alter many different bilayer properties, which all may contribute toward the changes in channel function observed in **Figs. 5** and **6**, it is useful to have a simple, descriptive term to summarize the changes in bilayer properties. Operationally, we therefore define changes in bilayer properties that at a constant bilayer thickness alter the disjoining force, as being changes in *bilayer stiffness* (**93**). A decrease in stiffness will decrease the disjoining force and increase gA channel appearance rate and lifetime and vice versa.

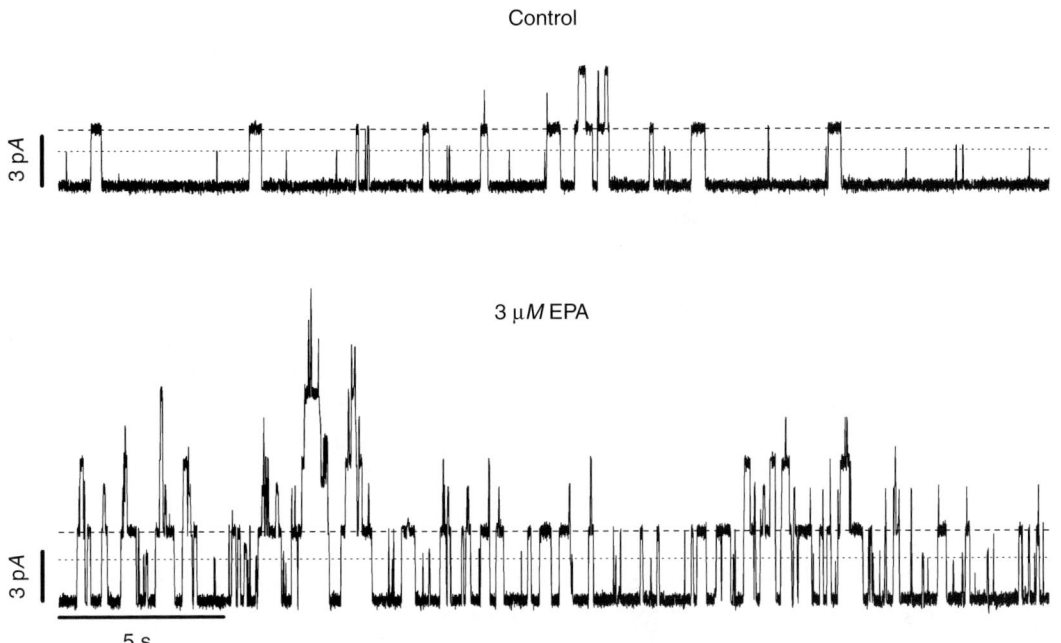

Fig. 5. Single-channel current traces obtained before and after the addition of 3 μM EPA to both sides of the bilayer that has been doped with des-Val^1-Gly^2-gA$^-$ and [Ala^1]gA. Two different gramicidins were present on both sides of the bilayer. The des-Val^1-Gly^2-gA$^-$ and [Ala^1]gA channels can be distinguished by virtue of their different current transition amplitudes, as indicated by the two horizontal lines in the top trace: des-Val^1-Gly^2-gA$^-$ · · · · ·; and [Ala^1]gA - - - - -. Bilayer made with dioleoylphosphatidylcholine; records obtained in 1 M NaCl, 10 mM HEPES buffered to pH 7.0, at 25°C; transbilayer voltage = 200 mV; filter: 500 Hz.

4. Using Gramicidin Channels as Molecular Force Transducers

In this section, the analytical framework is developed that is needed to fully exploit the use of gramicidin channels as molecular force transducers. Interested readers should also consult **ref. 108**.

4.1. Experiments With a Single Channel Type

First, the situation of a single channel type is discussed. The analysis can also be applied in a straightforward manner to the case of two different channels that do not form heterodimers.

4.1.1. Channel Lifetime as a Function of Hydrophobic Mismatch

The gramicidin channel dissociation rate constant (k_{-1}) is given by

$$k_{-1} = \frac{1}{\tau_0} \cdot \exp\left(\frac{-\Delta G_{tot}^{\ddagger,D \to M}}{k_B T}\right) \quad (10)$$

where $1/\tau_0$ is the frequency factor for the reaction and $\Delta G_{tot}^{\ddagger,D \to M}$ the transition state energy, such that the channel lifetime is given by

$$\ln(\tau) = -\ln(k_{-1}) = \ln(\tau_0) + \Delta G_{tot}^{\ddagger,D \to M}/k_B T \quad (11)$$

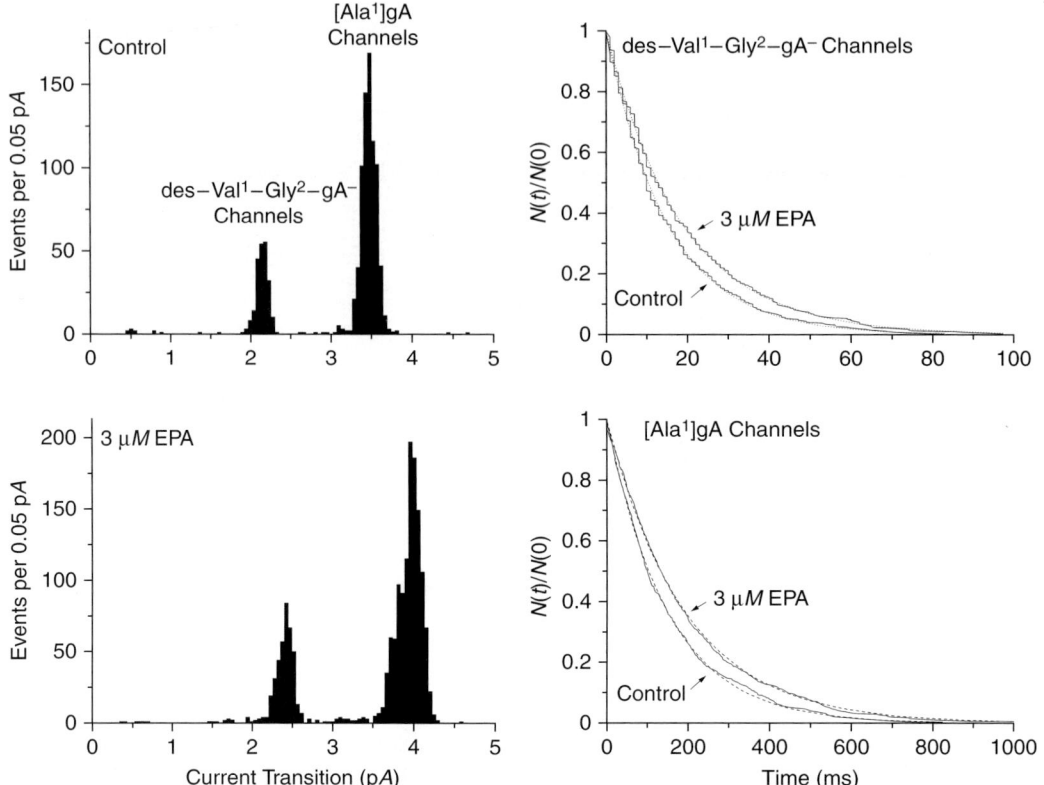

Fig. 6. Current transition amplitude histograms and lifetime distributions before and after the addition of 3 μM EPA to both sides of the bilayer that has been doped with des-Val^1-Gly^2-gA$^-$ and [Ala^1]gA. The two different channel types that were observed in the current traces (**Fig. 5**) are seen in the two peaks in the current transition amplitude histograms (**left panels**): the peaks at approx 2.1 pA (in the absence of EPA) and approx 2.5 pA (in the presence of EPA) represent the appearance and disappearance of des-Val^1-Gly^2-gA$^-$ channels; the peaks at approx 3.5 pA (in the absence of EPA) and approx 4.0 pA (in the presence of EPA) represent the appearance and disappearance of [Ala^1]gA channels. The lifetime distributions (**right panels**) were obtained based on analysis of the transitions underlying the two peaks in the histogram. Experimental details as in **Fig. 5**.

When a gA channel dissociates, the monomers separate a distance δ before the transition state is reached (*see* **Note 5**). $\Delta G_{\text{tot}}^{\ddagger,D\to M}$, thus includes contributions from subunit–subunit interactions as well as from channel–bilayer interactions, and can be decomposed into:

$$\Delta G_{\text{tot}}^{\ddagger,D\to M} = \Delta G_{\text{prot}}^{\ddagger,D\to M} + \Delta\Delta G_{\text{def}}^{\ddagger,D\to M} \qquad (12)$$

where $\Delta G_{\text{prot}}^{\ddagger,D\to M}$ is the intrinsic transition state energy and $\Delta\Delta G_{\text{def}}^{\ddagger,D\to M}$ is the difference in bilayer deformation energy for deformations of $2u_0$ ($= d_0 - l$) (the conducting channel) and $2u_0 - \delta$ ($= d_0 - l - \delta$) (the transition state), respectively:

$$\begin{aligned}\Delta\Delta G_{\text{def}}^{\ddagger,D\to M} &= H_B \cdot (2u_0 - \delta)^2 + H_X \cdot (2u_0 - \delta) \cdot c_0 + H_C \cdot c_0^2 - H_B \cdot (2u_0)^2 - H_X \cdot 2u_0 \cdot c_0 - H_C \cdot c_0^2 \\ &= -[H_B \cdot (4u_0 - \delta) + H_X \cdot c_0] \cdot \delta \\ &\approx -F_{\text{dis}} \cdot \delta \end{aligned} \qquad (13)$$

Combining **Eqs. 12** and **13**:

$$\Delta G_{\text{tot}}^{\ddagger,D\to M} = \Delta G_{\text{prot}}^{\ddagger,D\to M} - [H_B \cdot (4u_0 - \delta) + H_X \cdot c_0] \cdot \delta. \tag{14}$$

Combining **Eqs. 11** and **14**:

$$\ln(k_{-1}) = -\frac{\Delta G_{\text{prot}}^{\ddagger,D\to M} - [H_B \cdot (4u_0 - \delta) + H_X \cdot c_0] \cdot \delta}{k_B T} - \ln(\tau_0) \tag{15}$$

or, assuming that $\Delta G_{\text{prot}}^{\ddagger,D\to M}$ is invariant:

$$\frac{d[\ln(k_{-1})]}{du_0} = \frac{4 \cdot \delta}{k_B T} \cdot H_B \tag{16}$$

which allows for a determination of H_B—assuming that δ is known. But $u_0 = (d_0 - l)/2$, so **Eq. 16** can be rewritten in two different forms: (1) for fixed channel length and invariant subunit interface, such that $\Delta G_{\text{prot}}^{\ddagger,D\to M}$ indeed should be invariant:

$$\frac{d[\ln(k_{-1})]}{dd_0} = \frac{2 \cdot \delta}{k_B T} \cdot H_B \tag{17a}$$

(2) for fixed bilayer thickness, assuming no subunit-specific interactions, such that $\Delta G_{\text{prot}}^{\ddagger,D\to M}$ is invariant:

$$\frac{d[\ln(k_{-1})]}{dl} = -\frac{2 \cdot \delta}{k_B T} \cdot H_B \tag{17b}$$

Therefore, ln (k_{-1}) is a linear function of u_0 (of d_0 or l), which allows for determination of H_B from changes in τ as a function of d_0 (assuming δ is known), or from changes in τ as a function of l (assuming that δ is known, and that one can change l without altering $\Delta G_{\text{prot}}^{\ddagger,D\to M}$). In either case, the slope of the ln (k_{-1}) vs $(d_0 - l)$ relation does not depend on c_0. For the practical application of the method, δ is assumed to be 1.6 Å (*see* **Note 5**).

4.1.2. Channel Appearance Rate as a Function of Hydrophobic Mismatch

The gramicidin channel appearance rate (f) is given by

$$f = k_1 \cdot [M]^2 \tag{18}$$

where [M] is the monomer concentration in each bilayer leaflet (*see* **Note 6**) and the association rate constant (k_1) is given by:

$$k_1 = \frac{1}{\tau_0} \cdot \exp\left(\frac{-\Delta G_{\text{tot}}^{\ddagger,M\to D}}{k_B T}\right) \tag{19}$$

where $1/\tau_0$ is the frequency factor for the reaction, *see* **Eq. 10**, and $\Delta G_{\text{tot}}^{\ddagger,M\to D}$ the transition state energy. By analogy with **Subheading 4.1.1.**:

$$\Delta G_{\text{tot}}^{\ddagger,M\to D} = \Delta G_{\text{prot}}^{\ddagger,M\to D} + \Delta\Delta G_{\text{def}}^{\ddagger,M\to D} \tag{20}$$

where $\Delta G_{\text{prot}}^{\ddagger,M\to D}$ is the intrinsic transition state energy and $\Delta\Delta G_{\text{def}}^{\ddagger,M\to D}$ the difference in bilayer deformation energy for a deformation of 0 (the nonconducting monomeric subunits) and $2u_0 - \delta$ ($= d_0 - l - \delta$) (the transition state), respectively.

$$\Delta\Delta G_{\text{def}}^{\ddagger,M\to D} = H_B \cdot (2u_0 - \delta)^2 + H_X \cdot (2u_0 - \delta) \cdot c_0 + H_C \cdot c_0^2 - H_C \cdot c_0^2$$
$$= H_B \cdot (2u_0 - \delta)^2 + H_X \cdot (2u_0 - \delta) \cdot c_0 \qquad (21)$$

or

$$\ln(k_1) = -\ln(\tau) = -\frac{\Delta G_{\text{prot}}^{\ddagger,M\to D} + H_B \cdot (2u_0 - \delta)^2 + H_X \cdot (2u_0 - \delta) \cdot c_0}{k_B T} - \ln(\tau_0) \qquad (22)$$

The time-averaged number of conducting channels (the channel concentration) in the bilayer is given by:

$$[D] = k_1 \cdot [M]^2 / k_{-1} = f \cdot \tau \qquad (23)$$

Comparing **Eqs. 13** (or **15**) and **21** (or **22**), it is apparent that the association rate constant usually will vary as a stronger function of hydrophobic mismatch than the dissociation rate constants. Experimentally, however, the dissociation rate constants are more reproducible, because they do not depend on [M]. It is for that reason that the energetics of bilayer deformations usually is not evaluated using **Eq. 23**. The situation is not considered in detail here, but the reader is referred to **ref. 93** for an analysis of amphiphile-induced changes in bilayer elastic properties based on the above outlined arguments.

4.2. Experiments With Several Channel Types, Heterodimer Formation

This approach is based on the following three observations:

1. All bilayer-spanning $\beta^{6.3}$-helical gramicidin dimers are conducting channels *(105)*.
2. Gramicidin channel formation occurs through *trans*-bilayer association of two nonconducting $\beta^{6.3}$-helical monomers *(94)*.
3. Nonconducting $\beta^{6.3}$-helical gramicidin monomers are inserted into each of the two lipid bilayer leaflets *(106)*.

Further, although linear gramicidins can occur in several different conformations *(109–112)*, as long as the different conformers are in equilibrium among each other, one can describe gramicidin channel formation as a monomer ↔ dimer equilibrium between bilayer-embedded $\beta^{6.3}$-helical gramicidin monomers and dimers (as long as the different subunits have the same helix sense) (*see* **ref. 113**).

4.2.1. Basics of Heterodimer Formation

It is assumed that gramicidin channel appearances and disappearances can be equated with the association and dissociation of bilayer-spanning gramicidin dimers. If two different gramicidin analogs (A and B) are present in both monolayers of a membrane, one may observe not only symmetric, homodimeric AA and BB channels; but also asymmetric, heterodimeric AB and BA channels. The channel appearances and disappearances can be described as:

$$A_L + A_R \leftrightarrow AA, \quad K_{AA} = \frac{[AA]}{[A]_L \cdot [A]_R} \qquad (24)$$

$$B_L + B_R \leftrightarrow BB, \quad K_{BB} = \frac{[BB]}{[B]_L \cdot [B]_R} \tag{25}$$

$$A_L + B_R \leftrightarrow AB, \quad K_{AB} = \frac{[AB]}{[A]_L \cdot [B]_R} \tag{26}$$

$$B_L + A_R \leftrightarrow BA, \quad K_{BA} = \frac{[BA]}{[B]_L \cdot [A]_R} \tag{27}$$

where the subscripts "L" and "R" refer to the two bilayer halves, and the association constants are related to the standard free energies of dimerization for each different channel type (X) by

$$K_X = \exp(-\Delta G_X^0 / k_B T) = \exp\left[-(\Delta G_{X,\text{prot}}^0 + \Delta\Delta G_{X,\text{def}}^0)/k_B T\right] \tag{28}$$

where, as before, $\Delta G_{X,\text{prot}}^0$ denotes the energetic contributions because of the channel and sub unit–subunit interactions *per se*, and $\Delta\Delta G_{X,\text{def}}^0$ includes the energetic contributions that arise from the channel's interactions with the surrounding lipid bilayer. For simplicity, it is assumed that the monomers are symmetrically distributed: $[A]_L = [A]_R \equiv [A]$, and $[B]_L = [B]_R \equiv [B]$, but K_{AB} may differ from K_{BA} (because the relative stabilities of the two orientations may depend on the applied potential).

One cannot measure directly the bilayer concentrations of the different $\beta^{6.3}$-helical subunits, although they can be estimated from kinetic analysis of membrane conductance relaxations *(107)*; however, one can determine the dimer identities and concentrations, based on the different functional characteristics of the symmetric and asymmetric channels *(114)*. Consequently, even though it may not be possible to measure ΔG_X^0 itself, it *is* possible to measure the standard free energy *difference* for the heterodimers relative to the homodimers ($\Delta\Delta G^0$) by considering that the four kinetics schemes and equilibrium distributions in **Eqs. 24–27** formally can be represented by the following equilibrium distribution among conducting channels:

$$AA + BB \leftrightarrow AB + BA \tag{29}$$

with an equilibrium constant (K) given by:

$$K = \frac{[AB] \cdot [BA]}{[AA] \cdot [BB]} = \frac{K_{AB} \cdot K_{BA}}{K_{AA} \cdot K_{BB}} \tag{30}$$

with $\Delta\Delta G^0$ being defined as:

$$\Delta\Delta G^0 = -\frac{k_B T}{2} \ln(K) = \frac{(\Delta G_{AB}^0 + \Delta G_{BA}^0) - (\Delta G_{AA}^0 + \Delta G_{BB}^0)}{2} \tag{31a}$$

The factor 1/2 arises because it is convenient to measure $\Delta\Delta G^0$ per mole of monomeric subunit, i.e., per mole of A (or B) rather than AA (or BB). This convention is similar to that used previously *(114)*. (The presence of non-$\beta^{6.3}$-helical gramicidin conformers is immaterial for this argument, unless the folding of, and equilibrium distribution among, different gramicidin analogs somehow were influenced by the presence of other analogs.) If the two heterodimer orientations are indistinguishable, such that AB = BA = H, then **Eq. 31** can be rewritten as:

$$\Delta\Delta G^0 = -\frac{k_B T}{2}\ln(K) = \Delta G_H^0 - \frac{\Delta G_{AA}^0 + \Delta G_{BB}^0}{2} \qquad (31b)$$

an expression that often will be useful. For each dimer (channel) type the membrane concentration is given by:

$$[X] = f_X \cdot \tau_X \qquad (32)$$

where f_X is the channel's appearance rate and τ_X its average lifetime, so

$$\Delta\Delta G^0 = -\frac{k_B T}{2} \cdot \ln\left[\frac{(f_{AB}\cdot\tau_{AB})\cdot(f_{BA}\cdot\tau_{BA})}{(f_{AA}\cdot\tau_{AA})\cdot(f_{BB}\cdot\tau_{BB})}\right] \qquad (33)$$

or, in terms of measurable quantities:

$$\Delta\Delta G^0 = -\frac{k_B T}{2} \cdot \ln\left[\frac{(f_{hh}\cdot\tau_{hh})\cdot(f_{hl}\cdot\tau_{hl})}{(f_a\cdot\tau_a)\cdot(f_b\cdot\tau_b)}\right] \qquad (34a)$$

where "hh" and "hl" denote the low- and high-conductance heterodimers, whereas "a" and "b" denote the symmetrical channel types. If the two heterodimer orientations are indistinguishable, **Eq. 34a** becomes (*see* **Eq. 31b**) (*see also* **ref. *114***):

$$\Delta\Delta G^0 = -k_B T \cdot \ln\left(\frac{f_h \cdot \tau_h}{2\sqrt{f_a\cdot f_b}\cdot\sqrt{\tau_a\cdot\tau_b}}\right) \qquad (34b)$$

where f_h and τ_h are the heterodimer appearance rate and lifetime, respectively. The difference in activation energy for heterodimer formation relative to the symmetric channels ($\Delta\Delta G_f^\ddagger$) can be defined (and experimentally determined) by arguments that parallel those given above,

$$\Delta\Delta G_f^\ddagger = -\frac{k_B T}{2}\cdot\ln\left(\frac{f_{hh}\cdot f_{hl}}{f_a\cdot f_b}\right) \qquad (35a)$$

or, when the heterodimers are indistinguishable:

$$\Delta\Delta G_f^\ddagger = -k_B T \cdot \ln\left(\frac{f_h}{2\sqrt{f_a\cdot f_b}}\right). \qquad (35b)$$

Similarly, the difference in activation energy for heterodimer dissociation relative to the symmetric channels ($\Delta\Delta G_d^\ddagger$) can be defined, and determined, as:

$$\Delta\Delta G_d^\ddagger = \frac{k_B T}{2}\cdot\ln\left(\frac{\tau_{hh}\cdot\tau_{hl}}{\tau_a\cdot\tau_b}\right) \qquad (36a)$$

or, when the heterodimers are indistinguishable:

$$\Delta\Delta G_d^{\ddagger} = k_B T \cdot \ln\left(\frac{\tau_h}{\sqrt{\tau_a \cdot \tau_b}}\right). \quad (36b)$$

In either case, whether the heterodimers are distinguishable or indistinguishable:

$$\Delta\Delta G^0 = \Delta\Delta G_f^{\ddagger} - \Delta\Delta G_d^{\ddagger} \quad (37)$$

Whenever $\Delta\Delta G^0 \neq 0$, there are subunit-specific contributions to the energetics of channel formation *(114)*. To understand the molecular basis for $\Delta\Delta G^0 \neq 0$, it is helpful to recall that $\Delta G^0 = \Delta G_{prot}^0 + \Delta\Delta G_{def}^0$, such that

$$\Delta\Delta G^0 = \Delta\Delta G_{prot}^0 + \Delta\Delta\Delta G_{def}^0 \quad (38)$$

where:

$$\Delta\Delta G_{prot}^0 = \frac{\left(\Delta G_{AB,prot}^0 + \Delta G_{BA,prot}^0\right) - \left(\Delta G_{AA,prot}^0 + \Delta G_{BB,prot}^0\right)}{2} \quad (39)$$

and:

$$\Delta\Delta\Delta G_{def}^0 = \frac{\left(\Delta\Delta G_{AB,def}^0 + \Delta\Delta G_{BA,def}^0\right) - \left(\Delta\Delta G_{AA,def}^0 + \Delta\Delta G_{BB,def}^0\right)}{2}. \quad (40)$$

In the case that there are no subunit-specific interactions, $\Delta\Delta G_{prot}^0 = 0$ *(114)*. In case there are subunit-specific interactions, the $\Delta\Delta G_{prot}^0$ contribution to $\Delta\Delta G^0$ may be positive, because of an "engineered" instability at the subunit interface *(115)*, or negative, because of more favorable side chain interactions in the heterodimeric, as compared with the homodimeric, channels *(116)*. In either case, the major contribution to $\Delta\Delta G_{prot}^0$ is likely to be the $\Delta\Delta G_d^{\ddagger}$ (dissociation) contribution to $\Delta\Delta G^0$ (*see* **Eq. 37** and *see* **refs.** *114–116*). The $\Delta\Delta G_f^{\ddagger}$ contribution to $\Delta\Delta G_{prot}^0$ will be zero unless the bilayer spanning (and conducting) AA and BB homodimers have different folds *(114)*.

4.2.2. Effect of Bilayer Elastic Properties on the Relative Stability of Heterodimeric Channels

If the different subunits (A and B) have similar amino acid sequences, one would not expect that a channel-bilayer hydrophobic mismatch would have any impact on the relative heterodimer stability, on $\Delta\Delta G^0$. The situation becomes quite different in the case of heterodimeric channels formed between subunits of different length, as such experiments allow for another approach to measure H_B. In heterodimer experiments involving individual subunits that differ in length by two amino acid residues, the two homodimeric channels will differ in length by $2 \cdot \Delta$, where Δ is the length of an L–D pair of amino acids (*see* **Note 7**). The length of the heterodimeric channels will be the average of the lengths of the two homodimeric channels. For simplicity, the length of the heterodimeric channels (AB and BA) will be set to l, such that the lengths of the two homodimeric channels (AA and BB) are $(l - \Delta)$ and

$(l + \Delta)$. When individual subunits differ in length by two amino acid residues, the $\Delta\Delta G_{\text{def}}^{\text{M}\rightarrow\text{D}}$ contribution to ΔG^0 thus becomes (see **Eq. 8**):

$$\Delta\Delta G_{\text{AA,def}}^{\text{M}\rightarrow\text{D}} = H_{\text{B}} \cdot [d_0 - (l-\Delta)]^2 + H_{\text{X}} \cdot [d_0 - (l-\Delta)] \cdot c_0 \qquad (41\text{a})$$

$$\Delta\Delta G_{\text{AB,def}}^{\text{M}\rightarrow\text{D}} = \Delta G_{\text{BA,def}}^{0} = H_{\text{B}} \cdot (d_0 - l)^2 + H_{\text{X}} \cdot (d_0 - l) \cdot c_0 \qquad (41\text{b})$$

$$\Delta\Delta G_{\text{BB,def}}^{\text{M}\rightarrow\text{D}} = H_{\text{B}} \cdot [d_0 - (l+\Delta)]^2 + H_{\text{X}} \cdot [d_0 - (l+\Delta)] \cdot c_0 \qquad (41\text{c})$$

The $\Delta\Delta\Delta G_{\text{def}}^{0}$ contribution to $\Delta\Delta G^0$, see **Eq. 40**, thus becomes:

$$\Delta\Delta\Delta G_{\text{def}}^{0} = H_{\text{B}} \cdot (d_0 - l)^2 + H_{\text{X}} \cdot (d_0 - l) \cdot c_0$$
$$- H_{\text{B}} \cdot \frac{[(d_0 - l) - \Delta]^2 + [(d_0 - l) + \Delta]^2}{2} - H_{\text{X}} \cdot \frac{[(d_0 - l) - \Delta] + [(d_0 - l) + \Delta]}{2} \cdot c_0 = -H_{\text{B}} \cdot \Delta^2 \quad (42)$$

If $\Delta\Delta G_{\text{prot}}^{0} = 0$, one therefore can determine H_{B} by combining **Eqs. 34 and 42**:

$$H_{\text{B}} = \frac{k_{\text{B}}T}{2 \cdot \Delta^2} \cdot \ln\left[\frac{(f_{\text{hh}} \cdot \tau_{\text{hh}}) \cdot (f_{\text{hl}} \cdot \tau_{\text{hl}})}{(f_a \cdot \tau_a) \cdot (f_b \cdot \tau_b)}\right] = \frac{k_{\text{B}}T}{2 \cdot \Delta^2} \cdot \ln\left(\frac{f_{\text{hh}} \cdot f_{\text{hl}}}{f_a \cdot f_b} \cdot \frac{\tau_{\text{hh}} \cdot \tau_{\text{hl}}}{\tau_a \cdot \tau_b}\right) \qquad (43\text{a})$$

When the two heterodimer orientations are indistinguishable (see **Eqs. 31b and 34b**), **Eq. 43a** becomes:

$$H_{\text{B}} = -\frac{k_{\text{B}}T}{\Delta^2} \cdot \ln\left(\frac{f_{\text{h}} \cdot \tau_{\text{h}}}{2\sqrt{f_a \cdot f_b} \cdot \sqrt{\tau_a \cdot \tau_b}}\right). \qquad (43\text{b})$$

To proceed, it is helpful to examine the effects of bilayer elasticity on the relative appearance and lifetime terms (see **Eqs. 35 and 36**), as that becomes important for separating the $\Delta\Delta\Delta G_{\text{def}}^{0}$ and $\Delta\Delta G_{\text{prot}}^{0}$ contributions to $\Delta\Delta G^0$. To do so, it is helpful to rearrange **Eqs. 11 and 14** as

$$\tau = \tau_0 \cdot \exp\left\{\frac{\Delta G_{\text{prot}}^{\ddagger,\text{D}\rightarrow\text{M}} - [H_{\text{B}} \cdot (4u_0 - \delta) + H_{\text{X}} \cdot c_0] \cdot \delta}{k_{\text{B}}T}\right\}$$
$$= \tau_{\text{prot}} \cdot \exp\left\{\frac{-[H_{\text{B}} \cdot (4u_0 - \delta) + H_{\text{X}} \cdot c_0] \cdot \delta}{k_{\text{B}}T}\right\} \qquad (44)$$

where $\tau_{\text{prot}} = \tau_0 \cdot \exp(\Delta G_{\text{prot}}^{\ddagger,\text{D}\rightarrow\text{M}} / k_{\text{B}}T)$. It is then possible to express the lifetimes of the different channel types as:

$$\tau_{\text{AA}} = \tau_{\text{AA,prot}} \cdot \exp\left(\frac{-H_{\text{B}} \cdot \delta \cdot \{2 \cdot [d_0 - (l-\Delta)] - \delta\} + H_{\text{X}} \cdot \delta \cdot c_0}{k_{\text{B}}T}\right) \qquad (45\text{a})$$

$$\tau_{\text{AB}} = \tau_{\text{AB,prot}} \cdot \exp\left\{\frac{-H_{\text{B}} \cdot \delta \cdot [2 \cdot (d_0 - l) - \delta] + H_{\text{X}} \cdot \delta \cdot c_0}{k_{\text{B}}T}\right\} \qquad (45\text{b})$$

$$\tau_{BA} = \tau_{BA,prot} \cdot \exp\left\{\frac{-H_B \cdot \delta \cdot [2 \cdot (d_0 - l) - \delta] + H_X \cdot \delta \cdot c_0}{k_B T}\right\} \quad (45c)$$

$$\tau_{BB} = \tau_{BB,prot} \cdot \exp\left(\frac{-H_B \cdot \delta \cdot \{2 \cdot [d_0 - (l + \Delta)] - \delta\} + H_X \cdot \delta \cdot c_0}{k_B T}\right) \quad (45d)$$

and

$$\frac{\tau_{AB} \cdot \tau_{BA}}{\tau_{AA} \cdot \tau_{BB}} = \frac{\tau_{AB,prot} \cdot \tau_{BA,prot}}{\tau_{AA,prot} \cdot \tau_{BB,prot}} \quad (46)$$

because the exponential terms in **Eqs. 45a–d** cancel out in **Eq. 46**. That is, within the limits of the quadratic approximation (*see* **Eq. 4**), there is no bilayer contribution to the lifetime ratio. If $(\tau_{AB} \cdot \tau_{BA})/(\tau_{AA} \cdot \tau_{BB}) \neq 1$, then subunit specific interactions contribute to stabilization (or destabilization) of the heterodimeric channels relative to the two homodimeric channels. Similarly, one can express the bilayer contribution to the channel appearance rates, where

$$\Delta\Delta G_{def}^{\ddagger,M\to D} = H_B \cdot (d_0 - l - \delta)^2 + H_X \cdot (d_0 - l - \delta) \cdot c_0 \quad (47)$$

The channel appearance rates then are given by:

$$\begin{aligned}
f_{XY} &= [X]_L \cdot [Y]_R \cdot f_0 \cdot \exp\left(-\frac{\Delta G_{XY,prot}^{\ddagger,M\to D} + \Delta\Delta G_{XY,def}^{\ddagger,M\to D}}{k_B T}\right) \\
&= [X]_L \cdot [Y]_R \cdot f_{XY,prot} \cdot \exp\left[-\frac{H_B \cdot (d_0 - l - \delta)^2 + H_X \cdot (d_0 - l - \delta) \cdot c_0}{k_B T}\right]
\end{aligned} \quad (48)$$

where f_0 is a frequency factor and $f_{XY,prot} = f_0 \cdot \exp(-\Delta G_{XY,prot}^{\ddagger,M\to D} / k_B T)$. So, by analogy with the analysis for the lifetime ratio (and symmetric monomer distribution):

$$f_{AA} = [A]^2 \cdot f_{AA,prot} \cdot \exp\left\{-\frac{H_B \cdot [d_0 - (l - \Delta) - \delta]^2 + H_X \cdot [d_0 - (l - \Delta) - \delta] \cdot c_0}{k_B T}\right\} \quad (49a)$$

$$f_{AB} = [A] \cdot [B] \cdot f_{AB,prot} \cdot \exp\left[-\frac{H_B \cdot (d_0 - l - \delta)^2 + H_X \cdot (d_0 - l - \delta) \cdot c_0}{k_B T}\right] \quad (49b)$$

$$f_{BA} = [B] \cdot [A] \cdot f_{BA,prot} \cdot \exp\left[-\frac{H_B \cdot (d_0 - l - \delta)^2 + H_X \cdot (d_0 - l - \delta) \cdot c_0}{k_B T}\right] \quad (49c)$$

$$f_{BB} = [B]^2 \cdot f_{BB,prot} \cdot \exp\left\{-\frac{H_B \cdot [d_0 - (l + \Delta) - \delta]^2 + H_X \cdot [d_0 - (l + \Delta) - \delta] \cdot c_0}{k_B T}\right\} \quad (49d)$$

and

$$\frac{f_{AB} \cdot f_{BA}}{f_{AA} \cdot f_{BB}} = \frac{f_{AB,\text{prot}} \cdot f_{BA,\text{prot}}}{f_{AA,\text{prot}} \cdot f_{BB,\text{prot}}} \cdot \exp\left(\frac{H_B \cdot 2 \cdot \Delta^2}{k_B T}\right). \tag{50}$$

Assuming that $(f_{AB,\text{prot}} \cdot f_{BA,\text{prot}})/(f_{AA,\text{prot}} \cdot f_{BB,\text{prot}}) = 1$, which should be the case when the basic channel fold is not altered by the sequence alteration associated with the length change, H_B then can be expressed as:

$$H_B = \frac{k_B T}{2 \cdot \Delta^2} \cdot \ln\left(\frac{f_{AB} \cdot f_{BA}}{f_{AA} \cdot f_{BB}}\right) \tag{51a}$$

which should be compared with **Eq. 43a**. If the two heterodimer orientations are indistinguishable (*see* **Eqs. 31b** and **35b**), **Eq. 51a** becomes:

$$H_B = \frac{k_B T}{\Delta^2} \cdot \ln\left(\frac{f_h}{2\sqrt{f_a \cdot f_b}}\right). \tag{51b}$$

5. Experimental Materials and Methods
5.1. Synthesis of Gramicidin Analogs

Gramicidin analogs should be synthesized by solid-state peptide synthesis and twice purified by two-stage reversed-phase HPLC (Zorbax C8-80 Å, 4.6 mm × 25 cm, Agilent Technologies, Palo Alto, CA) with methanol/water as the mobile phase *(117)*. Samples collected from the HPLC column should be stored at –20°C as stock solutions. It is very important to have gramicidin analogs of the highest possible purity, as even minor amounts of impurities can produce aberrant channel activity, which will complicate the analysis and interpretation of the experimental results *(117)*. Before use, aliquots of the stock solutions are diluted in ethanol or dimethyl sulfoxide to a final concentration of about 1–100 n*M*. The gramicidin analogs are added to the aqueous solutions at each side of the bilayer, wherein they adsorb to bilayer/solution interface and fold into $\beta^{6.3}$-helical subunits *(106)*. (Because the linear gramicidins cross lipid bilayers poorly *(94)*, it is important to add the gramicidins to both sides of the bilayer.) The channel properties do vary as a function of the solvent used *(118)*; dimethyl sulfoxide and ethanol appear to be inert at the low concentrations used in single-channel measurements.

5.2. Electrophysiology

Single-channel measurements are done in planar bilayers, which can be formed from 2.5% w/v solution of lipids (phospholipids or phospholipid:cholesterol mixtures) in *n*-decane in a 100 µm thick Teflon partition (~1.5 mm diameter) separating two buffered or unbuffered electrolyte solutions. Experiments typically are done at 25 ± 1°C using the bilayer punch technique with a pipet tip of 30 µm diameter *(119)*, and a commercial patch-clamp amplifier. Small aliquots of the gramicidin(s) are added to both sides of a bilayer. The amount of gramicidin sample added in each experiment is adjusted to give about one channel event per second.

5.3. Data Acquisition and Analysis

Because many measurements are done using channels of different lengths, it is important that single-channel current transitions should be detected using a current transition-based algorithm, such as the one described in **ref. 119**. Single-channel current transition amplitude histograms and channel duration histograms can be constructed as shown in **Fig. 6**. These histograms differ from conventional current amplitude histograms by virtue that the histograms show the magnitude of the current transitions *per se*. The survivor plots of the channel durations should be fitted with single exponential distribution $N(t)/N(0) = \exp(-t/\tau)$, where $N(0)$ is the total number of channels, $N(t)$ the number of channels with a duration longer than time t, and τ the average channel lifetime.

6. Notes

1. Changes in lipid bilayer fluidity will change the kinetics of protein conformational transitions, but the forward and backward rates will change by the same factor *(123)*, meaning that the equilibrium distribution between different protein conformations will be impervious to changes in fluidity.
2. The energy density for monolayer bending is an approximation, as it relates to the situation in which all planes within the monolayer are bent similarly. This is not the case here because the plane separating the two monolayers is not bent. Indeed, the bilayer itself is plane, not bent.
3. Proteins are not rigid bodies; therefore, one would expect that the proteins would adapt to the bilayer, as the bilayer adapts to the protein. However, in practice proteins have bulk compressibility moduli *(124)* that are much higher than those of lipids *(125)*, such that a bilayer-channel hydrophobic mismatch will cause the bilayer to adapt to the protein.
4. ΔG_{def}^0 varies as a function of the bilayer-protein hydrophobic mismatch (u_0) and the lipid intrinsic curvature (c_0), which means that one should be able to express ΔG_{def}^0 as a Taylor expansion:

$$\Delta G_{def}^0(d_0 - l, c_0) = \Delta G_{def}^0(0,0) + \frac{\partial(\Delta G_{def}^0)}{\partial(d_0 - l)} \cdot (d_0 - l) + \frac{\partial(\Delta G_{def}^0)}{\partial c_0} \cdot c_0$$

$$+ \frac{1}{2} \frac{\partial^2(\Delta G_{def}^0)}{\partial(d_0 - l)^2} \cdot (d_0 - l)^2 + \frac{\partial^2(\Delta G_{def}^0)}{\partial(d_0 - l)\partial c_0} \cdot (d_0 - l) \cdot c_0 + \frac{1}{2} \cdot \frac{\partial^2(\Delta G_{def}^0)}{\partial c_0^2} \cdot c_0^2 + \ldots$$

   where the two first-order terms, by virtue of symmetry, will be zero (as the deformation energy for a small compression should be equal to that for a small extension, with a similar argument holding for bending). That is, the biquadratic form (**Eq. 5**) should be valid under rather general conditions, with $H_B = 0.5 \cdot \partial^2(\Delta G_{def}^0)/\partial(d_0 - l)^2$, $H_X = \partial^2(\Delta G_{def}^0)/\partial(d_0 - l)\partial c_0$ and $H_C = 0.5 \cdot \partial^2(\Delta G_{def}^0)/\partial c_0^2$
5. Based on experiments with gramicidin channels that miss hydrogen bonding at the subunit interface, the dissociation step appears to involve an axial separation of the two subunits by ~0.16 nm, the change in length owing to an L–D pair *(115)*. Similar estimates have been obtained from molecular modeling studies *(126)*, which suggest that the actual separation is more complex than a simple axial separation, involving also a rotation and lateral displacement of the subunits relative to each other. For the time being, it is assumed that $\delta = 0.16$ nm.
6. The channel activity is measured in a bilayer of finite area, meaning that the channel appearance rate is a function of the number of monomers in that particular area rather than the number of monomers per unit area. This issue will not be considered further.
7. $\Delta \approx \delta$, the distance the two subunits move apart to reach the transition state for channel dissociation. However, Δ, is likely to be better defined than δ, because it relates to channel equilibrium properties, as opposed to the kinetics of channel dissociation, which may be complex *(126)*, especially in the case of small hydrophobic mismatches, where the disjoining force imposed by the bilayer is small.

7. Conclusions

Gramicidin channels continue to be powerful tools for understanding many general features of bilayer regulation of membrane function. They also turn out to be useful for the development of single-molecule approaches to measure changes in membrane protein–lipid bilayer interactions, as changes in channel lifetimes and appearance rates provide direct information about changes in the energetics of protein-induced bilayer perturbations.

Acknowledgments

This work was supported in part by National Institutes of Health grants GM21342 (OSA) and RR15569 (REK). We thank Pablo Artigas, Helgi Ingolfsson, Jens A. Lundbæk, and Jon Sack for stimulating discussions and comments on the manuscript.

Nomenclature

c_0	Intrinsic monolayer curvature
d_0	Average hydrophobic thickness of unperturbed bilayer
d	Local bilayer thickness
f	Channel appearance rate
F_{dis}	Disjoining force the bilayer exerts on a bilayer-spanning channel (**Eq.**)
H_B	Phenomenological spring coefficient (**Eqs. 4 and 5**)
H_X	Phenomenological spring coefficient (**Eq. 5**)
H_C	Phenomenological spring coefficient (**Eq. 5**)
k_1	Gramicidin channel disassociation rate constant
k_{-1}	Channel dissociation rate constant
k_B	Boltzmann's constant
K_a	Bilayer area compression-expansion modulus
K_c	Bilayer splay-distortion modulus
K_D	Gramicidin channel dimerization constant (k_1/k_{-1})
l	Membrane protein, or gramicidin channel, hydrophobic length
T	Temperature in Kelvin
$2u_0$	Hydrophobic length difference between bilayer and protein (d_0-l)
$2u$	Local bilayer thickness change
δ	The distance two subunits separate in dissociation before transition state
Δ	Length of a dipeptide unit in the gramicidin sequence
ΔG^0_{def}	Bilayer deformation energy
ΔG^0_{prot}	Standard free energy change for a protein conformational change, or gramicidin channel formation, *per se*.
ΔG^0_{tot}	Standard free energy change for a protein conformational change, or gramicidin channel formation.
$\Delta\Delta G^{\ddagger}_f$	Activation energy difference between heterodimer and homodimer channel formation
$\Delta\Delta G^{\ddagger}_d$	Activation energy difference between heterodimer and homodimer channel dissociation
τ	Single-channel lifetime

References

1. Singer, S. J. and Nicolson, G. L. (1972) The fluid mosaic model of the structure of cell membranes. *Science* **175**, 720–731.
2. Haydon, D. A. and Hladky, S. B. (1972) Ion transport across thin lipid membranes. Critical discussion of mechanisms in selected systems. *Q. Rev. Biophys.* **5**, 187–282.
3. Andersen, O. S. (1978) Permeability properties of unmodified lipid bilayer membranes, in: *Membrane Transport in Biology*, (Giebisch, G., Tosteson, D. C., and Using, H. H. eds.), Springer, Berlin, pp. 369–446.
4. Finkelstein, A. (1987) *Water Movement Through Lipid Bilayers, Pores, and Plasma Membranes. Theory and Reality*. John Wiley, New York.
5. Finkelstein, A. (1976) Water and nonelectrolyte permeability of lipid bilayer membranes. *J. Gen. Physiol.* **68**, 127–135.
6. Sandermann, H., Jr. (1978) Regulation of membrane enzymes by lipids. *Biochim. Biophys. Acta* **515**, 209–237.
7. Sackmann, E. (1984) Physical basis of trigger processes and membrane structures, in *Biological Membranes*, (Chapman, D. ed.), Academic Press Inc. Ltd., London, pp. 105–143.
8. Devaux, P. F. and Seigneuret, M. (1985) Specificity of lipid-protein interactions as determined by spectroscopic techniques. *Biochim. Biophys. Acta* **822**, 63–125.
9. Bienvenüe, A. and Marie, J. S (1994) Modulation of protein function by lipids. *Curr. Top. Membr.* **40**, 319–354.
10. Dowhan, W. (1997) Molecular basis for membrane phospholipid diversity: why are there so many lipids? *Annu. Rev. Biochem.* **66**, 199–232.
11. Hilgemann, D. W., Feng, S., and Nasuhoglu, C. (2001). The complex and intriguing lives of PIP2 with ion channels and transporters. *Sci. STKE* **2001**, RE19.
12. Lee, A. G. (2004) How lipids affect the activities of integral membrane proteins. *Biochim. Biophys. Acta* **1666**, 62–87.
13. Myher, J. J., Kuksis, A., and Pind, S. (1989) Molecular species of glycerophospholipids and sphingomyelins of human erythrocytes: improved method of analysis. *Lipids* **24**, 396–407.
14. Fridriksson, E. K., Shipkova, P. A., Sheets, E. D., Holowka, D., Baird, B., and McLafferty, F. W. (1999) Quantitative analysis of phospholipids in functionally important membrane domains from RBL-2H3 mast cells using tandem high-resolution mass spectrometry. *Biochemistry* **38**, 8056–8063.
15. Pike, L. J., Han, X., Chung, K., and Gross, R. W. (2002) Lipid rafts are enriched in arachidonic acid and plasmenylethanolamine and their compostion is independent of caveolin-1 expression: a quantitative electrospray ionization/mass spectrometric analysis. *Biochemistry* **41**, 2075–2088.
16. Lemmon, M. A. and Ferguson, K. M. (2000) Signal-dependent membrane targeting by pleckstrin homology (PH) domains. *Biochem. J.* **350(Pt 1)**, 1–18.
17. McLaughlin, S., Wang, J., Gambhir, A., and Murray, D. (2002) PIP(2) and proteins: interactions, organization, and information flow. *Annu. Rev. Biophys. Biomol. Struct.* **31**, 151–175.
18. McIntosh, T. J. and Simon, S. A. (2006) Roles of bilayer material properties in function and distribution of membrane proteins. *Annu. Rev. Biophys. Biomol. Struct.* **35**, 177–198.
19. Unwin, P. N. T. and Ennis, P. D. (1984) Two configurations of a channel-forming membrane protein. *Nature* **307**, 609–613.
20. Unwin, N., Toyoshima, C., and Kubalek, E. (1988) Arrangement of the acetylcholine receptor subunits in the resting and desensitized states, determined by cryoelectron microscopy of crystallized *Torpedo* postsynaptic membranes. *J. Cell Biol.* **107**, 1123–1138.
21. Doyle, D. A., Morais Cabral, J., Pfuetzner, R. A., et al. (1998) The structure of the potassium channel: molecular basis of K^+ conduction and selectivity. *Science* **280**, 69–77.

22. Jiang, Y., Lee, A., Chen, J., Cadene, M., Chait, B. T., and MacKinnon, R. (2002) Crystal structure and mechanism of a calcium-gated potassium channel. *Nature* **417,** 515–522.
23. Jiang, Y., Lee, A., Chen, J., et al. (2003) X-ray structure of a voltage-dependent K$^+$ channel. *Nature* **423,** 33–41.
24. Jiang, Y., Ruta, V., Chen, J., Lee, A., and MacKinnon, R. (2003) The principle of gating charge movement in a voltage-dependent K$^+$ channel. *Nature* **423,** 42–48.
25. Toyoshima, C., Nakasako, M., Nomura, H., and Ogawa, H. (2000) Crystal structure of the calcium pump of sarcoplasmic reticulum at 2.6Å resolution. *Nature* **405,** 647–655.
26. Toyoshima, C., Nomura, H., and Tsuda, T. (2004) Lumenal gating mechanism revealed in calcium pump crystal structures with phosphate analogues. *Nature* **432,** 361–368.
27. Chang, G., Spencer, R. H., Lee, A. T., Barclay, M. T., and Rees, D. C. (1998) Structure of the MscL homolog from *Mycobacterium tuberculosis*: a gated mechanosensitive ion channel. *Science* **282,** 2220–2226.
28. Sukharev, S., Betanzos, M., Chiang, C., and Guy, H. R. (2001) The gating mechanism of the large mechanosensitive channel MscL. *Nature* **409,** 720–724.
29. Perozo, E., Cortes, D. M., Sompornpisut, P., Kloda, A., and Martinac, B. (2002) Open channel structure of MscL and the gating mechanism of mechanosensitive channels. *Nature* **418,** 942–948.
30. Salamon, Z., Wang, Y., Brown, M. F., Macleod, H. A., and Tollin, G. (1994) Conformational changes in rhodopsin probed by surface plasmon resonance spectroscopy. *Biochemistry* **33,** 13,706–13,711.
31. Sakmar, T. P. (1998) Rhodopsin: a prototypical G protein-coupled receptor. *Prog. Nucl. Acid Res. Mol. Biol.* **59,** 1–34.
32. Brown, M. F. (1994) Modulation of rhodopsin function by properties of the membrane bilayer. *Chem. Phys. Lipids* **73,** 159–180.
33. Isele, J., Sakmar, T. P., and Siebert, F. (2000) Rhodopsin activation affects the environment of specific neighboring phospholipids: an FTIR spectroscopic study. *Biophys. J.* **79,** 3063–3071.
34. Damjanovich, S., Edidin, M., Szöllõsi, J., and Trón, L. (1994) Mobility and Proximity in Biological Membranes. CRC Press, Boca Raton, FL.
35. Barenholz, Y. (2002) Cholesterol and other membrane active sterols: from membrane evolution to "rafts". *Prog. Lipid Res.* **41,** 1–5.
36. Lee, A. G. (1991) Lipids and their effects on membrane proteins: Evidence against a role for fluidity. *Prog. Lipid Res.* **30,** 323–348.
37. Iwata, S., Ostermeier, C., Ludwig, B., and Michel, H. (1995) Structure at 2 Å resolution of cytochrome c oxidase from *Paracoccus denitrificans*. *Nature* **376,** 660–669.
38. McAuley, K., Fyfe, P. K., Ridge, J. P., Isaacs, N. W., Cogdell, R. J., and Jones, M. R. (1999) Structural details of an interaction between cardiolipin and an integral membrane protein. *Proc. Natl. Acad. Sci. USA* **96,** 14,706–14,711.
39. Valiyaveetil, F. I., Zhou, Y., and MacKinnon, R. (2002) Lipids in the structure, folding, and function of the KcsA K$^+$ channel. *Biochemistry* **41,** 10,771–10,777.
40. Lee, A. G. (2003) Lipid–protein interactions in biological membranes: a structural perspective. *Biochim. Biophys. Acta* **1612,** 1–40.
41. Tefft, R. E., Jr., Carruthers, A., and Melchior, D. L. (1986) Reconstituted human erythrocyte sugar transporter activity is determined by bilayer lipid head groups. *Biochemistry* **25,** 3709–3718.
42. Starling, A. P., East, J. M., and Lee, A. G. (1995) Evidence that the effects of phospholipids on the activity of the Ca^{2+}-ATPase do not involve aggregation. *Biochem. J.* **308,** 343–346.
43. Cornelius, F. (2001) Modulation of Na, K-ATPase and Na-ATPase activity by phospholipids and cholesterol. I. steady-state kinetics. *Biochemistry* **40,** 8842–8851.
44. Perozo, E., Kloda, A., Cortes, D. M., and Martinac, B. (2002) Physical principles underlying the transduction of bilayer deformation forces during mechanosensitive channel gating. *Nat. Struct. Biol.* **9,** 696–703.

45. Navarro, J., Toivio-Kinnucan, M., and Racker, E. (1984) Effect of lipid composition on the calcium/adenosine 5′-triphosphate coupling ratio of the Ca^{2+}-ATPase of sarcoplasmic reticulum. *Biochemistry* **23**, 130–135.
46. Hui, S. -W. and Sen, A. (1989) Effects of lipid packing on polymorphic phase behavior and membrane properties. *Proc. Natl. Acad. Sci. USA* **86**, 5825–5829.
47. McCallum, C. D. and Epand, R. M. (1995) Insulin receptor autophosphorylation and signalling is altered by modulation of membrane physical properties. *Biochemistry* **34**, 1815–1824.
48. Chang, H. M. and Reitstetter, R. G. R. (1995) Lipid-ion channel interactions: Increasing phospholipid headgroup size but not ordering acyl chains alters reconstituted channel behavior. *J. Membr. Biol.* **145**, 13–19.
49. Hazel, J. R. (1995) Thermal adaption in biological membranes: is homeoviscous adaption the explanation? *Annu. Rev. Physiol.* **57**, 19–42.
50. Lindblom, G., Oradd, G., Rilfors, L., and Morein, S. (2002) Regulation of lipid composition in *Acholeplasma laidlawii* and *Escherichia coli* membranes: NMR studies of lipid lateral diffusion at different growth temperatures. *Biochemistry* **41**, 11,512–11,515.
51. Gruner, S. M. (1991) Lipid membrane curvature elasticity and protein function, in *Biologically Inspired Physics*, (Peliti, L., ed.), Plenum Press, New York, pp. 127–135.
52. Andersen, O. S., Sawyer, D. B., and Koeppe, R. E., II. (1992) Modulation of channel function by the host bilayer, in *Biomembrane Structure and Function*, (Easwaran, K. R. K. and Gaber, B., eds.), Adenine Press, Schenectady, NY, pp. 227–244.
53. Huang, H. W. (1986) Deformation free energy of bilayer membrane and its effect on gramicidin channel lifetime. *Biophys. J.* **50**, 1061–1070.
54. Mouritsen, O. G. and Bloom, M. (1984) Mattress model of lipid-protein interactions in membranes. *Biophys. J.* **46**, 141–153.
55. Gruner, S. M. (1985) Intrinsic curvature hypothesis for biomembrane lipid composition: a role for nonbilayer lipids. *Proc. Natl. Acad. Sci. USA* **82**, 3665–3669.
56. Helfrich, P. and Jakobsson, E. (1990) Calculation of deformation energies and conformations in lipid membranes containing gramicidin channels. *Biophys. J.* **57**, 1075–1084.
57. Ring, A. (1996) Gramicidin channel-induced lipid membrane deformation energy: influence of chain length and boundary conditions. *Biochim. Biophys. Acta* **1278**, 147–159.
58. Dan, N. and Safran, S. A. (1998) Effect of lipid characteristics on the structure of transmembrane proteins. *Biophys. J.* **75**, 1410–1414.
59. Nielsen, C., Goulian, M., and Andersen, O. S. (1998) Energetics of inclusion-induced bilayer deformations. *Biophys. J.* **74**, 1966–1983.
60. Nielsen, C. and Andersen, O. S. (2000) Inclusion-induced bilayer deformations: effects of monolayer equilibrium curvature. *Biophys. J.* **79**, 2583–2604.
61. Helfrich, W. (1973) Elastic properties of lipid bilayers: theory and possible experiments. *Z. Naturforsch.* **28C**, 693–703.
62. Helfrich, W. (1981) Amphiphilic mesophases made of defects, in *Physique des défauts (Physics of defects)*, (Balian, R., Kléman, M., and Poirier, J. -P., eds.), North-Holland Publishing Company, New York, pp. 716–755.
63. Partenskii, M. B. and Jordan, P. C. (2002) Membrane deformation and the elastic energy of insertion: Perturbation of membrane elastic constants due to peptide insertion. *J. Chem. Phys.* **117**, 10,768–10,776.
64. Lundbæk, J. A. and Andersen, O. S. (1999) Spring constants for channel-induced lipid bilayer deformations—estimates using gramicidin channels. *Biophys. J.* **76**, 889–895.
65. Cantor, R. (1997) Lateral pressures in cell membranes: a mechanism for modulation of protein function. *J. Phys. Chem. B* **101**, 1723–1725.
66. Bezrukov, S. M. (2000) Functional consequences of lipid packing stress. *Curr. Opin. Coll. Interface Sci.* **5**, 237–243.

67. Mitchell, D. C., Straume, M., Miller, J. L., and Litman, B. J. (1990) Modulation of metarhodopsin formation by cholesterol-induced ordering of bilayer lipids. *Biochemistry* **29,** 9143–9149.
68. Booth, P. J., Templer, R. H., Meijberg, W., Allen, S. J., Curran, A. R., and Lorch, M. (2001) *In Vitro* studies of membrane protein folding. *Crit. Rev. Biochem. Mol. Biol.* **36,** 501–603.
69. Keller, S. L., Bezrukov, S. M., Gruner, S. M., Tate, M. W., Vodyanoy, I., and Parsegian, V. A. (1993) Probability of alamethicin conductance states varies with nonlamellar tendency of bilayer phospholipids. *Biophys. J.* **65,** 23–27.
70. Epand, R. M. (1998) Lipid polymorphism and protein-lipid interactions. *Biochim. Biophys. Acta* **1376,** 353–368.
71. Seddon, J. M. (1990) Structure of the inverted hexagonal (H_{II}) phase, and non-lamellar phase transitions of lipids. *Biochim. Biophys. Acta* **1031,** 1–69.
72. Tate, M. W., Eikenberry, E. F., Turner, D. C., Shyamsunder, E., and Gruner, S. M. (1991) Nonbilayer phases of membrane lipids. *Chem. Phys. Lipids* **57,** 147–164.
73. Evans, E., Rawicz, W., and Hofmann, A. F. (1995) Lipid bilayer expansion and mechanical disruption in solutions of water-soluble bile acid, in *Bile Acids in Gastroenterology Basic and Clinical Advances*, (Hofmann, A. F., Paumgartner, G., and Stiehl, A. eds.), Kluwer Academic Publishers, Dordrecht, pp. 59–68.
74. McIntosh, T. J., Advani, S., Burton, R. E., Zhelev, D. V., Needham, D., and Simon, S. A. (1995) Experimental tests for protrusion and undulation pressures in phospholipid bilayers. *Biochemistry* **34,** 8520–8532.
75. Santore, M. M., Discher, D. E., Won, Y. -Y., Bates, F. S., and Hammer, D. A. (2002) Effect of surfactant on unilamellar polymeric vesicles: Altered membrane properties and stability in the limit of weak surfactant partitioning. *Langmuir* **18,** 7299–7308.
76. Ly, H. V. and Longo, M. L. (2004) The influence of short-chain alcohols on interfacial tension, mechanical properties, area/molecule, and permeability of fluid lipid bilayers. *Biophys. J.* **87,** 1013–1033.
77. Zhou, Y. and Raphael, R. M. (2005) Effect of salicylate on the elasticity, bending stiffness, and strength of SOPC membranes. *Biophys. J.* **89,** 1789–1801.
78. Evans, E. and Needham, D. (1987) Physical properties of surfactant bilayer membranes: thermal transitions, elasticity, rigidity, cohesion, and colloidal interactions. *J. Phys. Chem.* **91,** 4219–4228.
79. Needham, D. (1995) Cohesion and permeability of lipid bilayer vesicles, in *Permeability and Stability of Lipid Bilayers*, (Disalvo, E. A. and Simon, S. A., eds.), CRC Press, Boca Raton, FL, pp. 49–76.
80. Rawicz, W., Olbrich, K. C., McIntosh, T., Needham, D., and Evans, E. (2000) Effect of chain length and unsaturation on elasticity of lipid bilayers. *Biophys. J.* **79,** 328–339.
81. Allende, D., Vidal, A., Simon, S. A., and McIntosh, T. J. (2003) Bilayer interfacial properties modulate the binding of amphipathic peptides. *Chem. Phys. Lipids* **122,** 65–76.
82. Parsegian, V. A., Rand, R. P., and Rau, D. C. (1995) Macromolecules and water: probing with osmotic stress. *Methods Enzymol.* **259,** 43–94.
83. Rostovtseva, T. K., Liu, T. -T., Colombini, M., Parsegian, V. A., and Bezrukov, S. M. (2000) Positive cooperativity without domains or subunits in a monomeric membrane channel. *PNAS* **97,** 7819–7822.
84. Sawyer, D. B., Koeppe, R. E., II., and Andersen, O. S. (1989) Induction of conductance heterogeneity in gramicidin channels. *Biochemistry* **28,** 6571–6583.
85. Lundbæk, J. A. and Andersen, O. S. (1994) Lysophospholipids modulate channel function by altering the mechanical properties of lipid bilayers. *J. Gen. Physiol.* **104,** 645–673.
86. Lundbæk, J. A., Birn, P., Girshman, J., Hansen, A. J., and Andersen, O. S. (1996) Membrane stiffness and channel function. *Biochemistry* **35,** 3825–3830.
87. Lundbæk, J. A., Maer, A. M., and Andersen, O. S. (1997) Lipid bilayer electrostatic energy, curvature stress, and assembly of gramicidin channels. *Biochemistry* **36,** 5695–5701.

88. Bezrukov, S. M., Rand, R. P., Vodyanoy, I., and Parsegian, V. A. (1998) Lipid packing stress and polypeptide aggregation: alamethicin channel probed by proton titration of lipid charge. *Faraday Discuss.* 173–183, discussion 225–246.
89. Andersen, O. S., Nielsen, C., Maer, A. M., Lundbæk, J. A., Goulian, M., and Koeppe, R. E., II. (1999) Ion channels as tools to monitor lipid bilayer-membrane protein interactions: gramicidin channels as molecular force transducers. *Methods Enzymol.* **294,** 208–224.
90. Sawyer, D. B. and Andersen, O. S. (1989) Platelet-activating factor is a general membrane perturbant. *Biochim. Biophys. Acta* **987,** 129–132.
91. Hwang, T. C., Koeppe, R. E., II., and Andersen, O. S. (2003) Genistein can modulate channel function by a phosphorylation-independent mechanism: importance of hydrophobic mismatch and bilayer mechanics. *Biochemistry* **42,** 13,646–13,658.
92. Lundbæk, J. A., Birn, P. H. A. J., Søgaard, R., et al. (2004) Regulation of sodium channel function by bilayer elasticity: the importance of hydrophobic coupling: effects of micelle-forming amphiphiles and cholesterol. *J. Gen. Physiol.* **123,** 599–621.
93. Lundbæk, J. A., Birn, P., Tape, S. E., et al. (2005) Capsaicin regulates voltage-dependent sodium channels by altering lipid bilayer elasticity. *Mol. Pharmacol.* **68,** 680–689.
94. O'Connell, A. M., Koeppe, R. E., II., and Andersen, O. S. (1990) Kinetics of gramicidin channel formation in lipid bilayers: transmembrane monomer association. *Science* **250,** 1256–1259.
95. Andersen, O. S. and Koeppe, R. E., II. (1992) Molecular determinants of channel function. *Physiol. Rev.* **72,** S89–S158.
96. Andersen, O. S., Apell, H. -J., Bamberg, E., et al. (1999) Gramicidin channel controversy—the structure in a lipid environment. *Nat. Struct. Biol.* **6,** 609.
97. Sarges, R., Witkop, B., and Gramicidin, A. V. (1965) The structure of valine- and isoleucine-gramicidin A. *J. Am. Chem. Soc.* **87,** 2011–2019.
98. Arseniev, A. S., Lomize, A. L., Barsukov, I. L., and Bystrov, V. F. (1986) Gramicidin A transmembrane ion-channel. Three-dimensional structure reconstruction based on NMR spectroscopy and energy refinement. *Biol. Membr.* **3,** 1077–1104.
99. Ketchem, R. R., Roux, B., and Cross, T. A. (1997) High-resolution polypeptide structure in a lamellar phase lipid environment from solid state NMR derived orientational constraints. *Structure* **5,** 1655–1669.
100. Townsley, L. E., Tucker, W. A., Sham, S., and Hinton, J. F. (2001) Structures of gramicidins A, B, and C incorporated into sodium dodecyl sulfate micelles. *Biochemistry* **40,** 11,676–11,686.
101. Andersen, O. S., Koeppe, R. E., II., and Roux, B. (2005) Gramicidin channels. *IEEE Trans. Nanobiosci.* **4,** 10–20.
102. Elliott, J. R., Needham, D., Dilger, J. P., and Haydon, D. A. (1983) The effects of bilayer thickness and tension on gramicidin single-channel lifetime. *Biochim. Biophys. Acta* **735,** 95–103.
103. Benz, R., Fröhlich, O., Läuger, P., and Montal, M. (1975) Electrical capacity of black lipid films and of lipid bilayers made from monolayers. *Biochim. Biophys. Acta* **394,** 323–334.
104. Lewis, B. A. and Engelman, D. M. (1983) Lipid bilayer thickness varies linearly with acyl chain length in fluid phosphatidylcholine vesicles. *J. Mol. Biol.* **166,** 211–217.
105. Veatch, W. R., Mathies, R., Eisenberg, M., and Stryer, L. (1975) Simultaneous fluorescence and conductance studies of planar bilayer membranes containing a highly active and fluorescent analog of gramicidin A. *J. Mol. Biol.* **99,** 75–92.
106. He, K., Ludtke, S. J., Wu, Y., et al. (1994) Closed state of gramicidin channel detected by X-ray in-plane scattering. *Biophys. Chem.* **49,** 83–89.
107. Bamberg, E. and Läuger, P. (1973) Channel formation kinetics of gramicidin A in lipid bilayer membranes. *J. Membr. Biol.* **11,** 177–194.
108. Lundbæk, J. A. (2006) Regulation of membrane protein function by lipid bilayer elasticity—a single molecule technology to measure the bilayer properties experienced by an embedded protein. *J. Physics: Cond. Matt.* **18,** S1305–S1344.

109. Veatch, W. R., Fossel, E. T., and Blout, E. R. (1974) The conformation of gramicidin A. *Biochemistry* **13,** 5249–5256.
110. Bystrov, V. F. and Arseniev, A. S. (1988) Diversity of the gramicidin A spatial structure: two-dimensional proton NMR study in solution. *Tetrahedron* **44,** 925–940.
111. Abdul-Manan, N. and Hinton, J. F. (1994) Conformational states of gramicidin A along the pathway to the formation of channels in model membranes determined by 2D NMR and circular dichroism spectroscopy. *Biochemistry* **33,** 6773–6783.
112. Salom, D., Perez-Paya, E., Pascal, J., and Abad, C. (1998) Environment- and sequence-dependent modulation of the double-stranded to single-stranded conformational transition of gramicidin A in membranes. *Biochemistry* **37,** 14,279–14,291.
113. Koeppe, R. E., II., Providence, L. L., Greathouse, D. V., et al. (1992) On the helix sense of gramicidin A single channels. *Proteins* **12,** 49–62.
114. Durkin, J. T., Koeppe, R. E., II., and Andersen, O. S. (1990) Energetics of gramicidin hybrid channel formation as a test for structural equivalence. Side-chain substitutions in the native sequence. *J. Mol. Biol.* **211,** 221–234.
115. Durkin, J. T., Providence, L. L., Koeppe, R. E., II., and Andersen, O. S. (1993) Energetics of heterodimer formation among gramicidin analogues with an NH_2-terminal addition or deletion. Consequences of a missing residue at the join in channel. *J. Mol. Biol.* **231,** 1102–1121.
116. Saberwal, G., Greathouse, D., Koeppe, R. E., II., and Andersen, O. S. (1994) Contacts that are good for you: side chain contacts in the gramicidin channel. *Biophys. J.* **66,** A219.
117. Greathouse, D. V., Koeppe, R. E., II., Providence, L. L., Shobana, S., and Andersen, O. S. (1999) Design and characterization of gramicidin channels. *Methods Enzymol.* **294,** 525–550.
118. Sawyer, D. B., Koeppe, R. E., II., and Andersen, O. S. (1990) Gramicidin single-channel properties show no solvent-history dependence. *Biophys. J.* **57,** 515–523.
119. Andersen, O. S. (1983) Ion movement through gramicidin A channels. Single-channel measurements at very high potentials. *Biophys. J.* **41,** 119–133.
120. Mouritsen, O. G. and Andersen, O. S. (1998) Do we need a new biomembrane model. *In*: In Search of a New Biomembrane Model. Mouritsen, O. G. and Anderson, O. S., editors. *Biol. Skr. Dan. Vid. Selsk.* **49,** 7–12. Munksgaard, Copenhagen.
121. Borisenko, V., Lougheed, T., Hesse, J., et al. (2003) Simultaneous Optical and Electrical Recording of Single Gramicidin Channels. *Biophys. J.* **84,** 612–622.
122. Harms, G. S., Orr, G., Montal, M., Thrall, B. D., Colson, S. D., and Lu, H. P. (2003) Probing conformational changes of gramicidin ion channels by single-molecule patch-clamp fluorescence microscopy. *Biophys. J.* **85,** 1826–1838.
123. Schurr, J. M. (1970) The role of diffusion in bimolecular solution kinetics. *Biophys. J.* **10,** 701–716.
124. Gekko, K. and Noguchi, H. (1979) Compressibility of globular proteins in water at 25°C. *J. Phys. Chem.* **83,** 2706–2714.
125. Liu, N. and Kay, R. L. (1977) Redetermination of the pressure dependence of the lipid bilayer phase transition. *Biochemistry* **16,** 3484–3486.
126. Miloshevsky, G. V. and Jordan, P. C. (2004) Gating gramicidin channels in lipid bilayers: reaction coordinates and the mechanism of dissociation. *Biophys. J.* **86,** 92–104.

38

Ion-Channel Reconstitution

Francisco J. Morera, Guillermo Vargas, Carlos González, Eduardo Rosenmann, and Ramon Latorre

Summary

In this chapter, a detailed protocol is given for ion-channel reconstitution in the two most used model membranes: planar bilayers and liposomes. In the planar bilayer section, methods are described for the expression of ion channels in *Xenopus laevis* oocytes, the isolation of their membranes, the insertion of ion channels into the bilayer by vesicle fusion, and the recording of single-ion channel current measurements at a constant applied voltage. The reconstitution of bacterial channels in liposomes is also given. It includes the expression and purification of bacterial channels in *E. Coli* host strain XL1-blue, the insertion of the channels in liposomes, and the recording of their currents by patch clamping.

Key Words: Artificial planar lipid bilayers; ion-channel reconstitution; liposomes; oocyte membrane preparation; patch-clamp; single-channel recording; voltage-clamp.

1. Introduction

Ion channels are integral membrane proteins that form hydrophilic pores, which pass a controlled flow of ions down the electrochemical gradient. Every living cell has many types of ion channels, and they take part in many cellular processes like nerve conduction and muscle contraction, as well as the tactile, auditory and visual senses among others (*1*).

One method applied to study ion-channel function and its properties is the classical biochemical approach of purifying and assaying the protein of interest in a defined medium. But, without a membrane in which to place the channel protein under study, there can be no assay of its ion-transporting function (*2*). Ion-channel reconstitution is the assembly of an ion channel, taken from a biological membrane, in a model or "artificial" membrane formed to define the two aqueous phases for transport. Two kinds of model membranes have been successfully used in reconstitution work: planar bilayers and liposomes. Ion fluxes mediated by these proteins are then measured, usually by electrical techniques. Reconstitution presents distinct experimental advantages: it is a very well-controlled experimental system, especially when the channel protein is biochemically pure. It makes possible to study ion channels from intracellular membranes, which are not easily accessible to recording electrodes. It is possible to study the influence of membrane lipid composition on channel function. Planar bilayers have the additional advantage that the solution on both sides of the membrane can be easily perfused.

In this chapter, a detailed protocol is given for ion-channel reconstitution in planar bilayers and liposomes. In the planar bilayers section, methods are described for the expression of ion channels in *Xenopus* oocytes, the isolation of their membranes, the insertion of ion channels into the bilayer by vesicle fusion, and the recording of single-ion channel current measurements

From: *Methods in Molecular Biology, vol. 400: Methods in Membrane Lipids*
Edited by: A. M. Dopico © Humana Press Inc., Totowa, NJ

at a constant applied voltage. The reconstitution of bacterial channels in liposomes is also given. It includes the expression and purification of bacterial channels in XL1 blue strain, the insertion of the channels in liposomes and the recording of their currents by patch clamping.

2. Materials
2.1. Reconstitution of Cloned Ion Channels in Planar Bilayers

1. cDNA coding for the ion channel of interest cloned in a plasmid designed for expression in *Xenopus* oocytes, such as pBSTA.
2. Transcription in vitro system T7Message Machine (Ambion, Austin, TX).
3. *X. laevis* frogs (Nasco, Fort Atkinson, WI).
4. Tricaine.
5. Collagenase type 4 (Worthington, Lakewood, NJ).
6. OR2 solution (calcium-free): 82.5 mM NaCl, 2 mM KCl, 1 mM MgCl$_2$, and 5 mM HEPES, pH 7.5 at 18°C.
7. ND96 solution: 96 mM NaCl, 2 mM KCl, 1.8 mM CaCl$_2$, 1 mM MgCl$_2$, and 5 mM HEPES, pH 7.5 at 18°C.
8. Gentamicin (store at 4°C).
9. Nanoliter injector (World Precision Instrument, Inc., Sarasota, FL).
10. RNAse-free mineral oil.
11. Glass replacement 3.5 Nanoliter (World Precision Instrument, Inc., Sarasota, FL).
12. 10, 20, and 50% w/v sucrose solutions in 0.6 M KCl, and 10 mM HEPES, pH 6.8 (store at 4°C).
13. Protease inhibitors: phenylmethylsulfonylfluoride (PMSF), pepstatin, aprotinin, leupeptin, and *p*-aminobenzamidine.
14. SIGMACOTE® (Sigma, Atlanta, GA).
15. Palmitoyl-oleoyl phosphatidylethanolamine (POPE) and palmitoyl-oleoyl phospha-tidylcholine (POPC) (Avanti Polar Lipids, Birmingham, AL).
16. *n*-decane.
17. Planar bilayer setup: Faraday cage and vibration isolation table, bilayer clamp amplifier, acquisition system, eight-pole bessel filter, bilayer stirplate, perfusion system, bilayer cups and chambers, and headstage holder.
18. Recording bath solution: 100 mM KCl, 10 mM MOPS (Sigma, Atlanta, GA), pH 7.0 (store at 4°C).
19. Electrode solution: 3 M KCl.
20. A couple of Ag/AgCl electrodes.
21. Agar bridges made with 0.1% agar–agar in 3 M KCl.
22. Glass rod made from glass capillary.

2.2. Reconstitution of Bacterial Channels in Liposomes

1. XL1 blue strain (Stratagene, La Jolla, CA).
2. Transfection storage solution: 10% (w/v) polyethylene glycol 8000 or 3350, 5% (v/v) dimethyl sulfoxide, and 40 mM MgCl$_2$ or MgSO$_4$ in Luria-Bertani (LB) media, pH 6.5 at room temperature.
3. Ampicillin or kanamicin according to vector.
4. Isopropyl-β-D-thio-galactopyranoside (IPTG).
5. Buffer A: 50 mM Tris-HCl buffer pH 8.0, 100 mM KCl, and 1 mM PMSF at room temperature.
6. Buffer B: 1 mM PMSF, 1 μM aprotinin, and 10 μM leupeptin.
7. *n*-Decyl-β-D-maltopyranoside (DM).
8. *n*-Dodecyl-β-D-maltopyranoside (DDM).
9. 5 mM and 500 mM imidazole.
10. Asolectin (also named phosphatidylcholine; Avanti Polar Lipids). Store at −70°C.
11. Bio-Beads, SM-2 adsorbent (20–50 mesh), Bio-Rad Labs (Hercules, CA).
12. Microcapillary pipettes (calibrated) size: 100 μL (Sigma-Aldrich, Drummond Scientific Company, Broomal, PA; cat. no. 2-000-100).

Ion-Channel Reconstitution

13. Patch-clamp setup: Faraday cage and antivibration table, patch clamp amplifier, inverted microscope, acquisition system, pulse generator/stimulator, and micromanipulator.
14. Superdex 200 (GE Healthcare Bio-Sciences Corp., Piscataway, NJ).

3. Methods

3.1. Reconstitution of Cloned Ion Channels in Planar Bilayers

The methods described below outline (**3.1.1.**) the heterologous expression of ion channels in *X. laevis* oocytes, (**3.1.2.**) the isolation of membranes from oocytes, and (**3.1.3.**) the reconstitution of ion channels into planar lipid bilayers.

3.1.1. Expression of Ion Channels in X. Laevis Oocytes

Xenopus oocytes have become a popular ion-channel recording and screening system for their faithful and high level expression, as well as relatively low endogenous background current. Its large size, together with its tolerance for being impaled by multiple microelectrodes, makes it easy to inject mRNA for expression cloning and drug screening (*see* **Note 1**).

1. Oocytes are obtained from *X. laevis* frogs the same day or 1 d before injection. Procedures for the maintenance of *Xenopus* frogs and the preparation and injection of oocytes have been described previously *(3,4)*. Briefly, the frog is anesthetized by inmersion in tricaine at a concentration ranging from 0.15 to 0.35%. The oocytes are extracted with forceps through a small diagonal incision about 1 cm long in the abdomen. Follicle cell layers are removed with collagenase type 4 (Worthington; Lakewood, NJ) treatment at a concentration of 1.3 mg/mL (418 U/mg). For the treatment, the oocytes are incubated in a 15 mL solution of collagenase in OR2 solution at 20°C. Incubating the oocytes in a Falcon tube on a rotator allows better mixing than using a tissue culture dish. After shaking for 30 min at approx 180 rpm in an orbital shaker at 20°C, the 15 mL collagenase solution is replaced with a fresh one and incubated with shaking at approx 100 rpm for 30 min. Replace the solution a third time with 15 mL fresh collagenase and incubate at approx 100 rpm for 30 min. Monitor the treatment carefully by removal of oocytes, and stop treatment by rinsing the oocytes in OR2 several times. After rinsing the oocytes several times in ND96, transfer the oocytes to a plate with ND96 and gentamicin 50 µg/mL for incubations, and manually select the oocytes to be used for injection.
2. Prepare cRNA for injection from cDNA encoding the ion channel using the T7 mMessage mMachine system (Ambion, Austin, TX). Before filling the injection needle with RNA, centrifuge the RNA solution in a microfuge tube for about 60 s to pellet insoluble debris. This reduces the likelihood of needle clogging.
3. Injections of RNA can be performed with a Nanoliter injector (World Precision Instrument). To do so, the oocytes are placed in a 35-mm tissue culture dish under a dissecting microscope. To hold the oocytes in place during injections, a polypropylene mesh can be glued to the bottom of the dish. Injection needles are made in a pipette puller to draw out the glass bores that are used with the nanoliter injector. The needles should have a tip diameter of 20–40 µm. To attach the needles to the dispenser, add about 5 mm of RNAse-free mineral oil to the end of the needle opposite from the injection tip and insert the bore into the dispenser plunger. The plunger will force the oil toward the tip of the needle, which should result in a complete oil seal between the metal plunger and the injection needle tip. About 2 µL of RNA solution is placed on a square of RNAse-free Parafilm® (Pechiney Plastic Packaging Chicago, IL) and the solution is drawn into the nanoliter injector. It is important to watch the RNA being drawn into the needle through the microscope to ensure that the needle has not clogged. Once the needle is filled with RNA, it is positioned over each oocyte and gently lowered until it pierces the oocyte. As much as 100 nL can be injected into each oocyte; it is possible to inject 20 oocytes with one sample in a few minutes. Using a new needle for each RNA sample prevents cross-contamination and decreases the risk of needle clogging.

4. Injected oocytes are kept at 18°C in an incubator with ND96 solution supplemented with gentamicin 50 µg/mL. This solution must be changed every 12 h. Oocytes are used for the membrane preparation 3–6 d after RNA injection, as long as they are still healthy.

3.1.2. Isolation of Membranes From Oocytes

Membrane vesicles from oocytes injected with ion channels are prepared using a modified version of the method used by Perez et al. *(5)*. The entire procedure has to be performed at 4°C:

1. Rinse 30–40 oocytes using a solution containing 10% sucrose (w/v), 0.6 M KCl, 10 mM HEPES, pH 6.8, supplemented with 100 µM PMSF, 1 µM pepstatin, 1 µg/mL aprotinin, 1 µg/mL leupeptin, and 1 µM *p*-aminobenzamidine.
2. Transfer the oocytes into a 1-mL ground glass tissue grinder (Kontes Duall, Fisher Scientific, Vineland, N.J.) and homogenize manually for 5 min with the same solution (~10 µL/oocyte).
3. The homogenate (~200 µL) has to be layered onto a discontinuous sucrose gradient (0.75 mL of each 50% and 20%, w/v in 0.6 M KCl, and 10 mM HEPES, pH 6.8 plus protease inhibitors), and centrifuged at 100,000g for 1 h.
4. Discard the top lipid layer, collect the 20:50% interface (visible band) and dilute about three times with 10% sucrose (w/v), 0.6 M KCl, and 10 mM HEPES, pH 6.8.
5. Membranes have to be pelleted at 100,000g for 1 h., and resuspended in a final volume of approx 8–10 µL (10% sucrose (w/v), 0.6 M KCl, and 10 mM HEPES, pH 6.8) with a micropipette previously treated with SIGMACOTE.
6. Freeze the membranes in liquid N_2 and store them at –80°C until use in planar bilayer experiments (*see* **Note 2**).

3.1.3. Reconstitution of Ion Channels Into Planar Lipid Bilayers

This section describes the experimental procedures used to insert channels into painted planar lipid bilayers by vesicle fusion, the so-called "painted bilayer" method, first described by Mueller et al. *(6)*. In this method, a bilayer is made on an aperture between two aqueous compartments, which are called *cis*- and *trans*-compartments. The *cis*-compartment is connected to a voltage generator through an Ag/AgCl electrode, to control the transbilayer potential. The *trans*-compartment (virtual ground) is connected to the input of the current-measuring amplifier through a second Ag/AgCl electrode (*see* **Figs. 1** and **2**).

1. Prepare a mixture of POPE and POPC (4:1), dissolved in 25 µL of *n*-decane *(concentration ranging from 15 to 25 mg/mL)*. This mixture should be kept at room temperature and must be prepared on the same day of the experiment (*see* **Notes 3–5**).
2. Spread a droplet of the lipid mixture with a glass rod (*see* **Note 6**) onto and around the partition hole from the *cis* side of the chamber and allow it to dry.
3. Fill the electrode compartments with 3 M KCl and place the Ag/AgCl electrodes in them (*see* **Note 7**).
4. Fill the *cis*- and *trans*-compartments with the recording bath solution and place the agar bridges between them and the electrode compartments (*see* **Notes 8** and **9**).
5. Turn on the filter and the bilayer clamp amplifier. As the clean hole (*see* **Fig. 3**) does not have a very large resistance, the current signal saturates.
6. To form the bilayer, spread a droplet of the lipid mixture onto the hole from the *cis* side and then clean the glass rod applicator with a paper towel. This process has to be done until the bilayer forms.
7. When the lipids have filled the hole, the resistance increases and the signal stops from saturating. To know if the bilayer has formed, one can apply voltage pulses across the bilayer and "capacitive currents" should be measured (*see* **Fig. 4**). For a hole of 100 µm the capacitance should be in the order of 50–100 pF. Check also that the electrical resistance is higher than 10^9 Ω (*see* **Notes 10–12**).

Ion-Channel Reconstitution

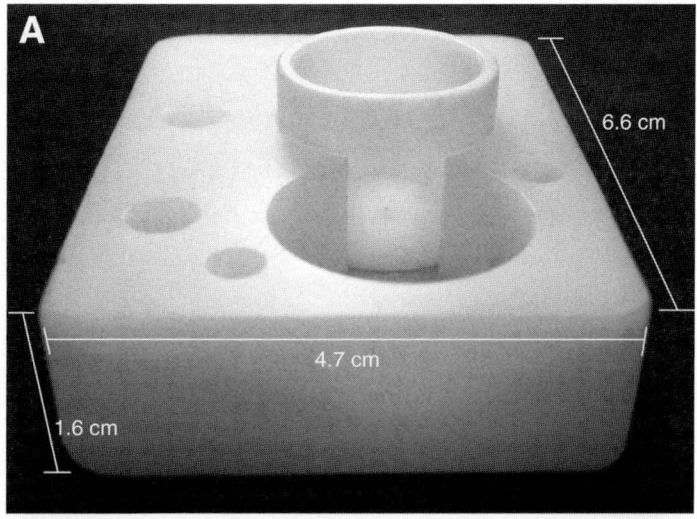

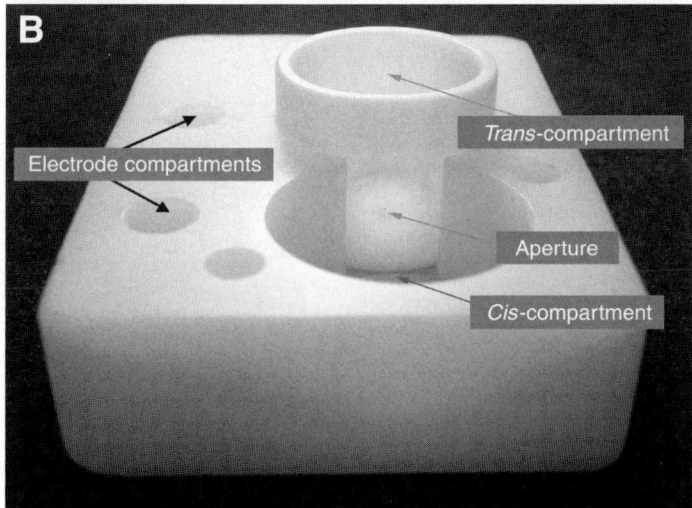

Fig. 1. A classical chamber for painting bilayers. Pits of different size are drilled into a plastic Delrin® (Dupont, Wilmington, DE) block. Two are for the insertion of the Ag/AgCl electrodes. Another consists of two intersecting holes; in one of them the cup is inserted. The cup is a cylinder made of Teflon® (Dupont, Wilmington, DE) having a hole of 300 µm in its center wherein the bilayer is formed. (A) Photograph of a chamber with its dimensions. (B) The same photograph with names of its essential parts. The two pits of the block without explanation are optional and can be used for perfusion; for this, they are connected with the *cis* pit by a small hole (not shown).

8. With the filter set at 2 kHz, the noise should be less than ±2 pA. If it is not, *see* **Note 13**.
9. Wait at least 2 min to make sure that the bilayer just formed is stable (i.e., it does not break).
10. Lightly touch the bilayer from the *cis* side with the glass rod with oocyte membrane preparation. One will know that the bilayer was touched because the capacitance decreases for a period of a few (~3) seconds and then goes back to its original value. If the bilayer breaks go back to **step 6** (*see* **Note 14**).
11. Wait at least 2 min for a channel to incorporate. One will know that the channel has been incorporated when the currents begin to change in steps (*see* **Fig. 5**). If after two or more minutes no

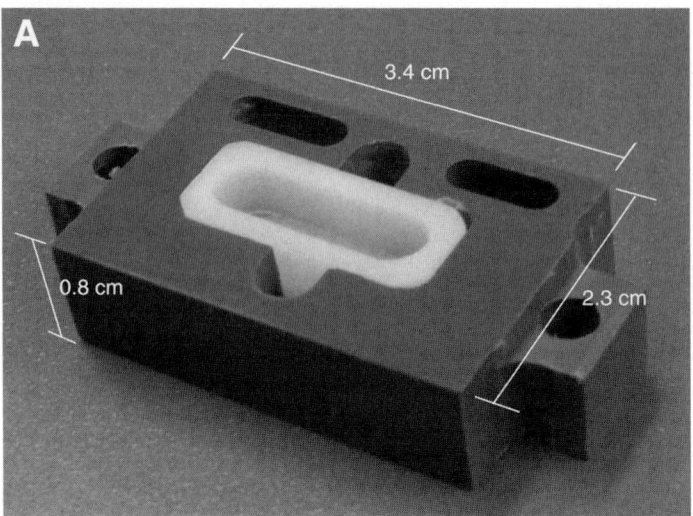

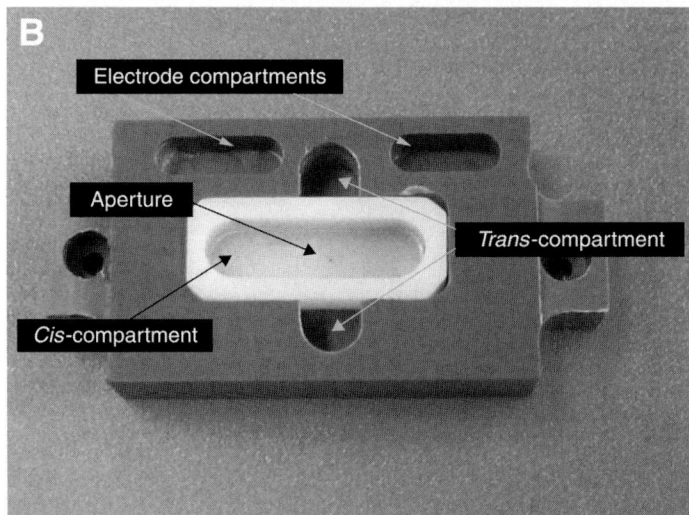

Fig. 2. Another chamber for painting bilayers. This chamber is smaller, and the bilayer is formed in a horizontal plane. The *trans*-compartment is below the *cis*-compartment. The block is made of PVC and the cup of Ertacetal-C® (Dotmar, Australia). (**A**) Photograph of a chamber with its dimensions. (**B**) Photograph of the same chamber with names of its essential parts.

channel has been incorporated, repeat **step 10**. To increase the chances of fusion, refer to **Note 15**; to decrease it, refer to **Note 16**.
12. One is ready to measure currents.

The procedure previously exposed is optimized for the large conductante calcium- and voltage-activated potassium channels *(5)*, but many ion channels have been expressed in *Xenopus* oocytes and reconstituted in planar artificial bilayers using a similar protocol. For example, a cloned epithelial Na^+ channel *(7,8)*, a brain Na^+ channel *(9)*, the calcium-activated chloride channel *(10)*, and Wolframin, a protein encoded by the gene associated with an autosomal recessive disease characterized by diabetic and optic atrophy disorder *(11)*. There are several other ways to obtain CFTR channels for incorporation into planar lipid bilayers, but

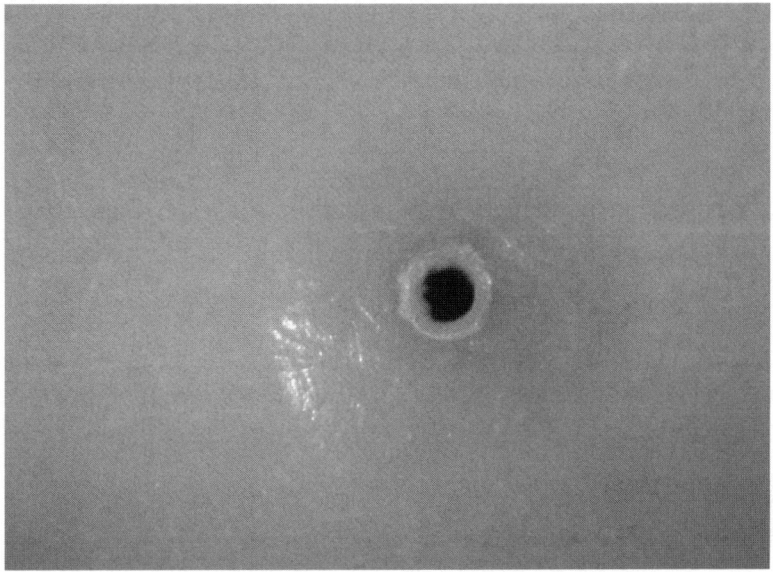

Fig. 3. A close up to the aperture of approx 250 μm made in the cup of the horizontal chamber.

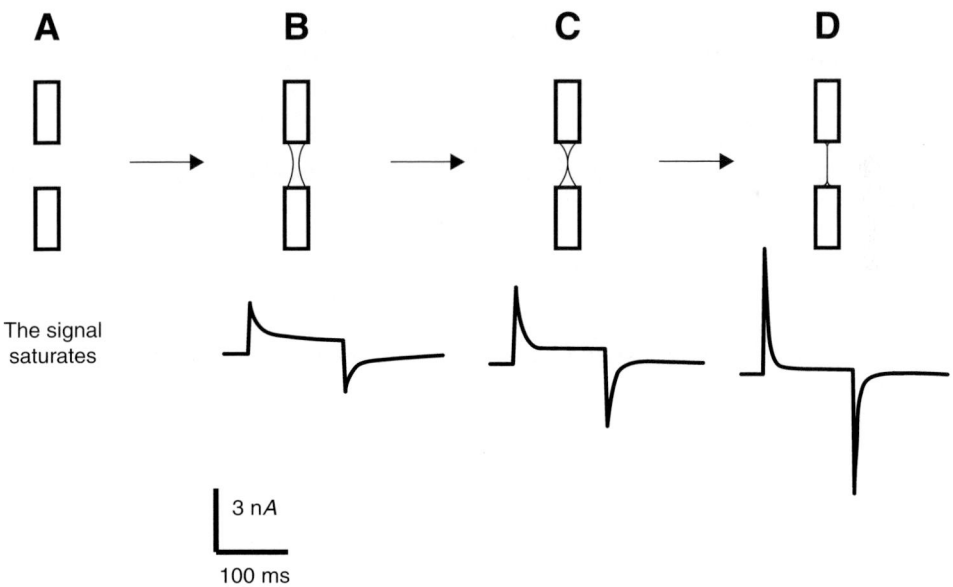

Fig. 4. Current across the lipid bilayer evoked by applied voltage pulses. (**A**) With no lipid in the hole; (**B**) when lipid is painted and fills the hole; (**C**) as the lipid thins and the bilayer starts forming; and (**D**) once the lipid bilayer has its maximum diameter.

expression in *Xenopus* oocytes and preparation of oocyte membrane vesicles is recommended as a method readily available in almost any laboratory *(12)*. It has been shown that endogenous Ca^{2+}-activated nonspecific channels incorporate into planar lipid bilayers of plasma membrane fractions from *X. laevis* oocytes without injecting *(13)*. Some new methods and advances for ion channel reconstitution in artificial bilayers are discussed in **Note 15**.

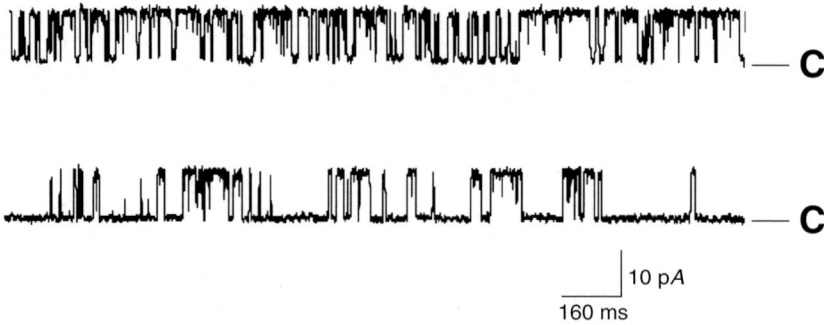

Fig. 5. Example of single-channel current recordings from a Ca^{2+}-activated potassium channel of large conductance (BK channel) reconstituted in lipid bilayers. Currents were evoked by a voltage pulse of 40 mV and contaminant Ca^{2+}. Both chambers are filled with 100 mM KCl, 10 mM MOPS-K, pH 7.0. The **C** indicates the closed state of the channel.

3.2. Reconstitution of Bacterial Channels in Liposomes

The methods described below outline: (**3.2.1.**) the expression and purification of bacterial channels in XL1 blue strain, (**3.2.2.**) the reconstitution of ion channels in liposomes, and (**3.2.3.**) the recording of ion channels by patch clamp in liposomes.

3.2.1. Expression and Purification of Bacterial Channels in Escherichia coli XL1 Blue Strain

1. Competent cell preparation (*E. coli* XL1 blue strain): in a sterile 50-mL tube add 5 mL of LB medium, 10 µL of antibiotic (ampicillin or kanamicine according to the vector) 200 µg/mL, and 50 µL of competent cell glycerol stock or a colony from the fresh plate (no more than 1 mo after its preparation) *(14)*. After 16 h, transfer 1 mL of the saturated culture to a sterile 500-mL flask containing 100 mL of LB medium. Incubate the cells at 37°C while shaking (~250 rpm) and measure absorbance (at 600nm) periodically until OD reaches 0.5. This takes approx 3 h and it is very important to avoid overgrowth. Next, chill the flask on ice (20 min) and collect the cells by centrifugation at 1200g for 5 min at 4°C. Then resuspend the cells in 10 mL of ice-cold transfection storage solution. Make 1-mL aliquots in prechilled tubes. Use 75–100 mL of the resuspended competent cells for each transformation.
2. From a fresh preparation of *E. coli* XL1 blue strain competent cells, transformed by thermal shock 50–100 µL of cells with ion channel DNA. After 1 h of incubation in a shaker at 37°C, cells are inoculated in a small flask containing 25 mL LB media, 200 µg/mL ampicillin, and 1% glucose. Allow them to grow overnight (O/N) for about 14–15 h. Do not let the preculture sit longer; if the preculture is too old, protein expression is impaired.
3. One liter of LB media is inoculated with 10 mL of the O/N culture (inoculation volume is 1% of the total volume) in the presence of 200 µg/mL ampicillin and 0.2% glucose (use a 3.8-L flask with special baffle at the bottom to improve culture aeration). The culture is started by placing those flasks in a shaker at 37°C with continuous agitation at about 228 rpm for 4 h. By the end of the 4-h period, the optic density (OD) in culture should be around 1 OD. When the OD is 1, make induction with IPTG so that the final concentration is 0.4 mM. Then, refresh the cells in 200 µg/mL ampicillin.
4. After IPTG induction, cells are incubated with continuous agitation for 5 h. Then, cells from the 2 L of culture are collected by centrifugation, and resuspended in 50 mL of buffer A (50 mM Tris-HCl buffer pH 8.0 and 100 mM KCl + 1 mM PMSF). Next, the cells are spun again in a small centrifuge and the pellet is weighted. Normally, cell pellets should weigh approx 3–3.5 g/L. The cells need to be processed immediately or frozen at –70°C until use. Do not use cells stored for more than 1 wk.

Ion-Channel Reconstitution

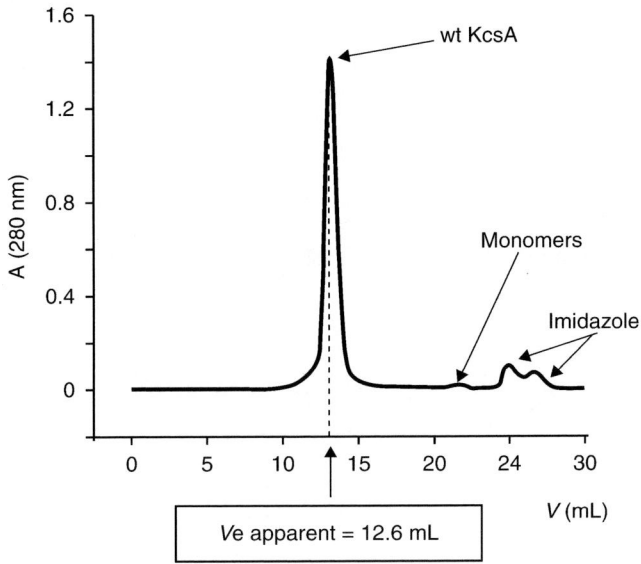

Fig. 6. Gel-filtration chromatography of *Streptomyces lividans* K$^+$ channel (KcsA) in DDM using a Superdex 200 column. KcsA is stable as a tetramer in DDM, and thus, can be maintained in detergent solutions for extended periods of time (1–2 wk).

5. To process cells for purification, a cell pellet weighing about 7 g is resuspended in 100 mL of cold buffer A, with the addition of protease inhibitor buffer B (1 mM PMSF, 1 μM aprotinin, and 10 μM leupeptin).
6. Cells are disrupted by passing twice through a cold cell homogenizer; the membrane suspension is spun down at 100,000 g for 1 h, and the pellet is resuspended in 100 mL buffer B; decyl maltoside solubilization grade is added to the membrane suspension (100 mL) to reach a final concentration of 40 mM. Membrane solubilization takes place for 2 h at room temperature in a rotating device that provides constant motion to the mix. Ending the 2 h period, spin down the sample at 100,000g for 45 min, recover supernatant, and incubate it with 5 mL of cobalt resin (50% w/v) per batch procedure for 2 h. The whole slurry is then loaded into a small column and washed extensively with buffer A and 1 mM DDM (or 5 mM DM) and 5 mM imidazole.
7. After an extensive column wash with 5 mM imidazole in buffer A and 1 mM DDM, elute the sample with 500 mM imidazole in buffer A supplemented with 1 mM DDM.
8. The yield for a wild-type channel is usually 4.5 ± 1.5 mg. Run a gel filtration to check the tetrameric state of the proteins (*see* **Note 21** and **Fig. 6**).

3.2.2. Reconstitution of Bacterial Ion Channels in Liposomes

1. Liposome generation: Clean a round bottom flask with water and chloroform (~100% purity). Put into the clean flash, 1–2 mL of the washed lipid asolectin. Rotate and evaporate for 30–40 min (25°C) until the formation of a thin layer of lipid in the flask is seen. Add 2–4 mL of buffer solution (100 mM KCl, 50 mM Tris-HCl, pH = 8.0) for a final concentration of 10 mg/mL (generally the initial concentration of lipids is around 20 mg/mL) and bath-sonicate for 3–5 min for liposome formation. This procedure guarantees relatively big yet not uniform liposomes *(2,15,16)*.
2. Reconstitution of membrane protein: add 2–4 mL of liposomes and the required volume of membrane protein (remember to always keep the protein in detergent) to a tube (15 mL). Generally, the proportion for single channel records is: 1:2000–5000 (protein mass:lipid mass), and for macroscopic currents 1:50–100 (protein mass:lipid mass). Add biobeads and wash with methanol and

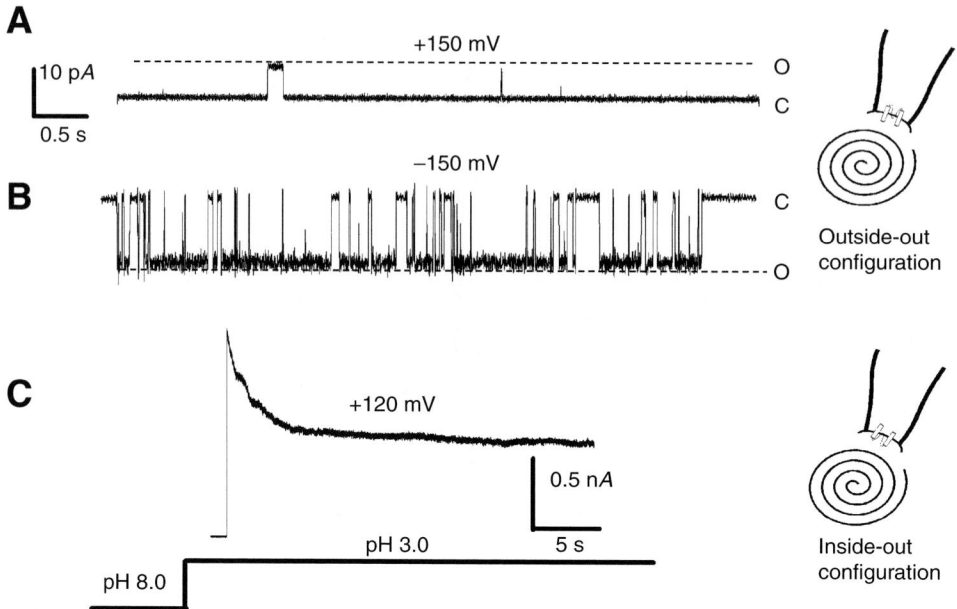

Fig. 7. *S. lividans* K$^+$ channel (KcsA) activity is preserved after reconstitution in liposomes. **(A)** Single-channel events recorded from an isolated proteoliposome patch made up of channel proteins and asolectin phospholipids (ratio 1:2000) at an applied voltage of +150 mV. The pipette solution was: 200 mM KCl, 5 mM MOPS, pH 3.0, and the bath solution was: 200 mM KCl, 5 mM MOPS, pH 3.0. The pipette resistance was approx 2.6 MΩ; **(B)** current events at −150 mV corresponding to the same patch in the outside-out patch configuration; and **(C)** macroscopic currents induced by KcsA channels. The ratio between proteins and asolectin was 1:100. The pipette solution was: 200 mM KCl, 5 mM MOPS, pH 7.0, with the bath solution having the same composition. The holding potential was +120 mV, and currents were elicited with a change of pH from 8.0 to 3.0.

water. Fill the tube with buffer solution (15 mL). Incubate at room temperature O/N (14–16 h). To remove the excess of detergent, quickly change the biobeads every 1–2 h (two to three times). Filter the biobeads and spin down the sample at 60,000g for 2 h. Resuspend the precipitate in the buffer solution (60 µL of solution/10 mg of lipids). Aliquote in 30 µL and store it at −70°C.

3.2.3. Patch Clamp Ion-Channel Recordings in Liposomes

1. To measure the current flux through ion channels, the liposome aliquots (30 µL) are spotted onto glass microslides and allowed to dehydrate under a vacuum at 4°C for 14–16 h (O/N), which is followed by rehydration (100 mM KCl, 50 mM Tris-HCl, pH 8.0) under humid conditions. The liposomes are ready 2 h after rehydration.
2. For liposome recording, standard patch-clamp techniques are used *(17)*. For single channel recording in the inside-out patch configuration, chamber and pipettes are filled with 100 mM KCl and 50 mM Tris-HCl, pH 8.0. A small aliquot (1–2 µL) of rehydrated liposomes is then placed in the 0.5-mL patch-clamp chamber. The patch pipette tip is gently touched against unilamellar blisters (which arise spontaneously from the liposomes), and suction is applied. A seal (>20 GΩ) should form either immediately or after application of a brief pulse of negative pressure (<50 mm Hg) applied to the interior of the patch pipette. To record single-channel currents, inside-out patches are formed by passing the pipette tip briefly through the solution–air interface. Channel activation is achieved by applying a voltage protocol. Single-channel currents can be filtered at 1 kHz and digitized at 5 kHz. Remember, the most important thing for the efficiency of the patch is the resistance (2.5–4 MΩ) and shape of the pipettes. It is recommended

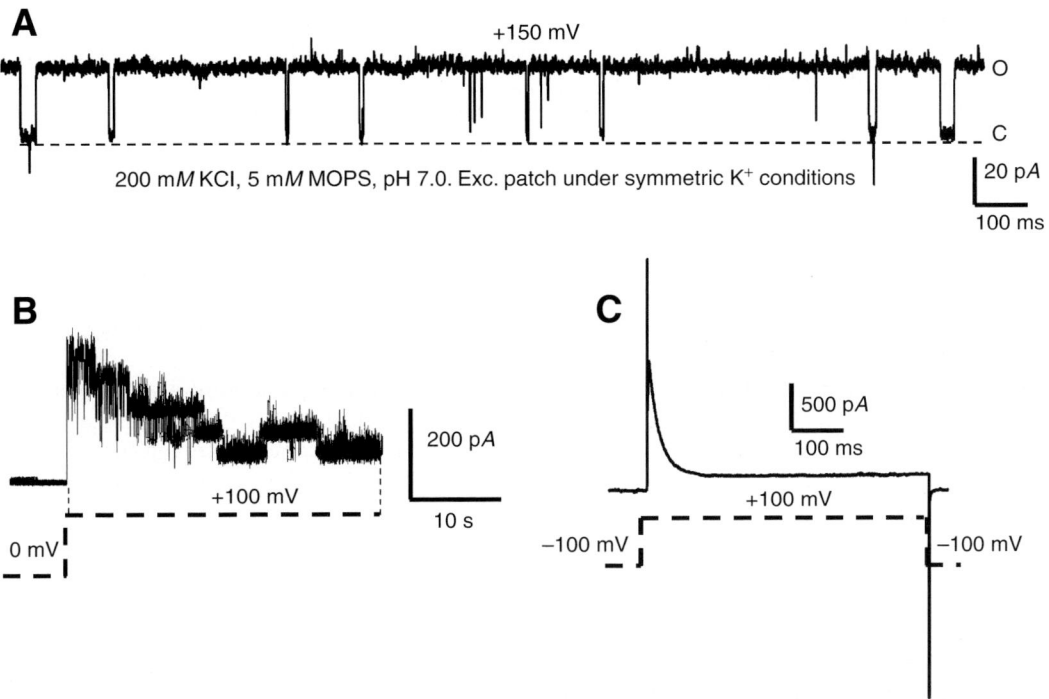

Fig. 8. *Aeropyrum pernix* K⁺ channels activate after dissolving channels in DM and reconstitution in asolectin-liposomes. Currents were elicited in cell-attached and inside-out patch configurations. The pipette solution was: 200 mM KCl, 5 mM MOPS, pH 7.0, and the bath solution was: 200 mM KCl, 5 mM MOPS, pH 7.0. **(A)** Single channel currents recorded at ⁺150 mV from an inside-out patch. The ratio protein: lipid was 1:500; **(B)** inactivation process following a potential step from 0 to +100 mV under the same conditions as the trace in A; **(C)** current corresponding to the activity of many *A. pernix* K⁺ channels KvAP channels in a patch made of proteins-lipid (ratio: 1:50). The potential protocol was from (100 to +100 mV and back to (100 mV, recorded in the cell-attached configuration. The pipette resistance was 1.5 MΩ.

that one split the patch manually. If more efficiency in the patch is needed, add to the recording solution 40 mM MgCl$_2$, provided that it does not interfere with the experiment (*see* **Fig. 8A**). For recording in outside-out patch configuration, the pipette needs to be slightly elongated. The first step toward outside-out patch is the whole-cell configuration, which is obtained by breaking the membrane with one potential pulse (40–50 mV of duration 1–2 s). Then, the pipette is slowly withdrawn to facilitate the ruptured membrane to reform in the outside-out configuration (*see* **Note 22** and **Fig. 7A**).
3. For macroscopic currents recordings, the resistance of pipette must be about 1.5–2.5 MΩ; pipette length is shorter than that of pipettes used in single channel recording. Otherwise, follow the same procedure as for single-channel recording. (*see* **Figs. 7B** and **8B,C**).

4. Notes
1. A major advantage of the oocyte model is that one can express well-defined and purified molecules (i.e., cRNA coding for the gene of interest) and measure their function in the membrane of a living cell. Hundreds of oocytes from the same female can be injected, insuring a well-controlled statistical analysis of the data. The main disadvantages of the system are twofold. First, there is a large biological variability from one set of oocytes to another independent of the traditional factors invoked, such as season, water quality, and so on. This intrinsic variability requires one to

perform a large number of independent experiments. The second pitfall is that the oocyte may lack or express endogenous components that are physiologically important to regulate the activity of the channel under study *(18)*.
2. The main criterion that makes a given membrane preparation suitable for fusion into planar lipid bilayers is its purity. The use of a highly purified membrane fraction provides the potential reproducibility that is critical in this kind of experimental work. However, crude membrane extracts have also been used in some cases.
3. Lipids usually come dissolved in chloroform. The chloroform or any other solvent present has to be completely evaporated before adding *n*-decane. This can be done by gently blowing a stream of nitrogen over the vial until the lipids are completely dried and stuck to the inner walls.
4. Before opening the stock of lipids from the refrigerator at −80°C, remember to be sure that it has reached room temperature, to avoid condensation of water inside the tube.
5. To preserve the lipid stock one has to avoid oxidation, whereas maintaining a constant concentration in chloroform. Thus, only expose the lipids to air when it is absolutely necessary. Before closing the stock solution, this should be put under a stream of nitrogen for 2 s (but not more!) to eliminate the oxygen present in the tube. Make the tube as air-tight as possible to avoid evaporation.
6. By "glass rod" is meant a glass capillary tube having a melted tip with the form of a sphere, which is used to spread the lipids. Some laboratories use a small brush or a plastic rod instead.
7. Check from time-to-time during the experiment that the electrodes have a uniform cover of AgCl. If the electrodes appear light or nonuniform they will need to be chlorinated again for at least 1 h. For this reason, it is advisable to have more than one pair of Ag/AgCl electrodes.
8. When measuring current across the bilayer one may see a slow and constant change in current called "drift." This happens because there is another electrical connection between the *cis*- and *trans*-compartments. If a drift is seen, check that the air surrounding the chamber is not too humid and that there is no water connecting the compartments. Sometimes there is a connection so small that it cannot be seen with the naked eye. To avoid this, a small quantity of silicone can be used to cover the problematic part of the chamber.
9. Because of adhesion and cohesion, water can go across the agar bridges contaminating the solution in the bath and causing noise. To avoid this, it is important that each bridge has a portion that locates completely over the chamber surface.
10. The presence of the solvent makes the bilayer thicker than that of a cell; thus, artificial bilayers have a smaller electrical capacitance per area. At higher voltages, the membrane with solvent thins, which causes an increase in the electrical capacitance.
11. If the lipids have filled the hole, but they do not thin enough to become a bilayer, apply a large voltage difference (e.g., 100 mV). One can also try painting the hole with a clean glass rod, to remove the excess lipids. If this does not work, clean the chamber and start the experiment again.
12. The thinning of the bilayer can be seen optically with a microscope. As a formed bilayer has a thickness of about 40 Å, it does not reflect light in the visible spectrum, so it appears black. This is why planar bilayers are also called "black membranes."
13. The noise in the bilayer recordings can be reduced by the following recommendations:
    a. The chamber and the measuring system must be enclosed in a Faraday cage.
    b. Floor and air vibrations can be attenuated by placing the chamber on a heavy platform seated on an inflated rubber.
    c. A smaller hole makes the bilayer more stable and lowers the electrical noise by reducing the electrical capacitance *(19)*.
14. For a more stable bilayer, use a larger fraction of POPC. This makes the bilayer more resistant, but also makes it more difficult to insert channel proteins.
15. To increase the chance of fusion one can:
    a. Add an aliquot of a concentrated solute to the *cis*-chamber to increase the osmolarity on the *cis* side with respect to the *trans*-side, which promotes the fusion of vesicles. The magnitude of the osmotic gradient necessary is variable, but a 3:1 gradient can be used as a start.

b. Include in the lipid mixture an acidic lipid such as palmitoyl-oleoyl phosphatidylserine (Avanti Polar Lipids) with a neutral lipid such as PE in, for example, a proportion of 1:3. A larger fraction of negative charged lipid increases the chances of fusion. However, there are two drawbacks of using PS: charged membranes can influence ion-channel properties *(20)* and make bilayers unstable.
   c. Use larger holes to form bilayers with greater surface areas, and thus, have a greater probability of fusion. However, realize that this increases the signal noise *(19)*.
   d. Use a magnetic stirring rod to stir the *cis*-chamber.
16. Sometimes too many channels incorporate. To keep channel incorporation low:
   a. Sonicate the vial with the preparation used on the day of the experiment for 5 s. This lowers the diameter of the vesicles, allowing fewer channels to incorporate into each vesicle.
   b. In **step 5**, **Subheading 3.1.2.**, resuspend the preparation with a lower concentration.
   c. Do not create an osmotic gradient.
   d. Do not use charged lipids, and use a fraction of POPC.
17. Many improvements have been done to study ion channels in planar bilayers since the method first described by Mueller in the early 1960s *(21)*. Montal and Mueller *(22)* devised a method for forming a bilayer that has virtually no solvent by folding two monolayers; they can be of different lipid composition. Most ion channels show no significant differences when inserted in these membranes compared with those studied using painted bilayers with *n*-decane (as described in this protocol). Another method is the formation of bilayers on the tip of patch-style pipettes, also called "tip–dip" bilayers *(23–25)*. Tip–dip bilayers are the combination of two other techniques: patch-clamping and planar lipid bilayers. In this technique, bilayers are made at the tip of patch-style pipettes by passing the tip of the pipette through a monolayer of lipid two times. The main advantage of tip–dip membranes is that they are very small (1–5 µm diameter) so the drawbacks associated with larger bilayers are avoided. Fast, small events can be detected with tip–dip membranes. Today, it is possible to create pores of any precise dimension in silanizated silicon wafers. In this material, bilayers are more stable than those formed on plastic. Single-ion channels can be incorporated when their currents have been measured *(26)*. There is a new method that forms bilayers easily and promptly (~10 s) by pressing them on an agarose layer, wherein fully functional ion channels incorporate *(27)*. This makes the first step toward the automatation of single ion-channel recordings.

    Although these methods form more stable and smaller bilayers, ion channels are still incorporated either by painting (as is presented in this protocol) or mixing the membrane preparation with the lipids before the formation of the artificial bilayer. With these methods, the insertion of an ion channel into a bilayer made in an aperture of less than 100 µm, while possible, has a low probability. The ability to insert a single ion channel in a controlled and fast way in a bilayer of any size is a major challenge for the progress of ion-channel reconstitution *(28,29)*.
18. In the past, it was thought that larger concentrations of a divalent cation, such as Ca^{2+} in the *cis* side of the bath solution resulted in increased channel incorporation; however, in the author's experience, the presence of Ca^{2+} does not influence the chance of fusion notoriously.
19. If one does not have a perfusion system, one can establish an osmotic gradient by adding an appropriate aliquot to the *cis* chamber, and then another in the *trans*-chamber to equalize the osmotic gradient.
20. When too much lipid or membrane preparation has been added (more than ~10 times each), the bilayer can get so thick that ion channels will not work. In this case, it is better to clean the chamber and start the protocol again.
21. If one cannot obtain a perfect peak for the gel filtration, use glycerol 10% during expression and purification of the protein.
22. If one has problem patching the liposomes (particularly in the outside-out configuration), try using the "Glup" preparation: mineral oil (mix of heavy:light oil, 1:4 mL) and Parafilm® (a square of about 10 cm^2 cut in small pieces). This cocktail, invented by Dr. Enrico Stefani, Department of Anesthesiology, University of California at Los Angeles, is heated until the Parafilm is dissolved.

After filling the patch pipette with solution, gently touch the pipette tip with the Glup solution. A Glup covered pipette usually improves the chances of successfully patching the liposome membrane.

Acknowledgments

We thank Dr. Cecilia Vergara for her advice and invaluable experimental tips and Mr. Charles Feinn for his helpful revision of the manuscript. C.G. thanks Dr. Luis Cuello for his invaluable input in the liposome section. This work was supported by Chilean Fondecyt Grant 103-0830. The Centro de Estudios Científicos is a Millennium Institute, and partially funded by a grant from the Fundación Andes.

References

1. Hille, B. (2001) *Ion Channels of Excitable Membranes.* Sinauer Associates Inc. Sunderland, MA.
2. Miller, C. (ed.) (1986) *Ion Channel Reconstitution.* Plenum Press, New York, NY.
3. Goldin, A. L. (1992) Manteinance of *Xenopus laevis* and oocyte injection. *Methods Enzymol.* **207,** 266–279.
4. Shih, T. M., Smith, R. D., Toro, L., and Goldin, A. L. (1998) High-level expression and detection of ion channels in *Xenopus* oocytes. *Methods Enzymol.* **293,** 529–556.
5. Perez, G., Lagrutta, A., Adelman, J. P., and Toro, L. (1994) Reconstitution of Expressed K_{Ca} Channels from *Xenopus* Oocytes to Lipid Bilayers. *Biophys. J.* **66,** 1022–1027.
6. Mueller, P., Rudin, D. O., Tien, H. T., and Wescott W. C. (1962) Reconstitution of cell membrane structure in vitro and its transformation into an excitable system. *Nature* **194,** 979–980.
7. Jovov, B., Tousson, A., Ji, H., et al. (1999) Regulation of Epithelial Na^+ Channels by Actin in Planar Lipid Bilayers and in the *Xenopus* Oocyte Expression System. *J. Biol. Chem.* **274(53),** 37,845–37,854.
8. Ismailov, I. I., Awayda, M. S., Berdiev, B. K., et al. (1996) Triple-barrel Organization of ENaC, a Cloned Epithelial Na^+ Channel. *J. Biol. Chem.* **271(2),** 807–816.
9. Berdiev, B. K., Mapstone, T. B., Markert, J. M., et al. (2001) pH Alterations "Reset" Ca^{2+} Sensitivity of Brain Na^+ Channel 2, a Degenerin/Epithelial Na^+ Ion Channel, in Planar Lipid Bilayers. *J. Biol. Chem.* **276(42),** 38,755–38,761.
10. Cunningham, S. A., Awayda, M. S., Bubien, J. K., et al. (1995) Cloning of an Epithelial Chloride Channel from Bovine Trachea. *J. Biol. Chem.* **270(52),** 31,016–31,026.
11. Osman, A. A., Saito, M., Makepeace, C., Permutt, M. A., Schlesinger, P., and Mueckler, M. (2003) Wolframin Expression Induces Novel Ion Channel Activity in Endoplasmic Reticulum Membranes and Increases Intracellular Calcium. *J. Biol. Chem.* **278(52),** 52,755–52,762.
12. Benos, D. J., Berdiev, B. K., and Ismailov, I. I. (2003) Measurement of CFTR chloride channel activity using the planar lipid bilayer technique *The European Working Group on CFTR Expression* http://pen2.ing.gulbenkian.pt/cftr/vr/physiology.html.
13. Young, G. P., Young, J. D., Deshpande, A. K., Goldstein, M., Koide, S. S., and Cohn, Z. A. (1984) A Ca^{2+}-activated channel from *Xenopus laevis* oocyte membranes reconstituted into planar bilayers. *Proc. Nati. Acad. Sci. USA* **81,** 5155–5159.
14. Cortes, D. M. and Perozo, E. (1997) Structural Dynamics of the *Streptomyces lividans* K^+ Channel (SKC1): Oligomeric Stoichiometry and Stability. *Biochemistry* **36,** 10,343–10,352.
15. Cuello, L. G., Cortes, M., and Perozo, E. (2004) Molecular Architecture of the KVAP Voltage-Dependent K^+ Channel in a Lipid Bilayer. *Science* **306(5695),** 491–495.
16. Heginbotham, L., Kolmakova-Partensky, L. and Miller, C. (1998) Functional Reconstitution of a Prokaryotic K^+ Channel. *J. Gen. Physiol.* **111(6),** 741–749.
17. Sukharev. S. I., Martinac. B., Arshavsky. V. Y., and Kung. C. (1993) Two types of mechanosensitive channels in the Escherichia coli cell envelope: solubilization and functional reconstitution. *Biophys. J.* **65,** 177–183.

18. Rossier, B. C. (1998) Mechanosensitivity of the epithelial sodium channel (ENaC): controversy or pseudocontroversy? *J. Gen. Physiol.* **112(2),** 95–96.
19. Alvarez, O., Benos, D., and Latorre, R. (1985) The Study of Ion Channels in Planar Lipid Bilayer Membranes. *J. Electrophysiol. Technol.* **12,** 159–177.
20. Latorre, R., Labarca, P., and Naranjo, D. (1992) Surface Charge Effects on Ion conduction in Ion Channels. *Methods Enzymol.* **207,** 471–501.
21. Montal, M. and Mueller, P. (1972) Formation of bimolecular membranes from lipid monolayers and a study of their electrical properties. *Proc. Natl. Acad. Sci. USA* **69(12),** 3561–3566.
22. Labarca, P. and Latorre, R. (1992) Insertion of Ion Channels into Planar Lipid Bilayers by Vesicle Fusion. *Methods Enzymol.* **207,** 447–463.
23. Coronado, R. and Latorre, R. (1983) Phospholipid bilayers made from monolayers on patch-clamp pipettes. *Biophys. J.* **43(2),** 231–236.
24. Suarez-Isla, B. A., Wan, K., Lindstrom, J., and Montal, M. (1983) Single-channel recordings from purified acetylcholine receptors reconstituted in bilayers formed at the tip of patch pipets. *Biochemistry* **22(10),** 2319–2323.
25. Ehrlich, B. E. (1992) Planar lipid bilayers on patch pipettes: Bilayer formation and ion channel incorporation. *Methods Enzymol.* **207,** 463–470.
26. Pantoja, R., Sigg, D., Blunck, R., Bezanilla, F., and Heath, J. R. (2001) Bilayer reconstitution of voltage-dependent ion channels using a microfabricated silicon chip. *Biophys. J.* **81(4),** 2389–2394.
27. Ide, T. and Ichikawa, T. (2005) A novel method for artificial lipid-bilayer formation. *Biosens. Bioelectron.* **21(4),** 672–677.
28. Williams, A. J. (1994) An introduction to the methods available for ion channel reconstitution, in *Microelectrode Techniques, The Plymouth workshop handbook*, (Ogden, D. C., ed.), Company of Biologist, Cambridge, UK, pp. 79–99.
29. Alvarez, O. (1986) How to set up a bilayers system, in *Ion Channel Reconstitution*, (Miller, C., ed.), Plenum Press, New York, NY.

39

The Use of Differential Scanning Calorimetry to Study Drug–Membrane Interactions

Thomas M. Mavromoustakos

Summary

Differential-scanning calorimetry is a thermodynamic technique widely used for studying drug–membrane interactions. This chapter provides practical examples on this topic, highlighting the caution to be taken in analyzing thermal data as well as scientific information that can be derived by the proper use of the technique. An example is given using model bilayers containing high concentration of the anesthetic steroid alphaxalone. It is shown that the breadth of the phase transitions and the maximum of the phase-transition temperature of the bilayer depend on the equilibration conditions before acquiring the thermal scan. In addition, the quality of the thermogram depends on its perturbation and incorporation effects; for dissecting these effects, a complementary technique such as solid-state nuclear magnetic resonance spectroscopy is necessary.

Differential-scanning calorimetry is a useful technique to study the interdigitation effect of a drug by monitoring ΔH changes. Cholesterol, a main constituent of membrane bilayers, appears to disrupt the interdigitating effect. In general, the thermal effects of the drug incorporated into a membrane bilayer depends on the drug stereoelectronic properties.

Key Words: Differential-scanning calorimetry; drugs; lipid bilayers; membranes; phosphatidylcholines; vinblastine.

1. Introduction

Differential-scanning calorimetry (DSC) is a thermodynamic technique suitable for studying phase transitions of membrane bilayers with and without inserted drug molecules. It has been extensively used to investigate the thermal changes caused by the incorporation of the drugs into the membrane bilayers. The technique is based on the differential heat flow between a sample that undergoes phase transition within a specific temperature range and the inert reference material. During a phase transition, when the sample undergoes a thermally induced event, the control system senses the resulting temperature difference between the sample and reference cells, and thus, supplies heat to the sample cell to hold its temperature equal to that of the reference. The temperature of the sample and the reference are varied at the same programmed rate and the excess specific or differential heat is recorded as a function of temperature. The following diagnostic parameters in an endothermic or exothermic event recorded during the phase transition are used for the study of drug–membrane interactions:

1. T_m (maximum of the temperature peak).
2. T_{onset} (the starting temperature of the phase transition).
3. $T_{m1/2}$ (the half-height width of the phase transition).
4. The area under the peak, which represents the enthalpy change during the transition (ΔH) (**Fig. 1**).

From: *Methods in Molecular Biology, vol. 400: Methods in Membrane Lipids*
Edited by: A. M. Dopico © Humana Press Inc., Totowa, NJ

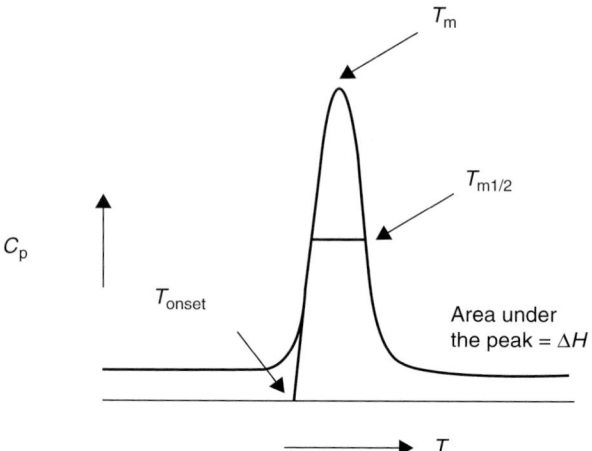

Fig. 1. An endothermic peak and the diagnostic parameters used for studying drug–membrane interactions.

Owing to the complexity of biological membranes and their instability, the study of their phase transitions is not an easy task. Therefore, lipid and especially, phospholipid bilayers (which share many of the conformational and dynamic properties of natural membranes) are used as model membranes. When hydrated, phosphatidylcholines spontaneously form multi-lamellar bilayers. Thus, these phospholipids have received a great deal of attention in terms of their physical properties in the presence and absence of drug molecules.

In particular, L-α-dipalmitoyl phosphatidylcholine bilayers have been studied extensively, as they show two endothermic transitions in a convenient temperature range: a broad low-enthalpy pretransition ($T_{m1/2}$ = 35.3°C) and a main phase transition (T_m = 41.2°C). Below the pretransition, phospholipid molecules are arranged in a one-dimensional lamellar gel phase (L_β'), whereas above the main transition they exist in the liquid-crystalline phase (L_α'). At temperatures between $T_{m1/2}$ and T_m, there is a ripple phase (P_β'), which on the basis of solid-state nuclear magnetic resonance (NMR) evidence, has been shown to be made up of coexisting gel and liquid-crystalline components (*1*).

In such complex systems the lineshapes of the thermograms depend on the equilibration time and temperature kept before running the experiment (*see* **Notes 10–13**). Generally, the lineshapes are broader when the sample is not kept in the freezer or run immediately after sealing. Longer equilibration times produce thermal profiles with shorter linewidths, whereas the enthalpy change remains constant. An example that confirms these principles is given with the potent anesthetic steroid alphaxalone (5α-pregnan-3α-ol-11,20-dione). Multilamellar vesicles (MLVs) of dipalmitoylphosphatidylcholine (DPPC)/alphaxalone preparation with a molar ratio of 80:20 (x = 0.20) were subjected to different equilibration conditions, before performing a series of DSC experiments.

Figure 2 shows a thermogram from a pure DPPC bilayer preparation and five thermal scans from a DPPC/alphaxalone preparation run at different stages of equilibration (*see* **Notes 1–7** and **9**). When the sample was run immediately following hydration, the phase-transition temperature was about 5–8 K less than that of DPPC. For example, Tm was 33°C in the first run, and increased to 35°C by the third run. Other subsequent scans gave identical thermograms to that obtained in the third run. These results are in agreement with the results obtained

DSC and Membrane–Drug Interactions

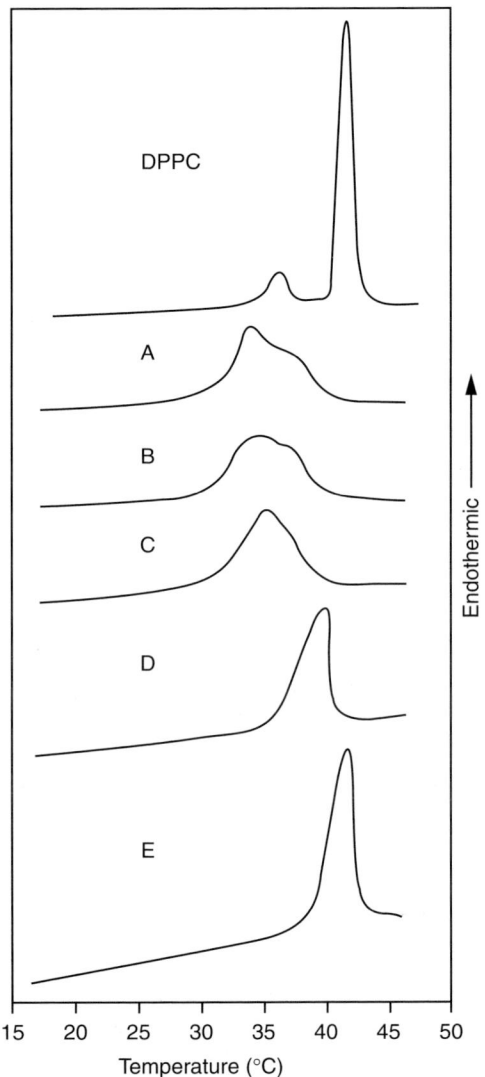

Fig. 2. Normalized thermogram of DPPC bilayers without steroid and thermograms of DPPC bilayers containing alphaxalone ($x = 0.20$) obtained (**A**) right after sealing; (**B**) second run; (**C**) third run; (**D**) after leaving at room temperature for 5 d; and (**E**) after leaving in the freezer for five additional days.

by Connor et al. *(2)* and Makriyannis et al. *(3)* in which it was reported that the presence of alphaxalone in DPPC bilayers reduced the phase-transition temperature by 8 and 4–5 K, respectively. When the sample was kept at room temperature for 5 d, the thermogram had a T_m at 40°C, only about 1 K less than that of pure DPPC bilayers. When the sample was left in the freezer for additional 5 d, it showed a T_m of 41°C, which possibly explains the previous results reported by O' Leary et al. *(4)*.

A plethora of publications examining the thermal effects of drugs in DPPC bilayers is available *(2–18)*. For example, Jain et al. *(14,15)* studied the influence of 34-adamantane, protoadamantane, and homoadamantane derivatives, as well as more than 100 other compounds on the thermotropic behavior of DPPC bilayers. These authors found that the T_m or half-height

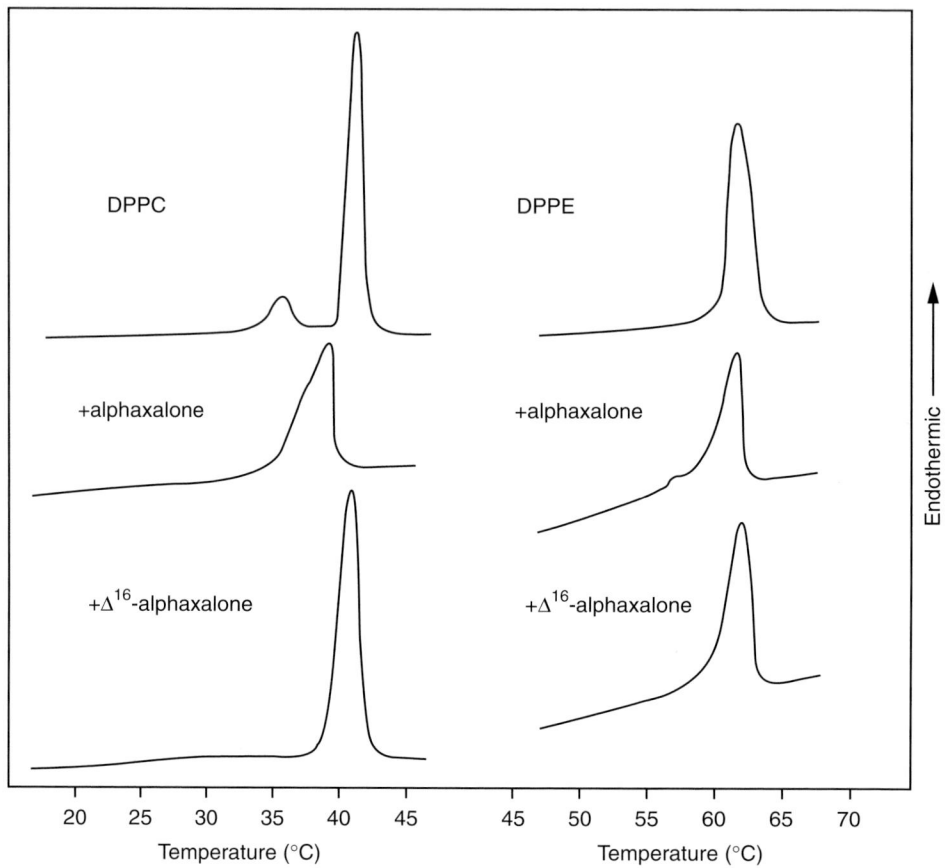

Fig. 3. Normalized thermograms of model membrane preparations of DPPC containing alphaxalone or Δ^{16}–alphaxalone at $x = 0.20$.

width of the DPPC bilayers was significantly affected with compound incorporation. They also found that there was a shift in the transition peak toward lower temperatures, and an increase in the half-height width of the transition. These parameters are related to the position and localization of the added drug along the thickness of the bilayer, and are sensitive to solute concentration.

The authors also stated that the degree to which these parameters were affected by the different compounds could not be accounted for by simple considerations of lipid/water partition coefficients, substitution constants based on free energy relationships, or the relative polarities or sizes of substitute groups. Their observations are consistent with the hypothesis that the position and orientation of a molecule within the bilayer are critical factors in determining the drug-relative potency. In turn, the localization of a molecule in a lipid bilayer was determined by the presence of the polar and apolar groups in the compound and by the geometric arrangements of these groups within the molecule.

Another issue that must be addressed is whether the observed different thermal effects of chemically similar compounds (i.e., alphaxalone and inactive congener Δ^{16}–alphaxalone) **(Fig. 3)** in MLVs are attributed to the compound abilities to incorporate in membrane bilayers or to pack properly within them. Bilayer incorporation is a matter of lipophilicity of the drug,

DSC and Membrane–Drug Interactions

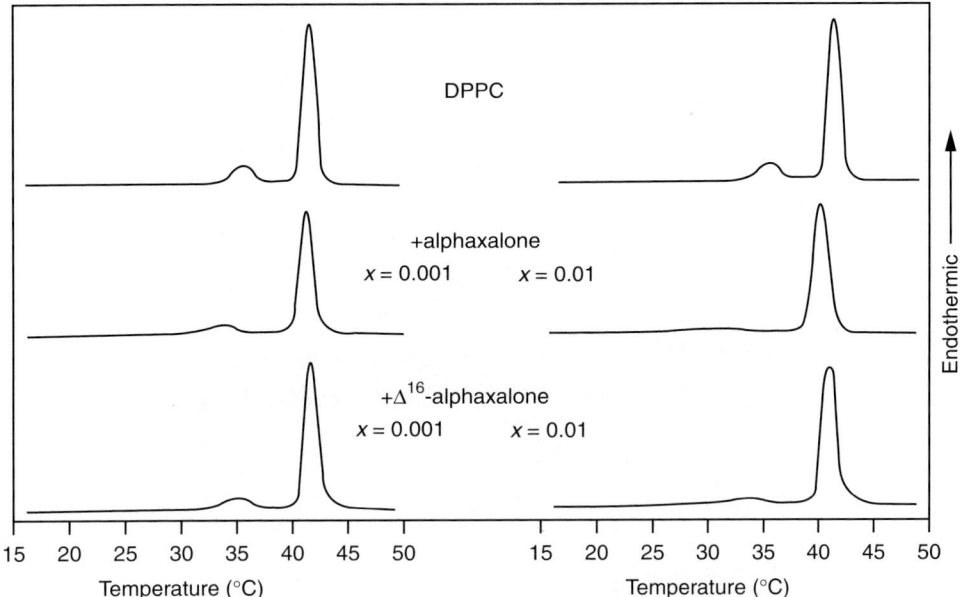

Fig. 4. Normalized thermograms of model membrane preparations of DPPC containing steroids with low concentrations of $x = 0.001$ (**left traces**) and $x = 0.01$ (**right traces**). (**Top traces**) lipid bilayer alone; middle traces: DPPC + alphaxalone; (**bottom traces**) DPPC + Δ^{16}–alphaxalone.

whereas drug packing in the bilayer is related to stereoelectronic features of the drug; imperfect packing of the lipid chains is associated with conformational and dynamic changes in the chain segments. These effects, which are observed as characteristic changes in the thermal scans are collectively described as "membrane perturbation."

Solid-state ^{2}H NMR is a useful technique to dissect between the two factors that are responsible for membrane perturbation by steroids or other membrane-active molecules, namely, degree of incorporation and packing characteristics *(19)*. The use of solid-state ^{2}H NMR shows that at $x = 0.01$ saturation has already occurred in the DPPC + Δ^{16}–alphaxalone, but not with DPPC + alphaxalone. Additionally, the spectral quantization allows to estimate that at a steroid concentration of $x = 0.01$, both steroids incorporated nearly equally. Having determined by an independent method that the concentration limit less than which both steroids incorporate equally into DPPC bilayers, DSC scans were obtained with $x = 0.01$ and $x = 0.001$ steroid concentrations. Changes in the main transition are not easily discernible at the low steroid concentrations. However, such differences can be observed clearly at the endothermic pretransition of the different DPPC preparations.

At the higher steroid concentration of $x = 0.01$, the differences between the thermal pretransitions of the two preparations are even starker, with the alphaxalone-containing preparation having a considerably broader pretransition. The aforementioned experiment demonstrates that when the two steroids are incorporated equally in a membrane system, alphaxalone perturbs the membrane bilayer system more effectively than its inactive Δ^{16}–alphaxalone. The experiment confirmed results by other authors reporting that the presence or absence of pretransition is related to the anesthetic activity of steroids *(11)* (**Fig. 4**).

Calorimetric and structural studies have demonstrated the formation of a thermodynamically stable, interdigitated gel phase. The most characteristic calorimetric feature associated with

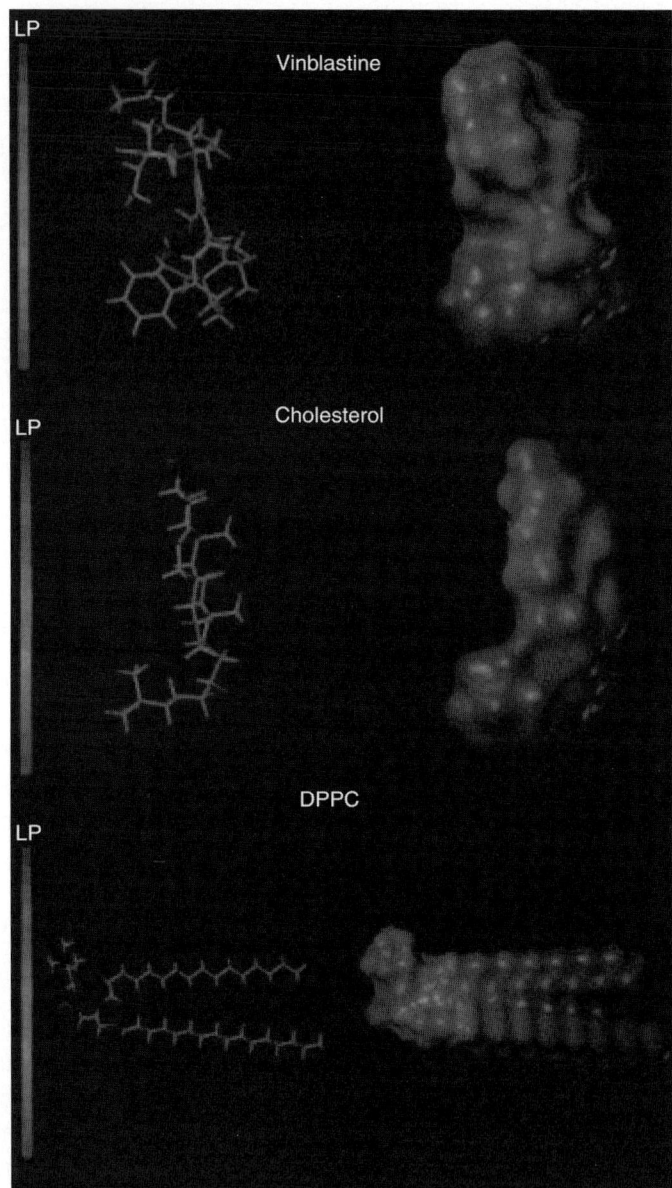

Fig. 5. Molecular structures of cholesterol, vinblastine, and DPPC and their lipophilic profiles. Blue colors represent the polar moiety and brown the hydrophobic segments.

this phase is a pronounced increase in the molar enthalpy of the main lipid-phase transition *(20–25)*. Such behavior was observed with the mitotic inhibitor vinblastine (**Fig. 5**), whose activity is exerted by entering the animal cells and diffusing through the plasma membranes *(26,27)*. The main transition at DPPC bilayers is associated with $\Delta H = 8.1 \pm 0.3$ kcal/mol, and the pretransition with $\Delta H = 1 \pm 0.3$ kcal/mol. (*see* **Note 8**). The encapsulation of vinblastine sulfate within the DPPC membranes at a molar ratio $x = 0.17$ causes significant changes in their thermotropic phase behavior. Specifically, the partitioning of vinblastine sulfate abolishes the pretransition, with an abrupt increase in $\Delta H = 10.4 \pm 0.4$ kcal/mol, a simultaneous broadening of the C_p peak, and the lowering of the transition temperature (**Fig. 6**).

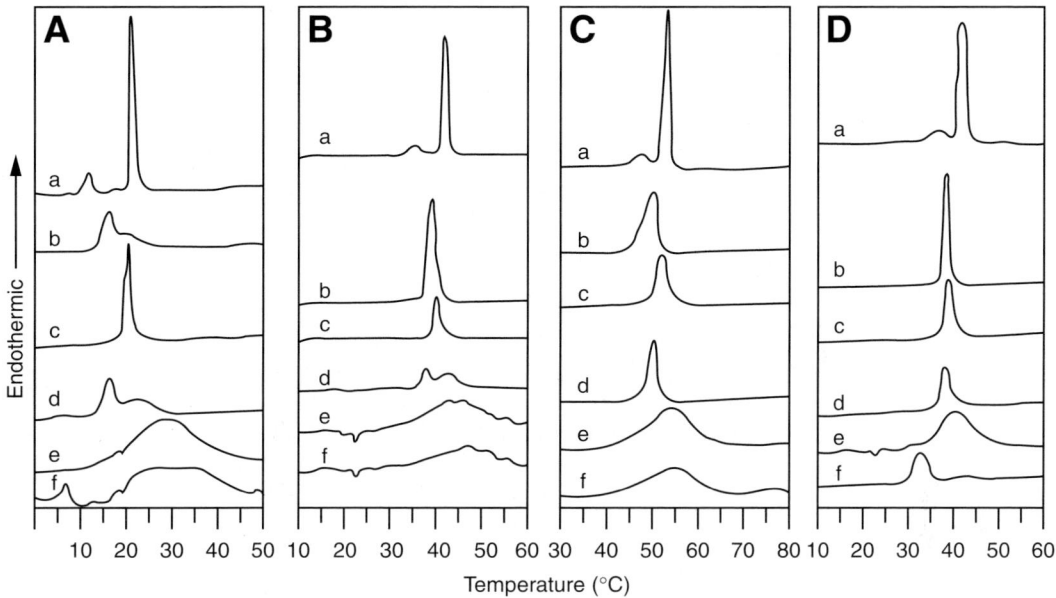

Fig. 6. DSC scans for lipid bilayers containing DMPC (**A**), DPPC (**B**), DSPC (**C**), and DPPG (**D**). (a) Phospholipid bilayers alone, (b) phospholipid bilayers containing $\chi = 0.17$ vinblastine, (c) phospholipid bilayers with incorporated $\chi = 0.10$ cholesterol, (d) Addition of $\chi = 0.17$ vinblastine in phospholipid bilayers with incorporated $\chi = 0.10$ cholesterol, (e) phospholipid bilayers with incorporated $\chi = 0.30$ cholesterol, and (f) addition of $\chi = 0.17$ vinblastine in phospholipid bilayers with incorporated $\chi = 0.30$ cholesterol.

Cholesterol has been reported to obstruct interdigitation of the alkyl chains. The interdigitated phase of diheptanoylphosphocholine (DHPC) is abolished in the DHPC: cholesterol membranes *(28)*. Membrane-partitioned short-chain alcohols are countered by cholesterol leading to the loss of interdigitization potential of membranes *(29,30)*. The presence of cholesterol primarily affects the fluidity of lipid membranes *(31)*. X-ray diffraction studies have shown that cholesterol positions parallel to the lipid bilayer acyl chains, with its hydrophobic steroidal and alkyl chain parts extending in the lipophilic region of the membrane, and the polar part locates in the vicinity of the esterified carbonyls. The hydroxyl group probably forms a hydrogen bond with the oxygen of the esterified carbonyl groups *(32)*.

The thermal effects of cholesterol on the phase transitions of mesomorphic states of phospholipid bilayers have been extensively studied by DSC. At low concentrations, cholesterol induces localized disorder in the gel phase. Cholesterol concentrations of approx 30% give rise to the formation of the liquid-ordered phase. Although the phase is liquid, there is a high degree of conformational ordering related to the lipid alkyl chains *(33–36)*. NMR experiments have demonstrated that cholesterol causes a conformational ordering effect above the phase-transition temperature of phosphatidylcholine bilayers and a positional disordering effect below the phase transition. This is termed the "buffer effect" *(37–42)*. These unique properties of cholesterol in lipid bilayers are all attributed to its molecular structure. More explicitly, it contains: (1) a β-OH group at position 3, which gives its amphipathic character, (2) a small relatively flexible tail, and (3) a flat fused-ring system.

The role of cholesterol (**Fig. 5**) in the lipid/vinblastine interdigitated bilayer system has been examined using DSC *(43)*. In addition, the role of lipid molecular characteristics in the vinblastine-induced interdigitation of the bilayer was examined. Such studies shed light on the molecular requirements for bilayer interdigitation induced by bulky amphoteric molecules. Moreover, the properties of cholesterol in phospholipid bilayers can be used to improve the drug-encapsulation capacity of liposomal carriers. The addition of cholesterol at $\chi = 0.10$ enhances the positional heterogeneity of DPPC bilayers in the gel phase rendering them more fluid-like. The C_p peak for the main lipid-phase transition of the MLVs broadens, the pretransitional peak disappears, and ΔH decreases (*see* **Fig. 6C**, second from the left).

When vinblastine at ($\chi = 0.17$) is incorporated in the DPPC/($\chi = 0.10$) cholesterol bilayer, the C_p peak sharpens, ΔH increases, and T_m decreases (**Fig. 6D**, second from the left). The vinblastine incorporation in MLVs prepared from DPPC/cholesterol ($\chi = 0.30$) has also been studied. The C_p peaks for $\chi = 0.30$ cholesterol are significantly broad. ΔH for MLV preparations containing DPPC/cholesterol ($\chi = 0.30$)/vinblastine ($\chi = 0.17$) is lower than that of the corresponding preparation without vinblastine. These calorimetric results provide strong evidence that vinblastine-induced interdigitation is obstructed by the presence of cholesterol in DPPC bilayers. Given, cholesterol concentration is high enough, interdigitation becomes impossible. **Figure 7** shows quantitatively ΔH changes vs cholesterol concentration for large unimellar vesicle (LUV) and MLV preparations *(43)*.

Vinblastine-induced interdigitation of lipid bilayers was studied separately in membranes containing phospholipids with shorter (DMPC) or longer alkyl chains (DSPC) than those of DPPC, and in bilayers containing phospholipids with a different head-group (DPPG). For dimyristoyl phosphatidylcholine (DMPC), distearoylphosphatidylcholine (DSPC), and dipalmitoylphosphatidylglycerol (DPPG), the modulation of the thermal properties induced by the addition of cholesterol was also studied. The DSC results for all these membrane bilayers are illustrated in **Fig. 6**, showing that the perturbing effect of vinblastine in these phospholipids is also specific.

Hydrated DMPC lipids exist in the gel phase for temperatures lower than 15°C, and in the liquid-crystalline phase for temperatures higher than 24°C. The presence of $\chi = 0.17$ vinblastine causes significant broadening of the phase-transition temperature range, and thus, the formation of lipid domains containing different vinblastine concentrations, as well as lowering of ΔH. The wide component at higher temperature may consist of pure DMPC bilayers, whereas the narrow component of DMPC bilayers may contain the vinblastine molecules intercalating into the bilayer.

The presence of $\chi = 0.10$ cholesterol in the DMPC bilayers results in abolishing the pretransition, marginal lowering of the phase-transition temperature, and decrease in ΔH. When a concentration of $\chi = 0.17$ of vinblastine is added, the profile resembles that of DPPC/$\chi = 0.17$ vinblastine. This indicates that the predominant concentration of drug over cholesterol governs the thermal effects of DPPC/cholesterol bilayers. Moreover, the presence of $\chi = 0.30$ cholesterol causes a more significant broadening of the phase transition and further lowering of ΔH. At this cholesterol concentration, the bilayers undergo a solid-disordered to a liquid-ordered phase transition. As shown in **Fig. 6**, the addition of $\chi = 0.17$ vinblastine furthers a disordering effect, almost entirely abolishing the main lipid-phase transition.

The results for DSPC, which has longer alkyl chains than DPPC (C-18 vs C-16), are analogous. DSPC bilayers exhibit the main lipid-phase transition around 54°C. The intercalation of $\chi = 0.17$ vinblastine into DSPC bilayers causes pronounced broadening of the temperature width of the phase transition, lowering the phase-transition temperature, and decreasing ΔH. The presence of $\chi = 0.10$ cholesterol in DSPC bilayers results in the abolishment of the pretransition, marginal lowering of the phase-transition temperature, and decrease of ΔH.

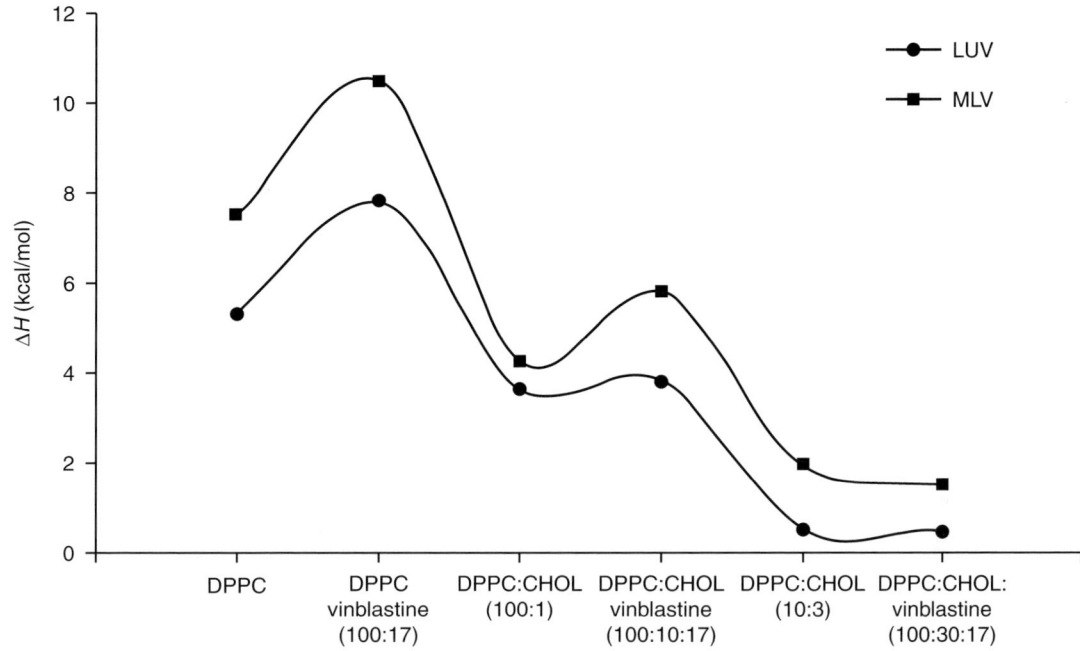

Fig. 7. ΔH changes vs cholesterol concentration for LUV (**bottom**) and MLV (**top**) preparations.

When $\chi = 0.17$ of vinblastine is added, a small additional decrease of the phase-transition temperature as well as of ΔH, are observed. The DSPC/cholesterol ($\chi = 0.30$) preparation shows a DSC thermogram exhibiting a broad peak that has a phase transition about at the same temperature as that of DSPC bilayers.

It is possible that phase separation effects are dominant and responsible for this C_p profile. When $\chi = 0.17$ vinblastine is added to the DPPC/cholesterol ($\chi = 0.30$) preparation, no significant differences from that of DPPC/cholesterol ($\chi = 0.30$) are observed, suggesting again that, as in DPPC bilayers, cholesterol determines the shape of the DSC profiles. DPPG bilayers show the pretransition and the main phase transitions at 37 and 42°C, respectively. The addition of $\chi = 0.17$ vinblastine results in lowering of the phase transition and abolishment of the pretransition as well as in decrease of ΔH. Similar effects are produced by $\chi = 0.10$ cholesterol. On the other hand, when $\chi = 0.17$ vinblastine is added to the DPPG/$\chi = 0.10$ cholesterol bilayer, an additional decrease in the main lipid-phase transition temperature is observed. The DPPG/$\chi = 0.30$ cholesterol preparation exhibits a very broad peak. The addition of $\chi = 0.17$ vinblastine results in a peak that ranges between 30 and 35°C, and a barely observable, very broad shoulder extended up to 55°C. Vinblastine fails to induce interdigitation in lipids other than DPPC. This is readily indicated by the ΔH trends recorded, which always show a decrease in ΔH in the presence of vinblastine in the bilayer (**Fig. 8**). The lowering of ΔH means that the interaction of the bulky vinblastine with the various phospholipids induces disorder (*8*). As expected, the obtained results also indicate that cholesterol augments vinblastine-induced disordering. This is an interesting observation and shows that partial interdigitation induced by vinblastine is only specific to DPPC bilayers. Thus, this interdigitation depends on the size of phospholipid alkyl chain and the specifics of its headgroup.

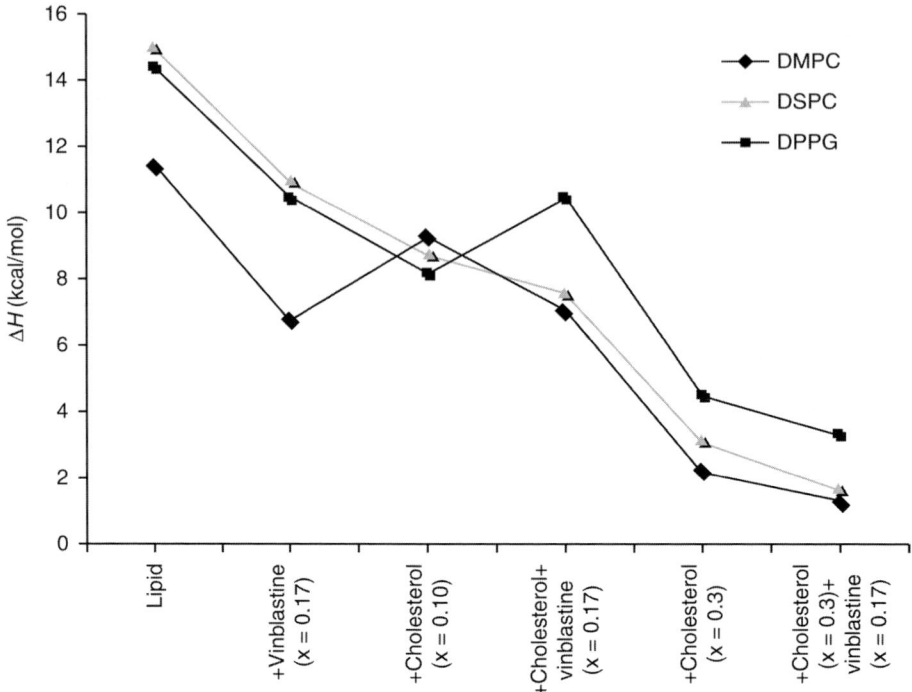

Fig. 8. ΔH changes vs cholesterol concentration for MLV preparations in DMPC, DSPC, and DPPG bilayers.

2. Materials

1. DPPC was purchased from Avanti Polar Lipids Inc. (Alabaster, AL).
2. Cholesterol and $CHCl_3$ were purchased from Sigma Aldrich (St. Louis, MO).
3. Vinblastine sulfate was obtained from Eli Lilly Co. (Indianapolis, Indiana).
4. Solution for pre-equilibration of LUVs (see **Subheading 3.1.2.**): 100 mM N-Tris (hydroxymethyl) methyl-2-aminoethanesulfonic acid (TES) and 100 mM NaCl, pH = 7.5, 300 mOs.

3. Methods

3.1. Differential-Scanning Calorimetry

3.1.1. Conventional DSC

1. Conventional DSC technique is applied for the study of MLV samples using a Perkin Elmer (Norwalk, CT) DSC-2 or DSC-7 calorimeters.
2. The lipid with or without drug is dissolved in chloroform. The solvent is then evaporated by rotavapore under vacuum (0.1 mmHg) at a temperature more than the transition temperature of the phospholipid.
3. For measurements, this dry residue is dispersed in appropriate amounts of bidistilled water by vortexing.
4. An aliquot of the samples (~5 mg) is sealed into stainless steel capsules obtained from Perkin Elmer (Norwalk).
5. Before scanning, the samples are held above their phase-transition temperature for 1–2 min to ensure equilibration.
6. All samples are scanned at least twice until identical thermograms are obtained using a scanning rate of 2.5°C/min.

7. The temperature scale of the calorimeter is calibrated using indium ($Tm = 156.6°C$) as the standard reference sample.

3.1.2. High-Precision Microcalorimetry

1. The Valerian Plotnikov Differential Scanning Calorimeter (VP-DSC) microcalorimeter (Microcal Inc. Northampton, USA) is used for the study of LUVs (*see* **ref. 43** and citations therein).
2. The lipid mixture is dissolved in chloroform.
3. The solvent is subsequently slowly evaporated in a rotary evaporator.
4. MLVs are formed by adding ammonium sulfate 150 mM (pH = 5.3, 300 mOs).
5. The preparation is then frozen–thawed 15 times, and is extruded 10 times through two polycarbonate membranes of 200–nm pore size diameter using an extruder (Lipex Biomembranes Inc., Vancouver, Canada) heated at 50°C.
6. The final lipid concentration of the formulations is 10 mg/mL.
7. The vesicle size distribution for the DPPC liposomes is determined using the Mastersizer (Malvern Instruments Ltd., Worcestershire, UK)
8. Repetitive extrusion of MLVs through two-stacked polycarbonate filters with 100-nm pore size after 10 freeze–thaw cycles results in LUVs with a size distribution of 180 nm in diameter.
9. Subsequently, the LUVs are passed through Sephadex G-75, pre-equilibrated with 100 mM TES and 100 mM NaCl (pH = 7.5, 300 mOs); thus, a transmembrane pH gradient was created.
10. LUVs are incubated with the drug at 60°C for 5 min.
11. The mixed lipid/drug vesicles with encapsulated drug are separated from untrapped drug by filtration through a Sephadex G-75 (Merck, Darmstadt, Germany) column.
12. Drugs usually are assayed by ultraviolet spectroscopy at 262 nm, and the drug-phospholipid molar ratio is determined by phosphate assay.
13. Heating and cooling scans are carried out at 15 K/h in order to study thermal history.

4. Notes

1. Follow precisely the calibration procedure described in the manual.
2. If you use indium, tin, lead, or zinc as references, encapsulate them in aluminium pans because they will alloy with the gold and platinum pans.
3. It is advisable to use the phospholipid bilayer alone as a second reference, (i.e., if you study the effect of a drug in DPPC bilayers, use DPPC itself as a second reference).
4. Optimize the baseline and substract it from the sample.
5. Make sure that N_2 is passed through the chamber, especially if you run experiments at low temperatures. If N_2 is not passed, the cells will be frozen.
6. Make sure that the gloves you wear are in good condition. Holes will cause the escape of N_2.
7. Use the proper reference sample during the experiment. Water or buffer is good enough for lipid bilayers.
8. Calculate ΔH at least three times. Find the mean value and report it with the standard deviation.
9. Check the carrier gas for time to time in order to make sure that your N_2 cylinder is not empty.
10. It is wise to run the samples containing lipid bilayers with or without drug molecule at high scanning rates (10°C/min); start from a low temperature of a few degrees below the phase transition and run up to a few degrees above the phase transition temperature. Leave the sample above the phase-transition for some time (it is advised for at least 15 min) and then return it to ambient temperature. Repeat this procedure more than three times (three cycles).
11. Make sure that your results are repeatable.
12. Store the samples in the refrigerator; run the samples after a few days or a week (or even a month) later.
13. Examine whether the phase transitions are stable, or some phase separation is observed. Keep in mind that you always have to follow the procedure described in **step 10**.

References

1. Janiak, M. J., Small, D. M., and Shipley, G. G. (1976) Nature of the thermal pretransition of synthetic phospholipids: dimyristoyl- and dipalmitoyllecithin. *Biochemistry* **15,** 4575–4580.
2. Connor, P., Mangat, B. S., and Rao, L. S. (1974) The labilization of lecithin liposomes by steroidal anesthetics: a correlation with anesthetic activity. *J. Pharm. Pharmacol.* **26(Suppl),** 120P.
3. Makriyannis, A., Siminovitch, D. J., Das Gupta, S. K., and Griffin, R. G. (1986) Studies on the interaction of anaesthetic steroids with phosphatidylcholine using ^{2}H and ^{13}C solid state NMR. *Biochim. Biophys. Acta* **849,** 49–55.
4. O' Leary, J. T., Ross, P. D., and Levin, W. I. (1984) Effects of Anesthetic and Nonanesthetic Steroids on Dipalmitoylphosphatidylcholine Liposomes: A Calorimetric and Raman Spectroscopic Investigation. *Biochemistry* **23,** 4636–4641.
5. Jorgensen, K., Ipsen, J. H., Mouritsen, O. G., Bennett, D., and Zuckermann, M. J. (1991) A general model for the interaction of foreign molecules with lipid membranes: drugs and anaesthetics. *Biochim. Biophys. Acta* **1062,** 227–238.
6. Ali, S., Minchey, S., Janoff, A., and Mayhew, E. A. (2000) Differential Scanning Calorimetry Study of Phosphocholines Mixed with Paclitaxdel and Its Bromoacylated Taxanes. *Biophys. J.* **78,** 246–256.
7. Huh, N. W., Porter, N. A., McIntosh, T. J., and Simon, S. A. (1996) The Interaction of Polyphenols with Bilayers: Conditions for Increasing Bilayer Adhesion. *Biophys. J.* **71,** 3261–3277.
8. Mavromoustakos, T., Theodoropoulou, E., Papahatjis, D., et al. (1996) Studies on the thermotropic effects of cannabinoids on phosphatidylcholine bilayers using differential scanning calorimetry and small angle X–ray diffraction. *Biochim. Biophys. Acta* **1281,** 235–244.
9. Kyrikou, I., Georgopoulos, A., Hatziantoniou, S., Mavromoustakos, T., and Demetzos, C. (2005) A comparative study of the effects of cholesterol and sclareol, a bioactive labdane type diterpene, on phospholipid bilayers. *Chem. Phys. Lipids* **133,** 125–134.
10. Rowat, A. C., Keller, D., and Ipsen, J. H. (2005) Effects of farnesol on the physical properties of DMPC membranes. *Biochim. Biophys. Acta* **1713,** 29–39.
11. Ryhanen, S. J., Alakoskela, J. M. I., and Kinnunen, P. K. J. (2005) Increasing surface charge density induces interdigitation in vesicles of cationic amphiphile and phosphatidylcholine. *Langmuir* **21,** 5707–5715.
12. Lohner, K. and Prenner, E. J. (1999) Differential scanning calorimetry and X-ray diffraction studies of the specificity of the interaction of antimicrobial peptides with membrane-mimetic systems. *Biochim. Biophys. Acta* **1462,** 141–156.
13. Kyrikou, I., Hadjikakou, S. K., Kovala-Demertzi, D., Viras, K., and Mavromoustakos, T. (2004) Effects of non-steroid anti-inflammatory drugs in membrane bilayers. *Chem. Phys. Lipids* **132,** 157–169.
14. Pajeva, I., Todorov, D. K., and Seydel, J. (2004) Membrane effects of the antitumor drugs doxorubicin and thaliblastine: comparison to multidrug resistance modulators verapamil and trans-flupentixol. *Eur. J. Pharm. Sci.* **21,** 243–250.
15. Mavromoustakos, T., Papadopoulos, A., Theodoropoulou, E., Dimitriou, C., and Antoniadou-Vyza, E. (1998) Thermal properties of adamantanol derivatives and their beta-cyclodextrin complexes in phosphatidylcholine bilayers. *Life Sci.* **62,** 1901–1910.
16. Zhao, L. Y. and Feng, S. S. (2005) Effects of lipid chain unsaturation and headgroup type on molecular interactions between paclitaxel and phospholipid within model biomembrane. *J. Col. Int. Sci.* **285,** 326–335.
17. Jain, M. K. and Wu, N. (1977) Effect of Small Molecules on the Dipalmitoyl Lecithin Liposomal Bilayer. *J. Membr. Biol.* **34,** 157–201.
18. Jain, M. K., Wu, N., Yen, -Min. W., Morgan, T. K., Briggs, M. S., and Murray, R. K. (1976) Phase Transition in a Lipid Bilayer. II. Influence off Adamantane Derivatives. *Chem. Phys. Lipids* **17,** 71–78.

19. Mavromoustakos, T., De-Ping Yang, A., and Makriyannis, A. (1995) Effects of the anesthetic steroid alphaxalone and its inactive Δ^{16}-analog on the thermotropic properties of membrane bilayers. A model for membrane perturbation. *Biochim. Biophys. Acta* **1239,** 257–264.
20. Cunningham, B. A. and Lis, L. J. (1986) Thiocynate and bromide ions influence the bilayer structural parameter of phosphatidylcholine bilayers. *Biochim. Biophys. Acta* **861,** 237–242.
21. Zhu, T. and Caffrey, M. (1994) Thermodynamic, thermomechanical and structural properties of a hydrated asymmetric phosphatidylcholine. *Biophys. J.* **65,** 939–954.
22. Rowe, E. S. (1985) Thermodynamic reversibility of phase transitions. Specific effects of alcohols on phosphatidylcholines. *Biochim. Biophys. Acta* **813,** 321–330.
23. McDaniel, R. V., McIntosh T. J., and Simon, S. A. (1983) Nonelectrolyte substitution for water in phosphatidylcholine bilayers. *Biochim. Biophys. Acta* **731,** 97–108.
24. McIntosh, T. J., McDaniel, R. V., and Simon, S. A. (1983) Induction of an interdigitated gel phase in fully hydrated phosphatidylcholine bilayers. *Biochim. Biophys. Acta* **731,** 109–114.
25. Schneider, M. F., Marsh, D., Jahn, W., Kloesgen, B., and Heimburg, T. (1999) Network formation of lipid membranes: triggering structural transitions by chain melting. *Proc. Natl. Acad. Sci. USA* **96,** 14,312–14,317.
26. Saraga, T. M. and Madelmont, G. (1983) Enhanced hydration of dipalmitoylphosphatidylcholine multibilayer by vinblastine sulfate. *Biochim. Biophys. Acta* **728,** 394–402.
27. Maswadeh, H., Demetzos, C., Daliani, I., et al. (2002) A molecular basis explanation of the dynamic and thermal effects of vinblastine sulfate upon dipalmitoylphosphatidylcholine bilayer membranes. *Biochim. Biophys. Acta* **1567,** 49–55.
28. Siminovitch, D. J., Ruocco, M. J., Makriyannis, A., and Griffin, R. G. (1987) The effect of cholesterol on lipid dynamics and packing in diether phosphatidylcholine bilayers. X-ray diffraction and ^{2}H-NMR study. *Biochim. Biophys. Acta* **901,** 191–200.
29. Rosser, M. F. N. (1999) Effects of alcohols on lipid bilayers with and without cholesterol: the dipalmitoylphosphatidylcholine system. *Biophys. Chem.* **81,** 33–44.
30. Bondar, O. P. and Rowe, E. S. (1998) Role of cholesterol in the modulation of interdigitation in phosphatidylethanols. *Biochim. Biophys. Acta* **1370,** 207–217.
31. Yeagle, P. L. (1985) Cholesterol and the cell membrane. *Biochim. Biophys. Acta* **822,** 267–287.
32. Franks, N. P. (1976) Structural analysis of hydrated egg lecithin and cholesterol bilayers: I. X-ray diffraction. *J. Mol. Biol.* **100,** 345–359.
33. McMullen, T. P. W., Lewis, R. N. A. H., and McElhaney, R. N. (1993) Differential scanning calorimetric study of the effect of cholesterol on the thermotropic phase behavior of a homologous series of linear saturated phosphatidylcholines. *Biochemistry* **32,** 516–522.
34. Estep, T. N., Mountcastle, D. B., Biltonen, R. L., and Thompson, T. E. (1978) Studies on the anomalous thermotropic behavior of aqueous dispersions of dipalmitoylphosphatidylcholine–cholesteol mixtures. *Biochemistry* **17,** 1984–1989.
35. Guo, W. and Hamilton, A. J. (1995) A multinuclear solid–state NMR study of phospholipid–cholesterol interaction. Dipalmitoylphosphatidylcholine–cholesterol binary system. *Biochemistry* **34,** 14,174–14,184.
36. Genz, A., Holzwarth, J. F., and Tsong, T. Y. (1986) The influence of cholesterol on the main phase transition of unilamellar dipalmytoylphosphatidylcholine vesicles. *Biophys. J.* **50,** 1043–1051.
37. Guo, W. and Hamilton, J. A. (1995) A multinuclear solid-state NMR study of phospholipid–cholesterol interactions. Dipalmitoyphosphatidylcholine–cholesterol binary system. *Biochemistry* **34,** 14,174–14,184.
38. Taylor, M., Akiyama, G., and Smith, I. C. P. (1981) The molecular dynamics of cholesterol in bilayer membranes: a deuterium NMR study. *Chem. Phys. Lipids* **29,** 327–339.
39. Huang, T. H., Lee, W. B., Gupta, D. S. K., Blume, A., and Griffin, R. G. (1993) A ^{13}C and ^{2}H nuclear magnetic resonance study of phosphatidylcholine/cholesterol interactions: characterization of liquid–gel phases. *Biochemistry* **32,** 13,277–13,287.

40. Murari, R., Murari, P. M., and Baumann, W. J. (1986) Sterol orientations in phosphatidylcholine liposomes as determined by deuterium NMR. *Biochemistry* **25,** 1062–1067.
41. Yeagle, P. L., Hutton, W. C., Huang, C. H., and Martin, R. B. (1977) Phospholipid head-group conformations; intermolecular interactions and cholesterol effects. *Biochemistry* **16,** 4344–4349.
42. Wu, W. -G. and Chin, L. -M. (1990) Comparison of lipid dynamics and packing in fully interdigitated monoarachidoylphosphatidylcholine and non-interdigitated dipalmitoylphosphatidylcholine bilayers: cross-polarization/magic angle spinning ^{13}C-NMR studies. *Biochim. Biophys. Acta* **1026,** 225–235.
43. Kyrikou, I., Daliani, I., Mavromoustakos, T., et al. (2004) The modulation of thermal properties of vinblastine by cholesterol in membrane bilayers. *Biochim. Biophys. Acta* **1661,** 1–8.

40

Atomic Force Microscopy to Study Interacting Forces in Phospholipid Bilayers Containing General Anesthetics

Zoya V. Leonenko, Eric Finot, and David T. Cramb

Summary

Atomic force microscopy (AFM) can be used to reveal intimate details about the effect of anesthetics on phospholipid bilayers. In AFM, surfaces are probed using a tip revealing lateral structural features at 10–20-nm resolution and height features at 0.5-nm resolution. Additionally, information on the viscoelasticity of the surface can be gained by examining the forces of tip–surface interactions. This is also known as force spectroscopy. In this chapter, the use of AFM to observe and quantify anesthetic-induced changes in phospholipid bilayers is detailed. The procedures developed to create supported phospholipid bilayers are described and the techniques developed to generate the best AFM images and force spectroscopy results have been revealed.

Key Words: Atomic force microscopy; force spectroscopy; halothane; phospholipid bilayer; DPPC; phase transition.

1. Introduction

The changes in the physical and chemical properties of biological membranes owing to incorporation of anesthetics are of great interest for understanding the mechanism of anesthetic action. The molecular theory of anesthetic action includes the hypothesis that anesthetics alter the lipid membrane structure, and therefore, its biophysical properties. Changes in lateral pressure within bilayers *(1)*, lipid packing in membranes *(2–5)*, polarization of the membrane, and the degree of motional disorder in lipid chains *(5,6)*, which leads to bilayer thinning *(6–8)* could all change the ability of cells to communicate with each other. Atomic force microscopy (AFM) is a valuable technique to address the structural changes in lipid bilayer membranes. AFM has proven to be advantageous not only for imaging biological samples in liquid media *(9,10)*, but also for allowing the study of sample physical properties by measuring force of interactions between the AFM tip and the sample surface *(11–13)*.

Previously, it was demonstrated that incorporation of general anesthetics into the bilayer produces structural alterations in a supported phospholipid bilayer *(8)*, such as bilayer area and height (i.e., domain formation), each of which depends on the time of incubation and concentration of anesthetics. Similar domain formation was observed when the membrane underwent a melting transition *(7)*. In spite of the visual similarity between an anesthetic-induced domain formation and the heat induced gel–liquid phase transition, when observed by AFM imaging, the mechanism of anesthetic action is likely to be different from the effect of membrane melting. It is assumed that the physical properties of the domains produced in lipid bilayers by halothane and by temperature are also different. To address these physical properties, the atomic force microscope can be used as a force apparatus; it has proven to be an advantageous tool for the measurements of interaction forces between AFM probe and the surface, giving information about the physical properties of the sample surface *(14–16)*. In this chapter, the procedures used to generate planar phospholipid bilayers supported on a mica substrate are described in detail. Also the sample conditions, which provide AFM images of the highest quality are detailed. Finally, the approach to force spectroscopy is conveyed.

2. Materials
2.1. Chemicals and Substrates

1,2-Dipalmitoyl phosphatidylcholine (DPPC) (Avanti Polar-lipids Inc., Alabaster, AL) is used without further purification. Pure halothane (Sigma Chemical Co., Saint Louis, MO) is used throughout. Freshly cleaved ASTMV-2 quality, scratch-free ruby mica (Asheville-Schoonmaker Mica Co., Newport News, VA) is used throughout the study as a substrate. Micro Cleaning Solution from ArrayIt® (Sunnyvale, CA 94089 USA) is used. Distilled, deionized, nanopure quality water is used in the generation of all vesicles.

2.2. AFM Accessories

Magnetically, (Au–Cr-coated) type I Maclevers® (Agilent Technologies AFM [Tempe, Arizona] are used for magnetic A/C (MAC) mode imaging and for force measurements. The nominal spring constants of Au–Cr-coated Maclevers used are 0.6 N/m. The tip radius of curvature is quoted as being typically 25 nm. Colloidal gold spheres of 5 and 14 nm in diameter from Ted Pella Inc. (Redding, CA) *(17)* are used for calibration of the AFM scanner's height resolution. A calibration grid from Molecular Imaging is used for calibration in the lateral direction.

3. Methods
3.1. Vesicle Solution Preparation

Multilamellar vesicles are prepared by ultrasonication of phospholipid solutions. First, a controlled volume of phospholipid dissolved in chloroform is added to a 10-mL round bottom flask. The chloroform is then removed using a gentle stream of dry nitrogen. Following this, the phospholipid is suspended in low-p*I* buffer or pure water, as appropriate, and stirred for 30 min at room temperature. This produces a cloudy suspension. The suspension is then subjected to 10 rounds of 10 min ultrasonication, followed by 15 min of stirring in a cold water bath. After this regimen, the DPPC solution is found to become clear, suggesting the dissolution of large particles.

3.2. Supported Bilayer Formation

The liquid sample cell is washed thoroughly. First, it is subjected to a 10-min sonication in a diluted microcleaning solution (MCS). The MCS is diluted by mixing 50 mL of 20X MCS and 950 mL ultrapure H_2O to make 1 L of 1X MCS. After sonication, the cell is rinsed with ultrapure water, and then sonicated 10 min further in ultrapure water. This procedure removes all debris from the cell.

To create the supported planar bilayer, mica is freshly cleaved with double-sided adhesive tape before each experiment. Mica consists of negatively charged layers that are bound together by large, positively charged interlayer cations (K^+ in the case of muscovite mica). Each stratum consists of two hexagonal layers of SiO_4, which are cross-linked by aluminum atoms having OH_3 groups incorporated. The electrostatic bonds between K^+ and O atoms from the layer are weak and easily broken. This layer is disrupted after a cleavage procedure, exposing a basal plane covered by K^+. Thus, freshly cleaved mica in water carries a slightly negative charge, with a density of 0.57 ions/nm *(18)*. It is possible to adsorb a DPPC lipid bilayer on mica without any further modification of the mica surface. Supported planar bilayers are prepared for AFM imaging by the vesicle fusion method *(8)* as follows. Aliquots of 20 µL (lipid concentration: 0.5 mg/mL; incubation time: 10 min) liposome solutions are deposited on freshly cleaved mica. After a well-defined period of time (usually 10 min) the mica is gently rinsed with ultrapure water and the cell is filled with ultrapure water. After this, the samples of supported bilayers are ready for imaging. Then, the sample chamber is mounted onto the AFM scanner. Supported bilayers are always kept in water, never dried or exposed to air.

To prepare bilayer samples with halothane, two methods are put forward. First, addition of halothane into the vesicle solution, followed by an incubation of 1.5 h. A supported bilayer could then be formed from the halothane-loaded vesicles method of vesicle fusion as described earlier for pure vesicles. Second, addition of small volumes (60 µL) of pure anesthetic into the AFM liquid cell containing the established supported bilayer, which is covered with 1 mL of water. The sample is incubated for 1.5 h after the addition of halothane. Then, the nonpartitioned halothane is carefully removed by pipeting with all the water filling the cell. The cell is then gently rinsed by slowly adding water by pipet and removing it again. Following this, the cell is filled with nanopure water. Fluorescence experiments show that an incubation time of 1.5 h is optimal for halothane partitioning between water and bilayer and reach equilibrium, whether one method is used or the other *(19)*. Both ways of adding halothane give similar results. For the sake of simplicity and reproducibility, the second method is used for all subsequent experiments.

3.3. AFM Imaging

AFM is a surface imaging technique with nanometer-scale lateral resolution and 0.1 nm resolution normal to the sample, which operates by monitoring the forces of interaction acting between a probe and a sample, while the sharp probe scans the sample surface. For imaging, MAC mode AFM is used; this is a tapping mode available through Molecular Imaging AFM in which a magnetically coated probe oscillates near its resonant frequency driven by an alternating magnetic field. All images are taken using a Pico SPM microscope with AFMS-165 scanner (Molecular Imaging Inc.) in a liquid cell. All samples are imaged in ultrapure water because degradation was observed in image quality with increasing p*I* *(8)*. The highest contrast images are always obtained by imaging in ultrapure water *(21)*.

In an attempt to minimize the osmotic shock to vesicles during rinsing, the original solutions are thus prepared with a low-p*I*. This rinsing process occurs before the AFM tip encounters the solution. Therefore, the tip is never exposed to a liposome solution, and should remain free of lipid contamination. Magnetically, (Au–Cr-coated) type I Maclevers (Molecular Imaging Inc.) are used for MAC mode imaging and force measurements. The scan rate is fixed to 20 mm/s. The standard MAC mode fluid cell (Molecular Imaging) is used throughout.

A supported DPPC bilayer is formed on mica from vesicle solution, rinsed with water, and covered with 1 mL of water in the AFM liquid cell; pure halothane (60 µL) is added into the liquid cell and the cell is incubated for 1.5 h. After this, nonpartitioned halothane is removed by gently rinsing and refilling the cell with water. The cell is mounted into the microscope and after an equilibration of 10–15 min, the sample is imaged in water in tapping mode (MAC mode). The amplitude set point for imaging is chosen at the near maximum of the resonance frequency (80–90% of the maximum amplitude).

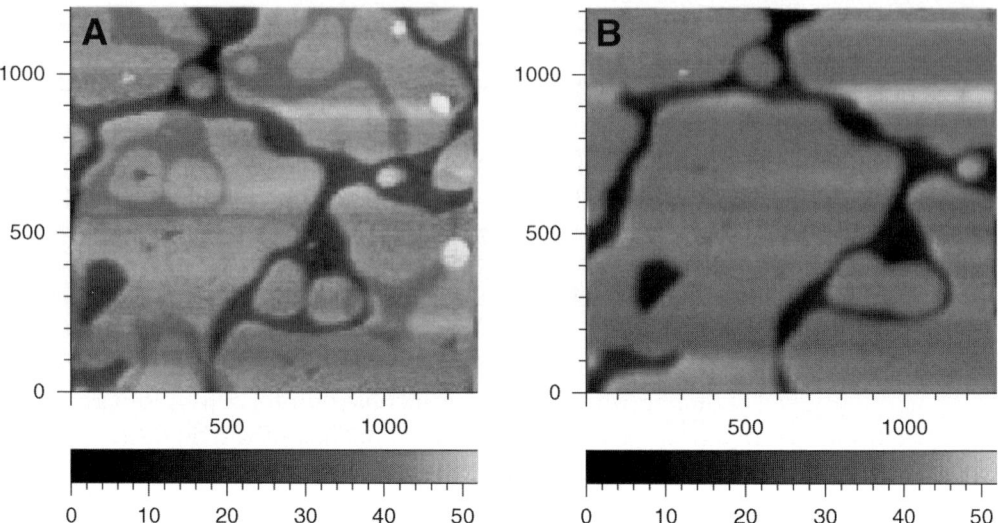

Fig. 1. A typical AFM image of DPPC bilayer with halothane incorporated. Supported bilayer was formed on mica from vesicle solution, rinsed with water and covered with 1 mL of water in AFM liquid cell, excess of halothane (60 µL) was added into the liquid cell, and the cell was incubated for 1.5 h. After this, nonpartitioned halothane was removed with water, AFM cell was gently rinsed and filled with water, Bilayer was imaged in taping mode (MAC mode). (**A**) imaged immediately after sample preparation and (**B**) after 30 min of imaging bilayer converts into thin domains.

A typical image of the DPPC bilayer with halothane incorporated is shown in **Fig. 1**. Halothane incorporation results in domain formation (*see* **Fig. 1**) *(20)*. The thickness of the higher and lower domains is measured as 3.5 nm.

3.4. AFM Force Measurements

In AFM force measurements, the probe is moved toward the sample, and the cantilever deflection is measured as a function of the extent of the piezo electric tube. During approach to the surface the attractive and repulsive forces being measured are characterized by van der Waals and electrostatic interactions, as well as solvation, hydration, and compression-related steric forces *(11)*. The probe is then retracted from the surface. The retraction force curves often show a hysteresis referred to as an adhesion pull-off event, which can be used to determine adhesion forces between the probe and the sample.

In the experiments, first the supported bilayer in tapping MAC mode is imaged. After the image is completed and saved, the regime is switched into the contact mode and then, using the same cantilever, the interacting forces are measured by positioning the AFM tip on different locations using the image obtained in tapping mode. Force curves are collected at 10 locations on the bilayer and then averaged. The force spectroscopy experiment consists of monitoring the interaction between the AFM tip and the substrate by sensing the cantilever deflection (Z_c) as a function of the piezo elongation (Z_p) and as the tip is moved toward and away from the substrate (i.e., raw data). The raw data averaged for each sample are shown on **Fig. 2**.

Force measurements were performed on DPPC bilayers formed from water solutions with and without halothane. It is a well-known fact that the force interactions depend on the velocity of the surface approach *(11)*, especially for soft, viscoelastic materials. The frequency sweep has been varied over about four decades, from 0.01 to 10 Hz. The Z-range is 50 nm. Note that

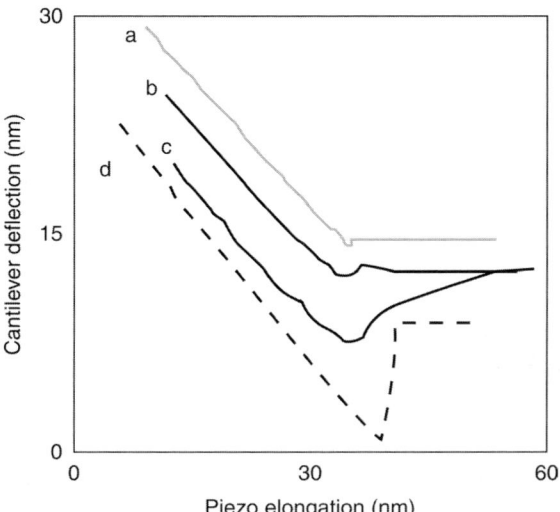

Fig. 2. Cantilever deflection dZ_c as a function of the piezotube elongation when the tip is retracted from the sample. DPPC bilayers deposited on mica were measured under three conditions: (b) the DPPC bilayer at $T = 20°C$ (c) the DPPC bilayer at $T = 50°C$ (d) the DPPC bilayer with halothane incorporated. The adhesion curve on bare mica (a) is shown for comparison.

varying the scan frequency affects simultaneously the velocity at which the tip is withdrawn from the bilayer and the contact time between the tip and the sample. At a fixed applied load, increasing the frequency sweep rate increases the tip velocity yet decreases the contact time.

Statistical analysis reveals that 10 measurements at different locations of the same area are required to determine parameters such as the adhesion and the long-range forces with 10% accuracy. Data are collected over a time period of 2 h, and a total of 100 force curves are analyzed for each sample. Force curves are saved as plots and then converted into text files for analysis. After the raw data are collected, they are converted into the "force vs separation plots," averaged and analyzed.

3.5. AFM Force Analysis

During AFM force measurements, the probe is moved toward the sample and the cantilever deflection is measured as a function of the extent of the piezo electric tube. Force measurements using the atomic force microscope (described in **Subheading 3.4.**) give the raw data of cantilever deflection (δV) vs piezo movement (z_{piezo}). Raw data are transformed numerically under a Matlab environment. As the tip–sample separation cannot be independently measured, a systematic procedure for calculating the sensitivity of the apparatus is developed (7). This measurement assumes that tip and the mica surface are brought into a nondeformable contact in the higher loading region. Raw data (Z_c vs Z_p) are then converted into force (F) vs surface-tip separation (D) using Hooke's law: $F = k Z_c$ where k is the spring constant of the cantilever, with the geometric relationship $D = Z_c - Z_p$ for incremental changes.

Statistical data are extracted from a large set of measurements. The authors use the Gaussian distribution in the following form:

$$P(x) = C \frac{1}{\sigma\sqrt{2\pi}} e^{-(x-\mu)^2/(2\pi^2)} \tag{1}$$

where $P(x)$ is the number of samples at the particular x point in the histogram, μ is the mean, σ is the standard deviation, and C is the normalization constant. The bin size of the histogram is determined statistically by dividing the range of measurements by the square root of the number of measurements. The average of the adhesion force and the standard deviation are obtained from the analysis of the whole set of the force curves, which are represented in histogram form ($P[x]$) and then adjusted using a Gaussian law.

3.5.1. Adhesion Forces

Surface force profiles are measured among the silicon nitride tip, mica, and a pure DPPC bilayer vs the profiles in a DPPC bilayer with halothane partitioned into it. To extract information about adhesive forces, the approach part of the force curve was analyzed. Raw data of the cantilever deflection corresponding to the retraction of the tip from the surface (after contact with the tip), are shown in **Fig. 2**, plots a–d. Data corresponding to the mica surface immersed in water are presented for comparison. Mica is characterized by an abrupt jump out of contact (plot a in **Fig. 2**).

To analyze adhesive forces, the same statistical analysis was conducted that was used in the previous work *(7)*. It has been demonstrated that the force of a unit interaction between an AFM tip and a surface can be determined from statistical analysis of a series of detachment force measurements. For a statistical analysis based on adhesive forces originating from a discrete number n, individual interactions or bonds have been used. The total force distribution follows Poisson statistics, where both the adhesion force (F_{adh}) and the variance (σ) originates from a number of individual bonds n:

$$F_{adh} = nF_s$$

and

$$\sigma^2 = nF_s^2$$

The force of one bond (F_s) is therefore given by the square of the variance of the force divided by F_{adh}. The value n is the ratio between F_{adh} and F_s. This analysis has the advantage that the knowledge of the mean radius of curvature is not required, and gives information on the nature of the bilayer. **Table 1** presents the standard deviation of the adhesion force in the gel phase DPPC bilayer and bilayer with halothane incorporated, as measured by the authors. Data for gel phase DPPC bilayer at room temperature and fluid phase DPPC bilayer at elevated temperature from the previous paper *(7)* are shown for comparison.

This analysis allows to correlate microscale changes in the bilayer physical properties with molecular structure and mobility of individual lipid molecules. Tail mobility can be seen as the number of molecules n in contact with the AFM probe: the larger the n, the higher the mobility. A change in mobility increases the number of bonding (n) and, thus, F_{adh} and σ.

3.5.2. Repulsive Forces

Repulsive forces are extracted by analyzing the approach part of experimental force curves. **Figure 3A** shows four patterns of repulsive forces, measured through the AFM tip when approaching the sample surface of pure mica (plot a), DPPC bilayer at room temperature (plot b), DPPC bilayer at 50°C (plot c) (*see also* **ref. 7**), and DPPC bilayer with an excess of halothane (plot d). The zero force intercept obtained from **Fig. 3B** is used to determine the "effective bilayer thickness" corresponding to the related tip–sample distance.

Table 1
Experimental Adhesion Force (F_{adh}) and Its Standard Error Sigma on DPPC Bilayer and DPPC With Halothane, Measured at Room Temperature; Calculation of the Number of Bindings n and the Mean Force of a Single Bond F_s

	DPPC[a] 22°C	DPPC	DPPC + halothane
F_{adh}	1 nN	1.5 nN	7 nN
σ	0.6 nN	0.8 nN	2 nN
F_s	0.3 nN	0.4 nN	0.5 nN
n	3	4	14

[a]Data on DPPC bilayer at room and elevated temperature, measured earlier during melting transition (*11*) are shown for comparison.

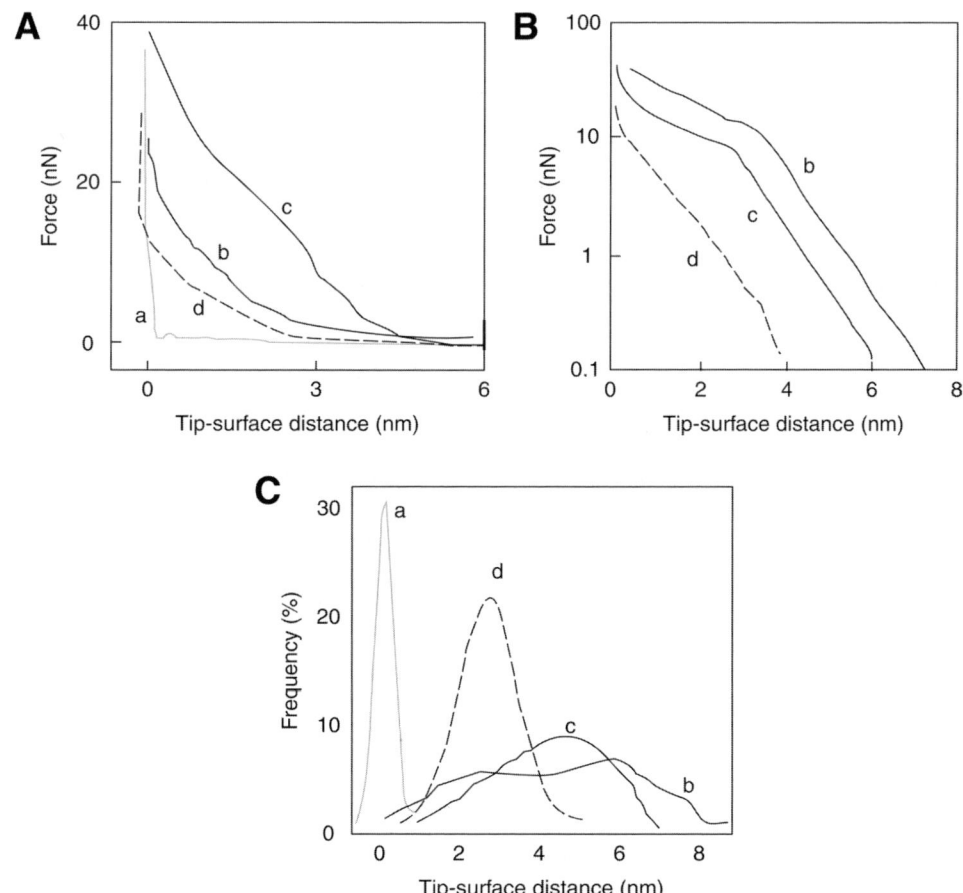

Fig. 3. **(A)** Surface force as a function of a tip–surface distance in liquid cell when the tip is approaching the sample; **(B)** logarithmic representation of data in panel **(A)**; **(C)** Distribution of the "effective thickness" determined from force analysis **(A)** on mica, DPPC and DPPC with halothane incorporated. Plots on **3A–C** are ascribed as: (a) bare mica (is shown for comparison), (b) the DPPC layer at $T = 22°C$, (c) the DPPC layer at $T = 50°C$, and (d) the DPPC layer with

4. Notes

1. All liposome solutions are sonicated at controlled temperature between 25° and 30°C, not to exceed the melting temperature of the DPPC bilayer (42°C) by addition of ice or cold water into the sonication bath.
2. When preparing supported bilayers on mica, the time of incubation and amount of solution added has to be adjusted, in order to obtain good coverage of mica by the bilayer. To form a supported DPPC bilayer on mica small aliquots (20–30 µL of 0.5 mg/mL) DPPC solution in water were used. The incubation time can vary from 5 to 30 min.
3. Sometimes it is necessary to produce holes in the bilayer in order to measure the bilayer thickness. One way to do so is using the AFM tip and scratch a hole in the bilayer by imaging a very small area at high scanning rate and high load force in contact mode. Although this method produces a hole in the bilayer, it may also produce mechanical defects in the tip, and contaminate the lipids adsorbing to the tip from the sample, which reduces the resolution and quality of the image. Another, less-invasive method is used to produce holes. First, one needs to scan a prepared, supported bilayer, because in general, holes are naturally present in the bilayer. If not, one can rinse the cell with supported bilayer with a stream of water from a plastic bottle; this procedure readily produces the required holes, and does not contaminate the cantilever.

Acknowledgment

Financial support from NSERC (DTC) is gratefully acknowledged.

References

1. Cantor, R. S. (1994) Breaking the Meyer-Overton Rule: Predicted Effects of Varying Stiffness and Interfacial Activity on the Intrinsic Potency of Anesthetics. *Biophys. J.* **80**, 2284.
2. Rowe, E. S. (1987) Induction of lateral phase separations in binary lipid mixtures by alcohol. *Biochemistry* **26**, 46.
3. Simon, S. A. and McIntosh, T. J. (1984) Surface ripples cause the large fluid spaces between gel phase bilayers containing small amounts of cholesterol. *Biochim. Biophys. Acta* **773**, 169.
4. Nambi, P., Rowe, E. S., and McIntosh, T. J. (1988) Studies of the ethanol-induced interdigitated gel phase in phosphatidylcholines using the fluorophore 1,6-diphenyl-1,3,5-hexatriene. *Biochemistry* **27**, 9175.
5. Mou, J., Yang, J., Huang, C., and Shao, Z. (1994) Alcohol Induces Interdigitated Domains in Unilamellar Phosphatidylcholine Bilayers. *Biochemistry* **33**, 9981.
6. McClain, R. L. and Breen, J. J. (2001) Phase Transition in Solid-Supported Lipid Bilayers. *Langmuir* **17**, 5121.
7. Leonenko, Z. V., Finot, E., Ma, H., Dahms, T., and Cramb, D. T. (2004) Investigation of temperature induced phase transitions in DOPC and DPPC phospholipid bilayers using temperature-controlled scanning force microscopy. *Biophys. J.* **86**, 3783.
8. Leonenko, Z. V., Carnini, A., and Cramb, D. T. (2000) Supported planar bilayer formation by vesicle fusion: the interaction of phospholipid vesicles with surfaces and the effect of gramicidin on bilayer properties using atomic force microscopy. *Biochim. Biophys. Acta* **1509**, 131.
9. Han, W. and Lindsay, S. M. (1998) Probing molecular ordering at a liquidsolid interface with a magnetically oscillated atomic force microscope. *Appl. Phys. Lett.* **72**, 1656.
10. Milhiet, P. E., Vie, V., Gioconde, M. C., and Le Grimellec, C. (2001) AFM Characterization of Model Rafts in Supported Bilayers. *Single Molecules* **2**, 109.
11. Cappella, B. and Dietler, G. (1999) Force-distance curves by atomic force microscopy. *Surf. Sci. Rep.* **34**, 1.
12. Franz, V., Loi, S., Muller, H., Bamberg, E., and Butt, H. -J. (2002) Tip penetration through lipid bilayers in atomic force microscopy. *Coll. Surf. B* **23**, 191.

13. Loi, S., Sun, G., Franz, V., and Butt, H. J. (2002) Rupture of molecular thin films observed in atomic force microscopy. II. Experiment. *Phys. Rev. E* **66(3),** 1602.
14. Fang, H. H. P., Chan, K. -Y., and Xu, L. -C. (2000) Quantification of bacterial adhesion forces using atomic force microscopy (AFM). *J. Microbiol. Methods* **40,** 89.
15. Beech, I. B., Smith, J. R., Steele, A. A., Penegar, I. and Campbell, S. A. (2002) The use of atomic force microscopy for studying interactions of bacterial biofilms with surfaces. *Colloids Surf. B Biointerface* **23,** 231.
16. Engler, A. J., Richert, L., Wong, J. Y., Picart, C., and Discher, D. E. (2004) Surface probe measurements of the elasticity of sectioned tissue, thin gels and polyelectrolyte multilayer films: Correlations between substrate stiffness and cell adhesion. *Surf. Sci.* **570,** 142.
17. Vesenka, J., Manne, S., Giberson, R., Marsh, T., and Henderson, E. (1993) Colloid gold particles as an incompressible atomic force microscope imaging standard for assessing the compressibility of biomolecules. *Biophys. J.* **65,** 992–997.
18. Muller, D. J., Amrein, M., and Engel, A. (1997) Adsorption of biological molecules to a solid support for scanning probe microscopy. *J. Struct. Biol.* **119,** 172–188.
19. Carnini, A., Phillips, H. A., Shamrakov, L. G., and Cramb, D. T. (2004) Revisiting lipid-general anesthetic interactions. II. Halothane location and changes in lipid bilayer microenvironment monitored by fluorescence. *Can. J. Chem.* **82,** 1139.
20. Leonenko, Z. and Cramb, D. (2004) Thinned bilayer domain formation caused by incorporation of volatile anesthetics into supported phospholipid bilayers. *Can. J. Chem.* **82,** 1.
21. Binnig, G., Quate, C. F., and Gerber, C. (1986) Atomis Force Microscope. *Phys. Rev. Lett.* **56,** 930.

Index

A

AChR, *see* Nicotinic acetylcholine receptor
Acrylodan-labeled intestinal fatty acid-binding protein (ADIFAB)
   fatty acid measurements
      adipocyte preparation, 32, 39
      assay, 33, 34, 39
      buffer sensitivity, 35–37, 39–41
      calibration, 33, 39
      fatty acid stock solution preparation, 31, 37
      large unilamellar vesicle preparation, 32, 39
      materials, 29, 30, 37
      principles, 30
      small unilamellar vesicle preparation, 31, 32, 37, 39
   fatty acid membrane binding and transport assay, 245
   overview, 27, 28
Adamantine, differential scanning calorimetry of drug–membrane interactions, 589, 590
ADIFAB, *see* Acrylodan-labeled intestinal fatty acid-binding protein
AFM, *see* Atomic force microscopy
Alphaxalone, differential scanning calorimetry of drug–membrane interactions, 590, 591
AMBER, molecular dynamics simulations, 91
Anesthetics, *see* Atomic force microscopy
Area strain, *see* Micropipet aspiration
Atomic force microscopy (AFM)
   anesthetic effects on lipid bilayers
      force
         analysis, 605, 606
         measurements, 604, 605
      imaging, 603, 604
      materials, 602
      overview, 601, 602
      supported bilayer formation, 602, 603
      vesicle preparation, 602
   near-field scanning optical microscopy samples, 471–473, 478, 479
   rock and roll method film assessment, 69, 70
   supported lipid bilayer analysis
      bilayer deposition, 509–511
      image bilayer-protein binding, 510, 511
      image phase-separated bilayers, 510, 511
      materials, 504, 505
      multilamellar vesicle preparation, 507, 508
      overview, 503–506
      small unilamellar vesicle preparation, 508–511

B

BAM, *see* Brewster angle microscopy
Bending modulus
   inverted hexagonal phase determination, 363
   micropipet aspiration measurements, 434
   X-ray scattering of inverted hexagonal phase, 363
Bilayer elasticity, *see* Elasticity
Bilayer phase, overview, 16, 17
Brewster angle microscopy (BAM), Langmuir films, 449
Brownian dynamics, *see* Diffusion

C

Caillé-de Gennes model, liquid-crystalline membranes and intermolecular correlation functions, 344, 349
Calorimetry, *see* Differential scanning calorimetry; Pressure perturbation calorimetry
CARS, *see* Coherent anti-stokes Raman scattering
Caveolae
   caveolin in biogenesis, 460
   disruption, 465
   isolation
      detergent isolation, 462, 466
      immunoisolation, 464–466
      materials, 461
      nondetergent isolation, 462, 465, 466
      plasma membranes, 462, 465, 466
      sucrose density centrifugation, 463, 464, 466
   microscopy, 460
   overview, 5, 6, 459, 460
Chain order, *see* Coherent anti-stokes Raman scattering
Chain orientation, *see* Coherent anti-stokes Raman scattering

CHARMM, molecular dynamics simulations, 91–95
CHIF
  function, 516
  nuclear magnetic resonance
    materials, 516–518
    overview, 515, 516, 522, 523
    protein expression and purification, 516–518
    solid-state nuclear magnetic resonance in bilayers
      nitrogen-15 studies, 523–525
      phosphorous-31 studies of lipids, 525
      sample preparation, 523
      two-dimensional PISEMA studies, 525, 526
    solution studies in micelles
      detergents, 519
      protein structure determination, 519, 520, 522
Cholesterol
  assay for sterol superlattice analysis, 164
  drug–membrane interaction effects, 593
  lipid rafts, *see* Lipid raft
  membrane effects, 489
Coherent anti-stokes Raman scattering (CARS)
  acyl chain
    order, 54, 55
    orientation, 55–57
  congested spectral bands, 58, 59
  data analysis, 53
  instrumentation, 50, 51
  lipid sample preparation, 51, 52
  picosecond CARS
    multiplex CARS, 50
    nonresonant background suppression, 49
    overview, 48, 49
  principles, 46–48
  sensitivity, 53, 54
  vibrational microscopy, 45, 46
Confocal microscopy
  fluorescence recovery after photobleaching, 271, 272, 274
  giant unilamellar vesicle shape studies
    dye partitioning analysis, 378
    least squares algorithm, 370, 375, 377
    maximum brightness method for shape tracing
      limitations, 375
      overview, 370, 371
    uncertainty estimation by tracing ideal images, 371, 372, 374, 375
    overview, 368–370
    phase area fraction determination, 377, 378
    shape fitting analysis
      first integral of shape equations, 379–381
      full integration of shape equations, 381–385
Cubic phase, overview, 17
Curvature strain
  enzyme activity effects, 22, 23
  overview, 20, 21
β-Cyclodextrin, cholesterol depletion for lipid raft disruption, 465

D

Dehydroergosterol
  concentration determination, 148, 149
  sterol oxidation assay, *see* Sterol oxidation
  structure, 146, 155
De-Pakeing, nuclear magnetic resonance data analysis, 114, 115
Differential scanning calorimetry (DSC)
  baseline establishment, 188–190
  data analysis, 190–193
  drug–membrane interactions
    conventional calorimetry, 596, 597
    high-precision microcalorimetry, 597
    materials, 596
    overview, 587–595, 597
  instrumentation, 179–181, 183
  principles, 171
  sample
    equilibration
      pretransitions of linear saturated diacylphosphatidylcholines, 184
      thermotropic transitions involving cubic phases, 187, 188
      thermotropic transitions involving lamellar-crystalline lipid phases, 184–187
    preparation, 183
    size, 183
  thermotropic processes at nonequilibrum conditions, 177–179
  thermotropic processes at thermodynamic equilibrium
    phase transition types, 172, 173
    theory, 173–177

Diffusion
  modeling
    Brownian dynamics algorithms, 299, 300
    caveats, 312, 313
    data collection and analysis, 305
    diffusive steps for normal diffusion, 301, 302
    dissipative particle dynamics, 300
    first passage time method, 311
    geometry, 301
    mean-square displacement, 295, 296, 304
    Monte Carlo simulations, 299, 300
    motion types, 296, 302, 303
    n-fold way, 310, 311
    particles, 303
    periodic boundary conditions, 301
    potentials and traps, 303, 304
    probability distribution function, 298
    programming and computation speed, 308–310
    propagators, 297, 298, 313, 314
    random number generator
      correlations, 307, 308
      requirements, 305
      seeds, 308
      selection, 306
      testing, 306, 307
      transformations, 308
      types, 305, 306
    resources, 312
    software packages, 311, 312
  rates, see Fluorescence recovery after photobleaching; Magic angle spinning nuclear magnetic resonance; Single-molecule detection; Stopped-flow fluorometry
Dissipative particle dynamics, diffusion modeling, 300
Dividing surface
  definition, 357, 358
  elastic moduli, 359, 360
  neutral surface, 360, 361
DSC, see Differential scanning calorimetry

E

Elasticity
  bending modulus, see Bending modulus
  bilayer barrier function, 543, 544
  dividing surface elastic moduli, 359, 360
  energetics of elastic bilayer deformations, 546–549, 563
  gramcidin channel studies
    bilayer stiffness modulation, 551, 552
    electrophysiology, 562, 563
    equation nomenclature, 564
    force transduction
      channel appearance as function of hydrophobic mismatch, 555, 556, 563
      channel lifetime as function of hydrophobic mismatch, 553–555, 563
      heterodimer studies, 556–563
      overview, 550, 551
    gramcidin analog synthesis, 562
  membrane proteins
    bilayer deformation induction, 546
    bilayer interactions, 545, 546, 563
    function and conformational changes, 544, 545
    ion channel regulation, 549
  micropipet aspiration studies, see Micropipet aspiration
  stresses and energy, 358, 359
Electrophysiology
  gramcidin channel studies of membrane elasticity, see Gramcidin channel
  reconstituted ion channels, see Ion channel reconstitution
Ewald sum, phospholipid interactions in membranes analysis, 131

F

Fatty acids
  fatty acid-binding protein acrylodan-labeled probe, see Acrylodan-labeled intestinal fatty acid-binding protein
  membrane binding and transport assays
    fluorescent probes and assay types, 241, 242
    materials, 240, 241, 249
    multiple probe assays, 245
    principles, 238–240
    single probe assays
      acrylodan-labeled intestinal fatty acid-binding protein assay, 245
      cis-PA assay, 244, 252, 253
      FPE assay, 244
      pyranine assay, 243, 251
    small unilamellar vesicle preparation
      FPE surface probe, 243, 251

nigericin calibration of pH, 243, 251
pH probe, 242, 250, 251
stopped-flow fluorescence assays, 246–248, 253, 254
metabolism, 237
quantification, *see* Acrylodan-labeled intestinal fatty acid-binding protein
Fluctuations, *see* Molecular fluctuations
Fluorescence recovery after photobleaching (FRAP)
advantages, 267
confocal microscopy, 271, 272, 274
data analysis, 273–275
fluorophore staining, 269
materials, 268, 269
principles, 267, 268
sample immobilization
bacteria, 269, 270
mammalian cells, 270
membrane fragments or vesicles, 270, 271
single-molecule detection comparison, 282
Fluorescence resonance energy transfer (FRET)
lipid rafts
materials, 492
phosphatidylcholine–cholesterol mixtures
acceptor partition coefficient, 493, 499
donor decay measurement and analysis, 494, 500
donor partition coefficient, 493
vesicle preparation, 493, 499
ternary phosphatidylcholine/sphingomyelin/cholesterol
domain analysis, 496–499
phase diagrams, 494–496, 500
nicotinic acetylcholine receptor Laurdan probing and fluorescence resonance energy transfer
data analysis, 535–537
fluorescence resonance energy transfer, 534, 535–537, 539
Laurdan generalized polarization, 534, 535, 537, 539, 540
materials, 533, 534, 539
overview, 532
receptor-rich membrane preparation, 534, 539
titration, 535, 540
theory, 491, 492
Force fields
molecular dynamics simulations, 91–95
phospholipid interactions in membranes
all-atom representation, 128, 129
simplified lipid representation, 129, 130
Fourier transform infrared spectroscopy (FTIR)
applications, 207, 208, 223
chain melting phase transitions, 212–214
data acquisition, 219, 220
data analysis, 220–223
instrumentation, 217
lipid phase transitions
crystalline/quasi-crystalline lipid structures, 214, 215
marker groups, 209–212
nonlamellar lipid phases, 215–217
overview, 208, 209
sample preparation, 217–219
FPE, *see* Fatty acids
FRAP, *see* Fluorescence recovery after photobleaching
FRET, *see* Fluorescence resonance energy transfer
FTIR, *see* Fourier transform infrared spectroscopy
FXYD2
function, 516
nuclear magnetic resonance
materials, 516–518
overview, 515, 516, 522, 523
protein expression and purification, 516–518
solid-state nuclear magnetic resonance in bilayers
nitrogen-15 studies, 523–525
phosphorous-31 studies of lipids, 525
sample preparation, 523
two-dimensional PISEMA studies, 525, 526
solution studies in micelles
detergents, 519
protein structure determination, 519, 520, 522

G

Giant unilamellar vesicle (GUV)
deformation studies with laser tweezers
deformation
bead utilization, 396, 398, 403
direct interactions with vesicle, 401
electroformation of vesicles, 393, 394, 402

Fourier analysis, 391, 392
holographic optical tweezer
  instrumentation, 390
image extraction of vessel boundary,
  401–403
materials, 392, 393, 402
microsphere incorporation into vesicles,
  394, 402
particle tracking, 391, 399–401, 403
sample preparation
  low index of refraction, 401, 402
  vesicles with internal beads, 395,
    396, 402
features and advantages of study, 333,
  368, 389
membrane pressure fluorescence microscopy
  studies
  data collection and analysis, 334, 335
  materials, 334
  overview, 333, 334
micropipet aspiration, see Micropipet
  aspiration
shape studies with confocal microscopy
  dye partitioning analysis, 378
  least squares algorithm, 370, 375, 377
  maximum brightness method for shape
    tracing
    limitations, 375
    overview, 370, 371
    uncertainty estimation by tracing ideal
      images, 371, 372, 374, 375
  overview, 368–370
  phase area fraction determination, 377, 378
  shape fitting analysis
    first integral of shape equations,
      379–381
    full integration of shape equations,
      381–385
Gramcidin channel
  bilayer elasticity studies
    bilayer stiffness modulation, 551, 552
    electrophysiology, 562, 563
    equation nomenclature, 564
    force transduction
      channel appearance as function
        of hydrophobic mismatch, 555,
        556, 563
      channel lifetime as function
        of hydrophobic mismatch,
        553–555, 563

heterodimer studies, 556–563
overview, 550, 551
gramcidin analog synthesis, 562
heterodimer formation, 556–559
regulation, 550
structure, 550
GROMOS, molecular dynamics simulations, 91
GUV, see Giant unilamellar vesicle

H

Hexagonal phase, overview, 17, 18
Holographic optical tweezers, see Laser tweezers

I

Immunoprecipitation, caveolae, 464–466
Inverted hexagonal phase
  features, 356, 357
  structure
    bending modulus determination, 363
    Luzzati plane, 361, 362
    measurable characteristics, 361
    pivotal plane as approximation for neutral
      surface, 362, 363
    prospects for study, 363, 364
Ion channel reconstitution
  applications, 571
  cloned ion channels in planar bilayers
    bacterial channels in liposomes
      incubation conditions, 579, 580
      patch clamp recordings, 580, 581,
        583, 584
      recombinant channel expression, 578,
        579, 583
    incubation conditions, 574–577, 582, 583
    membrane isolation, 574, 582
    Xenopus oocyte heterologous channel
      expression, 573, 574, 581, 582
  materials, 572, 573

L

Langmuir films
  lateral segregation of lipids
    fluorescence microscopy
      monolayers in situ, 451, 452
      overview, 449
      probes, 449–451, 455
      transferred monolayers, 452
    image analysis
      compression-driven lipid segregation
        quantification, 453–455

lateral film reorganization qualitative evaluation, 452, 453, 455
interfacial deposition
films from aqueous suspensions, 443, 444
films from organic solvent solutions, 442, 443, 454
Langmuir—Blodgett film solid supports
selection and cleaning, 446, 447
transfer, 447–449, 455
layer material and solvent selection, 440, 441
materials, 440
overview, 439, 440
surface balance
trough cleaning, 442
trough types, 441, 442
surface pressure versus area isotherms, 444–446, 455
near-field scanning optical microscopy membrane preparation by Langmuir—Blodgett transfer, 471, 472, 478
Large unilamellar vesicle (LUV), sterol superlattice assay preparation, 162–164, 167, 168
Laser scanning microscopy, *see* Confocal microscopy
Laser tweezers
giant unilamellar vesicle deformation studies
deformation
bead utilization, 396, 398, 403
direct interactions with vesicle, 401
electroformation of vesicles, 393, 394, 402
Fourier analysis, 391, 392
holographic optical tweezer instrumentation, 390
image extraction of vessel boundary, 401–403
materials, 392, 393, 402
microsphere incorporation into vesicles, 394, 402
particle tracking, 391, 399–401, 403
sample preparation
low index of refraction, 401, 402
vesicles with internal beads, 395, 396, 402
principles, 390
Lateral pressure profile, overview, 21

LAURDAN, *see* Giant unilamellar vesicle; Nicotinic acetylcholine receptor; Sterol superlattice
Lipid
assemblies, 3–5
definition, 1
structural diversity, 1–3
Lipid phase
bilayer/lamellar phase, 16, 17
cubic phase, 17
differential scanning calorimetry of transitions, *see* Differential scanning calorimetry
factors affecting preference
hydration, 19
structure, 18, 19
temperature, 19, 20
Fourier transform infrared spectroscopy of transitions, *see* Fourier transform infrared spectroscopy
hexagonal phase, 17, 18
normal versus inverted, 16
optical dynamometry of transitions, *see* Optical dynamometry
Lipid raft
cholesterol role, 489, 490
disruption, 465
fluorescence resonance energy transfer, *see* Fluorescence resonance energy transfer
isolation
detergent isolation, 462, 466
materials, 461
nondetergent isolation, 462, 465, 466
plasma membranes, 462, 465, 466
sucrose density centrifugation, 463, 464, 466
microscopy
fluorescence microscopy
bilayer preparation, 484–487
imaging, 486
material, 482, 483, 486
principles, 481–484
sample preparation, 484, 486
near-field scanning optical microscopy, *see* Near-field scanning optical microscopy
principles, 460
overview, 5, 459, 481
LUV, *see* Large unilamellar vesicle
Lysophospholipid, detergent properties, 8

Index

M

Magic angle spinning nuclear magnetic resonance (MAS-NMR)
  diffusion rate measurements
    bilayer curvature correction, 261
    materials, 258
    sample preparation, 258, 261
    signal attenuation in powder samples, 260, 261
    spectra acquisition and processing, 258, 259, 261, 262, 264
  pulsed magnetic field gradient combination, 257
MAS-NMR, *see* Magic angle spinning nuclear magnetic resonance
MathCad
  osmotic water permeability, 327–329
  solute permeability calculations, 330, 331
MD, *see* Molecular dynamics
Membrane bending stiffness, *see* Optical dynamometry
Membrane curvature, biological roles
  enzyme activity, 22, 23
  homeostasis, 22
  membrane fusion, 22
Membrane pressure, *see* Giant unilamellar vesicle; Osmotic stress technique
Microdomain, *see* Lipid raft
Micropipet aspiration (MPA)
  applications, 421, 422
  area compressibility modulus measurement, 434
  area strain equations, 428, 429, 435
  bending modulus measurement, 434
  chambers, 423, 424
  comparison with other techniques, 421
  data analysis, 434, 435
  experiment set-up, 433–435
  giant unilamellar vesicle preparation by electroformation, 422, 423, 432, 433
  glass cover slip coating, 431, 432
  instrumentation and software, 426, 427
  lysis tension measurement, 434
  manometer, 424, 425
  needle probe fabrication, 431
  pipet fabrication and characterization, 429–431
MLV, *see* Multilamellar vesicle
Molecular dynamics (MD)
  experimental observables, 95–97, 99
  force fields, 91–95
  optimization, 99
  overview, 89–91
  phospholipid interactions in membranes
    Boltzmann-weighted ensemble
      density profile analysis, 134, 135
      order parameter analysis, 135, 136
      sampling, 132, 133
    force fields
      all-atom representation, 128, 129
      simplified lipid representation, 129, 130
    headgroup region structure
      electrostatic potential, 137
      interfacial water structure, 138
      lipid structure, 137, 138
      overview, 136
    long-range interactions
      cutoffs, 141
      Ewald sum, 131
    overview, 127, 128
    periodic boundary conditions, 130
    solvation analysis
      large molecule solvation, 139
      small molecules and membrane permeability, 140
    void analysis
      test point approach, 138, 139
      Voronoi diagram method, 139
Molecular fluctuations
  fluctuation—dissipation theorem and ergodicity, 342
  liquid-crystalline membranes and intermolecular correlation functions, 343, 344
  membrane deformation and elasticity
    angular fluctuations, 350–352
    positional fluctuations, 349, 350
  molecular motions in membranes, 342, 343
  motional models of collective excitations of bilayers, 348
  overview, 341
  solid-state deuterium nuclear magnetic resonance of lipid dynamics, 344–349
  X-ray diffraction and intermembrane correlations, 344, 349
Monte Carlo simulations
  diffusion modeling, 299, 300
  sampling for phospholipid interactions in membranes
    configurational bias Monte Carlo, 133

extension-bias Monte Carlo, 133
initial configuration, 134
overview, 133
MPA, *see* Micropipet aspiration
Multilamellar vesicle (MLV)
differential scanning calorimetry of drug–membrane interactions, 588–595
interlamellar forces
measurement, 408
overview, 405, 406
preparation for atomic force microscopy, 507, 508
X-ray scattering
data acquisition and analysis, 414
instrumentation, 414
sample holder
loading, 413, 414
mylar windows, 413
sample cell, 412
teflon holders, 413
sample preparation, 411, 412
Near-field scanning optical microscopy (NSOM)
atomic force microscopy of samples, 471–473, 478, 479
data analysis, 477
dye-labeled lipid for phase separation visualization, 471, 478
imaging
bilayers in water, 476, 477, 479
monolayers, 475, 476
scanning conditions, 474
test samples, 474, 479
two-color imaging, 474
instrumentation, 473, 474
materials, 470
membrane preparation
Langmuir–Blodgett transfer, 471, 472, 478
vesicle fusion, 472, 478
principles, 469, 470
probe fabrication and installation, 472, 473
sample handling, 474, 479

N

Neutral surface
definition, 360, 361
pivotal plane as approximation, 362, 363
Nicotinic acetylcholine receptor (AChR)
Laurdan probing and fluorescence resonance energy transfer
data analysis, 535–537
fluorescence resonance energy transfer, 534, 535–537, 539
Laurdan generalized polarization, 534, 535, 537, 539, 540
materials, 533, 534, 539
overview, 532
receptor-rich membrane preparation, 534, 539
titration, 535, 540
signal transduction, 531
structure, 532
NMR, *see* Nuclear magnetic resonance
NSOM, *see* Near-field scanning optical microscopy
Nuclear magnetic resonance (NMR)
membrane proteins in micelles
detergents, 519
materials, 516–518
overview, 515, 516
protein expression and purification, 516–518
protein structure determination, 519, 520, 522
diffusion rates, *see* Magic angle spinning nuclear magnetic resonance
oriented membrane studies, *see* Solid-state nuclear magnetic resonance
inverse theory algorithms in analysis
alignment and structural organization, 119, 120
chain orientational order, 118, 119
de-Pakeing, 114, 115
indirectly observed data
existence, 104
stability, 105, 106
uniqueness, 104, 105
magnetic alignment, 115–117
overview, 103, 104
regularization
discretization, 106
L-curve, 107, 108
relaxation rate distribution, 110–112
self-consistency, 112, 113
singular-value decomposition truncation, 108, 109
software, 122–124
Tikhonov regularization, 109, 110
two dimensions, 120–122

Index

O

Optical dynamometry
  calibration, 233
  materials, 231, 235
  membrane bending stiffness measurement close to fluid–gel phase transition, 230, 234, 235
  optical trapping overview, 227, 228
  particle size determination, 233
  shear surface viscosity measurements
    Brownian motion dynamics, 323
    optical trapping dynamics, 234
    principles, 228–230, 233
    sedimentation velocity dynamics, 233, 234
  vesicle preparation, 231–233, 235
Osmotic stress technique
  imposition and measurement, 409, 410
  interlamellar force measurement, 408, 409
  overview, 407
  sterol-containing bilayer studies, 416, 417
  units of osmotic stress, 410
  work performed by osmotic stress, 409
Osmotic water permeability, *see* Stopped-flow fluorometry

P

Patch clamp, *see* Ion channel reconstitution
PDF, *see* Probability distribution function
Phase, *see* Lipid phase
Phosphocholine cytidyltransferase, curvature strain effects, 23
Phosphorous, assay for sterol superlattice analysis, 164, 168
PKC, *see* Protein kinase C
PPC, *see* Pressure perturbation calorimetry
Pressure perturbation calorimetry (PPC)
  applications, 197
  data acquisition, 199, 200, 204
  data evaluation, 200–204
  instrumentation, 198, 202–205
  principles, 197, 198
  sample preparation, 198, 203
Probability distribution function (PDF), diffusion modeling, 298
Propagator, diffusion modeling, 297, 298, 313, 314

Protein kinase C (PKC), curvature strain effects, 23
Pyranine, fatty acid membrane binding and transport assays, 243, 247, 248, 251, 254

R

Raman scattering, *see* Coherent anti-stokes Raman scattering
Random number generator (RNG), diffusion modeling
  correlations, 307, 308
  requirements, 305
  seeds, 308
  selection, 306
  testing, 306, 307
  transformations, 308
  types, 305, 306
Regularization, *see* Nuclear magnetic resonance
RNG, *see* Random number generator
Rock and roll method
  atomic force microscopy assessment, 69, 70
  film formation, 66, 67, 74
  hydration through vapor phase, 72, 73
  materials, 64
  orientation assessment, 67, 68
  overview, 63, 64, 73, 75
  packing samples, 71
  sample preparation, 64–66, 74
  substrate preparation, 66

S

Shape analysis, *see* Giant unilamellar vesicle
Shear surface viscosity, *see* Optical dynamometry
Signaling lipids, overview, 7, 8
Single-molecule detection (SMD)
  fluorescence recovery after photobleaching comparison, 282
  fluorescent labels, 285, 286
  historical perspective, 278
  materials, 282, 283
  principles, 278, 291
  sample preparation
    monolayer, 289, 291
    solid-supported bilayer
      film deposition, 288, 289, 291
      vesicle fusion, 287, 291
  tracer diffusion models
    heterogeneous membrane environment, 280–282
    homogeneous membrane environment, 279, 280

tracking analysis, 289–291
wide-field single-molecule fluorescence
   microscopy, 283–285, 291
Singular-value decomposition (SVD),
   nuclear magnetic resonance data analysis
   with truncation, 108, 109
Small unilamellar vesicle (SUV)
   fatty acid measurement assay preparation,
      31, 32, 37, 39
   preparation for atomic force microscopy,
      508–511
SMD, see Single-molecule detection
Solid-state nuclear magnetic resonance,
   see also Magic angle spinning nuclear
   magnetic resonance
   deuterium studies of molecular fluctuations
      lipid dynamics, 344–349
      membrane deformation and elasticity
         angular fluctuations, 350–352
         positional fluctuations, 349, 350
      motional models of collective excitations
         of bilayers, 348
   drug–membrane interactions, 591
   membrane proteins in bilayers
      materials, 516–518
      nitrogen-15 studies, 523–525
      overview, 515, 516, 522, 523
      phosphorous-31 studies of lipids, 525
      protein expression and purification,
         516–518
      sample preparation, 523
      two-dimensional PISEMA studies,
         525, 526
   oriented membranes
      deuterium studies
         acquisition and processing, 83, 84
         interpretation, 84, 85, 87
      glass slide preparation, 78
      materials, 78
      overview, 77
      phosphorous-31 studies
         acquisition and processing, 80, 81
         interpretation, 81, 83, 87
      preparation from
         lipid water dispersions, 79, 80
         liposomes, 80
         organic solvents, 78, 79, 87
   inverse theory algorithms in analysis,
      see Nuclear magnetic resonance
Spontaneous curvature

overview, 358
X-ray scattering studies, see X-ray scattering
Sterol oxidation, fluorescence assay in
   liposomal membranes
   applications, 153, 154
   data analysis
      apparent rate constant, 151
      initial rate, 152, 156
   dehydroergosterol
      concentration determination, 148, 149
      structure, 146, 155
   incubation conditions, 150, 151, 156
   materials, 147, 148, 155
   overview, 145–147, 155
   phospholipid concentration determination, 149
   sample preparation, 149, 155, 156
Sterol superlattice
   fluorescence assay
      cholesterol assay, 164
      large unilamellar vesicle preparation,
         162–164, 167, 168
      LAURDAN measurement, 164–166,
         168, 169
      materials, 161, 162, 166, 167
      overview, 161
      phosphorous assay, 164, 168
   regular distribution model, 159, 160
Stopped-flow fluorometry
   fatty acid binding and transport assays,
      246–248, 253, 254
   osmotic water permeability measurement
      data acquisition, 327
      equations, 329, 330
      liposome/vesicle preparation, 326
      materials, 325, 326
      MathCad calculations, 327–329
      overview, 323–325
      vesicle integrity test, 326, 327
   solute permeability measurement
      data acquisition, 327
      liposome/vesicle preparation, 326
      materials, 326
      MathCad calculations, 330, 331
      overview, 325
      vesicle integrity test, 326, 327
Superlattice, see Sterol superlattice
Supported lipid bilayer, see Atomic force
   microscopy
Surface pressure versus area isotherm,
   see Langmuir film

Index

SUV, *see* Small unilamellar vesicle
SVD, *see* Singular-value decomposition

T

Tikhonov regularization, nuclear magnetic resonance data analysis, 109, 110
Tilt modulus, overview, 21

V

Vibrational microscopy, *see* Coherent anti-stokes Raman scattering
Vinblastine, differential scanning calorimetry of drug–membrane interactions, 594, 595
Voronoi diagram, phospholipid interactions in membranes analysis, 139

W

Wide-field single-molecule fluorescence microscopy, *see* Single-molecule detection

X

Xenopus oocyte, *see* Ion channel reconstitution
X-ray scattering
  intermembrane correlations of molecular fluctuations, 344, 349
  inverted hexagonal phase structure
    bending modulus determination, 363
    Luzzati plane, 361, 362
    measurable characteristics, 361
    pivotal plane as approximation for neutral surface, 362, 363
    prospects for study, 363, 364
  multilamellar vesicle studies
    sample preparation, 411, 412
    sample holder
      loading, 413, 414
      mylar windows, 413
      sample cell, 412
      teflon holders, 413
    instrumentation, 414
    data acquisition and analysis, 414
  oriented, hydrated sample preparation, *see* Rock and roll method
  overview of lipid phase studies, 406, 407